Surveys in Differential Geometry

Vol. 1: Lectures given in 1990
edited by S.-T. Yau and H. Blaine Lawson

Vol. 2: Lectures given in 1993
edited by C.C. Hsiung and S.-T. Yau

Vol. 3: Lectures given in 1996
edited by C.C. Hsiung and S.-T. Yau

Vol. 4: Integrable systems
edited by Chuu Lian Terng and Karen Uhlenbeck

Vol. 5: Differential geometry inspired by string theory
edited by S.-T. Yau

Vol. 6: Essay on Einstein manifolds
edited by Claude LeBrun and McKenzie Wang

Vol. 7: Papers dedicated to Atiyah, Bott, Hirzebruch, and Singer
edited by S.-T. Yau

Vol. 8: Papers in honor of Calabi, Lawson, Siu, and Uhlenbeck
edited by S.-T. Yau

Vol. 9: Eigenvalues of Laplacians and other geometric operators
edited by A. Grigor'yan and S-T. Yau

Vol. 10: Essays in geometry in memory of S.-S. Chern
edited by S.-T. Yau

Vol. 11: Metric and comparison geometry
edited by Jeffrey Cheeger and Karsten Grove

Vol. 12: Geometric flows
edited by Huai-Dong Cao and S.-T. Yau

Vol. 13: Geometry, analysis, and algebraic geometry
edited by Huai-Dong Cao and S.-T.Yau

Vol. 14: Geometry of Riemann surfaces and their moduli spaces
edited by Lizhen Ji, Scott A. Wolpert, and S.-T. Yau

Volume VI

Surveys in Differential Geometry

Essays on Einstein manifolds

edited by

Claude LeBrun and McKenzie Wang

International Press
www.intlpress.com

Surveys in Differential Geometry, Vol. 6

Editors:
Claude LeBrun, Stony Brook University, New York
McKenzie Wang, McMaster University, Ontario

ISBN 978-1-57146-177-3

Paperback reissue 2010. Previously published in 1999 under ISBN 1-57146-068-3 (clothbound).

Typeset using the LaTeX system.

Preface

A dozen years have elapsed since Besse's *Einstein Manifolds* first offered the mathematical public a comprehensive overview of this vital area of differential-geometric research. The book's unqualified success is witnessed not only by the hundreds of research articles which cite it, but also by the entire generation of geometers for whom the book served as a catechism. A concomitant measure of Besse's success, however, is the degree to which the book has consequently become outdated; for every research problem discussed there has inevitably been attacked by scores of geometers, sometimes to considerable effect. Thus, while Besse still remains the best all-around introduction to the subject, its utility as a self-contained guide for the aspiring researcher has palpably declined in recent years.

With this in mind, Professor S.-T. Yau suggested to us, in early 1997, that the time was ripe for a broad new survey of the field. When he went on to invite us to oversee the project, however, we could only feel thunder-struck: deeply honored, of course, but also staggered by the prospect. The monumental task of compiling a comprehensive replacement for Besse, with uniform notation, a unified bibliography, and a common style of exposition, simply seemed too much to contemplate. On reflection, though, we eventually came to realize that what the mathematical community actually needed was, in any case, something rather different.

Thus, our aim here is merely to offer a photo-mosaic of current research: a collage of snapshots, each capturing some major trend, each taken from the vantage point of a major contributor to the field. The participating authors were given free rein to adopt whatever style seemed best adapted to their topics; and whatever success the book may have must merely reflect the quality of their work. We are therefore deeply grateful to these friends and colleagues for the time and energy they have so willingly sacrificed in order to bring these essays into the world.

We have organized our montage into a triptych, the first panel of which depicts the Einstein manifolds of special holonomy. Since geometries of this sort are naturally classified according to the parallel tensor fields they entail, this part of the book necessarily runs parallel to portions of Besse, although our treatment focuses almost entirely on recent developments. Topics covered include Kähler-Einstein manifolds (essays by G. Tian and S. T. Yau), hyper-Kähler manifolds (essay by A. Dancer), manifolds with holonomy G_2 or Spin(7) (essay by D. Joyce), and quaternion-Kähler manifolds (essay by S. Salamon). We are also delighted to be able to offer an authoritative treatment (essay by C. Boyer and K. Galicki) of 3-Sasakian manifolds, the importance of which was quite unforeseen by Besse; for while these manifolds *do not* have special holonomy, their fundamental ties to the Kähler, hyper-Kähler, and quaternion-Kähler cases allow one to generate them in profusion, providing a dazzling new source of explicit compact Einstein manifolds.

The second panel of our triptych concerns techniques applicable to the study of general Einstein manifolds. Topics covered include Ricci flow (essay by B. Chow), convergence and compactness (essay by P. Petersen), and the use of symmetry or bundle constructions (essay by M. Wang). This section also surveys the 4-dimensional case (C. LeBrun) and the asymptotically hyperbolic case (O. Biquard) in essays which compare metrics of special holonomy with general Einstein metrics.

The last panel of our triptych harkens back to the physical inspirations of our subject. We are particularly happy to be able to feature two essays on the Lorentzian case, written for an audience of Riemannian geometers. The first of these, by K. P. Tod, offers a broad survey of recent research on General Relativity; and the second, by D. Christodoulou, focuses on the author's joint work with Klainermann on the stability of Minkowski space. Finally, this section includes an essay by D. Calderbank and H. Pedersen on the Riemannian analog of Hermann Weyl's unified theory of gravitation and electro-magnetism.

Obviously, we have not been able to cover every active area with ties to the theory of Einstein manifolds. Sometimes, as with the study of self-dual 4-manifolds or extremal Kähler metrics, we have chosen to omit a topic entirely because the main thrust of current work has little bearing on our present subject. In other cases, we have been forced to give only a limited treatment of a major recent development because the experts we approached felt disinclined to duplicate up-to-date surveys they had written for publication elsewhere. Thus, while LeBrun's essay includes a brief introduction to entropy and minimal volume, the interested reader is urged to read the article by Besson, Courtois and Gallot entitled *Minimal Entropy and Mostow's Rigidity*, in *Ergodic Theory and Dynamical Systems* **16** (1996) 623–649. Similarly, while the articles of Joyce and Salamon draw attention to the importance of harmonic and Killing spinors on certain Einstein manifolds, the reader should refer to J.-P. Bourguignon's forthcoming survey in the *Bulletin of the American Mathematical Society* for a systematic development of these ideas.

Despite the notable progress chronicled in this volume, the study of Einstein manifolds remains in its infancy. For example, we still haven't a clue as to whether there are obstructions to the existence of Einstein metrics on compact manifolds of dimension > 4. It is our fondest hope that this book may play an indirect rôle in settling fundamental questions of this kind by refocusing the attention of the mathematical community on the many open problems in this fascinating and important area of Riemannian geometry.

Claude LeBrun
McKenzie Wang

TABLE OF CONTENTS

Part I : Einstein Manifolds and Special Holonomy

Part II: Towards a General Theory of Einstein Manifolds

Part III: Relativity Revisited

Part 1: Einstein Manifolds and Special Holonomy

Einstein Manifolds with Zero Ricci Curvature

S.-T. Yau

The original motivation for solving of the Einstein equation is to understand space-time in the absence of matter; see the essay by Tod in this volume for an overview of the mathematics of general relativity. The equation governs the manner in which space-time is influenced by the sole force of gravity. As is well known, singularities such as black holes can occur. The study of the vacuum Einstein equations is a difficult problem in nonlinear hyperbolic systems; see the essay by Christodoulou in this volume. Complete space-times with nontrivial gravitational radiation are not well understood.

Many exact solutions of the stationary vacuum Einstein equation have been found in the past, including the Schwarzschild solutions and Taub-NUT solutions. It is remarkable that a so-called Wick rotation [**29**] can be performed on these solutions to transform them into complete (positive-definite) Einstein manifolds without singularities. The (Euclidean) Schwarzschild metric is particularly significant as it is a complete non-singular Ricci-flat metric which is not Kähler. (The (Euclidean) Taub-NUT metric is Kähler. For more about these metrics, consult the paper of Lapedes [**48**].) Wick rotation is obtained by replacing t by it in suitable coordinates. Unfortunately there is no systematic theory of Wick rotations on Einstein metrics.

Wick rotation does not work for compact Einstein manifolds. It is not hard to show that up to torus fibrations, Ricci-flat compact manifolds have no continuous group of symmetries. It is therefore difficult to construct such manifolds by exploiting the symmetry group. In fact, for a long time, geometers believed that no compact Ricci-flat manifolds existed beyond these flat space-forms. In particular, it was a general belief that the conjecture of Calabi on Kähler manifolds was too good to be true, as it implied that the only obstruction for the existence of a Ricci-flat Kähler metric comes from the cohomological condition of the vanishing of the first Chern class. This conjecture of Calabi was in fact proved by the author in [**90**] in 1976.

On a compact Kähler manifold, the space of Kähler metrics with zero Ricci curvature is then parameterized by those complex structures with zero first Chern class and their Kähler classes. (It is not known whether every compact Ricci-flat metric on a Kähler manifold is Kähler or not.) This moduli space has played an important role in the effort to compactify string theory.

Until now, nobody has been able to find an explicit formula for any nontrivial Ricci-flat metric on a compact Kähler manifold. It is possible that the recent conjecture of Strominger-Yau-Zaslow [**78**] may solve this question. There are many more explicit examples when the manifold is non-compact. Besides the Taub-NUT space mentioned above, there are also metrics constructed by the twistor method (see the essay by A. Dancer in this volume). Right after I settled the compact case, I was interested in working out the general existence theorem for complete Ricci flat Kähler manifolds. In the International Congress in 1978 in Helsinki, I described a basic conjecture on such manifolds. It is believed that if such a manifold has finite topology, it must be the complement of a divisor D in a compact Kähler manifold M, and D is the anti-canonical divisor of M. (Anderson-Kronheimer-Lebrun [**1**] have found examples of such manifolds with infinite topology.) I could prove some special cases of this conjecture, for example the case of total space of holomorphic bundles over Fano manifolds. (Such examples were worked out at the same time by Calabi [**12**] using more explicit calculations.) More general cases were obtained by Tian and myself, where we assume D supports a non-singular anti-canonical divisor in a Fano manifold [**83**]. The geometry of the metric depends a lot on the multiplicity of the anti-canonical divisor along D. If the multiplicity is one, D is automatically a compact Kähler manifold with trivial first Chern class. No extra assumption is needed in this case and the volume growth of the metric is less than quadratic growth. When the multiplicity is greater than one, we need to make the assumption that there is a Kähler-Einstein metric on D (see the essay by Tian in this volume for criteria for the existence of such a metric). The volume growth of the Ricci-flat metric on the complement of D in this case behaves like standard Euclidean space.

Our construction of such metrics on the complement of D goes as follows.

Let Ω be a holomorphic n-form defined on the complement of D so that $\Omega \wedge \overline{\Omega}$ blows up along D. Then we want to find a complete Kähler metric $g_{i\bar{j}}$ with bounded curvature so that

$$\frac{\det(g_{i\bar{j}})dz^1 \wedge \cdots \wedge dz^n \wedge \overline{dz^1} \wedge \cdots \wedge \overline{dz^n}}{\Omega \wedge \overline{\Omega}}$$

is bounded from above and below by a constant. The idea is to perturb such a metric $g_{i\bar{j}}$ by the Hessian $\frac{\partial^2 u}{\partial z_i \partial \overline{z_j}}$ of some function u so that $g_{i\bar{j}} + \frac{\partial^2 u}{\partial z_i \partial \overline{z_j}}$ is bounded from above and below by a constant multiple of $g_{i\bar{j}}$ and its volume form is $\Omega \wedge \overline{\Omega}$. In this way, we know quite a bit about the Ricci-flat metric near D. We construct $g_{i\bar{j}}$ by taking a Kähler-Einstein metric on D and extending it in a neighborhood of D to satisfy the above conditions. The existence of u is proved by carefully estimating its C^0 norm, which in turn depends on a suitable estimate of Sobolev inequality constants. It will be interesting to find precise asymptotic behavior of the Ricci-flat metric near D. The parameterization of Kähler Ricci-flat metrics is much more difficult than in the compact case. One has to put constraints on the uniform behavior of the metrics near D. It has been known to experts for some time that the Taub-NUT metric provides a family of complete Ricci-flat Kähler metrics on $\mathbb{C}^2$. (See LeBrun [**49**] for an explication and also the essay by A. Dancer in this volume.) Perhaps they are related to various compactifications of $\mathbb{C}^2$ (and the same may hold for more general quasi-projective manifolds).

There have been many attempts to find explicit constructions of Ricci-flat manifolds. See the essays by M. Wang and A. Dancer in this volume for example. These constructions either use group actions to reduce the number of variables (including the concept of a moment map) or build bundles over existing Kähler-Einstein manifolds.

The existence of Ricci-flat metrics gives rise to a new tool to study Kähler manifolds with zero first Chern class (Calabi-Yau manifolds). For example, an immediate corollary is that if for some element ω in the Kähler cone of an n-dimensional Calabi-Yau manifold M and if $C_2(M) \cup \omega^{n-2} = 0$, then M is covered by the flat torus. This was observed by the author at the time when the Calabi Conjecture was solved. In the above statement, if we only assume ω to be on the boundary of the Kähler cone and $C_2(M) \cup \omega^{n-2} = 0$, then depending on the degeneracy of ω, one should be able to make use of differential geometry to classify those manifolds to be fiber spaces with generic fibers as lower-dimensional Calabi-Yau manifolds. P. Wilson was able to study this problem using algebraic methods in [**87**].

A very important consequence of the Ricci-flat metric is that any holomorphic n-form is parallel with respect to the metric. Therefore if it is non-degenerate at one point, it must be non-degenerate everywhere. In particular, a holomorphic symplectic Kähler manifold is automatically hyper-Kähler, and it therefore admits three compatible integrable complex structures. When the manifold has complex two dimensions, a non-vanishing holomorphic two-form Ω will then be parallel and a linear combination of Ω, $\overline{\Omega}$ and the Kähler form then gives rise to a family of parallel two-forms each of which in turn gives rise to an integrable complex structure compatible with the same metric. This family of complex structures is a very powerful tool in the study of hyper-kähler manifolds. For example, a K3 surface may not have rational curves in general. But (see Bryan-Leung [**10**]) this deformation will give rise to a complex structure which does contain rational curves.

The first general theory of Calabi-Yau manifolds was the study of two-dimensional surfaces due to Piatetski-Shapiro and Shafarevich [**67**] (Burns-Rapoport [**11**] for the case of Kähler manifolds). They found that the period map must be injective for the moduli space of K3 surfaces. The question of surjectivity was done much later and was due to Kulikov [**47**] and Pinkham-Persson [**66**]. Both of these papers are deep works and require a great deal of algebraic machinery.

These theorems were drastically simplified by the observations of Todorov [**84**], that the author's existence theorem for Ricci-flat metric can be applied. The key point is an observation of Hitchin [**38**] that the metric provides an S^2 family of complex structures. This rational curve of complex structures provides a way to move in the moduli space. (Much more rigorous and detailed treatments were then given by Siu [**76**].) There were expectations to generalize these methods to higher-dimensional Calabi-Yau manifolds. While this has not been carried out, the famous theorem of Bogomolov [**8**] on the unobstructedness of holomorphic symplectic Kähler manifolds was generalized to general Calabi-Yau manifolds by Tian [**80**] in his thesis and by Todorov [**85**] independently. This basic theorem played an important role on the later development of Calabi-Yau manifolds. (The analog of the formula for proving unobstructedness is being using by Kontsevich, Fukaya, and others to construct higher products in their attempts to work out the algebraic formulation of mirror symmetry [**5, 25**].)

Since the revolutions of string theory in theoretical physics, the theory of Kähler manifolds with zero Ricci curvature (i.e. Calabi-Yau manifolds) has gone through a vigorous change. The fundamental paper of Candelas, Horowitz, Strominger and Witten [14] studied the Kaluza-Klein model, where one wants to compactify a ten-dimensional space-time to a four-dimensional space-time by using compact six-dimensional manifolds with nontrivial parallel spinors. The final analysis shows that the compactification is given by a Calabi-Yau manifold of three complex dimensions. This famous paper immediately called for a great deal of work on constructing such manifolds, especially those with Euler number equal to ± 6 and with nontrivial fundamental group. At the beginning, physicists thought that there are only a couple of Calabi-Yau manifolds with three dimensions. During the first major conference on string theory [91], the author described many ways to construct these manifolds, and the physicists were rather surprised to find out that there should be at least on the order of ten thousand such manifolds. The author proposed to construct a large class of these manifolds by taking complete intersections of hypersurfaces in products of weighted projective spaces. The first important example is the complete intersection of two cubics in $\mathbb{CP}^3 \times \mathbb{CP}^3$ and a bidegree $(1,1)$ hypersurface. This manifold has Euler number equal to -18. I was able to find a group of order three which acts on it with no fixed point. The quotient manifold then has Euler number -6 and non-trivial fundamental group. Tian and the author [82] then found more examples in a similar way. The idea of taking complete intersections in products of weighted projective spaces was soon taken up by Candelas and his group to produce many examples for later study. It was observed by B. Greene that all these constructions lead to manifolds with the same topology. Greene and his coauthors even discussed the phenomenological implication of those manifolds [2].

String theory demands extensive calculations on the moduli space of Calabi-Yau manifolds. Since the local Torelli theorem holds, the period of the top-dimensional holomorphic form determines the local geometry of the moduli space. It was observed by Tian [80] and the physicists that the Kähler potential can be written as $\log ||\Omega||^2$ where Ω is a local holomorphic family of top-dimensional holomorphic forms. The fact that the holomorphic n-form defines a sub-line bundle of the (flat) bundle of n-dimensional cohomology classes gives a way to calculate the Weil-Petersson geometry with extra data. The quotient of this flat bundle by the line bundle describes the infinitesimal deformation of complex structures and hence gives the tangent bundle of the moduli space.

Two groups studied this kind of geometry (Candelas et al [13] and Strominger [77]). Strominger coined the name special geometry for it (he originally called it Kähler geometry of restricted type and the author suggested changing it to special geometry). Special geometry turns out to play an important role in later calculations of mirror symmetry.

The works of Gepner [28] and Greene-Vafa-Warner [34] show heuristically how to attach a conformal field theory and a path integration to certain Calabi-Yau manifolds. Soon after, Dixon [21] and Lerche-Vafa-Warner [50] made the prediction of mirror symmetry, which asserts that for any Calabi-Yau manifold M, one can associate another Calabi-Yau manifold M' so that by going from M to the mirror M', two three-point correlation functions (one associated to the complex deformations and the other associated to the Kähler deformations) are mapped to one another. The correlation function for complex deformations of M is simply the natural triple product of $H^1(T_M)$ (this works since $\Lambda^3 T$ is trivial). The correlation function for Kähler deformations is much more complicated. Besides the classical topological cup product on $H^1(T_M^*)$, one needs to add corrections due to integration over rational curves! B. Greene and the author called the last triple product the quantum cup product during the first conference on mirror manifolds in 1990 in Berkeley. Vafa called the cohomology arising from such a ring structure quantum cohomology.

For the important example of the quintic in $\mathbb{CP}^4$, Greene-Plesser **[33]** demonstrated the existence of the mirror based on arguments from conformal field theory. Immediately afterwards, Candelas et al **[15]** carried out the complete detailed calculation of the correlation functions based on the mirror statements. The identification of the special geometry on both the Kähler and the complex sides plays an important role. The calculation of such an identification is a spectacular piece of work in mathematics. It depends on studying the periods of holomorphic three-forms which satisfy a Picard-Fuchs equation and on understanding the monodromy associated to the degeneration of complex structure. This work of Candelas et al has greatly influenced the development of Calabi-Yau manifolds in the past ten years. In particular, it provides a beautiful formula to calculate the number of rational curves (which needs to be defined suitably) on the quintic. Even the existence of this formula was not expected in mathematics literature. Later developments due to many mathematicians are all basically reinterpretations of Candelas' formula in various forms.

Candelas' method of calculation was immediately carried out by many groups of mathematicians when the complex deformation space is one-dimensional. When the deformation space is multidimensional, the calculation requires a new method and this was carried out independently by Hosono-Klemm-Thiesen-Yau **[40]** and by Candelas-de la Ossa-Font-Katz-Morrison **[16]**. A further generalization was also done by Hosono-Lian-Yau **[41]**. In the former paper the Frobenius method and the hypergeometric system of Gelfand-Kapranov-Zelevinsky **[26, 27]** were extensively used. The formal parameter in the Frobenius method was later replaced by the hyperplane class in equivariant geometry. This gives the right interpretation of Candelas' formula in terms of equivariant geometry.

It makes sense to talk about the quantum cohomology ring structure for any Kähler manifolds. For manifolds with positive first Chern class, the associativity of quantum cohomology is sometimes enough to determine the instanton sum. This statement comes from the WDVV equations, which are due to a group of physicists (see **[88, 23]**). For these manifolds, mathematicians were able to exploit the associativity of the quantum cohomology to calculate the instanton sums. The concept of a Frobenius manifold was developed to understand these calculations, which in turn led to formulas for counting curves in homogeneous manifolds. On the other hand, it took a much longer time to actually prove the associativity of quantum cohomology.

The first attempt to prove this associativity was due to Ruan-Tian **[71]**. First of all, one needs to define the meaning of the instanton sum. Ruan **[70]** defined it for symplectic manifolds when the curve has genus zero. Then Ruan-Tian **[72]** generalized it to curves of arbitrary genus. The definition is modeled after Donaldson's definition of his gauge invariants for four-dimensional manifolds. A basic ingredient is the compactness argument for pseudo-holomorphic curves essentially due to Sacks-Uhlenbeck **[73]**. It was observed by Gromov **[35]** that pseudo-holomorphic curves can be used to study the rigidity of symplectic manifolds. Ruan-Tian's definition and proof of associativity for quantum cohomology works only for pseudo-holomorphic curves with respect to a generic choice of almost complex structure.

However integrable complex structures are far away from being generic, and therefore the instanton sum needs to be defined differently if we restrict ourselves to projective manifolds only.

Based on the works of Sacks-Uhlenbeck [**73**], Gromov [**35**], Parker-Wolfson [**65**] and others, Kontsevich [**46**] defined the concept of the compactification of the moduli space of rational maps from pointed rational curves to a projective manifold. When the projective manifold is a complete intersection in a certain homogeneous space, there is a way to define a certain obstruction bundle over the above compactified space. If the obstruction bundle has the same rank as the moduli space of maps, we can take the Euler number of the bundle. In general, however, one has to use the construction of the virtual cycle first done by Li-Tian [**51**] to define such a number. For a generic choice of projective hypersurface, the "number" of curves in a fixed topology can be defined in terms of these Euler numbers. For the quintic, we get

$$K_d = \sum_{k|d} n_{d/k} k^{-3},$$

where K_d is the Euler number and $n_{d/k}$ is the expected number of rational curves. This formula, called the covering formula, was discovered by Candelas et al [**17**] and rigorously justified by Aspinwall-Morrison [**4**] and Manin [**59**]. The number n_i is a projective invariant and should be called differently from the symplectic invariant mentioned above. A natural name should be the Schubert invariant to honor the fundamental work begun by Schubert a century ago.

In many important cases, Li-Tian [**52**] and Siebert [**75**] were able to prove these Schubert invariants are the same as the one defined by Ruan-Tian-Gromov. In particular, this demonstrates that the associativity law is valid for these Schubert invariants.

Candelas' formula for the quintic threefold is the following equation of formal power series in T:

$$\frac{5T^3}{6} + \sum_{d>0} K_d e^{dT} = \frac{5}{2}\left(\frac{f_1}{f_0}\frac{f_2}{f_0} - \frac{f_3}{f_0}\right),$$

where $T = \frac{f_1}{f_0}$, K_d is the Euler number as above, and for $i = 0, 1, 2, 3$,

$$f_i = \frac{1}{i!}\left(\frac{d}{dH}\right)^i\Big|_{H=0} \sum_{d\geq 0} e^{d(t+H)} \frac{\prod_{m=1}^{5d}(5H+m)}{\prod_{m=1}^{d}(H+m)^5}.$$

The f_i form a basis for the solution space of $L(f) = 0$, where L is the hypergeometric differential operator

$$L = \left(\frac{d}{dt}\right)^4 - 5e^t\left(5\frac{d}{dt}+1\right)\cdots\left(5\frac{d}{dt}+4\right).$$

Many people have made serious attempts to prove this formula. Witten [**88**] defined the concept of a linear sigma model, and Plesser-Morrison [**62**] made an (unsuccessful) attempt to use this concept to justify Candelas' formula. However, they did demonstrate the importance of the linear sigma model. Soon after, Kontsevich made a serious attempt to apply the Atiyah-Bott localization to prove Candelas' formula [**46**]. While he succeeded in computing the degree-four invariant for the quintic, his formulation is too complicated to be carried out in general. It is important to note that the above H used in the Frobenius method (see [**40**]) is interpreted as the equivariant hyperplane class. Following Kontsevich, Givental [**30**] made another attempt, using ideas of Witten and others and introducing quantum differential equations (these are just equations for determining a flat section of a certain canonically defined connection). However, his claimed proof is not complete. Finally, based on the works of Witten, Kontsevich, Li-Tian, and some new ideas on the concept of Euler data, Lian-Liu-Yau [**53**] gave the first complete proof of Candelas' formula in 1997. Some six months after the publication of [**53**], two works attempting to complete Givental's program appeared. The first one was due to Procesi et al [**7**] and the other one to Pandharipande [**64**]. The first paper did not claim to prove Candelas' formula in its final form, and the second used some ideas of Lian-Liu-Yau.

While the work of Lian-Liu-Yau does not give a construction of the mirror manifold, it does raise many interesting mathematical questions. One should interpret this theory as a theory of characteristic classes or K-theory over a mapping-space sigma model of algebraic manifolds. One advantage of such sigma models is that they allow us to restrict the maps to those from curves of a fixed topology, resulting in a finite-dimensional mapping space.

An important question involved in the theory of Lian-Liu-Yau is the following: Given an algebraic bundle V over an algebraic manifold M and the stable moduli space of maps $\mathcal{M}(g,k)$ from curves of genus g to M with homology class in $k \in H_2(M,\mathbb{Z})$, one can form a virtual bundle $\widetilde{V}$ over $\mathcal{M}(g,k)$ by looking at $H^0(C,f^*V) - H^1(C,f^*V)$, where $f: C \to M$ is a map in $\mathcal{M}(g,k)$. Given a theory of characteristic classes, i.e. a map b from the ring of holomorphic vector bundles to homology classes (which can be refined to algebraic cycles), one can then consider $b(\widetilde{V})$ and consider several numbers related to $b(\widetilde{V})$. For example, we can evaluate $b(\widetilde{V})$ over the Li-Tian class [**51**], which was defined by Li-Tian as a virtual moduli cycle (and subsequently understood by Behrend-Fantechi [**6**] using a somewhat different method), or we can consider a product of $b(\widetilde{V})$ with the Chern class of the tautological line bundle of $\mathcal{M}(g,k)$ and then evaluate this product over the Li-Tian cycle. The method of Lian-Liu-Yau can be used to compute these numbers for a large class of bundles V and M. This class includes, for example, convex and concave bundles over toric varieties or balloon manifolds. The computation of $b(\widetilde{V})$ can be considered as part of the K-theory over sigma models of algebraic manifolds.

It is important to carry out the computations of Lian-Liu-Yau in the most general possible setting. Equally important is to interpret the geometric meaning of the numbers computed. When b is the Euler class and $H^1(C,f^*V) = 0$, the number is interpreted to be related to the counting of the "number" of curves of genus g. This is how one computes the number of rational curves in a generic quintic in $\mathbb{CP}^4$. In that case, one takes V to be the line bundle $\mathcal{O}(5)$ over $\mathbb{CP}^4$. When V is $\mathcal{O}(-3)$ over $\mathbb{CP}^2$, we are dealing with numbers which arise in "local mirror symmetry," i.e. the "number" of rational curves in $\mathbb{CP}^2$ embedded as a hypersurface in a Calabi-Yau manifold (see the works of Vafa et al, e.g. [**45**], and the recent work of Chiang-Klemm-Lian-Roth-Yau-Zaslow [**18**]). The set of all these characteristic numbers over sigma models is very much related to the hypergeometric series of Gelfand-Kapranov-Zelevinsky [**26, 27**]. It would be very interesting to understand the internal structure of these numbers as a map from the K-groups of M.

When b is the Euler class, it is a remarkable theorem of Li-Tian that it is the same as counting the number (up to sign) of pseudo-holomorphic curves of a generic almost-complex structure compatible with the given symplectic structure. Using the proof of Lian-Liu-Yau, one should be able to extend the methods of Li, Ruan and Tian to show that the coefficients n_d of the generating function are integers. This should have deep interest for both combinatoricists and number theorists. The transformation from the hypergeometric series to the generating function is called the mirror transformation. It is also a remarkable fact that by choosing the right coordinates, the mirror transformation has a good q-expansion whose coefficients are integers (as was computed experimentally by Hosono-Klemm-Thiesen-Yau [**40**] and publicized by the authors). When the deformation of the mirror manifold is one-dimensional, this integral condition was verified by Lian-Yau [**57**]. This is a very important fact, as it was used by Lian-Yau to prove divisibility properties of the number of rational curves. For example, it was proved that n_i, the number of rational curves in a quintic, is divisible by 125 in the case i is not divisible by 5. However, such integral properties of the mirror map are not known for the multi-variable case and pose a challenging problem. Note that when the Calabi-Yau manifold has one or two dimensions, the mirror map is related to the j-function. In fact, Lian-Yau [**54, 55, 56**] observed that when the Calabi-Yau manifold is the K3 surface or when the Calabi-Yau manifold contains a pencil of K3 surfaces, the mirror map should be related to the automorphic form which appears in the moonshine conjecture related to the monster group. In his Harvard thesis, Chuck Doran made remarkable progress on this question, as he studied the Painlevé VI equation and its algebraic solutions extensively [**22**].

Duality.conjectures in the recent progress of string theory have clear implications in number theory as was indicated by works of Moore-Witten [**61**]. Also G. Moore has questions on the values of the mirror maps on certain special points on the moduli space determined by a variational principle [**60**]. All these questions imply that a very rich structure of number theory is hidden in the theory of mirror symmetries. Klemm-Lian-Roan-Yau [**44**] developed a generalization of the Schwarzian equation for the mirror map. It was based on such equations that the divisibility properties of number of rational curves were found.

While the theory of Lian-Liu-Yau is able to tackle many important questions in enumerative geometry, it does not explain the geometric meaning of mirror manifolds. The construction of Strominger-Yau-Zaslow [**78**], however, does provide such a framework. In a Calabi-Yau threefold, we look at the space of special Lagrangian (real) tori in the manifold. The moduli space of pairs of such a torus coupled with a $U(1)$ connection over it has a natural complex structure. This is conjectured to be the mirror manifold. Vafa [**86**] has recently extend the SYZ conjecture to include vector bundles in the picture. While Gross [**36, 37**] and Hitchin [**39**] have made significant progress on the SYZ conjecture, a full understanding of this theory is still far away. Key missing ingredients are explicit constructions of special Lagrangian submanifolds in general Calabi-Yau manifolds and holomorphic disks whose boundaries lie on given Lagrangian submanifolds. In any case, the SYZ picture is likely to be correct and it will be very interesting to combine the rigorous treatment of Lian-Liu-Yau with the picture of SYZ. It predicts a construction of a Ricci-flat metric and hopefully can be carried out by understanding the instanton corrections to the semi-flat metric.

Many years ago, Mukai [**63**] observed that the moduli space of $SU(n)$ bundles over a K3 surface has natural hyper-kähler structure. (This can be generalized to other hyper-kähler manifolds.) He introduced the concept of the Mukai transform, which is clearly related to the above theories. Hopefully, a complete mathematical theory encompassing all these ideas can be found soon.

Another important problem is to classify all three-dimensional Calabi-Yau manifolds and those four-dimensional ones that are elliptic fiber spaces. A very much related question is the understanding of construction of manifolds with G_2 and Spin(7) holonomy groups.

Only recently Joyce [**42, 43**] was able to construct non-trivial examples of such manifolds (see Joyce's essay in this volume). They were obtained by singular perturbation which is similar to the construction of C. Taubes on self-dual $SU(2)$ connections over four-manifolds [**79**]. While these manifolds clearly play an important role in the recent progress of string theory, their global structure is still hard to be understood. How do we parameterize them? Are they related to Kähler manifolds in a systematic way? How can we understand the moduli space of bundles with special holonomy groups over these manifolds or Calabi-Yau manifolds?

A recent development of string theory demands that a given Calabi-Yau manifold can be deformed to another. Since these manifolds may have different topology, one must go through singular manifolds to achieve such a goal. One is also allowed to identify manifolds which give rise to the same conformal field theory. Aspinwall-Greene-Morrison [**3**] has studied the deformation of conformal field theory of Calabi-Yau manifolds when these is a "flop" construction which changes the topology of the manifolds. Greene-Morrison-Strominger [**32**] also discussed how the quantum field theory changes when the manifold is deformed to acquire conifold points. These theories demonstrate the possibility of good physical theories even when the target space has singularities. This should mean that we can develop a good geometric theory even when the manifolds acquire singularities. This includes a good metric, a good Hodge theory, a good bundle theory, and a good enumerative geometry on such singular manifolds. Such geometries should reflect the quantum field theory mentioned above. In particular, one would like to see new geometric quantities to capture the limit of the "quantum" geometry when a smooth manifold approaches a singular one. Super-symmetric cycles which represent cycles collapsing to the singularities should play an important role in all these discussions.

In the discussion of connecting different Calabi-Yau manifolds, a particularly important process was suggested by M. Reid [**68**] (some initial ideas date back to Clemens [**20**]). We can destroy the second cohomology of a Calabi-Yau manifold by blowing down rational curves with negative normal bundle. There are theorems by Clemens [**19**], Friedman [**24**] and Tian [**81**] on how to deform the complex structure of the resulting singular manifold to that of a smooth complex manifold. These manifolds need not be algebraic (although they are birational to such manifolds). By passing through this kind of process, Reid suggests to connect all Calabi-Yau three-folds together. It is a rather tempting conjecture. However, since the manifold obtained by smoothing is not Kähler, a canonical Hermitian metric has to be defined to account for properties similar to those given by the Ricci-flat metric. A Weil-Petersson metric on the moduli space based on such canonical metrics would be important because it should help to identify the mirror map.

A few years ago Zaslow and I [**92**] demonstrated the relation between counting singular rational curves with nodes in a K3 surface and automorphic forms. Motivated by the formula, Göttsche [**31**] made the following conjecture for a more general Kähler surface X:

Let C be a sufficiently ample divisor on X, and K be the canonical divisor. Then the number of curves of genus g in $|C|$ passing through $r = -KC + g - \chi(\mathcal{O}_X)$ points is given by the coefficient of $q^{\frac{1}{2}C(C-K)}$ in the following power series in q:

$$B_1^{K^2} B_2^{CK} (DG_2)^r \frac{D^2 G_2}{(\Delta(D^2 G_2))^{\chi(\mathcal{O}_X)/2}},$$

where $D = q\frac{d}{dq}$, G_2 is the Eisenstein series

$$G_2(q) = -\frac{1}{24} + \sum_{k>0} (\sum_{d|k} d) q^k,$$

Δ is the discriminant

$$\Delta(q) = q \prod_{k>0} (1 - q^k)^{24},$$

and the $B_i(q)$ are certain universal power series.

Bryan-Leung [**10**] made the first step to give rigorous proof of the Yau-Zaslow formula for K3 surfaces when the cohomology class is primitive. It is remarkable that A.-K. Liu [**58**] was recently able to obtain the formula for general Kähler surfaces. (Some special cases were obtained with T.-J. Li jointly.) Using the idea of a family of Seiberg-Witten invariants, he is also able to study a similar question for algebraic three-folds which are elliptic fiber spaces. It is rather mysterious that the generating function for counting curves is related to automorphic forms. Perhaps some generalized theory of those forms will be developed in the near future.

Coming back to the classification of Calabi-Yau manifolds, it may be interesting to understand geometric cobordism among such manifolds. When do two Calabi-Yau three-folds bound a seven-dimensional manifold with G_2 holonomy? For G_2-manifolds, one can of course look for a Spin(7) manifold to be the total space.

Recent developments in string theory give rise to the following interesting question. If a manifold M is a metric cone over a compact manifold N such that M has special holonomy group, what conditions does this place on N? (See the chapter in this volume by Boyer-Galicki for details.) In the interesting case when M is Calabi-Yau, this is equivalent to N having an Einstein-Sasaki structure. Recently Boyer-Galicki [**9**] have studied such structures, but do not seem to construct new examples in real dimension 5.

Another interesting development in due to Sawon [**74**]. He uses techniques of Rozansky-Witten [**69**] to develop relations between the Chern numbers of hyperkähler manifolds. In particular, if the real dimension is 8, Sawon finds an upper bound of c_4. The Chern numbers of any Kähler-Einstein manifold satisfy the inequality [**89**]

$$(-1)^n 2(n+1) c_1^{n-2} c_2 \geq (-1)^n n c_1^n.$$

Of course, for a Calabi-Yau n-fold ($n \geq 3$), this inequality is trivial, but perhaps some of the relations developed by Sawon can be extended to more general Calabi-Yau manifolds.

References

[1] M. Anderson, P. Kronheimer and C. LeBrun, *Complete Ricci-flat Kähler manifolds of infinite topological type*, Comm. Math. Phys., **125** (1989), 637-642.

[2] P.S. Aspinwall, B.R. Greene, K.H. Kirklin and P.J. Miron, *Searching for three-generation Calabi-Yau manifolds*, Nucl. Phys. B, **294** (1987), 193-222.

[3] P.S. Aspinwall, B.R. Greene and D. Morrison, *Calabi-Yau moduli space, mirror manifolds and spacetime topology change in string theory*, Nuclear Phys. B, **416**(**2**) (1994), 414-480.

[4] P.S. Aspinwall and D. Morrison, *Topological field theory and rational curves*, Comm. Math. Phys., **151** (1993), 245-262.

[5] S. Barannikov and M. Kontsevich, *Frobenius manifolds and formality of Lie algebras of polyvector fields*, Internat. Math. Res. Notices, **1998** (**4**), 201-215, alg-geom/9710032.

[6] K. Behrend and B. Fantechi, *The intrinsic normal cone*, Invent. Math., **128**(**1**) (1997), 45-88.

[7] G. Bini, C. De Concini, M. Polino and C. Procesi, *On the work of Givental relative to mirror symmetry*, math.AG/9805097.

[8] F.A. Bogomolov, *Surfaces of class* VII_0 *and affine geometry* Math. USSR-Izv., **21**(**1**) (1983), 31-73.

[9] C.P. Boyer and K. Galicki *On Sasakian-Einstein geometry*, University of New Mexico preprint, September 1998.

[10] J. Bryan and N.C. Leung, *The enumerative geometry of* $K3$ *surfaces and modular forms*, alg-geom/9711031, 1997.

[11] D. Burns and M. Rapoport, *On the Torelli problem for Kählerian* $K3$ *surfaces*, Ann. Sci. École Norm. Sup., (4) **8**(**2**) (1975), 235-273.

[12] Calabi, *Métrique kählériennes et fibrés holomorphes*, Ann. Sci. Éc. Norm. Sup. (4), **12**(**2**) (1979), 269-294.

[13] P. Candelas, P.S. Green and T. Hübsch, *Rolling among Calabi-Yau vacua*, Nucl. Phys. B, **330** (1990), 49-102.

[14] P. Candelas, G. Horowitz, A. Strominger and E. Witten, *Vacuum configuration for superstrings*, Nucl. Phys. B, **258** (1985), 46-74.

[15] P. Candelas, M. Lynker and R. Schimmrigk, *Calabi-Yau manifolds in weighted* $\mathbb{P}^4$, Nucl. Phys. B, **341** (1990), 383-402.

[16] P. Candelas, X. de la Ossa, A. Font, S. Katz and D.R. Morrison, *Mirror symmetry for two-parameter models*, I, Nucl. Phys. B, **416**(**2**) (1994), 481-538.

[17] P. Candelas, X. de la Ossa, P. Green and L. Parkes, *A pair of Calabi-Yau manifolds as an exactly soluble superconformal theory*, Nucl. Phys. B, **359** (1991), 21-74.

[18] T.-M. Chiang, A. Klemm, B. Lian, M. Roth, S.-T. Yau and E. Zaslow, *Local mirror symmetry: calculations and interpretations*, in preparation.

[19] C.H. Clemens, *Degeneration techniques in the study of three-folds*, in 'Algebraic three-folds' (Varenna, 1981), 93-154; Lecture Notes in Math., **947**, Springer, Berlin-New York, 1982.

[20] C.H. Clemens, *Homological equivalence modulo algebraic equivalence is not finitely generated*, Publ. Math. IHES, **58** (1983), 19-38.

[21] L.J. Dixon, *Some world-sheet properties of superstring compactifications, on orbifolds and otherwise*, in 'Superstrings, unified theories and cosmology 1987' (Trieste, 1987), 67-126, ICTP Ser. Theoret. Phys., **4**, World Sci. Publishing, Teaneck, NJ, 1988.

[22] C. Doran, *Picard-Fuchs uniformization and geometric isomonodromic deformations: Modularity and variation of the mirror map*, Harvard thesis, 1999.

[23] B. Dubrovin, *Integrable systems in topological field theory*, Nucl. Phys. B, **379** (1992), 627-689.

[24] R. Friedman, *Simultaneous resolution of threefold double points*, Math. Ann., **274**(**4**) (1986), 671-689.

[25] K. Fukaya and Y.-G. Oh, *Zero-loop open strings in the cotangent bundle and Morse homotopy*, Asian J. Math., **1**(**1**) (1997), 96-180.

[26] I.M. Gelfand, A.V. Kapranov and A.V. Zelevinsky, *Hypergeometric functions and toric varieties*, Func. Anal. Appl., **23** (1989), 94-106; Correction in Func. Anal. Appl., **27** (1993), 295.

[27] I.M. Gelfand, A.V. Kapranov and A.V. Zelevinsky, *Generalized Euler integrals and A-hypergeometric functions*, Adv. Math., **84**(**2**) (1990), 255-271.

[28] D. Gepner, *Exactly solvable string compactifications on manifolds of* SU(N) *holonomy*, Phys. Lett. B, **199** (1987), 380-388.

[29] G.W. Gibbons, S.W. Hawking and M.J. Perry, *Path integrals and the indefiniteness of the gravitational action*, Nuclear Phys. B, **138(1)** (1978), 141-150.

[30] A. Givental, *Equivariant Gromov-Witten invariants*, Int. Math. Res. Notices, **1996(13)**, 613-663.

[31] L. Göttsche, *A conjectural generating function for numbers of curves on surfaces*, alg-geom/9711012.

[32] B.R. Greene, D.R. Morrison and A. Strominger, *Black hole condensation and the unification of string vacua*, Nuclear Phys. B, **451**(bf 1-**2**) (1995), 109-120.

[33] B.R. Greene and M.R. Plesser, *Duality in Calabi-Yau moduli space*, Nucl. Phys. B, **338** (1990), 15-37.

[34] B.R. Greene, C. Vafa and N.P. Warner, *Calabi-Yau manifolds and renormalization group flows*, Nucl. Phys. B, **324** (1989), 371-390.

[35] M. Gromov, *Pseudo-holomorphic curves in symplectic manifolds*, Invent. Math., **82(2)** (1985), 307-347.

[36] M. Gross, *Special Lagrangian fibrations* I: *Topology*, alg-geom/9710006.

[37] M. Gross, *Special Lagrangian fibrations* II: *Geometry*, math.AG/9809072.

[38] N.J. Hitchin, *Compact four-dimensional Einstein manifolds*, J. Diff. Geom., **9** (1974), 435-441.

[39] N.J. Hitchin, *The moduli space of special Lagrangian submanifolds*, Ann. Scuola Norm. Sup. Pisa Cl. Sci., **25** (1998), 503-515.

[40] S. Hosono, A. Klemm, S. Theisen and S.-T. Yau, *Mirror symmetry, mirror map and applications to Calabi-Yau hypersurfaces*, Comm. Math. Phys., **167(2)** (1995), 301-350; *Mirror symmetry, mirror map and applications to complete intersection Calabi-Yau spaces*, Nucl. Phys. B, **433(3)** (1995), 501-522.

[41] S. Hosono, B. Lian and S.-T. Yau, *Maximal degeneracy points of GKZ systems*, J.A.M.S., **10(2)** (1997), 427-443.

[42] D.D. Joyce, *Compact Riemannian* 7*-manifolds with holonomy* G_2, I, II, J. Diff. Geom., **43(2)** (1996), 291-328, 329-375.

[43] D.D. Joyce, *Compact* 8*-manifolds with holonomy* Spin(7), Invent. Math., **123(3)** (1996), 507-552.

[44] A. Klemm, B.H. Lian, S.S. Roan and S.-T. Yau, *A note on ODEs and mirror symmetry*, Functional Analysis on the eve of the 21st Century, Vol. II, Progr. Math., **132** (1993), 301-323.

[45] A. Klemm, P. Mayr and C. Vafa, *BPS states of exceptional non-critical strings*, hep-th/9607139.

[46] M. Kontsevich, *Enumeration of rational curves via torus actions*, in 'The moduli space of curves' (R. Dijkgraaf, C. Faber and G. van der Geer, eds.), Birkhäuser, (1995), 335-368.

[47] V.S. Kulikov, *Degenerations of* K3 *surfaces and Enriques surfaces*, Math. USSR-Izv., **11(5)** (1978), 957-989.

[48] A.S. Lapedes, *Black hole uniqueness theorems in classical and quantum gravity*, Seminar on Differential Geometry, 539-602; Ann. of Math. Stud., **102**, Princeton Univ. Press, Princeton, NJ, 1982.

[49] C. LeBrun, *Complete Ricci-flat Kähler metrics on* C^n *need not be flat*, Several complex variables and complex geometry, Part 2 (Santa Cruz, CA, 1989), 297-304; Proc. Sympos. Pure Math., **52**, Part 2, Amer. Math. Soc., Providence, RI, 1991.

[50] W. Lerche, C. Vafa and N.P. Warner, *Chiral rings in* $N = 2$ *superconformal theories*, Nucl. Phys. B, **324** (1989), 427-474.

[51] J. Li and G. Tian, *Virtual moduli cycles and Gromov-Witten invariants of algebraic varieties*, J. Amer. Math. Soc., **11(1)** (1998), 119-174.

[52] J. Li and G. Tian, *Comparison of the algebraic and the symplectic Gromov-Witten invariants*, alg-geom/9712035.

[53] B. Lian, K. Liu and S.-T. Yau, *Mirror principle*, I, Asian J. Math., **1(4)** (1997), 729-763.

[54] B.H. Lian and S.-T. Yau, *Mirror symmetry, rational curves on algebraic manifolds and hypergeometric series*, in 'ICMP Proceedings Paris 1994', 163-184, IP, 1995.

[55] B.H. Lian and S.-T. Yau, *Mirror maps, modular relations and hypergeometric series* I.

[56] B.H. Lian and S.-T. Yau, *Mirror maps, modular relations and hypergeometric series* II. *S*-duality and mirror symmetry (Trieste, 1995), Nucl. Phys. B Proc. Suppl., **46** (1996), 248-262.

[57] B.H. Lian and S.-T. Yau, *Integrality of certain exponential series*, in 'Algebra and Geometry, Proceedings of the International Conferance of Algebra and Geometry', 215-227, IP, 1998.

[58] A.-K. Liu, *Family blow up formula, admissable graphs and the counting of nodal curves*, preprint.

[59] Yu.I. Manin, *Generating functions in algebraic geometry and sums over trees*, in 'The Moduli Space of Curves', eds. R. Dijkgraaf, C. Faber, G. van der Geer, Progress in Math., Vol 129, Birkhäuser, (1995), 401-418.

[60] G. Moore, *Attractors and arithmetic*, hep-th/9807056; *Arithmetic and attractors*, hep-th/9807087.

[61] G. Moore and E. Witten, *Integration over the u-plane in Donaldson theory*, Adv. Theor. Math. Phys., **1**(**2**) (1997), 298-387.

[62] D. Morrison and R. Plesser, *Summing the instantons: quantum cohomology and mirror symmetry in toric varieties*, alg-geom/9412236.

[63] S. Mukai, *Symplectic structure of the moduli space of sheaves on an abelian or* $K3$ *surface*, Invent. Math., **77**(**1**) (1984), 101-116.

[64] R. Pandharipande, *Rational curves on hypersurfaces (after A. Givental)*, math.AG/9806133.

[65] T. Parker and J. Wolfson, *A compactness theorem for Gromov's moduli space*, J. Geom. Anal., **3** (1993), 63-98.

[66] U. Persson and H. Pinkham, *Degeneration of surfaces with trivial canonical bundle*, Ann. of Math., **113**(**1**) (1981), 45-66.

[67] I.I. Piatetski-Shapiro and I.R. Shafarevich, *Torelli's theorem for algebraic surfaces of type* K3, (Russian), Izv. Akad. Nauk SSSR Ser. Mat., **35** (1971), 530-572.

[68] M. Reid, *The moduli space of* 3*-folds with* $K = 0$ *may nevertheless be irreducible*, Math. Ann., **278**(**1-4**) (1987), 329-334.

[69] L. Rozansky and E. Witten, *Hyper-kähler geometry and invariants of three-manifolds*, Selecta Math. **3** (1997), 401-458.

[70] Y. Ruan, *Topological sigma model and Donaldson-type invariants in Gromov theory*, Duke Math. J., **83**(**2**) (1996), 461-500.

[71] Y. Ruan and G. Tian, *A mathematical theory of quantum cohomology*, J. Diff. Geom., **42**(**2**) (1995), 259-367.

[72] Y. Ruan and G. Tian, *Higher genus symplectic invariants and sigma madels coupled with gravity*, Invent. Math., **130**(**3**) (1997), 455-516.

[73] J. Sacks and K. Uhlenbeck, *The existence of minimal immersions of 2-spheres*, Ann. Math., **113** (1981), 1-24.

[74] J. Sawon, *The Rozansky-Witten invariants of hyper-kähler manifolds*, preprint.

[75] B. Siebert, *Algebraic and symplectic Gromov-Witten invariants coincide*, math.AG/9804108.

[76] Y.T. Siu, *Every* K3 *surface is Kähler*, Invent. Math., **73**(**1**) (1983), 139-150.

[77] A. Strominger, *Special geometry*, Comm. Math. Phys., **133**(**1**) (1990), 163-180.

[78] A. Strominger, S.-T. Yau and E. Zaslow, *Mirror symmetry is T-duality*, Nuclear Phys. B, **479** (1996), no. 1-2, 243-259.

[79] C.H. Taubes, *Self-dual Yang-Mills connections on non-self-dual* 4*-manifolds*, J. Diff. Geom., **17** (1982), 139-170.

[80] G. Tian, *Smoothness of the universal deformation space of compact Calabi Yau manifolds and its Petersson-Weil metric*, in 'Mathematical aspects of string theory' (San Diego, Calif., 1986), 629-646, Adv. Ser. Math. Phys., **1**, World Sci. Publishing, Singapore, 1987.

[81] G. Tian, *Smoothing* 3*-folds with trivial canonical bundle and ordinary double points*, in 'Essays on mirror manifolds', 458-479, Internat. Press, Hong Kong, 1992.

[82] G. Tian and S.-T. Yau, *Three-dimensional algebraic manifolds with* $C_1 = 0$ *and* $\chi = \Gamma 6$, in 'Math. Aspects of String Theory' (San Diego, Calif., 1986), 543-559, Adv. Ser. Math. Phys., **1**, World Sci. Publ., Singapore.

[83] G. Tian and S.-T. Yau, *Complete Kähler manifolds with zero Ricci curvature* I, II, J.A.M.S., **3** (1990), 579-610; Invent. Math., **106** (1991), 27-60.

[84] A.N. Todorov, *Applications of the Kähler-Einstein-Calabi-Yau metric to moduli of* K3 *surfaces*, Invent. Math., **61**(**3**) (1980), 251-265.

[85] A.N. Todorov, *The Weil-Petersson geometry of the moduli space of* SU(n ˘ 3) *(Calabi-Yau) manifolds*,I, Comm. Math. Phys., **126**(**2**) (1989), 325-346.

[86] C. Vafa, *Extending mirror conjecture to Calabi-Yau with bundles*, hep-th/9804131.
[87] P.M.H. Wilson, *The existence of elliptic fibre space structures on Calabi-Yau three-folds*, II, Math. Proc. Cambridge Philos. Soc., **123(2)** (1998), 259-262; Math. Ann., **300(4)** (1994), 693-703.
[88] E. Witten, *Two-dimensional gravity and intersection theory on moduli space*, Surveys in Diff. Geom., **1** (1991), 243-310.
[89] S.-T. Yau, *Calabi's conjecture and some new results in algebraic geometry*, Proc. Nat. Acad. Sci. USA, **74(5)** (1977), 1798-1799.
[90] S.-T. Yau, *On the Ricci curvature of a compact Kähler manifold and the Monge-Ampere equation*, I, Comm. Pure Appl. Math., **31** (1978), 339-411.
[91] S.-T. Yau, *Compact three dimensional Kähler manifolds with zero Ricci curvature*, Proc. of the Symposium on Anomalies, Geometry and Topology, Argonne, World Scientific, (1985), 395-406.
[92] S.-T. Yau and E. Zaslow, *BPS States, String Duality, and Nodal Curves on* $K3$, Nuclear Physics B, **471** (1996), 503-512; hep-th/9512121.

Harvard University, Department of Mathematics, Cambridge, MA 02138-2901
E-mail address: yau@math.harvard.edu

Hyper-Kähler Manifolds

Andrew S. Dancer

Department of Mathematics and Statistics
McMaster University
Hamilton, Ontario L8S 4K1
Canada
dancer@icarus.math.mcmaster.ca

A Riemannian manifold (M, g) is hyper-Kähler if it possesses complex structures I and J such that

(i) the metric g is Kähler with respect to both I and J; and
(ii) the given complex structures anti-commute: $IJ = -JI$

Defining $K = IJ$, it then follows that g is also Kähler with respect to K. Moreover, I, J, and K automatically satisfy the usual quaternionic multiplication table. Each tangent space of M therefore becomes a quaternionic vector space, and the dimension of M is therefore $4n$ for some integer n. Condition (i) implies that parallel transport commutes with I and J, and hence with K. Thus the holonomy group is contained in $O(4n) \cap GL(n, \mathbb{H}) = Sp(n)$. Hyper-Kähler manifolds therefore correspond to one of the possibilities on Berger's list of holonomy groups.

In fact $aI + bJ + cK$ is also a Kähler structure on M whenever $a^2 + b^2 + c^2 = 1$, so a hyper-Kähler manifold actually carries a two-sphere of Kähler structures. It is this fact which lies at the heart of the twistor approach to hyper-Kähler geometry, outlined later in the article. Note in particular that a hyper-Kähler manifold is a *symplectic* manifold in many different ways.

The existence of this large family of parallel two-forms gives us more rigidity than in the Kähler case. For example, any Kähler metric admits an infinite-dimensional space of Kähler perturbations, for we can just add $\partial\bar{\partial}f$ to ω, where f is a function with $\partial\bar{\partial}f$ sufficiently small. There is no analogue of this in hyper-Kähler geometry. Another contrast with the Kähler situation is that for a hyper-Kähler manifold the complex structures, and hence the metric, are determined by knowledge of the triple of two-forms $\omega_I, \omega_J, \omega_K$. For, viewing these forms as living in $\mathrm{Hom}(TM, T^*M)$, we find that $K = -\omega_I^{-1}\omega_J$ et cetera.

It is often useful to know that if, for some metric, I, J, K are almost-Hermitian structures satisfying (ii), and if the associated two-forms are all closed, then I, J, K are in fact integrable and we have a hyper-Kähler structure. Again, the analogous result for Kähler manifolds is false.

Supported in part by NSERC grant OPG0184235.

Let I be an arbitrary element of the two-sphere of complex structures, and let J, K be complex structures such that $IJ = K$ etc. Then the closed 2-form $\omega = \omega_J + i\omega_K$ is holomorphic with respect to I, and nondegenerate. That is, ω defines a *complex-symplectic structure* on (M, I).

Taking the $2n$th power of ω gives a trivialisation of the canonical bundle, so a hyper-Kähler manifold has $c_1 = 0$ (for any one of the two-sphere of complex structures). Moreover ω is parallel, so in fact we obtain a parallel trivialisation of the canonical bundle, and the Ricci tensor of M is zero. Alternatively, one may view this in terms of holonomy groups and observe that $Sp(n)$ is contained in $SU(2n)$. Note that in real dimension four, the hyper-Kähler condition is equivalent to Ricci-flat Kähler, as $Sp(1) = SU(2)$. In higher dimensions, however, the hyper-Kähler property is much stronger.

If a *compact* complex manifold M is complex-symplectic and Kähler, then it is hyper-Kähler [**4**]. For the complex-symplectic structure implies, as above, that c_1 vanishes, and so Yau's theorem [**91, 92**] tells us that M admits a Ricci-flat Kähler metric. A Bochner argument now shows that the holomorphic symplectic form is actually parallel with respect to the Yau metric. The Kähler form, together with the real and imaginary parts of the holomorphic symplectic form, spans the space of parallel two-forms which reduces the holonomy to $Sp(n)$.

It is still unclear in general when a complex-symplectic manifold admits a compatible hyper-Kähler structure. There are examples [**33, 34**] in all dimensions of compact complex-symplectic manifolds which do not support any hyper-Kähler metric (see §6). However, in this article we shall encounter many cases which *are* known to be hyper-Kähler; examples include coadjoint orbits of complex semsimple groups, spaces of representations of fundamental groups of surfaces into complex semisimple groups, and open subsets of cotangent bundles of Kähler manifolds. Looking for complex-symplectic manifolds and attempting to find a compatible hyper-Kähler structure has proved a very fruitful way of producing examples of hyper-Kähler manifolds.

1. Early Examples

In 1978 Eguchi and Hanson [**25**] found an example of a four-dimensional complete Ricci-flat metric which was of Bianchi IX type; that is, it admitted an isometric action of $SU(2)$ with generically three-dimensional orbits. In fact, their example is hyper-Kähler. Explicitly, the metric is

$$W^{-1}dr^2 + \frac{1}{4}r^2\left(\sigma_1^2 + \sigma_2^2 + W\sigma_3^2\right), \tag{1.1}$$

where

$$W = 1 - \frac{a^4}{r^4},$$

$\sigma_1, \sigma_2, \sigma_3$ are left-invariant one-forms, and a is a parameter.

If we denote by X_i the vector fields dual to σ_i, then one of the complex structures, I say, maps X_1 to X_2, and sends the normal $\partial/\partial r$ to $\frac{2}{rW}X_3$. We can pick an anticommuting complex structure J which maps the normal to $\frac{2}{r\sqrt{W}}X_1$ and X_2 to $\frac{1}{\sqrt{W}}X_3$.

If a is zero then (1.1) is just the Euclidean metric on $\mathbb{R}^4$. For nonzero a we obtain a smooth complete metric by letting r range over $[a, \infty)$. The apparent degeneracy at $r = a$ is just a coordinate singularity, where the three-dimensional orbits of $SU(2)$

collapse down to a special orbit which is a two-sphere. The underlying manifold of the Eguchi-Hanson metric is the cotangent bundle of this sphere, so the principal orbits are copies of $\mathbb{RP}^3$ rather than S^3. As the coefficients of σ_1^2 and σ_2^2 are equal, the full isometry group of the metric is actually $U(2)/\{\pm 1\}$, acting on T^*S^2 in the natural way. With respect to the complex structure I, the manifold is biholomorphic to the cotangent bundle of $\mathbb{CP}^1$, but for the other complex structures it is an affine algebraic variety, the nonsingular quadric in $\mathbb{C}^3$. We shall see a similar phenomenon for many other hyper-Kähler manifolds. Note that I is the only complex structure preserved by the full isometry group–the other complex structures are preserved by $SU(2)$ but rotated by the $U(2)$ action.

The Eguchi-Hanson metric was soon generalised in several ways. Calabi [**15**] constructed hyper-Kähler metrics on the cotangent bundle of $\mathbb{CP}^n$ for any n. One of the complex structures is just that induced from complex projective space. Indeed, Calabi proceeds by explicitly calculating a Kähler potential on this complex manifold such that the associated Kähler form and the natural complex-symplectic form on the cotangent bundle together define a hyper-Kähler structure. With respect to the other complex structures, however, M is biholomorphic to an affine variety, the coadjoint orbit $SL(n+1,\mathbb{C})/GL(n,\mathbb{C})$. Of course, this complex manifold has completely different properties from $T^*\mathbb{CP}^n$. For example, it has no compact complex submanifolds.

On another front, Belinskii, Gibbons, Page and Pope [**5**] found the general Bianchi IX hyper-Kähler metric in the case when the $SU(2)$ action preserves each Kähler structure. The metric may be written as

$$(abc)^2\,dt^2 + a^2\sigma_1^2 + b^2\sigma_2^2 + c^2\sigma_3^2, \tag{1.2}$$

where a, b, c must satisfy

$$2\frac{a'}{a} = b^2 + c^2 - a^2 \tag{1.3}$$

and the two equations obtained from (1.3) by cyclically permuting a, b, c. These equations are equivalent, after making the change of variables $f_1 = bc, f_2 = ac, f_3 = ab$, to the spinning top equations $\dot{f}_i = f_j f_k$, where (ijk) ranges over cyclic permutations of (123). These may be solved in terms of Jacobi elliptic functions, and the metric written down explicitly. The resulting metrics are generically *triaxial*, that is, the coefficients of σ_i^2 are all distinct so the isometry group is $SU(2)$ rather then $U(2)$. However the triaxial metrics are never complete. In fact, the only complete examples in the family considered by Belinskii et al. are Eguchi-Hanson and flat space. The former corresponds to the solution $f_1 = -\coth t, f_2 = f_3 = -\operatorname{csch} t$, while the latter comes from the solution $f_1 = f_2 = f_3 = -t^{-1}$. The other non-triaxial solution, corresponding to $f_1 = -\csc t, f_2 = f_3 = -\cot t$, gives an incomplete metric.

Also in four dimensions, Gibbons and Hawking [**29**] found the family of *multi-instanton metrics*. These admit a circle action preserving the hyper-Kähler structure, and can be written as

$$V d\underline{x}.d\underline{x} + V^{-1}(d\theta + \underline{\omega}.d\underline{x})^2. \tag{1.4}$$

Here θ is a coordinate on the circle fiber, $\underline{x}$ a coordinate on the three-dimensional quotient, $\underline{\omega}$ is the connexion form, and V is a real-valued function on the quotient. In order to obtain a hyper-Kähler metric we need $dV = *d\underline{\omega}$, so V is harmonic.

The multi-instanton metrics are obtained by setting

$$V(\underline{x}) = \sum_{i=1}^{m} \mid \underline{x} - p_i \mid^{-1}, \tag{1.5}$$

for some points $p_1, \dots, p_m$ in $\mathbb{R}^3$. With this choice of V we get a *complete* metric, provided the p_i are all distinct.

The underlying manifold fibres over $\mathbb{R}^3$ with generic fiber S^1, but at the points p_i the circle collapses to a point (metrically these are the points where V^{-1}, the length of the circle, is zero). The topology of the manifold is generated by the $m-1$ two-spheres which fiber over the closed line segment from p_i to p_{i+1}. If $m = 1$ we just get Euclidean space, and if $m = 2$ we recover the Eguchi-Hanson metric. The multi-instanton metrics were also obtained, using twistor methods, by Hitchin [**41**].

At about the same time, progress was made in a different, less explicit direction. Yau's proof of the Calabi conjecture [**91, 92**] showed that the $K3$ surface admitted a hyper-Kähler metric. This was the first *compact* hyper-Kähler manifold known, except for flat tori. We shall discuss the compact case further in the last section of this article.

2. Hyper-Kähler Quotients

The hyper-Kähler quotient construction, due to Hitchin, Karlhede, Lindström and Roček [**45**], is at present the most useful method of constructing complete hyper-Kähler manifolds. The beauty of the technique is that, starting from flat quaternionic space, it produces highly nontrivial manifolds. We shall see, for example, that the Calabi and multi-instanton spaces can be produced in this way.

The construction is inspired by the symplectic quotient of Marsden and Weinstein, a good account of which is given in the book of Guillemin and Sternberg [**36**]. If a group G acts on a symplectic manifold M preserving the symplectic form, then in good cases (for example, if G is semisimple), one obtains a *moment map* $\mu : M \to \mathfrak{g}^*$. The moment map is a G-equivariant map such that

$$d\mu(v)(X) = \omega(X^*, v) \tag{2.1}$$

for all tangent vectors v to M and all elements X of $\mathfrak{g}$. Here X^* denotes the vector field on M associated to X by the group action.

The equivariance of μ shows that if λ lies in the center of $\mathfrak{g}^*$ then G acts on the level set $\mu^{-1}(\lambda)$. The key result is that if λ is in addition a regular value of μ and G acts freely and properly on the level set, then the quotient $\mu^{-1}(\lambda)/G$ is a symplectic manifold.

The starting point for the hyper-Kähler quotient is a hyper-Kähler manifold M admitting an action of a Lie group G preserving the hyper-Kähler structure. This means that the action of G is isometric and holomorphic with respect to all the complex structures. In particular, the action of G preserves the Kähler forms $\omega_I, \omega_J, \omega_K$. If the associated three symplectic moment maps exist, then we can combine them into a single hyper-Kähler moment map $\mu : M \to \mathbb{R}^3 \otimes \mathfrak{g}^*$. It is useful to observe that if $IJ = K$ etc. then $\mu_J + i\mu_K$ is holomorphic with respect to the complex structure I. The level sets of this map are therefore Kähler with respect to I.

The main theorem of [**45**] says that if we pick λ_i in the center of $\mathfrak{g}^*$, and if $\underline{\lambda} = (\lambda_1, \lambda_2, \lambda_3)$ is a regular value of μ and G acts freely and properly on $\mu^{-1}(\underline{\lambda})$, then the quotient $\mu^{-1}(\underline{\lambda})/G$ is hyper-Kähler. To see this, observe that the quotient can be viewed as the symplectic quotient by G of $(\mu_J + i\mu_K)^{-1}(\lambda_2 + i\lambda_3)$. Now, the symplectic quotient of a Kähler manifold by an action preserving the Kähler structure is in fact Kähler, so $\mu^{-1}(\underline{\lambda})/G$ has a Kähler structure induced by I. The other Kähler structures are produced in the same way, by considering different combinations of μ_I, μ_J, μ_K.

Before we mention some examples, let us make a few remarks about the construction.

(i) It is sufficient that G should be semisimple for the moment maps to exist. This condition is definitely not necessary, as some of our later examples will show.
(ii) If the action of G on $\mu^{-1}(\underline{\lambda})$ is free, then $\underline{\lambda}$ is automatically a regular value of the moment map. The dimension of the quotient in this case is $\dim M - 4\dim G$.
(iii) If M is complete and G is compact, then the quotient, when smooth, is also complete.
(iv) If G has a nontrivial center, then, typically, the action of G on the level set is free for generic values of $\underline{\lambda}$. At special values, we get singularities.
(v) Although the quotient construction is a powerful method for showing existence of hyper-Kähler metrics, finding the metric *explicitly* may be difficult.

The hyper-Kähler quotient procedure is well-adapted to finding complex structures on hyper-Kähler manifolds. The tool used is the equivalence between Kähler and algebro-geometric quotients [**52**]. More precisely, if N is a Kähler manifold with an action of G preserving the Kähler structure, then the symplectic quotient of N by G can be identified with the quotient by $G_{\mathbb{C}}$ of the set of *stable* points–those points whose $G_{\mathbb{C}}$ orbits meet the zero set of the moment map for the G action.

In many concrete situations this notion of stability coincides with some more natural notion, for example Mumford stability of vector bundles. In any event, the crucial point is always to ascertain which orbits of the complex group meet the zero set of the moment map.

In the hyper-Kähler case, we just observe, as above, that the hyper-Kähler quotient $\mu^{-1}(0, \lambda_2, \lambda_3)/G$ is the symplectic quotient by G of the variety $(\mu_J + i\mu_K)^{-1}(\lambda_2 + i\lambda_3)$ and apply the previous discussion.

As an example of the construction, let $M = \mathbb{H}^n = \mathbb{C}^n \times \mathbb{C}^n$, and let $G = U(1)$ act by $e^{it}.(z,w) = (e^{it}z, e^{-it}w)$. The moment map is $\mu : (z,w) \mapsto (\frac{1}{2}(|z|^2 - |w|^2), \mathrm{Re}\langle z,w\rangle, \mathrm{Im}\langle z,w\rangle)$. Taking $\underline{\lambda} = (0, \mathrm{Re}\,\tau, \mathrm{Im}\,\tau)$, we find that the hyper-Kähler quotient is biholomorphic to the algebro-geometric quotient $\{(z,w) : \langle z,w\rangle = \tau\}//\mathbb{C}^*$, where $\mathbb{C}^*$ acts by $s.(z,w) = (sz, s^{-1}w)$. If we take τ to be nonzero then all points (z,w) with $\langle z,w\rangle = \tau$ are stable for this action. The quotient can be identified as a complex manifold with the cotangent bundle of $\mathbb{CP}^{n-1}$, and the hyper-Kähler metric is that found by Calabi.

There are some particularly interesting four-dimensional examples of hyper-Kähler quotients. The best understood case is that of asymptotically locally Euclidean (ALE) hyper-Kähler four-manifolds, studied by Kronheimer [**57, 58**]. The ALE condition means that the metric is asymptotically a quotient by a finite group of the Euclidean metric

$$dr^2 + r^2(\sigma_1^2 + \sigma_2^2 + \sigma_3^2), \tag{2.2}$$

on $\mathbb{R}^4$. More precisely, we require that the metric should agree with the quotient of the Euclidean metric modulo terms whose p^{th} derivatives decay at least as fast as r^{-4-p}.

The construction is begun by choosing a finite subgroup Γ of $SU(2)$. There are two infinite families, the cyclic and binary dihedral groups, together with the binary tetrahedral, octahedral and icosahedral groups. We can rephrase this by saying that the finite subgroups of $SU(2)$ have an A-D-E classification, that is, they correspond to the simply-laced Dynkin diagrams. Moreover the nodes of the extended Dynkin diagram correspond to the irreducible representations of Γ.

For each such Γ, let V denote the regular representation of Γ and let M be the Γ-invariant part of $\mathrm{End}(V) \otimes \mathbb{C}^2$, where Γ acts on $\mathbb{C}^2$ via inclusion into $SU(2)$. Let G be the subgroup of $PU(V)$ commuting with the action of Γ on $\mathrm{End}(V)$. Then the hyper-Kähler quotient of M by G is, for generic choices of level set, a smooth ALE hyper-Kähler four-manifold.

The zero level set yields the Kleinian singular space $\mathbb{C}^2/\Gamma$, with the flat metric, and the smooth quotients are desingularisations of this space. Their topology is generated by embedded two-spheres of self-intersection -2, intersecting according to the non-extended Dynkin diagram corresponding to Γ.

Asymptotically, the metric approaches the flat metric on $\mathbb{R}^4/\Gamma$. The analysis of the asymptotics relies on the homogeneity of the moment map with respect to dilations of M. Roughly speaking, this means that going out to infinity in the hyper-Kähler manifold corresponds to scaling the level set $\underline{\lambda}$ towards zero. The flatness of the space corresponding to the zero level set, together with the flatness of the connexion controlling the first variation of the metric about $\underline{\lambda} = \underline{0}$, yields the ALE property.

As explained in [**57**], one can use the work of McKay [**70**] to give a more concrete description of the quotient construction. Let Δ be the extended Dynkin diagram of Γ, with vertices $v_1, \dots, v_m$, and when $i \leq j$ and v_i is adjacent to v_j, let p_{ij} denote the edge of Δ from v_i to v_j. To each vertex v_i we attach an integer n_i, the dimension of the irreducible representation of Γ corresponding to v_i. Then M is the direct sum over the edges of $\mathbb{H}^{n_i n_j}$, while G is the product over the vertices of $U(n_i)$, factored out by the circle subgroup which acts trivially on M. To define the action, for each edge p_{ij} we view $\mathbb{H}^{n_i n_j}$ as $\mathrm{Hom}(\mathbb{C}^{n_i}, \mathbb{C}^{n_j}) \oplus \mathrm{Hom}(\mathbb{C}^{n_j}, \mathbb{C}^{n_i})$ and let $U(n_i)$ and $U(n_j)$ act in the natural way. In this picture the relation $A + C = 2I$ between the adjacency and Cartan matrices of the Dynkin diagram shows that the real dimension of M is $4|\Gamma|$. On the other hand, the dimension of G is $(\sum n_i^2) - 1$, which is $|\Gamma| - 1$ from elementary representation theory, so the hyper-Kähler quotients are four-dimensional, as required.

In the A_n case, corresponding to $\Gamma = \mathbb{Z}_{n+1}$, all the n_i are equal to 1, since Γ is abelian. Hence M is just $\mathbb{H}^{n+1}$ and $G = U(1)^n$. The resulting hyper-Kähler metrics are the Gibbons-Hawking multi-instanton series. Note that the description of the topology of the ALE spaces above agrees with our discussion of the multi-instanton spaces in the preceding section.

The multi-instanton metrics may be written down explicitly as in (1.4) because they admit a circle action preserving each Kähler structure. In terms of the algorithm for constructing the ALE spaces as a quotient, this is due to the fact that in the A_n case there are as many edges as vertices in the extended Dynkin diagram. This means that there is still a residual circle action after we perform the quotient. In the other cases the number of vertices is one greater than the number of edges, and we do not obtain a hyper-Kähler action on the quotient. No explicit form of the metric is known in these cases.

For each choice of Γ we get a family of hyper-Kähler metrics, depending on the choice of level set. The action of the isometry group of M means that certain choices of level set will give equivalent quotients, so not all these parameters are effective. For A_n, we can parametrise the metrics by the locations of the points $p_1, \cdots, p_{n+1}$ of (1.5), modulo translations and rotations of $\mathbb{R}^3$. The points must be distinct if we are to obtain a smooth metric. For $n = 1$ (the Eguchi-Hanson metric) we have a single parameter. Another way of viewing the parametrisation is via the *periods*, the numbers obtained by pairing the cohomology classes of the three Kähler forms with the homology classes represented by the embedded two-spheres. The periods determine the hyper-Kähler metric [**58**].

In fact, Kronheimer shows that these examples are the only ALE hyper-Kähler four-manifolds. His proof exploits the fact that ALE metrics admit an orbifold compactification. Hyper-Kähler metrics in dimension four are always anti-self-dual, so this compactification carries an anti-self-dual conformal structure and the methods of Penrose's twistor theory [**78**] may be applied. If the ALE condition is dropped, one may get other examples, the simplest of which are the *multi Taub-NUT* metrics. This family, discovered by Hawking [**39**], is given as in (1.4), but now we have $V = 1+\sum_{i=1}^{m} | \underline{x} - p_i |^{-1}$.

If $m = 1$, we obtain a $U(2)$-invariant metric, the celebrated Taub-NUT metric. This may be written in $SU(2)$-invariant form as

$$\frac{r+1}{4r}\, dr^2 + r(r+1)(\sigma_1^2 + \sigma_2^2) + \frac{r}{r+1}\sigma_3^2, \tag{2.3}$$

where $r \in [0, \infty)$. At $r = 0$ the $SU(2)$ orbit collapses to a point (a *nut* in the terminology of general relativity). The change of variables $r = s^2$ shows that the metric extends smoothly over the nut, and we obtain a metric defined on $\mathbb{R}^4$. Although complete, the Taub-NUT metric is not ALE, because the coefficient of σ_3^2 approaches a constant as r tends to infinity rather than growing like r^2. This kind of asymptotic behaviour, which is shared by the other multi-Taub-NUT metrics, is called Asymptotically Locally Flat (ALF). There is a detailed discussion of the Taub-NUT metric by LeBrun [**62**], who shows that all the complex structures are equivalent to that of $\mathbb{C}^2$. There is a circle subgroup of $U(2)$ preserving each complex structure, but the action of $SU(2)$ permutes them, making it hard to write them down explicitly in the coordinates of (2.3). The Taub-NUT example shows that a complex manifold ($\mathbb{C}^2$ in this case) may admit non-homothetic complete Ricci-flat Kähler metrics, in contrast to the situation for negative Einstein constant [**17, 93**].

One can also produce the multi-Taub-NUT metrics by replacing an $\mathbb{H}$ factor in the Kronheimer construction by an $\mathbb{R}^3 \times S^1$. It is interesting to note that these metrics may also be obtained as hyper-Kähler quotients by a noncompact group. For example, if $\mathbb{R}$ acts on $\mathbb{C}^2 \times \mathbb{C}^2$ by $t.(z_1, z_2, w_1, w_2) = (e^{it}z_1, e^{-it}z_2, w_1 + t, w_2)$, then the hyper-Kähler quotient is the Taub-NUT metric [**6**].

There are many higher-dimensional examples produced as hyper-Kähler quotients of $\mathbb{H}^n$ by subgroups of $Sp(n)$. For example, Lindström and Roček found hyper-Kähler metrics defined on the cotangent bundles of Grassmannians $SU(m+n)/S(U(m) \times U(n))$ by taking a hyper-Kähler quotient of $\mathbb{H}^{m(n+m)}$ by $U(m)$. The case $m = 1$ corresponds to the Calabi metric. We shall see later that cotangent bundles of arbitrary generalised flag manifolds also carry a hyper-Kähler structure.

One class of quotients whose properties can be calculated in detail is motivated by Delzant's construction of compact Kähler toric varieties as Kähler quotients [**22, 35**]. The starting point for Delzant's work is a polytope Δ in $\mathbb{R}^n$ defined by inequalities $\langle x, u_k \rangle \geq \lambda_k$ for vectors $u_1, \ldots, u_d$ and scalars $\lambda_1, \ldots, \lambda_d$. Here d is the number of faces of Δ. This data defines a subtorus N of T^d, for we can take the Lie algebra of N to be the kernel of the map from $\mathbb{R}^d$ to $\mathbb{R}^n$ given by sending the kth element of the standard basis to u_k. Delzant now considers the Kähler quotient of $\mathbb{C}^d$ by N, using the scalars λ_k to fix the level set. If the polytope satisfies certain constraints, then the resulting Kähler quotient is a smooth compact toric variety X of real dimension $2n$. Moreover X admits a Kähler action of $T^n \cong T^d/N$, and the image of the associated moment map is just Δ.

By performing a hyper-Kähler quotient of $\mathbb{H}^d$ by N in an analogous fashion, one obtains [**30, 8**] families of hyper-Kähler manifolds associated to hyperplane arrangements in $\mathbb{R}^n$. (We do not necessarily assume these define a polytope). The hyper-Kähler manifolds have real dimension $4n$ and have an action of $T^n \cong T^d/N$ preserving the hyper-Kähler structure. The methods of [**45, 77**] may therefore be applied to give rather explicit formulae for the metric and Kähler form. The manifolds are noncompact, but their topology is generated by a union of compact toric varieties. The Calabi spaces correspond to the hyperplanes defining the standard n-simplex, while the multi-instanton metrics correspond to a set of points in $\mathbb{R}$. There are many other complete examples given by more complicated hyperplane arrangments. One also obtains, by restriction, hyper-Kähler metrics on the cotangent bundles of toric varieties, but these metrics are usually incomplete. Taking the zero level set yields a singular quotient, but, in good cases, this is a cone over one of the *smooth* 3-Sasakian manifolds of [**12**]. Bielawski has investigated the interesting topological properties of these examples [**7**].

This may be a good point to mention a difference between symplectic and hyper-Kähler quotients. When performing symplectic quotients by a group with nontrivial center Z, the usual picture is that the set of critical values of the moment map is of real codimension one in $\mathfrak{z}^*$. We therefore have a wall-crossing phenomenon; as the level set crosses a critical value the symplectic quotient can change topology by a sequence of symplectic blowups and blowdowns. This situation has been extensively analysed in the beautiful paper of Guillemin and Sternberg [**38**].

In the hyper-Kähler case, the set of critical values is typically of codimension three or more in $\mathbb{R}^3 \otimes \mathfrak{z}^*$, so one expects the smooth quotients to all be diffeomorphic (although not isometric). For example, the ALE spaces of [**57**] are, when smooth, all diffeomorphic to the minimal resolution of the appropriate Kleinian singularity.

3. Moduli Spaces

One of the most important features about hyper-Kähler structures is that there are many naturally occurring examples. In particular, the moduli spaces arising in gauge theory often carry a hyper-Kähler metric. The key fact is that such moduli spaces may often be viewed as finite-dimensional hyper-Kähler quotients of an infinite-dimensional space of connexions by an infinite-dimensional gauge group.

As an example, let us consider connexions on a principal G-bundle over $\mathbb{R}^4$. The space $\mathcal{A}$ of connexions is an infinite-dimensional affine space modelled on $\Omega^1(\mathbb{R}^4, \mathfrak{g})$. We put a quaternionic structure on $\mathcal{A}$ by writing connexions as $(\partial_0 + A_0) + i(\partial_1 + A_1) + j(\partial_2 + A_2) + k(\partial_3 + A_3)$, where ∂_a denotes the ath partial derivative and

$\sum_{a=0}^{3} A_a dx_a$ is a $\mathfrak{g}$-valued one-form. The gauge group $\mathcal{G}$ may be identified with $C^\infty(\mathbb{R}^4, G)$.

If we restrict to a space $\mathcal{A}_0$ of connexions which are standardised at infinity, in some suitable sense, then we can put an L^2 metric on $\mathcal{A}_0$ which makes it into an infinite-dimensional hyper-Kähler manifold. Furthermore, if we consider the subgroup $\mathcal{G}_0$ of gauge transformations tending to the identity at infinity, then the action preserves the hyper-Kähler structure and admits a moment map. An integration by parts argument, together with the conditions at infinity, shows that the equations for the vanishing of the moment map are

$$[\partial_0 + A_0, \partial_i + A_i] = [\partial_j + A_j, \partial_k + A_k], \tag{3.1}$$

where (ijk) ranges over cyclic permutations of (123). The commutator terms are of course components of the curvature form of the connexion, and (3.1) is the self-dual Yang-Mills equation. Hence the moduli space of based instantons on $\mathbb{R}^4$ is a hyper-Kähler quotient of $\mathcal{A}_0$ by $\mathcal{G}_0$. Although $\mathcal{A}_0$ and $\mathcal{G}_0$ are infinite-dimensional, M is finite-dimensional, as alluded to above.

There are many variations on this theme. For example, one may replace Euclidean $\mathbb{R}^4$ by some other hyper-Kähler base manifold. Itoh [**48**] has considered moduli spaces of instantons on the two compact hyper-Kähler four-manifolds, the four-torus and the K3 surface. Instantons on ALE hyper-Kähler four-manifolds have been studied by Kronheimer and Nakajima [**61**]. In subsequent work, Nakajima has shown that the middle cohomology of these moduli spaces provides weight spaces for representations of affine Kac-Moody algebras [**74**].

Another modification, which has proved extremely fruitful, is to consider self-dual connexions invariant under some group of transformations of $\mathbb{R}^4$ preserving the hyper-Kähler structure. Some examples follow.

3.1. Nahm's Equations. These are the self-duality equations with $\mathbb{R}^3$ translation-invariance imposed. The equations then become the system of ordinary differential equations

$$\frac{dT_i}{dt} + [T_0, T_i] = [T_j, T_k], \tag{3.2}$$

for quadruples (T_0, T_1, T_2, T_3) of $\mathfrak{g}$-valued functions. As in (3.1), the triple (ijk) ranges over cyclic permutations of (123).

The simplest case is when T_i are smooth maps defined on a finite interval, say $[0, 1]$. The group of maps from $[0, 1]$ to G acts on solutions to (3.2) by gauge transformations

$$T_0 \mapsto \mathrm{Ad}(g)T_0 - \frac{dg}{dt}g^{-1}, \quad T_i \mapsto \mathrm{Ad}(g)T_i \quad (i = 1, 2, 3). \tag{3.3}$$

The quotient of the space of solutions by the group of gauge transformations vanishing at the endpoints is hyper-Kähler. We may see this by viewing the equations as the vanishing of a moment map for the action of this group on the space of maps $[0, 1] \to \mathfrak{g}^4$.

There are many modifications of this construction. One can have the Nahm matrices defined on a union of intervals, and impose boundary conditions at the endpoints. In particular, one often requires the Nahm matrices to have poles of prescribed residues at the endpoints of the intervals. In this case it is necessary to fix the residues for there to be a finite L^2 hyper-Kähler metric on the moduli space.

3.2. Self-Duality Equations on a Riemann Surface. If we look at self-dual G-connexions with $\mathbb{R}^2$ translation invariance, we get a system of equations involving a connexion A and a complex Higgs field Φ on $\mathbb{R}^2$. The Higgs field is regarded as a $(1,0)$-form taking values in the complex adjoint bundle associated to the principal G-bundle. These equations are conformally invariant, so also make sense on an arbitrary Riemann surface Σ. Hitchin [**43**] has carried out an intensive study of the resulting moduli space for $G = SU(2)$. If g is the genus of the Riemann surface, the moduli space M is a hyper-Kähler manifold of real dimension $12(g-1)$. For a special choice of complex structure, I say, M can be identified with a moduli space of stable Higgs pairs (V, Φ). Here V is a rank two holomorphic vector bundle on Σ, and Φ is a holomorphic section of $\mathrm{End}(V) \otimes K$, where K denotes the canonical bundle. The stability condition for such pairs is that the degree of any Φ-invariant rank one subbundle L of V should be less than half the degree of V. The cotangent bundle of the moduli space N of stable rank two bundles on Σ sits inside M as an open dense subset. We can regard N itself as the subset of M consisting of configurations with zero Higgs field. The stability condition for pairs reduces at such configurations to the ordinary stability condition for bundles.

There is an isometric circle action preserving I, defined by scalar multiplication of the Higgs field. The moment map for this action is just the L^2 norm squared of the Higgs field, and is a perfect Morse function on M. Hitchin uses this to get formulae for the Betti numbers of the moduli space. Note that the critical points for the Morse function, or equivalently, fixed points for the circle action, need not occur only when the Higgs field is zero. This is because we get a fixed point on the moduli space whenever (A, Φ) is *gauge equivalent* to $(A, e^{i\theta}\Phi)$ for all θ. The upshot is that as well as the critical submanifold N we also get critical submanifolds of higher index, and the Poincaré polynomial of M becomes quite complicated.

As with the Calabi examples, the other complex structures, which are all equivalent, make M into a Stein manifold. With respect to such a complex structure, the moduli space may be regarded as a space of representations of a central extension of $\pi_1(\Sigma)$ into $G_{\mathbb{C}}$ modulo conjugation. This is an example of a space with a natural complex-symplectic structure turning out to admit a hyper-Kähler metric.

Several authors have generalised these results to moduli spaces of parabolic Higgs bundles on Riemann surfaces (or equivalently orbifold Higgs bundles on orbifold Riemann surfaces) [**55, 75, 11**]. The dimension of the hyper-Kähler moduli spaces depends not just on the genus but also on the marked points and the parabolic data there. One obtains, by judicious choice of the various pieces of data, hyper-Kähler manifolds of dimension $4n$ for every n. In particular, taking Σ to be the Riemann sphere with four marked points gives a hyper-Kähler four-manifold fibering over $\mathbb{C}$. The generic fiber is an elliptic curve, and the fiber over the origin is the union of five rational curves intersecting like the extended Dynkin diagram of D_4. Hitchin has also constructed this manifold by twistor methods [**42**].

3.3. Monopoles. If instead we require $\mathbb{R}$-translation invariance, we obtain from the self-dual Yang-Mills equations the Bogomolny equations

$$*F_A = D_A \Phi, \tag{3.4}$$

for a connexion A and Higgs field Φ on $\mathbb{R}^3$. The Higgs field is a section of the real adjoint bundle associated to the principal G-bundle on which the connexion lives. Solutions to (3.4) with appropriate behaviour at infinity are called *monopoles*, and as with instantons we can study moduli spaces of these objects. Monopole moduli spaces are hyper-Kähler and admit an action of $SO(3)$, induced from spatial rotation in $\mathbb{R}^3$, which permutes the complex structures. Each complex structure identifies the moduli space with a space of based rational maps from the Riemann sphere into some flag manifold [**23, 47**].

In all these cases the moduli space has a nice description as a complex manifold. Formally, this is shown by the same kind of argument as in the finite-dimensional case, discussed in §2. One splits up the equations defining the moduli space into a real and a complex equation; in terms of the moment maps these are just $\mu_I = 0$ and $\mu_J + i\mu_K = 0$. The real equation is invariant under the action of the gauge group, but the complex equation is preserved by the action of the complexified gauge group. Then one shows that, subject to a stability condition, every orbit of the complexified gauge group in the set of solutions to the complex equations contains a solution to the real equations, unique up to the action of the real gauge group. The moduli space is thus identified with the quotient of $(\mu_J + i\mu_K)^{-1}(0)$ by the action of the complexified gauge group. In this infinite-dimensional setting, establishing the existence of a solution to the real equation in each orbit of stable points involves some analytical work.

For example, in the case of Nahm moduli spaces we may introduce the complex Nahm matrices

$$\alpha = T_0 - iT_1, \quad \beta = T_2 + iT_3.$$

Then the Nahm equations are equivalent to the real equation

$$\frac{d}{dt}(\alpha + \alpha^*) + [\alpha, \alpha*] + [\beta, \beta^*] = 0, \tag{3.5}$$

together with the complex equation

$$\frac{d\beta}{dt} + [\alpha, \beta] = 0. \tag{3.6}$$

In the case $G = SU(2)$ Donaldson showed by a variational argument that every solution to (3.6) can be transformed by a complex gauge transformation, unique up to the action of the real gauge group, into a configuration which also solves the real equation.

The next step, therefore, is to describe the solutions to the complex equation modulo (complex) gauge equivalence. Now, on the set where the Nahm matrices are smooth the complex equation may be trivialised in the sense that α may be gauged to zero, and hence β is gauged to a constant element of $\mathfrak{g}_{\mathbb{C}}$. The moduli space is in fact parametrised by an element of $\mathfrak{g}_{\mathbb{C}}$ and some further data which describes how the transformation gaugeing α to zero behaves at an endpoint of the interval. Donaldson uses this parametrisation to describe the moduli space as a set of rational maps. This result has been generalised to other groups by Hurtubise [**47**].

Moduli spaces of instantons on $\mathbb{R}^4$ are isometric, via the ADHM transform [**2, 66**], to moduli spaces of solutions to a system of matrix equations satisfying certain nondegeneracy conditions. These equations are in fact obtained by imposing $\mathbb{R}^4$-translation invariance on the self-dual Yang-Mills equations (3.1). Alternatively, they can be interpreted as the vanishing of a hyper-Kähler moment map for the action of a finite-dimensional group on a finite-dimensional vector space.

Similarly, the Nahm transform relates moduli spaces of monopoles and solutions to Nahm's equations [**72, 73**]. Given a monopole (A, Φ), one defines a family of Dirac operators coupled to (A, Φ), parametrised by a real number. This gives rise to an index bundle over a subset of $\mathbb{R}$, and projection from the trivial bundle defines a connexion which, written out in components, gives a solution to the Nahm equations. A similar procedure allows one to go in the reverse direction and build a monopole from Nahm matrices. The intervals on which the Nahm matrices are defined, and the boundary conditions, determine the gauge group and charge of the monopole. For example, in the case of $SU(2)$ monopoles the Nahm matrices are defined on a single interval and take values in $\mathfrak{u}(k)$, where k is the charge of the monopole. Moreover T_1, T_2, T_3 should have poles at the endpoints of the interval and their residues at each endpoint define the irreducible k-dimensional representation of $SU(2)$. Unlike the situation with instantons, as yet no finite-dimensional procedure for obtaining monopole spaces is known.

Monopole moduli spaces have been the focus of particular attention, because of the Manton programme for deducing information about monopole dynamics from geodesic flow on the moduli spaces [**67**]. In the simplest case, when $G = SU(2)$, the moduli space of charge k monopoles is $4k$-dimensional and is biholomorphic to the space of based rational maps of degree k from the Riemann sphere to itself. If $k = 1$ we just get the flat hyper-Kähler manifold $\mathbb{R}^3 \times S^1$, but for higher k the metric is nontrivial. After taking a cover, the moduli space splits as a product of $\mathbb{R}^3 \times S^1$ and a $(4k-4)$-dimensional hyper-Kähler manifold $\tilde{M}_k^0$ which is the universal cover of the moduli space M_k^0 of monopoles with fixed centre. Of course, this factor contains the interesting geometry.

Atiyah and Hitchin [**3**] studied the $k = 2$ case in detail. This example is particularly interesting because the metric can be found completely explicitly. This is possible because M_2^0 is now four-dimensional and, as mentioned earlier, admits an isometric action of $SO(3)$ permuting the Kähler structures. The orbits are generically three-dimensional, so the metric is of Bianchi IX type. Writing the metric as

$$(abc)^2 dt^2 + a^2\sigma_1^2 + b^2\sigma_2^2 + c^2\sigma_3^2 \tag{3.7}$$

it turns out that the coefficients a, b, c must satisfy the equation

$$2\frac{a'}{a} = (b-c)^2 - a^2 \tag{3.8}$$

and its cyclic permutations. (We can compare this with the equations (1.3) considered by Belinskii, Gibbons, Page and Pope [**5**] in their study of Bianchi IX hyper-Kähler metrics where the action preserves each complex structure individually).

The equations (3.8) are more difficult than those of [**5**], but they can be solved in terms of elliptic functions. There are three solutions corresponding to complete metrics; one gives flat space, one the Taub-NUT metric (2.3), and the last and most complicated gives the Atiyah-Hitchin moduli space metric. This metric is triaxial, but in the asymptotic region of M_2^0 approaches a $U(2)$-invariant metric. This is the so-called negative-mass version of the Taub-NUT metric. Although well-behaved in the asymptotic region, negative-mass Taub-NUT becomes singular in the core region, in contrast both to the Atiyah-Hitchin metric and the version of Taub-NUT given in (2.3). Of course, in the core region, negative mass Taub-NUT is no longer a good approximation to the Atiyah-Hitchin metric, which is smooth and complete.

In fact, Manton has shown that the asymptotic behaviour of the Atiyah-Hitchin metric can be deduced from the dynamics of well-separated monopoles [**68**]. This is a reversal of the usual procedure, in which geometric information is used to study the dynamics.

The underlying manifold of the Atiyah-Hitchin metric is diffeomorphic to $S^4 - \mathbb{RP}^2$. In fact, this identification can be made $SO(3)$-equivariant. We regard S^4 as the sphere in the real irreducible five-dimensional representation of $SO(3)$, which we can take concretely to be the set of traceless symmetric real three-by-three matrices with $SO(3)$ acting by conjugation. The generic stabiliser of the induced $SO(3)$ action on S^4 is just conjugate to the viergruppe of diagonal matrices in $SO(3)$, but there are two special orbits, obtained by considering matrices with two equal positive or two equal negative eigenvalues. Each special orbit is a copy of $\mathbb{RP}^2$. One of these is the $\mathbb{RP}^2$ we remove to obtain the Atiyah-Hitchin manifold, while the other is the unique special orbit in the Atiyah-Hitchin space. Physically, this orbit represents the set of axisymmetric monopoles; there is a unique such monopole for each choice of axis in $\mathbb{R}^3$.

Many other examples of hyper-Kähler manifolds may be obtained by considering Nahm or monopole moduli spaces. A particularly useful example is found by considering $\mathfrak{g}$-valued Nahm matrices smooth on a closed interval. Then, as observed by Kronheimer, the hyper-Kähler moduli space is biholomorphic to the cotangent bundle of $G_{\mathbb{C}}$. The complex description is especially simple here, because the smoothness of the Nahm matrices on the whole interval means that we can globally trivialise the complex Nahm equation–after a gauge transformation, its solutions are constant. The moduli space admits two commuting actions of G, obtained by considering gauge transformations which are the identity at one end of the interval but not the other. The moment maps are just evaluation of the Nahm matrices at the respective ends of the interval. Taking hyper-Kähler quotients of $T^*G_{\mathbb{C}}$ by a subgroup H of one copy of G leads [**20**] to a complete hyper-Kähler metric on a neighbourhood of the zero section in the vector bundle $G \times_H (\mathfrak{m}^* \otimes \mathbb{R}^3)$ over G/H, where $\mathfrak{m}$ is an $\mathrm{Ad}\, H$-invariant complement for $\mathfrak{h}$ in $\mathfrak{g}$. These metrics are hyper-Kähler analogues of the symplectic examples of Guillemin and Sternberg [**37**].

One can obtain a one-parameter family of deformations of the double cover $\tilde{M}_2^0$ of the Atiyah-Hitchin space by taking a hyper-Kähler quotient of a moduli space of $SU(3)$ monopoles by a circle action [**19**]. Making the deformation breaks the $SO(3)$ symmetry of $\tilde{M}_2^0$ down to a circle action. This action, although isometric, does not preserve all the complex structures, so the metric is not one of the Gibbons-Hawking family. The infinitesimal variation of the Kähler forms of the deformation is controlled by an $SO(3)$-invariant anti-self-dual two-form on $\tilde{M}_2^0$. This two-form can also be regarded as the curvature form on the pullback to $\tilde{M}_2^0$ of the index bundle on M_2^0, whose fiber over a monopole is the kernel of the Dirac operator associated to that monopole as in the Nahm transform [**72**]. The index bundle plays a key role in the study by Manton and Schroers of monopoles coupled to fermions [**69**]. In their work on the quantum dynamics of such particles, the wavefunction is interpreted as a section of the index bundle.

One can also construct a series of metrics on deformations of the Kleinian singularity corresponding to the dihedral group D_n for $n \geq 2$. R. Kobayashi showed the existence of such metrics by analytical methods [**54**]. Alternatively [**18**], one may replace an $\mathbb{H}^4$ factor in Kronheimer's construction for dihedral ALE spaces by a copy of $T^*GL(2,\mathbb{C})$ with the hyper-Kähler structure mentioned above. This is analogous to obtaining the multi-Taub-NUT family by replacing an $\mathbb{H}$ factor by an $\mathbb{R}^3 \times S^1$ in the construction of the multi-instanton series. This dihedral series has been further studied by Houghton [**46**], from the Nahm viewpoint, and by Chalmers [**16**], using twistor methods.

In dimension four, one may also mention examples of infinite topological type constructed in [**1, 31**] and [**32**]. These may be viewed as A_∞ and D_∞ limits of the ALE spaces, but, although complete, are not themselves ALE. They can be obtained using an infinite-dimensional hyper-Kähler quotient, or, in the A_∞ case, by a modification of the Gibbons-Hawking method to allow infinitely many points p_i in (1.5). In order to obtain a smooth, complete metric, the sequence of points must go off to infinity fast enough to ensure that the series $\sum \mid \underline{x} - p_i \mid^{-1}$ converges. The D_∞ examples can be realised as desingularisations of the quotient of the A_∞ spaces by an involution. The examples of [**1**] were the first examples of complete Ricci-flat Kähler manifolds of infinite topological type. The last property means that they cannot live on the complement of a divisor in a projective variety, answering in the negative a question posed by Yau.

As a final illustration of gauge theory methods, let us mention an example involving loop spaces. The space ΩG of based loops in a compact Lie group can be made into an infinite-dimensional symplectic (in fact Kähler) manifold [**79**]. Donaldson has shown that the corresponding complex-symplectic structure on $\Omega G_{\mathbb{C}}$ in fact comes from a hyper-Kähler structure on this infinite-dimensional space. The proof relies on interpreting $\Omega G_{\mathbb{C}}$ as a hyper-Kähler quotient of the space of $G_{\mathbb{C}}$ connexions on the trivial $G_{\mathbb{C}}$-bundle over the Poincaré disc by G-valued gauge transformations equal to the identity on the boundary. There is a detailed discussion in Hitchin's survey article [**44**].

4. Coadjoint Orbits

One particularly interesting family of higher-dimensional hyper-Kähler spaces is that of coadjoint orbits for complex semisimple Lie groups. It is well known that a coadjoint orbit of a compact Lie group G carries a symplectic structure. Explicitly, this is defined by the Kirillov-Kostant-Souriau form, given at $\beta \in \mathfrak{g}^*$ by

$$\omega(X^*(\beta), Y^*(\beta)) = \beta([X,Y]) \tag{4.1}$$

where X^*, Y^* are the vector fields on the orbit associated by the action to $X, Y \in \mathfrak{g}$. Similarly, a coadjoint orbit of the complexification $G_{\mathbb{C}}$ carries a complex-symplectic form. But more is true, it in fact admits a compatible hyper-Kähler metric.

The first work on this subject was done by Kronheimer [**59, 60**]. The central idea is to realise the coadjoint orbit as a moduli space of solutions to Nahm's equations. This space carries a hyper-Kähler structure via an infinite-dimensional quotient construction, as discussed in the preceding section.

As an example, let us consider the case of regular semsimple orbits. Let (τ_1, τ_2, τ_3) be a triple of elements in a Cartan subalgebra $\mathfrak{t}$ of $\mathfrak{g}$, and suppose that the intersection of their stabilisers is precisely the corresponding maximal torus. Kronheimer considers the Nahm equations for quadruples (T_0, T_1, T_2, T_3) of functions from $(-\infty, 0]$ to $\mathfrak{g}$. The gauge group of functions from $(-\infty, 0]$ to G acts on solutions as in (3.3). Clearly $(0, \tau_1, \tau_2, \tau_3)$ is a constant solution to the equations. We can obtain other constant solutions by applying the action of a constant element of G to this configuration. Kronheimer now looks at a moduli space M of solutions which, after conjugation by a constant element of G, approach $(0, \tau_1, \tau_2, \tau_3)$ exponentially fast as $t \to -\infty$. The group by which we quotient consists of gauge transformations which are the identity at 0 and have suitable decay at $-\infty$. The equations can then be interpreted, as usual, as the vanishing condition for the hyper-Kähler moment map for this gauge group. The analysis by Donaldson of the Nahm equations, as discussed in §3, carries over to give a description of a complex structure on the moduli space. It turns out that for a suitable choice of τ_2, τ_3, the moduli space M is biholomorphic (for some complex structure) to the orbit of $G_{\mathbb{C}}$ through $\tau_2 + i\tau_3$. This orbit may be viewed as a homogeneous space $G_{\mathbb{C}}/T_{\mathbb{C}}$ and the subset of constant Nahm matrices forms a copy of G/T inside the orbit. The Eguchi-Hanson space is the special case $G = SU(2)$. Burns [**14**], using twistor methods, had earlier shown the existence of a hyper-Kähler metric on a neighbourhood of G/T inside $G_{\mathbb{C}}/T_{\mathbb{C}}$ but Kronheimer's method also tells us the metric is complete.

Kronheimer also showed that nilpotent orbits admit a hyper-Kähler structure. More recently, Biquard and Kovalev [**9, 56**] have independently shown that *any* coadjoint orbit of $G_{\mathbb{C}}$ is hyper-Kähler. The basic strategy is the same as before–to identify the orbit with a suitable Nahm moduli space. One now considers triples (τ_1, τ_2, τ_3) in the Cartan algebra $\mathfrak{t}$ such that the Lie algebra $\mathfrak{h}$ of their common centraliser is allowed to strictly include $\mathfrak{t}$. The correct moduli space to work with consists of solutions to the Nahm equations with T_i asymptotic (possibly after applying the action of G) to $\tau_i + \rho(e_i)t^{-1}$ for $i = 1, 2, 3$. Here e_1, e_2, e_3 form a standard basis for $\mathfrak{su}(2)$, and ρ is a homomorphism from $\mathfrak{su}(2)$ to $\mathfrak{h}$. If ρ is zero we get a semisimple orbit. The nilpotent orbits, on the other hand, are obtained by setting each τ_i to zero. In this case the orbit we get is the one containing $\rho(e_2 + ie_3)$, where we have extended ρ to a homomorphism defined on $\mathfrak{sl}(2, \mathbb{C})$. This illustrates the principle that nilpotent orbits in $\mathfrak{g}_{\mathbb{C}}$ correspond to conjugacy classes of homomorphisms $\rho : \mathfrak{su}(2) \to \mathfrak{g}$.

The hyper-Kähler metrics obtained are complete if and only if the orbit is semisimple. These orbits are diffeomorphic to the cotangent bundles of flag manifolds for G. In fact for a suitable choice of τ_i they are even biholomorphic, though of course the complex structure here is not the same as that which comes from viewing the manifold as a $G_{\mathbb{C}}$ orbit. Donaldson and Kronheimer have an alternative construction of these hyper-Kähler structures, using gauge theory in two dimensions. In the special case when the flag manifold is a hermitian symmetric space, it is possible to construct the metric more explicitly without the use of gauge-theoretic or twistor methods [**10, 21**]. One also obtains in this way an *incomplete* hyper-Kähler metric on a neighbourhood of the zero section in the cotangent bundle of *noncompact* hermitian symmetric spaces.

If G is a classical group, the nilpotent orbits of $G_{\mathbb{C}}$ may also be obtained by a finite-dimensional hyper-Kähler quotient construction [**53**]. The relevant quotients are singular, but they have a smooth open dense set, which may be identified with a nilpotent orbit.

Nilpotent orbits have a particularly nice interpretation in terms of quaternionic Kähler geometry. Swann [**82**] has shown that given a quaternionic Kähler manifold N with positive scalar curvature, one can define a hyper-Kähler manifold $\mathcal{U}(N)$ whose underlying space is the total space of a $\mathbb{H}^*$- or $(\mathbb{H}^*/\mathbb{Z}_2)$-bundle over N. Now $\mathcal{U}(N)$ admits a free isometric action of $Sp(1)$ or $SO(3)$ permuting the complex structures. Moreover, if A is a complex structure and X_A the Killing field of the circle subgroup of $Sp(1)$ or $SO(3)$ fixing A, then AX_A is independent of A. Conversely, any hyper-Kähler manifold with such an action is $\mathcal{U}(N)$ for some quaternionic Kähler space N. As well as the $Sp(1)$ action, $\mathcal{U}(N)$ also admits a homothetic action of $\mathbb{R}^*$ whose conformal Killing field is AX_A. The manifolds $\mathcal{U}(N)$ may also be characterised as the hyper-Kähler manifolds which admit a hyper-Kähler potential, that is, a function $f : M \to \mathbb{R}$ which is a Kähler potential for each Kähler structure simultaneously.

It turns out that all nilpotent orbits are associated to a quaternionic Kähler manifold via the Swann construction. In terms of the Nahm data description of the nilpotent orbit, the $Sp(1)$ action is just the natural action on triples (T_1, T_2, T_3), while the homothetic action is the rescaling which replaces $T_i(t)$ by $cT_i(ct)$. Note that the boundary conditions at $-\infty$ prevent such actions from occurring in the non-nilpotent case. We can partially order the nontrivial nilpotent orbits of $G\mathbb{C}$ by saying $\mathcal{O}_1 \leq \mathcal{O}_2$ if and only if $\mathcal{O}_1$ is contained in the closure of $\mathcal{O}_2$. There is a unique nontrivial nilpotent orbit minimal with respect to this ordering, and it may be identified with $\mathcal{U}(N)$ where N is the Wolf space [**90**] associated to G. In general, the quaternionic Kähler manifold associated to a nilpotent orbit has a description due to Kobak and Swann [**53**], [**83**]. They define a functional on the Grassmannian $\mathrm{Gr}_3(\mathfrak{g})$ of oriented 3-planes in $\mathfrak{g}$ by $\psi(V) = \langle e_1, [e_2, e_3] \rangle$, where e_1, e_2, e_3 is an oriented orthonormal basis for V. As mentioned above, a nilpotent orbit in $\mathfrak{g}\mathbb{C}$ corresponds to a conjugacy class in G of homomorphisms $\rho : \mathfrak{su}(2) \to \mathfrak{g}$. The associated quaternionic Kähler manifold can then be identified with the stable manifold for ψ of the critical set containing the image of ρ.

Monopole moduli spaces, even though they have a Nahm description and hence admit an $Sp(1)$ action permuting the complex structures, do not fit into the above picture. In fact they do not admit the homothetic $\mathbb{R}^*$-action, as the intervals on which the Nahm matrices are defined are now finite. However, moduli spaces of instantons on $\mathbb{R}^4$ do arise from the Swann construction [**66**].

5. Twistors

The Penrose twistor construction [**78**] has been very successful in translating problems concerning self-dual four-manifolds into questions about the complex geometry of the twistor space. More recently it was realised [**80**] that the twistor construction could be generalised to higher dimensions for manifolds with a quaternionic structure.

A manifold M is *quaternionic* if there is a rank three subbundle $\mathcal{G}$ of $\mathrm{End}(TM)$, preserved by a torsion-free connexion, and locally spanned by sections I, J, K satisfying the relations $I^2 = J^2 = K^2 = -1$, $IJ = K$ etc. The twistor space is now the sphere bundle of $\mathcal{G}$. This space is a complex manifold in a natural way, and the fibers of the twistor projection $\pi : Z \to M$ (the *twistor lines*) are rational curves with normal bundle $2n\mathcal{O}(1)$, where $4n$ is the real dimension of M. There is also a *real structure* τ, an antiholomorphic involution defined by taking the antipodal map on the twistor lines.

If M is hyper-Kähler, then the almost complex structures exist globally, and are integrable. This means that the twistor fibration π is trivial, and that I, J, K define divisors in Z. We may regard Z as the smooth manifold $M \times \mathbb{CP}^1$, with complex structure given at (m, a, b, c) by $(aI + bJ + cK, I_0)$ where I_0 is the standard complex structure on S^2. As well as the twistor fibration, we have a holomorphic fibration $p : Z \to \mathbb{CP}^1$, whose fiber over $(a, b, c) \in \mathbb{CP}^1$ is just M endowed with the complex structure $aI + bJ + cK$. Each point of M defines a section of this fibration.

Of course this picture will arise even if we just have a hyperhermitian structure, that is, a triple of integrable hermitian structures multiplying like the quaternions. The extra information of a hyper-Kähler metric is encoded in a holomorphic section of the second exterior power of the cotangent bundle to the fibers of p, twisted by $p^*\mathcal{O}(2)$. Explicitly, this is given by

$$\omega = \omega_J + i\omega_K - 2\zeta\omega_I - \zeta^2(\omega_J - i\omega_K),$$

where ζ is an affine coordinate on $\mathbb{CP}^1$.

THEOREM 5.1. **[45]** *Let Z be the twistor space of a hyper-Kähler $4n$-manifold M. Then*

(i) *there is a holomorphic fibration $p : Z \to \mathbb{CP}^1$;*
(ii) *the fibration has a family of rational sections with normal bundle $2n\mathcal{O}(1)$;*
(iii) *there is a holomorphic section ω of $\Lambda^2 T_F^* \otimes p^*\mathcal{O}(2)$, defining a symplectic form on each fiber (T_F denotes the tangent bundle to the fibers); and*
(iv) *there is an antiholomorphic involution τ of Z, compatible with* (i–iii) *and inducing the antipodal map on $\mathbb{CP}^1$.*

Conversely any such Z is the twistor space of a hyper-Kähler $4n$-manifold M. This manifold may be identified with the set of real sections of the fibration p.

The procedure for recovering the hyper-Kähler metric from the twistor data goes as follows. Elements of the complexified tangent space to M at m may be regarded as sections of the normal bundle to the twistor line P_m over m. But $H^0(P_m, N)$ can be identified with $H^0(P_m, T_F)$, which in turn may be identified with $H^0(P_m, T_F(-1)) \otimes H^0(P_m, \mathcal{O}(1))$, since $T_F \cong 2n\mathcal{O}(1)$. The $\mathcal{O}(2)$-valued form ω defines a symplectic structure on the first factor, while we define a symplectic form on the second by $\langle a_1 + b_1\zeta, a_2 + b_2\zeta\rangle = a_1 b_2 - a_2 b_1$. The upshot is that we have a symmetric complex bilinear form on $T_m M \otimes \mathbb{C}$, and using the real structure we obtain a positive definite inner product on the real tangent space.

The twistor construction is compatible with the hyper-Kähler quotient in the following way. If G acts on M preserving the hyper-Kähler structure, then we can lift to a holomorphic action of G on Z preserving the twisted complex-symplectic form ω. In good cases, this action extends to a global action of the complexified group $G_{\mathbb{C}}$ and we obtain a moment map μ, which is a section of $\mathcal{O}(2) \otimes \mathfrak{g}_{\mathbb{C}}^*$. The quotient $\mu^{-1}(0)/G_{\mathbb{C}}$ is the twistor space of the hyper-Kähler quotient of M by G.

The simplest example of a twistor space is when $M = \mathbb{H}^n$. Then Z is the total space of $2n\mathcal{O}(1) \to \mathbb{CP}^1$. Many other examples, including the twistor space of the Taub-NUT metric, are worked out in **[42, 6, 45]**. As twistor spaces of hyper-Kähler manifolds fiber over $\mathbb{CP}^1$, they may locally be constructed by patching together copies of $\mathbb{C}^{2n} \times U$ and $\mathbb{C}^{2n} \times \tilde{U}$, where $U, \tilde{U}$ are the standard patches on the Riemann sphere. Often it is convenient to work with a Hamiltonian function for a vector field which exponentiates to give the patching diffeomorphism. A particularly nice case is that of hyper-Kähler $4n$-manifolds with an isometric action of T^n preserving each Kähler structure. The metric is then determined by a solution to a system of linear equations on $\mathbb{R}^{3n}$, and these functions are given in terms of a contour integral involving the Hamiltonian function. Alternatively, one can express the Kähler potential of the metric in terms of this integral. The Gibbons-Hawking metrics are the $n = 1$ case of this theory. As mentioned in §2, the toric hyper-Kähler manifolds considered in [**8**] provide a large class of complete examples where these techniques may be applied.

The twistor method is more general than that of hyper-Kähler quotients, although it has the disadvantage that global properties of the metric are difficult to check in the twistor picture. A recent use of twistor techniques is the result of B. Feix [**26**] that, given a real-analytic Kähler manifold N, a hyper-Kähler metric will exist on some neighbourhood of the zero section in T^*N. This has also been proved independently by Kaledin, using different methods [**51**]. The metric is compatible with the standard complex-symplectic structure on T^*N, and is preserved by the natural circle action obtained by scalar multiplication in each cotangent space. Again, we have an example of an obviously complex-symplectic manifold in fact admitting a hyper-Kähler structure. This metric will not, however, always be complete. Indeed, the Milnor-Wolf theorem shows that a necessary condition for completeness is that finitely generated subgroups of $\pi_1(N)$ must have polynomial growth.

6. Compact Examples

The preceding techniques have usually lead to non-compact manifolds. In this section, we shall consider the compact case, where fewer examples are known.

The two basic examples are the four-torus with a flat metric and a $K3$ surface with the Yau metric. These are, in fact, the only compact four-dimensional hyper-Kähler manifolds. Note that any such manifold is Ricci-flat and anti-self-dual, so gives equality in the Hitchin-Thorpe inequality $3|\tau| \leq 2\chi$ for compact Einstein four-manifolds [**40**]. Hitchin' s list of examples where equality is attained also includes quotients of the above examples, but these are only *locally* hyper-Kähler.

Of course, the existence of a hyper-Kähler metric on $K3$ follows from Yau's proof of the Calabi conjecture, together with the fact that in dimension four a Ricci-flat Kähler metric is hyper-Kähler. There are other interesting ways of viewing this hyper-Kähler metric, however.

The basic idea goes back to Page's 1978 article [**76**]. A classical construction due to Kummer shows that some $K3$ surfaces may be obtained by quotienting the four-torus by an involution and resolving the singularities. Each of the 16 singularities gives rise to a (-2) curve in the $K3$. Page suggested that this picture could be made to work on the Riemannian level. The idea was to glue copies of the Eguchi-Hanson metric on $T^*\mathbb{CP}^1$ to the flat metric on the complement of the singularities in $T^4/\mathbb{Z}_2$, and then smooth out so as to obtain a hyper-Kähler metric on $K3$.

More recently LeBrun and Singer [**64**] gave a rigorous treatment of this idea (see also [**84**]). They use a generalisation to the orbifold case of the Donaldson-Friedman technique for producing self-dual metrics on connected sums [**24**]. In the case of the $K3$ the orbifolds are $T^4/\mathbb{Z}_2$ and 16 copies of the one-point compactification of Eguchi-Hanson, and the resulting singular twistor space can be deformed to a smooth twistor space corresponding to a hyper-Kähler metric on the $K3$. Geometrically, one obtains sequences of $K3$ metrics which converge on an open dense set to the flat metric on $T^4/\mathbb{Z}_2$. Suitable rescalings of the metric at the fixed points converge to Eguchi-Hanson in a bubbling-off phenomenon.

A Kummer-type construction was used by Joyce to construct compact manifolds with exceptional holonomy in dimension 7 and 8 (see [**49, 50**] and the article in this volume). In this case twistor arguments are replaced by direct analytical methods. Joyce has also given a proof along these lines of the existence of a hyper-Kähler metric on the K3 surface.

Compact hyper-Kähler manifolds cannot arise from a finite-dimensional hyper-Kähler quotient [**44**]. It remains an open question whether the $K3$ surface can be produced as a quotient in an infinite-dimensional setting, for example as a gauge theory moduli space. In this connection, it is interesting to note that Braam, Maciocia and Todorov [**13**] have shown that some hyper-Kähler structures on $K3$ surfaces may be obtained by realising $K3$ as a desingularisation of a compactification of a moduli space of instantons on the four-torus. More precisely, they consider the moduli space of $SO(3)$-instantons with $p_1 = -4$ and $(w_2)^2 \equiv 0 \bmod 4$. This is an eight-dimensional space, but it admits an action of the four-torus induced by translations on the base manifold. The four-dimensional quotient $\mathcal{M}$ has some $\mathbb{Z}_2$ orbifold singularities arising from finite stabilisers of the action. There may also be orbifold singularities due to reducible instantons. Finally, compactifying $\mathcal{M}$ introduces a further orbifold singularity corresponding to the quaternion group of order eight. Resolving these singularities yields a K3 surface, which is not in general the Kummer surface of the torus.

Higher dimensional irreducible hyper-Kähler manifolds have been constructed from the four-dimensional examples by Fujiki [**27**] in dimension 8, and later in all dimensions by Mukai [**71**] and Beauville [**4**]. We describe Beauville's method here.

If S is a compact complex manifold, we let $S^{[r]}$ denote the Hilbert scheme whose points are finite analytic subschemes of S of length r. There is a birational map between $S^{[r]}$ and the rth symmetric product of S. If S has dimension one, the symmetric powers are nonsingular, and the map is an isomorphism. If S has dimension two, the symmetric powers are no longer smooth, but $S^{[r]}$ is a desingularisation. We shall take S to be two-dimensional from now on.

Beauville shows that if S admits a complex-symplectic form, so does $S^{[r]}$. The corresponding statement is also true for Kähler structures, so, as S is assumed to be compact, it follows that whenever S is hyper-Kähler, so is $S^{[r]}$. Applying this construction to the $K3$ surface, Beauville obtains compact simply-connected irreducible hyper-Kähler manifolds of dimension $4r$ for each r. The case $r = 2$ was also found by Fujiki, who obtained it as a quotient by an involution of the manifold obtained by blowing up the diagonal in a product of two copies of $K3$. If instead we start with the four-torus, we get a family of spaces which are locally reducible. However, they admit a map onto the four-torus, defined by the group law, and the fiber is a compact, simply-connected irreducible hyper-Kähler manifold of dimension $4r-4$. If $r = 2$ this is a $K3$ surface. There are also hyper-Kähler deformations of the members of these two families of irreducible hyper-Kähler manifolds.

It is interesting to note that the moduli space of charge r $SU(2)$ monopoles may be viewed as $S^{[r]}$ where S is the flat *noncompact* hyper-Kähler manifold $\mathbb{R}^3 \times S^1$, the charge one moduli space. The physical interpretation is that monopoles are not pure point particles, so the charge r moduli space is not the rth symmetric power of the charge one moduli space, but rather a smoothed out version of this space.

Guan's construction of compact complex-symplectic manifolds admitting no hyper-Kähler metric starts by applying Beauville's method to a Kodaira surface [**33**]. The resulting manifolds inherit their complex-symplectic structure from the Kodaira surface, but, like this surface, have odd first Betti number so admit no Kähler metric. Guan constructs further examples by resolving singularities of complex-symplectic quotients of his initial examples, and by taking deformations [**34**].

A fair amount is now known about the topology of compact hyper-Kähler manifolds. One early observation [**4**] was that compact hyper-Kähler manifolds of dimension $4n$ with holonomy *equal* to $Sp(n)$ are simply connected. Ricci-flatness, and an argument involving the Cheeger-Gromoll splitting theorem, implies that the universal cover is compact. The Bochner argument shows that every holomorphic form on M is parallel, so the holonomy assumption implies that the only holomorphic forms are scalar multiples of powers of the complex-symplectic form ω. The same assertion holds on $\tilde{M}$, so M and $\tilde{M}$ have the same holomorphic Euler characteristic. As this quantity is multiplicative under covers, the assertion follows.

Wakakuwa and Fujiki studied the Hodge and Betti numbers of compact hyper-Kähler manifolds [**89, 28**]. In particular, Fujiki observed that the Hodge numbers satisfy the relation $h^{p,q} = h^{2m-p,q}$, $(0 \leq p, q \leq 2m)$, where $2m$ is the complex dimension of the manifold. In fact, wedging with the $(m-p)^{\text{th}}$ power of the complex-symplectic form provides an isomorphism between the relevant spaces.

Recently Salamon [**81**] has discovered some interesting relations between the Betti numbers of a compact hyper-Kähler manifold. Riemann-Roch techniques yield identities relating Hodge and Chern numbers on a compact Kähler manifold [**65**], and in the Ricci-flat case we have $c_1 = 0$ so a relation on the Hodge numbers follows. For hyper-Kähler manifolds, Fujiki's result allows this to be rewritten as the following equation satisfied by the *Betti* numbers of M:

$$2\sum_{j=1}^{2m}(-1)^j(3j^2 - m)b_{2m-j} = mb_{2m}.$$

Salamon deduces that the Euler characteristic, signature and middle Betti number are even unless the real dimension of M is divisible by 32. Another consequence is that m times the Euler characteristic is divisible by 24. A familiar illustration of these relations is the $K3$ surface, when $m = 1$, $b_1 = 0$, $b_2 = 22$, $\chi = 24$ and $\tau = 16$. As an example in higher dimensions, we may consider the eight-dimensional manifold with holonomy $Sp(2)$ obtained from the four-torus by the Beauville construction. The Poincaré polynomial of this space is $1 + 7t^2 + 8t^3 + 108t^4 + 8t^5 + 7t^6 + t^8$, and the signature is 84.

Verbitsky has also studied the cohomology of compact hyper-Kähler manifolds [**85, 86, 87, 88**]. As the complex structures are parallel they define an action of $\mathfrak{su}(2)$ on $H^*(M)$, which one may integrate to an action of $SU(2)$. Classes which are invariant under this action are of particular importance. If N is a complex submanifold of M with respect to one complex structure I, and if the class represented by the Poincaré dual of (N, I) is invariant, then in fact N is a hyper-Kähler submanifold. Verbitsky uses this to show that for generic elements I of the two-sphere of complex structures, any I-complex submanifold is hyper-Kähler. An interesting consequence is that for such a complex structure M has no divisors, so (M, I) is not projective algebraic,

It is interesting to contrast compact hyper-Kähler manifolds with compact quaternionic Kähler manifolds of positive scalar curvature. In the latter case considerable progress has been made in classification using twistor methods, exploiting the Fano property of the twistor space [**63**]. For hyper-Kähler manifolds, whose twistor spaces do not have such nice properties, much less is currently known about the classification problem.

References

[1] M. Anderson, P. B. Kronheimer and C. LeBrun. *Complete Ricci-flat Kähler manifolds of infinite topological type*, Commun. Math. Phys. **125** (1989) 637–642

[2] M. F. Atiyah, V. Drinfeld, N. J. Hitchin and Y. Manin. *Construction of instantons*, Phys. Lett. A **65** (1978) 185–187.

[3] M. F. Atiyah and N.J. Hitchin. *The geometry and dynamics of magnetic monopoles.* Princeton University Press, Princeton, 1988.

[4] A. Beauville. *Variétés Kähleriennes dont la première classe de Chern est nulle*, J. Diff. Geometry **18** (1983) 755–782

[5] V. Belinskii, G.W. Gibbons, D. Page and C. Pope. *Asymptotically Euclidean Bianchi IX metrics in quantum gravity*, Phys. Lett. B **76** (1978) 433–435

[6] A. Besse. *Einstein manifolds*, Springer-Verlag, Berlin, 1987.

[7] R. Bielawski. *Betti numbers of 3-Sasakian quotients of spheres by tori*, Bull. Lond. Math. Soc. **29** (1997) 731–736

[8] R. Bielawski and A.S. Dancer. *The geometry and topology of toric hyper-Kähler manifolds*, Communications in Analysis and Geometry, to appear.

[9] O. Biquard. *Sur les équations de Nahm et la structure de Poisson des algèbres de Lie semisimples complexes*, Math. Ann. **304** (1996) 253–276

[10] O. Biquard and P. Gauduchon. *Hyper-Kahler metrics on cotangent bundles of hermitian symmetric spaces*, in J.E. Andersen et al, ed. "Geometry and Physics" (Aarhus 1995), Lecture Notes in Pure and Applied Mathematics vol. 184, Marcel Dekker, New York, 1997, pp. 287–298.

[11] H. Boden and K. Yokogawa. *Moduli spaces of parabolic Higgs bundles and parabolic $K(D)$ pairs over smooth curves I*, Int. J. Math. **7** (1996) 573–598

[12] C. P. Boyer, K. Galicki and B. M. Mann. *Quaternionic reduction and Einstein manifolds*, Communications in Analysis and Geometry **1** (1993) 229–279

[13] P. Braam, A. Maciocia, A. Todorov. *Instanton moduli as a novel map from tori to K3 surfaces*, Invent. Math. **108** (1992) 419–451

[14] D. Burns. *Some examples of the twistor construction*, in A. Howard and P.-M. Wong ed. "Contributions to several complex variables: in honour of Wilhelm Stoll" Vieweg, Braunschweig, 1986, pp. 51–65

[15] E. Calabi. *Métriques kählériennes et fibrés holomorphes*, Ann. Sci. Éc. Norm. Sup. **12** (1979) 269–294

[16] G. Chalmers. *The implicit metric on a deformation of the Atiyah-Hitchin manifold*, Phys. Rev. D **58** (1998) 125011

[17] S-Y. Cheng and S-T. Yau. *On the existence of a complete Kähler metric on non-compact complex manifolds and the regularity of Fefferman's equation*, Comm. Pure and Appl. Math. **33** (1980) 507–544

[18] A. S. Dancer. *Dihedral singularities and gravitational instantons*, J. Geom. Phys. **12** (1993) 77–91

[19] A. S. Dancer. *A family of hyper-Kähler manifolds*, Quarterly Journal of Mathematics **45** (1994) 463–478

[20] A.S. Dancer and A. F. Swann. *Hyper-Kähler metrics associated to compact Lie groups*, Math. Proc. Cam. Phil. Soc. **120** (1996) 61–69

[21] A. S. Dancer and R. Szőke. *Symmetric spaces, adapted complex structures and hyper-Kähler structures*, Quarterly Journal of Mathematics **48** (1997) 27–38

[22] T. Delzant. *Hamiltoniens périodiques et images convexes de l'application moment*, Bull. Soc. Math. France **116** (1988) 315–339

[23] S. K. Donaldson. *Nahm's equations and the classification of monopoles*, Commun. Math. Phys. **96** (1984) 387–407

[24] S. K. Donaldson and R. Friedman. *Connected sums of self-dual manifolds and deformations of singular spaces* Nonlinearity **2** (1989) 197–239

[25] T. Eguchi and A.J. Hanson. *Asymptotically flat self-dual solutions to Euclidean gravity*, Phys. Lett. B **74** (1978) 249–251

[26] B. Feix, Cambridge dissertation (1997)

[27] A. Fujiki. *On primitively symplectic compact Kähler V-manifolds of dimension four*, in K. Ueno, ed. "Classification of algebraic and analytic manifolds", Birkhäuser, 1983, 71–250.

[28] A. Fujiki. *On the de Rham cohomology group of a compact Kähler symplectic manifold*, Algebraic Geometry, Sendai 1985, Adv. Stud. Pure Math. **10** (1987) 105–165

[29] G. W. Gibbons and S.W. Hawking. *Gravitational multi-instantons*, Phys. Lett. B **78** (1978) 430–432

[30] R. Goto. *On toric Hyper-Kähler manifolds given by the Hyper-Kähler quotient method*, in "Infinite Analysis" Part A,B (Kyoto 1991) Adv. Ser. Math. Phys. **16** World Scientific Publishing, River Edge NJ, 1992, pp. 317–338

[31] R. Goto. *On Hyper-Kähler manifolds of type A_∞*, G.A.F.A. **4** (1994) 424–454

[32] R. Goto. *On Hyper-Kähler manifolds of type A_∞ and D_∞*, Commun. Math. Phys. **198** (1998) 469–491

[33] D. Guan. *Examples of compact holomorphic symplectic manifolds which are not Kählerian II*, Invent. Math. **121** (1995) 135–145

[34] D. Guan. *Examples of compact holomorphic symplectic manifolds which are not Kählerian III*, Int. J. Math. **6** (1995) 709–718

[35] V. Guillemin. *Kähler structures on toric varieties*, J. Diff. Geometry **40** (1994) 285–309

[36] V. Guillemin and S. Sternberg. *Symplectic techniques in physics*, Cambridge University Press, Cambridge, 1984.

[37] V. Guillemin and S. Sternberg. *A normal form for the moment map*, in S. Sternberg ed. "Differential geometric methods in mathematical physics", Reidel Publishing Company, 1984, pp. 161–175.

[38] V. Guillemin and S. Sternberg. *Birational equivalence in the symplectic category*, Invent. Math. **97** (1989) 485–522

[39] S. W. Hawking. *Gravitational instantons*, Phys. Lett. A **60** (1977) 81–83.

[40] N.J. Hitchin. *Compact four-dimensional Einstein manifolds*, J. Diff. Geometry **9** (1974) 435–441

[41] N.J. Hitchin. *Polygons and gravitons*, Math. Proc. Camb. Phil. Soc. **85** (1979) 465–476

[42] N.J. Hitchin. *Twistor construction of Einstein metrics*, in N.J. Hitchin, T. Willmore ed. "Global Riemannian Geometry" (Durham 1983), Ellis Horwood, Chichester, 1984, pp. 115–125.

[43] N. J. Hitchin. *The self-duality equations on a Riemann surface*, Proc. Lond. Math. Soc. **55** (1987) 59–126

[44] N. J. Hitchin. *Hyper-Kähler manifolds*, Seminaire Bourbaki Vol 1991/2 Asterisque **206** (1992) 137–166

[45] N. J. Hitchin, A. Karlhede, U. Lindström and M. Roček. *Hyper-Kähler metrics and supersymmetry*, Commun. Math. Phys. **108** (1987) 535–589

[46] C. J. Houghton. *New hyper-Kähler manifolds by fixing monopoles*, Phys. Rev. D. **56** (1997) 1220–1227

[47] J. Hurtubise. *The classification of monopoles for the classical groups*, Commun. Math. Phys. **120** (1989) 613–641

[48] M. Itoh. *Quaternion structure on the moduli space of Yang-Mills connections*, Math. Ann. **276** (1987) 581–593

[49] D. Joyce. *Compact Riemannian 7-manifolds with holonomy G_2 I,II*, J. Diff. Geometry **43** (1996) 291–328, 329–375

[50] D. Joyce. *Compact 8-manifolds with holonomy Spin(7)*, Invent. Math. **123** (1996) 507–552

[51] D. Kaledin. *Hyper-Kähler structures on total spaces of holomorphic cotangent bundles*, preprint, 1997.

[52] F. Kirwan. *Cohomology of quotients in symplectic and algebraic geometry*, Princeton University Press, Princeton, 1985.

[53] P. Kobak and A.F. Swann. *Classical nilpotent orbits as hyper-Kähler quotients*, Int. J. Math. **7** (1996) 193–210

[54] R. Kobayashi. *Ricci-flat Kähler metrics on affine algebraic manifolds and degeneration of Kähler-Einstein K3 surfaces*, II, Adv. Stud. Pure Math. **18** (1990) 137–228

[55] H. Konno. *Construction of the moduli space of stable parabolic Higgs bundles on a Riemann surface*, J. Math. Soc. Japan **45** (1993) 253–276

[56] A. Kovalev. *Nahm's equations and complex adjoint orbits*, Quarterly Journal of Mathematics **47** (1996) 41–58

[57] P. B. Kronheimer. *The construction of ALE spaces as hyper-Kähler quotients*, J. Diff. Geometry **29** (1989) 665–683 .

[58] P. B. Kronheimer. *A Torelli-type theorem for gravitational instantons.* J. Diff. Geometry **29** (1989) 685–697

[59] P. B. Kronheimer. *A hyper-Kählerian structure on coadjoint orbits of a semisimple complex group*, J. Lond. Math. Soc. **42** (1990) 193–208

[60] P. B. Kronheimer. *Instantons and the geometry of the nilpotent variety*, J. Diff. Geometry **32** (1990) 473–490

[61] P. B. Kronheimer and H. Nakajima. *Yang-Mills instantons on ALE gravitational instantons*, Math. Ann. **288** (1990) 263–307

[62] C. LeBrun. *Complete Ricci-flat Kähler metrics on* $\mathbb{C}^n$ *need not be flat*, Proc. Symp. Pure Math. **52, 2** (1991) 297–304

[63] C. LeBrun and S. M. Salamon. *Strong rigidity of positive quaternion Kähler manifolds*, Invent. Math. **118** (1994) 109–132

[64] C. LeBrun and M. A. Singer. *A Kummer-type construction of self-dual 4-manifolds*, Math. Ann. **300** (1994) 165–180

[65] A. Libgober and J.W. Wood. *Uniqueness of the complex structure on Kähler manifolds of certain homotopy types*, J. Diff. Geometry **32** (1990) 139–154

[66] A. Maciocia. *Metrics on the moduli space of instantons over Euclidean 4-space*, Commun. Math. Phys. **135** (1991) 467–482

[67] N. S. Manton. *A remark on the scattering of BPS monopoles*, Phys. Lett. B **110** (1982) 54–56

[68] N. S. Manton. *Monopole interactions at long range*, Phys. Lett. B **154** (1985) 397–400

[69] N. S. Manton and B. Schroers. *Bundles over moduli spaces and the quantisation of BPS monopoles*, Ann. Phys. **225** (1993) 290–338

[70] J. McKay. *Graphs, singularities and finite groups*, Proc. Symp. Pure Math. **37** (1980) 183–186

[71] S. Mukai. *Symplectic structure of the moduli space of sheaves on an abelian or K3 surface*, Invent. Math **77** (1984) 101–116.

[72] W. Nahm. *The construction of all selfdual monopoles by the ADHM method*, in N. Craigie, P. Goddard, W. Nahm ed. "Monopoles in quantum field theory" (Trieste 1981) World Scientific Publishing, Singapore, 1982, pp. 87–94.

[73] H. Nakajima. *Monopoles and Nahm's equations*, in T. Mabuchi, S. Mukai ed. "Einstein metrics and Yang-Mills connections" Marcel Dekker, New York, 1993, pp. 193–211

[74] H. Nakajima. *Instantons on ALE spaces, quiver varieties, and Kac-Moody algebras*, Duke Math. J. **76** (1994) 365–416

[75] E. B. Nasatyr and B. Steer. *Orbifold Riemann surfaces and the Yang-Mills-Higgs equations*, Ann. Sc. Norm. Sup. Pisa **22** (1995) 596–643

[76] D. N. Page. *A physical picture of the K3 gravitational instanton*, Phys. Lett. B **80** (1978) 55–57

[77] H. Pedersen and Y. S. Poon. *Hyper-Kähler metrics and a generalisation of the Bogomolny equations*, Commun. Math. Phys. **117** (1988) 569-580

[78] R. Penrose. *Nonlinear gravitons and curved twistor theory*, Gen. Rel. Grav. **7** (1976) 31–52

[79] A. Pressley and G. B. Segal. *Loop Groups*, Oxford University Press, Oxford, 1986.

[80] S. M. Salamon. *Differential geometry of quaternionic manifolds* Ann. Sci. Éc. Norm. Sup. **19** (1986) 31–55

[81] S. M. Salamon. *On the topology of Kähler and hyper-Kähler manifolds*, Topology **35** (1996) 137–155

[82] A. F. Swann. *Hyper-Kähler and quaternionic Kähler geometry*, Math. Ann. **289** (1991) 421–450

[83] A. F. Swann. *Homogeneous twistor spaces and nilpotent orbits*, Math. Ann. **313** (1999) 161-188

[84] P. Topiwala. *A new proof of the existence of Kähler-Einstein metrics on K3, I,II*, Invent. Math. **89** (1987) 425–448, 449–454

[85] M. Verbitsky. *Tri-analytic subvarieties of Hyper-Kähler manifolds*, G.A.F.A. **5** (1995) 92–104

[86] M. Verbitsky. *Hyper-Kähler embeddings and holomorphic symplectic geometry I*, J. Alg. Geometry **5** (1996) 401–413

[87] M. Verbitsky. *Cohomology of compact Hyper-Kähler manifolds and its applications*, G.A.F.A. **6** (1996) 601–611

[88] M. Verbitsky. *Algebraic structures on hyper-Kähler manifolds*, Math. Res. Lett. **3** (1996) 763–767

[89] H. Wakakuwa. *On Riemannian manifolds with homogeneous holonomy group Sp(n)*, Tohoku. Math. J. (2), **10** (1958) 274–303

[90] J. Wolf. *Complex homogeneous contact manifolds and quaternionic symmetric spaces*, J. Math. and Mech., **14** (1965) 1033–1047

[91] S-T. Yau. *On Calabi's conjecture and some new results in algebraic geometry*, Proc. Nat. Acad. Sci. U.S.A. **74** (1977) 1798–9

[92] S-T. Yau. *On the Ricci curvature of a compact Kähler manifold and the complex Monge-Ampère equation I* (1978) Comm. Pure and Appl. Math **31**, 339–411

[93] S-T. Yau. *A general Schwarz lemma for Kähler manifolds*, Amer. J. Math. **100** (1978) 197–203

Current address: Jesus College, Oxford, OX1 3DW, **United Kingdom**

Compact Riemannian Manifolds with Exceptional Holonomy

Dominic Joyce

Suppose that M is an orientable n-dimensional manifold, and g a Riemannian metric on M. Then the *holonomy group* $\mathrm{Hol}(g)$ of g is an important invariant of g. It is a subgroup of $SO(n)$. For generic metrics g on M the holonomy group $\mathrm{Hol}(g)$ is $SO(n)$, but for some special g the holonomy group may be a proper Lie subgroup of $SO(n)$. When this happens the metric g is compatible with some extra geometric structure on M, such as a complex structure.

The possibilities for $\mathrm{Hol}(g)$ were classified in 1955 by Berger. Under conditions on M and g given in §1, Berger found that $\mathrm{Hol}(g)$ must be one of $SO(n)$, $U(m)$, $SU(m)$, $Sp(m)$, $Sp(m)Sp(1)$, G_2 or $Spin(7)$. His methods showed that $\mathrm{Hol}(g)$ is intimately related to the Riemann curvature R of g. One consequence of this is that metrics with holonomy $Sp(m)Sp(1)$ for $m > 1$ are automatically *Einstein*, and metrics with holonomy $SU(m)$, $Sp(m)$, G_2 or $Spin(7)$ are *Ricci-flat*.

Now, people have found many different ways of producing examples of metrics with these holonomy groups, by exploiting the extra geometric structure – for example, quotient constructions, twistor geometry, homogeneous and cohomogeneity one examples, and analytic approaches such as Yau's solution of the Calabi conjecture. Naturally, these methods yield examples of Einstein and Ricci-flat manifolds. In fact, metrics with special holonomy groups provide the *only* examples of compact, Ricci-flat Riemannian manifolds that are known (or known to the author).

The holonomy groups G_2 and $Spin(7)$ are known as the *exceptional holonomy groups*, since they are the exceptional cases in Berger's classification. Here G_2 is a holonomy group in dimension 7, and $Spin(7)$ is a holonomy group in dimension 8. Thus, metrics with holonomy G_2 and $Spin(7)$ are examples of Ricci-flat metrics on 7- and 8-manifolds. The exceptional holonomy groups are the most mysterious of the groups on Berger's list, and have taken longest to reveal their secrets – it was not even known until 1985 that metrics with these holonomy groups existed.

The purpose of this chapter is to describe the construction of compact Riemannian manifolds with holonomy G_2 and $Spin(7)$. These constructions were found in 1994-5 by the present author, and appear in [**16, 17**] for the case of G_2, and in [**15**] for the case of $Spin(7)$. They are also summarized in a short survey paper [**18**], and will be discussed at much greater length in the author's forthcoming book [**19**].

These constructions are interesting because they provide new examples of Ricci-flat 7- and 8-manifolds – in fact, the G_2 case provides the only known examples of compact, simply-connected Ricci-flat manifolds of odd dimension. They are also important to physicists working in String Theory, who need compact 7-manifolds with holonomy G_2 to explain why the universe apparently has only 4 dimensions, rather than the 11 dimensions it really ought to have.

We begin in §1 with an introduction to holonomy groups of Riemannian metrics, and Berger's classification. Sections 2 and 3 define the holonomy groups G_2 and $Spin(7)$, and give the background material we shall need. Section 4 is an aside on metrics with holonomy $SU(2)$, and the Kummer construction for such metrics on the $K3$ surface. The G_2 and $Spin(7)$ constructions are in fact motivated by and modelled on the Kummer construction, so we describe this first as a simple model.

Sections 5-9 explain the construction of metrics of holonomy G_2 and $Spin(7)$ on compact 7- and 8-manifolds. The ideas are first summarized in §5, which divides the proof into four steps. These steps are then covered in more detail in §6-§9 respectively. The most difficult part is Step 3, which uses analysis to construct a solution of a nonlinear elliptic partial differential equation. Finally, in §10 we suggest some areas for future research.

1. Riemannian Holonomy Groups

Section 1.1 introduces the theory of Riemannian holonomy groups, §1.2 discusses G-structures and their torsion, and §1.3 describes the classification of possible holonomy groups of Riemannian metrics. Some good references on this material are Salamon [**30**], Joyce [**19**], and Kobayashi and Nomizu [**20**, Chapters 2-4].

1.1. Basic Notions. Throughout this section, let M be a connected manifold of dimension n and g a Riemannian metric on M, and let ∇ be the Levi-Civita connection of g, regarded as a connection on the tangent bundle TM of M. Suppose that $\gamma : [0,1] \to M$ is a smooth path, with $\gamma(0) = p$ and $\gamma(1) = q$. Let s be a smooth section of $\gamma^*(TM)$, so that $s : [0,1] \to TM$ with $s(t) \in T_{\gamma(t)}M$ for each $t \in [0,1]$. Then we say that s is *parallel* if $\nabla_{\dot\gamma(t)}s(t) = 0$ for all $t \in [0,1]$, where $\dot\gamma(t)$ is $\frac{d}{dt}\gamma(t) \in T_{\gamma(t)}M$.

It turns out that for each $v \in T_pM$, there is a unique parallel section s of $\gamma^*(TM)$ with $s(0) = v$. Define a map $P_\gamma : T_pM \to T_qM$ by $P_\gamma(v) = s(1)$. Then P_γ is well-defined and linear, and is called the *parallel transport map* along γ. This definition easily generalizes to piecewise-smooth paths γ. Since ∇ is the Levi-Civita connection of g, we have $\nabla g = 0$. Using this one can show that $P_\gamma : T_pM \to T_qM$ is *orthogonal* with respect to the metric g on T_pM and T_qM.

Here is the definition of holonomy group

DEFINITION 1.1. Fix a point $p \in M$. We say that γ is a *loop based at* p if $\gamma : [0,1] \to M$ is a piecewise-smooth path with $\gamma(0) = \gamma(1) = p$. If γ is a loop based at p, then the parallel transport map P_γ lies in $O(T_pM)$, the group of orthogonal linear transformations of T_pM. Define the *(Riemannian) holonomy group* $\mathrm{Hol}_p(g)$ *of* g *based at* p to be

$$\mathrm{Hol}_p(g) = \{P_\gamma : \gamma \text{ is a loop based at } p\} \subset O(T_pM). \tag{1.1}$$

A loop γ based at p is called *null-homotopic* if it can be deformed to the constant loop at p. Define the *restricted (Riemannian) holonomy group* $\mathrm{Hol}^0_p(g)$ *of* g to be

$$\mathrm{Hol}^0_p(g) = \{P_\gamma : \gamma \text{ is a null-homotopic loop based at } p\}. \tag{1.2}$$

The following properties are elementary, and easy to prove.

PROPOSITION 1.2. *Both* $\mathrm{Hol}_p(g)$ *and* $\mathrm{Hol}^0_p(g)$ *are subgroups of* $O(T_pM)$. *Suppose that* $p,q \in M$. *Since* M *is connected, we can find a piecewise-smooth path* $\gamma : [0,1] \to M$ *with* $\gamma(0) = p$ *and* $\gamma(1) = q$, *so that* $P_\gamma : T_pM \to T_qM$. *Then*

$$P_\gamma \,\mathrm{Hol}_p(g)\, P_\gamma^{-1} = \mathrm{Hol}_q(g) \qquad \text{and} \qquad P_\gamma \,\mathrm{Hol}^0_p(g)\, P_\gamma^{-1} = \mathrm{Hol}^0_q(g). \tag{1.3}$$

By choosing an orthonormal basis for T_pM we can identify $O(T_pM)$ with the Lie group $O(n)$. Thus we may regard $\mathrm{Hol}_p(g)$ and $\mathrm{Hol}^0_p(g)$ as subgroups of $O(n)$. Changing the choice of basis changes the subgroups by conjugation by an element of $O(n)$. Thus, $\mathrm{Hol}_p(g)$ and $\mathrm{Hol}^0_p(g)$ may be regarded as subgroups of $O(n)$ defined up to conjugation. Moreover, equation (1.3) shows that in this sense, $\mathrm{Hol}_p(g)$ and $\mathrm{Hol}^0_p(g)$ are *independent of the base point* p. Therefore, we omit the subscript p, and write $\mathrm{Hol}(g)$ for the holonomy group of g and $\mathrm{Hol}^0(g)$ for the reduced holonomy group of g, both of which are subgroups of $O(n)$ defined up to conjugation.

Our next result, taken from [**20**, p. 73, p. 186], is rather more difficult.

THEOREM 1.3. *The reduced holonomy group* $\mathrm{Hol}^0(g)$ *is a closed, connected, Lie subgroup of* $SO(n)$. *It is the connected component of* $\mathrm{Hol}(g)$ *containing the identity, and is normal in* $\mathrm{Hol}(g)$. *There is a surjective group homomorphism* $\phi : \pi_1(M) \to \mathrm{Hol}(g)/\mathrm{Hol}^0(g)$. *Thus* $\mathrm{Hol}(g)/\mathrm{Hol}^0(g)$ *is countable, and if* M *is simply-connected then* $\mathrm{Hol}(g) = \mathrm{Hol}^0(g)$.

Because $\mathrm{Hol}^0(g)$ is a Lie group, it has a Lie algebra. We define the *holonomy algebra* $\mathfrak{hol}(g)$ of g to be the Lie algebra of $\mathrm{Hol}^0(g)$. Then $\mathfrak{hol}(g)$ is a Lie subalgebra of $\mathfrak{o}(n)$, defined up to the adjoint action of $O(n)$. Similarly, define $\mathfrak{hol}_p(g)$ to be the Lie algebra of $\mathrm{Hol}^0_p(g)$, which is a Lie subalgebra of $\mathfrak{o}(T_pM)$. Using g we may identify $\mathfrak{o}(T_pM)$ with $\Lambda^2T^*_pM$, so that $\mathfrak{hol}_p(g)$ becomes a vector subspace of $\Lambda^2T^*_pM$.

Now the holonomy algebra $\mathfrak{hol}(g)$ is intimately connected with the Riemann curvature tensor $R^a{}_{bcd}$ of g. Actually, we find it more convenient to lower the index a and work with the tensor $R_{abcd} = g_{ae}R^e{}_{bcd}$, which we also call the Riemann curvature. Here are two results relating R_{abcd} and $\mathfrak{hol}_p(g)$.

THEOREM 1.4. *The Riemann curvature tensor* R_{abcd} *lies in* $S^2\,\mathfrak{hol}_p(g)$ *at* p, *where* $\mathfrak{hol}_p(g)$ *is regarded as a subspace of* $\Lambda^2T^*_pM$. *It also satisfies the first and second Bianchi identities*

$$R_{abcd} + R_{adbc} + R_{acdb} = 0, \tag{1.4}$$

$$\text{and} \qquad \nabla_e R_{abcd} + \nabla_c R_{abde} + \nabla_d R_{abec} = 0. \tag{1.5}$$

The second result is the *Ambrose-Singer Holonomy Theorem* [**2**].

THEOREM 1.5. *Let* p *and* q *be points in* M, *let* $\gamma : [0,1] \to M$ *be piecewise-smooth with* $\gamma(0) = p$ *and* $\gamma(1) = q$, *and let* $P_\gamma : T_pM \to T_qM$ *be the parallel transport map. Write tensors at* p *using the tensor indices* a, b, *and tensors at* q *using the tensor indices* i, j, k, l. *Let* $v, w \in T_qM$, *so that*

$$(P_\gamma)^i_a (P_\gamma)^j_b R_{ijkl} v^k w^l \quad \text{lies in} \quad \Lambda^2T^*_pM, \tag{1.6}$$

where R_{ijkl} *is the Riemann curvature at* q. *Then* $\mathfrak{hol}_p(g)$ *is the vector subspace of* $\Lambda^2T^*_pM$ *spanned by all elements of the form* (1.6), *for all* $q \in M$.

Let (M_1, g_1) and (M_2, g_2) be Riemannian manifolds, of positive dimension. Then the product $M_1 \times M_2$ has tangent spaces $T_{(p,q)}(M_1 \times M_2) \cong T_pM_1 \oplus T_qM_2$. Thus we may define the *product metric* $g_1 \times g_2$ to be $g_1|_p + g_2|_q$ at each (p, q) in $M_1 \times M_2$. This makes $M_1 \times M_2$ into a Riemannian manifold $(M_1 \times M_2, g_1 \times g_2)$, called a *Riemannian product*. The holonomy group of a product metric is the product of the corresponding holonomy groups.

PROPOSITION 1.6. *Let* (M_1, g_1) *and* (M_2, g_2) *be Riemannian manifolds, with Riemannian product* $(M_1 \times M_2, g_1 \times g_2)$. *Then* $\mathrm{Hol}(g_1 \times g_2) = \mathrm{Hol}(g_1) \times \mathrm{Hol}(g_2)$.

We call a Riemannian manifold (M, g) *reducible* if every point has an open neighbourhood isometric to a Riemannian product, and *irreducible* if it is not reducible. Here is a kind of converse to Proposition 1.6.

THEOREM 1.7. *Let* (M, g) *be an irreducible Riemannian manifold of dimension* n. *Then the natural representation of* $\mathrm{Hol}(g)$ *on* $\mathbb{R}^n$ *is irreducible.*

There is a class of Riemannian manifolds called the *Riemannian symmetric spaces* which are important in the theory of Riemannian holonomy groups. A Riemannian symmetric space is a special kind of Riemannian manifold with a transitive isometry group. The theory of symmetric spaces was worked out by Élie Cartan in the 1920's, who classified them completely, using his own classification of Lie groups and their representations.

A Riemannian metric g is called *locally symmetric* if $\nabla R = 0$, and *nonsymmetric* if it is not locally symmetric. It turns out that every locally symmetric metric is locally isometric to a Riemannian symmetric space. The relevance of symmetric spaces to holonomy groups is that many possible holonomy groups are the holonomy group of a Riemannian symmetric space, but are not realized by any nonsymmetric metric. Therefore, by restricting attention to nonsymmetric metrics one considerably reduces the number of possible Riemannian holonomy groups. For more information about symmetric spaces, see Kobayashi and Nomizu [**21**, §XI] or Helgason [**14**].

1.2. Holonomy Groups and Torsion-Free G-Structures. Now we explain a useful mathematical tool for studying holonomy groups.

DEFINITION 1.8. Let M be a manifold of dimension n, and F the frame bundle of M. Then F is a principal bundle over M with fibre $GL(n, \mathbb{R})$. Let G be a Lie subgroup of $GL(n, \mathbb{R})$. Then a *G-structure* on M is a principal subbundle P of F, with fibre G.

Let (M, g) be a riemannian n-manifold, with frame bundle F. Then each point of F is a basis $(e_1, \ldots, e_n)$ for one of the tangent spaces T_pM of M. Define P to be the subset of F of bases $(e_1, \ldots, e_n)$ which are *orthonormal* with respect to g. Then P is a principal subbundle of F with fibre $O(n)$, and so P is an *$O(n)$-structure* on M. This gives a 1-1 correspondence between $O(n)$-structures and Riemannian metrics on M.

Now let M be an n-manifold, G a Lie subgroup of $O(n)$, and Q a G-structure on M. Then $P = O(n) \cdot Q$ is an $O(n)$-structure on M containing Q, which corresponds to a Riemannian metric g on M. Let ∇ be the Levi-Civita connection of g. Then ∇ is a connection on P. We say that Q is *torsion-free* if ∇ preserves the subbundle Q of P. To each G-structure Q on M we can associate a tensor $T(Q)$ called the *torsion* of Q, which measures the failure of ∇ to preserve Q, and Q is torsion-free if and only if $T(Q) = 0$.

The relationship between torsion-free G-structures and holonomy groups is given by the following Proposition.

PROPOSITION 1.9. *Let (M, g) be a manifold of dimension n, and let P be the $O(n)$-structure on M associated to g. Suppose G is a Lie subgroup of $O(n)$. Then there exists a torsion-free G-structure Q contained in P if and only if* $\mathrm{Hol}(g) \subseteq G$. *More generally, there is a 1-1 correspondence between the set of torsion-free G-structures on M contained in P, and the homogeneous space*

$$\{a \in O(n) : a^{-1}\mathrm{Hol}(g)a \subseteq G\}/G. \tag{1.7}$$

Let Q be a G-structure on M. Then G is a subgroup of $GL(n, \mathbb{R})$, and so it has a natural representation on $\mathbb{R}^n$, and therefore also on $V_{k,l} = \bigotimes^k \mathbb{R}^n \otimes \bigotimes^l (\mathbb{R}^n)^*$. Now the tensor bundle $\bigotimes^k TM \otimes \bigotimes^l T^*M$ over M is canonically isomorphic to $Q \times_G V_{k,l}$. This has two important consequences. Firstly, if the representation $V_{k,l}$ of G splits into a direct sum of subrepresentations, then the tensor bundle $\bigotimes^k TM \otimes \bigotimes^l T^*M$ on M has a corresponding splitting into subbundles. Thus, a G-structure decomposes tensors into components.

Secondly, if an element $t \in V_{k,l}$ is fixed by the action of G, there is a corresponding tensor T in $C^\infty(\bigotimes^k TM \otimes \bigotimes^l T^*M)$. Moreover, if Q is torsion-free, then we have $\nabla T = 0$, where ∇ is the Levi-Civita connection of the metric g associated to Q. Thus, to each torsion-free G-structure we can associate a number of *constant tensors*, that is, tensors T on M with $\nabla T = 0$.

PROPOSITION 1.10. *Let M be a manifold, g a Riemannian metric on M, and ∇ the Levi-Civita connection of g. Then there is a 1-1 correspondence between elements t of $V_{k,l}$ fixed by* $\mathrm{Hol}(g)$, *and constant tensors $T \in C^\infty(\bigotimes^k TM \otimes \bigotimes^l T^*M)$.*

1.3. Classification of Riemannian Holonomies. The following very important result was proved by Berger [**5**, Thm. 3, p. 318] in 1955.

THEOREM 1.11 (Berger). *Suppose that M is a simply-connected manifold of dimension n, and that g is a Riemannian metric on M, that is irreducible and nonsymmetric. Then exactly one of the following seven cases holds.*

(*i*) $\mathrm{Hol}(g) = SO(n)$,
(*ii*) $n = 2m$ *with* $m \geq 2$, *and* $\mathrm{Hol}(g) = U(m)$ *in* $SO(2m)$,
(*iii*) $n = 2m$ *with* $m \geq 2$, *and* $\mathrm{Hol}(g) = SU(m)$ *in* $SO(2m)$,
(*iv*) $n = 4m$ *with* $m \geq 2$, *and* $\mathrm{Hol}(g) = Sp(m)$ *in* $SO(4m)$,
(*v*) $n = 4m$ *with* $m \geq 2$, *and* $\mathrm{Hol}(g) = Sp(m)Sp(1)$ *in* $SO(4m)$,
(*vi*) $n = 7$ *and* $\mathrm{Hol}(g) = G_2$ *in* $SO(7)$, *or*
(*vii*) $n = 8$ *and* $\mathrm{Hol}(g) = Spin(7)$ *in* $SO(8)$.

In fact, Berger also included the possibility $n = 16$ and $\mathrm{Hol}(g) = Spin(9)$, but this was shown not to occur by Alekseevskii [**1**] and Brown and Gray [**8**]. To simplify the classification, Berger made three assumptions: that M is *simply-connected*, that g is *irreducible*, and that g is *nonsymmetric*. If we work with $\mathrm{Hol}^0(g)$ instead of

Hol(g), then we need not suppose M is simply-connected. The holonomy group of a reducible metric is a product of holonomy groups of irreducible metrics, and the holonomy groups of locally symmetric metrics follow from Cartan's classification of Riemannian symmetric spaces. Thus, these three assumptions can easily be removed.

Here is a sketch of Berger's proof of Theorem 1.11. As M is simply-connected, Theorem 1.3 shows that Hol(g) is a closed, connected Lie subgroup of $SO(n)$, and since g is irreducible, Theorem 1.7 shows that the representation of Hol(g) on $\mathbb{R}^n$ is irreducible. So, suppose that H is a closed, connected subgroup of $SO(n)$ with irreducible representation on $\mathbb{R}^n$, and Lie algebra $\mathfrak{h}$. The classification of all such groups H follows from the classification of Lie groups (and is of considerable complexity). Berger's method was to take the list of all such groups H, and to apply two tests to each possibility to find out if it could be a holonomy group. The only groups H which passed both tests are those in the Theorem.

Berger's tests are algebraic and involve the curvature tensor. Suppose that R_{abcd} is the Riemann curvature of a metric g with $\mathrm{Hol}(g) = H$. Then Theorem 1.4 shows that $R_{abcd} \in S^2\mathfrak{h}$, and the first Bianchi identity (1.4) applies. But if $\mathfrak{h}$ has large codimension in $\mathfrak{o}(n)$, then the vector space $\mathfrak{R}^H$ of elements of $S^2\mathfrak{h}$ satisfying (1.4) will be small, or even zero. However, Theorem 1.5 shows that $\mathfrak{R}^H$ must be big enough to generate $\mathfrak{h}$. For many of the candidate groups H this does not hold, and so H cannot be a holonomy group. This is the first test.

Now $\nabla_e R_{abcd}$ lies in $(\mathbb{R}^n)^* \otimes \mathfrak{R}^H$, and also satisfies the second Bianchi identity (1.5). Frequently these imply that $\nabla R = 0$, so that g is locally symmetric. Therefore we may exclude such H, and this is Berger's second test. Later, Simons [**31**] found a rather shorter proof of Theorem 1.11 based on showing that Hol(g) must act transitively on the unit sphere in $\mathbb{R}^n$.

The holonomy groups G_2 and *Spin*(7), cases (*vi*) and (*vii*) of Theorem 1.11, are known as the *exceptional holonomy groups* because they are the exceptional cases in the classification. The existence of metrics with holonomy G_2 and *Spin*(7) was first shown in 1985 by Bryant [**9**], using the theory of exterior differential systems. Explicit examples of *complete* metrics with holonomy G_2 and *Spin*(7) on noncompact manifolds were found in 1989 by Bryant and Salamon [**10**]. Then in 1994-5, the present author constructed examples of metrics with holonomy G_2 on compact 7-manifolds [**16**], [**17**], and of metrics with holonomy *Spin*(7) on compact 8-manifolds [**15**]. In the rest of this chapter we will explain these examples and the methods used to contruct them.

2. The Holonomy Group G_2

In this section we give a brief introduction to the geometry of metrics with holonomy G_2, beginning with their local properties, and then moving on to discuss the topology of compact Riemannian manifolds with holonomy G_2. All the results below can be found in Salamon [**30**, Chapter 11], or Joyce [**16, 17**]. Here is a definition of G_2 as a subgroup of $GL(7,\mathbb{R})$.

DEFINITION 2.1. Let $(x_1, \ldots, x_7)$ be coordinates on $\mathbb{R}^7$. Write ω_{ijk} for the 3-form $dx_i \wedge dx_j \wedge dx_k$ on $\mathbb{R}^7$, and ω_{ijkl} for the 4-form $dx_i \wedge dx_j \wedge dx_k \wedge dx_l$. Define a 3-form φ_0 on $\mathbb{R}^7$ by

$$\varphi_0 = \omega_{127} + \omega_{136} + \omega_{145} + \omega_{235} - \omega_{246} + \omega_{347} + \omega_{567}. \tag{2.1}$$

The subgroup of $GL(7,\mathbb{R})$ preserving φ_0 is called the *exceptional Lie group* G_2. It is a compact, connected, simply-connected, semisimple, 14-dimensional Lie group. It also preserves the Euclidean metric

$$g_0 = dx_1^2 + \cdots + dx_7^2 \tag{2.2}$$

on $\mathbb{R}^7$, the orientation on $\mathbb{R}^7$, and the 4-form

$$*\varphi_0 = \omega_{1234} + \omega_{1256} - \omega_{1357} + \omega_{1467} + \omega_{2367} + \omega_{2457} + \omega_{3456}. \tag{2.3}$$

Since G_2 is a subgroup of $SO(7)$, a G_2 structure on a 7-manifold M induces a metric g and an orientation on M. Combining g and the orientation gives the *Hodge star*, a linear map $* : \Lambda^k T^*M \to \Lambda^{7-k}T^*M$. The forms φ_0 and $*\varphi_0$ of (2.1) and (2.3) are related by the Hodge star on $\mathbb{R}^7$, which is why we use this notation.

Let M be a 7-manifold, and φ a 3-form on M. We call φ a *positive 3-form* if for every $p \in M$, there exists an isomorphism between T_pM and $\mathbb{R}^7$ that identifies $\varphi|_p$ and the 3-form φ_0 of (2.1). Since G_2 is the subgroup of $GL(7,\mathbb{R})$ preserving φ_0, it follows that there is a 1-1 correspondence between positive 3-forms φ on M, and G_2-structures Q on M. Moreover, as in §1.2, to each G_2-structure Q on M we may associate a 3-form φ, a metric g, and a 4-form $*\varphi$, corresponding to the tensors (2.1), (2.2) and (2.3) on $\mathbb{R}^7$.

Thus, to each G_2-structure Q on M there corresponds a unique pair (φ, g), where φ is a positive 3-form and g a compatible Riemannian metric. For the rest of this chapter, we will adopt the following abuse of notation: we shall refer to the pair (φ, g) as a *G_2-structure*. Of course it is not, exactly, a G_2-structure, but it does at least define a unique G_2-structure.

PROPOSITION 2.2. *Let M be a 7-manifold and (φ, g) a G_2-structure on M. Then the following are equivalent:*

(*i*) $\mathrm{Hol}(g) \subseteq G_2$, *and φ is the induced 3-form,*

(*ii*) $\nabla\varphi = 0$ *on M, where ∇ is the Levi-Civita connection of g, and*

(*iii*) $d\varphi = d^*\varphi = 0$ *on M.*

The tensor $\nabla\varphi$ is called the *torsion* of the G_2-structure (φ, g). If $\nabla\varphi = 0$ then the G_2-structure is called *torsion-free*. In §1.2 we explained that a G-structure on M induces a splitting of the bundles of tensors on M into irreducible components. Here is the decomposition of the exterior forms on a 7-manifold with a G_2-structure.

PROPOSITION 2.3. *Let M be a 7-manifold and (φ, g) a G_2-structure on M. Then $\Lambda^k T^*M$ splits orthogonally into components as follows, where Λ^k_l is a vector subbundle of dimension l corresponding to an irreducible representation of G_2:*

(*i*) $\Lambda^1 T^*M = \Lambda^1_7$, (*ii*) $\Lambda^2 T^*M = \Lambda^2_7 \oplus \Lambda^2_{14}$,
(*iii*) $\Lambda^3 T^*M = \Lambda^3_1 \oplus \Lambda^3_7 \oplus \Lambda^3_{27}$, (*iv*) $\Lambda^4 T^*M = \Lambda^4_1 \oplus \Lambda^4_7 \oplus \Lambda^4_{27}$,
(*v*) $\Lambda^5 T^*M = \Lambda^5_7 \oplus \Lambda^5_{14}$, *and* (*vi*) $\Lambda^6 T^*M = \Lambda^6_7$.

The Hodge star $*$ of g gives an isometry between Λ^k_l and Λ^{7-k}_l.

Let the orthogonal projection from $\Lambda^k T^*M$ to Λ^k_l be denoted π_l. So, for instance, if $\xi \in C^\infty(\Lambda^2 T^*M)$, then $\xi = \pi_7(\xi) + \pi_{14}(\xi)$. We saw from Theorem 1.4 that the holonomy group of a Riemannian metric g constrains its Riemann curvature. Using this one can show:

PROPOSITION 2.4. *Let g be a Riemannian metric on a 7-manifold. If* $\mathrm{Hol}(g) \subseteq G_2$*, then g is Ricci-flat.*

Now suppose that M is a *compact* manifold, and that g is a Riemannian metric on M with $\mathrm{Hol}(g) = G_2$. Then, from the Proposition, g is Ricci-flat. Consider the de Rham cohomology group $H^1(M, \mathbb{R})$. By Hodge theory, each class in $H^1(M, \mathbb{R})$ is represented by a unique 1-form α with $d\alpha = d^*\alpha = 0$. However, since M is compact and g is Ricci-flat, one can prove by a well-known argument of Bochner [**7**] that any such 1-form satisfies $\nabla\alpha = 0$.

But by Proposition 1.10, since $\mathrm{Hol}(g) = G_2$ fixes no nonzero vectors in $(\mathbb{R}^7)^*$, there are no nonzero constant 1-forms on M. Thus $H^1(M, \mathbb{R}) = 0$. One can then use the Cheeger-Gromoll splitting Theorem [**6**, Cor. 6.67] to show that the fundamental group $\pi_1(M)$ of M is *finite*. Thus we prove the following result, which is [**17**, Prop. 1.1.1].

PROPOSITION 2.5. *Let M be a compact 7-manifold, and suppose that (φ, g) is a torsion-free G_2-structure on M. Then* $\mathrm{Hol}(g) = G_2$ *if and only if $\pi_1(M)$ is finite.*

The next result is deduced from [**16**, Theorem C].

THEOREM 2.6. *Let M be a compact 7-manifold, let $\mathcal{X}$ be the set of torsion-free G_2-structures on M, and let $\mathcal{D}$ be the group of diffeomorphisms of M isotopic to the identity. Then $\mathcal{X}/\mathcal{D}$ is a smooth manifold of dimension $b^3(M)$. Moreover, the map from $\mathcal{X}/\mathcal{D}$ to $H^3(M, \mathbb{R})$ taking (φ, g) to the cohomology class $[\varphi]$ of φ is a local diffeomorphism.*

This Theorem is proved by studying small deformations of a fixed torsion-free G_2-structure, and is in that sense a purely local result. In fact, at present we know very little about the global geometry of the moduli space of metrics with holonomy G_2 on any 7-manifold.

3. The Holonomy Group *Spin*(7)

We shall now give a very similar treatment of the holonomy group *Spin*(7). The material in this section can be found in Salamon [**30**, Chap. 12] or Joyce [**15**]. First we define *Spin*(7) as a subgroup of $GL(8, \mathbb{R})$.

DEFINITION 3.1. Let $\mathbb{R}^8$ have coordinates $(x_1, \ldots, x_8)$. Write ω_{ijkl} for the 4-form $dx_i \wedge dx_j \wedge dx_k \wedge dx_l$ on $\mathbb{R}^8$. Define a 4-form Ω_0 on $\mathbb{R}^8$ by

$$\begin{aligned}\Omega_0 = {}& \omega_{1256} + \omega_{1278} + \omega_{3456} + \omega_{3478} + \omega_{1357} - \omega_{1368} - \omega_{2457} \\ & + \omega_{2468} - \omega_{1458} - \omega_{1467} - \omega_{2358} - \omega_{2367} + \omega_{1234} + \omega_{5678}.\end{aligned} \tag{3.1}$$

The subgroup of $GL(8, \mathbb{R})$ preserving Ω_0 is the holonomy group *Spin*(7). It is a compact, connected, simply-connected, semisimple, 21-dimensional Lie group, which is isomorphic as a Lie group to the double cover of $SO(7)$. This group also preserves the orientation on $\mathbb{R}^8$ and the Euclidean metric $g_0 = dx_1^2 + \cdots + dx_8^2$ on $\mathbb{R}^8$.

Let M be an 8-manifold and Ω a 4-form on M. We call Ω an *admissible 4-form* if for every $p \in M$, there is an isomorphism between T_pM and $\mathbb{R}^8$ that identifies $\Omega|_p$ and the 4-form Ω_0 of (3.1). Then there is a 1-1 correspondence between *Spin*(7)-structures Q and admissible 4-forms Ω on M. Each *Spin*(7)-structure Q induces a

4-form Ω on M and a metric g on M, corresponding to Ω_0 and g_0 on $\mathbb{R}^8$. As with G_2, we shall abuse notation by referring to the pair (Ω, g) as a *Spin*(7)-*structure*. The next three results correspond to Propositions 2.2-2.4.

PROPOSITION 3.2. *Let M be a compact 8-manifold and (Ω, g) a Spin(7)-structure on M. Then the following are equivalent:*

(*i*) $\mathrm{Hol}(g) \subseteq \mathit{Spin}(7)$, *and Ω is the induced 4-form,*

(*ii*) $\nabla\Omega = 0$ *on M, where ∇ is the Levi-Civita connection of g, and*

(*iii*) $d\Omega = 0$ *on M.*

Again, $\nabla\Omega$ is the *torsion* of the *Spin*(7)-structure (Ω, g), and (Ω, g) is *torsion-free* if $\nabla\Omega = 0$. Since *Spin*(7) lies in $SO(8)$, a *Spin*(7)-structure on an 8-manifold M induces a natural orientation on M, and so we have the *Hodge star* $* : \Lambda^k T^*M \to \Lambda^{8-k}T^*M$.

PROPOSITION 3.3. *Let M be an 8-manifold and (Ω, g) a Spin(7)-structure on M. Then $\Lambda^k T^*M$ splits orthogonally into components as follows, where Λ^k_l is a vector subbundle of dimension l corresponding to an irreducible representation of Spin(7):*

$$(i)\ \Lambda^1T^*M = \Lambda^1_8, \qquad (ii)\ \Lambda^2T^*M = \Lambda^2_7 \oplus \Lambda^2_{21}, \qquad (iii)\ \Lambda^3T^*M = \Lambda^3_8 \oplus \Lambda^3_{48},$$

$$(iv)\ \Lambda^4T^*M = \Lambda^4_+T^*M \oplus \Lambda^4_-T^*M,\ \ \Lambda^4_+T^*M = \Lambda^4_1 \oplus \Lambda^4_7 \oplus \Lambda^4_{27},\ \ \Lambda^4_-T^*M = \Lambda^4_{35},$$

$$(v)\ \Lambda^5T^*M = \Lambda^5_8 \oplus \Lambda^5_{48}, \qquad (vi)\ \Lambda^6T^*M = \Lambda^6_7 \oplus \Lambda^6_{21}, \qquad (vii)\ \Lambda^7T^*M = \Lambda^7_8.$$

The Hodge star $*$ gives an isometry between Λ^k_l and Λ^{8-k}_l. In part (iv), $\Lambda^4_+T^*M$ and $\Lambda^4_-T^*M$ are the $+1$- and -1- eigenspaces of $*$ on Λ^4T^*M respectively.

The orthogonal projection from Λ^kT^*M to Λ^k_l will be written π_l.

PROPOSITION 3.4. *Let g be a Riemannian metric on a 8-manifold. If* $\mathrm{Hol}(g) \subseteq$ *Spin*(7), *then g is Ricci-flat.*

Now suppose that M is a *compact* 8-manifold, and (Ω, g) a torsion-free *Spin*(7)-structure on M. Since *Spin*(7) is simply-connected, we deduce that M is spin. Therefore there are positive and negative spin bundles S_+, S_- over M, with fibre $\mathbb{R}^8$, and the Dirac operator D acts by $D : C^\infty(S_+) \to C^\infty(S_-)$. As M is Ricci-flat, it has zero scalar curvature. Thus, by a well-known argument of Lichnerowicz [**26**], all the spinors in $\mathrm{Ker}\, D$ and $\mathrm{Ker}\, D^*$ are constant.

However, as for the case of tensors in Proposition 1.10, the constant spinors on a spin Riemannian manifold (M, g) are determined entirely by the holonomy group $\mathrm{Hol}(g)$. Therefore, $\mathrm{Hol}(g)$ determines $\mathrm{Ker}\, D$ and $\mathrm{Ker}\, D^*$, and thus it determines the index $\mathrm{ind}\, D = \dim \mathrm{Ker}\, D - \dim \mathrm{Ker}\, D^*$. But D is an elliptic operator, so the Atiyah-Singer Index Theorem shows [**3**, Thm. 5.3] that $\mathrm{ind}\, D$ is a topological invariant of M known as the *$\hat{A}$-genus $\hat{A}(M)$ of M.*

In this way, the holonomy group $\mathrm{Hol}(g)$ determines a topological invariant $\hat{A}(M)$ of M, which can be written as a linear combination of Betti numbers of M. Conversely, if we know the Betti numbers of M, we can determine $\hat{A}(M)$ and hence $\mathrm{Hol}(g)$. Thus we prove the following result, which is [**15**, Thm. C].

THEOREM 3.5. *Suppose that M is a compact, simply-connected 8-manifold and that (Ω, g) is a torsion-free Spin(7)-structure on M. Then M is spin, and the* *volume form $\Omega \wedge \bar{\Omega}$ gives a natural orientation on M. Define the $\hat{A}$-genus $\hat{A}(M)$ of M by*

$$24\hat{A}(M) = -1 + b^1 - b^2 + b^3 + b^4_+ - 2b^4_-, \tag{3.2}$$

where b^i are the Betti numbers of M, and $b^4_\pm$ are the dimensions of the spaces of self-dual and anti-self-dual 4-forms in $H^4(M, \mathbb{R})$. Then $\hat{A}(M)$ is equal to 1, 2, 3 or 4, and the holonomy group $\mathrm{Hol}(g)$ *of g is determined by $\hat{A}(M)$ as follows:*

(*i*) $\mathrm{Hol}(g) = Spin(7)$ *if and only if* $\hat{A}(M) = 1$,

(*ii*) $\mathrm{Hol}(g) = SU(4)$ *if and only if* $\hat{A}(M) = 2$,

(*iii*) $\mathrm{Hol}(g) = Sp(2)$ *if and only if* $\hat{A}(M) = 3$, *and*

(*iv*) $\mathrm{Hol}(g) = SU(2) \times SU(2)$ *if and only if* $\hat{A}(M) = 4$.

Every compact, Riemannian 8-manifold with holonomy group $Spin(7)$ is simply-connected.

Finally, here is [**15**, Thm. D], which describes the moduli space of torsion-free $Spin(7)$-structures on a compact 8-manifold.

THEOREM 3.6. *Let M be a simply-connected, compact 8-manifold admitting torsion-free $Spin(7)$-structures, let $\mathcal{X}$ be the set of torsion-free $Spin(7)$-structures on M, and let $\mathcal{D}$ be the group of diffeomorphisms of M isotopic to the identity. Then $\mathcal{X}/\mathcal{D}$ is a smooth manifold of dimension $\hat{A}(M) + b^4_-(M)$.*

4. Metrics with Holonomy $SU(2)$ on the $K3$ surface

Before discussing the construction of metrics with holonomy G_2 and $Spin(7)$ on compact 7- and 8-manifolds, we will first explain a simpler construction of the same type: that of metrics of holonomy $SU(2)$ on a particular compact 4-manifold, the $K3$ surface. This is a good illustration of the general plan used for G_2 and $Spin(7)$, but the details are easier because the dimension is lower. Section §4.1 describes the *Eguchi-Hanson space*, an explicit Riemannian manifold with holonony $SU(2)$. Then §4.2 covers the *Kummer construction* for metrics with holonomy $SU(2)$ on the $K3$ surface, in which the Eguchi-Hanson space appears as an important ingredient.

Let $(x_1, \ldots, x_4)$ be the usual coordinates on $\mathbb{R}^4$. Then $SU(2)$ acts on $\mathbb{R}^4$ preserving the Euclidean metric $g_0 = dx_1^2 + dx_2^2 + dx_3^2 + dx_4^2$ and the three 2-forms

$$\begin{aligned} \omega_0^I = dx_1 \wedge dx_2 + dx_3 \wedge dx_4, \qquad \omega_0^J = dx_1 \wedge dx_3 - dx_2 \wedge dx_4 \\ \text{and} \qquad \omega_0^K = dx_1 \wedge dx_4 + dx_2 \wedge dx_3. \end{aligned} \tag{4.1}$$

Moreover, the subgroup of $GL(4, \mathbb{R})$ that preserves ω_0^I, ω_0^J and ω_0^K is exactly $SU(2)$. Therefore, if X is a 4-manifold, there is a 1-1 correspondence between $SU(2)$-structures on X and triples $(\omega^I, \omega^J, \omega^K)$ of 2-forms on X, such that for each point $p \in X$ there exists an isomorphism $T_pX \cong \mathbb{R}^4$ that identifies ω^I, ω^J and ω^K with ω_0^I, ω_0^J and ω_0^K respectively.

By an abuse of notation, we shall refer to a triple $(\omega^I, \omega^J, \omega^K)$ of 2-forms on X with this property as an $SU(2)$*-structure* on X. To each $SU(2)$-structure $(\omega^I, \omega^J, \omega^K)$ on X, we may associate a Riemannian metric g and almost complex structures I, J, K on X in a natural way, which satisfy the quaternion relations $IJ = -JI = K$.

It can be shown that an $SU(2)$-structure $(\omega^I, \omega^J, \omega^K)$ is *torsion-free* if and only if ω^I, ω^J and ω^K are *closed* 2-forms. If $(\omega^I, \omega^J, \omega^K)$ is torsion-free, then $\mathrm{Hol}(g) \subseteq$

$SU(2)$, the almost complex structures I, J and K are integrable, the metric g is Kähler with respect to each of I, J and K, and the Kähler forms of the Kähler structures (I, g), (J, g) and (K, g) are ω^I, ω^J and ω^K respectively.

As all of this structure shows, Riemannian manifolds with holonomy $SU(2)$ have a very rich and interesting geometry. In particular, metrics with holonomy $SU(2)$ are *Kähler metrics,* in three different ways. Given a complex manifold M, it is often easy to write down a large number of explicit Kähler metrics on M using Kähler potentials, and other devices from Kähler geometry. If g is a Kähler metric on a simply-connected complex surface, then $\mathrm{Hol}(g) \subseteq SU(2)$ if and only if g is Ricci-flat. Thus, one method of constructing metrics with holonomy $SU(2)$ is to find solutions to the equation $\mathrm{Ric}(g) = 0$ for a Kähler metric g. This can be done explicitly in examples with symmetry, or using analysis as in Yau's solution to the Calabi conjecture [**34**].

4.1. The Eguchi-Hanson Space. The simplest nontrivial example of a Riemannian manifold with holonomy $SU(2)$ is the Eguchi-Hanson space [**11**], which is a family of complete metrics on the noncompact 4-manifold $T^*\mathbb{CP}^1$. We will write down this metric explicitly in coordinates. Let $\mathbb{C}^2$ be equipped with complex coordinates (z_1, z_2), and the standard flat Kähler metric $g_0 = |dz_1|^2 + |dz_2|^2$. The involution $-1 : (z_1, z_2) \mapsto (-z_1, -z_2)$ acts on $\mathbb{C}^2$, preserving g_0 and fixing 0. Thus $\mathbb{C}^2/\{\pm 1\}$ is a singular complex manifold with one singular point at 0, and the metric g_0 pushes down to $\mathbb{C}^2/\{\pm 1\}$.

Let X be the blow-up of $\mathbb{C}^2/\{\pm 1\}$ at 0, and let $\pi : X \to \mathbb{C}^2/\{\pm 1\}$ be the blow-down map. Then X is a nonsingular complex manifold biholomorphic to $T^*\mathbb{CP}^1$, with $\pi_1(X) = \{1\}$ and $H^2(X, \mathbb{R}) = \mathbb{R}$. The *radius function* r given by $r^2 = |z_1|^2 + |z_2|^2$ on $\mathbb{C}^2$ pushes down to $\mathbb{C}^2/\{\pm 1\}$, and so lifts to X. Let $t \geq 0$, and define a function f_t on X by

$$f_t = \sqrt{r^4 + t^4} + 2t^2 \log r - t^2 \log\left(\sqrt{r^4 + t^4} + t^2\right). \tag{4.2}$$

This is the Kähler potential for the Eguchi-Hanson metric, and is taken from [**24**, p. 593]. For each $t > 0$, define 2-forms ω_t^I, ω_t^J and ω_t^K on X by

$$\omega_t^I = \tfrac{1}{2} i \partial\bar{\partial} f_t, \quad \omega_t^J = \mathrm{Re}(dz_1 \wedge dz_2), \quad \text{and} \quad \omega_t^K = \mathrm{Im}(dz_1 \wedge dz_2). \tag{4.3}$$

Then $(\omega_t^I, \omega_t^J, \omega_t^K)$ is a torsion-free $SU(2)$-structure on X. The associated metric g_t is the *Eguchi-Hanson metric*, and has holonomy $SU(2)$.

From (4.2) we find that the asymptotic behaviour of g_t near infinity in X is

$$g_t = \pi^*(g_0) + O(t^4 r^{-4}). \tag{4.4}$$

Thus, the Eguchi-Hanson metric is asymptotic to the flat metric g_0 on $\mathbb{C}^2/\{\pm 1\}$ at infinity. Metrics with this property are called *Asymptotically Locally Euclidean*, or *ALE* for short. Similarly, the 2-forms ω_t^I, ω_t^J and ω_t^K have the asymptotic behaviour

$$\begin{aligned} \omega_t^I = \pi^*(\omega_0^I) + O(t^4 r^{-4}), \qquad & \omega_t^J = \pi^*(\omega_0^J) + O(t^4 r^{-4}) \\ \text{and} \qquad & \omega_t^K = \pi^*(\omega_0^K) + O(t^4 r^{-4}). \end{aligned} \tag{4.5}$$

4.2. The Kummer Construction. The $K3$ *surface* is a compact 4-manifold which has a family of complex structures, each making the $K3$ into a complex surface. These complex surfaces are of particular interest to algebraic geometers. From Yau's proof of the Calabi conjecture [**34**], it is known that the $K3$ surface possesses a 58-parameter family of metrics of holonomy $SU(2)$. An approximate description of some of these metrics was given by Page [**28**], which employs an idea known as the *Kummer construction.*

Proofs of the existence of metrics of holonomy $SU(2)$ on $K3$ using Page's idea have been given by Topiwala [**33**] and LeBrun and Singer [**25**] using twistor theory, and by the author [**17**, Ex. 1] using analysis. Here is a brief sketch of the Kummer construction for the $K3$ surface, and the metrics of holonomy $SU(2)$ upon it.

Let $\mathbb{C}^2$ have complex coordinates (z_1, z_2) and metric g_0 as above. Define a subset Λ of $\mathbb{C}^2$ by $\Lambda = \{(a+ib, c+id) : a,b,c,d \in \mathbb{Z}\}$. Then Λ is a *lattice* in $\mathbb{C}^2$, and $\mathbb{C}^2/\Lambda$ is the 4-torus T^4. It is a complex manifold, with flat Kähler metric g_0. Define an involution $\sigma : T^4 \to T^4$ by

$$\sigma : (z_1, z_2) + \Lambda \longmapsto (-z_1, -z_2) + \Lambda. \tag{4.6}$$

Then σ has 16 fixed points, the points $(z_1, z_2)+\Lambda$ with $z_j \in \{0, \frac{1}{2}, \frac{1}{2}i, \frac{1}{2}+\frac{1}{2}i\}$. Thus $T^4/\langle\sigma\rangle$ is a singular complex manifold with a flat Kähler metric g_0, and 16 singular points, $s_1, \ldots, s_{16}$ say, modelled on 0 in $\mathbb{C}^2/\{\pm 1\}$.

Let Y be the blow-up of $T^4/\langle\sigma\rangle$ at each singular point s_j. Then Y is a compact, nonsingular complex surface called a $K3$ surface. The blow-ups replace each singular point with a complex curve $\mathbb{CP}^1$. These blow-ups are modelled on the construction of the Eguchi-Hanson space X. Hence, we may regard Y as the result of gluing 16 copies of X into the orbifold $T^4/\langle\sigma\rangle$.

Define subsets $A_j = \{y \in Y : d(\pi(y), s_j) < \frac{1}{5}\}$ in Y for $j = 1, \ldots, 16$, and let $B = \{y \in Y : d(\pi(y), s_j) > \frac{1}{6} \text{ for } j = 1, \ldots, 16\}$, where $d(,)$ is the metric on $T^4/\langle\sigma\rangle$ induced by g_0. Then Y is the union of the A_j and B. The intersections $A_j \cap A_k$ are empty for $j \neq k$, but $A_j \cap B$ is an 'annulus' diffeomorphic to $(\frac{1}{6}, \frac{1}{5}) \times S^3/\langle\sigma\rangle$.

Now A_j is naturally isomorphic to an open subset of the Eguchi-Hanson space X. Therefore on each A_j we have a family $(\omega_t^I, \omega_t^J, \omega_t^K)$ of torsion-free $SU(2)$-structures, depending on $t > 0$. Similarly, B is naturally isomorphic to an open subset of $T^4/\langle\sigma\rangle$, so on B we have the flat $SU(2)$-structure $(\omega_0^I, \omega_0^J, \omega_0^K)$. On the overlap $A_j \cap B$, we can compare these two $SU(2)$-structures using (4.5). Since $r \in (\frac{1}{6}, \frac{1}{5})$ on $A_j \cap B$, we have $r^{-4} < 6^4$, and so

$$\omega_t^I = \omega_0^I + O(t^4), \quad \omega_t^J = \omega_0^J + O(t^4), \quad \text{and} \quad \omega_t^K = \omega_0^K + O(t^4) \quad \text{on } A_j \cap B. \tag{4.7}$$

This equation shows that when $t > 0$ is small, the $SU(2)$-structures $(\omega_t^I, \omega_t^J, \omega_t^K)$ defined on A_j, and $(\omega_0^I, \omega_0^J, \omega_0^K)$ defined on B, are close to each other on $A_j \cap B$. We can exploit this fact to construct torsion-free $SU(2)$-structures on Y. The argument runs as follows. First one uses a partition of unity to write down an $SU(2)$-structure $(\hat\omega_t^I, \hat\omega_t^J, \hat\omega_t^K)$ on Y, that equals $(\omega_t^I, \omega_t^J, \omega_t^K)$ on $A_j \setminus B$ and $(\omega_0^I, \omega_0^J, \omega_0^K)$ on $B \setminus \bigcup_j A_j$, and interpolates smoothly between the two on the overlaps $A_j \cap B$.

Then $(\hat\omega_t^I, \hat\omega_t^J, \hat\omega_t^K)$ is torsion-free on $A_j \setminus B$ and $B \setminus \bigcup_j A_j$, but on $A_j \cap B$ it has nonzero torsion. However, using (4.7) one can ensure that the torsion is $O(t^4)$, and so when t is small, the torsion is also small. Finally, using analysis one proves that an $SU(2)$-structure on Y with sufficiently small torsion, can be deformed to a nearby, torsion-free $SU(2)$-structure. Therefore, for all small enough $t > 0$ we construct a new, torsion-free $SU(2)$-structure $(\tilde\omega_t^I, \tilde\omega_t^J, \tilde\omega_t^K)$ on Y that is close to $(\omega_t^I, \omega_t^J, \omega_t^K)$ on A_j, and close to $(\omega_0^I, \omega_0^J, \omega_0^K)$ on B. The metric $\tilde g_t$ associated to this $SU(2)$-structure is a metric on the $K3$ surface with holonomy $SU(2)$.

5. Compact Manifolds with Holonomy G_2 and $Spin(7)$

In the rest of this chapter we will describe a construction for compact Riemannian 7-manifolds with holonomy G_2, and a very similar construction for compact Riemannian 8-manifolds with holonomy $Spin(7)$. Both constructions are motivated by and modelled on the Kummer construction of §4 for metrics of holonomy $SU(2)$ on the $K3$ surface. They can be divided into four steps. Here is a summary of each. For simplicity we will describe the G_2 case only, but the $Spin(7)$ case is very similar.

Step 1. Let T^7 be the 7-torus. Let (φ_0, g_0) be a flat G_2-structure on T^7. Choose a finite group Γ of isometries of T^7 preserving (φ_0, g_0). Then the quotient T^7/Γ is a singular, compact 7-manifold.

For certain special groups Γ there is a method to resolve the singularities of T^7/Γ in a natural way, using complex geometry. We get a nonsingular, compact 7-manifold M, together with a map $\pi : M \to T^7/\Gamma$, the resolving map.

Step 2. On M, we explicitly write down a 1-parameter family of G_2-structures (φ_t, g_t) depending on a real variable $t \in (0, \epsilon)$. These G_2-structures are not torsion-free, but when t is small, they have small torsion. As $t \to 0$, the G_2-structure (φ_t, g_t) converges to the singular G_2-structure $\pi^*(\varphi_0, g_0)$.

Step 3. We prove using analysis that for all sufficiently small t, the G_2-structure (φ_t, g_t) on M, with small torsion, can be deformed to a G_2-structure $(\tilde{\varphi}_t, \tilde{g}_t)$, with zero torsion.

Step 4. Finally, we show that $\tilde{g}_t$ is a metric with holonomy G_2 on the compact 7-manifold M, using topological invariants of M.

We shall explain Steps 1-4 at much greater length in sections 6-9 respectively. By considering different groups Γ acting on T^7 and T^8, we are able to find metrics with holonomy G_2 and $Spin(7)$ on many topologically distinct 7- and 8-manifolds. It also happens that the same orbifold T^k/Γ can admit several topologically distinct resolutions, and this increases the number of examples.

In **[17]**, the author gave examples of 68 distinct, compact 7-manifolds with holonomy G_2, and in **[15]**, examples of 95 distinct, compact 8-manifolds with holonomy $Spin(7)$. The forthcoming book **[19]** will provide many more examples, by using more powerful mathematical tools to resolve singularities, and by studying the possibilities for the finite group Γ in a systematic way.

6. Orbifolds and Resolutions

This section explains Step 1 of §5 in greater detail. For simplicity, we will mostly confine our attention to the case of 7-manifolds and holonomy G_2, but the case of 8-manifolds with holonomy $Spin(7)$ is very similar. Section 6.1 introduces orbifolds T^7/Γ with flat G_2-structures, and their singular points. Then §6.2 discusses resolutions of T^7/Γ, and describes a special way of resolving orbifolds T^7/Γ using complex geometry. We shall see later that a resolution constructed in this way admits a family of torsion-free G_2-structures. Section 6.3 gives a simple example of an orbifold T^7/Γ, and how to resolve it, and §6.4 gives a similar example in 8 dimensions.

6.1. Orbifolds of T^7 with Flat G_2-Structures. Let (φ_0, g_0) be the standard, flat G_2-structure on $\mathbb{R}^7$, given in Definition 2.1. Let Λ be a *lattice* in $\mathbb{R}^7$, so that $\Lambda \cong \mathbb{Z}^7$. Then Λ acts as a group of translations on $\mathbb{R}^7$, and the quotient $\mathbb{R}^7/\Lambda$ is the 7-torus T^7. Moreover, this action of Λ on $\mathbb{R}^7$ preserves φ_0 and g_0, and thus there is a flat G_2-structure (φ_0, g_0) on T^7.

Now the group of linear transformations of $\mathbb{R}^7$ preserving (φ_0, g_0) is $G_2 \subset GL(7,\mathbb{R})$, and the group of linear transformations of $\mathbb{R}^7$ preserving Λ is $GL(7,\mathbb{Z}) \subset GL(7,\mathbb{R})$. Thus, the subgroup of $GL(7,\mathbb{R})$ preserving both (φ_0, g_0) and Λ is $F = G_2 \cap GL(7,\mathbb{Z})$. Here F is a finite group, as it is both discrete and compact. Note that the embedding of $GL(7,\mathbb{Z})$ in $GL(7,\mathbb{R})$, and thus the finite group F, is not fixed but depends on the choice of lattice Λ in $\mathbb{R}^7$. The group F acts on T^7 preserving the G_2-structure (φ_0, g_0). But T^7 acts on itself by translations, and this action also preserves (φ_0, g_0). Together, these actions of F and T^7 on T^7 generate a group $F \ltimes T^7$, which turns out to be the full group of automorphisms of T^7 that preserve (φ_0, g_0).

Let Γ be a finite subgroup of $F \ltimes T^7$. Then Γ acts on T^7 preserving (φ_0, g_0), and thus T^7/Γ is an *orbifold*, equipped with a flat G_2-structure (φ_0, g_0). The singular points of T^7/Γ are easy to describe. Let $x \in T^7$, so that $x\Gamma \in T^7/\Gamma$. Let $\Gamma_x = \{\gamma \in \Gamma : \gamma(x) = x\}$ be the *stabilizer* of x. If $\Gamma_x = \{1\}$ then $x\Gamma$ is a *nonsingular point* of T^7/Γ. If $\Gamma_x \neq \{1\}$ then $x\Gamma$ is a *singular point* of T^7/Γ. Moreover, Γ_x acts naturally on $\mathbb{R}^7$, and the singularity at $x\Gamma$ is modelled on the singularity at 0 of $\mathbb{R}^7/\Gamma_x$. This action of Γ_x on $\mathbb{R}^7$ makes Γ_x into a subgroup of G_2.

Thus, $x\Gamma$ is a singular point of T^7/Γ if and only if x is fixed by some nonidentity element of Γ. It is convenient to adopt the following notation: for each nonidentity element $\gamma \in \Gamma$, define S_γ to be the set of points $x\Gamma \in T^7/\Gamma$ for $x \in T^7$ with $\gamma(x) = x$. Define S to be the set of singular points of T^7/Γ. Then S is the union (not necessarily a disjoint union) of the S_γ for $\gamma \neq 1$ in Γ. Suppose that $\alpha \neq 1$ is an element of G_2. Then the subset of $\mathbb{R}^7$ fixed by α is either $\mathbb{R}$ or $\mathbb{R}^3$. Therefore, if $\gamma \neq 1$ lies in Γ then S_γ is either empty, or has dimension 1 or 3. Hence the singular set S is a finite union of (singular) 1-manifolds and 3-manifolds in T^7/Γ.

6.2. Resolutions of T^7/Γ with Torsion-Free G_2-Structures. Our goal is to resolve the singularities of the orbifold T^7/Γ to get a compact 7-manifold M, and to construct a family of torsion-free G_2-structures on M that are in some sense close to the singular G_2-structure (φ_0, g_0) on T^7/Γ. By a *resolution* of T^7/Γ, we mean a pair (M, π), where M is a compact 7-manifold and $\pi : M \to T^7/\Gamma$ a continuous map, such that the restriction $\pi : M \setminus \pi^{-1}(S) \to (T^7/\Gamma) \setminus S$ is a diffeomorphism, and for each $s \in S$, the subset $\pi^{-1}(s)$ is a finite union of compact submanifolds of M.

In general T^7/Γ has not just one, but infinitely many resolutions (M, π). However, nearly all of these resolutions are unsuitable for our purposes, and most of them do not even admit G_2-structures. To be able to construct torsion-free G_2-structures on M, we must restrict our attention to orbifolds T^7/Γ with a particular kind of singularity, and then resolve these singularities in a special way.

Let $x\Gamma$ be a *generic* point of S. Then the singularity of T^7/Γ at $x\Gamma$ is modelled on that of $\mathbb{R}^7/\Gamma_x$ at 0. Now we saw above that the singular set S of T^7/Γ is a finite union of (singular) 1-manifolds and 3-manifolds. As $x\Gamma$ is generic, near $x\Gamma$ we see that S is nonsingular and of dimension 1 or 3. Suppose first that S is of dimension 1 near $x\Gamma$. Then there is a natural splitting $\mathbb{R}^7 \cong \mathbb{R} \oplus \mathbb{R}^6$ preserved by Γ_x, and Γ_x

acts trivially on $\mathbb{R}$ and freely on $\mathbb{R}^6 \setminus \{0\}$. As Γ_x preserves the G_2-structure (φ_0, g_0) on $\mathbb{R}^7$, it follows that Γ_x is a subgroup of G_2. But the subgroup of G_2 fixing a subspace $\mathbb{R} \subset \mathbb{R}^7$ is $SU(3)$, so that Γ_x is a subgroup of $SU(3)$.

Similarly, if S has dimension 3 near $x\Gamma$, we can show that Γ_x lies in a subgroup $SU(2)$ of G_2. Thus, if $x\Gamma$ is a generic singular point of T^7/Γ, then one of two possibilities holds:

(*i*) There is a natural splitting $\mathbb{R}^7 \cong \mathbb{R}^3 \oplus \mathbb{C}^2$, and $SU(2)$ acts trivially on $\mathbb{R}^3$ and in the usual way on $\mathbb{C}^2$. The group Γ_x is a finite subgroup of $SU(2)$ which acts freely on $\mathbb{C}^2 \setminus \{0\}$, and $\mathbb{R}^7/\Gamma_x \cong \mathbb{R}^3 \times (\mathbb{C}^2/\Gamma_x)$.

(*ii*) There is a natural splitting $\mathbb{R}^7 \cong \mathbb{R} \oplus \mathbb{C}^3$, and $SU(3)$ acts trivially on $\mathbb{R}$ and in the usual way on $\mathbb{C}^3$. The group Γ_x is a finite subgroup of $SU(3)$ which acts freely on $\mathbb{C}^3 \setminus \{0\}$, and $\mathbb{R}^7/\Gamma_x \cong \mathbb{R} \times (\mathbb{C}^3/\Gamma_x)$.

The key observation is that the singularities $\mathbb{C}^2/\Gamma_x$ and $\mathbb{C}^3/\Gamma_x$ occurring here are *complex singularities.* Now in the field of complex algebraic geometry, the problem of resolving singularities of complex manifolds has been studied for many years, and is very well understood in complex dimensions 2 and 3. In particular, if F is a nontrivial finite subgroup of $SU(2)$, it is known that the quotient singularity $\mathbb{C}^2/F$ can be resolved in a unique way to give a complex manifold X, which has a family of ALE metrics with holonomy $SU(2)$. These metrics were explicitly constructed and classified by Kronheimer [**22, 23**]. In the simplest case $F = \{\pm 1\}$, we get the Eguchi-Hanson space of §4.1.

Similar results are known for the case of $SU(3)$. If F is any finite subgroup of $SU(3)$, Roan [**29**] has shown that $\mathbb{C}^3/\Gamma$ admits a suitable complex resolution Y. The family of ALE metrics with holonomy $SU(3)$ on Y are not known explicitly except in special cases, but the author [**19**] has proved that such metrics exist in every case, by following Yau's proof of the Calabi conjecture. Tian and Yau [**32**] have also proved some related results.

Suppose X is a resolution of $\mathbb{C}^2/F$, and g_X is an ALE metric on X with holonomy $SU(2)$. Let $g_{\mathbb{R}^3}$ be a flat metric on $\mathbb{R}^3$. Then $g = g_{\mathbb{R}^3} \times g_X$ is a metric on $\mathbb{R}^3 \times X$ which has holonomy $\mathrm{Hol}(g) = \{1\} \times SU(2)$. Since $\{1\} \times SU(2) \subset G_2$, the metric g extends to a torsion-free G_2-structure (φ, g) on $\mathbb{R}^3 \times X$, which is asymptotic to the flat G_2-structure (φ_0, g_0) on $\mathbb{R}^7/F$ as one approaches infinity in X. In the same way, if Y is a resolution of $\mathbb{C}^3/F$ and g_Y an ALE metric on Y with holonomy $SU(3)$, we may construct torsion-free G_2-structures (φ, g) on $\mathbb{R} \times Y$ that are asymptotic to (φ_0, g_0) on $\mathbb{R}^7/F$.

Thus, if $x\Gamma$ is a singular point of T^7/Γ and Γ_x lies in some subgroup of G_2 conjugate to $SU(2)$ or $SU(3)$, then one can resolve the quotient singularity $\mathbb{R}^7/\Gamma_x$ using complex geometry, in such a way that the resolution carries a family of torsion-free G_2-structures (φ, g) asymptotic to (φ_0, g_0).

For our construction, we first choose an orbifold T^7/Γ for which Γ_x lies in $SU(2)$ or $SU(3)$ for every singular point $x\Gamma$ of T^7/Γ. Next, we build a resolution (M, π) of T^7/Γ, which is modelled at each singular point on the resolution from complex geometry described above. Then, by gluing together torsion-free G_2-structures on the different regions of M, we can write down a family of G_2-structures on M with small torsion. This will be explained in §7, but first we will give examples of orbifolds T^7/Γ and T^8/Γ, and how to resolve them.

6.3. An Example of an Orbifold T^7/Γ and its Resolution. We begin with an example of a suitable group Γ. Let $(x_1, \ldots, x_7)$ be coordinates on $T^7 =$

$\mathbb{R}^7/\mathbb{Z}^7$, where $x_i \in \mathbb{R}/\mathbb{Z}$. Let (φ_0, g_0) be the flat G_2-structure on T^7 defined by (2.1) and (2.2). Let α, β and γ be the involutions of T^7 defined by

$$\alpha\big((x_1,\ldots,x_7)\big) = (-x_1,-x_2,-x_3,-x_4,x_5,x_6,x_7), \tag{6.1}$$

$$\beta\big((x_1,\ldots,x_7)\big) = (-x_1,\tfrac{1}{2}-x_2,x_3,x_4,-x_5,-x_6,x_7), \tag{6.2}$$

$$\gamma\big((x_1,\ldots,x_7)\big) = (\tfrac{1}{2}-x_1,x_2,\tfrac{1}{2}-x_3,x_4,-x_5,x_6,-x_7). \tag{6.3}$$

By inspection, α, β and γ preserve (φ_0, g_0), because of the careful choice of exactly which signs to change. Also, $\alpha^2 = \beta^2 = \gamma^2 = 1$, and α, β and γ commute. Thus they generate a group $\Gamma = \langle \alpha, \beta, \gamma \rangle \cong \mathbb{Z}_2^3$ of isometries of T^7 preserving the flat G_2-structure (φ_0, g_0). The following Lemma is proved in [**16**, §2.1].

LEMMA 6.1. *The only nonidentity elements of Γ with fixed points are α, β and γ. Each of S_α, S_β and S_γ are 4 copies of T^3, so that the singular set S of T^7/Γ is a disjoint union of 12 copies of T^3. Every component of S is a singularity modelled on that of $T^3 \times \mathbb{C}^2/\{\pm 1\}$.*

Now the natural resolution of the complex singularity $\mathbb{C}^2/\{\pm 1\}$ is the Eguchi-Hanson space X of §4.1, with its resolving map $\pi : X \to \mathbb{C}^2/\{\pm 1\}$. Therefore, $T^3 \times X$ is a resolution of the singularity $T^3 \times \mathbb{C}^2/\{\pm 1\}$, with resolving map $\pi : T^3 \times X \to T^3 \times \mathbb{C}^2/\{\pm 1\}$. Each component of S is modelled on $T^3 \times \mathbb{C}^2/\{\pm 1\}$, and we resolve T^7/Γ by replacing this with $T^3 \times X$, using the resolving map π. In this way we construct a compact, nonsingular 7-manifold M with a map $\pi : M \to T^7/\Gamma$, making (M, π) into a resolution of T^7/Γ. Later we will construct a family of metrics with holonomy G_2 on M.

6.4. An Example of an Orbifold T^8/Γ and Its Resolution. Let $(x_1, \ldots, x_8)$ be coordinates on $T^8 = \mathbb{R}^8/\mathbb{Z}^8$, where $x_i \in \mathbb{R}/\mathbb{Z}$. Define a flat *Spin*(7)-structure (Ω_0, g_0) on T^8 as in Definition 3.1. Let α, β, γ and δ be the involutions of T^8 defined by

$$\alpha\big((x_1,\ldots,x_8)\big) = (-x_1,-x_2,-x_3,-x_4,x_5,x_6,x_7,x_8), \tag{6.4}$$

$$\beta\big((x_1,\ldots,x_8)\big) = (x_1,x_2,x_3,x_4,-x_5,-x_6,-x_7,-x_8), \tag{6.5}$$

$$\gamma\big((x_1,\ldots,x_8)\big) = (\tfrac{1}{2}-x_1,\tfrac{1}{2}-x_2,x_3,x_4,\tfrac{1}{2}-x_5,\tfrac{1}{2}-x_6,x_7,x_8), \tag{6.6}$$

$$\delta\big((x_1,\ldots,x_8)\big) = (-x_1,x_2,\tfrac{1}{2}-x_3,x_4,\tfrac{1}{2}-x_5,x_6,\tfrac{1}{2}-x_7,x_8). \tag{6.7}$$

By inspection, α, β, γ and δ preserve Ω_0 and g_0. It is easy to see that $\alpha^2 = \beta^2 = \gamma^2 = \delta^2 = 1$, and that $\alpha, \beta, \gamma, \delta$ all commute. Define Γ to be the group $\langle \alpha, \beta, \gamma, \delta \rangle$. Then $\Gamma \cong (\mathbb{Z}_2)^4$ is a group of automorphisms of T^8 preserving (Ω_0, g_0).

The following Lemma is proved in [**15**, §3.2].

LEMMA 6.2. *The only nonidentity elements of Γ with fixed points in T^8 are $\alpha, \beta, \gamma, \delta$ and $\alpha\beta$. The corresponding singular sets are as follows:*

(*i*) *S_α is 4 copies of $T^4/\{\pm 1\}$,*
(*ii*) *S_β is 4 copies of $T^4/\{\pm 1\}$,*
(*iii*) *S_γ is 2 copies of T^4,*
(*iv*) *S_δ is 2 copies of T^4, and*
(*v*) *$S_{\alpha\beta}$ is 64 points.*

Here S_α and S_β intersect in $S_{\alpha\beta}$. Each point $x\Gamma \in S_{\alpha\beta}$ has stabilizer $\Gamma_x = \{1, \alpha, \beta, \alpha\beta\}$.

Thus, the singular set of T^8/Γ is rather more complex than that of the previous example. However, we can still resolve T^8/Γ to get a compact 8-manifold M, using only the Eguchi-Hanson space X. First consider S_γ, which is two disjoint copies of T^4, each modelled on the singularity of $T^4 \times \mathbb{C}^2/\{\pm 1\}$. The resolution of this singularity is $T^4 \times X$, and we may use this to resolve both S_γ and S_δ.

Now the singular sets S_α and S_β are not disjoint, but rather, each component $T^4/\{\pm 1\}$ in S_α meets each component $T^4/\{\pm 1\}$ in S_β in 4 points, which lie in $S_{\alpha\beta}$. The singularity at each point in $S_{\alpha\beta}$ is modelled on $\mathbb{C}^2/\{\pm 1\} \times \mathbb{C}^2/\{\pm 1\}$. Here S_α corresponds locally to the subset $\mathbb{C}^2/\{\pm 1\} \times \{0\}$, and S_β corresponds locally to the subset $\{0\} \times \mathbb{C}^2/\{\pm 1\}$. Now the natural resolution of $\mathbb{C}^2/\{\pm 1\} \times \mathbb{C}^2/\{\pm 1\}$ is $X \times X$, and this is how we resolve near each point in $S_{\alpha\beta}$.

Each component $T^4/\{\pm 1\}$ of S_α and S_β is modelled locally on $T^4/\{\pm 1\} \times \mathbb{C}^2/\{\pm 1\}$, and the resolution of this is $K3 \times X$, where $T^4/\{\pm 1\}$ is resolved to give the $K3$ surface using the Eguchi-Hanson space X, as in §4.2. Combining these resolutions gives a compact 8-manifold M with a map $\pi : M \to T^8/\Gamma$, making (M, π) a resolution of T^8/Γ. In [**15**], the author constructs a family of metrics with holonomy $Spin(7)$ on M.

7. G_2- and $Spin(7)$-Structures with Small Torsion

Now we will explain Step 2 of §5, concentrating on the G_2 case. Suppose that we are given an orbifold T^7/Γ equipped with a flat G_2-structure (φ_0, g_0), and a resolution (M, π) of T^7/Γ, constructed in the way explained in §6.2. Let S be the singular set of T^7/Γ, and let $S_1, \ldots, S_l$ be the connected components of S. Now, for simplicity, and in order to be as explicit as possible, we shall suppose that each component S_j is a copy of T^3, and the corresponding singularity is modelled on $T^3 \times \mathbb{C}^2/\Gamma_j$, where Γ_j is a finite subgroup of $SU(2)$. Then each S_j is desingularized using $T^3 \times X_j$, where X_j is a complex resolution of $\mathbb{C}^2/\Gamma_j$. This is the case in the example of §6.3. However, the methods we use also work with more complicated singularities, as explained in [**17**, §2.2].

Let ζ be a positive constant, and define $A_j = \{m \in M : d(\pi(m), S_j) < 2\zeta\}$ for $j = 1, \ldots, l$, and $B = \{m \in M : d(\pi(m), S_j) > \zeta \text{ for } j = 1, \ldots, l\}$, where $d(\,,\,)$ is the metric on T^7/Γ induced by g_0. Let ζ be chosen sufficiently small that the A_j are all disjoint, and each A_j is of the form $A_j \cong T^3 \times Y_j$, where Y_j is an open subset in X_j with boundary S^3/Γ_j. Then, as in §4.2, our manifold M is the union of open sets A_j and B, where the A_j are disjoint, but $A_j \cap B$ is an 'annulus' diffeomorphic to the product $T^3 \times (\zeta, 2\zeta) \times S^3/\Gamma_j$.

The subset B is naturally isomorphic to an open subset of T^7/Γ, and therefore it carries a flat G_2-structure (φ_0, g_0). Consider the restriction of (φ_0, g_0) to $A_j \cap B$. We may identify $A_j \cap B$ with an open subset of $T^3 \times \mathbb{C}^2/\Gamma_j$, and under this identification, φ_0 has the form

$$\varphi_0 = \nu^I \wedge \omega_0^I + \nu^J \wedge \omega_0^J + \nu^K \wedge \omega_0^K + \nu^I \wedge \nu^J \wedge \nu^K, \tag{7.1}$$

where ν^I, ν^J and ν^K are constant, linearly independent 1-forms on T^3, and $(\omega_0^I, \omega_0^J, \omega_0^K)$ is the flat $SU(2)$-structure on $\mathbb{C}^2/\Gamma_j$, as in §4.

Now, as we explained in §6.2, it is known that the manifold X_j admits a family of ALE metrics with holonomy $SU(2)$. Therefore, as in §4.1, for each $t > 0$ we can find a torsion-free $SU(2)$-structure $(\omega_t^I, \omega_t^J, \omega_t^K)$ on X_j, satisfying the asymptotic conditions

$$(7.2) \quad \omega_t^I = \omega_0^I + O(t^4 r^{-4}), \quad \omega_t^J = \omega_0^J + O(t^4 r^{-4}) \quad \text{and} \quad \omega_t^K = \omega_0^K + O(t^4 r^{-4})$$

near infinity in X_j. Motivated by (7.1), define a 3-form φ_t' on A_j by

$$(7.3) \qquad \varphi_t' = \nu^I \wedge \omega_t^I + \nu^J \wedge \omega_t^J + \nu^K \wedge \omega_t^K + \nu^I \wedge \nu^J \wedge \nu^K.$$

Then it turns out that φ_t' induces a *torsion-free G_2-structure* (φ_t', g_t') on A_j. Moreover, since $r \in (\zeta, 2\zeta)$ on $A_j \cap B$, equations (7.1) and (7.2) imply that

$$(7.4) \qquad \varphi_t' = \varphi_0 + O(t^4) \qquad \text{on } A_j \cap B.$$

Thus, on the subset A_j of M we have a torsion-free G_2-structure (φ_t', g_t') for each $t > 0$, and on the subset B of M we have a torsion-free G_2-structure (φ_0, g_0). On the overlaps $A_j \cap B$, the difference between the G_2-structures is $O(t^4)$ by (7.4), and so when t is small, the two G_2-structures are close together. It is easy to use a partition of unity to write down a G_2-structure (φ_t, g_t) on M, which equals (φ_t', g_t') on $A_j \setminus B$ and equals (φ_0, g_0) on $B \setminus \bigcup_j A_j$, and interpolates smoothly between the two on the intersections $A_j \cap B$. This G_2-structure will be torsion-free on $A_j \setminus B$ and $B \setminus \bigcup_j A_j$, but will have nonzero torsion on $A_j \cap B$.

Now, for the purposes of the analysis in the next section, we need to estimate three geometric invariants of this G_2-structure (φ_t, g_t). These are the *torsion* $\nabla\varphi_t$, the *injectivity radius* $\delta(g_t)$ of g_t, and the *Riemann curvature* $R(g_t)$ of g_t. Using (7.4), one can ensure that $\|\nabla\varphi_t\|_{C^0} = O(t^4)$, and this is an estimate for the torsion. Let $g_{j,t}$ be the ALE metric with holonomy $SU(2)$ on X_j used in the construction. One can choose the metrics $g_{j,t}$ to be *homothetic* for all t, so that after applying an automorphism of X_j depending on t, we may choose $g_{j,t}$ to be isometric to $t^2 g_{j,1}$ for each $t > 0$.

It immediately follows that $\delta(g_{j,t}) = t\,\delta(g_{j,1})$ and $\|R(g_{j,t})\|_{C^0} = t^{-2}\|R(g_{j,1})\|_{C^0}$. When t is small, it is easy to see that the injectivity radius and curvature of the ALE metrics $g_{j,t}$ make the dominant contribution to the injectivity radius and curvature of the g_t on M. Therefore, we expect $\delta(g_t)$ to be $O(t)$, and $\|R(g_t)\|_{C^0}$ to be $O(t^{-2})$ for small t. Arguing in this way, one may estimate $\delta(g_t)$ and $\|R(g_t)\|_{C^0}$.

We state the existence and some important estimates for the G_2-structures (φ_t, g_t) in the following Theorem, which summarizes the results of this stage of the construction.

THEOREM 7.1. *On the compact 7-manifold M described in §6.3, and on many other compact 7-manifolds constructed in a similar fashion, one can write down the following data explicitly in coordinates:*

- *Positive constants A_1, A_2, A_3 and ϵ,*
- *A G_2-structure (φ_t, g_t) on M with $d\varphi_t = 0$ for each $t \in (0, \epsilon)$, and*
- *A 3-form ψ_t on M with $d^*\psi_t = d^*\varphi_t$ for each $t \in (0, \epsilon)$.*

These satisfy the three conditions:

(*i*) $\|\psi_t\|_{L^2} \le A_1 t^4$ *and* $\|d^*\psi_t\|_{L^{14}} \le A_1 t^4$,

(*ii*) *the injectivity radius $\delta(g_t)$ satisfies $\delta(g_t) \ge A_2 t$, and*

(*iii*) *the Riemann curvature $R(g_t)$ satisfies $\|R(g_t)\|_{C^0} \le A_3 t^{-2}$.*

Here the operator d^* and the norms $\|\,.\,\|_{L^2}$, $\|\,.\,\|_{L^{14}}$ and $\|\,.\,\|_{C^0}$ depend on g_t.

For a proof of this result, see [**16**, §2.2] and [**17**, §2.2]. Here is a brief explanation. From Proposition 2.2 we see that $\nabla\varphi_t = 0$ if and only if $d\varphi_t = d^*\varphi_t = 0$.

It turns out to be more convenient to work with $d\varphi_t$ and $d^*\varphi_t$, rather than $\nabla\varphi_t$. Also, it is possible to choose the G_2-structure (φ_t, g_t) on M to satisfy $d\varphi_t = 0$, which means that the only nonzero component of the torsion left is $d^*\varphi_t$.

The 3-form ψ_t appearing in Theorem 7.1 should be interpreted as a *first integral* of $d^*\varphi_t$. Since $d^*\varphi_t = d^*\psi_t$, part (i) of the Theorem implies that $\|d^*\varphi_t\|_{L^{14}} \leq A_1 t^4$. Thus, part (i) gives an $O(t^4)$ estimate on the torsion $\nabla\varphi_t$ of (φ_t, g_t). The reasons for choosing φ_t closed, and introducing ψ_t in the way we have, will become clear in the next section.

For the $Spin(7)$ construction, the result corresponding to Theorem 7.1 is the following, which is proved in [**15**, §4].

THEOREM 7.2. *On the compact 8-manifold M given in §6.4, and on many other compact 8-manifolds, one can write down the following data explicitly in coordinates:*

- *Positive constants A_1, A_2, A_3 and ϵ,*
- *A $Spin(7)$-structure (Ω_t, g_t) on M for each $t \in (0,\epsilon)$, and*
- *A 4-form ϕ_t on M for each $t \in (0,\epsilon)$ satisfying $d\Omega_t + d\phi_t = 0$.*

These satisfy the three conditions:

(i) $\|\phi_t\|_{L^2} \leq A_1 t^{9/2}$ *and* $\|d\phi_t\|_{L^{10}} \leq A_1 t$,

(ii) *the injectivity radius $\delta(g_t)$ of g_t satisfies $\delta(g_t) \geq A_2 t$, and*

(iii) *the Riemann curvature $R(g_t)$ of g_t satisfies $\big\|R(g_t)\big\|_{C^0} \leq A_3 t^{-2}$.*

Here all norms are taken w.r.t. the metric g_t on M.

8. Deforming to Torsion-Free G_2- and $Spin(7)$-Structures

Now we explain Step 3 of §5, which in the G_2 case is accomplished by the following Theorem.

THEOREM 8.1. *In the situation of Theorem 7.1, there are positive constants κ, K depending only on A_1, A_2, A_3 and ϵ, such that for every t with $0 < t \leq \kappa$, there exists a smooth, torsion-free G_2-structure $(\tilde\varphi_t, \tilde g_t)$ on M with $\|\tilde\varphi_t - \varphi_t\|_{C^0} \leq K t^{1/2}$.*

This result is proved in [**16**, §3]. The proof is not easy, and it represents most of the hard work in [**16**]. Note that the proof given there also involves estimates on the volume and diameter of the Riemannian manifold (M, g_t), but it turns out that these are unnecessary and can be removed. An improved proof will be given in [**19**].

The rest of this section gives a sketch of the proof of Theorem 8.1, ignoring several technical points. The treatment follows that given in [**19**], which differs a little from that of [**16**]. We begin in §8.1 with an aside on G_2-structures and exterior forms, which leads in §8.2 to a way to write the deformation problem as a nonlinear elliptic partial differential equation. Section 8.3 then explains how to construct a smooth solution using analytic methods.

8.1. G_2-Structures and Forms on 7-Manifolds. Let φ_0 and $*\varphi_0$ be the 3- and 4-forms defined on $\mathbb{R}^7$ in (2.1) and (2.3). Now $GL(7,\mathbb{R})$ acts linearly on $\mathbb{R}^7$, and this induces an action of $GL(7,\mathbb{R})$ on $\Lambda^3(\mathbb{R}^7)^*$. Let $\mathcal{P}^3$ be the orbit of φ_0 in $\Lambda^3(\mathbb{R}^7)^*$ under this action. From §2, the stabilizer of φ_0 in $GL(7,\mathbb{R})$ is G_2, and therefore $\mathcal{P}^3$ is isomorphic to $GL(7,\mathbb{R})/G_2$. Since $\dim GL(7,\mathbb{R}) = 49$ and $\dim G_2 = 14$, it follows that $\dim \mathcal{P}^3 = 49 - 14 = 35$. But $\dim \Lambda^3(\mathbb{R}^7)^*$ is also $\binom{7}{3} = 35$. Thus, $\mathcal{P}^3$ is an *open set* in $\Lambda^3(\mathbb{R}^7)^*$. This means that the 3-form φ_0 used to define G_2-structures is a generic 3-form on $\mathbb{R}^7$.

In the same way, let $\mathcal{P}^4$ be the orbit of the 4-form $*\varphi_0$ under the action of $GL(7,\mathbb{R})$ on $\Lambda^4(\mathbb{R}^7)^*$. The stabilizer of $*\varphi_0$ is $G_2 \times \{\pm 1\}$, giving $\dim \mathcal{P}^4 = 35 = \dim \Lambda^4(\mathbb{R}^7)^*$, and so $\mathcal{P}^4$ is an open set in $\Lambda^4(\mathbb{R}^7)^*$. Now each element φ of $\mathcal{P}^3$ determines a unique G_2-structure (φ, g) on $\mathbb{R}^7$, and this in turn determines a 4-form $*\varphi$ in $\mathcal{P}^4$. Define a map $\Theta : \mathcal{P}^3 \to \mathcal{P}^4$ by $\Theta(\varphi) = *\varphi$. It is important to note that Θ is a *nonlinear* map. This is because the Hodge star $*$ used to define $*\varphi$ depends on the metric g, but g itself depends on φ.

Next, we extend these ideas from $\mathbb{R}^7$ to a general 7-manifold M. Let $\mathcal{F}$ be the *frame bundle* of M, which is a principal bundle with fibre $GL(7,\mathbb{R})$. Then we may write $\Lambda^3 T^*M = \mathcal{F} \times_{GL(7,\mathbb{R})} \Lambda^3(\mathbb{R}^7)^*$. Now $\mathcal{P}^3$ is a subset of $\Lambda^3(\mathbb{R}^7)^*$ invariant under $GL(7,\mathbb{R})$. Therefore we may define $\mathcal{P}^3 M = \mathcal{F} \times_{GL(7,\mathbb{R})} \mathcal{P}^3$, which is a subbundle of $\Lambda^3 T^*M$ with fibre $\mathcal{P}^3$. Similarly, define $\mathcal{P}^4 M = \mathcal{F} \times_{GL(7,\mathbb{R})} \mathcal{P}^4$, which is a subbundle of $\Lambda^4 T^*M$ with fibre $\mathcal{P}^4$. And as $\Theta : \mathcal{P}^3 \to \mathcal{P}^4$ commutes with the $GL(7,\mathbb{R})$-actions, it induces a map of bundles $\Theta : \mathcal{P}^3 M \to \mathcal{P}^4 M$.

A 3-form φ on M which lies in $\mathcal{P}^3 M$ at every point is called *positive*. Clearly, there is a 1-1 correspondence between G_2-structures on M, and positive 3-forms. By Proposition 2.2, if φ is a positive 3-form and (φ, g) the corresponding G_2-structure, then (φ, g) is *torsion-free* if and only if $d\varphi = d^*\varphi = 0$. But $d^*\varphi = 0$ if and only if $d(*\varphi) = 0$, and $*\varphi = \Theta(\varphi)$. Thus, (φ, g) is torsion-free if and only if $d\varphi = 0$ and $d\,\Theta(\varphi) = 0$.

Therefore, to construct torsion-free G_2-structures on a 7-manifold, we must look for sections φ of $\mathcal{P}^3 M$ satisfying the equations $d\varphi = d\,\Theta(\varphi) = 0$. Because Θ is nonlinear, we have to solve a *nonlinear partial differential equation* on the 3-form φ. The reason it helps to write the equations in this form is that the nonlinearity of the equations is obvious: it is packaged up in the nonlinear function Θ. If, instead, we wrote the condition as $\nabla\varphi = 0$ or $d\varphi = d^*\varphi = 0$, it would still be nonlinear, because the operators ∇ and d^* depend on φ in a nontrivial way, but the nonlinearity would be hidden.

8.2. Reformulating the Problem as a Non-Linear Elliptic PDE. Now consider the situation of Theorem 7.1. We are given a 7-manifold M, a family of positive 3-forms φ_t and a family of 3-forms ψ_t for $t \in (0, \epsilon)$, which satisfy $d\varphi_t = 0$ and $d\,\Theta(\varphi_t) = d(*\psi_t)$. We regard t as being small and fixed. Our goal is to deform φ_t to a positive 3-form $\tilde{\varphi}_t$ satisfying $d\tilde{\varphi}_t = 0$ and $d\,\Theta(\tilde{\varphi}_t) = 0$. Let η be a 2-form on M, and put $\tilde{\varphi}_t = \varphi_t + d\eta$. Since $\mathcal{P}^3 M$ is an open subset of $\Lambda^3 T^*M$, it follows that if $d\eta$ is small in C^0, then $\tilde{\varphi}_t$ is a positive 3-form, and defines a G_2-structure on M. Also, $d\tilde{\varphi}_t = 0$ holds automatically.

Thus, the condition for $\tilde{\varphi}_t$ to define a torsion-free G_2-structure is that $d\eta$ should be small in C^0, and $d\,\Theta(\varphi_t + d\eta) = 0$. The function Θ can be expanded about φ_t to give

$$\Theta(\varphi_t + d\eta) = *\varphi_t + \tfrac{7}{3} * \pi_1(d\eta) + 2 * \pi_7(d\eta) - *d\eta - F(d\eta). \tag{8.1}$$

Here $F : \Lambda^3 T^*M \to \Lambda^4 T^*M$ is a smooth function, such that $F(\chi)$ is defined when $|\chi|$ is small, and satisfies $F(\chi) = O(|\chi|^2)$. Also, in (8.1) and in the rest of the section, the Hodge star $*$, the projections π_k and the operator d^* depend on the G_2-structure (φ_t, g_t).

Equation (8.1) expresses $\Theta(\varphi_t + d\eta)$ as the sum of a constant term $*\varphi_t$, a term $\frac{7}{3} * \pi_1(d\eta) + 2 * \pi_7(d\eta) - *d\eta$ that is linear in $d\eta$, and a remainder $F(d\eta)$ that is at least quadratic in $d\eta$. We shall use (8.1) to rewrite the equation $d\,\Theta(\varphi_t + d\eta) = 0$ in a form that we are able to solve. Here is the first stage in this.

LEMMA 8.2. *Let η be a 2-form on M, and define a real function f on M by*

$$\pi_1(d\eta) = f\varphi_t. \tag{8.2}$$

Then the following two equations are equivalent:

$$d\,\Theta(\varphi_t + d\eta) = \tfrac{7}{3}df \wedge (*\varphi_t - *\psi_t) + 2d(*\pi_7(d\eta)), \tag{8.3}$$

and

$$d^* d\eta = d^*\psi_t + \tfrac{7}{3}d^*(f\psi_t) + *dF(d\eta). \tag{8.4}$$

PROOF. Substituting (8.2) into (8.1) and applying d, we find that

$$d\,\Theta(\varphi_t + d\eta) = d * \varphi_t + \tfrac{7}{3}d(f * \varphi_t) + 2d * \pi_7(d\eta) - d * d\eta - dF(d\eta). \tag{8.5}$$

Now $d * \varphi_t = d * \psi_t$, and thus $df \wedge (*\varphi_t - *\psi_t) = d(f * \varphi_t) - d(f * \psi_t)$. Putting these into (8.5) and rearranging gives

$$\begin{aligned} d\,\Theta(\varphi_t + d\eta) =& \tfrac{7}{3}df \wedge (*\varphi_t - *\psi_t) + 2d * \pi_7(d\eta) \\ & - d * d\eta + d * \psi_t + \tfrac{7}{3}d(f * \psi_t) - dF(d\eta). \end{aligned} \tag{8.6}$$

Therefore (8.3) holds if and only if

$$d * d\eta = d * \psi_t + \tfrac{7}{3}d(f * \psi_t) - dF(d\eta). \tag{8.7}$$

Applying the Hodge star to this equation and using the fact that $*d* = -d^*$ on 3-forms gives (8.4), as we have to prove. □

Next, one proves the following Proposition.

PROPOSITION 8.3. *Suppose that η is a smooth 2-form on M with $|d\eta|$ small, that f is a real function on M, and that* (8.3) *holds. Then $\tilde{\varphi}_t = \varphi_t + d\eta$ is a closed, positive 3-form and*

$$d\,\Theta(\tilde{\varphi}_t) = 0, \qquad df = 0 \qquad \text{and} \qquad d(*\pi_7(d\eta)) = 0. \tag{8.8}$$

Thus, $\tilde{\varphi}_t$ defines a torsion-free G_2-structure on M.

This rather curious result is proved using the special geometry of G_2-structures, and it shows that if (8.3) holds, then all three terms in the equation must actually be zero. Combining the Proposition and the previous Lemma, we arrive at the following formulation of the problem.

PROPOSITION 8.4. *Suppose η is a smooth 2-form on M with $|d\eta|$ small, and f a real function on M, that satisfy*

$$(dd^* + d^*d)\eta = d^*\psi_t + \tfrac{7}{3}d^*(f\psi_t) + *dF(d\eta) \tag{8.9}$$

and

$$f\varphi_t = \pi_1(d\eta). \tag{8.10}$$

Then $\tilde{\varphi}_t = \varphi_t + d\eta$ defines a torsion-free G_2-structure on M.

PROOF. Since M is compact, the subsets $\operatorname{Im} d$ and $\operatorname{Im} d^*$ in $C^\infty(\Lambda^2 T^*M)$ are L^2-orthogonal, and so have zero intersection. Now the l.h.s. of (8.9) is the sum of $dd^*\eta \in \operatorname{Im} d$ and $d^*d\eta \in \operatorname{Im} d^*$, but the r.h.s. lies wholly in $\operatorname{Im} d^*$. Therefore $dd^*\eta = 0$, implying that $d^*\eta = 0$ by integration by parts. Thus, (8.9) implies (8.4), and any solutions η, f of (8.9) and (8.10) also satisfy (8.2) and (8.4). But Lemma 8.2 then shows that (8.3) holds, and finally Proposition 8.3 proves that if in addition $|d\eta|$ is small, then $\tilde{\varphi}_t$ defines a torsion-free G_2-structure, as we want. □

The operator $dd^* + d^*d$ appearing on the left hand side of (8.9) is a *second order linear elliptic operator.* Of course, the terms $d^*(f\psi_t)$ and $*dF(d\eta)$ on the right hand side are also of second order in η, since f is a component of $d\eta$. But when t is small, ψ_t is small, and thus $d^*(f\psi_t)$ is small compared to f. Also, since $F(\chi) = O(|\chi|^2)$, if $d\eta$ is small, then $F(d\eta)$ is even smaller.

Therefore, when ψ_t and $d\eta$ are both small, the second and third terms on the right hand side of (8.9) are small compared to the left hand side. In this case, since ellipticity is an open condition, (8.9) is a *nonlinear elliptic partial differential equation* for η. Now, a great deal is known about the properties of linear and nonlinear elliptic equations, and there is a body of well understood techniques for studying their solutions. For an introduction to this area, see Aubin [**4**], or Gilbarg and Trüdinger [**12**]. We shall use these techniques to show that (8.9) and (8.10) have a smooth solution.

8.3. Constructing a Solution to the Equation. Theorem 8.1 now follows from Proposition 8.4 and the next Theorem.

THEOREM 8.5. *There exist positive constants κ, K depending on the constants A_1, A_2, A_3 and ϵ of Theorem 7.1, such that for each t with $0 < t \le \kappa$, there exists a smooth 2-form η on M with $\|d\eta\|_{C^0} \le Kt^{1/2}$ satisfying* (8.9) *and* (8.10).

Here is a brief sketch of the proof of this result. We solve (8.9) by iteration, introducing sequences $\{\eta_j\}_{j=0}^\infty$ and $\{f_j\}_{j=0}^\infty$ with $\eta_0 = f_0 = 0$, satisfying the inductive relations

$$(dd^* + d^*d)\eta_{j+1} = d^*\psi_t + \tfrac{7}{3}d^*(f_j\psi_t) + *dF(d\eta_j) \tag{8.11}$$

$$\text{and} \qquad f_{j+1}\varphi_t = \pi_1(d\eta_{j+1}). \tag{8.12}$$

Suppose by induction that smooth $\eta_0, \dots, \eta_k$ and $f_0, \dots, f_k$ exist and satisfy (8.11) and (8.12) for $j < k$. Now, provided $F(d\eta_k)$ is well-defined, which happens if $|d\eta_k|$ is small, the r.h.s. of (8.11) for $j = k$ is well-defined and lies in $\operatorname{Im} d^*$. Therefore, by Hodge theory, there exists a smooth 2-form η_{k+1} satisfying (8.11). If in addition we ask that η_{k+1} be L^2-orthogonal to the Hodge forms representing $H^2(M, \mathbb{R})$, then η_{k+1} is unique. The function f_{k+1} is then defined uniquely by (8.12) for $j = k$.

Thus by induction, provided $|d\eta_j|$ remains sufficiently small for $F(d\eta_j)$ to be well-defined, the sequences $\{\eta_j\}_{j=0}^\infty$ and $\{f_j\}_{j=0}^\infty$ exist and can even be chosen uniquely. If these sequences converge to limits η and f, then taking limits in (8.11) and (8.12) shows that η and f satisfy (8.9) and (8.10), giving us the solution we want. The key to proving this is an inductive estimate: one shows that there are positive constants C_1, C_2, K and κ depending only on the constants A_1, A_2, A_3 and ϵ of Theorem 7.1, such that if η_j, η_{j+1}, f_j and f_{j+1} satisfy (8.11) and (8.12) and the inequalities

$$\|d\eta_j\|_{L^2} \le C_1 t^4, \qquad \|\nabla d\eta_j\|_{L^{14}} \le C_2 \quad \text{and} \quad \|d\eta_j\|_{C^0} \le Kt^{1/2}, \tag{8.13}$$

and if $t \leq \kappa$, then

$$\|d\eta_{j+1}\|_{L^2} \leq C_1 t^4, \quad \|\nabla d\eta_{j+1}\|_{L^{14}} \leq C_2 \quad \text{and} \quad \|d\eta_{j+1}\|_{C^0} \leq K t^{1/2}. \tag{8.14}$$

Here is how the inductive estimates (8.14) are proved. For the first one, we take the L^2-inner product of (8.11) with η_{j+1} and integrate by parts. A short calculation shows that

$$\|d\eta_{j+1}\|_{L^2} \leq \|\psi_t\|_{L^2} + \tfrac{7}{3}\|\psi_t\|_{L^2}\|d\eta_j\|_{C^0} + C_3\|d\eta_j\|_{L^2}\|d\eta_j\|_{C^0}, \tag{8.15}$$

where C_3 is a constant such that $|F(\chi)| \leq C_3|\chi|^2$, for small 3-forms χ.

Equation (8.15) gives an *a priori estimate* for $\|d\eta_{j+1}\|_{L^2}$ in terms of $\|d\eta_j\|_{L^2}$ and $\|d\eta_j\|_{C^0}$, which are bounded by (8.13), and $\|\psi_t\|_{L^2}$, which is bounded by $A_1 t^4$. When t is small enough (depending on C_1, K, C_3 and A_1), we can show that $\|d\eta_{j+1}\|_{L^2} \leq C_1 t^4$, as we have to prove. The 3-form ψ_t was introduced solely to achieve this inequality.

Next we prove the second inequality of (8.14). Using parts (*ii*) and (*iii*) of Theorem 7.1 we may show that if χ is a closed 3-form on M then

$$\|\nabla\chi\|_{L^{14}} \leq C_4\big(\|d^*\chi\|_{L^{14}} + t^{-4}\|\chi\|_{L^2}\big), \tag{8.16}$$

where C_4 is a positive constant depending on A_2 and A_3. This is an *elliptic regularity result* for the elliptic operator $d+d^*$ acting on exterior forms on M. We substitute $\chi = d\eta_{j+1}$ in (8.16). The term $\|d^*\chi\|_{L^{14}}$ can be estimated in terms of norms of ψ_t and $d\eta_j$ using (8.11), and the term $\|\chi\|_{L^2}$ is $\|d\eta_{j+1}\|_{L^2}$, which we have already bounded. Again, when t is small enough we can show that $\|\nabla d\eta_{j+1}\|_{L^{14}} \leq C_2$, as we want.

Lastly, we prove the third inequality of (8.14). If χ is a 3-form on M, then one can use parts (*ii*) and (*iii*) of Theorem 7.1 to show that

$$\|\chi\|_{C^0} \leq C_5\big(t^{1/2}\|\nabla\chi\|_{L^{14}} + t^{-7/2}\|\chi\|_{L^2}\big), \tag{8.17}$$

where C_5 depends on A_2 and A_3. This is a *Sobolev embedding result.* The third inequality of (8.14) follows from the first two and (8.17), provided we take $K = C_5(C_2 + C_1)$.

The remainder of the proof is comparatively straightforward. By induction on j, the estimates (8.13) hold for all j. It soon follows that the sequences $\{\eta_j\}_{j=0}^{\infty}$ and $\{f_j\}_{j=0}^{\infty}$ exist, and converge in the appropriate Sobolev spaces to limits η, f, for which (8.9) and (8.10) hold. Taking the limit in (8.13) shows that $\|d\eta\|_{C^0} \leq K t^{1/2}$. Since (8.9) is elliptic for small t and $|d\eta|$, one can then show that η is smooth using standard analytic techniques, and the proof of Theorem 8.1 is complete.

For the $Spin(7)$ case, the result corresponding to Theorem 8.1 is the following.

THEOREM 8.6. *In the situation of Theorem 7.2, there are positive constants κ, K depending only on A_1, A_2, A_3 and ϵ, such that for every t with $0 < t < \kappa$, there exists a smooth, torsion-free $Spin(7)$-structure $(\tilde\Omega_t, \tilde g_t)$ on M with $\|\tilde\Omega_t - \Omega_t\|_{C^0} \leq K t^{1/2}$.*

The proof of this Theorem is given in [**15**, §5]. It is somewhat different to the G_2 case above.

9. Finishing the Proof

Having proved Theorem 8.1, we have found examples of compact 7-manifolds M admitting torsion-free G_2-structures. Now if (φ, g) is a torsion-free G_2-structure, then the holonomy group $\mathrm{Hol}(g)$ must be a subgroup of G_2. The final part of the construction, Step 4 of §5, is to show that $\mathrm{Hol}(g) = G_2$, rather than some proper subgroup. From Proposition 2.5, we see that $\mathrm{Hol}(g) = G_2$ if and only if the fundamental group $\pi_1(M)$ of M is finite. Therefore, to show that there exist metrics with holonomy G_2 on M, we just have to compute $\pi_1(M)$ and verify that it is finite.

EXAMPLE 9.1. We shall calculate the Betti numbers $b^k(M)$ and the fundamental group of the compact 7-manifold M of §6.3. Since M is compact and connected, we have $b^0(M) = 1$ and $b^k(M) = b^{7-k}(M)$ by Poincaré duality. Thus, it is enough to work out $b^1(M)$, $b^2(M)$ and $b^3(M)$. Now, the cohomology $H^*(T^7/\Gamma)$ is simply the Γ-invariant part of $H^*(T^7)$. Thus we can easily work out the Betti numbers of T^7/Γ, which are

$$b^1(T^7/\Gamma) = 0, \quad b^2(T^7/\Gamma) = 0 \quad \text{and} \quad b^3(T^7/\Gamma) = 7. \tag{9.1}$$

To make M we glue in 12 patches of the form $T^3 \times X$, where X is the Eguchi-Hanson space. Each of these patches increases b^k by $b^k(T^3 \times X) - b^k(T^3 \times \mathbb{C}^2/\{\pm 1\})$. Using the Künneth theorem and the Betti numbers $b^1(T^3) = 3$, $b^1(X) = 0$ and $b^2(X) = 1$, one can show that each patch adds 0 to b^1, 1 to b^2 and 3 to b^3. Together with (9.1) this gives

$$b^1(M) = 0, \quad b^2(M) = 12 \quad \text{and} \quad b^3(M) = 43. \tag{9.2}$$

The fundamental group of M is also easy to work out: it turns out that T^7/Γ is simply-connected, and the process of resolving does not change the fundamental group, and so M is also simply-connected. Now, Theorem 8.1 constructs torsion-free G_2-structures on M. But the fundamental group of M is finite, so by Proposition 2.5, these torsion-free G_2-structures come from metrics on M with holonomy G_2. Thus, M admits metrics with holonomy G_2. By (9.2) and Theorem 2.6, the moduli space of metrics with holonomy G_2 on M is a smooth, 43-dimensional manifold.

In the case of compact 8-manifolds M with torsion-free $Spin(7)$-structures (Ω, g), the topological test is more complicated: by Theorem 3.5, we have $\mathrm{Hol}(g) = Spin(7)$ if and only if M is simply-connected and $\hat{A}(M) = 1$, where $\hat{A}(M)$ is a linear combination of the Betti numbers $b^k(M)$ and $b^4_\pm(M)$. Thus, to prove that $\mathrm{Hol}(g) = Spin(7)$, we evaluate the fundamental group and Betti numbers of M and verify they satisfy the right conditions. Here is an example.

EXAMPLE 9.2. We shall work out the Betti numbers and fundamental group of the compact 8-manifold M of §6.4. Since M is compact and connected we have $b^0(M) = 1$ and $b^k(M) = b^{8-k}(M)$, so it is enough to find $b^1(M), b^2(M), b^3(M)$ and $b^4(M)$. But M is oriented, and thus $b^4(M)$ splits into the sum of $b^4_+(M)$ and $b^4_-(M)$. Working out the Γ-invariant part of $H^*(T^8)$ shows that

$$b^1(T^8/\Gamma) = b^2(T^8/\Gamma) = b^3(T^8/\Gamma) = 0 \quad \text{and} \quad b^4_+(T^8/\Gamma) = b^4_-(T^8/\Gamma) = 7. \tag{9.3}$$

To find the Betti numbers of the resolution M, we must add contributions from the resolution of each component of the singular set. These are more difficult to work out than in the previous example. In brief, each copy of $T^4/\{\pm 1\}$ in S_α and

S_β fixes b^1 and b^3, adds 1 to b^2, and adds 3 to each of b^4_+ and b^4_-. Each copy of T^4 in S_γ and S_δ fixes b^1 and adds 1 to b^2, 4 to b^3 and 3 to each of b^4_+ and b^4_-. Each point in $S_{\alpha\beta}$ fixes b^1, b^2, b^3 and b^4_-, and adds 1 to b^4_+. Combining these with (9.3), we find that M has Betti numbers

$$(9.4)\quad b^1(M) = 0, \quad b^2(M) = 12, \quad b^3(M) = 16, \quad b^4_+(M) = 107, \quad b^4_-(M) = 43.$$

As in Example 9.1, it turns out that T^8/Γ is simply-connected, and the resolution does not change the fundamental group, and so M is simply-connected. From (9.4) and the definition (3.2) of $\hat{A}(M)$, we see that $\hat{A}(M) = 1$. Therefore Theorem 3.5 applies, so M admits metrics with holonomy $Spin(7)$. By (9.4) and Theorem 3.6, the moduli space of metrics of holonomy $Spin(7)$ on M is a smooth, 44-dimensional manifold.

10. Interesting Questions

We finish by suggesting some questions and open problems for future research. Here is our first question.

- Which compact 7- and 8-manifolds M admit metrics with holonomy G_2 and $Spin(7)$?

The construction described above yields many examples of compact manifolds with holonomy G_2 and $Spin(7)$ metrics. The construction described above yields many examples of compact manifolds with holonomy G_2 and $Spin(7)$ metrics. However, there may be many compact manifolds with holonomy G_2 and $Spin(7)$ that cannot be constructed by this method, and we know almost nothing about them. It is not even sure whether there are finitely or infinitely many compact manifolds admitting metrics with holonomy G_2 and $Spin(7)$. The author guesses that there are only finitely many.

Our second group of questions concerns the moduli space of metrics.

- On a given 7- or 8-manifold M, what does the moduli space of metrics with holonomy G_2 or $Spin(7)$ look like?
- Can this moduli space be compactified by adding extra 'ideal' points, corresponding to singular metrics?
- What kinds of singularities occur in these 'ideal' singular manifolds, and how are they resolved?

These seem to be difficult problems, but ones on which some progress can be made. The simplest sort of singularities to consider are orbifold singularities. When the orbifold group locally lies in some $SU(m)$ in G_2 or $Spin(7)$, we have a good theory of when and how the singularities can be resolved, which will be explained in **[19]**. Some things can also be proved for more general orbifold groups.

Thirdly, we discuss special submanifolds of G_2 and $Spin(7)$ manifolds. There is a beautiful theory called *calibrated geometry*, which was introduced by Harvey and Lawson **[13]**. The idea is that in a Riemannian manifold with an extra geometric structure such as a Kähler structure, there is a special type of minimal submanifold called a *calibrated submanifold.* For example, complex submanifolds of Kähler manifolds are calibrated submanifolds.

Riemannian 7-manifolds with holonomy G_2 have two types of calibrated submanifold, *associative 3-manifolds*, and *coassociative 4-manifolds*. Riemannian 8-manifolds with holonomy $Spin(7)$ have just one type, *Cayley 4-manifolds*. The deformation theory of all three was worked out by McLean **[27]**. Examples of compact associative and coassociative submanifolds in compact 7-manifolds with holonomy G_2 are given in [**17**, §4.2], and similar methods yield examples of Cayley 4-manifolds in $Spin(7)$-manifolds. Calibrated submanifolds seem to play a similar rôle in exceptional geometry to holomorphic curves in complex manifolds.

- Describe the calibrated submanifolds N in a given compact Riemannian manifold M with exceptional holonomy.
- Both coassociative 4-manifolds and Cayley 4-manifolds can occur in families of positive dimension. What do the singular elements of these families look like?
- What happens to the calibrated submanifolds in M as we deform the metric on M?
- One can define invariants of compact manifolds with exceptional holonomy, by counting calibrated submanifolds with a fixed homology class, as the Gromov invariant counts pseudo-holomorphic curves in a symplectic manifold. What is the theory of these invariants? Are there any connections with physics?

Finally, we note that compact manifolds with holonomy G_2 (and to a lesser extent $Spin(7)$) are arousing interest in the branch of theoretical physics known as 'string theory'. The most popular version of string theory predicts that the universe should have 10 dimensions. But we can only see four dimensions, including time.

To account for the discrepancy, it is supposed that the universe is locally a product of 4-dimensional Minkowski space with a compact 6-manifold with very small radius, of order 10^{-33}cm. The metric on this 6-manifold must have holonomy $SU(3)$, that is, it is a Calabi-Yau 3-fold. So string theorists have studied Calabi-Yau 3-folds from their own point of view, and have come up with some extraordinary mathematical predictions, known as Mirror Symmetry.

However, a new theoretical model called *M-theory* predicts that the universe should actually have 11 dimensions. To get from 4 dimensions to 11 dimensions we take the product with a compact 7-manifold with very small radius, and at the classical level this 7-manifold must have holonomy G_2. It seems very likely that string theory will throw new light on the geometry of compact manifolds with holonomy G_2 and $Spin(7)$, and their calibrated submanifolds.

References

[1] D. Alekseevskii, *Riemannian spaces with exceptional holonomy*, Functional Anal. Appl., 2 (1968), pp. 97–105.

[2] W. Ambrose and I. Singer, *A theorem on holonomy*, Trans. Amer. Math. Soc., 75 (1953), pp. 428–443.

[3] M. Atiyah and I. Singer, *The index of elliptic operators. III*, Annals of mathematics, 87 (1968), pp. 546–604.

[4] T. Aubin, *Nonlinear analysis on manifolds. Monge-Ampère equations*, vol. 252 of Grundlehren der math. Wissenschaften, Springer-Verlag, New York, 1982.

[5] M. Berger, *Sur les groupes d'holonomie homogène des variétés à connexion affines et des variétés riemanniennes*, Bull. Soc. Math. France, 83 (1955), pp. 279–330.

[6] A. Besse, *Einstein Manifolds*, Springer-Verlag, New York, 1987.

[7] S. Bochner, *Vector fields and Ricci curvature*, Bull. A.M.S., 52 (1946), pp. 776–797.

[8] R. Brown and A. Gray, *Riemannian manifolds with holonomy group Spin(9)*, in Differential Geometry (in honour of Kentaro Yano), S. Kobayashi et al., eds., Tokyo, 1972, Kinokuniya, pp. 41–59.

[9] R. Bryant, *Metrics with exceptional holonomy*, Annals of mathematics, 126 (1987), pp. 525–576.

[10] R. Bryant and S. Salamon, *On the construction of some complete metrics with exceptional holonomy*, Duke Mathematical Journal, 58 (1989), pp. 829–850.

[11] T. Eguchi and A. Hanson, *Asymptotically flat solutions to Euclidean gravity*, Physics letters, 74B (1978), pp. 249–251.

[12] D. Gilbarg and N. Trudinger, *Elliptic partial differential equations of second order*, vol. 224 of Grundlehren der math. Wissenschaften, Springer-Verlag, Berlin, 1977.

[13] R. Harvey and H. Lawson, *Calibrated geometries*, Acta Math., 148 (1982), pp. 47–157.

[14] S. Helgason, *Differential geometry and symmetric spaces*, Academic Press, New York, 1962.

[15] D. D. Joyce, *Compact 8-manifolds with holonomy* $Spin(7)$, Inventiones mathematicae, 123 (1996), pp. 507–552.

[16] ———, *Compact Riemannian 7-manifolds with holonomy* G_2. *I*, Journal of Differential Geometry, 43 (1996), pp. 291–328.

[17] ———, *Compact Riemannian 7-manifolds with holonomy* G_2. *II*, Journal of Differential Geometry, 43 (1996), pp. 329–375.

[18] ———, *Compact manifolds with exceptional holonomy*, in Geometry and physics, J. Andersen, J. Dupont, H. Pedersen, and A. Swann, eds., vol. 184 of Lecture notes in pure and applied mathematics, New York, 1997, Marcel Dekker.

[19] ———, *Compact Riemannian manifolds with special holonomy groups*, Oxford University Press, 2000 (anticipated).

[20] S. Kobayashi and K. Nomizu, *Foundations of Differential Geometry*, vol. 1, Wiley, New York, 1963.

[21] ———, *Foundations of Differential Geometry*, vol. 2, Wiley, New York, 1963.

[22] P. Kronheimer, *The construction of ALE spaces as hyperkähler quotients*, Journal of Differential Geometry, 29 (1989), pp. 665–683.

[23] ———, *A Torelli-type theorem for gravitational instantons*, Journal of Differential Geometry, 29 (1989), pp. 685–697.

[24] C. LeBrun, *Counterexamples to the generalized positive action conjecture*, Communications in mathematical physics, 118 (1988), pp. 591–596.

[25] C. LeBrun and M. Singer, *A Kummer-type construction of self-dual 4-manifolds*, Mathematische Annalen, 300 (1994), pp. 165–180.

[26] A. Lichnerowicz, *Spineurs harmoniques*, C. R. Acad. Sci. Paris, 257 (1963), pp. 7–9.

[27] R. McLean, *Deformations and moduli of calibrated submanifolds*, PhD thesis, Duke University, 1990.

[28] D. Page, *A physical picture of the K3 gravitational instanton*, Physics letters, 80B (1978), pp. 55–57.

[29] S.-S. Roan, *Minimal resolution of Gorenstein orbifolds*, Topology, 35 (1996), pp. 489–508.

[30] S. Salamon, *Riemannian geometry and holonomy groups*, vol. 201 of Pitman Res. Notes in Math., Longman, Harlow, 1989.

[31] J. Simons, *On transitivity of holonomy systems*, Annals of mathematics, 76 (1962), pp. 213–234.

[32] G. Tian and S.-T. Yau, *Complete Kähler manifolds with zero Ricci curvature. II*, Inventiones mathematicae, 106 (1991), pp. 27–60.

[33] P. Topiwala, *A new proof of the existence of Kähler-Einstein metrics on K3. I*, Inventiones mathematicae, 89 (1987), pp. 425–448.

[34] S. T. Yau, *On the Ricci curvature of a compact Kähler manifold and the complex Monge-Ampère equations. I*, Communications on pure and applied mathematics, 31 (1978), pp. 339–411.

Lincoln College, Oxford, OX1 3DR, England

Kähler-Einstein Manifolds of Positive Scalar Curvature

Gang Tian

1. Introduction

A Riemannian metric on M is Kähler-Einstein if it is Kähler and its Ricci curvature form is a constant multiple of its Kähler form. Such a metric provides a special solution of the Einstein equation on Riemannian manifolds. In 50's, E. Calabi asked when a compact Kähler manifold M admits any Kähler-Einstein metrics.

Since the Ricci form represents the first Chern class $c_1(M)$, a necessary condition for the existence of Kähler-Einstein metrics is that $c_1(M)$ is definite. In fact, Calabi conjectured that any (1,1)-form representing $c_1(M)$ is the Ricci form of some Kähler metric on M (the Calabi conjecture). In particular, the conjecture implies the existence of Ricci-flat Kähler metrics in case $c_1(M) = 0$. The Calabi conjecture was solved by Yau in 1977 [**28**]. Around the same time, Aubin and Yau proved independently the existence of Kähler-Einstein metrics on Kähler manifolds with negative first Chern class [**1, 28**]. Therefore, it had been known by the middle of 70's that $c_1(M)$ being zero or negative is also sufficient for the existence of Kähler-Einstein metrics on the underlying manifold.

Back to early 50's, using the maximum principle, Calabi had proved the uniqueness of Kähler-Einstein metrics within a fixed Kähler class for Kähler manifolds with non-positive first Chern class. In 1986, Bando and Mabuchi proved the uniqueness of Kähler-Einstein metrics on compact Kähler manifolds with positive first Chern class.

We remain to study the existence problem of Kähler-Einstein metrics on a compact Kähler manifold M with $c_1(M) > 0$. Now Kähler-Einstein metric, if they exist, must have positive scalar curvature.

In algebraic geometry, a Kähler manifold M with $c_1(M) > 0$ is called a Fano manifold.

In these notes, we summarize and discuss basic results on Kähler-Einstein manifolds with positive scalar curvature. We will emphasize on recent progress, particularly, the existence problem. The materials chosen here may depend on the author's interest.

¿From now on, unless specified, M always denotes a compact Kähler manifold with positive first Chern class. We will identify a Kähler metric g with its Kähler

1991 *Mathematics Subject Classification.* Primary 53C20.

Supported in part by a grant from the National Science Foundation.

form ω_g, or simply ω if there is no confusion. In local coordinates, if g is given by hermitian matrices $\{g_{i\bar{j}}\}$, then

$$\omega = \omega_g = \sum_{i,=1}^{n} g_{i\bar{j}} dz_i \wedge d\bar{z}_j,$$

where n is the complex dimension of M.

2. Holomorphic Obstructions

There are holomorphic obstructions to existence of Kähler-Einstein metrics with positive scalar curvature.

We denote by $\eta(M)$ the Lie algebra of all holomorphic vector fields on M. In 1957, Matsushima proved

THEOREM 2.1. **[16]** *If M admits a Kähler-Einstein metric with positive scalar curvature, then $\eta(M)$ is reductive.*

Its proof can be found in [**16**] or [**11**]. It follows from this theorem that there are compact Kähler manifolds with the first Chern class positive and which do not admit Kähler-Einstein metrics. For instance, if M is the blow-up of $\mathbb{CP}^2$ at one or two points, then $\eta(M)$ is not reductive, consequently, such an M does not have any Kähler-Einstein metrics.

In 1983, Futaki introduced another holomorphic obstruction: let M be as above. Since $c_1(M) > 0$, one can find a Kähler metric ω such that $\frac{\sqrt{-1}}{2\pi}\omega = \omega_g$ represents $c_1(M)$. If $\mathrm{Ric}(\omega)$ denotes the Ricci form of ω, then $\frac{\sqrt{-1}}{2\pi}\mathrm{Ric}(g)$ also represents $c_1(M)$, therefore, there is a real-valued function h_ω such that

$$\mathrm{Ric}(\omega) - \omega = \partial\bar{\partial} h_\omega, \quad \int_M (e^{h_\omega} - 1)\omega^n = 0.$$

Futaki proved in [**10**] that for any X in $\eta(M)$, the integral

$$\int_M X(h_\omega)\omega^n$$

is independent of choices of ω. Hence, one can define a holomorphic invariant $f_M : \eta(M) \mapsto \mathbb{C}$ by

$$f_M(X) = \int_M X(h_\omega)\omega^n.$$

In fact, f_M is a Lie algebra character, so it descends to $\eta(M)/[\eta(M), \eta(M)]$. Moreover, we have

THEOREM 2.2. **[10]***If M admits a Kähler-Einstein metric with positive scalar curvature, then $f_M \equiv 0$.*

In [**10**], Futaki constructed an example of 3-dimensional M with $c_1(M) > 0$ such that $\eta(M)$ is reductive and $f_M \neq 0$. Hence, such an M does not admit any Kähler-Einstein metrics.

The Futaki invariant can be computed by formulas similar to the Bott residue formula for Chern numbers.

DEFINITION 2.3. X is non-degenerate if the zero set of X is the disjoint union of smooth connected complex submanifolds $\{Z_\lambda\}_{\lambda\in\Lambda}$ and if at each $z \in Z_\lambda$, $DX : T_zM/T_zZ_\lambda \mapsto T_zM/T_zZ_\lambda$ is non-degenerate, i.e., $\det(DX|_{T_zM/T_zZ_\lambda}) \neq 0$.

We denote by $L_\lambda(X)$ the induced homomorphism $DX : T_zM/T_zZ_\lambda \mapsto T_zM/T_zZ_\lambda$. The following was proved in [**11**].

THEOREM 2.4. *Let M be a compact Kähler manifold with $c_1(M) > 0$. Then for any $X \in \eta(M)$,*

$$f_M(X) = \frac{1}{n+1}\sum_{\lambda\in\Lambda}\int_{Z_\lambda}\frac{\mathrm{tr}(L_\lambda(X)+c_1(M))^{n+1}}{\det(L_\lambda(X)+\frac{\sqrt{-1}}{2\pi}K_\lambda)},$$

where K_λ is the curvature form of the induced metric on TM/TZ_λ by ω.

REMARK 2.5. The Futaki invariant is only a special case of the Calabi-Futaki invariant defined for any Kähler class on any compact Kähler manifolds. The Calabi-Futaki invariant can be defined in a way similar to what we did above (cf. [**11**]). It is an obstruction to existence of Kähler metrics with constant scalar curvature. There is also a residue formula computing the Calabi-Futaki invariant analogous to the above one in general cases [**25**]. Also we refer readers to [**5, 14**] for some references on Kähler metrics with constant scalar curvature.

The residue formula in Theorem 2.4 can be expressed more explicitly in the following two cases:

1. If X only has isolated zeroes so $\dim Z_\lambda = 0$ and hence the only terms that will contribute to the integral are the degree 0 terms and therefore
$$f_M(X) = \frac{1}{n+1}\sum_{\lambda}\frac{\mathrm{tr}(L_\lambda(X))^{n+1}}{\det(L_\lambda(X))}.$$
In particular if M has a Kähler-Einstein metric then we know that $f_M = 0$ and therefore this puts a constraint on the zero set of the vector field.
2. If M is a complex surface, then $\Lambda = \Lambda_0 \cup \Lambda_1$ where $\Lambda_i = \{\lambda \in \Lambda; \dim Z_\lambda = i\}$, and
$$f_M(X) = \frac{1}{3}\sum_{\lambda\in\Lambda_0}\frac{\mathrm{tr}(L_\lambda(X))^3}{\det(L_\lambda(X))} + \frac{2}{3}\sum_{\lambda\in\Lambda_1}L_\lambda(X)(c_1(M)(Z_1)+1).$$

Let us apply this to the blow-up M of $\mathbb{CP}^2$ at one point, say $[1,0,0]$. Choose the vector field X obtained by differentiating automorphisms:

$$\phi(\lambda)([z_0,z_1,z_2]) = [z_0,\lambda z_1,\lambda z_2]$$

at $\lambda = 1$. The zero set of X consists of two connected components E_0 and E_∞, where E_0 is the exceptional divisor of M and E_∞ is the pull-back of the divisor $\{z_0 = 0\}$ in $\mathbb{CP}^2$. It is easy to see that $L_\lambda(X) = 1$ along E_0 and -1 along E_∞. Hence, we have

$$f_M(X) = \frac{2}{3}(2-8) = -4.$$

The Futaki invariant of M is not zero. In a similar way, one can show that the blow-up of $\mathbb{CP}^2$ at two points has non-vanishing Futaki invariant.

In [**8**], Ding and Tian defined the generalized Futaki invariant for any almost Fano variety (possibly singular). They used this new invariant to construct new obstructions to existence of Kähler-Einstein metrics. Such obstructions were refined and used to define K-stability in [**24**]. The simplest case of these obstructions can be described as follows:

Let $\pi : N \mapsto D$ is a holomorphic submersion, such that M is biholomorphic to a fiber $N_z = \pi^{-1}(z)$ for some $z \in D$, where D is the unit disk in $\mathbb{C}^1$. We further assume that there is a family of automorphisms $\sigma(\lambda)$ of N which can descend to the dilations $z \mapsto \lambda z$ on D ($\lambda \le 1$). Usually, we denote by X_N the holomorphic vector field on N_0 induced by those automorphisms. More precisely, $X_N = -\sigma'(1)$. If N is holomorphically different from $M \times D$, we say that $\pi : N \mapsto D$ is non-trivial. Otherwise, we say that N is trivial.

THEOREM 2.6. *If M admits a Kähler-Einstein metric with positive scalar curvature, then for any fibration $\pi : N \mapsto D$ as above, we have* $\mathrm{Re}(f_{N_0}(X_N)) \ge 0$, *the equality holds if and only if N is trivial, i.e., X_N does not preserve M.*

Theorem 2.6 can be used to disprove a long-standing conjecture in the case of complex dimensions higher than two. The conjecture claims that *any compact Kähler manifold M with $c_1(M) > 0$ and $\eta(M) = \{0\}$ admits a Kähler-Einstein metric.* A counterexample can be briefly described as follows (see section 6 for details): let $G(4,7)$ be the complex Grassmannian manifold consisting of all 4-dimensional subspaces in $\mathbb{C}^7$, for any 3-dimensional subspace $P \subset \wedge^2\mathbb{C}^7$, one can define a sub-variety X_P in $G(4,7)$ by

$$X_P = \{U \in G(4,7) \mid P \text{ projects to zero in } \wedge^2 (C^7/U)\}$$

For a generic P, X_P is a smooth 3-fold with $c_1(X_P) > 0$. These manifolds were first constructed by Iskovskih (cf. [**12**], [**17**]).

Take P_a to be the subspace spanned by bi-vectors

$$\begin{aligned} &3e_1 \wedge e_6 - 5e_2 \wedge e_5 + 6e_3 \wedge e_4 + \textstyle\sum_{j+k\ge 8} a_{1jk}e_j, \\ &3e_1 \wedge e_7 - 2e_2 \wedge e_6 + e_3 \wedge e_5 + \textstyle\sum_{j+k\ge 9} a_{2jk}e_j, \\ &e_2 \wedge e_7 - e_3 \wedge e_6 + e_4 \wedge e_5 + \textstyle\sum_{j+k\ge 10} a_{3jk}e_j, \end{aligned}$$

where e_i are euclidean basis of $\mathbb{C}^7$ and $a = \{a_{ijk}\}$.

Then we can deduce the following from Theorem 2.6

COROLLARY 2.7. *For generic a, X_{P_a} has neither nontrivial holomorphic vector fields nor Kähler-Einstein metrics.*

3. Kähler-Einstein Metrics and Complex Monge-Ampere Equations

Before 1987, only known Kähler-Einstein metrics were either homogeneous or of cohomogeneity one. They can be reduced to solving an either algebraic or ODE equation (cf. [**19**]). Kähler-Einstein metrics were first constructed on manifolds without any holomorphic vector fields by Tian [**21**], Tian-Yau [**26**] and Siu [**20**] by solving certain complex Monge-Ampere equations. Those manifolds have only finite automorphism group. They include Fermat hypersurfaces in $\mathbb{CP}^{n+1}$ of degree n and $n+1$, complex surfaces with symmetries. The method in [**21**] and [**26**] was based on the following holomorphic invariant introduced in [**21**]. More examples of Kähler-Einstein manifolds were later constructed by A. Nadel in [**18**]. In particular, Nadel proved that there are Kähler-Einstein metrics on Fermat hypersurfaces in $\mathbb{CP}^{n+1}$ of degree $d \ge 1 + n/2$.

In this section, we give an analytic criterion for the existence of Kähler-Einstein metrics with positive scalar curvature. We will also show that every Fermat hypersurface admits a Kähler-Einstein metric.

Let M be a compact Kähler manifold with $c_1(M) > 0$, and ω be a Kähler metric with the Kähler class $c_1(M)$. We define $P(M,\omega)$ to be the set of smooth functions φ satisfying: $\omega + \partial\bar{\partial}\varphi > 0$. It is essentially the set of Kähler metrics with the Kähler class $c_1(M)$.

If G is a maximal compact subgroup of the automorphism group of M, then we may take ω to be G-invariant and define

$$P_G(M,\omega) = \{\varphi \in P(M,\omega) \mid \sigma^*\varphi = \varphi, \forall \sigma \in G\}.$$

For any $\varphi \in P(M,\omega)$, we put $\omega_\varphi = \omega + \partial\bar{\partial}\varphi$. If ω_φ is a Kähler-Einstein metric, then

$$-\partial\bar{\partial}\log\left(\frac{\omega_\varphi^n}{\omega^n}\right) = \omega + \partial\bar{\partial}\varphi - \mathrm{Ric}(\omega),$$

since $\mathrm{Ric}(\omega) - \omega = \partial\bar{\partial}h_\omega$, we can derive from the above

$$(\omega + \partial\bar{\partial}\varphi)^n = e^{h_\omega - \varphi}\omega^n, \ \omega + \partial\bar{\partial}\varphi > 0. \tag{3.1}$$

This is a complex Monge-Ampere equation. The existence of Kähler-Einstein metrics on M is equivalent to solvability of (3.1).

In order to solve (3.1), we used the continuity method and consider

$$(\omega + \partial\bar{\partial}\varphi)^n = e^{h_\omega - t\varphi}\omega^n, \ \omega + \partial\bar{\partial}\varphi > 0, \tag{3.2}$$

where $t \in [0,1]$; for a given t, we will refer to this as equation $(3.2)_t$. Let S be the set of t in $[0,1]$ such that $(3.2)_t$ is solvable. Then by Yau's solution for the Calabi conjecture [**28**], $0 \in S$.

LEMMA 3.1. *Let φ be a solution of $(3.2)_t$, and $\omega_t = \omega_\varphi$. Then $\mathrm{Ric}(\omega_t) \geq t\omega_t$, and the equality holds if and only if $t = 1$.*

PROOF. . Taking $\partial\bar{\partial}$ on both sides of $(3.2)_t$, we obtain

$$\mathrm{Ric}(\omega_t) = t\omega_t + (1-t)\omega.$$

Then the lemma follows. □

Combining this with a Bochner identity, Aubin [**2**] proved in that the first nonzero eigenvalue of ω_t is bigger than t for $t < 1$. On the other hand, the linearization of $(3.2)_t$ at φ is given by $\Delta_t + t$, where Δ_t denotes the Laplacian operator of ω_t. Therefore, the linearization is invertible if $t < 1$. So by using the Implicit Function Theorem, Aubin concluded in [**2**] that S is open in $[0,1]$.

It remains to show that S is closed. It amounts to establishing an a priori C^3-estimate for solutions of $(3.2)_t$. Repeating Yau's computations for C^2 in [**28**], one can show

LEMMA 3.2. *There is a uniform constant $c > 0$, such that for any solution φ of 3.2, we have*

$$||\varphi||_{C^2} \leq c e^{c \sup_M |\varphi|}.$$

Then, using Calabi's computations for C^3 in [**28**], one can further show that $||\varphi||_{C^3}$ can be bounded uniformly by an a prior C^0-estimate for any solution φ of $(3.2)_t$.

However, such a C^0-estimate does not exist due to those obstructions in Section 2. Therefore, it is of fundamental importance to find a correct geometric condition for validity of such a C^0-estimate.

A sufficient condition can be described as follows: we define

$$\alpha(M) = \sup\{\alpha \mid \text{there is a } C(\alpha) > 0, s.t. \frac{1}{V}\int_M e^{\alpha(\varphi - \sup_M \varphi)}\omega^n \leq C(\alpha), \tag{3.3}$$

One can show that it is always positive and independent of G and ω, so it is a holomorphic invariant [**21**]. It was shown in [**21**] (cf. [**7**]) that *M has a Kähler-Einstein metric whenever $\alpha(M) > \frac{n}{n+1}$*. The existence of Kähler-Einstein metrics was shown in [**21**], [**26**] and [**18**] by establishing $\alpha(M) > \frac{n}{n+1}$. In particular, Nadel proved elegant vanishing theorems involving multiplier sheaves if $\alpha(M)$ is less than 1.

Unfortunately, $\alpha(M)$ may not exceed $\frac{n}{n+1}$ even if M has a Kähler-Einstein metric.

It turns out that (3.1) is the Euler-Lagrange equation of a functional F_ω, which will be defined as follows: for any $\varphi \in P(M,\omega)$,

$$\begin{aligned} F_\omega(\varphi) &= J_\omega(\varphi) - \frac{1}{V}\int_M \varphi\omega^n - \log\left(\frac{1}{V}\int_M e^{h_\omega - \varphi}\omega^n\right) \\ J_\omega(\varphi) &= \sum_{i=0}^{n-1} \frac{i+1}{n+1}\frac{1}{V}\int_M \partial\varphi \wedge \bar{\partial}\varphi \wedge \omega^i \wedge (\omega + \partial\bar{\partial}\varphi)^{n-i-1}, \end{aligned}$$

where $V = \int_M \omega^n = c_1(M)^n$. An easy computation shows that (3.1) is just the first variation of F_ω. Clearly, $J_\omega(\varphi) \geq 0$ whenever $\varphi \in P(M,\omega)$. So we may consider J_ω as a generalized energy.

DEFINITION 3.3. We say that F_ω is proper on $P_G(M,\omega)$, if there is an increasing function $\mu : \mathbb{R} \mapsto [c(\omega), \infty)$, such that $\lim_{t\to\infty} \mu(t) = \infty$ and $F_\omega \geq \mu(J_\omega)$ on $P_G(M,\omega)$, where $c(\omega)$ is some constant.

An easy computation shows that F_ω satisfies the following cocycle condition:

$$F_\omega(\varphi) - F_{\omega_\psi}(\varphi - \psi) = F_\omega(\psi).$$

It follows that for any $\psi \in P_G(M,\omega)$, if F_ω is proper, so is F_{ω_ψ}. Clearly, it is also independent of choices of the maximal compact subgroup G. Therefore, the properness of F_ω is an intrinsic property of the underlying manifold M.

THEOREM 3.4. [**24**] *Let M be a compact Kähler manifold with $c_1(M) > 0$. Then M has a Kähler-Einstein metric if and only if F_ω is proper on $P_G(M,\omega)$.*

This gives an analytic criterion for the existence of Kähler-Einstein metrics. The properness can often be checked. In general, we believe that the properness is equivalent to certain stability of underlying manifolds in Geometry Invariant Theory.

REMARK 3.5. The arguments in [**7**] shows that F_ω is essentially proper if $\alpha(M) > \frac{n}{n+1}$.

Now let us apply this theorem to proving that any Fermat hypersurfaces admit Kähler-Einstein metrics.

A Fermat hypersurface of degree d is of the form

$$M = \{z_0^d + \dots + z_{n+1}^d = 0\} \subset \mathbb{CP}^{n+1}.$$

Since $c_1(M) > 0$, we have $d < n+2$. Note that for other d, M has non-positive $c_1(M)$, so it has a Kähler-Einstein metric with non-positive scalar curvature.

Consider the group G_0 generated by

$$\sigma_i : [z_0, \dots, z_{n+1}] \mapsto [z_0, \dots, e_d z_i, \dots, z_{n+1}]$$

where $e_d = e^{\frac{2\pi\sqrt{-1}}{d}}$. Clearly, M is invariant under G_0. Let G be a maximal compact subgroup of $\mathrm{Aut}(M)$ containing G_0 and ω be a G-invariant metric. We will show that F_ω is proper on $P_G(M,\omega)$. Let

$$\pi_i : M \to \mathbb{CP}^n_i$$

be the projection onto $\mathbb{CP}_i^n = \{[z_0, ..., z_{i-1}, 0, z_{i+1}, ..., z_{n+1}]\} \simeq \mathbb{CP}^n$. Note that this map is well-defined because $[0, ..., 0, 1, 0, ..., 0] \notin M$. So any $\phi \in P_G(M, \omega)$ is of the form $(n+2-p)\pi_i^*\varphi$ for some $\varphi \in P(M, \omega_{FS})$, where ω_{FS} is the Fubini-Study metric on $\mathbb{CP}^n$. Note that $c_1(M) = (n+2-p)[\pi^*\omega_{FS}]$. We may normalize $\int_M e^{h_\omega - \phi}\omega^n = V$.

For any Kähler metric ω and $\phi \in P(M, \omega)$, we put

$$F_\omega^0(\phi) = J_\omega(\phi) - \frac{1}{V}\int_M \phi\omega^n.$$

This functional also satisfies the cocycle condition, so

$$F_\omega(\phi) = F_\omega^0(\phi) = F^0_{(n+2-d)\pi_i^*\omega_{FS}}(\phi) - F_\omega^0(u),$$

where $(n+2-d)\pi_i^*\omega_{FS} = \omega + \partial\bar{\partial}u$ may not be a Kähler metric, but u and consequently, $F_\omega^0(u)$ are bounded. Furthermore, one can check that

$$F^0_{(n+2-d)\pi_i^*\omega_{FS}}(\phi) = \frac{n+2-d}{n+1} F^0_{(n+1)\omega_{FS}}\left(\frac{n+1}{n+2-d}\varphi\right).$$

Since $\mathbb{CP}^n$ has a canonical Kähler-Einstein metric $(n+1)\omega_{FS}$, by Theorem 2.1, we know that $F_{(n+1)\omega_{FS}}$ is bounded from below on $P(\mathbb{CP}^n, (n+1)\omega_{FS})$. This implies that

$$F_\omega(\phi) \geq \frac{n+2-d}{n+1}\log\left(\frac{1}{V}\int_M e^{-\frac{n+1}{n+2-d}\phi}(n+2-d)^n\pi_i^*\omega_{FS}^n\right) - C.$$

Note that C always denotes a uniform constant. Consequently,

$$F_\omega(\phi) \geq \frac{n+2-p}{n+1}\log\left(\frac{1}{V}\int_M e^{-\frac{n+1}{n+2-d}\phi}(n+2-d)^n\sum_{i=0}^{n+1}\pi_i^*\omega_{FS}^n\right) - C,$$

since $\sum_{i=0}^{n+1}\pi_i^*\omega_{FS}^n \geq c\omega$ for some $c > 0$, we have

$$F_\omega(\phi) \geq \frac{n+2-d}{n+1}\log\left(\frac{1}{V}\int_M e^{-\frac{d-1}{n+2-d}\phi}e^{h_\omega-\phi}\omega^n\right) - C.$$

We may assume that $d > 1$.

CLAIM 1. There is a uniform $C > 0$ such that

$$\sup_M \phi \leq C\left(1 + \log\left(\frac{1}{V}\int_M e^{-\frac{d-1}{n+2-d}\phi}e^{h_\omega-\phi}\omega^n\right)\right).$$

PROOF. Fix any α in $(0, \alpha(M))$, then for any $\phi \in P(M, \omega)$,

$$\frac{1}{V}\int_M e^{-\alpha(\phi - \sup_M\phi)}\omega^n \leq C(\alpha).$$

Choose $\delta > 0$ such that

$$\alpha = \frac{\delta(n+1)}{\delta(n+2-d)+d-1}.$$

Then by the Hölder inequality, we have

$$\begin{aligned} 1 &= \tfrac{1}{V}\textstyle\int_M e^{h_\omega-\phi}\omega^n \\ &\leq e^{-\delta sup_M\phi}\left(\tfrac{1}{V}\textstyle\int_M e^{\alpha(h_\omega-\phi+\sup_M\phi)}\omega^n\right)^{\frac{\delta}{\alpha}}\left(\tfrac{1}{V}\textstyle\int_M e^{\frac{d-1}{n+2-d}(h_\omega-\phi)}e^{h_\omega-\phi}\omega^n\right)^{\frac{(1-\delta)(n+2-d)}{n+1}}. \end{aligned}$$

This shows the claim. □

It follows from this claim that F_ω is proper on $P_G(M,\omega)$, and consequently, M admits a Kähler-Einstein metric.

In fact, if M has a Kähler-Einstein metric, the functional F_ω is not only proper, but satisfies the following nonlinear inequality.

THEOREM 3.6. *Let (M,ω) be a Kähler-Einstein manifold with $Ric(\omega) = \omega$. Then there are constants $\delta = \delta(n)$ and $c = c(n, \lambda_2(\omega) - 1)$, such that for any $\phi \in P(M,\omega)$ which satisfies $\phi \perp \Lambda_1$, we have*

$$F_\omega(\phi) \geq J_\omega(\phi)^\delta - c,$$

which is the same as

$$\frac{1}{V}\int_M e^{-\phi}\omega^n \leq e^c e^{J_\omega(\phi) - \frac{1}{V}\int_M \phi\omega^n - J_\omega(\phi)^\delta},$$

where Λ_1 is the space of eigenfunctions with eigenvalue 1, and $\lambda_2(\omega)$ is the next eigenvalue of ω after 1.

This was proved in [**24**] under a further condition on ϕ. This condition was removed later by Zhu and myself.

Indeed, it was conjectured in [**24**] that *for any $\phi \in P(M,\omega)$ which satisfies $\phi \perp \Lambda_1$, we have*

$$F_\omega(\phi) \geq \delta J_\omega(\phi) - c,$$

where $\delta = \delta(n) > 0$ and $c = c(n, \lambda_2(\omega) - 1) > 0$ are as above.

Finally, we would like to cite the uniqueness theorem of Bando and Mabuchi.

THEOREM 3.7. [**4**] *The solution of (3.1) is unique modulo the connected component $Aut_0(M)$ of $Aut(M)$ containing the identity if it exists, where $Aut(M)$ denotes the group of all holomorphic automorphisms of M. In particular, it implies the uniqueness of the Kähler-Einstein metric on M if it exists.*

4. The Case of Complex Surfaces

In this section, we assume that $n = 2$, i.e., M is a complex surface with positive first Chern class. By the classification theory of complex surfaces, M is either $\mathbb{CP}^1 \times \mathbb{CP}^1$ or the blow-up of $\mathbb{CP}^2$ at k points $(0 \leq k \leq 8)$.

THEOREM 4.1. [**22**] *Let M be a complex surface with $c_1(M) > 0$. Then M admits a Kähler-Einstein metric if and only if its Lie algebra $\eta(M)$ of holomorphic fields is reductive.*

Theorem 4.1 gives the complete solution to the existence of Kähler-Einstein metrics on complex surfaces.

Now we outline the proof of Theorem 4.1 in [**22**]. First, by Theorem 1.1, if M has a Kähler-Einstein metric, then $\eta(M)$ is reductive. So we need to only prove the sufficient part. By the classification theory of complex surfaces, we may assume that M is a blow-up of $\mathbb{CP}^2$ at k points $(1 \leq k \leq 8)$, since $\mathbb{CP}^2$ and $\mathbb{CP}^1 \times \mathbb{CP}^1$ have homogeneous Kähler-Einstein metrics. A direct computation shows that $\eta(M)$ is not reductive if $k = 1$ or 2. It was shown in [**26**] that $\alpha(M) \geq 1$ when $k = 3$ and $\alpha(M) \geq 3/4$ when $k = 4$, so M has a Kähler-Einstein metric when $k = 3, 4$. The case that $k = 3$ was also proved independently by Siu [**20**].

Now, we may assume that $k \geq 5$. Then the moduli space of complex structures on the differentiable manifold underlying M is connected and $(k-3)$-dimensional. We denote by $\mathcal{M}_k$ such a moduli space.

Let M be a complex surface with positive first Chern class. By the main theorem in **[26]**, each $\mathcal{M}_k$ contains at least one Kähler-Einstein surface M_k'. Choose a smooth family of Kähler surfaces $\{M_t\}_{0\le t\le 1}$ with $c_1(M_t) > 0$, such that $M_1 = M$ and $M_0 = M_k'$ admits a Kähler-Einstein metric g_0.

Now we define

$$E = \{t \in [0,1] \mid M_{t'} \text{ has a K} - \text{E metric for any } t' \le t\}.$$

Then E is nonempty since $0 \in E$.

Since each M_t has no non-trivial holomorphic vector fields, by the Implicit Function Theorem, one can easily show that E is open in $[0,1]$.

Hence, in order for M to have a Kähler-Einstein metric, we suffice to prove that E is closed. The closure of E is the most difficult part in the proof of Theorem 4.1. We accomplish this in the following two steps.

Without loss of generality, we may assume that $E = [0,1)$. Let ω_t $(0 \le t < 1)$ be the Kähler-Einstein metric on M_t. We need to show that ω_t converges to a metric ω_1 on M in C^2-topology. Since $\{M_t\}_{0\le t\le 1}$ is a smooth family of Kähler surfaces with $c_1(M) > 0$, there are Kähler metrics $\tilde{\omega}_t$ on M_t $(0 \le t \le 1)$ such that their Kähler class is $C_1(M)$. We may further assume that for each $t < 1$, there is a unique function φ_t such that $\omega_t = \tilde{\omega}_t + \partial\bar{\partial}\varphi_t$ and

$$\omega_t^2 \;=\; e^{\tilde{h}_{\tilde{\omega}_t} - \varphi_t}\,\tilde{\omega}_t^2 \text{ on } M_{t\cdot t} \tag{4.1}$$

As we argued in last section, if there is an a prior C^0-estimate of φ_t $(0 \le t < 1)$, then by taking a subsequence if necessary, we may have that ω_t converges to a metric on M.

Instead of deriving such a C^0-estimate directly, we will first give a partial C^0-estimate for φ_t.

First we introduce a notion of height functions for line bundles. Let L be a positive Hermitian line bundle over M. Let ω be a Kähler metric with $c_1(L)$ as its Kähler class. We define a height function $h_L(\cdot,\omega)$ by

$$h_L(x,\omega) = \sup_{0\ne s\in H^0(M,L)} \sqrt{\frac{a(s,\bar{s})(x)}{\frac{1}{V}\int_M a(s,\bar{s})\omega^n}},$$

where $x \in M$, $V = \int_M \omega^n$ and $a(\cdot,\cdot)$ is a Hermitian metric of L with the curvature form ω. Though such a hermitian metric $a(\cdot,\cdot)$ is not unique, the above h_L will remains the same. If L is very ample, then h_L is strictly positive. In fact, it was proved in **[23]** that $m^{-1}h_{L^m}(\cdot, m\omega)^{\frac{1}{m}}$ converges to 1 uniformly on M as m tends to infinity.

In our situation, L is the anti-canonical bundle K_M^{-1}. We simply write h_m for $h_{K_M^{-m}}$. The following was first proposed in **[25]**.

CONJECTURE 1. For any $\epsilon > 0$, there are $m = m(n,\epsilon)$ and $c = c(m,n,\epsilon) > 0$ such that $h_m(\cdot, m\omega) \ge c$ for any Kähler metric ω with the Kähler class $c_1(M)$ on any compact Kähler manifold M of dimension n, provided that $\mathrm{Ric}(\omega) \ge \epsilon\omega$.

It is possible that the conclusion in the above conjecture is true for any sufficiently large m. This conjecture should be also true for a general ample line bundle L over a much larger class of Kähler manifolds. For example, it is possible that *for any* $d > 0$, *there are* $m = m(n,d)$ *and* $c = c(m,n,d) > 0$ *such that* $h_{L^m}(\cdot, m\omega) \ge c$ *for any Kähler metric* ω *with the Kähler class* $c_1(L)$ *on any compact Kähler manifold* M *of dimension* n, *provided that* ω *has constant scalar curvature and its diameter* diam(ω) *is bounded from above by d.* Recently, Zhiqin Lu was able to verify this for hyperbolic Riemann surfaces.

The following was proved in **[22]** (though it was stated differently)

THEOREM 4.2. *The above conjecture holds for any Kähler-Einstein surfaces of complex dimension two. In fact, for* $m > 0$ *and* $m \equiv 0 \bmod 6$, *there is a constant* $C(m)$ *such that* $h_m(\cdot, m\omega) \geq C(m)$ *for any Kähler-Einstein surfaces.*

The proof of this theorem used the Cheeger-Gromov's compactness Theorem, Uhlenbeck's estimates for Yang-Mills connections and L^2-estimates for $\bar\partial$-operators.

Now we explain what this means to φ_t. Let m be given in the above theorem. Choose any orthonormal basis $\{S_i(t)\}_{0\leq i\leq N_m}$ of $H^0(M, K_M^{-m})$ with respect to the inner product induced by ω_t. Then

$$\log(\sum_{i=0}^{N_m} a_t(S_i, \bar S_i)) \geq \log C(m), \tag{4.2}$$

where $a_t(\cdot,\cdot)$ denotes the Hermitian metric on K_M^{-1} defined by ω_t^n.

LEMMA 4.3. *For* $S \in H^0(M_t, K_M^{-m})$ $(t < 1)$, *we have*

$$\sup_{M_t} a_t(S, \bar S)(x) \leq \frac{C(m)}{V} \int_{M_t} a_t(S, \bar S)\omega_t^n.$$

Note that we always denote by $C(m)$ *a constant depending only on* m.

PROOF. An easy computation shows that

$$-\Delta_t \sqrt{a_t(S, \bar S)} \leq m \sqrt{a_t(S, \bar S)},$$

where Δ_t denotes the Laplacian operator of ω_t. Since $\mathrm{Ric}(\omega_t) = \omega_t$, the Sobolev constant can be bounded independent of t. So the lemma follows from the above differential inequality and a standard iteration. □

It follows from this lemma and (4.2) that

$$|\log(\sum_{i=0}^{N_m} a_t(S_i, \bar S_i))| \leq \log C(m). \tag{4.3}$$

Let $\tilde a_t$ be the Hermitian metric with the curvature $\tilde\omega_t$ with

$$\sup_{M_t} \log(\sum_{i=0}^{N_m} \tilde a_t(S_i, \bar S_i)) = 0.$$

Then we have

$$\varphi_t - \frac{1}{m}\log(\sum_{i=0}^{N_m} \tilde a_t(S_i, \bar S_i)) = \frac{1}{m}\log(\sum_{i=0}^{N_m} a_t(S_i, \bar S_i)).$$

It follows from this and (4.3) that

$$\|\varphi_t - \sup_M \varphi_t - \frac{1}{m}\log(\sum_{i=0}^{N_m} \tilde a_t(S_i, \bar S_i))\|_{C^0(M_t)} \leq C(m) \tag{4.4}$$

By using orthogonal transformations if necessary, we may assume that $S_i = \lambda_i \tilde S_i$ for an orthonormal basis $\{\tilde S_i\}$ of $H^0(M, K_M^{-m})$ with respect to the inner product induced by $\tilde{\omega}_t$. Moreover, we may arrange $0 < \lambda_0 < \cdots < \lambda_{N_m}$. Then (4.4) implies that each $\varphi_t - \sup_{M_t} \varphi_t$ is uniformly bounded in any compact subset outside the zero locus of $S_{N_m}(t)$. For this reason, we call (4.4) a partial C^0-estimate of φ_t.

To finish the proof of Theorem 4.1, we need a numerical criterion for the existence of a Kähler-Einstein metric on M.

Let $P_{G,m,k}(M, \tilde{\omega}_1)$ be the collection of all G-invariant functions of the form $\frac{1}{m}\log(\sum_{i=1}^k \tilde{a}_1(S_i, \bar{S}_i))$, where $\{S_i\}_{0\le i\le N_m}$ is an orthonormal basis of $H^0(M, K_M^{-m})$ with respect to the metric $\tilde{\omega}_1$.

Define

$$\alpha_{m,k}(M) = \sup\{\alpha \mid \exists C_\alpha > 0, \text{ s.t. } \frac{1}{V}\int_M e^{-\alpha(\varphi - \sup_M \varphi)}\tilde{\omega}_1^2 \le C_\alpha, \forall \varphi \in P_{G,m,k}(M, \tilde{\omega}_1)\}$$

THEOREM 4.4. *Let m be given in Theorem 4.2. If either $\alpha_{m,1}(M) > 2/3$ or $\alpha_{m,2}(M) > 2/3$ and $\alpha_{m,1}(M) > l_0/(l_0+1)$ where $l_0 = \max\{0, 2 - \frac{3\alpha_{m,2}(M)-2}{\alpha_{m,2}(M)}\}$, then M admits a Kähler-Einstein metric.*

Indeed, the conditions in Theorem 4.4 assures the properness of $F_{\tilde{\omega}_t}$ discussed in last section.

By direct computations, we proved in [**24**] that $\alpha_{6,1}(M) \ge 2/3$, $\alpha_{6,2}(M) > 2/3$ for any complex surface M with $c_1(M) > 0$. Thus Theorem 4.1 is proved.

The proof of Theorem 4.1 can be simplified by using Theorem 3.4 in the cases that M is a blow-up of $\mathbb{CP}^2$ at k points with $k = 5, 7, 8$. The case that $k = 6$ still needs to use Theorem 4.2. Let us show this by establishing directly existence of Kähler-Einstein metrics on any blow-up M of $\mathbb{CP}^2$ at 7 points in general position (i.e., $c_1(M) > 0$).

Let M be a blow-up of $\mathbb{CP}^2$ at 7 points. Then it is a double covering $\pi : M \mapsto \mathbb{CP}^2$ with branch locus along a smooth quartic curve. Assume that G contains the deck transformation of π. As before, ω denotes a fixed, G-invariant Kähler metric with the Kähler class $c_1(M)$. Then for any $\varphi \in P_G(M, \omega)$, $\pi^*\pi_*\varphi = 2\varphi$. We normalize φ such that

$$\frac{1}{V}\int_M e^{h_\omega - \varphi}\omega^2 = 1.$$

Put $f = \pi^*\omega_{FS}^2/\omega^2$. It is non-negative and vanishes along the branch locus of π. Moreover, we have

$$\frac{1}{V}\int_M |f|^{-\frac{4}{5}}\omega^2 \le c < \infty.$$

Using the Hölder inequality, we deduce from this

$$\frac{1}{V}\int_M e^{-\frac{4}{3}\varphi}\omega^2 \le c\left(\frac{1}{V}\int_M e^{-3\varphi}\pi^*\omega_{FS}^2\right)^{\frac{4}{9}}.$$

Note that c always denotes a uniform constant. By using Theorem 3.4 for $\mathbb{CP}^2$, we can show (cf. the example in last section) that

$$F_\omega(\varphi) \ge \frac{1}{3}\log\left(\frac{1}{V}\int_M e^{-3\varphi}\pi^*\omega_{FS}^2\right) - c.$$

It follows from the above inequalities that

$$F_\omega(\varphi) \ge \frac{1}{4}\sup_M \varphi - c.$$

Therefore, F_ω is proper on $P_G(M,\omega)$, and consequently, M admits a Kähler-Einstein metric.

5. Kähler-Ricci Solitons

Let M be a compact Kähler manifold. A Kähler-Ricci soliton is a Kähler metric ω such that $\mathrm{Ric}(\omega) - \omega = L_X\omega$, where X is a complex vector field and L_X is its Lie derivative. Since both the Ricci curvature and the metric are real and of type $(1,1)$, the field X has to be holomorphic and its imaginary part is a Killing field. Since ω is d-closed, $L_X\omega = \partial i_X\omega$ and $\overline{\partial} i_x\omega = 0$. It follows that there is a function θ such that $L_X\omega = \partial\overline{\partial}\theta$. This implies that $c_1(M) > 0$, i.e., Kähler-Ricci solitons can exist only on a compact Kähler manifold with positive first Chern class. We also notice that a Kähler-Ricci soliton is Kähler-Einstein if and only if the Futaki invariant vanishes.

The soliton-type solutions have been widely studied in the regularity theory of nonlinear parabolic equations. Ricci-solitons were first studied by R. Hamilton in his works on the Ricci flow. Its Kähler version was studied by H. Cao in his works on Kähler-Ricci flow. Interesting examples of Kähler-Ricci solitons were constructed by Koiso [**13**], H. Cao [**6**].

As in the case of Kähler-Einstein metrics, we can also reduce the existence of Kähler-Ricci solitons to solving a complex Monge-Ampere equation as follows.

We may assume that $c_1(M) > 0$, in particular, M is simply-connected. Choose a Kähler metric ω with the Kähler class $c_1(M)$. Let θ_X be the function defined by

$$i_X\omega = \overline{\partial}\theta_X, \quad \frac{1}{V}\int_M e^{\theta_X}\omega^n = 1.$$

Then for any $\varphi \in P(M,\omega)$, we have

$$i_X\omega_\varphi = \overline{\partial}(\theta_X + X(\varphi)).$$

Now if ω_φ is a Kähler-Ricci soliton, then

$$\mathrm{Ric}(\omega_\varphi) - \omega_\varphi = \partial\overline{\partial}(\theta_X + X(\varphi)).$$

Hence,

$$(\omega + \partial\overline{\partial}\varphi)^n = e^{h_\omega - \varphi - \theta_X - X(\varphi)}\omega^n. \tag{5.1}$$

If M has no nontrivial holomorphic fields, then this equation reduces to (3.1).

Recently, it was proved by X.H.Zhu and the author that the solution of (5.1) is unique modulo automorphisms of M. This extends clearly the uniqueness theorem of Bando-Mabuchi.

The equation (5.1) bears considerable resemblance to (3.1), and is, moreover, the Euler-Lagrange equation of a functional similar to F_ω. However, few results are known pertaining to the solvability of (5.1). If M has sufficiently many symmetries, then one can reduce it to an ordinary differential equation and solve it (cf. [**13**]). On the other hand, the counterexample in Section 2 shows that (5.1) is not always solvable. A plausible conjecture is the following: *Given any compact Kähler manifold M with $c_1(M) > 0$, either M has a unique Kähler-Ricci soliton or there is a smooth holomorphic fibration $\pi : N \mapsto D$, where D is the unit disc in $\mathbb{C}$, such that M is biholomorphic to each $\pi^{-1}(z)$ with $z \neq 0$ and $\pi^{-1}(0)$ admits a unique Kähler-Ricci soliton.*

Note that M can be separated from the central fiber $\pi^{-1}(0)$ in the moduli space of complex structures. Also unless N is a trivial fibration, N and consequently, $\pi^{-1}(0)$, has a nontrivial holomorphic field.

6. CM-Stability

In this section, we show a connection between the existence of Kähler-Einstein metrics and the stability of underlying manifolds. It was proved by Donaldson-Uhlenbeck-Yau [9], [**27**] hat the existence of Hermitian-Yang-Mills connections is equivalent to the stability of the underlying holomorphic bundle. Inspired by this, in late 80's, S.T. Yau proposed the following

CONJECTURE 2 (Yau). *The existence of Kähler-Einstein metrics should correspond to certain stability of the underlying manifold in the geometric invariant theory.*

This stability of algebraic manifolds should be described in terms of an action of a linear group G on a Hilbert scheme together with a G-linearized line bundle L. There is often a clear choice of such a Hilbert scheme, which parameterizes all manifolds of type similar to that of M. However, there may be possibly many stability conditions which are specified by G-linearized line bundles. Therefore, to solve the above conjecture, we first need to identify the stability corresponding to Kähler-Einstein metrics, or equivalently, the G-linearized line bundle L defining the stability. Next we need to show how Kähler-Einstein metrics are related to the stability. Now let us briefly describe how these can be done.

In fact, our arguments also show a connection between Kähler metrics with constant scalar curvature and the stability on any compact Kähler manifolds.

Let $G = SL(N+1, \mathbb{C})$. Consider a G-equivariant holomorphic fibration $\pi : \mathcal{X} \mapsto Z$ with equi-dimensional fibers. We further assume that there is a G-equivariant line bundle $\mathcal{L}$ on $\mathcal{X}$ with a hermitian metric h, such that its curvature $R(h)$ restricts to a Kähler metric over each fiber. We let Z_0 be the sub-variety of Z consisting of those smooth fibers. Clearly, Z_0 is G-invariant.

Consider the virtual bundle

$$(6.1)\qquad \mathcal{E} = (n+1)(\mathcal{K}^{-1} - \mathcal{K}) \otimes (\mathcal{L} - \mathcal{L}^{-1})^n - n\mu(\mathcal{L} - \mathcal{L}^{-1})^{n+1},$$

where $\mathcal{K} = K_{\mathcal{X}} \otimes K_Z^{-1}$ is the relative canonical bundle, and μ is equal to

$$\frac{c_1(M) \wedge c_1(L)^{n-1}}{c_1(L)^n},$$

where $M = \pi^{-1}(z)$ for a fixed $z \in Z_0$ and $L = \mathcal{L}|_M$.

We define L_Z to be the inverse of the determinant line bundle $\det(\mathcal{E}, \pi)$.

A straightforward computation shows that

$$(6.2)\qquad \begin{aligned} & ch_{n+1}((n+1)(\mathcal{K}^{-1} - \mathcal{K}) \otimes (\mathcal{L} - \mathcal{L}^{-1})^n - n\mu(\mathcal{L} - \mathcal{L}^{-1})^{n+1} \\ = \; & 2^{n+1}\left((n+1)c_1(\mathcal{K}^{-1})c_1(\mathcal{L})^n - n\mu c_1(\mathcal{L})^{n+1})\right) \end{aligned}$$

Therefore, by the Grothendieck-Riemann-Roch Theorem,

$$(6.3)\qquad c_1(L_Z) = 2^{n+1}\pi_*\left((n+1)c_1(\mathcal{K})c_1(\mathcal{L})^n + n\mu c_1(\mathcal{L})^{n+1}\right)$$

We also denote by $\mathcal{L}_Z^{-1}$ the total space of the line bundle L_Z^{-1} over Z. Then G acts naturally on $\mathcal{L}_Z^{-1}$. Recall that $M = \pi^{-1}(z)$ $(z \in Z_0)$ is weakly CM-stable with respect to L, if the orbit $G \cdot \tilde{z}$ in $\mathcal{L}_Z^{-1}$ is closed, where $\tilde{z}$ is any nonzero vector in the fiber of L_Z^{-1} over z; If, in addition, the stabilizer G_z of z is finite, then M is CM-stable. We also recall that M is CM-semistable, if the 0-section is not in the closure of $G \cdot \tilde{z}$. Clearly, this G-stability (resp. G-semistability) is independent of choices of $\tilde{z}$.

THEOREM 6.1. **[24]** *Let $\pi : \mathcal{X} \mapsto Z$ be as above. Assume that M has a Kähler-Einstein metric and $c_1(M) = \mu c_1(L)$. Then M is weakly CM-stable. If X_z has no nontrivial holomorphic vector fields, it is actually CM-stable with respect to L.*

EXAMPLE 6.2. Let us apply Theorem 6.1 to proving Corollary 2.7.

Let Q be the universal quotient bundle over $G(4,7)$, which consists of all 4-subspaces in $\mathbb{C}^7$.

Let π_i $(i = 1,2)$ be the projection from $G(4,7) \times G(3, H^0(G(4,7), \wedge^2 Q))$ onto its ith-factor, and let S be the universal bundle over $G(3, H^0(G(4,7), \wedge^2 Q))$. Then there is a natural endomorphism over $G(4,7) \times G(3, H^0(G(4,7), \wedge^2 Q))$

$$\Phi : \pi_2^* S \mapsto \pi_1^* \wedge^2 Q, \ \Phi(v)|_{(x,P)} = v_x \in \wedge^2 Q. \tag{6.4}$$

Naturally, one can regard Φ as a section in $\pi_2^* S^* \otimes \pi_1^*(\wedge^2 Q)$.

We define

$$\mathcal{X} = \{(x,P) \in G(4,7) \times G(3, H^0(G(4,7), \wedge^2 Q))$$

One can show that $\mathcal{X}$ is smooth.

If $L = \det(Q)$, then $c_1(L)$ is the positive generator of $H^2(W, \mathbb{Z})$.

Consider the G-equivariant fibration $\pi = \pi_2|_{\mathcal{X}} : \mathcal{X} \mapsto Z$, where $G = SL(7, \mathbb{C})$ and $Z = G(3, H^0(G(4,7), \wedge^2 Q))$. Its generic fibers are smooth and of dimension 3. Then Z_0 parameterizes all Fano 3-folds X_P.

Using the Adjunction Formula, one can show

$$c_1(\mathcal{K}) = -\pi_1^* c_1(L) - 3\pi_2^* c_1(S).$$

Therefore, it follows that

$$c_1(L_Z) = 16\pi_* \left(12\pi_2^* c_1(S^*) \pi_1^* c_1(L)^3 - \pi_1^* c_1(L)^4\right).$$

So L_Z is ample.

By the definition of P_a, one can show that none of $G \cdot P_a$ is closed. It implies that any generic X_{P_a} admits no Kähler-Einstein metrics, Furthermore, any generic X_{P_a} admits no nonzero holomorphic fields. In particular, we have proved Corollary 2.7.

The ideas in the proof of Theorem 6.1 can be outlined as follows. For simplicity, we assume that M has no nontrivial holomorphic vector fields.

We will start with a simple criterion for stability.

LEMMA 6.3. *Let $||\cdot||_Z$ be any fixed hermitian metric on L_Z^{-1}. Given any z in Z_0, define a function F_0 on G by*

$$F_0(\sigma) = \log\left(||\sigma(\tilde{z})||_Z\right), \ \sigma \in G.$$

Then M is CM-stable if and only if F_0 is proper on G.

Next we recall **[3, 15]** the definition of the K-energy. Let ω be any Kähler metric representing a positive multiple of $c_1(L)$. Then for any $\varphi \in P(M, \omega)$, the K-energy ν_ω is defined by

$$\nu_\omega(\varphi) = -\frac{1}{V}\int_0^1 \int_M \dot{\varphi}_t(\mathrm{Ric}(\omega_t) - \omega_t) \wedge \omega_t^{n-1} \wedge dt$$

where $\{\varphi_t\}_{0\le t\le 1} \subset P(M,\omega)$ is any path with $\varphi_0 = 0$ and $\varphi_1 = \varphi$, and $\omega_t = \omega + \partial\bar\partial\varphi_t$.

Let h be a hermitian metric on $\mathcal{L}$ over $\mathcal{X}$ such that its curvature $R(h)$ restricts to Kähler metrics on fibers. Fix a function θ such that $R(h)|_M = \omega + \partial\bar\partial\theta$. Define φ_σ by

$$\varphi_\sigma(x) = \theta(x) + \log\left(\frac{h(x)}{h(\sigma(x))}\right).$$

Then we have a function D on G given by

$$D_\omega(\sigma) = \nu_\omega(\varphi_\sigma).$$

Next, we compare D_ω with F_0 on G.

PROPOSITION 6.4. *There is a uniform constant c such that*

$$F_0(\sigma) \ge (n+1)2^{n+1}D_\omega(\sigma) - c.$$

This is the most technical part in the proof of Theorem 6.1 (cf. [**24**]). In fact, the conclusion holds for any ample line bundles over any compact Kähler manifolds.

Now, in order to complete the proof of Theorem 6.1, we suffice to prove that ν_ω is proper. If $\mu \le 0$, it follows from the inequality

$$\nu_\omega(\varphi) \ge \frac{1}{V}\int_M \left(\log\left(\frac{\omega_1^n}{\omega^n}\right)\omega_1^n + h_\omega(\omega^n - \omega_1^n)\right).$$

If $\mu > 0$, then we have

$$\nu_\omega(\varphi) \ge F_\omega(\varphi) + \frac{1}{V}\int_M h_\omega\omega^n.$$

Then the properness of ν_ω follows from this and Theorem 3.6.

We conjecture that *the K-energy is proper on any compact Kähler manifold with constant scalar curvature.* This is true for any Kähler class which admits a Kähler-Einstein metric.

References

[1] Aubin, T.: Equations du type de Monge-Ampère sur les variétés Kähleriennes compactes. C. R. Acad. Sci. Paris 283 (1976), 119 - 121.

[2] Aubin, T.: Réduction du Cas Positif de l'Equation de Monge-Ampére sur les Varietes Kählerinnes Compactes á la Demonstration d'une Inégalité. J. Func. Anal., 57 (1984), 143-153.

[3] Bando, S.: The K-energy Map, Almost Einstein-Kähler Metrics and an Inequality of the Miyaoka-Yau Type. Tohoku Math. J., 39 (1987), 231-235.

[4] Bando, S. and Mabuchi, T.: Uniqueness of Einstein Kähler metrics modulo connected group actions. Algebraic Geometry, Adv. Studies in Pure Math., 10 (1987).

[5] Calabi, E.: Extremal Kähler metrics. Seminar on Diff. Geom., Ann. of math. Stud., 102, Princeton Univ. Press, 1982.

[6] Cao, H.D.: Existence of gradient Kähler-Ricci solitons. Elliptic and parabolic methods in geometry, Minneopolis, MN, 1994, edited by B. Chow, R. Gulliver, S. Levy, J. Sullivan. AK Peters, 1996.

[7] Ding, W.: Remarks on the existence problem of positive Kähler-Einstein metrics. Math. Ann 282 (1988), 463-471.

[8] Ding, W. and Tian, G.: Kähler-Einstein metrics and the generalized Futaki invariants. Invent. Math., 110 (1992), 315-335.

[9] Donaldson, S.: Anti-self-dual Yang-Mills Connections Over Complex Algebraic Surfaces and Stable Vector Bundles. Proc. London Math. Soc., 50 (1985), 1-26.

[10] Futaki, A.: An obstruction to the existence of Einstein-Kähler metrics. Inv. Math., 73 (1983), 437-443.

[11] Futaki, A.: Kähler-Einstein Metrics and Integral Invariants. Lecture Notes in Mathematics, 1314, Springer-Verlag.

[12] Iskovskih, V.A.: Fano threefolds I. Math. USSR Izv., vol. 11 (1977), 485-527; II. Math. USSR Izv., vol. 12 (1978), 469-506.

[13] Koiso, N.: On rationally symmetric Hamilton's equation for Kähler-Einstein metrics. Algebraic Geometry. Adv. Studies in Pure Math., vol. 18-1 (1990), Sendei.

[14] LeBrun, C.: Scalar-flat Kähler metrics on blown-up ruled surfaces. J. Reine Angew. Math. 420 (1991), 161-177.

[15] Mabuchi, T.: K-energy maps integrating Futaki invariants. Tohoku Math. J., 38 (1986), 245-257.

[16] Matsushima, Y.: Sur la structure du group homéomorphismes analytiques d'une certaine variété Kaehlérienne. Nagoya Math. J., 11 (1957), 145-150.

[17] Mukai, S.: Fano 3-folds. Complex Projective Geometry, London Math. Soc. Lec. Notes, ser. 179, Cambridge Univ. Press, 1992.

[18] Nadel, A.: Multiplier ideal sheaves and existence of Kähler-Einstein metrics of positive scalar curvature. Proc. Nat. Acad. Sci. USA, Vol. 86, No. 19 (1989).

[19] Sakane, Y.: Examples of Compact Kähler-Einstein Manifolds with Positive Ricci Curvature. Osaka J. Math., 23 (1986), 586-617.

[20] Siu, Y.T.: The Existence of Kähler-Einstein Metrics on Manifolds with Positive Anti-canonical Line Bundle and a Suitable Finite Symmetry Group. Ann. Math., 127, 585-627 (1988).

[21] Tian, G.: On Kähler-Einstein metrics on certain Kähler Manifolds with $C_1(M) > 0$. Invent. Math., 89 (1987), 225-246.

[22] Tian, G.: On Calabi's conjecture for complex surfaces with positive first Chern class. Inv. Math. Vol.101, No.1 (1990), 101-172.

[23] Tian, G.: On a set of polarized Kähler metrics on algebraic manifolds. J. Diff. Geom., 32 (1990), 99-130.

[24] Tian, G.: Kähler-Einstein metrics with positive scalar curvature. Invent. Math., 130 (1997), 1-39.

[25] Tian, G.: Kähler-Einstein metrics on algebraic manifolds. Proceeding of the C.I.M.E. 1994 conference on "Transcendental Methods in Algebraic Geometry", edited by F. Catanese and C. Ciliberto. Lecture Notes in Math., vol. 1646.

[26] Tian, G. and Yau, S.T.: Kähler-Einstein metrics on complex surfaces with $C_1(M)$ positive. Comm. Math. Phys., 112 (1987).

[27] Uhlenbeck, K. and Yau, S.T.: On the Existence of Hermitian-Yang-Mills Connections on Stable Vector Bundles. Comm. Pure Appl. Math., 1986.

[28] Yau, S.T.: On the Ricci curvature of a compact Kähler manifold and the complex Monge-Ampére equation, I^*. Comm. Pure Appl. Math., 31 (1978), 339-441.

MIT, Building 2, Cambridge, MA 02139
E-mail address: tian@math.mit.edu

Quaternion-Kähler Geometry

S.M. Salamon

Introduction

Interest in quaternion-Kähler manifolds and metrics has developed during the past decades from at least four separate, originally unrelated, sources: (i) the classification of holonomy groups, (ii) the theory of quaternionic manifolds, (iii) self-duality in 4-dimensions, (iv) σ-models in theoretical physics.

An understanding of (i) leads to the holonomy definition of a quaternion-Kähler manifold that dates back to the 1960's. It needs clarification when the dimension is 4, for which curvature conditions enter explicitly into the definition. This aspect of the theory regards (iii), and involves the decomposition of the Weyl tensor, the significance of which was not fully understood (at least in the Riemannian context) until the end of the 1970's. The net result is that quaternion-Kähler manifolds are always Einstein, though their nature depends very much on the sign (positive, negative or zero) of the scalar curvature s.

One aspect of quaternion-Kähler geometry that is inherently higher-dimensional arises from the representation theory of the simplest non-abelian group $SU(2)$. This is most evident in the description, first given by Wolf, of a class of symmetric spaces that include the only complete examples known for $s > 0$. Despite this limitation in the global theory, there is a surprisingly rich geometry underlying the definitions, and the title and contents of this chapter are meant to reflect that. Some of the underlying constructions that we present are valid for the broader theory of quaternionic manifolds, described in §4, that was developed independently by Bérard Bergery and the author, building on earlier work of many others [**17, 105**].

For the uninitiated, the most important point to be made is that a quaternion-Kähler manifold M is not necessarily Kähler in the usual complex sense. Indeed, locally M has a compatible Kähler metric if and only if $s = 0$, in which case it is hyper-Kähler, and amenable to techniques described elsewhere in this volume. On the other hand, the quaternionic structure of M can always be 'untwisted' by passing to the total space of a suitable fibre bundle, and a quaternion-Kähler manifold does always possess associated higher-dimensional complex manifolds, which are Kähler if $s > 0$. Of particular importance in this case is the twistor space, a contact Fano manifold, whose algebraic geometry provides the best hope for tackling the classification problem. Results of LeBrun and others in this direction are presented in §5.

Curvature conditions follow automatically from torsion ones in dimensions greater than 4, and it is sometimes possible to conclude that a given manifold M is quaternion-Kähler without identifying the Einstein metric explicitly. Each point of M corresponds to a rational curve in its twistor space, and one approach is to reconstruct M from the identification of such rational curves and their deformations. We shall illustrate this process in §6. Many techniques for the construction of quaternion-Kähler metrics also have their origins in the 4-dimensional theory. Important constructions of quaternion-Kähler (equivalently, self-dual Einstein) 4-manifolds that we do not discuss can be found in the works of Hitchin [**57**], Pedersen [**93**] and Tod [**115**].

The fundamental role played by isometry groups in the theory of quaternion-Kähler manifolds pervades the whole chapter, though §6 and §7 are specifically devoted to properties of group actions. This includes Swann's generalization of Wolf spaces and their relationship to complex nilpotent orbits and work of Kronheimer, and the quotient construction of Galicki-Lawson. Morse theory turns up in an essential way in these topics, and leads to a number of open problems. An important tool is that of a moment mapping and, whilst this can be interpreted entirely within the realm of symplectic geometry, it comes in other flavours that are peculiar to the quaternionic setting. The quotient construction produces an abundance of local quaternion-Kähler metrics, though the appearance of orbifolds is inherent in the theory. Four-dimensional quaternion-Kähler metrics themselves give rise to Ricci-flat metrics with holonomy equal to G_2 or $Spin(7)$ [**29**], and are therefore especially relevant to the general search for Einstein metrics.

It would be true to say that this chapter represents only a selection of topics in what is a very active field. A final section is devoted to the topology associated to quaternion-Kähler structures, and reflects the author's own interest. The main applications are to the case of positive scalar curvature and link in with §5. The general philosophy is to try to duplicate results known to hold for the Wolf spaces to arbitrary quaternion-Kähler manifolds with $s > 0$. The theory in §8 also allows one to pose a number of related questions for compact quaternion-Kähler manifolds with $s \leq 0$, or indeed the more general class of quaternionic manifolds.

We conclude by mentioning some other topics that we do not pursue further. In a direction related to (iv) above, there is a large class of solvable Lie groups with quaternion-Kähler metrics with $s < 0$ that were discovered by Alekseevsky [**3**], and have recently been re-classified within the framework of supergravity [**38, 35**]. In this set-up, one considers mappings from a space-time of dimension d into a complete target manifold with a specific geometrical structure, such as Kähler, special Kähler, quaternion-Kähler. In the latter case, topological considerations place one in the realm of negative scalar curvature, though this point of view leads to constructions uniting the various geometries. The theory of special Kähler metrics (Kähler ones admitting a certain type of flat symplectic connection) has recently captured the imagination of mathematicians [**42**], and is likely to lead to further developments in the quaternionic field.

An early result of Gray to the effect that any quaternionic submanifold of a quaternion-kähler manifold is totally geodesic [**51**] puts a stop to any naïve theory of submanifolds. Methods in [**33**] effectively classify quaternionic submanifolds of symmetric spaces, and other types of submanifolds have been considered in [**83**]. There is also a vast literature concerned with the classification of various types of submanifolds of quaternionic projective space. But the most effective direction is the study of certain holomorphic submanifolds of twistor space, and there is an extensive theory of harmonic mappings of surfaces into quaternion-Kähler manifolds [**32, 67**] that generalizes the more familiar situation of mappings into 4-manifolds.

1. Almost-Complex Structures and the Canonical 4-Form

A hyper-Kähler manifold ('HK manifold' for short) can be defined as a Riemannian manifold of dimension $4n \geq 4$ admitting an anti-commuting pair I, J of almost-complex structures, relative to both of which the metric is Kähler. This implies that I, J and therefore $K = IJ$ are parallel or 'constant' relative to the Levi-Civita connection, and define integrable complex structures. The triple of endomorphisms I, J, K behaves like the imaginary quaternions, and if (a,b,c) is a unit vector in $\mathbb{R}^3$ then $aI + bJ + cK$ is also a parallel complex structure. Thus, an HK manifold is endowed with a set of complex structures parametrized by the 2-sphere.

Quaternion-kähler manifolds form a more general class of Riemannian manifolds that incorporate not just hyper-Kähler ones, but also the quaternionic projective space space $\mathbb{HP}^n$. Actually, one can define a Riemannian symmetric space that is quaternion-Kähler with an arbitrary compact simple isometry group, and $\mathbb{HP}^n$ corresponds to the case $Sp(n+1) = C_{n+1}$. On a general quaternion-Kähler manifold ('QK manifold' for short), it is not possible to find individual structures I, J, K that are parallel, but only a bundle V of endomorphisms with fibre isomorphic to the imaginary quaternions which as a whole is preserved by the Levi-Civita connection ∇. Locally, one can therefore find I_1, I_2, I_3 satisfying

$$I_1 I_2 = I_3 = -I_2 I_1, \tag{1.1}$$

and 1-forms α_i such that

$$\begin{aligned} \nabla I_1 &= \qquad\qquad\; - \alpha_3 \otimes I_2 + \alpha_2 \otimes I_3, \\ \nabla I_2 &= \alpha_3 \otimes I_1 \qquad\qquad\;\; - \alpha_1 \otimes I_3, \\ \nabla I_3 &= -\alpha_2 \otimes I_1 + \alpha_1 \otimes I_2. \end{aligned} \tag{1.2}$$

These equations were considered by Ishihara [**59**].

Identify $\mathbb{R}^{4n}$ with the space $\mathbb{H}^n$ of quaternion column vectors, so that the Euclidean inner product is given by $\langle v, w\rangle = \mathrm{Re}(v^* w)$, where * is the operation of transposing and quaternionically conjugating entries. The group $Sp(n)$ of quaternion $n \times n$ matrices for which A^*A equals the identity then acts isometrically on $\mathbb{R}^{4n}$ by left multiplication. An HK manifold is then the same as a Riemannian manifold whose holonomy reduces to this group. The parallel complex structures I, J, K arise from the right action of the corresponding unit quaternions.

The group $Sp(n)$ is a subgroup of $SO(4n)$, but not a maximal one since it commutes with the action of the group $Sp(1)$ of unit quaternions on $\mathbb{R}^{4n}$ by right multiplication. The enlarged group of transformations

$$v \longmapsto Avq^*, \qquad A \in Sp(n),\ q \in Sp(1) \tag{1.3}$$

is denoted by $Sp(n)Sp(1)$. It is a subgroup of $SO(4n)$ isomorphic to the quotient $Sp(n) \times_{\mathbb{Z}_2} Sp(1)$, where $\mathbb{Z}_2$ is generated by $(-I, -1)$.

DEFINITION 1.1. A QK manifold is a Riemannian manifold of dimension $4n$ whose holonomy group is contained in the group $Sp(n)Sp(1)$.

Since $Sp(1)Sp(1) = SO(4)$, the geometry resulting from this definition generalizes that of oriented Riemannian 4-manifolds. For the moment though, we shall assume that $n \geq 2$.

It is an immediate consequence of the above definition that the frame bundle of a QK manifold reduces to a principal bundle with structure group $Sp(n)Sp(1)$. The bundle V defined above is none other than that associated to the homomorphism

$$Sp(n)Sp(1) \to Sp(1)/\mathbb{Z}_2 \cong SO(3),$$

and a local basis $\{I_1, I_2, I_3\}$ of V satisfying (1.1) is determined up to the action of $SO(3)$ at each point. Each almost-complex structure I_i determines a 2-form ω_i by the usual identity $\omega(X,Y) = g(I_i X, Y)$, and the ω_i are analogues of self-dual 2-forms in 4 dimensions. It is easy to see that the 4-form

$$\Omega = \omega_1 \wedge \omega_1 + \omega_2 \wedge \omega_2 + \omega_3 \wedge \omega_3 \tag{1.4}$$

is independent of the choice of basis, and Ω^n is nowhere zero. This form was introduced by Bonan in [**20**]. An orientation of M can be defined by decreeing that the volume form is a constant positive multiple of Ω^n.

Replacing the I_i's in (1.2) by ω_i's, we see that Ω is parallel, and therefore closed. Since the stabilizer of Ω in $GL(4n, \mathbb{R})$ is exactly $Sp(n)Sp(1)$, the holonomy reduction of a QK manifold is characterized by the existence of a 4-form which is

(i) in the same $GL(4n, \mathbb{R})$-orbit as Ω at each point, and

(ii) parallel.

If M has dimension at least 12, it turns out surprisingly that (ii) is equivalent to requiring that the 4-form be closed [**112**]. The case of 8 dimensions is in a certain sense richer, as there exist metrics at least locally admitting closed but non-parallel 4-forms satisfying (i). This contrasts with the case of $Spin(7)$ holonomy which, as observed in [**28**], is defined by a closed 4-form linearly equivalent to

$$\tilde{\Omega} = \omega_1 \wedge \omega_1 + \omega_2 \wedge \omega_2 - \omega_3 \wedge \omega_3.$$

On its own, condition (i) defines the class of 'almost QK manifolds'. Let $\mathfrak{h}^\perp$ denote the orthogonal complement of $\mathfrak{h} = \mathfrak{sp}(n) + \mathfrak{sp}(1)$ in $\mathfrak{so}(4n)$; this is in fact an irreducible representation of $Sp(n)Sp(1)$ (denoted $\Lambda_0^2 \otimes \Sigma^2$ below). General principles imply that, on any almost QK manifold, the covariant derivative $\nabla_X \Omega$ (for any tangent vector X) belongs to a subspace of $\bigwedge^4 T^*$ isomorphic to $\mathfrak{h}^\perp$. Numerous classes of almost quaternionic manifolds can be defined by decomposing the space $T^* \otimes \mathfrak{h}^\perp$ and imposing corresponding conditions on Ω [**106, 111**], though we shall adopt a different approach in §4. Analogues of quaternion-Kähler geometry with torsion have been studied by theoretical physicists (see e.g. [**58**]).

Remark. The fact that 4-forms arise automatically from a holonomy reduction may be deduced from the following purely algebraic observation. Let $\mathfrak{h}$ denote the Lie algebra of a compact holonomy group H, regarded as a subspace of $\bigwedge^2 T^*$. Then there is an H-equivariant mapping

$$b \colon S^2\mathfrak{h} \longrightarrow \bigwedge^4 T^* \tag{1.5}$$

defined simply by skewing 2-forms together. The symmetric product $S^2\mathfrak{h}$ contains at least a 1-dimensional space of H-invariant elements, so let ρ be such an element. If $b(\rho) = 0$ then ρ is a curvature-like tensor satisfying the first Bianchi identity, and the fundamental work of É. Cartan implies that ρ is the curvature tensor of a Riemannian symmetric space. Otherwise, $b(\rho)$ will determine a non-zero parallel

4-form. In the case of $\mathfrak{h} = \mathfrak{sp}(n) + \mathfrak{sp}(1)$, both these possibilities occur as there is a 2-dimensional space of invariants in $S^2\mathfrak{h}$, spanned by elements ρ_1, ρ_2 with $\rho_1 \in \ker b$. The corresponding symmetric space is $\mathbb{HP}^n$ with curvature tensor ρ_1, and the 4-form $b(\rho_2)$ is proportional to Ω.

A more careful analysis of the 'Bianchi map' (1.5) for $\mathfrak{h} = \mathfrak{sp}(n) + \mathfrak{sp}(1)$ shows that, provided $n \geq 2$, its kernel is spanned only by ρ_1 and the highest-weight component W in the tensor product $\mathfrak{sp}(n) \otimes \mathfrak{sp}(n)$. The summand W contains the curvature tensor of an HK manifold, and since it has no components in common with the space S^2T^* of Ricci tensors, one deduces

COROLLARY 1.2. *Any QK manifold of dimension $4n \geq 8$ is Einstein, and its scalar curvature s vanishes if and only if it is locally HK, i.e. its restricted holonomy group H_0 is a subgroup of $Sp(n)$.*

In this case, the classification of possible holonomy groups H having connected component $H_0 = Sp(n)$ has been carried out by McInnes [**85**], and a related theory of 'locally quaternion-Kähler' manifolds is developed in [**96**]. It also makes sense to talk about QK manifolds with zero curvature.

Definition 1.1 is the traditional one, though we are now able to point out drawbacks in the terminology.

(i) As it stands, a 4-dimensional QK manifold is none other than an oriented Riemannian one. The curvature restrictions that apply from (1.5) in higher dimensions disappear because $\bigwedge^4 T^*$ is just 1-dimensional. However, we may re-impose them by redefining a quaternion-Kähler 4-manifold to be an oriented Riemannian one whose curvature tensor belongs to

$$S^2_0(\textstyle\bigwedge^2_+ T^*) \oplus \langle \rho_1 \rangle,$$

relative to (3.3), where ρ_1 spans $\ker b$. The first component is the 'positive half' W_+ of the Weyl tensor, and the second is a multiple of the constant curvature tensor of $S^4 = \mathbb{HP}^1$. This condition is equivalent to asserting that M is 'self-dual' (meaning that $W_- = 0$) and Einstein.

(ii) A quaternion-Kähler manifold is not in general Kähler, since $Sp(n)Sp(1)$ is not a subgroup of $U(2n)$. (For this reason, the author sees the abbreviation 'QK' as avoiding a certain amount of embarassment.) In fact, if H is a subgroup of

$$(Sp(n)Sp(1)) \cap U(2n) = Sp(n)U(1),$$

then H_0 is a subgroup of $Sp(n)$ [**119**] and M is locally HK. As we have remarked, the two situations are distinguished locally by the Ricci tensor.

(iii) There is a natural tendency to restrict the terminology 'quaternion-Kähler' to the case of non-zero scalar curvature, a situation significantly different from the hyper-Kähler one. This is indeed the approach we adopt in this chapter, although there are further links between the two classes of manifolds that transcend the definitions. For example, we shall see that any QK manifold with positive scalar curvature can be realized as a type of quotient of an HK manifold of 4 dimensions greater. This is disconcerting as it implies that the theory of QK manifolds can logically (at least in the $s > 0$ regime) be subsumed into the theory of HK ones!

(iv) One cannot restrict to the case of non-zero scalar curvature by demanding that the holonomy group should equal $Sp(n)Sp(1)$. For this would exclude most of the symmetric space examples, which have holonomy of the form $K\,Sp(1)$ where K is a subgroup of $Sp(n)$. We study these in the next section.

On a compact Kähler manifold, wedging with the fundamental 2-form ω induces a non-singular map on cohomology in appropriate dimensions; this is the well-known Lefschetz property that relates to formality properties of the de Rham algebra of a Kähler manifold. One of the earliest results for a compact QK manifold M of dimension $4n$ was the analogous result found by Kraines [**73**]. With a slight improvement of Fujiki [**44**], this states that wedging with the 4-form Ω determines an injection

$$H^k(M,\mathbb{R}) \hookrightarrow H^{k+4}(M,\mathbb{R}), \qquad k \leq n-1.$$

Refinements were also made in [**21**]. Of course, given that Ω is a closed 4-form, with $\Omega^n \neq 0$, it is also true that $b_{4i} > 0$ for $0 \leq i \leq n$.

Much more can be said when the scalar curvature is positive. A complete quaternion-Kähler manifold with $s > 0$ is called a 'positive QK manifold', and because of the Einstein condition, completeness is equivalent to compactness. The author proved that a positive QK manifold has vanishing 'odd' Betti numbers $b_{2i+1} = 0$ [**102**]. Building on this, one can show that the differences

$$\beta_{2i} = b_{2i} - b_{2i-4}, \qquad i \leq n, \tag{1.6}$$

are all non-negative [**44**]. They are in fact the Betti numbers of an associated 3-Sasakian manifold defined in §5, and feature again after Theorem 8.2(ii).

To conclude this section we quote a result that is relevant to the remarks in (ii) above. Its proof exploits Theorem 5.5(ii) below, and the well-known fact that there is no almost complex structure on $\mathbb{HP}^1 \cong S^4$.

THEOREM 1.3. [**5**] *No positive QK manifold admits a compatible almost complex structure.*

An almost-complex structure is said to be compatible if it is a section of V, so that it can be expressed locally as $aI_1 + bI_2 + cI_3$ with the relations (1.1) and a, b, c functions satisfying $a^2 + b^2 + c^2 = 1$.

2. Symmetric Spaces and Grassmannian Geometry

The essence of quaternion-Kähler geometry is captured by the 1964 paper of Wolf [**119**]. In it, he characterizes the quaternionic structure of a class of symmetric spaces, and discusses what is now known as their twistor fibrations. In a paper [**2**] which coined the term 'Wolf space', Alekseevsky went on to show that any homogeneous QK manifold with $s > 0$ must in fact be one of these symmetric spaces. In this section we begin by listing the spaces in question, and then explain how their existence is related to the theory of 3-dimensional subalgebras.

By a 'Wolf space' we mean a quaternion-Kähler symmetric space with $s > 0$. There are two Wolf spaces of real dimension 4, namely

$$S^4 = \frac{SO(5)}{SO(4)}, \quad \mathbb{CP}^2 = \frac{SU(3)}{S(U(2)\times U(1))}.$$

The curvature of both of these spaces is self-dual and Einstein, in accordance with the revised definition in 4 dimensions.

In general, there is a Wolf space corresponding to each simple compact Lie algebra. In dimension $4n$, there are three families

$$\mathbb{HP}^n = \frac{Sp(n+1)}{Sp(n)\times Sp(1)}, \quad \mathrm{Gr}_2(\mathbb{C}^{n+2}) = \frac{U(n+2)}{U(n)\times U(2)}, \quad \mathrm{Gr}_4(\mathbb{R}^{n+4}) = \frac{SO(n+4)}{SO(n)\times SO(4)}.$$

There are coincidences $\mathbb{HP}^1 = S^4 = \mathbb{G}\mathrm{r}_4(\mathbb{R}^5)$ and $\mathbb{G}\mathrm{r}_2(\mathbb{C}^3) = \mathbb{CP}^2$ for $n = 1$, and $\mathbb{G}\mathrm{r}_2(\mathbb{C}^4) = \mathbb{G}\mathrm{r}_4(\mathbb{R}^6)$ for $n = 2$. In addition, there are the exceptional spaces

$$\frac{G_2}{SO(4)}, \qquad \frac{F_4}{Sp(3)Sp(1)},$$

$$\frac{E_6}{SU(6)Sp(1)}, \quad \frac{E_7}{Spin(12)Sp(1)}, \quad \frac{E_8}{E_7Sp(1)},$$

corresponding to $n = 2, 7, 10, 16, 28$ respectively.

Needless to say, these homogeneous spaces arise as orbits for the action of the isometry group G on a suitable linear space. Any adjoint orbit of G on its Lie algebra $\mathfrak{g}$ is a complex homogeneous space, and the only Wolf space that arises in this way is the complex Grassmannian. Indeed, if we identify $\mathfrak{su}(n+2)$ with the space $\Lambda_0^{1,1}(\mathbb{C}^{n+2})$ of primitive $(1,1)$-forms on $\mathbb{C}^{n+2}$ then the orbit of an element $\alpha\wedge\overline{\beta}$ (with $\langle\alpha,\beta\rangle = 0$) is isomorphic to $\mathbb{G}\mathrm{r}_2(\mathbb{C}^{n+2})$. In the same vein, $\mathbb{HP}^n$ arises as the $Sp(n+1)$-orbit of a suitable element in $\Lambda_0^2(\mathbb{C}^{2n+2})$, and the Plücker embedding exhibits $\mathbb{G}\mathrm{r}_4(\mathbb{R}^{n+4})$ as a $SO(n+4)$-orbit of a simple 4-form inside $\Lambda^4(\mathbb{R}^{n+4})$.

Remark. The following topological properties of the Wolf spaces reflect general results on QK manifolds that we shall discuss below:

(i) The list consists of exactly those irreducible Riemannian symmetric spaces for which $H^2(M,\mathbb{Z}) \cong \mathbb{Z}_2$, together with $\mathbb{HP}^n$ and $\mathbb{G}\mathrm{r}_2(\mathbb{C}^{n+2})$ (which of course have $H^2(M,\mathbb{Z})$ equal to 0 and $\mathbb{Z}$ respectively) [**31**].

(ii) The only Wolf spaces with 4th Betti number $b_4 > 1$ are the Grassmannians $\mathbb{G}\mathrm{r}_2(\mathbb{C}^{n+2})$ $(n \geq 2)$ and $\mathbb{G}\mathrm{r}_4(\mathbb{R}^{n+4})$ $(n \geq 2)$. The only space with $b_4 > 2$ is $\mathbb{G}\mathrm{r}_4(\mathbb{R}^8)$, which has two distinct quaternion-Kähler structures related by an outer automorphism of $SO(8)$, and $b_4 = 3$ [**96**].

(iii) Those with isometry group of type A, D, E have

$$\dim G = 2\chi + b_{2n-2} + b_{2n}, \tag{2.1}$$

where χ denotes the Euler characteristic.

It is conjectured that any positive QK manifold is a Wolf space. The following compilation of results in this direction represents the state of play at the time this article was written.

THEOREM 2.1. *A positive QK manifold M of dimension $4n$ is necessarily symmetric if one of the following is true:*
(i) $n \leq 2$,
(ii) $n \leq 4$ and $b_4 = 1$,
(iii) the isometry group of M has rank at least $n+1$.

Part (i) for $n = 1$ is Hitchin's theorem asserting that S^4 and $\mathbb{CP}^2$ are the only two compact 4-manifolds with a metric which is self-dual, Einstein, with positive scalar curvature [**55, 17**]. The corresponding result for $n = 2$ was proved in [**98**], and by a different method in [**80**]. A key starting point is the existence of a space of Killing vector fields of sufficiently positive dimension.

The biggest gap in a potential proof of the conjecture is the unknown answer to the conundrum of whether a QK manifold of dimension greater than 16 has any non-zero Killing vector fields. Estimates on the size of the isometry group G are known for $n = 3$ or $n = 4$, and in these cases it is proved in [**49**] that if $b_4 = 1$

then M is isometric to $\mathbb{HP}^n$, whence (ii). The author is confident that a complete analysis will soon be possible in dimension 16 and less.

A valuable bound is that the rank of G cannot exceed $n+1$ [**98**]. Part (iii) is a recent result of Bielawski that is proved by reconstructing the spaces in question as quotients using the techniques hinted at in §7. The only positive QK manifolds with an isometry group of rank $n+1$ are in fact $\mathbb{HP}^n$ and $\mathbb{G}\mathrm{r}_2(\mathbb{C}^{n+2})$.

The special nature of the isotropy groups of the Wolf spaces is emphasized by the next result. To formulate it, suppose that M is a QK manifold with isometry group G, and let $x \in M$. If $\mathfrak{g}_x$ denotes the Lie algebra of the isotropy subgroup at x, then $\mathfrak{g}_x \subseteq \mathfrak{sp}(n) + \mathfrak{sp}(1)$, and

PROPOSITION 2.2. *If $\mathfrak{g}_x$ contains the summand $\mathfrak{sp}(1)$ of $\mathfrak{sp}(n) + \mathfrak{sp}(1)$ for all x, then M is locally symmetric.*

This follows because ∇R lies in a space ($S^5E \otimes \Sigma^1$ in notation below) that has no non-zero elements invariant by $Sp(1)$.

A non-trivial isotropy subgroup at a single point can even impose quite severe constraints on the curvature tensor [**98**], and an algebraic classification of curvature jets is feasible in certain circumstances. The study of possible isotropy subgroups is important in the classification of QK manifolds with a cohomomogeneity-one group action, that is in progress by the authors of [**36, 18, 6**].

Remark. The symmetric spaces listed above have non-compact duals with $s < 0$, and the embedding of these in their compact partners is discussed in [**119**]. By a theorem of Borel [**22**], the dual of any compact QK symmetric space admits compact quotients. Constraints on the fundamental group of such a quotient arise from work of Corlette [**34**]. He proved that any Riemannian metric with non-positive curvature operator on a compact quotient of quaternionic hyperbolic space $\mathbb{H}H^n$, $n \geq 2$, is the induced one. In this connection, it is known that the homogeneous non-symmetric spaces of [**3**] do not admit compact quotients, and the only known compact examples of QK manifolds with $s < 0$ are locally symmetric.

The tangent space at any point to a Riemannian symmetric space $M = G/H$ can be identified with an orthogonal complement $\mathfrak{m}$ of the Lie algebra $\mathfrak{h}$ in $\mathfrak{g}$. The holonomy group of M then coincides with the linear isotropy group, that is the image of the natural homomorphism $\ell\colon H \to \mathrm{Aut}(\mathfrak{m})$, and is isomorphic to H if G has trivial centre. For each of the Wolf spaces, it is easy to check explicitly that $\ell(H)$ has the form $K \times_{\mathbb{Z}_2} Sp(1)$ where $K \subseteq Sp(n)$, in accordance with (1.3). In fact, H is the normalizer of a 3-dimensional subgroup of G that is generated by a highest root vector of the Lie algegra $\mathfrak{g}$ [**119**]. It follows that G/H parameterizes a 'conjugacy class' of 3-dimensional subalgebras of $\mathfrak{g}$, a fact that we now proceed to generalize.

Let G be a compact semisimple Lie group of dimension d, with Lie algebra $\mathfrak{g}$, and let $\mathfrak{su}(2)$ denote the Lie algebra generated by the matrices

$$A_1 = \begin{pmatrix} i & 0 \\ 0 & -i \end{pmatrix}, \quad A_2 = \begin{pmatrix} 0 & -1 \\ 1 & 0 \end{pmatrix}, \quad A_3 = \begin{pmatrix} 0 & i \\ i & 0 \end{pmatrix} \tag{2.2}$$

over $\mathbb{R}$, with Lie bracket given by $[A_1, A_2] = A_1A_2 - A_2A_1 = -2A_3$ etc. A non-zero homomorphism of Lie algebras $\rho\colon \mathfrak{su}(2) \to \mathfrak{g}$ is necessarily injective, and we may regard $\rho(\mathfrak{su}(2))$ is an element of the Grassmannian $\mathbb{G}\mathrm{r}_3(\mathfrak{g})$ of real oriented

3-dimensional subspaces of $\mathfrak{g}$. There is a canonical 3-form on the Lie algebra $\mathfrak{g}$ that defines a real-valued G-invariant function ψ on $\mathbb{Gr}_3(\mathfrak{g})$ defined by setting

$$\psi(V) = -\langle v_1, [v_2, v_3]\rangle,$$

where $V \in \mathbb{Gr}_3(\mathfrak{g})$ is a subspace with an oriented basis $\{v_1, v_2, v_3\}$, orthonormal relative to the Killing form of $\mathfrak{g}$.

LEMMA 2.3. *V is a critical point of ψ with $\psi(V) > 0$ if and only if $V = \rho(\mathfrak{su}(2))$ for some homomorphism ρ.*

To understand this result, recall that the tangent space $T_V\mathbb{Gr}_3(\mathfrak{g})$ can be identified with

$$\mathrm{Hom}(V, V^\perp) \cong V^* \otimes V^\perp, \tag{2.3}$$

where $V^\perp$ is the orthogonal complement of V in $\mathfrak{g}$; if $\alpha \in \mathrm{Hom}(V, V^\perp)$, then the corresponding vector is the one tangent to the curve $t \longmapsto V_t = \mathrm{span}\{v + t\alpha(v) : v \in V\}$ in $\mathbb{Gr}_3(\mathfrak{g})$ at $t = 0$. The gradient of ψ at $V \in \mathbb{Gr}_3(\mathfrak{g})$ is the linear mapping characterized by

$$v_1 \longmapsto -[v_2, v_3] - \psi(V)v_1 \in V^\perp, \tag{2.4}$$

whenever $\{v_1, v_2, v_3\}$ is an oriented orthonormal basis of V.

The orbit of $\rho(\mathfrak{su}(2))$ under the adjoint action of G forms the critical manifold L_ρ, and

$$L_\rho \cong G/N_\rho,$$

where N_ρ denotes the normalizer of $\rho(\mathfrak{su}(2))$. A trajectory or flow line of the vector field gradψ is a curve in $\mathbb{Gr}_3(\mathfrak{g})$ satisfying

$$V'(t) = \mathrm{grad}\,\psi(V(t)). \tag{2.5}$$

It was verified by Burstall that the Hessian of ψ is non-degenerate in normal directions to the critical submanifolds L_ρ, which means that Morse-Bott theory can be applied to the flow lines as in [**66**]. The union of L_ρ and those points on trajectories $V(t)$ with $\lim_{t\to-\infty} V(t) \in L_\rho$ is the so-called unstable manifold M_ρ associated to L_ρ. There are inclusions

$$L_\rho \subseteq M_\rho \subset \mathbb{Gr}_3(\mathfrak{g}),$$

and M_ρ is G-equivariantly diffeomorphic to the total space of the normal bundle to L_ρ in M_ρ. In this way, M_ρ parametrizes a distinguished family of 3-dimensional subspaces of $\mathfrak{g}$ including the subalgebras conjugate to $\rho(\mathfrak{su}(2))$.

THEOREM 2.4. [**114**] *Let $\rho\colon \mathfrak{su}(2) \to \mathfrak{g}$ be any non-zero Lie algebra homomorphism. Then M_ρ has a G-invariant QK metric with $s > 0$.*

This theorem was proved by Swann by relating (2.5) to Nahm's equations, and then via twistor theory to complex nilpotent codajoint orbits and work of Kronheimer [**75**]. We shall explain in §6 that if $V \in M_\rho$ then the isotropic elements of its complexification V_c are nilpotent, a property that generalizes that enjoyed by 3-dimensional subalgebras. Indeed, the theorem belongs to a select class of results in which a Morse flow is used to classify a family of objects, the most obvious of which correspond to critical points. An analogous example of this in an infinite-dimensional setting appears in the paper [**30**] on harmonic maps.

The infinitesimal quaternionic structure is readily identified at points of L_ρ, and this will help to introduce the algebraic structure of the tangent space of an arbitrary QK manifold. Let us denote the complexification $\rho(\mathfrak{su}(2))_c$ by $\mathfrak{sl}(2,\mathbb{C})$. The latter has a complex $(k+1)$-dimensional irreducible representation that we denote by Σ^k, isomorphic to $S^k\mathbb{C}^2$ and the space of homogeneous polynomials of degree k in two variables. There is a decomposition

$$\mathfrak{g}_c \cong \mathfrak{sl}(2,\mathbb{C}) \oplus \bigoplus_{k\geq 0} \mu_k\Sigma^k = \mathfrak{n}_c \oplus \bigoplus_{k>0} \mu_k\Sigma^k, \tag{2.6}$$

where $\mu_k\Sigma^k$ denotes $\Sigma^k \oplus \cdots \oplus \Sigma^k$, with $\mu_k \geq 0$ summands, and $\mathfrak{n}$ is the Lie algebra of N_ρ. The adjoint representation $\mathfrak{sl}(2,\mathbb{C})$ is itself isomorphic to the space Σ^2 of homogeneous quadratic polynomials.

Fix $V \in L_\rho$, so that $V^\perp \cong \bigoplus_{k\geq 0} \mu_k\Sigma^k$. Using (2.3) and the isomorphism $\Sigma^2 \otimes \Sigma^k \cong \Sigma^{k-2} \oplus \Sigma^k \oplus \Sigma^{k+2}$ (with $k>0$ and $\Sigma^{-1} = \{0\}$), we obtain

$$T_V\mathbb{G}\mathrm{r}_3(\mathfrak{g}) = T_+ \oplus T_0 \oplus T_-,$$

where

$$\begin{aligned}(T_+)_c &\cong \bigoplus_{k\geq 2} \mu_k\Sigma^{k-2}\\ (T_0)_c &\cong \bigoplus_{k\geq 1} \mu_k\Sigma^{k}\\ (T_-)_c &\cong \bigoplus_{k\geq 1} \mu_k\Sigma^{k+2} \oplus \mu_0\Sigma^2.\end{aligned}$$

Then T_0 coincides with the tangent space T_VL_ρ to the orbit through V, and $T_0 \oplus T_+$ is the tangent space to M_ρ. It has complexification

$$(T_0 \oplus T_+)_c \cong E \otimes_c \Sigma^1, \tag{2.7}$$

where $E = \bigoplus_{k\geq 1} \mu_k\Sigma^{k-1}$. The quaternionic structure of E originates from each summand Σ^k when k is odd, and pairs $\Sigma^k \oplus \Sigma^k$ when k is even. Therefore the action of $SU(2)$ on $T_0 \oplus T_+$ factors through $Sp(n)Sp(1)$, where $2n = \dim_{\mathbb{C}} E = \sum_{k\geq 1} k\mu_k$.

Any root space $\mathfrak{g}_\alpha$ generates such a subalgebra $\rho(\mathfrak{su}(2))$, but a general homomorphism ρ is determined up to conjugacy by assigning an integer in the set $\{0,1,2\}$ to each simple root of $\mathfrak{g}$ according to rules prescribed by Dynkin (this is explained by [**67**] in a useful context). The dimension of N_ρ is as small as possible when $\mathfrak{su}(2)$ is the span of an orthonormal basis $\{v_1, v_2, v_3\}$ of $\mathfrak{g}$ where $v_1 + iv_2$ belongs to a highest root space $\mathfrak{g}_\gamma$ of $\mathfrak{g}_c$. We shall call such a subalgebra minimal. The functional ψ attains its maximum value on the Wolf space G/N_ρ, where ρ arises from a highest root. In this case $T_+ = 0$ (equivalently $\mu_k = 0$ for all $k \geq 2$).

3. Representations and the Dirac Operator

The representation of the structure group $Sp(n)Sp(1)$ on the complexified tangent space $(T_x)_c$ of an arbitrary QK manifold is determined by (1.3) and coincides with the right-hand side of (2.7). The structure group of a QK manifold lifts globally to $Sp(n) \times Sp(1)$ if and only if $\varepsilon = 0$, where $\varepsilon \in H^2(M, \mathbb{Z}_2)$ is the class induced by the short exact sequence

$$1 \to \mathbb{Z}_2 \to Sp(n) \times Sp(1) \to Sp(n)Sp(1) \to 1,$$

and introduced explicitly in [**84**]. The significance of this lifting condition was first realized by Sakamoto [**101**], in a study of sectional curvature and pinching.

Over an open set on which the obstruction ε vanishes, it is conventional to write

$$T_c = E \otimes H, \tag{3.1}$$

where E and H now represent complex vector bundles of rank $2n$ and 2 respectively, underlying the standard representations of $Sp(n)$ and $Sp(1)$ on $\mathbb{H}^n$ and $\mathbb{H}$ respectively. Since the latter are self-dual, one can also replace T by T^* in (3.1) without affecting its validity.

Given the well-known isomorphism $Sp(1) \times Sp(1) \cong Spin(4)$ over an oriented Riemannian 4-manifold, E and H are in this case the same as the spin bundles, denoted V_+ and V_- in [**8**]. Thus E and H exist globally over S^4, and indeed over $\mathbb{HP}^n$ for all $n \geq 1$ since $H^2(\mathbb{HP}^n, \mathbb{Z}_2) = 0$. Regarded as a quaternionic line bundle, H is simply the tautological bundle whose fibre H_x at a point $x \in \mathbb{HP}^n$ is the line represented by that point, and E_x can be identified with the complement $H^\perp$ in $\mathbb{H}^{n+1}$. The resulting decomposition

$$\mathbb{H}^{n+1} = E_x \oplus H_x$$

characterizes the action of the isotropy group $Sp(n) \times Sp(1)$ on $\mathbb{H}^{n+1}$.

Example. The algebra underlying the 4-dimensional situation is also relevant when one examines the 8-dimensional space $G_2/SO(4)$. Identifying $SO(4) = Sp(1)Sp(1)$, its inclusion in G_2 is described by the decomposition $\mathbb{C}^7 = S^2V_- \oplus (V_- \otimes V_+)$ of the standard representation of G_2. This equation provides the well-known link between self-duality in dimension 4 and G_2-structures in 7. Furthermore,

$$(\mathfrak{g}_2)_c \cong S^2V_- \oplus S^2V_+ \oplus (S^3V_- \otimes V_+),$$

and the last summand is effectively the isotropy representation $\mathfrak{m}$. So we may take $E = S^3V_+$ and $H = V_-$; these are not globally defined bundles as $H^2(G_2/SO(4), \mathbb{Z}) \cong \mathbb{Z}_2$ is generated by ε.

The use of the locally-defined bundles E and H is a very convenient tool in describing exterior forms and other natural tensors on a QK manifold. For example, anticipating the notation below, the bundle of 2-forms can be written

$$\begin{aligned} \textstyle\bigwedge^2 T_c^* &\cong \textstyle\bigwedge^2(E \otimes H) \\ &\cong (S^2E \otimes \textstyle\bigwedge^2 H) \oplus (\textstyle\bigwedge^2 E \otimes S^2H) \\ &\cong S^2E \oplus S^2H \oplus (\textstyle\bigwedge_0^2 E \otimes S^2H) \\ &\cong \Lambda_1^2 \oplus \Sigma^2 \oplus (\Lambda_0^2 \otimes \Sigma^2). \end{aligned} \tag{3.2}$$

It has three irreducible real components, corresponding to $\mathfrak{h} = \mathfrak{sp}(n) \oplus \mathfrak{sp}(1)$ and $\mathfrak{h}^\perp$. This generalizes the celebrated decomposition

$$\textstyle\bigwedge^2 T^* = \bigwedge_+^2 \oplus \bigwedge_-^2 = \mathfrak{sp}(1)_+ \oplus \mathfrak{sp}(1)_- \tag{3.3}$$

on an oriented Riemannian 4-manifold, and leads to extensions of Yang-Mills theory (see §4).

It is natural to ask how the standard representation Δ of $Spin(4n)$ of dimension 2^{2n} decomposes relative to the natural homomorphism $Sp(n) \times Sp(1) \to Spin(4n)$

for $n \geq 2$. To proceed, one needs to distinguish certain representations of $Sp(n)$. Choosing standard coordinates on the Lie algebra of a maximal torus, we may identify irreducible $Sp(n)$-modules with n-tuples of integers

$$(a_1, a_2, \ldots, a_n), \qquad a_1 \geq a_2 \geq \cdots \geq a_n \geq 0,$$

corresponding to dominant weights (this is explained in [**106**]). In particular, E corresponds to $(1, 0, \ldots, 0)$, and the symmetric power $S^m E$ corresponds to $(m, 0, \ldots, 0)$ and is irreducible.

We shall be more interested in the summands

$$\Lambda_q^m = (\underbrace{2, \ldots, 2}_{q}, \underbrace{1, \ldots, 1}_{m-2q}, 0 \ldots, 0), \qquad 0 \leq q \leq \lfloor m/2 \rfloor. \tag{3.4}$$

of the m-fold tensor product $\bigotimes^m E$. The space Λ_0^m is isomorphic to the so-called primitive or effective summand of $\bigwedge^m E$, and the representations (3.4) all arise from tensor products of primitive ones:

LEMMA 3.1. $\Lambda_0^m \otimes \Lambda_0^n \cong \bigoplus_{k=0}^{\min(m,n)} \Lambda_k^{m+n}$.

Recall that the irreducible complex representations of $Sp(1)$ are merely the symmetric powers $S^q H$ which we denote by Σ^q. The spin representation of M is given by combining the primitive summands with these symmetric powers.

PROPOSITION 3.2. [**10, 116**] $\Delta \cong \bigoplus_{q=0}^{n} \Lambda_0^{n-q} \otimes \Sigma^q$.

The summands of Δ arise from representations of $Sp(n)Sp(1)$ (rather than just $Sp(n) \times Sp(1)$) if and only if n is even, and in this case they determine vector bundles defined globally on M. (We usually denote these associated bundles by the same symbol as the representation, relying on the context to make the meaning clear.) In the case in which M is hyper-Kähler, Σ^q becomes a trivial bundle of dimension $q+1$. Moreover, the vector bundle E exists globally and is isomorphic to the holomorphic tangent bundle $T^{1,0}$ relative to any compatible complex structure. ($T^{1,0}$ is also isomorphic to its dual $\Lambda^{1,0}$ by means of the appropriate holomorphic symplectic form.) It is well known that in this case Δ is the full exterior algebra on E, and this is consistent with the above proposition.

COROLLARIES 3.3. *(i) A QK manifold of even quaternionic dimension n is always spin.*
(ii) An HK manifold of quaternionic dimension n has a complex $(n+1)$-dimensional space of harmonic spinors.

On an HK manifold, the Dirac operator can be identified with $\overline{\partial} + \overline{\partial}^*$ acting on the full exterior algebra of E, and the relevant operators can be 'strung out' into the usual Dolbeault complex. A similar phenomenon occurs on a QK $4n$-manifold for which n is even or $\varepsilon = 0$. Namely there is an elliptic complex of the form

$$0 \to \Lambda_0^n \xrightarrow{D} \Lambda_0^{n-1} \otimes \Sigma^1 \xrightarrow{D} \Lambda_0^{n-2} \otimes \Sigma^2 \to \cdots \to \Sigma^n \to 0 \tag{3.5}$$

(of course, to make sense of the notation, the objects between the arrows are now sheaves or sections of the corresponding vector bundles). This complex has the remarkable property that it can be coupled to any vector bundle V with a connection whose curvature lies in the space $\mathfrak{h} = \mathfrak{sp}(n) \oplus \mathfrak{sp}(1)$ without destroying the property that $D^2 = 0$. This fact leads to one possible generalization of the Sieberg-Witten equations to a quaternionic context.

Analogues of the Dirac complex (3.5) that do not require a Riemannian metric for their definition are studied in [**105, 11**]. The most obvious such complexes are those obtained by tensoring (3.5) by Σ^{k+n} and rearranging the pieces to give

$$0 \to \Sigma^k \xrightarrow{\overline{D}_k} E \otimes \Sigma^{k+1} \to \textstyle\bigwedge^2 E \otimes \Sigma^{k+2} \to \cdots \to \bigwedge^n E \otimes \Sigma^{k+n} \to 0. \tag{3.6}$$

The reappearance of full exterior powers of E ensures that they can indeed be defined relative to the G-structure used in the definition of a quaternionic manifold (see §4). A recent application of (3.6) for $k = 0$ is in the definition of a quaternionic version of analytic torsion [**81**].

For $k \geq 1$, the $\overline{D}_k$ are 'twistor operators'; each is overdetermined and has locally a finite-dimensional space of solutions that have special geometrical significance. For example, a solution of $\overline{D}_1\zeta = 0$ determines a hypercomplex structure, and a solution of $\overline{D}_2\zeta = 0$ a 'quaternionic complex structure' [**60**].

Remark. The work of Friedrich and others [**15**] on eigenvalues of the Dirac operator on a compact spin manifold has motivated work on the problem of finding a lower bound λ of the Dirac operator on a QK manifold of dimension $4n$ with n even [**54, 74**]. The result is that

$$\lambda^2 \geq \frac{n+3}{n+2}\frac{s}{4},$$

with equality occurs if and only if $M = \mathbb{HP}^n$ and the eigenvector lies in the summand $\Lambda_0^n \oplus (\Lambda_0^{n-1} \otimes \Sigma^1)$. An analogous sharp lower bound is known for the Laplacian acting on functions on a QK manifold [**79, 4**].

The 'Fueter complex'

$$0 \to H \xrightarrow{D} E \xrightarrow{\Delta} \textstyle\bigwedge^3 E \to \bigwedge^4 E \otimes \Sigma^1 \to \cdots \to \bigwedge^{n+3} E \otimes \Sigma^n \to 0 \tag{3.7}$$

is a version of (3.6) for $k = -3$, and incorporates a natural analogue D of the $\overline{\partial}$-operator in complex analysis. All the operators in this complex are first order, except for Δ which is second order. Local sections of H solving the equation $Df = 0$ in flat space $\mathbb{H}^n$ are the so-called quaternionic regular functions defined by Fueter, and twistor theory can be used to show that (3.7) is a resolution of the associated sheaf, so that for example $g = Df$ is locally solvable if and only if $\Delta g = 0$. Explicit expressions for the operators can be found in [**1**]. Much of what can be done in flat space extends to the class of hypercomplex manifolds, and Joyce has developed an extensive programme aimed, amongst other things, at reconstructing HK metrics from their function theory [**63, 99**].

A decomposition of the exterior forms on a QK manifold can in theory be determined by the formula

$$\Delta \otimes \Delta \cong \bigoplus_{k=0}^{4n} \textstyle\bigwedge^k T^*.$$

Lemma 3.1 implies that each summand has the form $\Lambda_q^p \otimes \Sigma^r$. To illustrate this, we quote without proof that

$$\textstyle\bigwedge^4 T^* \cong (\Lambda_2^4 \oplus \Lambda_0^2 \oplus \mathbb{R}) \oplus ((\Lambda_1^4 \oplus \Lambda_1^2 \oplus \Lambda_0^2) \otimes \Sigma^2) \oplus ((\Lambda_0^4 \oplus \Lambda_0^2 \oplus \mathbb{R}) \otimes \Sigma^4).$$

On the other hand, for $\mathfrak{h} = \mathfrak{sp}(n) + \mathfrak{sp}(1)$,

$$\begin{aligned} S^2\mathfrak{h} &\cong S^2(S^2E) \oplus (S^2E \otimes \Sigma^2) \oplus S^2(\Sigma^2) \\ &\cong (S^4E \oplus \Lambda_2^4 \oplus \Lambda_0^2 \oplus \mathbb{R}) \oplus (\Lambda_1^2 \otimes \Sigma^2) \oplus \Sigma^4 \oplus \mathbb{R}. \end{aligned}$$

All these summands occur in $\bigwedge^4 T^*$ apart from a trivial summand and $W = S^4E$, and this can be used to justify the remarks before Corollary 1.2.

COROLLARY 3.4. *The curvature tensor of a QK manifold equals* $R_Q + s\rho_1$, *where* R_Q *takes values in* S^4E *and* ρ_1 *is* $Sp(n)Sp(1)$*-invariant.*

On a HK manifold, the space of curvature tensors is isomorphic to S^4E. A choice of complex structure on an HK manifold yields an identification between E and the holomorphic cotangent bundle T^*, so we may regard R as a smooth section of $T^* \otimes S^3E$. The second Bianchi identity can be used show that R actually determines an element of the sheaf cohomology group $H^1(M, \mathcal{O}(S^3T^*))$. This last result is relevant to theory resulting from the so-called Witten-Rosansky invariants [**100, 64**], and some of the relevant representation theory appears in [**50**].

Corollary 3.4 leads to the idea due originally to Roček that, in certain circumstances, HK metrics can be constructed as the limit of a sequence of QK metrics with scalar curvature tending to zero. Although no general theory for such a phenomenon as yet exists, this idea has led to the whole programme relating hyper-Kähler to quaternion-Kähler described later in this chapter.

4. Quaternionic Manifolds and Bundles

The inability to choose a global basis of complex structures on a quaternion-Kähler manifold M can be overcome by passing to the total space of an associated bundle. This approach is however best viewed within the wider context of quaternionic manifolds, which we now describe.

DEFINITION 4.1. A quaternionic manifold is a smooth manifold of dimension $4n \geq 8$ admitting a G-structure and a torsion-free G-connection, where G denotes the subgroup $GL(n,\mathbb{H})GL(1,\mathbb{H})$ of $GL(4n,\mathbb{R})$.

The group G is defined as in (1.3), but without reference to an inner product on $\mathbb{R}^{4n}$. Thus, $GL(n,\mathbb{H})$ is the commutator of the group $GL(1,\mathbb{H})$ of transformations $v \longmapsto vq^*$, q a non-zero quaternion, and

$$G = GL(n,\mathbb{H})Sp(1) \cong GL(n,\mathbb{H}) \times_{\mathbb{Z}_2} Sp(1).$$

To complete the definition, it is logical to define a quaternionic manifold of real dimension 4 to be one with a self-dual conformal structure.

Let M be a quaternionic manifold. The homomorphism

$$G \to Sp(1)/\mathbb{Z}_2 \cong SO(3)$$

given by projection to the second factor allows one to define bundles over M associated to various representations of $SO(3)$. First, let $SO(V)$ denote the principal $SO(3)$ bundle parametrizing triples $\{I_1, I_2, I_3\}$ of almost-complex structures satisfying (1.1), whose existence does not require a Riemannian metric.

Let F be any space (linear or otherwise) on which $SO(3)$ acts, and let $\mathbf{F}$ denote the fibre bundle associated to $SO(V)$ with fibre F. Here are some obvious candidates for F:

(i) $SO(3)$, acted on by itself by left translation;

(ii) the standard representations $\mathbb{R}^3$ and $\mathbb{C}^3$;
(iii) the 2-sphere S^2 in $\mathbb{R}^3$;
(iv) $(\mathbb{C}^2 \setminus \{0\})/\mathbb{Z}_2$, where $\mathbb{C}^2$ is the standard representation of $Sp(1)$;
(v) $S^3/\mathbb{Z}_2 \cong \mathbb{RP}^3$, where S^3 is the set of unit vectors in the above $\mathbb{C}^2$.

In each case the bundle $\mathbf{F}$ has been well studied. In (i) it is simply $SO(V)$ itself, and in (ii) we recover the bundle V of endomorphisms defining the quaternionic structure and its complexification V_c. In (iii) $\mathbf{F}$ is the subset of unit vectors in V; it is denoted by Z and called the twistor space of M. In (iv) we shall see that $\mathbf{F}$ can be identified with the total space, minus its zero section, of a complex line bundle L^* over Z. Finally, (v) coincides with (i) as $SO(V)$ may be also be identified with the set of 'unit' vectors in L^*.

The geometry of M is simplified to a greater or lesser extent when passing to the total space of each of the above bundles, and there are pros and cons to focussing on each case. However, it is Z that encodes the underlying quaternionic structure of M most directly into complex geometry. We shall always denote the projection $Z \to M$ by π, and a fibre $\pi^{-1}(x)$ by Z_x. Then each point $z \in Z_x$ is an almost-complex structure on T_xM of the form $a_1I_1 + a_2I_2 + a_3I_3$, where $\{I_1, I_2, I_3\}$ is a local orthonormal basis of V. Thus, a section s of Z over an open set M' of M can itself be regarded as an almost-complex structure I_s on M'. Let $\sigma\colon Z \to Z$ denote the 'antipodal mapping' $I \longmapsto -I$ defined on each S^2 fibre, and with no fixed points.

THEOREM 4.2. **[17, 104]** *Over a quaternionic manifold M, the total space Z admits a complex structure with the property that (i) its fibres are rational curves with normal bundle $2n\mathcal{O}(1)$, (ii) σ is anti-holomorphic, and (iii) a local section $s(M')$ is a complex submanifold if and only if I_s is an integrable complex structure.*

Here, $2n\mathcal{O}(1)$ is short for $\mathcal{O}(1) \otimes \mathbb{C}^{2n}$, where $\mathcal{O}(1)$ denotes the hyperplane line bundle; more generally $\mathcal{O}(k)$ will denote the tensor power $\mathcal{O}(1)^{\otimes k}$. It is a corollary that the quaternionic structure of M is always generated locally by a complex structure I_1 and an almost-complex structure I_2 anti-commuting with I_1. If I_2 is also integrable then the resulting structure is hypercomplex (see below).

We shall often call the fibres of Z over M the 'twistor lines'. By identifying a vector $I_1 \in Z_x$ with the projective class of $I_2 + iI_3$, one may also regard $C = Z_x$ as the conic of null lines in the projective plane $\mathbb{P}((V_x)_c)$. In many situations the vector spaces V_x are explicitly realized as subspaces of a 'universal' vector space $\mathbb{V}$. In any case, since

$$H^0(C, 2n\mathcal{O}(1)) \cong \mathbb{C}^{4n}, \quad H^1(C, 2n\mathcal{O}(1)) = \{0\}, \tag{4.1}$$

Kodaira's theory implies that C belongs to a complex $4n$-dimensional family of rational curves. The existence of such a curve C with the given normal bundle thus captures the essential geometry of a twistor space.

The complex structure J on Z characterized by Theorem 4.2(iii) may be defined by first identifying Z locally with the complex projective bundle

$$\mathbb{P}(H) = \mathbb{P}(H \otimes A^\lambda), \tag{4.2}$$

where A is the real line bundle arising from the standard representation of the centre $\mathbb{R}^*$ of G. The point is that the the twistor operator $\overline{D}_1$ defined in (3.6) is only invariantly defined if Σ^1 is replaced by $\tilde{H} = H \otimes A^\lambda$ for an appropriate value of the 'weight' λ (computed in **[95]** to equal $n/(n+1)$). The integrability of J may then be deduced by applying the proof of [**8**, Theorem 4.1] and results on the curvature of quaternionic manifolds from [**105**]. Because Z is now complex analytically a projective bundle, it is a corollary of this approach that there exists a holomorphic line bundle L over Z which restricts to $\mathcal{O}(2)$ on each fibre. The existence of a torsion-free connection is precisely the condition that guarantees the integrability of (Z, J).

A special case of a quaternionic manifold is a manifold with a torsion-free connection $\overline{\nabla}$ preserving a $GL(n, \mathbb{H})$-structure. In this case V is trivial and there exist globally-defined triples of parallel complex structures $\{I_1, I_2, I_3\}$. Such a manifold is called hypercomplex, and bears the same relationship to quaternionic that hyper-Kähler bears to quaternion-Kähler (see [**106**] and references therein). In fact, a hypercomplex structure is uniquely specified by two anti-commuting complex structures I_1, I_2, for in this case $I_3 = I_1 I_2$ is also complex, and the ('Obata') connection $\overline{\nabla}$ is uniquely determined.

If M is a hypercomplex manifold then the twistor space Z is trivial as a smooth bundle, and the projection $Z \to \mathbb{CP}^1$ is holomorphic. Points of M correspond to sections of π with normal bundle $2n\mathcal{O}(1)$. This point of view has proved particularly valuable in the construction of non-compact HK manifolds [**17**], and in classifying deformations of hypercomplex manifolds [**94**].

One reason for including Definition 4.1 in this chapter is that curvature can be defined in this more general context.

PROPOSITION 4.3. [**105**] *(i) A quaternionic manifold has a tensor R_Q that is the component of the curvature of a torsion-free G-connection ∇ independent of the choice of ∇.*
(ii) On a hypercomplex manifold, the curvature of $\overline{\nabla}$ equals $R_Q + R_\kappa$, where $R_\kappa \in \Lambda^2_1$ represents the curvature 2-form of $\kappa = \bigwedge^{2n,0}$.

The canonical bundle κ in (ii) is the complexification of a real bundle arising from a homomorphism $GL(n, \mathbb{H}) \to \mathbb{R}^*$, and is therefore independent of the complex structure chosen to define $\bigwedge^{2n,0}$. Observe that $\Lambda^2_1 \cong S^2 E$ is a subspace of $\bigwedge^2 T^* M$ defined by (3.2) by the G-structure. It can be identified with the intersection of the spaces $\bigwedge^{1,1}$ of $(1,1)$-forms relative to each almost-complex structure $I \in Z_x$. A 2-form is called self-dual if it takes values in this subspace, which coincides with $\bigwedge^2_+$ when $n = 1$.

As the notation implies, on a QK manifold R_Q can be identified with the non-trivial component of the Riemann tensor defined by Corollary 3.4, and in 4 dimensions, it would just be the non-vanishing half W_+ of the Weyl tensor. A compact simply-connected quaternionic manifold with $R_Q \equiv 0$ is necessarily isomorphic to $\mathbb{HP}^n$.

A hypercomplex manifold for which $R_Q = 0 = R_\kappa$ is covered by coordinate charts with constant quaternionic linear transition functions. This class of manifolds was considered by Sommese in the paper [**109**], which contains one of the earliest references to the concept of the twistor space. Such affine flat examples include $S^{4n-1} \times S^1$, and the abelian hypercomplex nilmanifolds considered in [**40**] which are quotients of $\mathbb{H}^n$. The tensor R_κ is a type of skew-symmetric Ricci tensor, and less trivial examples with $R_\kappa = 0$ include metrics which are conformally HK. A compact hypercomplex 4-manifold M necessarily has $R_\kappa = 0$, and Boyer effectively used this to show that either M admits a HK metric (and is therefore a torus or K3 surface), or else is diffeomorphic to a Hopf surface [**25, 65**].

Example. Suppose that M is a QK manifold with a compatible hypercomplex structure. Let ∇ represent its Levi-Civita connection, R the Riemann tensor, and $\overline{R}$ the curvature of $\overline{\nabla}$. The difference $\nabla - \overline{\nabla}$ may be regarded as a 1-form α with the property that $\alpha_i = I_i\alpha$ in (1.2), and

$$S = R - \overline{R} = \nabla\alpha + \tfrac{1}{2}\alpha\wedge\alpha.$$

Since R_Q must coincide with that component of R in S^4E, it follows that the symmetric part of S is proportional to the Riemannian metric g, and its skew part $d\alpha$ a self-dual 2-form. This approach is used in **[5]** to show that with certain additional assumptions M must be quaternionic hyperbolic space.

Let F be a complex vector bundle over a quaternionic manifold M, and suppose that ∇ is a connection on F. The curvature R_∇ of ∇ is a 2-form with values in $\mathrm{End}F$, and referring to (3.2) we record the

DEFINITION 4.4. **[105, 82, 90, 48]** The connection ∇ is called quaternionic, of type B_2 or c_2-self-dual if R_∇ is self-dual as a 2-form so that $R_\nabla \in \mathrm{End}F\otimes\Lambda^2_1$.

These connections satisfy the Yang-Mills equations. Although their moduli spaces are known in some special cases with $\dim M \geq 8$ **[82, 88]**, 'quaternionic Yang-Mills theory' is still in its infancy. The self-duality condition on the curvature of ∇ enables the complexes of differential operators described in §3 to be extended by tensoring by F, and a number of cohomological results are known **[89]**.

On a hypercomplex manifold, the connection $\overline{\nabla}$ induces covariant derivatives on all vector bundles associated (even locally) to the $GL(n,\mathbb{H})$-structure. The same is true on a QK manifold equipped with its Levi-Civita connection ∇. The following result is related to Proposition 4.3(ii).

LEMMA 4.5. *On a hypercomplex or QK manifold, the connection induced on E is c_2-self-dual.*

Now suppose that M is a quaternionic manifold with $\varepsilon = 0$, and that ∇ is self-dual. Then there exists a twistor operator

$$\overline{D}_1 : F\otimes\tilde{H} \to F\otimes\tilde{E}\otimes S^2\tilde{H},$$

where the tildes represent appropriate weights. If F has complex rank $2r$ and ∇ preserves a $GL(r,\mathbb{H})$-structure on F then $\overline{D}_1$ is an operator between real vector bundles of rank $4r$ and $12nr$ respectively.

THEOREM 4.6. **[105]** *With the above hypotheses, the real $(4n+4r)$-dimensional total space M_F of $F\otimes\tilde{H}$ is a quaternionic manifold.*

The twistor space of M_F can be identified with the total space of the complex vector bundle $(\pi^{-1}F)\otimes L^{1/2}$ over Z, where $L^{1/2}$ is a holomorphic square root of L determined by the smooth splitting $\pi^{-1}H = L^{1/2}\oplus\overline{L^{1/2}}$. Indeed, the fact that $\pi^{-1}F$ is a holomorphic vector bundle over Z follows from the fact that the curvature of the pulled-back connection $\pi^{-1}\nabla$ has no $(0,2)$-component, and a celebrated integrability theorem of Atiyah **[8, 39]**. For example, if M is hypercomplex, then ∇ is c_2-self-dual if and only if ∇ is 'triholomorphic'.

Example. Let M be a QK manifold. Lemma 4.5 implies that the total space M_E of the tangent bundle TM of a QK manifold is itself quaternionic. This applies in particular to S^4 and $\mathbb{CP}^2$. In the latter case $E \cong \kappa^{1/2}\oplus\overline{\kappa}^{-1/2}$ where κ denotes the canonical line bundle whose curvature is a multiple of the Kähler-form $\omega \in \Lambda^+$.

Over an arbitrary quaternionic manifold one can take $F = \mathbb{C}^2 = \mathbb{H}$, and ∇ the trivial connection. Then M_F can be identified with the total space of the quaternionic line bundle $\tilde{H}$. Let $\tilde{H}^*$ denote $\tilde{H}$ with its zero section removed; this may be regarded as a principal $\mathbb{H}^*$-bundle over M. The proof of Theorem 4.2 implies that $\tilde{H}^*$ admits a complex structure J, and the orbit of J under $\mathbb{H}^*$ is a 2-sphere $\{aI + bJ + cK : a^2 + b^2 + c^2 = 1\}$ of complex structures satisfying (1.1). This endows $\tilde{H}^*$ and

$$U = \tilde{H}^*/\mathbb{Z}_2 \tag{4.3}$$

with a hypercomplex structure. Natural though the definition of the twistor space Z is, there is a sense in which it involves the choice of a complex structure, and the construction of U overcomes this objection.

Example. The homogeneous space $SU(n+2)/SU(n)$ that fibres over the Wolf space $M = \mathbb{G}\mathrm{r}_2(\mathbb{C}^{n+2})$ is closely related to U, although of course one has compact fibre and the other not. Let $U' \cong U/\mathbb{Z}$ denote the bundle associated to U with fibre $U(1) \times SO(3)$. Then there is a principal $U(1)$-bundle P over M such that $SU(n+2)/SU(n)$ double covers the quotient of $U' \times P$ by the diagonal action of $U(1)$. This action preserves the hypercomplex structure, and the latter persists because P has a c_2-self-dual connection which enables its complexififcation to be viewed as a holomorphic bundle over the twistor space Z [**60, 12**]. The case $n = 1$ relates to the fact that $SU(3)$ is itself hypercomplex.

There are many ways in which the last example can be generalized to construct compact hypercomplex structures on Stiefel manifolds [**23**], and Lie groups [**110, 61**]. We conclude this section by summarizing the latter, which proceeds by extending the decomposition (2.6) by a sequence of minimal 3-dimensional subalgebras.

Given a 3-dimensional subalgebra $\mathfrak{sp}(1) = \mathfrak{sp}(1)_1$ of $\mathfrak{g} = \mathfrak{h}_0$ generated by a highest root, one may regard its centralizer $\mathfrak{h} = \mathfrak{h}_1$ as a Lie algebra in its own right. If this is not abelian it will contain a minimal 3-dimensional subalgebra $\mathfrak{sp}(1)_2$ (see the end of §2), and we may write $\mathfrak{h}_{i-1} = \mathfrak{sp}(1)_i \oplus \mathfrak{h}_i \oplus \mathfrak{m}_i$ for $i \geq 1$. Hence,

$$\mathfrak{g} = \mathfrak{h}_k \oplus \bigoplus_{i=1}^{k} (\mathfrak{sp}(1)_i \oplus \mathfrak{m}_i),$$

where $\mathfrak{m}_i \cong \mathbb{C}^{2m_i} \otimes \Sigma^1$, as an $\mathfrak{sp}(1)_i$-module, but lies in the centralizer of $\mathfrak{sp}(1)_j$ when $j > i$. The subalgebra $\mathfrak{h}_i$ is the centralizer of $\mathfrak{sp}(1)_i$ in $\mathfrak{h}_{i-1}$, and the process can be continued unless $\mathfrak{h}_k$ is abelian. Whether or not this is the case, $U(1)^k \times (G/H_k)$ has a hypercomplex structure with tangent space isomorphic to

$$\mathbb{R}^k \oplus (\mathfrak{g}/\mathfrak{h}_k) \cong \bigoplus_{i=1}^{k} (\mathbb{R} \oplus \mathrm{Im}\,\mathbb{H} \oplus \mathbb{H}^{m_i}).$$

One can also replace H_j by its semisimple part by adjusting the number of extra $U(1)$ factors required. Taking k maximal then yields a hypercomplex structure on $U(1)^k \times G$ for some integer k no greater than the rank of G (or 3 if G has rank 2).

5. Fano Twistor Spaces

The Levi-Civita connection on M determines a horizontal distribution on all the associated bundles considered in the last section. By general principles [**17, 117**] that are discussed elsewhere in this volume, all these total spaces admit families of Einstein metrics. The situation is particularly important when the scalar curvature is positive.

THEOREM 5.1. [**26, 102, 113**] *If M is a QK manifold with $s>0$ then*
(i) Z has a Kähler-Einstein metric,
(ii) U has an HK metric, and distinct QK metrics, and
(iii) $SO(V)$ has a 3-Sasakian metric.

We might add that a Ricci-flat metric can be defined on V using the techniques of [**29**]. This metric will be irreducible and (it almost follows) will have holonomy group equal to $SO(4n+3)$, a fact that is significant as there are few known examples of Ricci-flat metrics without reduced holonomy.

We shall now comment on additional structures that exist on these manifolds, spending most time on Z. The horizontal space at a point $z\in Z_x$ can be identified with T_xM and therefore has a natural complex structure determined by z. We let D denote the corresponding bundle of complexified horizontal vectors of type $(1,0)$. The isomorphism (4.2) leads to an interpretation of the relevant structures in terms of the 'EH' formalism. The almost-complex structure determined by the projective line $[h]$ with $h\in H$ has $E\otimes[h]$ as its subspace of $(1,0)$-vectors in $(T_xM)_c$. The quotient TZ/D can be identified with the holomorphic line bundle L introduced after (4.2). It follows that

$$D\cong\pi^*E\otimes L^{1/2},$$

and this bundle acquires a holomorphic structure over Z, by Proposition 4.5, reflecting the Einstein curvature of M.

The following result appears in [**102**] (though what is here called L is there called L^2).

PROPOSITION 5.2. *If $s\neq 0$, D is a holomorphic contact distribution on Z.*

To explain this, observe that the exact sequence

$$0\to D\to TZ\to L\to 0 \tag{5.1}$$

determines a holomorphic 1-form $\theta\in H^0(Z,\mathcal{O}(T^*Z\otimes L))$ with values in L. Although $d\theta$ itself can only be calculated by choosing a local section of L, its restriction to $\bigwedge^2 D$ is independent of this choice and gives rise to an element of $H^0(\bigwedge^2 D^*\otimes L)$ that is non-degenerate provided $s\neq 0$. This is the contact condition, and corresponds to the distribution D being 'maximally non-integrable'.

Theorem 4.2 with Proposition 5.2 combine to give a powerful encryption of QK metrics. LeBrun has shown that a complex contact manifold (Z,D) with a fixed-point free anti-involution σ and a family of rational curves transverse to D with normal bundle $2n\mathcal{O}(1)$ is the twistor space of a pseudo-Riemannian metric with holonomy in $Sp(p,n-p)Sp(1)$ for some p [**77**]. This inversion theorem also led him to prove that the moduli space of complete such metrics on $\mathbb{R}^{4n}$ is infinite-dimensional [**78**]. We shall exemplify a family of rational curves with the stated properties in §6.

If $\kappa = \bigwedge^{2n+1} T^* Z$ denotes the canonical bundle of Z, the well-defined section

$$\theta \wedge (d\theta)^n \in H^0(Z, \mathcal{O}(\kappa \otimes L^{n+1}))$$

yields an isomorphism $\kappa^* \cong L^{n+1}$. If $s > 0$, the 2-form defining the Kähler-Einstein metric of Z is proportional to the curvature of a natural connection on L [**102**], and L is an ample line bundle. Since the same is true of the anticanonical bundle κ^*, Z is by definition a Fano manifold. The algebraic geometry of Fano 3-folds from the twistor space point of view can be found in [**56**] and [**92**].

It is an open problem to determine conditions on a contact Fano manifold to ensure that it is the twistor space of a positive QK manifold. In this direction,

THEOREM 5.3. [**79, 86**] *If Z is a compact Kähler-Einstein manifold with a holomorphic contact structure then Z is the twistor space of some QK manifold M.*

The space U of (4.3) can be identified with the total space of L^* over Z with its zero section removed. The contact form θ pulls back (and then evaluates) to a genuine 1-form on U. The contact condition ensures that the exterior derivative

$$\omega = d\theta \tag{5.2}$$

is in fact a holomorphic symplectic form on U. Indeed, any contact manifold has a 'symplectification', and given Z, U is it. In the case in which $s > 0$, the total space of L^* has a natural Kähler metric, and the HK structure on U is generated by an action of $Sp(1)/\mathbb{Z}_2 = SO(3)$.

Conversely, suppose that N is a hyper-Kähler manifold with a free action of $SO(3)$ inducing a transitive action on the 2-sphere S^2 of complex structures. If, furthermore, IX_I is independent of $I \in S^2$ (where X_I is the vector field generated by the circle subgroup preserving I) then N is locally isometric to the bundle U of some QK manifold. This fact allows one to construct the join of QK manifolds. If M_1 and M_2 are both QK, then the product $U_1 \times U_2$ has a HK structure. It follows that the manifold $M_1 * M_2 = (U_1 \times U_2)/\mathbb{H}^*$ (locally isomorphic to an open set of the quaternionic projective bundle $\mathbb{P}(H_1 \oplus H_2)$) is quaternion-Kähler. Taking $M = M_1$ to have $s > 0$ and $M_2 = \{x\}$ to be a point establishes the existence of a QK metric on $M * \{x\} \cong U$ with positive scalar curvature. More details, as well as a related discussion of HK potentials, can be found in [**113**].

The bundle U can also be viewed as a cone over $SO(V)$, and its HK structure reflects the 3-Sasakian structure of $SO(V)$ in accordance with the theory of Killing spinors [**9**]. When M is a self-dual Einstein 4-manifold then $SO(V)$ actually carries an Einstein metric with so-called 'weak holonomy G_2', and the quaternionic line bundle H associated metrics with holonomy $Spin(7)$ [**43, 49**]. The great significance of the bundle $SO(V)$ is that it may be a manifold even in situations in which M has orbifold singularities; this has led to some surprisingly rich classification questions [**27**], that are presented elsewhere in this volume. The manifold $SO(V)$ also has an underlying 'quaternionic contact structure', a notion exploited in [**19**] for the local construction of QK metrics.

There are many general results that apply to a Fano contact manifold Z without the assumption that it fibres over a QK manifold. The exterior powers of the 'DTL' sequence (5.1) provide important information. Associated long exact sequences relate the Dolbeault cohomology spaces $H^q(Z, \mathcal{O}(\bigwedge^p D \otimes L^p))$ and

$H^{p,q}(Z,\mathcal{O})$. This allows one to deduce that the Hodge numbers $h^{p,q}$ of Z vanish if $p \neq q$, and derive the following formula for holomorphic Euler characteristics.

LEMMA 5.4. **[102]** $\chi(Z, \mathcal{O}(\bigwedge^p D \otimes L^{-p-r})) = \begin{cases} 0, & 1 \leq r \leq n-p, \\ (-1)^p h^{p,p}, & r = 0. \end{cases}$

The index of a Fano manifold is by definition the largest root of κ that can be extracted, and it follows from a well-known characterization of Kobayashi-Ochiai **[69]** that if the index $2n+2$, the Fano manifold is biholomorphically equivalent to $\mathbb{CP}^{2n+1}$. The index of a twistor space is $n+1$ unless L itself has a square root which occurs if and only if $\varepsilon = 0$. Two simply-connected complex contact manifolds are contact-isomorphic if and only if they are biholomorphic **[80]**. A fuller discussion of automorphism groups will be postponed until §7, but part (i) of the next theorem now follows.

THEOREM 5.5. **[102, 80]** *Let M be a positive QK manifold of dimension $4n$.*
(i) If $\varepsilon = 0$ then M is isometric to $\mathbb{HP}^n$.
(ii) If $b_2(M) > 0$ then M is isometric to $\mathbb{G}\mathrm{r}_2(\mathbb{C}^{n+2})$.

As first pointed out by LeBrun, the characterization of QK manifolds with $b_2 \geq 1$ is a spin-off of results of Wiśniewski **[118]** within the context of Mori's programme. The crucial property of the twistor space Z of such a manifold is the existence of a rational curve C with $C \cdot L = 1$ whose homology class is not proportional to that of a fibre of π. Through each point the family of such rational curves actually spans out a projective space and Z can be identified with the total space of the projectivization of a vector bundle over a variety X. The mapping $f \colon Z \to X$ is a so-called Fano contraction, and its fibres are tangent to the contact distribution. The key point here is that if C is any rational curve satisfying $L \cdot C = 1$, then the pullback of θ to C is zero since $H^1(\mathbb{CP}^1, \mathcal{O}(\Omega^1(1))) = 0$. It turns out that the existence of a contact structure on Z allows one to deduce that X is isomorphic to $\mathbb{CP}^{n+1}$, and $Z \cong \mathbb{P}(T^*\mathbb{CP}^{n+1})$.

A study of Fano manifolds Z with $b_2(Z) = 1$ (corresponding to $b_2(M) = 0$) is accomplished in the papers **[70, 87]**. A key theorem asserts that in each fixed dimension, the top power of $c_1(Z)$ is bounded, and this implies that there are only finitely many deformation types. On the other hand, under appropriate hypotheses, a Fano contact structure is rigid under defomation, whence

THEOREM 5.6. **[80]** *Up to homothety, there are only finitely many positive QK manifolds of dimension $4n$.*

The general theory of polarized varieties, as described by Fujita **[45]**, is especially relevant to the study of low-dimensional Fano manifolds. We set

$$R_k = H^0(Z, \mathcal{O}(L^k)), \quad r_k = \dim R_k, \tag{5.3}$$

and omit the subscripts when $k = 1$. The fact that L is ample implies that the natural map

$$\nu_k \colon Z \to \mathbb{P}(R_k^*)$$

is an embedding for k sufficiently large. However, we shall be more concerned with $\nu = \nu_1$, which is a well-defined mapping only if the base locus B of the linear system $|L|$ is empty. Relative to the 'polarization' defined by L, the Δ-genus is defined by

$$\Delta(Z) = \deg(Z) - r + 2n + 1,$$

where $\deg(Z) = \langle \ell^{2n+1}, [Z] \rangle$ and $\ell = c_1(L)$. Then

$$\Delta(Z) \geq \dim B + 1 \tag{5.4}$$

(with the convention that $\dim \emptyset = -1$), and this equation limits the size of B.

Given that $c_1(TZ) = (n+1)\ell$, the Riemann-Roch theorem implies that

$$r_k = \langle e^{k\ell}\mathrm{td}(Z), [Z] \rangle = \left\langle e^{\ell(n+1+2k)/2} \hat{A}(Z), [Z] \right\rangle, \tag{5.5}$$

where $\hat{A}(Z)$ is defined by (8.4). It follows that there exists a polynomial

$$P(k) = \deg(Z) \frac{k^{2n+1}}{(2n+1)!} + \text{lower powers of } k,$$

such that $r_k = P(k)$ for $k \geq 0$. This is the so-called Hilbert polynomial of the polarized variety (Z, L). Geometrical properties of μ are encapsulated in the natural homomorphism

$$\bigoplus_{k=0}^{\infty} S^k R \to \bigoplus_{k=0}^{\infty} R_k \tag{5.6}$$

of coordinate rings. The space R_k is spanned by the pullbacks of homogeneous polynomials of degree k to Z. We shall see in §6 that, for a twistor space, $R = R_1$ is isomorphic to the complexification $\mathfrak{g}_c$ of the Lie algebra of the isometry group G of M. Indeed, the individual linear mappings $s_k \colon S^k R \to R_k$ of (5.6) are G-equivariant, and an understanding of the resulting representations leads to models for twistor spaces.

The dimension of the space of polynomials of degree k in $N+1$ variables equals $\binom{k+N}{N}$, and the Hilbert polynomial of $(\mathbb{CP}^N, \mathcal{O}(1))$ is

$$\binom{t+N}{N} = \frac{1}{N!}(t+N)(t+N-1)\cdots(t+1).$$

If X is an embedded hypersurface of $\mathbb{CP}^N$ of degree h, then the kernel of (5.6) is generated by the element of $S^k R$ whose zero set defines Z. It follows that the Hilbert polynomial of X is

$$\binom{t+N}{N} - \binom{t+N-h}{N}.$$

A dual situation occurs when Y is a covering of $\mathbb{CP}^N$ of degree d branched over a hypersurface of degree dh. In this case, the cokernel of (5.6) is generated by an element of R_h, and the Hilbert polynomial of Y is

$$\binom{t+N}{N} + \binom{t+N-h}{N}.$$

Example. These situations are combined when Z is a branched covering of a hypersurface of $\mathbb{CP}^N$. An analysis of the representations R_k for $n = 4$ shows that one of the many potential twistor spaces Z of a real 16-dimensional QK manifold has

$$P(t) = \binom{t+10}{10} - \binom{t+8}{10} + \binom{t+6}{10} - \binom{t+4}{10} = \frac{4t^9}{9!} + \text{lower terms},$$

and the values of $r_k = P(k)$ for $k \geq 1$ are

(5.7) $$11,\ 65,\ 275,\ 936,\ 2728,\ 7072,\ 16720,\ 36685, \dots .$$

This is consistent with Z being the double-covering of a hyperquadric H in $\mathbb{CP}^{10}$, branched over the intersection of H with an octic, although a positive identification of this sort requires more explicit knowledge of (5.6).

6. Isometry Groups and Moment Mappings

Let M be a manifold with a symplectic 2-form ω, and a vector field X which is an infinitesimal automorphism of ω. Thus,

$$0 = \mathcal{L}_X\omega = X \lrcorner\, d\omega + d(X \lrcorner\, \omega) = d(X \lrcorner\, \omega),$$

and there exists a real-valued function f (defined on at least an open set of M) such that $X \lrcorner\, \omega = df$. This basic observation underlies much of this section, though we shall see in due course that analogues of f can be constructed on manifolds with geometrical structures that are not obviously 'symplectic'.

Next, suppose that M is a hyper-Kähler manifold, so that we can choose symplectic 2-forms $\omega_1, \omega_2, \omega_3$ associated to a standard triple of complex structures. If X is a Killing vector field on M whose corresponding 1-parameter group of isometries preserves the HK structure, then the above observation shows that, locally, there exist functions f_1, f_2, f_3 such that $df_i = X \lrcorner\, \omega_i$. These functions constitute the 'hyper-Kähler moment mapping' for the 1-dimensional group action, but it is convenient to represent them by means of the 2-form $\zeta = \sum_{i=1}^{3} f_i\omega_i$, so that

$$d\zeta = \sum_{i=1}^{3} df_i \wedge \omega_i = \tfrac{1}{2} X \lrcorner\, \Omega$$

in terms of (1.4).

This process generalizes to the QK case as follows. First we identify V with the subbundle of $\bigwedge^2 T^*M$ with fibre isomorphic to $\mathfrak{sp}(1)$. A section ζ of the bundle V is called a 'twistor function' if

(6.1) $$d\zeta = \tfrac{1}{2} X_\zeta \lrcorner\, \Omega,$$

for some vector field X_ζ. The terminology is taken from **[60]**, and the equation (6.1) is equivalent to the assertion that $\overline{D}_2\zeta = 0$, where $\overline{D}_2$ is the operator described in (3.6) using the Levi-Civita connection.

LEMMA 6.1. **[102]** *Let M be a QK manifold of dimension $4n \geq 8$ with non-zero scalar curvature. The mapping $\zeta \longmapsto X_\zeta$ establishes a bijective correspondence between the space of twistor functions and the space of Killing vector fields.*

The inverse mapping is obtained as follows. If X is a Killing vector field then at each point ∇X belongs to the subspace of $\operatorname{End} T$ determined by the holonomy algebra $\mathfrak{sp}(n) \oplus \mathfrak{sp}(1)$ **[71]**. Then, up to a universal constant, $(\nabla X)^V = s\zeta$, where the left-hand side is the component of ∇X in V. This works because the relevant component of the derivative of $(\nabla X)^V$ is proportional to X, thanks to the Ricci identity and Einstein condition.

Remark. It follows that, given a Killing vector field X on a QK 8-manifold,

$$(\sqrt{s}\,\zeta, X) \in \Sigma^2 \oplus \Lambda^1_0 \subset \Delta$$

is an eigenvector for the Dirac operator. Other eigensections of Δ are generated by vector fields X for which ∇X belongs to the subspace of $\operatorname{End} T$ isomorphic to $\bigwedge^2 E \subset \mathfrak{gl}(2,\mathbb{H})$. Such vector fields are non-isometric automorphisms of the quaternionic structure, and in the compact case exist only on $\mathbb{HP}^2$ [**4, 79**].

The fibre of V at $x \in M$ is naturally isomorphic to the space $H^0(Z_x, \mathcal{O}(2))$ of holomorphic sections of the restriction of the holomorphic line bundle L to the twistor line Z_x. In this way we obtain a mapping f from sections of V to sections of L over Z, and the following is a well-known example of the 'twistor transform':

LEMMA 6.2. [**102**] *The mapping f induces an isomorphism between the space of twistor functions and the space $H^0(Z, \mathcal{O}(L))^\sigma$ of σ-invariant holomorphic sections of L over Z.*

Consider the beginning of the long exact sequence

$$(6.2) \qquad 0 \to H^0(Z, \mathcal{O}(D)) \to H^0(Z, \mathcal{O}(TZ)) \leftrightarrows H^0(Z, \mathcal{O}(L)) \to \dots$$

associated to (5.1). It is known that map $Y \longmapsto Y \lrcorner\, \theta$ induces an isomorphism between the space of infinitesimal automorphisms of the contact structure and $H^0(Z, \mathcal{O}(L))$, and it follows that any σ-invariant automorphism of the contact structure arises from an isometry of M [**91**]. Such an isometry will, in turn, induce a holomorphic vector field on Z, and this process provides the indicated splitting of the sequence (6.2).

Let Y be a holomorphic vector field on Z preserving the contact structure, and let $s = Y \lrcorner\, \theta$ be the corresponding section of L. The latter defines a genuine function on U, and the equation

$$0 = \mathcal{L}_Y \theta = d(Y \lrcorner\, \theta) + Y \lrcorner\, d\theta = ds + Y \lrcorner\, \omega$$

allows us to interpret $-s$ as a holomorphic moment mapping on U. More invariantly, given $s \in \mathfrak{g}_c$, a moment mapping

$$\mu\colon U \to \mathfrak{g}_c^*$$

is defined by $\mu(u)(s) = s(u)$ for $u \in U$ and $s \in \mathfrak{g}_c$.

The mapping

$$\nu\colon Z \to \mathbb{P}(H^0(Z, \mathcal{O}(L))^*) \cong \mathbb{P}(\mathfrak{g}_c^*)$$

discussed in §5 may now be regarded as the projectivization of μ. The image $\nu(Z_x)$ of each fibre is determined by the restriction

$$i_*\colon H^0(Z, \mathcal{O}(L)) \to H^0(Z_x, \mathcal{O}(2)),$$

which corresponds (after complexification) to the mapping

$$Y \lrcorner\, \theta \mapsto (\nabla Y)_x^V \in V.$$

Suppose that the image of i_* is 3-dimensional for every $x \in M$, so that ν maps the conic Z_x to a conic in $\mathbb{P}(\mathfrak{g}_c^*)$. After dualizing, one then obtains a mapping $M \to \mathbb{G}\mathrm{r}_3(\mathfrak{g})$, and the situation may be summarized schematically:

$$\begin{array}{ccc} U & \xrightarrow{\mu} & \mathfrak{g}_c^* \\ \downarrow & & \downarrow \\ Z & \overset{\nu}{\dashrightarrow} & \mathbb{P}(\mathfrak{g}_c^*) \\ \downarrow & & \\ M & \dashrightarrow & \mathbb{G}\mathrm{r}_3(\mathfrak{g}) \end{array} \tag{6.3}$$

Dotted arrows indicate mappings whose domain of definition may be a subset of that indicated.

We are now in a position to explain the fundamental link between quaternion-Kähler geometry and complex nilpotent orbits. Let G_c denote a complex semisimple Lie group and let G be a compact subgroup of G_c corresponding to a real form $\mathfrak{g}$ of its Lie algebra $\mathfrak{g}_c = \mathfrak{g} + i\mathfrak{g}$. An element of $\mathfrak{g}_c$ is nilpotent if and only if it lies in the intersection of all invariant polynomials on $\mathfrak{g}_c$, and the set of such elements forms the 'nilpotent variety' $\mathcal{N}$. Suppose that M admits a group of isometries G that does not preserve any almost-complex structure that arises as a local section of Z. Roughly speaking this is a 'fullness' assumption for the way the group interacts with the quaternionic structure. In this case, L^k cannot have any G-invariant divisors for any k, and so $H^0(Z, \mathcal{O}(L^k))$ cannot contain any G-invariant elements. If follows that $\nu(Z) \subset \mathcal{N}/\mathbb{C}^*$.

The fullness condition is certainly satisfied if ν is an embedding (i.e. L is very ample), or more generally if $\dim \nu(Z) = 2n+1$. In these cases, it follows that $\nu(Z)$ is a nilpotent coadjoint orbit, and this leads to

THEOREM 6.3. [**16**] *If Z is a compact Fano twistor space and $\nu(Z)$ has the same dimension as Z, then M must be a Wolf space.*

Let us now consider the general nilpotent orbit. The algebra of invariant polynomials on $\mathfrak{g}$ is generated by a finite set $p_i \in S^{k_i}\mathfrak{g}^*$ with degrees $k_1, \dots, k_r$, and we may take $k_1 = 2$ and p_1 to be the Killing form. Thus, any nilpotent element of $\mathfrak{g}_c$ must be isotropic or 'null' relative to the Killing form, and if $\mathfrak{g} = \mathfrak{su}(2)$ this condition is of course sufficient. More generally, fix a non-zero homomorphism $\rho\colon \mathfrak{su}(2) \to \mathfrak{g}$. Then for any $V \in L_\rho$, it is easy to see that isotropic elements of V_c are nilpotent not just in $V_c \cong \mathfrak{sl}(2,\mathbb{C})$, but also in $\mathfrak{g}_c$. Thus, isotropic elements of V_c belong to the G_c-orbit

$$U_\rho = \{\mathrm{Ad}(g)(\xi) : g \in G_c\}, \tag{6.4}$$

where $\xi = \rho(A_1 + iA_2)$ in the notation of (2.2). Conversely, any nilpotent orbit in $\mathfrak{g}_c$ may be written in the form (6.4) for some element ξ arising from a real homomorphism $\mathfrak{su}(2) \to \mathfrak{g}$ [**72**].

The fact that $\xi \in U_\rho$ if and only if $\lambda\xi \in U_\rho$ for any non-zero complex number λ (infinitesimally, it is actually true that $\xi \in (\mathrm{ad}(e))^2\mathfrak{g}_c$) ensures that the projectivized nilpotent orbit

$$Z_\rho = U_\rho/\mathbb{C}^*$$

is defined as a submanifold of $\mathbb{P}(\mathfrak{g}_c)$. Now, U_ρ is equipped with the Kostant-Kirillov (holomorphic) symplectic form ω, and has even complex dimension. A contact 1-form θ is induced on Z_ρ for which the quotient line bundle L over Z is the pullback of $\mathcal{O}(1)$ on $\mathbb{P}(\mathfrak{g}_c)$.

Consider the quadric

$$Q = \{[v] \in \mathbb{P}(\mathfrak{g}_c) : \langle v, v\rangle = 0\}$$

of isotropic elements, itself isomorphic to the Grassmannian $\mathbb{G}\mathrm{r}_2(\mathfrak{g})$. Given $V \in \mathbb{G}\mathrm{r}_3(\mathfrak{g})$, we let C_V denote the conic $\mathbb{P}(V_c) \cap Q$. Recall Theorem 2.4. We have already shown that if $V \in L_\rho$ then $C_V \subset Z_\rho$, and this is also valid for $V \in M_\rho$ for the following reasons. Let $\xi = v_1 + iv_2$ be an isotropic vector in $V \in M_\rho \setminus L_\rho$, where $\{v_1, v_2, v_3\}$ is an oriented orthonormal basis of V. It follows that

$$\begin{aligned} \nabla\psi\big|_V(\xi) &= (-[v_2, v_3] - \psi(V)v_1) + i(-[v_3, v_1] - \psi(V)A_v) \\ &= i[v_1 + iv_2, v_3] - \psi(V)(v_1 + iv_2), \end{aligned}$$

belongs to $T_\xi U_\rho$. This means that tangent vectors to the flow lines of ψ preserve the nilpotency property to first order at all points of $M_\rho \setminus L_\rho$. It is also easy to check that for all $\alpha \in T_V\mathbb{G}\mathrm{r}_3(\mathfrak{g})$, we have $\alpha \in T_0 \oplus T_+$ if and only if $\alpha(\xi) \in \mathrm{ad}(\xi)\mathfrak{g}_c$ for all isotropic elements $\xi \in V_c$.

Let $\tilde{Z}_\rho$ denote the tautological bundle over M_ρ whose fibre at a point $V \in M_\rho$ is the conic C_V, so that there is a natural mapping $f\colon \tilde{Z}_\rho \to Z_\rho$. Since $C = C_V$ satisfies (4.1), it belongs to a complex $4n$-dimensional family of projective lines all of the form C_V with $V \in \mathbb{G}\mathrm{r}_3(\mathfrak{g})$. We may restrict to a family of real dimension $4n$ by considering only those lines which are invariant by the real structure σ of $\mathfrak{g}_c$ which preserves Z_ρ. It follows that $f(Z_\rho)$ is an open subset of Z_ρ. On the other hand, the Morse theory implies that $f(\tilde{Z}_\rho)$ is closed, since if $\{V_n\}$ is a sequence in M_ρ converging to $V_\infty \in \mathbb{G}\mathrm{r}_3(\mathfrak{g})$, then the flow line containing V_∞ emanates from a critical manifold L'_ρ with $\psi|L'_\rho \geq \psi|L$. It is now easy to show that f is bijective, and this analysis leads to the following characterization of Z_ρ as a twistor space.

THEOREM 6.4. [**114**] *Let $\rho\colon \mathfrak{su}(2) \to \mathfrak{g}$ be a non-zero homomorphism. There is a fibration $\pi\colon Z_\rho \to M_\rho$ such that*
(i) $\pi^{-1}(V) = C_V$;
(ii) C_V has normal bundle $N_V \cong 2n\mathcal{O}(1)$ where $2n$ is its rank;
(iii) C_V is transverse to the contact distribution.

The only known proof of (ii) relies on Kronheimer's analysis of the nilpotent orbit U_ρ [**75**], though it is instructive to indicate its validity for $V \in L_\rho$. Suppose that Z_ρ has complex dimension $2n+1$. The fibre of N_V at $[\xi] \in C_V$ is obtained by applying ξ to the right-hand side of (2.6). Decomposing each space $\mathrm{ad}(\xi)\Sigma^k$ in (2.6) into weights, it follows that

$$\deg(N_V) = \sum_{k\geq 1} k\mu_k = 2n,$$

and we may write $N_V \cong \bigoplus_{i=1}^{2n} \mathcal{O}(k_i)$ with $\sum_{i=1}^{2n} k_i = 2n$. The existence of holomorphic vector fields pointing in all directions at each point of Z_ρ implies that $k_i \geq 0$. Since C_V is a conic,

$$\left.\frac{T\mathbb{P}(\mathfrak{g}\mathbb{C})}{T\mathbb{P}(V\mathbb{C})}\right|_Z \cong V^\perp \otimes \mathcal{O}(2),$$

and this vector bundle contains N_V as a subbundle. Because $\mathcal{O}(-k_i+2)$ contains no non-zero holomorphic sections if $-k_i+2<0$, it follows that $k_i \le 2$. Moreover $k_i = 2$ for some i if and only if N_V has a constant section determined by a non-zero element in $\bigcap_{[\xi]\in C_V} \mathrm{ad}(\xi)\mathfrak{g}_c$, but this intersection is easily seen to be zero.

Remark. Given ρ, there exists a complex flag manifold $F_\rho = G_c/P$ and a 'canonical fibration'

$$f_\rho \colon Z_\rho = U_\rho/\mathbb{C}^* \to F_\rho$$

with the property that the fibres of f are tangent to the contact distribution of Z_ρ. This mapping gives rise to a contact map $U_\rho/\mathbb{C}^* \to \mathbb{P}(T^*F)$ that was exploited by Kobak [**67**] in the study of harmonic maps. This construction is closely related to the Springer resolution of the nilpotent variety of $\mathfrak{g}_c$.

7. Divisors and Quotients

To begin this section, let M be a (not necessarily compact) 4-dimensional manifold with an anti-self-dual conformal structure, so $W_+ = 0$. This is equivalent to saying that M is a 1-dimensional quaternionic manifold, as the assumption imples that the twistor space Z is a complex manifold with a holomorphic line bundle L satisfying $L^2 \cong \kappa^*$.

Suppose that M has a complex structure I, and let D_I denote the divisor of Z formed from the disjoint union of the sections I and $-I$.

THEOREM 7.1. [**97**] *$D_I \in |L|$ if and only if the conformal class contains a Kähler metric.*

Such a metric has zero scalar curvature by standard curvature properties, and is therefore 'scalar-flat Kähler' (SFK). This theorem leads to a characterization of anti-self-dual Hermitian surfaces [**24**].

Now suppose that M is a QK manifold of dimension $4n \ge 4$. From the point of view of §6, the relevance of Theorem 7.1 is that a non-zero Killing vector field X gives rise to a such a divisor of L. The corresponding complex structure I_ζ is the one whose 2-form is $\zeta/|\zeta|$, and is defined away from the zero set

$$M_0 = \{x \in M : \zeta(x) = 0\}.$$

If $x \in M \setminus M_0$ then D intersects Z_x in two points corresponding to $\pm I_\zeta$. To check explicitly that I_ζ is integrable, extend $I_\zeta = I_1$ to an orthonormal basis $\{I_1, I_2, I_3\}$ of V. In terms of (1.2), the condition (6.1) tells us that $\alpha_i = I_i\alpha$ for $i = 1, 2$ and some 1-form α. This implies that

$$\nabla_{I_\zeta X} = I_\zeta \nabla_X I_\zeta,$$

which is indeed the condition that I_ζ be a complex structure.

Suppose that M is a QK manifold with an action by S^1 preserving the quaternionic structure. Let $X = X_\zeta$ denote the corresponding Killing vector field, and let f denote the S^1-equivariant function $||\zeta||^2$. It follows from (6.1) that $df = 2\,X \lrcorner\, \zeta$, or

$$\frac{1}{\|\zeta\|}\mathrm{grad}\, f = I_\zeta X. \tag{7.1}$$

The various moment equations can be neatly encapsulated in the statement that, if t is a formal variable that behaves like a closed 2-form with $X \lrcorner\, t = 0$, then

$$\hat{\Omega} = \Omega - 2\zeta t + f t^2$$

satisfies $d\hat{\Omega} + (X \lrcorner\, \hat{\Omega})t = 0$. This means that $\hat{\Omega}$ is an equivariantly closed extension of Ω, and the situation is analogous to that of Kähler geometry [**14, 7**].

From (7.1), the set of critical points of f consists of the union of the zero set M_0 and the set

$$M^{S^1} = \{x \in M : X(x) = 0\}$$

of fixed points of the S^1-action. If $M_0 \cap M^{S^1} = \emptyset$, so that S^1 acts freely on M_0, then the result of Galicki-Lawson (Theorem 7.3 below) implies that $Q = M_0/S^1$ is a QK manifold. The gradient flow therefore determines a diagram

$$Q \leftarrow M_c/S^1 \rightarrow M^{S^1}, \tag{7.2}$$

where $M_c = f^{-1}(c)$, c is less than the first critical value of f, and every connected component of the fixed point set M^{S^1} is a Kähler submanifold of M. The Hermitian manifold $(M \setminus M_0, g, I_z)$ has been studied in [**5**], and M_c/S^1 may be regarded as a Kähler quotient of it. The correspondence (7.2) realizes M as a 'vehicle' for the abstract geometry associated to the quotient Q, and can also be used to compute the topology of Q using the methods of [**66**]. This and other applications of Morse theory to quaternion-Kähler geometry discussed in this section, are due to Battaglia [**14**].

Example. The only complete example of this construction arises from the action of S^1 on $\mathbb{H}^{n+1}$ by left multiplication by e^{it} which we now examine in some detail. The commutator of S^1 in $Sp(n+1)$ is $U(n+1)$, and we may regard $\mathbb{H}^{n+1}$ is the real vector space underlying $\mathbb{C}^{n+1} \otimes \mathbb{C}^2$ by writing a quaternionic vector as $a + bj$ with $a, b \in \mathbb{C}^{n+1}$. The action of $\mathbb{H}^*$ on $\mathbb{C}^2$ extends to an action of $GL(2,\mathbb{C})$ commuting with $U(n+1)$, and this action is free outside the set S of simple elements (those with rank less than 2) in the tensor product. The mapping $a + bj \longmapsto \langle a, b \rangle$ defines an isomorphism

$$(\mathbb{H}^{n+1} \setminus S)/GL(2,\mathbb{C}) \cong \mathbb{G}\mathrm{r}_2(\mathbb{C}^{n+1}),$$

and it follows that $\mathbb{HP}^n \setminus \mathbb{CP}^n$ is the total space of a bundle over $\mathbb{G}\mathrm{r}_2(\mathbb{C}^{n+1})$ with fibres isomorphic to $GL(2,\mathbb{C})/\mathbb{H}^* \cong \mathbb{HP}^1 \setminus \mathbb{CP}^1$. The 2-form ζ on $\mathbb{HP}^n$ can be identified with the mapping

$$a + bj \longmapsto \overline{(a+bj)}i(a+bj) = (\|a\|^2 - \|b\|^2)i + 2\bar{a}b\,k,$$

and the fibre over a 2-plane $\Pi \in \mathbb{G}\mathrm{r}_2(\mathbb{C}^{n+1})$ intersects M_0 in the circle $U(2)/Sp(1)$ each point of which is represented by a unitary basis of Π. Thus M_0/S^1 is also isomorphic to $\mathbb{G}\mathrm{r}_2(\mathbb{C}^{n+1})$.

This example illustrates the way in which a quotient by a non-compact group may be replaced by a compact quotient. If X is the Killing vector field corresponding to the S^1 action on $\mathbb{HP}^n$, then the tangent vectors I_1X, I_2X, I_3X defined at any given point arise from a basis of $\mathfrak{sl}(2,\mathbb{C})/\mathfrak{su}(2)$, and are always orthogonal to M_0. In general one does not have the luxury of an $SL(2,\mathbb{C})$ action, though lifting to the twistor space does enable one to complexify an S^1 action. In the example above, $M^{S^1} = \mathbb{CP}^n$ and M_c/S^1 can be identified with the flag manifold

$U(n+1)/(U(1)\times U(1)\times U(n))$. In general one might conjecture that M_c/S^1 is isomorphic to the twistor space of Q.

THEOREM 7.2. [**14**] *If S^1 acts on a positive QK manifold M and the action is free on M_0, then M is isometric to $\mathbb{HP}^n$.*

Moreover, it is effectively a consequence of Theorem 5.5(ii) that $\mathbb{G}\mathrm{r}_2(\mathbb{C}^{n+2})$ is the only positive QK manifold that can be obtained as a QK quotient by a circle action [**13**].

Passing to the more general case of a group action, suppose that G is a connected Lie group acting on a QK manifold M as a group of isometries. For each point $x\in M$ we obtain a mapping $\mathfrak{g}\to V$. This gives rise to a 'moment section' $\gamma\in\Gamma(M,\mathfrak{g}^*\otimes V)$ which is G-equivariant.

THEOREM 7.3. [**47**] *If G acts freely on the zero set M_0 of γ then $Q=M_0/G$ is a QK manifold.*

This quotient construction commutes with the hyper-Kähler quotient construction on U, in the sense that the bundle associated to Q with fibre $\mathbb{H}^*/\mathbb{Z}_2$ can be realized as an HK quotient of U [**113**]. There is also a corresponding notion of 'contact' quotient construction on Z, enabling one to extend the commutativity of (6.3) to quotients as well as embeddings. Incidentally, the paper [**52**] describes an example of a holomorphic contact orbifold obtained as a Kähler S^1-quotient, and used to compactify a moduli space of Higgs bundles.

Remark. The quaternionic quotient construction arises naturally in relation to the decomposition of a tensor product into irreducible components. Let G be a compact Lie group with complexification G_c. Suppose that W is a complex $(2n+2)$-dimensional vector space upon which G acts linearly, commuting with an antilinear involution $j\colon W\to W$ and an identification $W^*\cong W$. The action $\mathfrak{g}\otimes W\to W$ induces a G-equivariant mapping

$$\mathfrak{p}\colon W\otimes W\to\mathfrak{g}^*,$$

which is best regarded as projection to a summand of the tensor product. Let

$$W_0=\{w\in W:\mathfrak{p}(w\otimes w)=0=\mathfrak{p}(w\otimes jw)\}$$

(the notation is taken from [**103**]). Then the quotient by G of the flat hyper-Kähler structure on W is W_0/G, and the QK quotient of $\mathbb{P}(W)\cong\mathbb{HP}^n$ is $W_0/(G\times\mathbb{H}^*)$. In this set-up, the function $f(w)=i\mathfrak{p}(w\otimes jw)$ is a real-valued moment mapping for the action of G on the Kähler submanifold $W_0'=\{w\in W:\mathfrak{p}(w\otimes w)=0\}$, and the curves $\exp(itA)$, with $A\in\mathfrak{g}$, are trajèctories the gradient flow of $\|f\|^2$. It follows from the theory of [**66**] that $W_0/G=f^{-1}(0)/G$ can be identified with the geometrical invariant theory quotient of W_0'.

Less standard actions of S^1 give rise to a host of quaternion-Kähler orbifolds. Starting from $\mathbb{HP}^2$, an analogous but more general procedure produces self-dual Einstein metrics, for example on the weighted projective planes

$$\mathbb{CP}^2_{1,1,k}=\frac{\mathbb{C}^3\setminus\{0\}}{(x,y,z)\longmapsto(\lambda x,\lambda y,\lambda^k z)},\qquad\lambda\in\mathbb{C}^*.$$

This orbifold has $(0,0,1)$ as a unique singular point, and may be viewed as a 1-point compactification of the total space of the line bundle L^{-k} over $\mathbb{CP}^1$. In [**60**], Joyce shows that it admits a $U(2)$-invariant twistor function characterizing the SFK metric constructed by LeBrun in [**76**]. He also shows that formal combinations of such metrics can be then deformed to self-dual metrics on connected sums $n\mathbb{CP}^2$.

Although complete metrics with negative scalar curvature exist [**78**], singularities seem inevitable in the positive case. From the quotient point of view, these arise when G does not act freely on M_0. An exhaustive study of the resulting singularities has been given in [**37**].

Example. The action of S^1 on the Wolf space $M = \mathbb{G}\mathrm{r}_4(\mathbb{R}^7)$ arising from inclusions $S^1 \subset U(3) \subset SO(7)$ has been described in [**68**]. The fixed point set M^{S^1} consists of two copies of $\mathbb{CP}^2$, one of which lies in M_0. In fact, the quotient M_0/S^1 can be identified with the locally-symmetric singular space $\mathbb{Z}_3\backslash G_2/SO(4)/\mathbb{Z}_3$ which is itself an $\mathbb{H}^*$-quotient of the nilpotent variety $\{A \in \mathfrak{sl}(3,\mathbb{C}) : A^3 = 0\}$.

Let M be an arbitrary QK manifold. The subgroup S^1 of $SO(3)$ fixing the complex structure I_1 on the associated bundle U gives rise to a moment mapping $\mu_1 : U \to \mathbb{R}$ which equals the square of radial distance measured on the fibres of U over M. Then

$$Z \cong \mu_1^{-1}(1)/S^1$$

may be regarded as a Kähler quotient of U. The Kähler form of Z can be used to construct a Kähler quotient of Z, which provides an unexplored link between quaternion-Kähler and Kähler geometry. For example the action of S^1 on an open set of a QK manifold M gives rise to a Kähler metric on a space of the same real dimension.

One can always define a 'quadratic moment mapping'

$$F : M \to S^2\mathfrak{g}^*$$

by $F(x)(A,B) = \langle \zeta_A(x), \zeta_B(x) \rangle$, for $A, B \in \mathfrak{g}$. This is the generalization of the function f defined for an S^1 action above, and can also be regarded as the $Sp(1)$-invariant component of the hyper-Kähler moment mapping on the associated bundle U. A computation of the derivative of F shows that it will be an immersion if, at each point x, γ has rank 3 and the quaternionic span of the tangent space to the orbit $G(x)$ equals T_xM. In general, $S^2\mathfrak{g}^*$ will decompose into a number of irreducible G-modules U_i, and we may write $F(x) = \sum F_i(x)$, with $F_i(x) \in U_i$. One of the U_i is the 1-dimensional space spanned by the Killing form of $\mathfrak{g}$, and there will be other trivial summands if and only if G is not simple. The orbit $G(x)$ will then fibre over each of the orbits $G(F_i(x))$ with $F_i(x) \neq 0$. For example, if G is not simple, a component of F determines a G-equivariant mapping $M \to \mathfrak{g}^*$, and the orbits of G will all fibre over coadjoint ones.

Example. Let $M = \mathbb{G}\mathrm{r}_4(\mathbb{R}^{n+4})$ and $G = SO(n+4)$. If $\mathfrak{g} = \mathfrak{so}(n+4)$ then the decomposition of $S^2\mathfrak{g}$ is well known from the theory of curvature tensors; we may write

$$S^2\mathfrak{g} \cong W \oplus \Lambda^4 \oplus S^2_0 \oplus \mathbb{R}.$$

The component of F in $\mathbb{R}$ is the norm of the moment mapping, and must be constant given that G acts transitively on M. Since the stabilizer of $F(m)$ must contain the stabilizer of m, we may deduce that $F(M) \subset \Lambda^4 \oplus \mathbb{R}$, and ignoring the $\mathbb{R}$-component, F is a Plücker embedding.

We remarked in §2 that the Wolf space $\mathbb{G}\mathrm{r}_4(\mathbb{R}^8)$ has some special properties. Each point of the twistor space Z of its non-compact dual defines a null line in the complex quadric Q^4 (in $\mathbb{CP}^5$), which is a compactification and complexification of Minkowski space. This is the basic example of a phenomenon discovered in **[77]** that associates to any real analytic conformal manifold N of signature $(3, n-1)$ a QK $4n$-manifold M with $s < 0$. The twistor space of M is an open set of the space of null geodesics of a complexification of N.

8. Characteristic Classes and Constraints

Suppose that M is a QK manifold for which (partly for convenience of exposition) $\varepsilon = 0$, so that M has a distinguished principal G-bundle P with $G = Sp(n) \times Sp(1)$. Then P may be regarded as the pullback of the universal bundle E_G by a suitable map f to the classifying space $B_G = E_G/G$. Characteristic classes on M arise by pulling back elements of the cohomology ring $H^*(B_G)$, which is identified with the subspace of W-invariant elements of $H^*(B_T)$ where T is a maximal torus of G and $W = N(T)/T$ is the Weyl group.

If $x_1, \dots, x_n$ are coordinates on the Lie algebra $\mathfrak{sp}(n)$ and y is a coordinate on $\mathfrak{sp}(1)$, then W permutes the former and changes an arbitrary number of signs, so that $|W| = 2^{n+1}n!$. We may interpret the symbols $x_1, \dots, x_n, y$ as cohomology classes on B_T by means of the natural isomorphism $H^1(T) \cong H^2(B_T)$, and $H^*(B_G)$ is generated by the elementary polynomials

$$c_{2k} = \sum_{r_1<r_2<\cdots<r_k} {x_{r_1}}^2 {x_{r_2}}^2 \cdots {x_{r_k}}^2, \qquad k = 1, \dots, n,$$
$$u = -y^2.$$

Then $f^*c_2, \dots, f^*c_{2n}$ are the Chern classes of the vector bundle E, and f^*u is minus the 2nd Chern class of H. Moreover, these classes are well defined as elements of $H^*(M, \mathbb{Q})$ even when the structure of M does not lift to G. For example, $4f^*u$ is always integral and represented in de Rham cohomology by a constant multiple of the 4-form Ω. In practice it is often convenient to replace the Chern classes of E by the Pontrjagin classes of M. The kth such class equals f^*p_k (for $1 \le k \le n$) where p_k is the polynomial of degree $2k$ defined by

$$\sum_{k=1}^{2n} (-1)^k p_k = \prod_{r=1}^{2n} (1 - {X_r}^2).$$

An irreducible G-module V decomposes under T as the direct sum of 1-dimensional spaces V_λ, where the weight $\lambda \in \mathfrak{t}^*$ describes the eigenvalues of $(e^{ix_1}, \dots, e^{ix_1}, e^{iy})$. The representation ring $R[T]$ is thus the polynomial algebra generated by $e^{\pm x_1}, \dots, e^{\pm x_n}$, $e^{\pm y}$, and restriction to T defines an injective ring homomorphism

$$\mathrm{ch}\colon R[G] \to \mathbb{C}[[x_1, \dots, x_n, y]],$$

whose image is contained in the space of W-invariant formal power series. The usual Chern character of the vector bundle $P \times_G V$ is simply $f^*\mathrm{ch}(V)$.

As an illustration, first consider

$$\begin{aligned}
\mathrm{ch}(E \otimes H) &= (2n - c_2 + \tfrac{1}{12}({c_2}^2 - 2c_4) + \cdots)(2 + u + \tfrac{1}{12}u^2 + \cdots) \\
&= 4n + 2(nu - c_2) + \tfrac{1}{6}({c_2}^2 - 2c_4 - 6c_2u + nu^2) + \cdots
\end{aligned}$$

and it follows that

$$\begin{aligned} p_1 &= 2(nu - c_2), \\ p_2 &= {c_2}^2 + 2c_4 + (6-4n)c_2u + (2n-1)nu^2. \end{aligned}$$

The Euler class of the oriented real vector bundle TM is the pullback of the class

(8.1) $$e = \prod_{r=1}^{2n} X_r = \prod_{r=1}^{n} ({x_r}^2 - y^2) = \sum_{r=0}^{n} c_{2r} u^{n-r}$$

on B_G. The last term coincides with the Chern class c_{2n} of the virtual vector bundle $E - H$. In this sense, the lemma is analogous to the fact that the Euler class e underlying a complex vector bundle V is its top Chern class.

Example. Over the projective space $\mathbb{HP}^n$, the direct sum $E \oplus H$ can be identified with the trivial bundle with fibre $\mathbb{H}^{n+1}$ with H the tautological line bundle and E its complement. Hence

$$(1 + c_2 + c_4 + \cdots + c_{2n})(1-u) = 1,$$

$c_{2k} = u^k$ and $e = (n+1)u^n$. Since $\chi(\mathbb{HP}^n) = n+1$, we deduce that $u^n = 1$. On an arbitrary compact QK manifold, with our choice of orientation, all one knows is that the 'quaternionic volume'

$$v(M) = \langle (4u)^n, [M] \rangle$$

is a positive integer, and $\deg(Z) = 2v(M)$. A known estimate for the Chern number c_1^{2n+1} on the twistor space Z [**80**] implies that $v(M) \leq 2v(\mathbb{HP}^n) = 2 \cdot 4^n$. It is also known that the integer

(8.2) $$v'(M) = (n-1)v(M) - 2\left\langle p_1 (4u)^{n-1}, [M] \right\rangle$$

is non-negative, with equality if and only if M is $\mathbb{HP}^n$ [**102**].

Let

(8.3) $$D \colon \Gamma(M, \Delta_+) \to \Gamma(M, \Delta_-)$$

denote the Dirac operator described in §3. If $V \in R[G]$, the Dirac operator may be coupled to the corresponding virtual vector bundle in order to define the index

$$\mathrm{ind}(V) = \dim \ker D - \dim \operatorname{coker} D.$$

The Atiyah-Singer index theorem implies that

$$\mathrm{ind}(V) = \left\langle \mathrm{ch}(V)\hat{A}(TM) \right\rangle,$$

where angular brackets indicate the evaluation or 'integration' of a cohomology class on the fundamental cycle $[M]$. The $\hat{A}$ class is defined by

(8.4) $$\begin{aligned} \hat{A} &= \textstyle\prod_{r=1}^{2n} \dfrac{\frac{1}{2}X_r}{\sinh(\frac{1}{2}X_r)} &= 1 + \hat{A}_1 + \hat{A}_2 + \cdots \\ & &= 1 - \tfrac{1}{24}p_1 + \tfrac{1}{45\Delta 2}(7{p_1}^2 - 4p_2) + \cdots \end{aligned}$$

To present the results of this section in a more general context, we shall first consider the case of a compact oriented 8-dimensional Riemannian 8-manifold M with a holonomy group K of rank less than 4. If M is neither locally reducible nor symmetric, then K has to be one of the groups

K	$Sp(2)Sp(1)$	$Spin(7)$	$SU(4)$	$Sp(2)$
k	0	1	2	3

In all cases M is a spin manifold. The last three groups correspond to Ricci-flat geometries, and the integer k is the dimension of the space of parallel spinors [**117**], which we may arrange to equal $\dim\ker D$ in (8.3).

The existence of the $*$ operator on 4-forms allows one to decompose $b_4 = b^+ - b^-$, where

$$b^\pm = \dim\{\sigma \in \gamma(M, \Lambda^4 T^*M) : *\sigma = \pm\sigma,\ d\sigma = 0\}.$$

The Hirzebruch signature theorem implies that

$$b^+ - b^- = \tfrac{1}{45}(7p_2 - {p_1}^2),$$

and the Atiyah-Singer index theorem asserts that k is equal to the $\hat{A}$ genus $\hat{A}_2$ given in (8.4). Using the extra equation

$$4p_2 - {p_1}^2 = 8\chi$$

that follows merely from the topological reduction (see for example (8.1)), we obtain

PROPOSITION 8.1. [**62**] $k = \frac{1}{24}(b^+ - 2b^- + b_3 - b_2 + b_1 - 1)$.

It follows that the Betti numbers are 'relatively large' in the Ricci-flat cases. Indeed, if M is an irreducible HK manifold so that $H = Sp(2)$ then $b_1 = 0$ and Hodge theory implies that $b^- = 3h_0^{1,1} = 3(b_2 - 3)$. It follows that

$$b_3 + b_4 = 10b_2 + 46, \tag{8.5}$$

and the right-hand side is at least 76. By contrast, if $H = Sp(2)Sp(1)$ and $s > 0$, then b_1, b_3, b^- all vanish, and

$$b_4 = 1 + b_2. \tag{8.6}$$

This equation plays a crucial role in proofs that M must in fact be one of the symmetric spaces $\mathbb{G}\mathrm{r}_2(\mathbb{C}^4)$, $G_2/SO(4)$, $\mathbb{HP}^2$ (recall Theorem 2.1). If M is not the complex Grassmannian, then $b_4 = 1$ which gives an extra relation between characteristic classes.

It is rather surprising that the equations (8.5),(8.6) have non-trivial generalizations to higher dimensions:

THEOREM 8.2. [**108, 80**] *Let M be a compact Riemannian manifold of dimension $4n$ with holonomy group K and scalar curvature s.*

(i) If $K \subseteq Sp(n)$ (so that $s = 0$) then $n\chi = 6\sum_{i=1}^{2n}(-1)^i i^2 b_{2n-i}$.

(ii) If $K \subseteq Sp(n)Sp(1)$ and $s > 0$ then $\sum_{i=0}^{n-1}[6i(n-1-i) - (n-1)(n-3)]b_{2i} = \frac{1}{2}n(n-1)b_{2n}$.

The more attractive form of the first equation reflects the fact that products of HK manifolds remain HK. Quaternion-kähler manifolds enjoy no such functorial properties, though the second equation can be expressed in the more memorable form

$$\sum_{i=1}^{\lfloor m/2 \rfloor} i(m-i)(m-2i)\gamma_i = 0, \tag{8.7}$$

by setting $m = n+1$, $\gamma_i = \beta_{2i} - \beta_{2m-2i}$ and $\beta_{2i} = b_{2i} - b_{2i-4}$ (see (1.6) and [**49**]).

Remark. Fix k with $1 \leq k \leq n$. There is a non-trivial action of $\mathbb{Z}_2$ on $\mathbb{HP}^n$ induced by changing signs of k of the coordinates of $\mathbb{H}^{n+1}$. Work on other manifolds with exceptional holonomy suggest that in assessing the relevance of topological constraints to orbifolds like $N = \mathbb{HP}^n/\mathbb{Z}_2$ it is more relevant to consider the 'string-theoretic' Poincaré polynomial $P_{st}(t)$ formed by adding contributions for the fixed point set. The latter is a disjoint union $\mathbb{HP}^{k-1} \sqcup \mathbb{HP}^{n+1-k}$, so

$$P_{st}(t) = Q_n + t^{2k}Q_{n-k} + t^{2n-2k+2}Q_{k-1},$$

where $Q_n = \sum_{i=0}^{n} t^{4i}$. It follows that, in the above notation, N has

$$\beta_{2i} = \begin{cases} 1, & i = k \text{ or } n+1-k, \\ 0, & \text{otherwise.} \end{cases}$$

and $\gamma_{2i} = 0$ for all i. This provides some justification for the symmetry in (8.7).

The complicated nature of $\hat{A}_k$ for $k \geq 2$ leads one to seek virtual representations V for which $\mathrm{ch}(V)$ has only terms of near top dimension. To make this precise, let $\mathcal{F}_k$ denote the space spanned by virtual representations V of $Sp(n)Sp(1)$ such that $\mathrm{ch}(V)$ has no terms of degree less than $2k$. The sequence of ideals $\cdots \supset \mathcal{F}_k \supset \mathcal{F}_{k+1} \supset \cdots$ is an example of the γ-filtration that is used to derive the graded structure of cohomology from K-theory [**46**]. It is easy to describe the restriction of this filtration to $R[Sp(1)]$. Since ch is a ring homomorphism and $\mathcal{F}_j\mathcal{F}_k \subset \mathcal{F}_{j+k}$, we have $(\Sigma^2 - 3)^k \in \mathcal{F}_k$ for any k, and in fact $(\Sigma^2-3)^k$ generates $\mathcal{F}_k \cap R[Sp(1)]$. Expanding both sides of the equation

$$\mathrm{ch}((\Sigma^2 - 3)^{n+k}) = (4u + \tfrac{4}{3}u^2 + \cdots)^{n+k}$$

and recalling (5.5) and (8.2) yields

PROPOSITION 8.3. [**53**] *Let M be a compact QK manifold of dimension $8m$. Then*

$$\sum_{i=0}^{m+k} (-1)^i \binom{4m+2k+1}{i} r_{m+k-i} = \begin{cases} \frac{1}{48}(v' + 3(n-1)v), & k = -1, \\ v, & k = 0, \\ 0, & k \geq 1. \end{cases}$$

An analogous result holds when M has odd quaternionic dimension.

Example. Let $n = 2m = 4$. The Hilbert polynomial is completely determined by the dimension $r = r_1$ of the isometry group and the quaternionic volume v. Combining the formula for $k = -1$ with (5.4) gives

$$7 + \tfrac{3}{16}v \leq r \leq 2v + 9.$$

For example, $r = 11$ and $v = 2$ gives the values (5.7) in §5, though the resulting polarized variety with $\Delta = 2$ must be smooth and has topology inconsistent with that of a twistor space.

A deeper analysis of the filtration $\mathcal{F}_k$ is used in the proof of Theorem 8.2. This relies on the existence of $W_n \in R[G]$ such that

$$\mathrm{ch}(W_n) = y(1 + \tfrac{1}{24}p_1),$$

where y has degree $4n-4$, ensuring that $\mathrm{ind}(W_n)$ vanishes [**107**]. The element W_n is generated from representations of the form $R^{p,q} = \Lambda_0^p \otimes \Sigma^q$, and the associated vector bundles may be coupled to the Dirac operator provided $p+q+n$ is even. The associated indices

$$i^{p,q} = \mathrm{ind} D(R^{p,q})$$

play a fundamental role.

The index of the Dirac operator on M coupled to Σ^q is equal to that of the Dolbeault complex on Z coupled to the line bundle $\mathcal{O}(q-n)$, whence

$$i^{p,q} = \chi(Z, \mathcal{O}(\pi^* \Lambda_0^p (q-n)).$$

Using $h^{p,p}(Z) = b_{2p-2} + b_{2p}$ (with $b_{-2} = 0$), Lemma 5.5 implies that on a positive QK manifold of dimension $4n$,

$$i^{p,q} = \begin{cases} 0, & n = p+q+2r,\ r > 0, \\ (-1)^p (b_{2p-2} + b_{2p}), & n = p+q, \\ r_{(q-n)/2}, & p = 0,\ q \geq n. \end{cases}$$

Observe that $i^{0,n+2} = r_1$ is the dimension of the isometry group. The next result is the generalization of (2.1) to the case of an abstract QK manifold, and is proved in the same way as Theorem 8.3:

PROPOSITION 8.4. **[41]** $\sum_{p=0}^{n} (-1)^p i^{p,n+2-p} = 2\chi + b_{2n-2} + b_{2n}.$

It follows from **[79]** that $i^{1,n+1} = 0$ unless $M = \mathbb{HP}^n$. It is an interesting problem to understand the indices $i^{p,n-p+1}$ for $p \geq 2$, and identify the cohomology spaces that might contribute to their non-vanishing.

ACKNOWLEDGEMENT. The author is grateful to D. Joyce for commenting on a draft of this essay.

References

[1] W. W. ADAMS, C. A. BERENSTEIN, P. LOUSTAUNAU, I. SABADINI, AND D. C. STRUPPA, *Regular functions of several quaternionic variables and the Cauchy-Fueter complex*. J. Geom. Anal., to appear.

[2] D. V. ALEKSEEVSKY, *Compact quaternion spaces*, Funktsional. Anal. i Prilozhen., 2 (1968), pp. 11–20. English translation: Functional Anal. Appl. **2** (1968), 106–114.

[3] ———, *Classification of quaternionic spaces with a transitive solvable group of motions*, Math. USSR Isvestija, 9 (1975), pp. 297–339.

[4] D. V. ALEKSEEVSKY AND S. MARCHIAFAVA, *Transformations of a quaternionic Kähler manifold*, C. R. Acad. Sci. Paris, Sér. I, 320 (1995), pp. 703–708.

[5] D. V. ALEKSEEVSKY, S. MARCHIAFAVA, AND M. PONTECORVO, *Compatible almost complex structures on quaternion Kähler manifolds*, Ann. Global Anal. and Geom., 16 (1998).

[6] D. V. ALEKSEEVSKY AND F. PODESTÀ, *Compact cohomogeneity one Riemannian manifolds of positive Euler characteristic and quaternionic Kähler manifolds*, in USA-Brazil Workshop on Geometry, Topology and Physics, 1997, 1998.

[7] M. F. ATIYAH AND R. BOTT, *The moment map and equivariant cohomology*, Topology, 23 (1984), pp. 1–28.

[8] M. F. ATIYAH, N. J. HITCHIN, AND I. M. SINGER, *Self-duality in four-dimensional Riemannian geometry*, Proc. Roy. Soc. London, A 362 (1978), pp. 425–461.

[9] C. BÄR, *Real Killing spinors and holonomy*, Comm. Math. Phys., 154 (1993), pp. 509–521.

[10] R. BARKER AND S. M. SALAMON, *Analysis on a generalized Heisenberg group*, J. London Math. Soc., 28 (1983), pp. 184–192.

[11] R. J. BASTON, *Quaternionic complexes*, J. Geom. Phys., 8 (1992), pp. 29–52.

[12] F. BATTAGLIA, *A hypercomplex Stiefel manifold*, Differential Geom. Appl., 6 (1996), pp. 121–128.

[13] ———, *S^1-quotients of quaternion-Kähler manifolds*, Proc. Amer. Math. Soc., 27 (1996), pp. 2185–2192.

[14] ———, *Circle actions and Morse theory on quaternion-Kähler maifolds*. J. London Math. Soc., to appear.

[15] H. BAUM, T. FRIEDRICH, R. GRUNEWALD, AND I. KATH, *Twistors and Killing spinors on Riemannian manifolds*, Teubner-Verlag, Stuttgart/Leipzig, 1991.

[16] A. BEAUVILLE, *Fano contact manifolds and nilpotent orbits*. preprint, 1997.

[17] A. L. BESSE, *Einstein manifolds*, vol. 10 of Ergebnisse der Mathematik und ihrer Grenzgebiete, 3. Folge, Springer, Berlin, Heidelberg and New York, 1987.

[18] R. BIELAWSKI, *On the hyperkähler metrics associated to singularities of nilpotent varieties*, Ann. Global Anal. and Geom., 14 (1996), pp. 177–191.

[19] O. BIQUARD, *Métriqes d'einstein asymptotiquement symétriques*. preprint, 1998.

[20] E. BONAN, *Tenseur de structure d'une variété presque quaternioniennes*, C. R. Acad. Sci. Paris, 259 (1964), pp. 45–48.

[21] ———, *Sur l'algèbre extérieure d'une variété presque hermitienne quaternionique*, C. R. Acad. Sci. Paris, 296 (1983), pp. 601–602.

[22] A. BOREL, *Compact Clifford-Klein forms of symmetric spaces*, Topology, 2 (1963), pp. 111–122.

[23] C. BOYER, K. GALICKI, AND B. MANN, *Hypercomplex structures on Stiefel manifolds*, Ann. Global Anal. and Geom., 14 (1996), pp. 81–105.

[24] C. P. BOYER, *Conformal duality and compact complex surfaces*, Math. Ann., 274 (1986), pp. 517–526.

[25] ———, *A note on hyperhermitian four-manifolds*, Proc. Amer. Math. Soc., 102 (1988), pp. 157–164.

[26] C. P. BOYER, K. GALICKI, AND B. M. MANN, *The geometry and topology of 3-Sasakian manifolds*, J. reine angew. Math., 455 (1994), pp. 183–220.

[27] C. P. BOYER, K. GALICKI, B. M. MANN, AND E. G. REES, *Compact 3-Sasakian 7-manifolds with arbitrary second Betti number*, Invent. Math., 131 (1996), pp. 321–344.

[28] R. BRYANT AND R. HARVEY, *Submanifolds in hyper-Kähler geometry*, J. Amer. Math. Soc., 2 (1989), pp. 1–31.

[29] R. L. BRYANT AND S. M. SALAMON, *On the construction of some complete metrics with exceptional holonomy*, Duke Math. J., 58 (1989), pp. 829–850.

[30] F. BURSTALL AND M. GUEST, *Harmonic two-spheres in compact symmetric spaces, revisited*, Math. Ann., 309 (1997), pp. 541–572.

[31] F. BURSTALL AND J. RAWNSLEY, *Twistor theory for Riemannian symmetric spaces, with applications to harmonic maps of Riemann surfaces*, vol. 1424 of Lecture Notes in Mathematics, Springer-Verlag, 1990.

[32] F. BURSTALL, J. RAWNSLEY, AND S. M. SALAMON, *Stable harmonic 2-spheres in symmetric spaces*, Bull. Amer. Math. Soc., 16 (1987), pp. 274–278.

[33] B.-Y. CHEN AND T. NAGANO, *Totally geodesic submanifolds of symmetric spaces II*, Duke J. Math., 45 (1978), pp. 405–425.

[34] K. CORLETTE, *Archimedian superrigidity and hyperbolic geometry*, Ann. of Math., 135 (1992), pp. 165–182.

[35] V. CORTÉS, *Alekseevskian spaces*, Differential Geom. Appl., 6 (1996), pp. 129–168.

[36] A. S. DANCER AND A. F. SWANN, *HyperKähler metrics of cohomogeneity one*, J. Geom. Phys., 21 (1997), pp. 218–230.

[37] ———, *The geometry of singular quaternionic Kähler quotients*. Internat. J. Math., to appear.

[38] B. DE WIT AND A. VAN PROEYEN, *Special geometry, cubic polynomials and homogeneous quaternionic spaces*, Comm. Math. Phys., 149 (1992), pp. 307–333.

[39] S. K. DONALDSON AND P. B. KRONHEIMER, *The geometry of four-manifolds*, Oxford Mathematical Monographs, Clarendon Press, Oxford, 1990.

[40] I. DOTTI AND A. FINO, *Abelian hypercomplex 8-dimensional nilmanifolds*. preprint, 1998.

[41] A. FINO AND S. M. SALAMON, *Observations on the topology of symmetric spaces*, in Proceedings on Geometry and Physics (Aarhus and Odense, 1995), J. E. Andersen et al., eds., vol. 184 of Lect. Notes in Pure and Appl. Math., Marcel Dekker, New York, 1996, pp. 275–286.

[42] D. FREED, *Special Kähler manifolds*. preprint, 1997.

[43] T. Friedrich, I. Kath, A. Moroianu, and U. Semmelmann, *On nearly parallel g_2-structures*. preprint, 1995.

[44] A. Fujiki, *On the de Rham cohomology group of a compact Kähler symplectic manifold*, in Algebraic Geometry, Sendai 1985, T. Oda, ed., vol. 10 of Adv. Studies in Pure Math., North-Holland, 1987.

[45] T. Fujita, *Classification theories of polarized varieties*, Cambridge University Press, Cambridge, 1990.

[46] W. Fulton and S. Lang, *Riemann-Roch Algebra*, Springer-Verlag, Berlin, 1985.

[47] K. Galicki and H. B. Lawson, *Quaternionic reduction and quaternionic orbifolds*, Math. Ann., 282 (1988), pp. 1–21.

[48] K. Galicki and Y. S. Poon, *Duality and Yang-Mills theory on quaternionic kähler manifolds*, J. Math. Phys., 32 (1991), pp. 1263–1268.

[49] K. Galicki and S. M. Salamon, *Betti numbers of 3-Sasakian manifolds*, Geom. Dedicata, 63 (1996), pp. 45–68.

[50] S. Garoufalidis and H. Nakamura, *Some IHX-type relations and symplectic representation theory*, Math. Res. Letters, 5 (1998), pp. 391–402.

[51] A. Gray, *A note on manifolds whose holonomy group is a subgroup of $Sp(n)\,\Delta Sp(1)$*, Mich. Math. J., 16 (1969), pp. 125–128.

[52] T. Hausel, *Compactification of moduli of Higgs bundles*. preprint, 1998.

[53] R. Herrera, *Topics in Geometry and Topology*, PhD thesis, Oxford University, 1997.

[54] O. Hijazi and J.-L. Milhorat, *Minoration des valeurs prorpres de l'opérateur de Dirac sur les variétés spin Kähler-quaternioniennes*, J. Math. Pures Appl., 74 (1995), pp. 387–414.

[55] N. Hitchin, *The self-duality equations on a Riemann surface*, Proc. Lond. Math. Soc., 55 (1987), pp. 59–126.

[56] N. J. Hitchin, *Kählerian twistor spaces*, Proc. London Math. Soc., 43 (1981), pp. 133–150.

[57] ———, *Twistor spaces, Einstein metrics and isomonodromic deformations*, J. Differential Geom., 42 (1995), pp. 30–112.

[58] P. S. Howe, A. Opfermann, and G. Papadopoulos, *Twistor spaces for qkt manifolds*. preprint, 1997.

[59] S. Ishihara, *Quaternion kählerian manifolds*, J. Differential Geom., 9 (1974), pp. 483–500.

[60] D. Joyce, *The hypercomplex quotient and the quaternionic quotient*, Math. Ann., 290 (1991), pp. 323–340.

[61] ———, *Compact hypercomplex and quaternionic manifolds*, J. Differential Geom., 35 (1992), pp. 743–761.

[62] ———, *Compact riemannian 8-manifolds with holonomy Spin(7)*, Invent. Math., 123 (1996), pp. 507–552.

[63] ———, *Hypercomplex algebraic geometry*, Quart. J. Math., 49 (1998), pp. 129–162.

[64] M. Kapranov, *Rozansky-Witten invariants via Atiyah classes*. preprint, 1997.

[65] M. Kato, *Compact differentiable 4-folds with quaternionic structure*, Math. Ann., 248 (1980), pp. 79–96.

[66] F. C. Kirwan, *Cohomology of quotients in algebraic geometry*, Princeton University Press, Princeton, 1984.

[67] P. Kobak, *Twistors, nilpotent orbits and harmonic maps*, in Harmonic maps and integrable systems, A. P. Fordy and J. C. Wood, eds., vol. 23 of Aspects of Math., Vieweg, Braunschweig, Wiesbaden, 1994, pp. 295–319.

[68] P. Kobak and A. F. Swann, *Quaternionic geometry of a nilpotent variety*, Math. Ann., 297 (1993), pp. 747–764.

[69] S. Kobayashi and T. Ochiai, *Characterizations of complex projective spaces and hyperquadrics*, J. Math. Kyoto Univ., 13 (1973), pp. 31–47.

[70] J. Kollár, Y. Miyaoka, and S. Mori, *Rational connectedness and boundedness of Fano manifolds*, J. Differential Geom., 36 (1992), pp. 765–779.

[71] B. Kostant, *Holonomy and the Lie algebra of infinitesimal motions of a Riemannian manifold*, Trans. Amer. Math. Soc., 80 (1955), pp. 528–542.

[72] ———, *The principal three-dimensional subgroup and the Betti numbers of a complex simple Lie group*, Amer. J. Math., 81 (1959), pp. 973–1032.

[73] V. Kraines, *Topology of quaternionic manifolds*, Trans. Amer. Math. Soc., 122 (1966), pp. 357–367.

[74] W. KRAMER, U. SEMMELMANN, AND G. WEINGART, *Eigenvalue estimates for the Dirac operator on quaternionic Kähler manifolds.* preliminary version, 1997.

[75] P. B. KRONHEIMER, *Instantons and the geometry of the nilpotent variety*, J. Differential Geom., 32 (1990), pp. 473–490.

[76] C. LEBRUN, *Counter-examples to the generalized positive action conjecture*, Commun. Math. Phys., 118 (1988), pp. 591–596.

[77] ———, *Quaternionic-Kähler manifolds and conformal geometry*, Math. Ann., 284 (1989), pp. 353–376.

[78] ———, *On complete quaternionic-Kähler manifolds*, Duke Math. J., 63 (1991), pp. 723–743.

[79] ———, *Fano manifolds, contact structures, and quaternionic geometry*, Int. J. Maths., 3 (1995), pp. 419–437.

[80] C. LEBRUN AND S. M. SALAMON, *Strong rigidity of positive quaternion-Kähler manifolds*, Invent. Math., 118 (1994), pp. 108–132.

[81] N. C. LEUNG AND S. YI, *Analytic torsion for quaternionic manifolds and related topics.* preprint, 1997.

[82] M. MAMONE CAPRIA AND S. M. SALAMON, *Yang-Mills fields on quaternionic spaces*, Nonlinearity, 1 (1988), pp. 517–530.

[83] S. MARCHIAFAVA, *Su alcune sottovarietà che ha interesse considerare in una varietà Kähleriana quaternionale*, Rendiconti di Matematica, Serie VIi, Roma, 10 (1990), pp. 493–529.

[84] S. MARCHIAFAVA AND G. ROMANI, *Sui fibrati con struttura quaternionale generalizzata*, Annali di Matematica pura ed applicata, 107 (1976), pp. 131–157.

[85] B. MCINNES, *Methods of holonomy theory for Ricci-flat riemannian manifolds*, J. Math. Phys., 32 (1991), pp. 888–896.

[86] A. MOROIANU AND U. SEMMELMANN, *Kählerian Killing spinors, complex contact structures and twistor spaces.* preprint.

[87] A. NADEL, *The boundedness of degree of Fano varieties with picard number one*, J. Amer. Math. Soc., 4 (1991), pp. 681–692.

[88] Y. NAGATOMO, *Vanishing theorem for cohomology groups of c_2-self-dual bundles on quaternionic Kähler manifolds*, Differential Geom. Appl., 5 (1995), pp. 79–97.

[89] Y. NAGATOMO AND T. NITTA, *Vanishing theorems for quaternionic complexes*, Bull. London Maths. Soc., 29 (1997), pp. 359–366.

[90] T. NITTA, *Connections for vector bundles over quaternionic Kähler manifolds*, Proc. Japan Acad., 63 (1987), pp. 23–25.

[91] T. NITTA AND M. TAKEUCHI, *Contact structures on twistor spaces*, J. Math. Soc. Japan, 39 (1987), pp. 139–162.

[92] W. OXBURY, *Twistor spaces and Fano threefolds*, Quart. J. Math., 45 (1994), pp. 343–366.

[93] H. PEDERSEN, *Eguchi-Hanson metrics with cosmological constant*, Classical Quantum Gravity, 2 (1985), pp. 579–587.

[94] H. PEDERSEN AND Y. S. POON, *Deformations of hypercomplex structures*, J. Reine Angew Math., 499 (1998), pp. 81–99.

[95] H. PEDERSEN, Y. S. POON, AND A. F. SWANN, *Hypercomplex structures associated to quaternionic manifolds.* Differential Geom. Appl. (to appear).

[96] P. PICCINI, *Local quaternion-kähler manifolds.* preprint.

[97] M. PONTECORVO, *On twistor spaces of anti-self-dual hermitian surfaces*, Trans. Amer. Math. Soc., 331 (1992), pp. 653–661.

[98] Y. S. POON AND S. M. SALAMON, *Quaternionic Kähler 8-manifolds with positive scalar curvature*, J. Differential Geom., 33 (1991), pp. 363–378.

[99] D. G. QUILLEN, *Quaternionic algebra and sheaves on the Riemann sphere*, Quart. J. Math., 49 (1998), pp. 163–198.

[100] L. ROZANSKY AND E. WITTEN, *Hyperkähler geometry and invariants of three-manifolds.* preprint, 1996.

[101] K. SAKAMOTO, *On the topology of quaternion Kähler manifolds*, Tôhoku Math. J., 26 (1974), pp. 389–405.

[102] S. M. SALAMON, *Quaternionic Kähler manifolds*, Invent. Math., 67 (1982), pp. 143–171.

[103] ———, *Harmonic 4-spaces*, Math. Ann., 269 (1984), pp. 169–178.

[104] ———, *Harmonic and holomorphic maps*, in Geometry Seminar "L. Bianchi" II, E. Vesentini, ed., vol. 1164 of Lecture Notes in Mathematics, Springer-Verlag, 1985, pp. 161–224.

[105] ——, *Differential geometry of quaternionic manifolds*, Ann. Scient. Éc. Norm. Sup., 19 (1986), pp. 31–55.

[106] ——, *Riemannian geometry and holonomy groups*, vol. 201 of Pitman Research Notes in Mathematics, Longman, Harlow, 1989.

[107] ——, *Index theory and quaternionic Kähler manifolds*, in Differential Geometry and its Applications, Proceedings, Opava, Czechoslovakia, 1992, O. Kowalski and D. Krupka, eds., Silesian University Opava, Czech Republic, 1993, pp. 387–404.

[108] ——, *On the cohomology of Kähler and hyper-Kähler manifolds*, Topology, 35 (1996), pp. 137–155.

[109] A. J. SOMMESE, *Quaternionic manifolds*, Math. Ann., 212 (1975), pp. 191–214.

[110] P. SPINDEL, W. SEVRIN, A. AND. TROOST, AND A. VAN PROEYEN, *Extended supersymmetric σ-models on group manifolds*, Nucl. Phys., B 308 (1988), pp. 662–698.

[111] A. F. SWANN, *Aspects symplectiques de la géométrie quaternionique*, C. R. Acad. Sci. Paris, 308 (1989), pp. 225–228.

[112] ——, *Quaternionic kähler geometry and the fundamental 4-form*, in Proceedings of the Curvature Geometry Workshop, Lancaster, 1989, C. T. J. Dodson, ed., ULDM Publ., 1989, pp. 165–174.

[113] ——, *HyperKähler and quaternionic Kähler geometry*, Math. Ann., 289 (1991), pp. 421–450.

[114] ——, *Homogeneous twistor spaces and nilpotent orbits.* preprint, 1997.

[115] K. P. TOD, *The su_∞ Toda field equation and special 4-dimensional metrics*, in Proceedings on Geometry and Physics (Aarhus and Odense, 1995), J. E. Andersen et al., eds., vol. 184 of Lect. Notes in Pure and Appl. Math., Marcel Dekker, New York, 1996, pp. 307–312.

[116] M. Y. WANG, *Parallel spinors and parallel forms*, Anal. Global Anal. Geom., 7 (1989), pp. 59–68.

[117] ——, *Einstein metrics and quaternionic Kähler manifolds*, Math. Z., 210 (1992), pp. 305–325.

[118] J. A. WIŚNIEWSKI, *On Fano manifolds of large index*, Manuscripta Math., 70 (1991), pp. 145–152.

[119] J. WOLF, *Complex homogeneous contact manifolds and quaternionic symmetric spaces*, J. Math. Mech., 14 (1965), pp. 1033–1047.

MATHEMATICAL INSTITUTE, 24–29 ST GILES, OXFORD, OX1 3LB, UK
E-mail address: salamon@maths.ox.ac.uk

3-Sasakian Manifolds

Charles Boyer and Krzysztof Galicki

CONTENTS

1. Introduction

We begin this essay with a brief history of the subject, for our exposition shall otherwise pay scant attention to the chronological incidentals. In 1960 Sasaki [**110**] introduced a geometric structure related to an almost contact structure. This geometry became known as Sasakian geometry and has been studied extensively ever since. In 1970 Kuo [**83**] refined this notion and introduced manifolds with Sasakian 3-structures (see also [**84**], [**117**]). Independently, the same concept was invented by Udrişte [**125**]. Between 1970 and 1975 this new kind of geometry was investigated almost exclusively by a group of Japanese geometers, including Ishihara, Kashiwada, Konishi, Kuo, Tachibana, Tanno, and Yu. Already in [**83**] we learn that the 3-Sasakian geometry has some interesting topological implications. Using earlier results of Tachibana about the harmonic forms on compact Sasakian spaces [**116**], Kuo showed that odd Betti numbers up to the middle dimension must

1991 *Mathematics Subject Classification.* Primary 53C20.

Supported in part by a grant from the National Science Foundation.

be divisible by 4. In 1971 Kashiwada observed that every 3-Sasakian manifold is Einstein with a positive Einstein constant [**69**]. In the same year Tanno proved an interesting theorem about the structure of the isometry group of every 3-Sasakian space [**118**]. In a related paper he studied a natural 3-dimensional foliation on such spaces showing that, if the foliation is regular, then the space of leaves is an Einstein manifold of positive scalar curvature [**119**]. Tanno clearly points to the importance of the analogy with the quaternionic Hopf fibration $S^3 \to S^7 \to S^4$, but does not go any further. In fact, Kashiwada's paper mentions a conjecture speculating that every 3-Sasakian manifold is of constant curvature [**69**]. She attributed this conjecture to Tanno and, at the time, these were the only known examples.

Very soon after, however, it became clear that such a conjecture could not possibly be true. This is due to a couple of papers by Ishihara and Konishi [**68**], [**66**]. They made a fundamental observation that the space of leaves of the natural 3-dimensional foliations mentioned above has a "quaternionic structure", part of which is the Einstein metric discovered by Tanno. This led Ishihara to an independent study of this "sister geometry": quaternionic Kähler manifolds [**67**]. His paper is very well-known and is almost always cited as the source of the explicit coordinate description of quaternionic Kähler geometry. Among other results Ishihara showed that his definition implies that the holonomy group of the metric is a subgroup of $Sp(n)\cdot Sp(1)$, thus providing an important connection with the earlier studies of such manifolds by Alekseevsky [**3**], Bonan [**21**], Gray [**58**], Kraines [**77**], and Wolf [**132**]. In 1975 Konishi [**76**] proved the existence of a Sasakian 3-structure on a natural principal $SO(3)$-bundle over any quaternionic Kähler manifold of positive scalar curvature. This, with the symmetric examples of Wolf, gives precisely all of the homogeneous 3-Sasakian spaces. Yet, at the time they did not appear explicitly and escaped any systematic study until much later.

In fact, 1975 seems to be the year when 3-Sasakian manifolds are relegated to an almost complete obscurity which lasted for about 15 years. From that point on the two "sisters" fair very differently. The extent of this can be best illustrated by the famous book on Einstein manifolds by Besse [**14**]. The book appeared in 1987 and provided the reader with an excellent, up-to-date, and very complete account of what was known about Einstein manifolds 10 years ago. But one is left in the dark when trying to find references to any of the papers on 3-Sasakian manifolds we have cited; 3-Sasakian manifolds are never mentioned in Besse. The other "sister", on the contrary, received a lot of space in a separate chapter. Actually Einstein metrics on Konishi's bundle do appear in Besse (see [**14**] 14.85, 14.86) precisely in the context of the $SO(3)$-bundles over positive quaternionic Kähler manifolds as a consequence of a theorem of Bérard-Bergery ([**14**], 9.73). Obviously, the absence of 3-Sasakian spaces in Besse's book was the result rather than the cause of this obscurity. One could even say it was justified by the lack of any interesting examples. The authors have puzzled over this phenomenon without any sound explanation. One can only speculate that it is the holonomy reduction that made quaternionic Kähler manifolds so much more attractive an object. Significantly, the holonomy group of a 3-Sasakian manifold never reduces to a proper subgroup of the special orthogonal group. And when in 1981 Salamon [**106, 107**], independently with Bérard-Bergery [**12**], generalized Penrose's twistor construction for self-dual 4-manifolds introducing the twistor space over an arbitrary quaternionic Kähler manifold, the research on quaternionic Kähler geometry flourished, fueled by powerful tools from complex algebraic geometry.

Finally, in the early nineties, 3-Sasakian manifolds start a comeback. They begin to appear in two completely different contexts. First, in the study of manifolds with real Killing spinors, Friedrich and Kath notice that the existence of one such spinor leads naturally to a Sasakian-Einstein structure while three of them give the manifold a 3-Sasakian structure [**11, 46**]. Assuming regularity they are able to combine the result of Hitchin [**63**] and Friedrich and Kurke [**49**] and obtain a classification of all regular complete 7-manifolds with 3-Sasakian structure [**47**]. This appears to be the first classification result about 3-Sasakian manifolds. In 1993 the classification problem for manifolds admitting Killing spinors found an elegant formulation in terms of holonomy groups [**10**]. Bär observes that if (M, g) is a simply connected spin manifold with a non-trivial real Killing spinor then the metric cone $(C(M), \bar{g})$ must admit a parallel spinor. In particular $(C(M), \bar{g})$ is Ricci-flat and $\mathrm{Hol}(\bar{g})$ is quite restricted so that only very few groups can occur. One such possibility is $\mathrm{Hol}(\bar{g}) = Sp(m+1)$ which gives the cone a hyper-Kähler structure. It easily follows that M must be 3-Sasakian.

Independently, the hyper-Kähler geometry of the cone $C(\mathcal{S})$ was the starting point of our research on 3-Sasakian manifold. In 1991 the authors, together with Ben Mann, discovered that 3-Sasakian manifolds appear naturally as levels sets of a certain moment map on a hyper-Kähler manifold with an isometric $SU(2)$-action rotating the triple of complex structures [**25**]. In fact, if some obstructions for the $SU(2)$-action vanish, then the hyper-Kähler manifold is precisely a cone on a 3-Sasakian space and, at the same time, it is the Swann's bundle over the associated quaternionic Kähler orbifold of positive scalar curvature [**115**]. We quickly realized that $\mathcal{S}$ is ultimately related to three other Einstein geometries: its hyper-Kähler cone $C(\mathcal{S})$, the associated twistor space $\mathcal{Z}$, and the associated quaternionic Kähler orbifold $\mathcal{O}$. In this review we call the collection of these four geometries together with all the relevant maps $\diamondsuit(\mathcal{S})$. Thus, every $\mathcal{S}$ comes together with a fundamental diagram

$$\begin{array}{ccccc} & & C(\mathcal{S}) & & \\ & \swarrow & & \searrow & \\ \mathcal{Z} & \longleftarrow & \downarrow & \longrightarrow\!\!\!\!- & \mathcal{S}. \\ & \searrow & & \swarrow & \\ & & \mathcal{O} & & \end{array}$$

More importantly we also realized that, even when $\mathcal{O}$ and $\mathcal{Z}$ are compact Riemannian orbifolds, $\mathcal{S}$ can be a smooth manifold. This moment marks the beginning of our efforts to understand the geometry and topology of 3-Sasakian manifolds.

They have led us through the classification of all 3-Sasakian homogeneous spaces and a discovery of a new quotient construction of infinitely many homotopy types of non-regular compact 3-Sasakian manifolds [**26**]. In dimension 7 these examples turned out to be certain Eschenburg bi-quotients of $U(3)$ by a 2-torus [**40**] and [**41**]. We gave a complete analysis of the geometry and topology of such spaces [**26**]. The next important step was the second author's work with Simon Salamon [**54**]. There we noticed that Kuo's theorem about odd Betti numbers of 3-Sasakian manifolds being divisible by 4 missed a crucial point. Because of the isometric $SU(2)$-action, all odd Betti numbers up to the middle dimension must actually vanish. In the regular case we were able to show that 3-Sasakian cohomology is just the primitive cohomology of both $\mathcal{Z}$ and $\mathcal{O}$. These results were then extended to the orbifold case in [**22**], where we also made a systematic study of the orbifold twistor spaces $\mathcal{Z}$ and gave an orbifold extension of the LeBrun's inversion theorem [**87**]. Finally, the Vanishing Theorem for Betti numbers provided us with the tools to study the geometry and topology of more complicated examples. This study [**33, 34**] used a rational spectral sequence and culminated in discovering that, in dimension 7, all rational homology types not excluded by the Vanishing Theorem do occur and can be constructed explicitly. These examples illustrate the richness of 3-Sasakian geometry in dimension 7. For example, there is an infinite family of 3-Sasakian 7-manifolds that admit metrics of positive sectional curvature, while there is another infinite family that can admit no metrics whose sectional curvature is bounded below by an arbitrary fixed negative number! Later in [**32**] we discovered how to handle the integral spectral sequence giving integral results for our 7-dimensional examples up through the second homology group. We also studied [**31**] the higher dimensional analogue showing that these meet with an entirely different fate.

This review chapter is intended to give the reader a self-contained account of everything we have learned about such spaces to date. We have tried to gather all the known results. In a chapter like this it would be impossible to present every proof so we do quote some theorems just referring to the literature. But we have tried to include as many proofs as possible so that the review is not simply a long dry list of theorems, propositions, and corollaries. When it comes to references we make no claim of completeness, though we have tried to do our best. We apologize for any omissions. At the end we hope to be able to convince our reader that the 3-Sasakian geometry is at least as fascinating as any other "sister" geometry of the fundamental diagram $\diamondsuit(\mathcal{S})$.

Our review is organized as follows: We begin by setting up definitions, notation, and describing elementary properties of Sasakian, Sasakian-Einstein, and 3-Sasakian manifolds in Section 1. Next we discuss fundamentals about the geometry of the associated foliations (arrows in the diagram $\diamondsuit(\mathcal{S})$). We then give a classification of homogeneous geometries in Section 3. Section 4 is all about Betti numbers of Sasakian and 3-Sasakian manifolds while Section 5 is a very brief look at the Killing spinors and G_2 structures. The following section describes the geometry of the 3-Sasakian quotient construction. After this we give a detailed study of "toric" 3-Sasakian manifolds. We conclude with a handful of open problems, questions, and some conjectures followed by an appendix on fundamental properties of orbifolds.

Acknowledgments: The authors would like thank Ben Mann who is a friend and has been a collaborator on much of our work. We also thank our other collaborators Simon Salamon and Elmer Rees. We thank Roger Bielawski, Alex Buium, Claude LeBrun, Liviu Ornea, and Uwe Semmelmann for discussions and valuable comments. Last, but not least, the second named author would like to than Max-Planck-Institute für Mathematik in Bonn for support and hospitality. This review was written during his stay in Bonn.

2. Definitions and Basic Properties

In this section we introduce notation, definitions, and discuss some elementary properties of Sasakian, Sasakian-Einstein, and 3-Sasakian manifolds. Traditionally Sasakian structures were defined via contact structures by adding a Riemannian metric with some additional conditions. We take a simpler and more geometric approach that uses the holonomy reduction of the associated metric cone.

2.1. Sasakian Manifolds.

DEFINITION 2.1.1. *Let $(\mathcal{S}, g)$ be a Riemannian manifold of real dimension m. We say that $(\mathcal{S}, g)$ is Sasakian if the holonomy group of the metric cone on $\mathcal{S}$ $(C(\mathcal{S}), \bar{g}) = (\mathbb{R}_+ \times \mathcal{S},\ dr^2 + r^2 g)$ reduces to a subgroup of $U(\frac{m+1}{2})$. In particular, $m = 2n+1, n \geq 1$ and $(C(\mathcal{S}), \bar{g})$ is Kähler.*

The following proposition provides three alternative characterizations of the Sasakian property, the first one, perhaps, most in the spirit of the the original definition of Sasaki [**110**]:

PROPOSITION 2.1.2. *Let $(\mathcal{S}, g)$ be a Riemannian manifold, ∇ the Levi-Civita connection of g, and let $R(X,Y) : \Gamma(T\mathcal{S}) \to \Gamma(T\mathcal{S})$ denote the Riemann curvature tensor of ∇. Then the following conditions are equivalent:*
(i) There exists a Killing vector field ξ of unit length on $\mathcal{S}$ so that the tensor field Φ of type $(1,1)$, defined by $\Phi(X) = \nabla_X \xi$, satisfies the condition

$$(\nabla_X \Phi)(Y) = g(\xi, Y)X - g(X,Y)\xi$$

for any pair of vector fields X and Y on $\mathcal{S}$.
(ii) There exists a Killing vector field ξ of unit length on $\mathcal{S}$ so that the Riemann curvature satisfies the condition

$$R(X,\xi)Y = g(\xi, Y)X - g(X,Y)\xi,$$

for any pair of vector fields X and Y on $\mathcal{S}$.
(iii)$(\mathcal{S}, g)$ is Sasakian.

PROOF. : We outline the proof of the equivalence of (i) and (iii). The equivalence of (i) and (ii) is a simple calculation relating $(\nabla_X \Phi)(Y)$ to $R(X,\xi)Y$ and is left to the reader (see [**134**]).

We first show how (iii) implies (i). Let X, Y be any two vector fields on $\mathcal{S}$ viewed as vector fields on $C(\mathcal{S})$ and $\bar{\nabla}$ be the Levi-Civita connection of $\bar{g}$. Then we have the following warped product formulas for the cone metric connection [**99**] pg. 206:

$$(2.1)\qquad \bar{\nabla}_{\partial_r}\partial_r = 0, \qquad \bar{\nabla}_{\partial_r} X = \bar{\nabla}_X \partial_r = \frac{1}{r} X, \qquad \bar{\nabla}_X Y = \nabla_X Y - r g(X,Y)\partial_r.$$

Since the holonomy group of the cone $(C(\mathcal{S}), \bar{g})$ reduces to a subgroup of $U(\frac{m+1}{2})$ there is a parallel complex structure I on $C(\mathcal{S})$, i.e., I commutes with $\bar{\nabla}$. We can identify $\mathcal{S}$ with $\mathcal{S} \times \{1\} \subset C(\mathcal{S})$ and define

$$(2.2)\qquad \xi = I(\partial_r), \qquad \eta(Y) = g(\xi, Y), \qquad \Phi(Y) = \nabla_Y \xi$$

for any vector field $Y \in \Gamma(T\mathcal{S})$. It is then a simple calculation to show that ξ is actually a unit Killing vector field on $\mathcal{S}$ and it satisfies the curvature condition in (i). Clearly, ξ is unit by definition and we have

$$g(\nabla_Y \xi, X) = \bar{g}(\bar{\nabla}_Y \xi + g(\xi, Y)\partial_r, X) = \bar{g}(\bar{\nabla}_Y I(\partial_r), X) =$$
$$= \bar{g}(I(\bar{\nabla}_Y \partial_r), X) = \bar{g}(I(Y), X)$$

which is skew-symmetric in X and Y. The second condition follows from $\bar{\nabla} I = 0$, definition of $\Phi(Y) = \nabla_Y(I\partial_r)$, and the formulas 2.1. Conversely, we can construct a Kähler structure on $C(\mathcal{S})$ as follows: Let $\Psi = r\partial_r$ denote the Euler field on $C(\mathcal{S})$ and define smooth section of End $TC(\mathcal{S})$ by the formula

$$IY = \Phi(Y) - \eta(Y)\Psi, \qquad I\Psi = \xi,$$

where $\eta(Y) = g(\xi, Y)$ is the dual 1-form of ξ. It is easy to see that I is an almost complex structure on $C(\mathcal{S})$ and the metric $\bar{g}$ is Hermitian. To show that $C(\mathcal{S})$ is Kähler it is enough to show that $\bar{\nabla}I = 0$. This is done by a direct calculation using the definition of I and equations 2.2. □

The above discussion shows that there is a natural splitting of the tangent bundle $TC(\mathcal{S})$ as $TC(\mathcal{S}) = \mathcal{L}_\Psi \oplus \mathcal{L}_\xi \oplus \mathcal{H}$ where $\mathcal{L}_X$ denotes the trivial line bundle generated by the nowhere vanishing vector field X, and $\mathcal{H}$ is a complement with respect to the metric $\bar{g}$. It follows immediately that the frame bundle of any Sasakian manifold of dimension $2n+1$ reduces to the group $1 \times U(n)$ [**110**]. It follows that every Sasakian manifold has a canonical Spin^c structure [**97**].

In view of the above proposition the triple $\{\xi, \eta, \Phi\}$ is called a *Sasakian structure* on $(\mathcal{S}, g)$, the Killing vector field ξ and the 1-form η are called the *characteristic vector field* and the *characteristic 1-form* of the Sasakian structure, respectively. We next give some elementary properties of Sasakian structures. All of them follow as an immediate consequence of the definition and Proposition 2.1.2.

PROPOSITION 2.1.3. *Let $(\mathcal{S}, g)$ be a Sasakian manifold, $\{\xi, \eta, \Phi\}$ its Sasakian structure, and X and Y any pair of vector fields on $\mathcal{S}$. Furthermore, let*

$$N_\Phi(Y, X) = [\Phi Y, \Phi X] + \Phi^2[Y, X] - \Phi[Y, \Phi X] - \Phi[\Phi Y, X]$$

be the Nijenhuis torsion tensor of Φ. Then

(i) $$\Phi\circ\Phi(Y) = -Y + \eta(Y)\xi,$$

(ii) $$\Phi\xi = 0, \qquad \eta(\Phi Y) = 0,$$

(iii) $$g(X, \Phi Y) + g(\Phi X, Y) = 0, \qquad g(\Phi Y, \Phi X) = g(Y, X) - \eta(Y)\eta(X),$$

(iv) $$d\eta(Y, X) = 2g(\Phi Y, X), \qquad N_\Phi(Y, X) = d\eta(Y, X) \otimes \xi.$$

A Sasakian manifold is not necessarily Einstein. As a simple consequence of the relation between Ricci curvature of $\mathcal{S}$ and its metric cone $C(\mathcal{S})$, the Einstein condition can be expressed in terms of Ricci-flatness of the cone metric $\bar{g}$ and we get

PROPOSITION 2.1.4. *Let $(\mathcal{S}, g)$ be a Sasakian manifold of dimension $2n+1$. Then the metric g is Einstein if and only if the cone metric $\bar{g}$ is Ricci-flat, i.e., $(C(\mathcal{S}), \bar{g})$ is Kähler Ricci-flat (Calabi-Yau). In particular, it follows that the restricted holonomy group $\mathrm{Hol}^0(\bar{g}) \subset SU(n+1)$ and that the Einstein constant of g is positive and equals $2n$.*

An immediate consequence of the this proposition and Myers' Theorem is:

COROLLARY 2.1.5. *A complete Sasakian-Einstein manifold is compact with diameter less than or equal to π and with finite fundamental group.*

Now $\mathrm{Hol}^0(\bar{g})$ is the normal subgroup of the full holonomy group $\mathrm{Hol}(\bar{g})$ that is the component connected to the identity. There is a canonical epimorphism

$$\pi_1(\mathcal{S}) = \pi_1(C(\mathcal{S})) \longrightarrow \mathrm{Hol}(\bar{g})/\mathrm{Hol}^0(\bar{g}),$$

so if $\mathcal{S}$ is simply-connected its structure group reduces to $1 \times SU(n)$ and it will admit a spin structure. We have

COROLLARY 2.1.6. *Let $\mathcal{S}$ be a Sasakian-Einstein manifold such that the full holonomy group of the cone metric $\mathrm{Hol}(\bar{g})$ is contained in $SU(m+1)$. Then $\mathcal{S}$ admits a spin structure. In particular, every simply-connected Sasakian-Einstein manifold admits a spin structure.*

We give some examples that illustrate the complications in the presence of fundamental group. The hypothesis of this corollary is not necessary as the second example shows.

EXAMPLE 2.1. The real projective space $\mathcal{S} = \mathbb{RP}^{2n+1}$ with its canonical metric is Sasakian-Einstein, and the cone $C(\mathcal{S}) = (\mathbb{C}^{n+1} - \{0\})/\mathbb{Z}_2$ with the usual antipodal identification. We have $\mathrm{Hol}(\bar{g}) \simeq \pi_1(\mathcal{S}) \simeq \mathbb{Z}_2$. When n is odd the antipodal map τ is in $SU(n+1)$, so $\mathcal{S} = \mathbb{RP}^{2n+1}$ admits a spin structure. But when n is even the antipodal map τ does not lie in $SU(n+1)$, which obstructs a further reduction of the structure group. In this case it is well-known that $\mathcal{S} = \mathbb{RP}^{2n+1}$ does not admit a spin structure. In fact the generator of $\mathrm{Hol}(\bar{g}) \simeq \mathbb{Z}_2$ is the obstruction. There are many other similar examples. An example that shows that the hypothesis in Corollary 2.1.6 is not necessary is the following: Consider the lens space $L(p;,q_1,\cdots,q_n) \simeq S^{2n+1}/\mathbb{Z}_p$ where the q_i's are relatively prime to p. The action on $\mathbb{C}^{n+1} - \{0\}$ is generated by $(z_0, z_1, \cdots, z_n) \mapsto (\eta z_0, \eta^{q_1} z_1, \cdots, \eta^{q_n} z_n)$ where η is a primitive pth root of unity. It is known [**44**] that if p is odd, $L(p;q_1,\cdots,q_n)$ admits a spin structure. However, if $\sum_i q_i + 1$ is not divisible by p, the holonomy group $\mathrm{Hol}(\bar{g}) \simeq \mathbb{Z}_p$ does not lie in $SU(n+1)$.

Let $\mathcal{S}$ be a Sasakian manifold, suppose that the characteristic vector field ξ is complete. Since ξ has unit norm, it defines a 1-dimensional foliation $\mathcal{F}$ on $\mathcal{S}$. We shall be interested in the case when all the leaves of $\mathcal{F}$ are compact.

DEFINITION 2.1.7. *Let $(\mathcal{S}, g)$ be be a compact Sasakian manifold and let $\mathcal{F}$ be the 1-dimensional foliation defined by ξ. We say that $\mathcal{S}$ is quasi-regular if the foliation $\mathcal{F}$ is quasi-regular, i.e., each point $p \in \mathcal{S}$ has a cubical neighborhood U such that any leaf $\mathcal{L}$ of $\mathcal{F}$ intersects a transversal through p at most a finite number of times $N(p)$. Furthermore, $\mathcal{S}$ is called regular if $N(p) = 1$ for all $p \in \mathcal{S}$.*

It is known that the quasi-regular property is equivalent to the condition that all the leaves of the foliation are compact. In the regular case, the foliation $\mathcal{F}$ is simple, and defines a global submersion. In fact it defines a principal S^1 bundle over its space of leaves. In the quasi-regular case it is well-known [**122, 94**] that ξ generates a locally free circle action on $\mathcal{S}$, and that the space of leaves is a compact orbifold (See the appendix for a brief review of orbifolds and their relation to foliations, in particular see Theorem A.1.2). We shall denote the space of leaves of the foliation $\mathcal{F}$ on $\mathcal{S}$ by $\mathcal{Z}$. Then the natural projection $\pi : \mathcal{S} \longrightarrow \mathcal{Z}$ is a Siefert fibration. It is an example of what we call a principal V-bundle over $\mathcal{Z}$. In Section 2 we shall study this foliation in detail.

2.2. 3-Sasakian Spaces. Using all the definitions of the previous section we now describe a more specialized situation. Again, this can be done by an additional holonomy reduction requirement.

DEFINITION 2.2.1. *Let $(\mathcal{S}, g)$ be a Riemannian manifold of real dimension m. We say that $(\mathcal{S}, g)$ is 3-Sasakian if the holonomy group of the metric cone on $\mathcal{S}$ $(C(\mathcal{S}), \bar{g}) = (\mathbb{R}_+ \times \mathcal{S},\ dr^2 + r^2 g)$ reduces to a subgroup of $Sp(\frac{m+1}{4})$. In particular, $m = 4n+3, n \geq 1$ and $(C(\mathcal{S}), \bar{g})$ is hyperkähler.*

Since $C(\mathcal{S})$ is hyperkähler it has a hypercomplex structure $\{I^1, I^2, I^3\}$. We can define $\xi^a = I^a(\partial_r)$ for each $a = 1,2,3$. Then using the well-known properties of a hypercomplex structure together with Proposition 2.1.2 gives:

PROPOSITION 2.2.2. *Let $(\mathcal{S}, g)$ be a Riemannian manifold and let ∇ denote the Levi-Civita connection of g. Then $\mathcal{S}$ is 3-Sasakian if and only if it admits three characteristic vector fields $\{\xi^1, \xi^2, \xi^3\}$ (that is, satisfying any of the corresponding conditions in Proposition 2.1.2) such that $g(\xi^a, \xi^b) = \delta_{ab}$ and $[\xi^a, \xi^b] = 2\epsilon_{abc}\xi^c$.*

REMARK 2.1. By using Proposition 2.2.2 we can easily generalize the definition of a 3-Sasakian structure to orbifolds. A Riemannian orbifold $\mathcal{S}$ is a *3-Sasakian orbifold* if it admits three characteristic vector fields satisfying the conditions of Proposition 2.2.2, and if the action of the local uniformizing groups leaves the characteristic vector fields $\{\xi^1, \xi^2, \xi^3\}$ invariant.

The triple $\{\xi^1, \xi^2, \xi^3\}$ defines $\eta^a(Y) = g(\xi^a, Y)$ and $\Phi^a(Y) = \nabla_Y \xi^a$ for each $a = 1,2,3$. We call $\{\xi^a, \eta^a, \Phi^a\}_{a=1,2,3}$ the *3-Sasakian structure* on $(\mathcal{S}, g)$. The hyperkähler geometry of the cone $C(\mathcal{S})$ gives $\mathcal{S}$ a "quaternionic structure" reflected by the composition laws of the (1,1) tensors Φ^a. The following proposition describes additional properties of $\{\xi^a, \eta^a, \Phi^a\}$ not listed in Proposition 2.1.3(i-iv).

PROPOSITION 2.2.3. *Let $(\mathcal{S}, g)$ be a 3-Sasakian manifold and let $\{\xi^a, \eta^a, \Phi^a\}_{a=1,2,3}$ be its 3-Sasakian structure. Then*

$$
\begin{aligned}
\eta^a(\xi^b) &= \delta^{ab}, \\
\Phi^a \xi^b &= -\epsilon^{abc}\xi^c, \\
\Phi^a \circ \Phi^b - \xi^a \otimes \eta^b &= -\epsilon^{abc}\Phi^c - \delta^{ab}\mathrm{id}.
\end{aligned}
$$

REMARK 2.2. For any $\tau = (\tau_1, \tau_2, \tau_3) \in \mathbb{R}^3$ such that $\tau_1^2 + \tau_3^2 + \tau_3^2 = 1$ the vector field $\xi(\tau) = \tau_1\xi^1 + \tau_2\xi^2 + \tau_3\xi^3$ has the Sasakian property. Therefore a 3-Sasakian manifold has not just 3 but an S^2 worth of Sasakian structures. This is in complete analogy with the hyperkähler case, and perhaps the name *hypersasakian* would have been more consistent. However, most of the existing literature uses the name *Sasakian 3-structure* or, as we do, *3-Sasakian structure.* Thus we have decided to stay with the latter.

Since a hyperkähler manifold is Ricci-flat, Proposition 2.1.4 and its corollary immediately imply:

COROLLARY 2.2.4. *Every 3-Sasakian manifold $(\mathcal{S}, g)$ of dimension $4n+3$ is Einstein with Einstein constant $\lambda = 2(2n+1)$. Moreover, if $\mathcal{S}$ is complete it is compact with finite fundamental group.*

The important result that every 3-Sasakian manifold is Einstein was first obtained by Kashiwada [**69**] using tensorial methods. One can also easily verify the structure group of any 3-Sasakian manifold is reducible to $Sp(n) \times \mathbb{I}_3$, where $\mathbb{I}_3$ denotes the three by three identity matrix [**83**]. It follows [**26**] that

COROLLARY 2.2.5. *Every 3-Sasakian manifold $(\mathcal{S}, g)$ is spin.*

If $(\mathcal{S}, g)$ is compact the characteristic vector fields $\{\xi^1, \xi^2, \xi^3\}$ are complete and define a 3-dimensional foliation $\mathcal{F}_3$ on $\mathcal{S}$. The leaves of this foliation are necessarily compact as $\{\xi^1, \xi^2, \xi^3\}$ defines a locally free $Sp(1)$ action on $\mathcal{S}$. Hence, the foliation $\mathcal{F}_3$ is automatically quasi-regular and the space of leaves is a compact orbifold. We shall denote it by $\mathcal{O}$.

DEFINITION 2.2.6. *Let $(\mathcal{S}, g)$ be be a compact 3-Sasakian manifold of dimension $4n+3$, $n \geq 1$, and let $\mathcal{F}_3$ be the 3-dimensional foliation defined by $\{\xi^1, \xi^2, \xi^3\}$. We say that $\mathcal{S}$ is regular if $\mathcal{F}_3$ is regular.*

REMARK 2.3. When $\dim(\mathcal{S}) = 3$ the leaf space of the foliation $\mathcal{F}_3$ is a single point so it makes no sense to talk about the regularity of $\mathcal{F}_3$. In this case we will say that $\mathcal{S}$ is *regular* if the foliation $\mathcal{F}_1$ defined by the characteristic vector field ξ^1 is regular.

For any $\tau \in S^2$ we can consider again the characteristic vector field $\xi(\tau)$ associated with the direction τ. This vector field defines a 1-dimensional foliation $\mathcal{F}_\tau \subset \mathcal{F}_3 \subset \mathcal{S}$. This foliation has compact leaves and defines a locally free circle action $U(1)_\tau \subset Sp(1)$ on $\mathcal{S}$. In the next section we will describe the geometry of these foliations. Here, we simply conclude by the following observation concerning regularity properties of the foliations $\mathcal{F}_\tau \subset \mathcal{F}_3$ [**119**]:

PROPOSITION 2.2.7. *Let $(\mathcal{S}, g)$ be a compact 3-Sasakian manifold. If $\mathcal{F}_3$ is regular then $\mathcal{F}_\tau$ is regular for all $\tau \in S^2$. Conversely, if $\mathcal{F}_\tau$ is regular for some $\tau = \tau_0 \in S^2$ then it is regular for all τ and, hence, $\mathcal{F}_3$ is regular. Furthermore, if $\mathcal{F}_3$ is regular then either all the leaves are diffeomorphic to $SO(3)$ or all the leaves are diffeomorphic to S^3.*

Actually in the regular case it follows from a deeper result of Simon Salamon [**106**] that all leaves are diffeomorphic to S^3 in precisely one case, namely when $\mathcal{S} = S^{4n+3}$. (See the next section for further discussion.)

REMARK 2.4. Note that every Sasakian-Einstein 3-manifold must also have a 3-Sasakian structure. This is because in dimension four Ricci-flat and Kähler is equivalent to hyperkähler. Every compact 3-Sasakian 3-manifold, by Proposition 2.1.2(iii), must be a space of constant curvature 1. Hence, $\mathcal{S}$ is covered by a unit round 3-sphere and , in fact, it is always the homogeneous spherical space form S^3/Γ, where Γ is a discrete subgroup of $Sp(1)$ [**111**]. The homogeneous spherical space forms in dimension 3 are well-known. They are $Sp(1)/\Gamma$ where Γ is one of the finite subgroups of $Sp(1)$, namely: $\Gamma = \mathbb{Z}_m$ the cyclic group of order m, $\Gamma = \mathbb{D}^*_m$ a binary dihedral group with m is an integer greater than 2, $\Gamma = \mathbb{T}^*$ the binary tetrahedral group, $\Gamma = \mathbb{O}^*$ the binary octahedral group, $\Gamma = \mathbb{I}^*$ the binary icosahedral group. The only regular 3-Sasakian manifolds in dimension 3 are S^3 and $SO(3)$. More generally, the diffeomorphism classification of compact Sasakian 3-manifolds was recently completed by Geiges [**55**]. In addition to S^3/Γ one gets compact quotients of the double cover of $PSL_2(\mathbb{R})$ and the 3-dimensional Heisenberg group.

REMARK 2.5. A Sasakian-Einstein structure on a 3-Sasakian manifold does not have to be a part of the 3-Sasakian structure. The simplest example when this is the case is the lens space $\mathbb{Z}_k \backslash S^3$. Consider the unit 3-sphere $S^3 \simeq Sp(1)$ as the unit quaternion $\sigma \in \mathbb{H}$. Such a sphere has two 3-Sasakian structures generated by the left and the right multiplication. Consider the homogeneous space $\mathbb{Z}_k \backslash S^3$, where the $\mathbb{Z}_k$-action is given by the multiplication from the left by $\rho \in Sp(1)$, $\rho^k = 1$. The quotient still has the "right" 3-Sasakian structure. But it also has a "left" Sasakian structure (the centralizer of $\mathbb{Z}_k$ in $Sp(1)$ is an S^1 and it acts on the coset from the left). This left Sasakian structure is actually regular while none of the Sasakian structures of the right 3-Sasakian structure can be regular unless $k = 1, 2$ [**120**].

3. The Fundamental Foliations

In this section we discuss the foliations associated with Sasakian and 3-Sasakian manifolds and describe their consequences.

3.1. The Sasakian Foliation. As mentioned in Section 2.1 a Sasakian manifold defines a Riemannian foliation of dimension 1. Using the basic properties described in Propositions 2.1.2 and 2.1.3. we have

PROPOSITION 3.1.1. *Let $(\mathcal{S}, g)$ be a Sasakian manifold, and let $\mathcal{F}$ denote the foliation defined by the characteristic vector field ξ. Then*
(i) The metric g is bundle-like.
(ii) The leaves of $\mathcal{F}$ are totally geodesic.
(iii) The complementary vector bundle $\mathcal{H}$ to the trivial line subbundle of $T\mathcal{S}$ generated by ξ defines a strictly pseudoconvex CR structure on $\mathcal{S}$ with vanishing Webster torsion.

In order to have a well behaved space of leaves we need a further assumption on the foliation. We have a generalization of the well-known Boothby-Wang fibration Theorem:

THEOREM 3.1.2. *Let $\mathcal{S}$ be a complete quasi-regular Sasakian manifold. Then*
(i) The leaves of $\mathcal{F}$ are all diffeomorphic to circles with cyclic leaf holonomy groups.
(ii) The space of leaves $\mathcal{Z} = \mathcal{S}/\mathcal{F}$ has the structure of a Kähler orbifold.
Suppose further that $(\mathcal{S}, g)$ is Sasakian-Einstein. Then
(iii) The leaf space $\mathcal{Z}$ is a simply-connected normal projective algebraic variety with a Kähler-Einstein metric h of positive scalar curvature $4n(n+1)$ in such a way that $\pi : (\mathcal{S}, g) \to (\mathcal{Z}, h)$ is an orbifold Riemannian submersion.
(iv) $\mathcal{Z}$ has the structure of a $\mathbb{Q}$-factorial Fano variety. Hence, it is uniruled with Kodaira dimension $\kappa(\mathcal{Z}) = -\infty$.

PROOF. Parts (i) and (ii) are straightforward generalizations of the Boothby-Wang fibration in the Sasakian setting [**19, 134**] to the quasi-regular case. The point is that the CR structure on $\mathcal{S}$ pushes down to give a complex structure on $\mathcal{Z}$ and the Sasakian nature of $\mathcal{S}$ guarantees that the complex structure will be Kähler. That $\mathcal{Z}$ is projective algebraic is a consequence of Baily's version [**9**] of the Kodaira Embedding Theorem. Simple connectivity follows essentially from Kobayashi's argument in the smooth case by using the singular version of the Riemann-Roch Theorem due to Baum, Fulton, and Macpherson. The uniruledness is a result of Miyaoka and Mori [**93**]. For details we refer the reader to [**22, 23**]. □

Let us recall that a complex variety X is $\mathbb{Q}$-factorial variety if for every Weil divisor D there exists a positive integer m such that mD is a Cartier divisor. The smallest such integer $m(D)$ is called the order of D. If X is compact the least common multiple taken over all Weil divisors on X is the order of X. Now on a compact complex orbifold Weil divisors coincide with Baily divisors **[22]** and Baily divisors correspond to line V-bundles. On X we have the group $\mathrm{Pic}^{orb}(X)$ of holomorphic line V-bundles on X and its subgroup $\mathrm{Pic}(X)$ of holomorphic line bundles or absolute line V-bundles in Baily's terminology **[8, 9]**. It is not difficult to prove **[23]**

PROPOSITION 3.1.3. *Let $\mathcal{S}$ be a complete Sasakian-Einstein manifold, and let $\mathcal{Z}$ be the space of leaves of the foliation $\mathcal{F}$ on $\mathcal{S}$. Then Pic($\mathcal{Z}$) is free, and a subgroup of $Pic^{orb}(\mathcal{Z})$ which satisfies*
(i) $Pic^{orb}(\mathcal{Z}) \otimes \mathbb{Q} \simeq Pic(\mathcal{Z}) \otimes \mathbb{Q}$.
(ii) If $\pi_1^{orb}(\mathcal{Z}) \simeq 0$, then $Pic^{orb}(\mathcal{Z}) \simeq Pic(\mathcal{Z})$.

For an inversion theorem to Theorem 3.1.2 in the Sasakian-Einstein case and the construction of many nontrivial examples the reader is referred to **[23]** and **[11]** in the regular case. In particular, in dimension 5 we have

THEOREM 3.1.4. **[11]** *Let $\mathcal{S}$ be a simply-connected regular Sasakian-Einstein manifold of dimension 5. Then $\mathcal{S}$ is one of the following: S^5, the Stiefel manifold $V_2(\mathbb{R}^4)$ of 2-frames in $\mathbb{R}^4$, or the total space S_k of the S^1 bundles $S_k \to P_k$ for $3 \leq k \leq 8$ where P_k is a Del Pezzo surface with a Kähler-Einstein metric* **[124]**. *It is known that S_k is diffeomorphic to the k-fold connected sum $S^2 \times S^3 \# \cdots \# S^2 \times S^3$.*

3.2. The One Dimensional 3-Sasakian Foliation. Fixing a Sasakian structure, say (ξ^1, Φ^1, η^1) in the 3-Sasakian structure, we notice that subbundle $\mathcal{H} = \ker \eta^1$ of $T\mathcal{S}$ together with $I = -\Phi^1|\mathcal{H}$ define the CR structure on $\mathcal{S}$. Actually a 3-Sasakian structure gives a special kind of CR structure, namely, a CR structure with a compatible holomorphic contact structure. Notice that the complex valued one form on $\mathcal{S}$ defined by $\eta^+ = \eta^2 + i\eta^3$ is type $(1,0)$ on $\mathcal{S}$. Moreover, one checks that η^+ is holomorphic with respect to the CR structure I. Although the 1-form η^+ is not invariant under the circle action generated by ξ^1, the trivial complex line bundle L^+ generated by η^+ is invariant. Thus, the complex line bundle L^+ pushes down to a nontrivial complex V-line bundle $\mathcal{L}$ on $\mathcal{Z}$. Let V denote the one dimensional complex vector space generated by L^+. Writing the circle action as exp $(i\phi\xi^1)$ shows that V is the representation with character $e^{-2i\phi}$, and since $\mathcal{S}$ is a principal S^1 V bundle over $\mathcal{Z}$, the twisted product $\mathcal{L} \simeq \mathcal{S} \times_{S^1} V$ is a complex line V-bundle on $\mathcal{Z}$. Now we can define a map of V-bundles $\theta : T^{(1,0)}\mathcal{Z} \longrightarrow \mathcal{L}$ by

$$\theta(X) = \eta^+(\hat{X}),$$

where $\hat{X}$ is the horizontal lift of the vector field X on $\mathcal{Z}$. Notice that $\theta(X)$ is not a function on $\mathcal{Z}$ but a section of $\mathcal{L}$. Now a straightforward computation shows that $\eta^+ \wedge (d\eta^+)^n$ is a nowhere vanishing section of $\Lambda^{(2n+1,0)}\mathcal{H}$ on $\mathcal{S}$, and thus $\theta \wedge (d\theta)^n$ is a nowhere vanishing section of $K \otimes \mathcal{L}^{n+1}$, where K is the canonical V-line bundle (see the Appendix) on $\mathcal{Z}$. Hence, in $\mathrm{Pic}^{orb}(\mathcal{Z})$ we have the relation $\mathcal{L}^{n+1} \otimes K = 1$. So the contact line V-bundle is $\mathcal{L} \simeq K^{-\frac{1}{n+1}}$ in $\mathrm{Pic}^{orb}(\mathcal{Z})$. Alternatively, the subbundle $\ker \theta$ is a holomorphic subbundle of $T^{(1,0)}\mathcal{Z}$ which is maximally non-integrable. This defines the complex contact structure on $\mathcal{Z}$. Of course, this construction depends on a choice of direction $\tau \in S^2$ in the 2-sphere of complex structures. However, the transitive action of $Sp(1)$ on S^2 guarantees that this structure is unique up to isomorphism as complex contact manifolds. We have [**25, 22**]:

THEOREM 3.2.1. *Let $\mathcal{S}$ be a complete 3-Sasakian manifold, choose a direction $\tau \in S^2$, and let $\mathcal{Z}_\tau$ denote the space of leaves of the corresponding foliation $\mathcal{F}_\tau$. Then $\mathcal{Z}_\tau$ is a compact $\mathbb{Q}$-factorial contact Fano variety with a Kähler-Einstein metric h of scalar curvature $8(2n+1)(n+1)$ such that the natural projection $\pi : \mathcal{S} \longrightarrow \mathcal{Z}_\tau$ is an orbifold Riemannian submersion with respect to the Riemannian metrics g on $\mathcal{S}$ and h on $\mathcal{Z}_\tau$.*

We call the space $\mathcal{Z}_\tau$, usually just written $\mathcal{Z}$, the *twistor space* associated to $\mathcal{S}$. Actually there is another object that could merit the name the twistor space of $\mathcal{S}$, namely the trivial 2-sphere bundle $S^2 \times \mathcal{S}$ with the structure induced from the twistor space $S^2 \times C(\mathcal{S})$ of the hyperkähler cone.

An important property of the twistor space in the case of quaternionic Kähler manifolds is that it is ruled by rational curves. The same is true in our case as long as one allows for singularities. We have

PROPOSITION 3.2.2. *$\mathcal{Z}$ is ruled by a real family of rational curves C with possible singularities on the singular locus of $\mathcal{Z}$. All the curves C are simply-connected, but $\pi_1^{orb}(C)$ can be a non-trivial cyclic group.*

For any line V-bundle $\mathcal{L}$ we let $\hat{\mathcal{L}}$ denote $\mathcal{L}$ minus its zero section.

PROPOSITION 3.2.3. *Let $\mathcal{Z}$ be the twistor space of a 3-Sasakian manifold $\mathcal{S}$ of dimension $4n+3$, and assume that $\pi_1^{orb}(\mathcal{Z}) = 0$. If the contact line V-bundle $\mathcal{L}$ (or equivalently its dual $\mathcal{L}^{-1}$) has a root in $Pic^{orb}(\mathcal{Z})$, then it must be a square root, namely $\mathcal{L}^{\frac{1}{2}}$. Moreover, in this case if both $\hat{\mathcal{L}}$ and $\hat{\mathcal{L}}^{\frac{1}{2}}$ are proper in the sense of Kawasaki, then we must have $\mathcal{Z} = \mathbb{P}^{2n+1}$. In particular, this holds if the total space of $\hat{\mathcal{L}}$ is smooth.*

PROOF. By Proposition 3.1.3 $\mathrm{Pic}^{orb}(\mathcal{Z})$ is torsion free. So the proof in [**22**] now goes through. By Proposition 3.2.2 $\mathcal{Z}$ is ruled by rational curves C which on the singular locus take the form $\Gamma\backslash\mathbb{P}^1$. Now the restriction $\mathcal{L}^{-1}|C$ is $\mathcal{O}(-2)$ which is a V-bundle if C is singular. In either case it has only a square root namely the tautological V-bundle $\mathcal{O}(-1)$. Since these curves C cover $\mathcal{Z}$ this proves the first statement. The second statement follows from a modification of an argument due to Kobayashi and Ochiai [**75**] and used by Salamon [**106**]. The main point is that since $\hat{\mathcal{L}}, \hat{\mathcal{L}}^{\frac{1}{2}}$ are proper and it follows that we can apply Kawasaki's Riemann-Roch Theorem [**70**] together with the Kodaira-Baily Vanishing Theorem [**9**] to arbitrary powers of the line V-bundle $\mathcal{L}^{\frac{1}{2}}$ to give $(n+1)(2n+3)$ infinitesimal automorphisms of the complex contact structure on $\mathcal{Z}$. Since $\pi_1^{orb}(\mathcal{Z}) = 0$, these integrate to global automorphisms on $\mathcal{Z}$ and the result follows. See the Appendix and [**22**] for details. □

REMARK 3.1. : There is an error in the statement of Proposition 4.3 of [**22**]. The error is in leaving out the assumptions that $\pi_1^{orb}(\mathcal{Z})$ is trivial and that the contact line bundle is proper. Example 3.1 below shows that the conclusion in Proposition 3.2.3 does not necessarily hold if the hypothesis $\pi_1^{orb}(\mathcal{Z}) = 0$ is omitted. Likewise, Example 3.2 below gives a counterexample when the condition that $\mathcal{L}$ be proper is omitted.

EXAMPLE 3.1. Consider the 3-Sasakian lens space $L(p;q) = \mathbb{Z}_p\backslash S^7$ constructed as follows: S^7 is the unit sphere in the quaternionic vector space $\mathbb{H}^2$ with quaternionic coordinates u_1, u_2. The action of $\mathbb{Z}_p$ is the left action defined by $(u_1,u_2) \mapsto (\tau u_1, \tau^q u_2)$, where $\tau^p = 1$ and p and q are relatively prime positive integers. If $p = 2m$ for some integer m then $-\mathrm{id}$ is an element of $\mathbb{Z}_{2m}$, so the 3-Sasakian manifolds $L(2m;q)$ and $L(m;q)$ both have the same twistor space, namely $\mathcal{Z} = \mathbb{Z}_m\backslash\mathbb{P}^3$, and $\pi_1^{orb}(\mathcal{Z}) \simeq \mathbb{Z}_m$. There are clearly many similar examples in all dimensions equal to 3 mod 4.

EXAMPLE 3.2. Consider the 3-Sasakian 7 manifolds $\mathcal{S}(p_1,p_2,p_3)$ described in Section 8.4 below, where the p_i's are pairwise relatively prime, and precisely one of the p_i's is even, say p_1. $\mathcal{S}(p_1,p_2,p_3)$ is simply-connected and its twistor space $\mathcal{Z}(p_1,p_2,p_3)$ has $\pi_1^{orb}(\mathcal{Z}(p_1,p_2,p_3)) = 0$. Now there is a $\mathbb{Z}_2$ acting on $\mathcal{S}(p_1,p_2,p_3)$, but not freely, which acts as the identity on $\mathcal{Z}(p_1,p_2,p_3)$. Thus, $\mathbb{Z}_2\backslash\mathcal{S}(p_1,p_2,p_3)$ has $\mathcal{Z}(p_1,p_2,p_3)$ as its twistor space, and as a V-bundle $\mathbb{Z}_2\backslash\mathcal{S}(p_1,p_2,p_3) \longrightarrow \mathcal{Z}(p_1,p_2,p_3)$ is not proper in the sense of Kawasaki [**71**]. Thus, the V-bundle $\hat{\mathcal{L}}$ is not proper, and Kawasaki's Riemann-Roch theorem [**70**] cannot be applied.

We now wish to formulate a converse to Theorem 3.2.1.

DEFINITION 3.2.4. *A complete $\mathbb{Q}$-factorial Fano contact variety $\mathcal{Z}$ is said to be good if the total space of the principal circle bundle $\mathcal{S}$ associated with the contact V-line bundle $\mathcal{L}$ is a smooth compact manifold.*

Thus, for good $\mathbb{Q}$-factorial Fano contact varieties, $\mathcal{S}$ desingularizes $\mathcal{Z}$. As discussed in the Appendix this happens precisely when all the leaf holonomy groups inject into the group S^1 of the bundle. Notice also that in this case $\mathcal{S}$ is necessarily compact. We now are ready for:

THEOREM 3.2.5. *A good $\mathbb{Q}$-factorial Fano contact variety $\mathcal{Z}$ is the twistor space associated to a compact 3-Sasakian manifold if and only if it admits a compatible Kähler-Einstein metric h.*

PROOF. Let $\mathcal{Z}$ be a good $\mathbb{Q}$-factorial Fano contact variety with a compatible Kähler-Einstein metric h. Choose the scale of h so that the scalar curvature is $8(2n+1)(n+1)$. Let $\pi : \mathcal{S} \to \mathcal{Z}$ denote the principal orbifold circle bundle associated to $\mathcal{L}$. It is a smooth compact submanifold embedded in the dual of the contact V-line bundle $\mathcal{L}^{-1}$. The Kähler-Einstein metric h has Ricci form $\rho = 4(n+1)\omega$, where ω is the Kähler form on $\mathcal{Z}$, and ρ represents the first Chern class of K^{-1}. Let η^1 be the connection in $\pi : \mathcal{S} \to \mathcal{Z}$ with curvature form $2\pi^*\omega$. Then the Riemannian metric $g_{\mathcal{S}}$ on $\mathcal{S}$ can be defined by $g_{\mathcal{S}} = \pi^*h + (\eta^1)^2$. It is standard (see the proof in Example 1 of Section 4.2 in [**11**]) that $g_{\mathcal{S}}$ is Sasakian-Einstein. As in Proposition 2.2.4 of [**115**] the V-bundle $\mathcal{L}\otimes\Lambda^{(1,0)}\mathcal{Z}$ has a section θ such that the Kähler-Einstein metric h decomposes as $h = |\theta|^2 + h_D$, where h_D is a metric in the V-bundle D. Let us write $\pi^*\theta = \eta^+$. Since $\mathcal{S}$ is a circle bundle in $\mathcal{L}^{-1}$, the contact bundle $\mathcal{L}$ trivializes when pulled back to $\mathcal{S}$. This together with the condition that $\theta \wedge (d\theta)^n$ is nowhere vanishing on $\mathcal{Z}$ implies that η^+ is a nowhere vanishing complex valued 1-form on $\mathcal{S}$. So the metric $g_{\mathcal{S}}$ on $\mathcal{S}$ can be written as

$$g_{\mathcal{S}} = (\eta^1)^2 + |\eta^+|^2 + \pi^*h_D.$$

We claim that this metric is 3-Sasakian. To see this consider the total space M of the dual of the contact V-line bundle minus its 0 section which is $\mathcal{S} \times \mathbb{R}^+$. Put the cone metric $dr^2 + r^2 g$ on M. The natural $\mathbb{C}^*$ action on M induces homotheties of this metric. Now using a standard Weitzenböck argument, LeBrun [**87**] shows that M has a parallel holomorphic symplectic structure and his argument works just as well in our case. Let ϑ denote the pullback of the contact form θ to M which is a holomorphic 1-form on M that is homogeneous of degree 1 with respect to the $\mathbb{C}^*$ action. Thus $\Upsilon = d\vartheta$ is a holomorphic symplectic form on M which is parallel with respect to the Levi-Civita connection of the cone metric. Hence, $(M, dr^2 + r^2 g)$ is hyperkähler. Furthermore, if $\{I^a\}_{a=1}^3$ denote hyperkähler endomorphisms on M, ϑ^2, ϑ^3 are the real and imaginary parts of ϑ, and ϑ^1 is the pullback of η^1 to M, then LeBrun shows that

$$\vartheta^1 I^1 = \vartheta^2 I^2 = \vartheta^3 I^3.$$

It then follows from our previous work [**25**] that g is 3-Sasakian. But by construction $\mathcal{Z}$ is the space of leaves of the foliation generated by ξ^1, so $\mathcal{Z}$ must be the twistor space of the compact 3-Sasakian manifold $\mathcal{S}$. □

3.3. The Three Dimensional 3-Sasakian Foliation. Next we consider the three dimensional foliation $\mathcal{F}_3$ discussed in Section 2.2.

PROPOSITION 3.3.1. *Let $(\mathcal{S}, g)$ be a 3-Sasakian manifold such that the characteristic vector fields ξ^a are complete. Let $\mathcal{F}_3$ denote the the canonical three dimensional foliation on $\mathcal{S}$. Then*
(i) The metric g is bundle-like.
(ii) The leaves of $\mathcal{F}_3$ are totally geodesic spherical space forms $\Gamma\backslash S^3$ of constant curvature one, where $\Gamma \subset Sp(1) = SU(2)$ is a finite subgroup.
(iii) The 3-Sasakian structure on $\mathcal{S}$ restricts to a 3-Sasakian structure on each leaf.
(iv) The generic leaves are either $SU(2)$ or $SO(3)$.

PROOF. The proof of (i), (ii), and (iii) follow from the basic relations for 3-Sasakian manifolds as in Proposition 2.1.1. To prove (iv) we notice that the foliation $\mathcal{F}_3$ is regular restricted to the generic stratum $\mathcal{S}_0$. By (ii) and regularity there is a finite subgroup $\Gamma \subset SU(2)$ such that the leaves of this restricted foliation are all diffeomorphic to $\Gamma\backslash S^3$, which is 3-Sasakian by (iii). Now the regularity of $\mathcal{F}_3$ on $\mathcal{S}_0$ implies that its leaves must all be regular with respect to the foliation generated by ξ^1. But a result of Tanno [**118**] says that the only regular 3-Sasakian 3-manifolds have $\Gamma = \mathrm{id}$ or $\mathbb{Z}_2$, in which case (iv) follows. □

EXAMPLE 3.3. Consider the 3-Sasakian lens space $L(p;q) = \mathbb{Z}_p\backslash S^7$ of Example 2.1. If p is odd then $-\mathrm{id}$ is not an element of $\mathbb{Z}_p$ so the generic leaf of the foliation $\mathcal{F}_3$ is S^3. The singular stratum consists of two leaves both of the form $\mathbb{Z}_p\backslash S^3$ with leaf holonomy group $\mathbb{Z}_p$. These two leaves are described by $u_2 = 0$ and $u_1 = 0$, respectively. If p is even then $-\mathrm{id}$ is an element of $\mathbb{Z}_p$, so the generic leaf is $SU(2)/\mathbb{Z}_2 = SO(3)$, and the leaf holonomy of the two singular leaves is $\mathbb{Z}_{\frac{p}{2}}$.

The next theorem was first proved by Ishihara [**67**] in the regular case using slightly different methods. First we need to describe our structures in the orbifold category. Recall that a quaternionic Kähler structure on a Riemannian manifold M is defined by

DEFINITION 3.3.2. *A Riemannian orbifold $\mathcal{O}$ is called a quaternionic Kähler orbifold if there is a rank 3 V-subbundle $\mathcal{G}$ of the endomorphism V-bundle $\mathrm{End}\, TM$ of TM which is preserved by the Levi-Civita connection and is locally generated by* *almost complex structures I, J, K that satisfy the algebra of the quaternions, and the action of the local uniformizing groups preserves the bundle $\mathcal{G}$. An alternative definition which works only in dimension greater than 4 is that $\mathcal{O}$ is a Riemannian orbifold whose holonomy group is a subgroup of $Sp(n)\cdot Sp(1)$.*

It is well-known that the strata of a quaternionic Kähler orbifold are not necessarily quaternionic Kähler [**52**]. The strata will be quaternionic Kähler if the local uniformizing groups act trivially on the fibres of $\mathcal{G}$ [**39**]. The group of the bundle $\mathcal{G}$ is $SO(3)$ with the adjoint representation. Thus, for each local uniformizing system on $\mathcal{O}$ there is a group homomorphism $\psi_i : \Gamma_i \longrightarrow SO(3)$.

THEOREM 3.3.3. *Let $(\mathcal{S}, g)$ be a 3-Sasakian manifold of dimension $4n+3$ such that the characteristic vector fields ξ^a are complete. Then the space of leaves $\mathcal{S}/\mathcal{F}_3$ has the structure of a quaternionic Kähler orbifold $(\mathcal{O}, g_{\mathcal{O}})$ of dimension $4n$ such that the natural projection $\pi : \mathcal{S} \longrightarrow \mathcal{O}$ is a principal V-bundle with group $SU(2)$ or $SO(3)$ and a Riemannian orbifold submersion such that the scalar curvature of $g_{\mathcal{O}}$ is $16n(n+2)$.*

PROOF. We can split $T\mathcal{S} = \mathcal{V}_3 \oplus \mathcal{H}$, where $\mathcal{V}_3$ is the subbundle spanned by the characteristic vector fields $\{\xi^1, \xi^2, \xi^3\}$ and the "horizontal" bundle is the orthogonal complement $\mathcal{H} = \mathcal{V}_3^{\perp}$. Let $h\Phi^a = \Phi^a \mid_{\mathcal{H}}$ be the restriction of characteristic endomorphisms. One can easily see that

$$h\Phi^a \circ h\Phi^b = -\delta^{ab}\mathbf{1} + \sum_c \epsilon^{abc} h\Phi^c.$$

It follows that $\mathcal{H}$ is pointwise a quaternionic vector space and $\mathcal{O}$ is a compact quaternionic orbifold. We must show that the metric $g_{\mathcal{O}}$ obtained from g by the orbifold Riemannian submersion $\pi : \mathcal{S} \to \mathcal{O}$ has its holonomy group reduced to a subgroup of $Sp(n)\cdot Sp(1)$. This can be done by constructing a parallel 4-form on $\mathcal{O}$. Consider $\phi^a = d\eta^a$ and

$$\bar{\Phi}^a = \phi^a + \sum_{b,c} \epsilon^{abc} \eta^b \wedge \eta^c.$$

It is easy to see that the 4-form $\Omega = \sum_a \bar{\Phi}^a \wedge \bar{\Phi}^a$ is horizontal and $Sp(1)$-invariant. It follows that there is a unique 4-form $\hat{\Omega}$ on the orbifold $\mathcal{O}$ invariant under the action of the local uniformizing groups such that $\pi^*\hat{\Omega} = \Omega$. One can show that $\hat{\Omega}$ is parallel on $\mathcal{O}$ using standard tensor computation with O'Neill formulas (see [**25, 52**] for details). In the case $n = 1$ the parallelism of the 4-form does not further restrict Riemannian geometry of $\mathcal{O}$. However, one can show that $(\mathcal{O}, g_{\mathcal{O}})$ is a compact self-dual Einstein orbifold. Self-duality follows easily from the fact that $\mathcal{O}$ is quaternionic. The fact that the metric is Einstein is a simple computation and in the regular case can be found in [**119**]. □

There is an important inversion theorem of Theorem 3.3.3 originally in the regular case due to Konishi [**76**]. By now there are several proofs of this, all of them related. Given a quaternionic Kähler orbifold $\mathcal{O}$ one can construct the Salamon twistor space $\mathcal{Z}$ and then get $\mathcal{S}$ from the inversion theorem of [**22**]. Another approach would be to construct the orbifold version of Swann's bundle [**115**] on $\mathcal{O}$ and then use the results of [**25**] to obtain $\mathcal{S}$. Here our proof is essentially that of Konishi's, only slightly modified to handle the orbifold situation.

THEOREM 3.3.4. *Let $(\mathcal{O}, g_{\mathcal{O}})$ be a quaternionic Kähler orbifold of dimension $4n$ with positive scalar curvature $16n(n+2)$. Then there is a principal $SO(3)$ V-bundle over $\mathcal{O}$ whose total space $\mathcal{S}$ admits a 3-Sasakian structure with scalar curvature $2(2n+1)(4n+3)$.*

PROOF. Let $\mathcal{G}$ denote the V-subbundle of End $T\mathcal{O}$ describing the quaternionic structure. Let $\{\tilde{U}_i\}$ be local uniformizing neighborhoods that cover $\mathcal{O}$ and $\mathcal{I}_i^a$ a local framing of $\mathcal{G}$ on $\tilde{U}_i$ that satisfies

$$\mathcal{I}_i^a \circ \mathcal{I}_i^b = -\delta^{ab}\mathrm{id} + \epsilon^{abc}\mathcal{I}_i^c.$$

Since $\mathcal{O}$ is quaternionic Kähler there are 1-forms τ_i^a on each $\tilde{U}_i$ such that

$$\nabla\mathcal{I}_i^a = \epsilon^{abc}\tau_i^b \otimes \mathcal{I}_i^c.$$

Now the structure group of the V-bundle End $T\mathcal{O}$ is $Sp(n)\cdot Sp(1)$, and that of the V-subbundle $\mathcal{G}$ is $SO(3)$. Let $\pi : \mathcal{S} \longrightarrow \mathcal{O}$ denote the principal $SO(3)$ V-bundle associated to $\mathcal{G}$. The local 1-forms τ_i^a are the components of an $\mathfrak{so}(3)$ connection $\tau_i = \sum_{a=1}^3 \tau_i^a e_a$, where $\{e_a\}$ denotes the standard basis of $\mathfrak{so}(3)$ which satisfies the Lie bracket relations $[e_a, e_b] = 2\epsilon^{abc}e_c$. The local connection forms satisfy the well-known relations

$$\tau_i = \mathrm{ad}_{g_{ij}}\tau_j + g_{ij}^{-1}dg_{ij}$$

in $\tilde{U}_i \cap \tilde{U}_j$ for some smooth map $g_{ij} : \tilde{U}_i \cap \tilde{U}_j \longrightarrow SO(3)$. Furthermore, from **[67]** one checks that the curvature forms

$$\omega_i^a = d\tau_i^a + \epsilon^{abc}\tau_i^b \wedge \tau_i^c \tag{3.1}$$

satisfy the relation $2g_{\mathcal{O}}(\mathcal{I}_i^a X, Y) = \omega_i^a(X, Y)$. Now on each $\tilde{U}_i$ there exists a smooth local section $\sigma_i : \tilde{U}_i \longrightarrow \mathcal{S}$ and on $\mathcal{S}$ there is a global 1-form η^a such that $\tau_i^a = \sigma_i^*\eta^a$. On $\mathcal{S}$ we define a Riemannian metric by

$$g = \pi^* g_{\mathcal{O}} + \sum_{a=1}^{3} \eta^a \otimes \eta^a.$$

By construction the vector fields ξ^a generating the $SO(3)$ action on $\mathcal{S}$ are dual to the forms η^a with respect to this metric, viz $g_{\mathcal{S}}(X, \xi^a) = \eta^a(X)$ for any vector field X on $\mathcal{S}$. Also by construction the vector fields ξ^a are Killing fields with respect to the metric g. Now define the $(1,1)$ tensor field $\Phi^a = \nabla\xi^a$. Since the ξ^a are mutually orthogonal vector fields of unit length on $\mathcal{S}$, one easily checks that $\Phi^a\xi^b = -\epsilon^{abc}\xi^c$. Thus Φ^a splits as

$$\Phi^a = h\Phi^a + \epsilon^{abc}\xi^b \otimes \eta^c.$$

One then checks using 3.1 that on each open set $\pi^{-1}(U_i)$, $h\Phi^a$ equals the horizontal component of $(\sigma_i)_*\mathcal{I}_i^a$. From this one then shows that

$$\Phi^a \circ \Phi^b - \xi^a \otimes \eta^b \;=\; -\epsilon^{abc}\Phi^c - \delta^{ab}\mathrm{id}.$$

The result then follows by Propositions 2.2.2 and 2.2.3. □

We mention that from the discussion in the Appendix it follows that if the homomorphisms $\psi_i : \Gamma_i \longrightarrow SO(3)$ are injective the total space $\mathcal{S}$ will be a smooth 3-Sasakian manifold.

Konishi's construction gives an $SO(3)$ bundle over $\mathcal{O}$. In the case that $\mathcal{O}$ is a smooth manifold there is a well-known obstruction **[106]** to lifting this bundle to an $Sp(1)$ bundle, the Marchiafava-Romani class ε. Actually ε is the obstruction to lifting the principal $Sp(n)\cdot Sp(1)$ frame bundle to an $Sp(n)\times Sp(1)$ bundle [**91, 106**]. This obstruction also occurs when $\mathcal{O}$ is an orbifold as long as one uses Haefliger's orbifold cohomology (see Appendix). The class ε is the image of the connecting homomorphism

$$\delta : H^1_{orb}(\mathcal{O},\mathcal{G}) \longrightarrow H^2_{orb}(\mathcal{O},\mathbb{Z}_2), \tag{3.2}$$

where $\mathcal{G}$ is the sheaf of germs of smooth orbifold maps from open sets of $\mathcal{O}$ to the group $Sp(n)\cdot Sp(1)$. If following Salamon [**106**] we write $T\mathcal{O}\otimes\mathbb{C}\simeq \mathbf{E}\otimes\mathbf{H}$, then ε is the second Stiefel-Whitney class w_2 of the bundle $S^2(\mathbf{H})$ over $\mathcal{O}$. We have

PROPOSITION 3.3.5. *The principal $SO(3)$ V-bundle constructed in Theorem 3.3.4 lifts to a principal $Sp(1)$ V-bundle if and only if $\varepsilon \in H^2_{orb}(\mathcal{O},\mathbb{Z}_2)$ vanishes. Moreover, when $\varepsilon = 0$ the 3-Sasakian structure on the total space $\mathcal{S}$ of the $SO(3)$ V-bundle lifts to the total space $\mathcal{S}'$ of the $Sp(1)$ V-bundle.*

Thus, in the case that $\varepsilon = 0$ there are precisely two 3-Sasakian orbifolds $\mathcal{S},\mathcal{S}'$ corresponding to the quaternionic Kähler orbifold $\mathcal{O}$. Let $\mathcal{Z}$ denote the twistor space of the orbifold $\mathcal{O}$. Then likewise, since $S^2\simeq SO(3)/S^1\simeq Sp(1)/S^1$ the two 3-Sasakian orbifolds $\mathcal{S}$ and $\mathcal{S}'$ have the same twistor space $\mathcal{Z}$. When $\mathcal{O}$ is a smooth manifold a result of Salamon [**106**] says that $\varepsilon = 0$ if and only if the quaternionic Kähler manifold $\mathcal{O}$ is quaternionic projective space. If we impose the condition that the orbifolds $\mathcal{S}$ and $\mathcal{S}'$ are smooth manifolds, there is a similar result.

THEOREM 3.3.6. *If two 3-Sasakian manifolds $\mathcal{S}$ and $\mathcal{S}'$ are associated to the same quaternionic Kähler orbifold $\mathcal{O}$ or equivalently the same twistor space $\mathcal{Z}$, then both $\mathcal{S}$ and $\mathcal{S}'$ have the same universal covering space $\tilde{\mathcal{S}}$ and $\tilde{\mathcal{S}}$ is a standard 3-Sasakian sphere.*

PROOF. We work with the twistor space $\mathcal{Z}$. Now $\mathcal{S}$ and $\mathcal{S}'$ are unit circle bundles in the line V-bundles $\mathcal{L}^{-1}$ and $(\mathcal{L}')^{-1}$ respectively. Moreover, since $\mathcal{S}'$ is a double cover of $\mathcal{S}$, it follows that $\mathcal{L}=(\mathcal{L}')^2$. Consider the universal orbifold cover $\tilde{\mathcal{Z}}$ of $\mathcal{Z}$ with $\pi_1^{orb}(\tilde{\mathcal{Z}})=0$. Pull back the V-bundles $\mathcal{L}$ and $\mathcal{L}'$ to V-bundles $\tilde{\mathcal{L}}$ and $\tilde{\mathcal{L}}'$ on $\tilde{\mathcal{Z}}$ respectively. These bundles satisfy $\tilde{\mathcal{L}}'=\tilde{\mathcal{L}}^{\frac{1}{2}}$. By construction and naturality of the covering maps $\tilde{\mathcal{L}}$ is the contact line bundle on $\tilde{\mathcal{Z}}$. Moreover, $\tilde{\mathcal{S}}$ and $\tilde{\mathcal{S}}'$ which are the total spaces of the pullbacks of $\mathcal{S}$ and $\mathcal{S}'$ to $\tilde{\mathcal{Z}}$ are both smooth manifolds since they cover smooth manifolds. Thus, by Proposition 3.2.3 $\tilde{\mathcal{Z}}\simeq\mathbb{P}^{2n+1}$. It follows that $\tilde{\mathcal{S}}'\simeq S^{4n+3}$. □

REMARK 3.2. Konishi also considers the case when the quaternionic Kähler manifold has negative scalar curvature. This gives a Sasakian 3-structure on $\mathcal{S}$ with indefinite signature $(3,4n)$.

Finally, we give some general results concerning the curvature of any 3-Sasakian manifold. Since the curvature of any Riemannian manifold is completely determined by its sectional curvature and the sectional curvature of any Sasakian manifold [**19, 134**] is completely determined by the Φ-sectional curvature, we essentially give the latter. This shows that the local geometry of any 3-Sasakian manifold determines and is determined by that of its associated quaternionic Kähler orbifold.

PROPOSITION 3.3.7. *Let $(\mathcal{S},g,\xi^a)$ be a 3-Sasakian manifold and let K and $\check{K}$ denote the sectional curvatures of g and its transverse component $\check{g}$, respectively. Then if X is any horizontal vector field of unit length on $\mathcal{S}$, we have $K(X,\Phi^aX)=\check{K}(X,\Phi^aX)-3$.*

PROOF. We first notice that [14], 9.29c gives $K(X,\Phi^aX) = \check{K}(X,\Phi^aX) - 3|A_X\Phi^aX|^2$, where A is O'Neill's tensor [14] which is essentially the curvature of the $\mathfrak{sp}(1)$-valued orbifold connection [25]. One then shows that for any horizontal vector fields X, Y on $\mathcal{S}$ we have

$$A_XY = \sum_{a=1}^{3} g(\Phi^aX,Y)\xi^a, \tag{3.3}$$

and the identity follows. □

3.4. The Second Einstein Metric. Of course, by an Einstein metric we actually mean a homothety class of Einstein metrics. In this section we shall show by using a theorem of Berard-Bergery [14] that every 3-Sasakian manifold has at least two distinct homothety classes of Einstein metrics. The method involves the canonical variation [14] associated with Riemannian submersions. Due to the local nature of the calculations involved this construction holds equally well for orbifold Riemannian submersions. The canonical variation is constructed as follows [14]: Let $\pi : M \longrightarrow B$ be an orbifold Riemannian submersion with g the Riemannian metric on M. Let $\mathcal{V}$ and $\mathcal{H}$ denote the vertical and horizontal subbundles of the tangent bundle TM. For each real number $t > 0$ we construct a one parameter family g_t of Riemannian metrics on M by defining

$$g_t|\mathcal{V} = tg|\mathcal{V}, \qquad g_t|\mathcal{H} = g|\mathcal{H}, \qquad g_t(\mathcal{V},\mathcal{H}) = 0. \tag{3.4}$$

So for each $t > 0$ we have an orbifold Riemannian submersion with the same base space. Furthermore, if the fibers of g are totally geodesic, so are the fibers of g_t. We apply the canonical variation to the orbifold Riemannian submersion $\pi : \mathcal{S} \longrightarrow \mathcal{O}$. The metric as well as other objects on $\mathcal{O}$ will be denoted with a check such as $\check{g}$.

THEOREM 3.4.1. *Every 3-Sasakian manifold admits a second Einstein metric of positive scalar curvature.*

PROOF. We apply the canonical variation to the orbifold Riemannian submersion $\pi : \mathcal{S} \longrightarrow \mathcal{O}$. According to the Bérard-Bergery Theorem [14], 9.73 there are several conditions to check. First, the connection $\mathcal{H}$ must be a Yang-Mills connection. The condition for this is [14]:

$$\sum_i g((\nabla_{X_i}A)_{X_i}X, \xi^a) = 0$$

for each $a = 1,2,3$ and where X_i is a local orthonormal frame of $\mathcal{H}$, X is any horizontal vector field, and A is O'Neill's tensor. Actually we can use standard computations together with Proposition 2.1.2 and Equation 3.3 to prove the stronger condition $(\nabla_{X_i}A)_{X_i}X = 0$. Second $|A|^2$ must be constant. To compute this notice that using [14] and Equation 3.3 we find

$$g(A_{X_i}, A_{X_j}) = 3\delta_{ij}, \qquad g(A\xi^a, A\xi^b) = 4n\delta^{ab}. \tag{3.5}$$

This gives $|A|^2 = 12n$. The final condition to be satisfied is $(\check{\lambda})^2 - \hat{\lambda}(12n+18) > 0$, where $\check{\lambda}$ and $\hat{\lambda}$ are the Einstein constants for $\mathcal{O}$ and the fibers, respectively, and we have made use of 3.5. Since in our case $\check{\lambda} = 4(n+2)$ and $\hat{\lambda} = 2$, we see that the inequality is satisfied. □

The scalar curvature of any metric g_t in the canonical variation of the metric g is given by the formula $s_t = 16n(n+2) + 6/t - 12nt$ [**14**]. Moreover, the value of t that gives the second Einstein metric is $t_0 = \frac{\lambda}{\bar{\lambda}-\lambda} = \frac{1}{2n+3}$. The ratio of the two metrics depends only on the homothety class and is given by

$$\frac{s_{\frac{1}{2n+3}}}{s_1} = 1 + \frac{6(n+1)}{(2n+3)(2n+1)}.$$

In the special case $n = 1$ that is dim $\mathcal{S} = 7$, both the 3-Sasakian metric and the second Einstein metric have weak G_2 holonomy [**48, 54**]. See Theorem 5.2.9 below.

3.5. Invariants and the Classification of 3-Sasakian Structures. We consider the question of equivalence of 3-Sasakian manifolds. A 3-Sasakian structure $\{\xi^a, \eta^a, \Phi^a\}_{a=1}^3$ on a manifold $(\mathcal{S}, g)$ is determined completely by the metric g and the characteristic vector fields ξ^a.

DEFINITION 3.5.1. *Two 3-Sasakian manifolds $(\mathcal{S}, g)$ and $(\mathcal{S}', g')$ are said to be isomorphic if there exist a diffeomorphism $F : \mathcal{S} \longrightarrow \mathcal{S}'$ and an $\phi \in Sp(1)$ such that $F^*g' = g$ and $\acute{\xi}^a = (Ad_\phi)_* F_* \xi^a$, where Ad denotes the adjoint action of $Sp(1)$ on its Lie algebra $\mathfrak{sp}(1)$.*

In practice we shall always choose a basis $\acute{\xi}^a$ of the 3-Sasakian structure on $\mathcal{S}'$ so that $F_*\xi^a = \acute{\xi}^a$. Now given such a diffeomorphism $F : \mathcal{S} \longrightarrow \mathcal{S}'$ it is clear that the corresponding foliations are F-related, that is that $F_*\mathcal{F}_1 = \mathcal{F}'_1$ and $F_*\mathcal{F}_3 = \mathcal{F}'_3$. This induces a commutative diagram of orbifold diffeomorphisms

$$(3.6)\qquad \begin{array}{ccc} \mathcal{S} & \xrightarrow{F} & \mathcal{S}' \\ \downarrow & & \downarrow \\ \mathcal{Z} & \xrightarrow{F_1} & \mathcal{Z}' \\ \downarrow & & \downarrow \\ \mathcal{O} & \xrightarrow{F_3} & \mathcal{O}'. \end{array}$$

This implies that if L_x is the leaf of $\mathcal{F}_3$ at $x \in \mathcal{O}$, then $F(L_x)$ is the leaf at $F_3(x) \in \mathcal{O}'$, that is, $L'_{F_3(x)} = F(L_x)$. Let $G(L)$ denote the leaf holonomy group of the leaf L. Then we have $G(L'_{F_3(x)}) \approx G(L_x)$. More generally let $\mathcal{G}(\mathcal{S})$ denote the holonomy groupoid [**95**] of the foliation $\mathcal{F}_3$, that is the set of triples $(x, y, [\alpha])$ where $x, y \in \mathcal{S}$ lie on the same leaf L_x of $\mathcal{F}_3$ and $[\alpha]$ is the holonomy equivalence class of all piecewise smooth paths from x to y lying in L_x. Multiplication in the groupoid $\mathcal{G}(\mathcal{S})$ is defined on pairs of triples $(x, y, [\alpha]), (x', y', [\alpha'])$ precisely when $y = x'$ and then $(x, y, [\alpha]) \cdot (x', y', [\alpha']) = (x, y', [\alpha'\alpha])$. Furthermore, the subgroup of triples $(x, x, [\alpha])$ with x fixed is identified with the holonomy group $G(L_x)$. With this structure, $\mathcal{G}(\mathcal{S})$ is a locally compact topological groupoid [**95**]. (Actually $\mathcal{G}(\mathcal{S})$ is a smooth manifold of dimension $4n+6$ but we do not use this here). We have

PROPOSITION 3.5.2. *Let $F : \mathcal{S} \longrightarrow \mathcal{S}'$ be an isomorphism of 3-Sasakian manifolds. Then F induces an isomorphism $F_* : \mathcal{G}(\mathcal{S}) \longrightarrow \mathcal{G}(\mathcal{S}')$ of topological groupoids.*

The groupoid $\mathcal{G}(\mathcal{S})$ will be studied in a forthcoming work. For now we are interested in the unordered list $(\Gamma_1, \Gamma_2, \cdots)$ of holonomy groups in $\mathcal{G}(\mathcal{S})$ up to abstract isomorphism. This list is finite if $\mathcal{S}$ is complete and it provides important invariants of a 3-Sasakian manifold. Since the leaves of the foliation $\mathcal{F}_3$ are all spherical space forms, the groups Γ_i are all either subgroups of $Sp(1)$ or all subgroups of $SO(3)$, depending on whether the Marchiafava-Romani class ε of the quaternionic Kähler orbifold $\mathcal{O}$ is 0 or 1, respectively. Notice that it follows from its definition and 3.6 above that the class ε is an invariant of the 3-Sasakian structure on the manifold $\mathcal{S}$. Indeed, ε can be identified with a certain secondary characteristic class of the foliation $\mathcal{F}_3$. Thus, the Marchiafava-Romani class splits the isomorphism classes $\mathbb{S}$ of 3-Sasakian manifolds into the disjoint union $\mathbb{S}_0 + \mathbb{S}_1$ depending on whether ε is 0 of 1. A further rough classification scheme is given by

DEFINITION 3.5.3. *$\mathcal{S}$ is said to be:*
(i) regular if all the Γ_i are the identity.
(ii) of cyclic type if all the Γ_i are cyclic.
(iii) of dihedral type if all the Γ_i are either cyclic or dihedral or binary dihedral with at least one Γ_i non-Abelian.
(iv) of polyhedral type if at least one of the Γ_i is one of the polyhedral groups, tetrahedral, octahedral, or icosahedral (or the corresponding binary double covers) groups.

The definition of regular here coincides with that of Definition 2.2.6. In general 3-Sasakian dimension $4n+3$ the only known examples of 3-Sasakian manifolds of polyhedral or dihedral type are the spherical space forms $\Gamma\backslash S^{4n+3}$ and $\Gamma\backslash\mathbb{RP}^{4n+3}$, where Γ is a binary polyhedral or a binary dihedral group in the first case and a polyhedral or a dihedral group in the second. The action is that induced by the diagonal action of Γ on the quaternionic vector space $\mathbb{H}^{n+1}$. However, in dimension 7 there exist 3-Sasakian manifolds of dihedral or polyhedral type [**24, 53**] which are not spherical space forms. All other known non-regular 3-Sasakian manifolds are of cyclic type and are discussed in detail in Section 8 The following is essentially due to Salamon:

THEOREM 3.5.4. *Let $\mathcal{S}$ be a complete regular 3-Sasakian manifold with $\varepsilon = 0$. Then $\mathcal{S} \simeq S^{4n+3}$ or $\mathbb{RP}^{4n+3}$.*

For more results about regular 3-Sasakian manifolds see Section 5.4 below. Next we consider an important infinitesimal rigidity result. In the regular case this rigidity is a simple consequence of the results of LeBrun [**86**] and Nagatomo [**98**] (see [**54**]). In the general case it was recently proved by Pedersen and Poon.

THEOREM 3.5.5. [**102**] *Complete 3-Sasakian manifolds are infinitesimally rigid.*

PROOF. The deformation theory of 3-Sasakian manifolds is tied to the deformation theory of hypercomplex manifolds studied previously in [**101**]. Let $\mathcal{S}$ be a complete 3-Sasakian manifold. Then the compact manifold $S^1 \times \mathcal{S}$ has a natural hypercomplex structure [**26**]. Thus, its twistor space W is compact and fibers holomorphically over $\mathbb{CP}^1$. Moreover, there is a holomorphic foliation on W whose leaves are elliptic Hopf surfaces, and whose space of leaves is the twistor space $\mathcal{Z}$ associated to $\mathcal{S}$. The geometry of the corresponding deformation theory is as follows. Deformations $(\mathcal{S}_t, g_t)$ of the 3-Sasakian structure $(\mathcal{S}_0, g_0)$ on $\mathcal{S}$ correspond to deformations of the hypercomplex structure on $S^1 \times \mathcal{S}$ of the form $S^1 \times \mathcal{S}_t$. In turn these deformations correspond to deformations of the holomorphic fibration $p: W \longrightarrow \mathbb{CP}^1$. Thus, there are natural projections:

$$
\begin{array}{ccc}
 & W & \\
p \swarrow & & \searrow \Phi \\
\mathbb{CP}^1 & & \mathcal{Z},
\end{array}
\tag{3.7}
$$

where each fiber of p is a divisor in W diffeomorphic to $S^1 \times \mathcal{S}$ and Φ is an orbifold submersion whose leaves are elliptic Hopf surfaces. Now the product map $p \times \Phi : W \longrightarrow \mathbb{CP}^1 \times \mathcal{Z}$ is an orbifold submersion whose leaves are elliptic curves. The differential of $p \times \Phi$ induces the exact sequence of sheaves

$$0 \longrightarrow \mathcal{O}_W \longrightarrow \Theta_W \longrightarrow \Phi^*\Theta_{\mathcal{Z}} \oplus p^*\Theta_{\mathbb{CP}^1} \longrightarrow 0,$$

where $\mathcal{O}_W$ denotes the structure sheaf of W and Θ denotes the holomorphic tangent sheaf. Then using standard techniques together with the Kodaira-Baily vanishing theorem and the orbifold version of the Akizuki-Nagano vanishing theorem, Pedersen and Poon show that the virtual parameter space for 3-Sasakian deformations lies in

$$
\begin{aligned}
&H^0(\mathcal{Z},\mathcal{O}_{\mathcal{Z}}) \otimes H^1(F,\mathcal{O}_F) \oplus H^0(\mathcal{Z},\Theta_{\mathcal{Z}}) \otimes H^1(F,\mathcal{O}_F) \\
&\oplus H^1(\mathcal{Z},\Theta_{\mathcal{Z}}) \otimes H^0(F,\mathcal{O}_F) \oplus H^1(W,p^*\Theta_{\mathbb{CP}^1}),
\end{aligned}
\tag{3.8}
$$

where F is the generic elliptic Hopf surface $S^1 \times Sp(1)$. One then analyzes each summand of 3.8 to show that there are no 3-Sasakian deformations. For example, possible deformations lying in the last summand vanish by results of Horikawa, while 3-Sasakian deformations lying in the second and third summands must preserve the complex contact structure on $\mathcal{Z}$. There are no such deformations in the third summand by the Kodaira-Baily vanishing theorem. Elements in the second summand correspond to complex contact transformations that are invariant under the $U(1) \times U(1)$ action coming from a discrete quotient of the $\mathbb{C}^*$ principal action on $\mathcal{L}$, and there are no such elements. Finally, elements of the first summand correspond to scale changes in the S^1 factor of $S^1 \times Sp(1)$ and these hypercomplex deformations do not come from 3-Sasakian ones. □

While this theorem says that there is no "infinitesimal moduli", there may well be discrete moduli of 3-Sasakian structures. Indeed, we believe that the work of Kruggel [**82**] can be used with the aid of a computer to construct distinct 3-Sasakian structures on the same manifold. See the last paragraph of secion 8.4 below.

4. Homogeneous Spaces

In this section we classify Sasakian-Einstein and 3-Sasakian homogeneous spaces. We begin with the Sasakian-Einstein case.

4.1. Homogeneous Sasakian-Einstein Manifolds. As a Sasakian vector field ξ is Killing, every Sasakian, and, hence, Sasakian-Einstein manifold $\mathcal{S}$ has non-trivial isometries. Recall the following well-known terminology. Let G be a complex semi-simple Lie group. A maximal solvable complex subgroup B is called a Borel subgroup, and B is unique up to conjugacy. Any complex subgroup P that contains B is called a parabolic subgroup. Then the homogeneous space G/P is called a generalized flag manifold. A well-known result of Wang [**2**] says that every simply-connected homogeneous Kähler manifold is a generalized flag manifold.

DEFINITION 4.1.1. *A compact Sasakian-Einstein manifold $\mathcal{S}$ is called a homogeneous Sasakian-Einstein manifold if there is a transitive group K of isometries on $\mathcal{S}$ that preserve the Sasakian structure, that is, if $\phi^k \in$ Diff $\mathcal{S}$ corresponds to $k \in K$, then $\phi^k_*\xi = \xi$. (This implies that both Φ and η are also invariant under the action of K.)*

Note that K is a compact Lie group by compactness of $\mathcal{S}$. The following is a result of [**23**].

THEOREM 4.1.2. *Let $\mathcal{S}$ be a compact quasi-regular homogeneous Sasakian-Einstein manifold. Then $\mathcal{S}$ is an S^1-bundle over a generalized flag manifold G/P. Conversely, given any generalized flag manifold G/P there is a circle bundle $\pi : \mathcal{S} \longrightarrow G/P$ whose total space $\mathcal{S}$ is a homogeneous Sasakian-Einstein manifold.*

PROOF. As in Proposition 4.6 of [**26**], $\mathcal{S}$ is regular. By Theorem 3.1.2 $\mathcal{S}$ fibers over a simply-connected Fano variety $\mathcal{Z}$ with a Kähler-Einstein metric of positive scalar curvature. Since the action of K commutes with ξ it sends fibers to fibers, and thus acts transitively on $\mathcal{Z}$. But by Wang's theorem [**2**], $\mathcal{Z} = G/P$ for some complex semi-simple Lie group G and some parabolic subgroup $P \subset G$. Now K preserves the Kähler-Einstein structure, and thus the complex structure. So $K \subset G$. In fact G is just the complexification of K its maximal compact subgroup [**126**].

Conversely, by a theorem of Matsushima [**14**] every G/P admits K invariant Kähler-Einstein metric, where K is the maximal compact subgroup of G. Moreover, there is a subgroup $U \subset K$ such that $G/P = K/U$. Then by the Kobayashi construction described in the previous section there is a circle bundle over G/P whose total space $\mathcal{S}$ admits a Sasakian-Einstein metric. By the construction one easily sees that this metric is homogeneous. □

The following corollary lists all the possible G/P in the first three dimensions:

COROLLARY 4.1.3. *Let $\mathcal{S}$ be a compact homogeneous Sasakian-Einstein manifold of dimension $2n+1$. Then $\mathcal{S}$ is a circle bundle over*
(i) $\mathbb{CP}^1$ when $n = 1$,
(ii) $\mathbb{CP}^2$ or $\mathbb{CP}^1 \times \mathbb{CP}^1$ when $n = 2$,
(iii) $\mathbb{CP}^3$, $\mathbb{CP}^2 \times \mathbb{CP}^1$, $\mathbb{CP}^1 \times \mathbb{CP}^1 \times \mathbb{CP}^1$, the complex flag $F_{3,2,1} = SU(3)/T^2$, and the real Grassmannian $Gr_2(\mathbb{R}^5)$ when $n = 3$.

REMARK 4.1. Note that $(\mathcal{S}, g)$ does not have to be simply-connected. For each G/P and each $k \in \mathbb{Z}^+$ we get a homogeneous S^1 bundle over G/P with fundamental group $\pi_1 = \mathbb{Z}_k$. It can be obtained as a discrete $\mathbb{Z}_k$-quotient of the unique simply-connected model of such $\mathcal{S}$.

4.2. Homogeneous 3-Sasakian Manifolds. Every 3-Sasakian manifold $(\mathcal{S}, g)$ has a nontrivial isometry group $I(\mathcal{S}, g)$ of dimension at least three. We first recall some of the results about $I(\mathcal{S}, g)$.

DEFINITION 4.2.1. *Let $I_0(\mathcal{S}, g) \subset I(\mathcal{S}, g)$ be the subgroup of the isometry group which preserves the 3-Sasakian structure, that is if $\phi^k \in$ Diff $\mathcal{S}$ corresponds to $k \in I_0(\mathcal{S}, g)$ then $\phi^k_* \xi^a = \xi^a$, for all $a = 1, 2, 3$. Then $I_0(\mathcal{S}, g)$ is called the group of 3-Sasakian isometries and when it acts transitively on $(\mathcal{S}, g)$ the space $\mathcal{S}$ is said to be a 3-Sasakian homogeneous space.*

LEMMA 4.2.2. *Let $(\mathcal{S}, g)$ be a 3-Sasakian manifold and $X \in \mathfrak{i}$ be a Killing vector field on $\mathcal{S}$. Let $\mathcal{L}_X$ denote the Lie derivative with respect to X. Then the following conditions are equivalent:*
(i) $\mathcal{L}_X \Phi^a = 0$, for $a = 1, 2, 3$.
(ii) $\mathcal{L}_X \eta^a = 0$, for $a = 1, 2, 3$.
(iii) $\mathcal{L}_X \xi^a = 0$, for $a = 1, 2, 3$.
Furthermore, if any (hence, all) of the conditions above is satisfied, then for any vector field Y on $\mathcal{S}$ we have $X\eta^a(Y) = \eta^a([X, Y])$.

The above lemma gives alternative characterizations of the Lie algebra of $I_0(\mathcal{S}, g)$ and it easily follows from the definition and properties of the 3-Sasakian structure. As its immediate consequence we get the following theorem [**118**]:

THEOREM 4.2.3. *Let $(\mathcal{S}, g)$ be a complete 3-Sasakian manifold which is not of constant curvature. $\mathfrak{i}$ and $\mathfrak{i}_0$ denote the Lie algebras of $I(\mathcal{S}, g)$ and $I_0(\mathcal{S}, g)$, respectively. Then as Lie algebras $\mathfrak{i} = \mathfrak{i}_0 \oplus \mathfrak{sp}(1)$, where $\mathfrak{sp}(1)$ is the Lie algebra generated by $\{\xi^1, \xi^2, \xi^3\}$.*

Notice that any of the first three conditions in Lemma 4.2.2 can be used to describe the Lie subalgebra $\mathfrak{i}_0 \subset \mathfrak{i}$. Moreover, the equivalence of conditions (iii) and (i) says that the Lie algebra $\mathfrak{c}(\mathfrak{sp}(1))$ of the centralizer of $Sp(1)$ in $I(\mathcal{S}, g)$ is precisely $\mathfrak{i}_0$. Globally, on the group level we obtain:

PROPOSITION 4.2.4. *Let $(\mathcal{S}, g)$ be a complete 3-Sasakian manifold. Then both the isometry groups $I(\mathcal{S}, g)$ and $I_0(\mathcal{S}, g)$ are compact. Furthermore, if $(\mathcal{S}, g)$ is not of constant curvature then either $I(\mathcal{S}, g) = I_0(\mathcal{S}, g) \times Sp(1)$ or $I(\mathcal{S}, g) = I_0(\mathcal{S}, g) \times SO(3)$. Finally, if $(\mathcal{S}, g)$ does have constant curvature then $I(\mathcal{S}, g)$ strictly contains either $I_0(\mathcal{S}, g) \times Sp(1)$ or $I_0(\mathcal{S}, g) \times SO(3)$ as a proper subgroup and $I_0(\mathcal{S}, g)$ is the centralizer of $Sp(1)$ or $SO(3)$.*

PROOF. The first assertion follows from Corollary 2.2.5 and a standard result of Myers and Steenrod (cf. [**14**]). Next, since $I_0(\mathcal{S}, g)$, $Sp(1)$, and $SO(3)$ are all compact, the direct sum on the Lie algebra level given in Theorem 4.2.3 also gives a direct product of Lie groups. The last assertion follows immediately from lemma 4.2.2. □

PROPOSITION 4.2.5. *Let $(\mathcal{S}, g)$ be a 3-Sasakian homogeneous space. Then all leaves are diffeomorphic and $\mathcal{S}/\mathcal{F}_3$ is a quaternionic Kähler manifold where the natural projection $\pi : \mathcal{S} \to \mathcal{S}/\mathcal{F}_3$ is a locally trivial Riemannian fibration. Furthermore, $I_0(\mathcal{S}, g)$ acts transitively on the space of leaves $\mathcal{S}/\mathcal{F}_3$.*

PROOF. Let $\psi : I_0(\mathcal{S}, g) \times \mathcal{S} \to \mathcal{S}$ denote the action map so that, for each $a \in I_0(\mathcal{S}, g)$, $\psi_a = \psi(a, \cdot)$ is a diffeomorphism of $\mathcal{S}$ to itself. Proposition 4.2.4 implies that the isometry group $I(\mathcal{S}, g)$ contains $I_0(\mathcal{S}, g) \times Sp(1)$ where either $Sp(1)$ acts effectively or its $\mathbb{Z}_2$ quotient $SO(3) \simeq Sp(1)/\mathbb{Z}_2$ acts effectively. Since the Killing vector fields ξ^a for $a = 1, 2, 3$ are both the infinitesimal generators of the group $Sp(1)$ and a basis for the vertical distribution $\mathcal{V}$, it follows that $Sp(1)$ acts transitively on each leaf with isotropy subgroup of a point some finite subgroup $\Gamma \subset Sp(1)$. Now let p_1 and p_2 be any two points of $\mathcal{S}$ and let $\mathcal{L}_1$ and $\mathcal{L}_2$ denote the corresponding leaves through p_1 and p_2, respectively. Since $I_0(\mathcal{S}, g)$ acts transitively on $\mathcal{S}$, there exists an $a \in I_0(\mathcal{S}, g)$ such that $\psi_a(p_1) = p_2$. Now ψ_a restricted to $\mathcal{L}_1$ maps $\mathcal{L}_1$ diffeomorphically onto its image, and, since the $Sp(1)$ factor acts transitively on each leaf and commutes with $I_0(\mathcal{S}, g)$, the image of ψ_a lies in $\mathcal{L}_2$. But the same holds for the inverse map $\psi_{a^{-1}}$ with $\mathcal{L}_1$ and $\mathcal{L}_2$ interchanged, so the leaves must be diffeomorphic. Thus, the leaf holonomy is trivial and $\pi : \mathcal{S} \rightarrow \mathcal{S}/\mathcal{F}_3 = \mathcal{O}$ is a locally trivial Riemannian fibration. The fact that the space of leaves $\mathcal{O}$ is a quaternionic Kähler manifold now follows from Ishihara's Theorem 3.3.4. Finally, the constructions above shows directly that $I_0(\mathcal{S}, g)$ acts transitively on $\mathcal{O}$. □

The following classification theorem is now immediate from Proposition 4.2.5, the result of Alekseevsky which states that all homogeneous quaternionic Kähler manifolds of positive scalar curvature are symmetric [**4**], and Proposition 2.2.7:

THEOREM 4.2.6. *Let $\mathcal{S}$ be a 3-Sasakian homogeneous space. Then $\mathcal{S} = G/H$ is precisely one of the following:*

$$\frac{Sp(n+1)}{Sp(n)}, \quad \frac{Sp(n+1)}{Sp(n)\times\mathbb{Z}_2}, \quad \frac{SU(m)}{S\big(U(m-2)\times U(1)\big)}, \quad \frac{SO(k)}{SO(k-4)\times Sp(1)},$$

$$\frac{G_2}{Sp(1)}, \quad \frac{F_4}{Sp(3)}, \quad \frac{E_6}{SU(6)}, \quad \frac{E_7}{\mathrm{Spin}(12)}, \quad \frac{E_8}{E_7}.$$

Here $n \geq 0$, $Sp(0)$ denotes the trivial group, $m \geq 3$, and $k \geq 7$. Hence, there is one-to-one correspondence between the simple Lie algebras and the simply-connected 3-Sasakian homogeneous manifolds.

Below we give the fundamental diagram $\diamondsuit(G/H)$ for each 3-Sasakian homogeneous space of Theorem 4.2.6:

$$\begin{array}{ccccc} & & \mathbb{R}_+ \times G/H & & \\ & \swarrow & & \searrow & \\ G/H\cdot U(1) & & \longleftarrow\!\!\!\!\!\!\!+\!\!\!\!\!\!\longrightarrow & & G/H, \\ & \searrow & & \swarrow & \\ & & G/H\cdot Sp(1) & & \end{array} \tag{4.1}$$

where $G/H\cdot Sp(1)$ are precisely the Wolf spaces [**132**].

REMARK 4.2. Note that a homogeneous 3-Sasakian manifold is necessarily simply-connected with the exception of the real projective space. This is in sharp contrast with the Sasakian-Einstein case. Also notice that a 3-Sasakian manifold can be Riemannian homogeneous (i.e., the full isometry group acts transitively) but not 3-Sasakian homogeneous. This is true for the lens spaces $\Gamma\backslash S^3$, with $|\Gamma| > 2$. Observe that $\mathbb{Z}_k\backslash S^3$, $k > 2$, is a homogeneous Sasakian-Einstein manifold but not 3-Sasakian homogeneous.

Theorem 4.2.6 does not specify what is the 3-Sasakian metric on the coset G/H. In Section 7.2 we will describe a quotient construction of the 3-Sasakian homogeneous spaces with $G = SU(n+1)$ and $G = SO(n+1)$. Here we quote a theorem of Bielawski [**15**], which gives an explicit description of these metrics in all cases.

THEOREM 4.2.7. *Let $\mathcal{S} = G/H$ be one of the spaces in Theorem 4.2.6. and let $\mathfrak{g} = \mathfrak{h} \oplus \mathfrak{m}$ be the corresponding decomposition of the Lie algebras. Then there is a natural decomposition $\mathfrak{m} = \mathfrak{sp}(1) \oplus \mathfrak{m}'$ and the metric g on $\mathcal{S}$ is given in terms of the scalar product on $\mathfrak{m}$*

$$\|m\|^2 = - < \sigma, \sigma > - \frac{1}{2} < m', m' >,$$

where $\sigma \in \mathfrak{sp}(1)$, $m' \in \mathfrak{m}'$, and $< \cdot, \cdot >$ is the Killing form on $\mathfrak{g}$. In particular, the metric g is not naturally reductive with respect to the homogeneous structure on $\mathcal{S}$.

REMARK 4.3. In the case when $\mathcal{S}$ is of constant curvature the canonical metric on S^{4n+3} (or $\mathbb{RP}^{4n+3}$) is *not* the standard homogeneous metric on the homogeneous space $Sp(n+1)/Sp(n)$ (or $Sp(n+1)/Sp(n) \times \mathbb{Z}_2$) with respect to the reductive decomposition $\mathfrak{sp}(n+1) \simeq \mathfrak{sp}(n) + \mathfrak{m}$. It is, of course, the standard homogeneous metric with respect to the naturally reductive decomposition $\mathfrak{o}(4n+4) \simeq \mathfrak{o}(4n+3) + \mathfrak{m}$. This is quite special to the sphere and orthogonal group. In general the 3-Sasakian homogeneous metrics are not naturally reductive with respect to any reductive decomposition.

5. 3-Sasakian Cohomology

In this section we will describe some cohomological properties of 3-Sasakian manifolds $\mathcal{S}$. We prove a vanishing theorem and then derive a relation between the Betti numbers of $\mathcal{S}$ and the Betti numbers of the associated orbifolds $\mathcal{Z}$ and $\mathcal{O}$. We conclude with various implications of these relations in the case $\mathcal{S}$ is regular.

5.1. Sasakian Manifolds and Harmonic Theory. We start by recalling some old results about harmonic forms on Sasakian manifolds due to Tachibana [**116**]. Let $(\mathcal{S}, g)$ be a compact Sasakian manifold of dimension $2m+1$ with Sasakian structure $\{\xi, \eta, \Phi\}$ and let $\Omega^p(\mathcal{S})$ be the space of smooth p-forms on $\mathcal{S}$. Furthermore, let $\mathcal{H}^p(\mathcal{S}) = \{u \in \Omega^p(\mathcal{S}) : du = 0 = d * u\}$ denote the finite-dimensional space of harmonic p-forms. By Hodge theory any harmonic form u is necessarily invariant under the isometry group $I(\mathcal{S}, g)$ The tensor Φ extends to an endomorphism of $\Omega^p(\mathcal{S})$ by setting

$$(\Phi u)(X_1, X_2, \ldots, X_p) = \sum_{i=1}^{p} u(X_1, \ldots, \Phi X_i, \ldots, X_p).$$

With this notation we have [**116**]

THEOREM 5.1.1. : *Let $u \in \mathcal{H}^p(\mathcal{S})$, $p \leq m$. Then $\xi \lfloor u = 0$, and $\Phi(u) \in \mathcal{H}^p(\mathcal{S})$.*

PROOF. The first statement is easy to prove in the case $p = 1$. Indeed, let $u = \alpha + f\eta$ be a closed invariant 1-form, where $\alpha(\xi) = 0$, and f is a function. The vanishing of the Lie derivative of u along ξ implies that $0 = d(\xi \lfloor u) = df$, so that f is a constant and $0 = d\alpha + fd\eta$. Then

$$0 = \int_{\mathcal{S}} d(\alpha \wedge (d\eta)^{m-1} \wedge \eta) = -\int_{\mathcal{S}} f(d\eta)^m \wedge \eta.$$

Since $(d\eta)^m \wedge \eta$ is a non-zero multiple of the volume form of $\mathcal{S}$, we obtain $f = 0$ and $\xi \lfloor u = 0$. The general case of the original proof uses an explicit computation in local coordinates and we omit it here. The second statement follows immediately from 5.1.1 and the fact that Φ preserves horizontal subspaces. □

Let us define the following $\mathcal{I} : \mathcal{H}^p(\mathcal{S}) \to \mathcal{H}^p(\mathcal{S})$ endomorphism for $p \leq m$:

$$(\mathcal{I}u)(X_1, \ldots, X_p) = u(\Phi X_1, \ldots, \Phi X_p)$$

The basic identity 2.1.3(i) together with Theorem 5.1.2 shows that $\mathcal{I}u$ is a linear combination of $(\Phi)^k u$ for $0 \leq k \leq p$. Thus $\mathcal{I}$ also maps $\mathcal{H}^p(\mathcal{S})$ into itself. The following proposition is now a simple consequence of the definition 5.1.3 and Theorem 5.1.2 [**19, 20**]:

PROPOSITION 5.1.2. *Let $\mathcal{I} : \mathcal{H}^p(\mathcal{S}) \to \mathcal{H}^p(\mathcal{S})$ and $p \leq m$. Then $\mathcal{I} \circ \mathcal{I} = (-1)^p$. In particular, when p is odd, $\mathcal{I}$ defines an almost complex structure on the vector space $\mathcal{H}^p(\mathcal{S})$.*

COROLLARY 5.1.3. *Let $(\mathcal{S}, g)$ be a compact Sasakian manifold of dimension $2m + 1$. Then the Betti numbers b_p for p odd and $p \leq m$ are even.*

In the case of compact Sasakian-Einstein manifolds this and the fact that the fundamental group is finite are the only known general topological restrictions on $\mathcal{S}$. Under some additional curvature conditions we can get further restrictions. For example, it is known [**19**] that a compact simply-connected Sasakian manifold of positive sectional curvature is isometric to a sphere. For other similar results see [**19, 57**] and references therein.

5.2. A Vanishing Theorem. Now, let $(\mathcal{S}, g)$ be a compact 3-Sasakian manifold of dimension $4n + 3$ and 3-Sasakian structure $\{\xi^a, \eta^a, \Phi^a\}$. Throughout this section we shall suppose that $p \leq 2n + 1$. Referring to the splitting of the tangent bundle of $\mathcal{S}$ into $T\mathcal{S} = \mathcal{V}_3 \oplus \mathcal{H}$, we shall say that a p-form $u \in \Omega^p(\mathcal{S})$ has bidegree $(i, p - i)$ if it is a section of the subbundle of $\bigwedge^p T^*\mathcal{S}$ isomorphic to the dual of $\bigwedge^i \mathcal{V}_3 \otimes \bigwedge^{p-i} \mathcal{H}$. In particular, u is called 3-horizontal if it has bidegree $(0, p)$, or equivalently if $\xi^a \lfloor u = 0$ for $a = 1, 2, 3$. An element $\omega \in \Omega^p(\mathcal{S})$ is called invariant if $h^*\omega = \omega$ for all $h \in Sp(1)$. In the regular case, there is a principal $Sp(1)$-bundle $\pi : \mathcal{S} \to \mathcal{O}$, and ω is both 3-horizontal and invariant if and only if it is the pullback $\pi^*\hat{\omega}$ of a form $\hat{\omega}$ on the quaternionic Kähler base $\mathcal{O}$. Now the curvature forms $\bar{\Phi}^a$ defined in the proof of Theorem 3.3.3 are horizontal with respect to the foliation $\mathcal{F}_3$. The Killing fields ξ^a transform according to the adjoint representation of $Sp(1)$, and the same is true of the associated triples η^a, $d\eta^a$, and Φ^a. For example, if $h \in Sp(1)$, we may write

$$h_*\Phi^a = \sum_b h^{ab}\Phi^b, \quad a = 1, 2, 3,$$

where h^{ab} are components of the image of h in $Sp(1)/\mathbb{Z}_2 \cong SO(3)$. The 3-forms

$$(5.1) \qquad \Upsilon = \eta^1 \wedge \eta^2 \wedge \eta^3, \qquad \Theta = \sum_a \eta^a \wedge \bar{\Phi}^a = \sum_a \eta^a \wedge d\eta^a + 6\Upsilon$$

have respective bidegrees $(3,0),(1,2)$, and are clearly invariant. Their exterior derivatives are

$$(5.2) \qquad d\Upsilon = \eta^1 \wedge \eta^2 \wedge \bar{\Phi}^3 + \eta^2 \wedge \eta^3 \wedge \bar{\Phi}^1 + \eta^3 \wedge \eta^1 \wedge \bar{\Phi}^2, \qquad d\Theta = \Omega + 2d\Upsilon,$$

where the 4-form Ω is defined in section 3.3. In fact, Ω is the canonical 4-form determined by the quaternionic structure of Proposition 2.2.3 of the subbundle $\mathcal{H}$, and is the pullback of the fundamental 4-form $\hat{\Omega}$ on the quaternionic Kähler orbifold $\mathcal{O}$ (see section 3.2).

Theorem 5.1.2(i) implies that any harmonic p-form with $p \leq 2n+1$ on the compact 3-Sasakian manifold $\mathcal{S}$ is 3-horizontal. Apply 5.1.1 so as to obtain $\Phi^a : \mathcal{H}^p(\mathcal{S}) \to \mathcal{H}^p(\mathcal{S}), \quad a = 1,2,3, \quad p \leq 2n+1$, and 5.1.3 to get

$$(\mathcal{I}^a u)(X_1, X_2, \ldots, X_p) = u(\Phi^a X_1, \Phi^a X_2, \ldots, \Phi^a X_p).$$

Now, using the basic identities of Proposition 2.2.3 we can generalize Proposition 5.1.4 to get the following result due to Kuo [**83**]:

PROPOSITION 5.2.1. *Let $\mathcal{I}^a : \mathcal{H}^p(\mathcal{S}) \to \mathcal{H}^p(\mathcal{S})$, $a = 1,2,3$, and $p \leq 2n+1$. Then*

$$\mathcal{I}^b \circ \mathcal{I}^a = (-\delta^{ab})^p \mathbf{I} + \sum_c (\epsilon^{abc})^p \mathcal{I}^c.$$

In particular, when p is odd, $\{\mathcal{I}^1, \mathcal{I}^2, \mathcal{I}^3\}$ defines an almost quaternionic structure on the vector space $\mathcal{H}^p(\mathcal{S})$.

We are now ready to prove the main theorem of this section:

THEOREM 5.2.2. *Let $u \in \mathcal{H}^p(\mathcal{S})$, $p \leq 2n+1$.*
(i) If p is odd then $u \equiv 0$.
(ii) If p is even then $\mathcal{I}^a u = u$ for $a = 1,2,3$.

PROOF. Let $u \in \mathcal{H}^p(\mathcal{S})$. We shall in fact show that $\mathcal{I}^1 u = \mathcal{I}^2 u$ irrespective of whether p is even or odd; the result then follows from the identities in Proposition 5.2.1 and symmetry between the indices 1,2,3. By 5.2, we may choose an isometry $h \in Sp(1)$ so that $h_* \Phi^1 = \Phi^2$. Both u and $\mathcal{I}^1 u$ are harmonic, so $h^* u = u$ and

$$(\mathcal{I}^1 u)(X_1, \ldots, X_p) = (h_*(\mathcal{I}^1 u))(X_1, \ldots, X_p) = u((h_*\Phi^1)(X_1), \ldots, (h_*\Phi^1)(X_p)) =$$
$$= u(\Phi^2 X_1, \ldots, \Phi^2 X_p) = (\mathcal{I}^2 u)(X_1, \ldots, X_p).$$

□

COROLLARY 5.2.3. *Let $(\mathcal{S}, g)$ be a compact 3-Sasakian manifold of dimension $4n+3$. Then the odd Betti numbers b_{2k+1} of $\mathcal{S}$ are all zero for $0 \leq k \leq n$.*

We should point out that Corollary 5.2.3 does not apply to compact Sasakian or even Sasakian-Einstein manifolds. In [**23**] the authors construct examples of Sasakian-Einstein manifolds with certain non-vanishing odd Betti numbers within the range given in Corollary 5.2.3. For example, in dimension 7 there are circle bundles over Fermat hypersurfaces in $\mathbb{CP}^3$, as well as circle bundles over certain complete intersections that admit Sasakian-Einstein structures and have $b_3 \neq 0$.

These are the only known examples of Sasakian-Einstein manifolds which cannot admit any 3-Sasakian structure.

5.3. 3-Sasakian Cohomology As Primitive Cohomology. We are going to consider connection between the cohomology of $\mathcal{S}$ and that of $\mathcal{Z}$ and $\mathcal{O}$. We will use the vanishing theorem and orbifold Gysin sequence arguments for the diagram of orbifold bundles of $\diamondsuit(\mathcal{S})$:

$$\begin{array}{ccc} \mathcal{S} & \longrightarrow & \mathcal{Z}. \\ \downarrow & \swarrow & \\ \mathcal{O} & & \end{array} \tag{5.3}$$

PROPOSITION 5.3.1. *Let $\mathcal{S}$ be a compact 3-Sasakian manifold of dimension $4n+3$ and $\mathcal{Z} = \mathcal{S}/S^1$ be the twistor space. Then $b_p(\mathcal{S}) = b_p(\mathcal{Z}) - b_{p-2}(\mathcal{Z})$, for $p \leq 2n+1$. In particular, all odd Betti numbers of $\mathcal{Z}$ vanish.*

PROOF. The result follows form the rational Gysin sequence applied to the orbifold fibration $S^1 \to \mathcal{S} \to \mathcal{Z}$. First, note that the bundle $S^1 \longrightarrow \mathcal{S} \longrightarrow \mathcal{Z}$ is a circle V-bundle over a compact Kähler-Einstein orbifold $\mathcal{Z}$. As explained in Section 3, up to a possible $\mathbb{Z}_2$ cover, $\mathcal{S}$ is the total space of the unit circle bundle in the dual of the contact line V-bundle on $\mathcal{Z}$, and the Kähler-Einstein metric of $\mathcal{Z}$ arises in accordance with the orbifold version of the Kobayashi's theorem [**74, 22**] . It follows that the connecting homomorphism δ is given by wedging with a non-zero multiple of the Kähler form of $\mathcal{Z}$. When $\mathcal{Z}$ is smooth this is well-known to be injective so long as $p \leq 2n+2$. However, the Lefschetz decomposition is equally true for compact orbifolds and the result still holds in this more general situation [**22**]. The Gysin sequence therefore reduces to a series of short exact sequences up to and including $H^{2n+1}(\mathcal{S})$, and the proposition follows. □

PROPOSITION 5.3.2. *Let $\mathcal{S}$ be a compact 3-Sasakian manifold of dimension $4n+3$ and let $\mathcal{O} = \mathcal{S}/\mathcal{F}_3$. Then $b_{2p}(\mathcal{S}) = b_{2p}(\mathcal{O}) - b_{2p-4}(\mathcal{O})$, for $p \leq 2n+1$.*

PROOF. The result follows form the Gysin sequence applied to the orbifold fibration $\mathcal{L} \to \mathcal{S} \to \mathcal{O}$. Since the principal orbit of the $Sp(1)$ action (or generic leaf $\mathcal{L}$) is either S^3 or $SO(3)$ the usual Gysin sequence argument applies as well in this situation (see the Appendix). We have

$$\cdots \to H^i(\mathcal{S},\mathbb{Q}) \to H^{i-3}(\mathcal{O},\mathbb{Q}) \xrightarrow{\delta} H^{i+1}(\mathcal{O},\mathbb{Q}) \to H^{i+1}(\mathcal{S},\mathbb{Q}) \to H^{i-2}(\mathcal{S},\mathbb{Q}) \to \cdots$$

and the statement of the proposition follows easily from the vanishing of the odd Betti numbers of $\mathcal{S}$. □

Recall that the vector space of primitive harmonic p-forms $\mathcal{H}^p_0(\mathcal{Z},\mathbb{Q})$ of the orbifold $\mathcal{Z}$ is isomorphic to the cokernel of the injective mapping $L_{\mathcal{Z}} : \mathcal{H}^{p-2}(\mathcal{Z}) \hookrightarrow \mathcal{H}^p(\mathcal{Z})$, $p \leq 2n$ defined by wedging with the Kähler 2-form. We define the primitive Betti numbers $b^0_p(\mathcal{Z})$ of $\mathcal{Z}$ as the dimension of $\mathcal{H}^p_0(\mathcal{Z})$. Proposition 5.3.1 says that the primitive Betti numbers of $\mathcal{Z}$ are the usual Betti numbers of $\mathcal{S}$ and it follows from the fact that for, $0 \leq r \leq 2n+1$, an r-form on $\mathcal{S}$ is harmonic if and only if it is the lift of a primitive harmonic form on $\mathcal{Z}$ [**22**]. Similarly, the vector space of primitive harmonic p-forms $\mathcal{H}_0^p(\mathcal{O}, \mathbb{Q})$ of the orbifold $\mathcal{O}$ is isomorphic to the cokernel of the injective mapping $L_{\mathcal{O}} : \mathcal{H}^{p-4}(\mathcal{O}) \hookrightarrow \mathcal{H}^p(\mathcal{O}), \quad p \leq 2n+2$ defined by wedging with the quaternionic Kähler 4-form Ω. The injectivity of this mapping is well-known in the smooth case [**21, 50, 77**] and it extends to the orbifold case. We define the primitive Betti numbers $b_p^0(\mathcal{O})$ of $\mathcal{O}$ as the dimension of $\mathcal{H}_0^p(\mathcal{O})$. Proposition 5.3.2 says that the primitive Betti numbers of $\mathcal{O}$ are the usual Betti numbers of $\mathcal{S}$. Again, Proposition 5.3.3 is a consequence of the fact that an r-form on $\mathcal{S}$ is harmonic if and only if it is the lift of a primitive harmonic form on $\mathcal{O}$, $0 \leq r \leq 2n+1$.

5.4. Regular 3-Sasakian Cohomology, Finiteness, and Rigidity. In this Section we shall assume that $\mathcal{S}$ is regular and, hence, both $\mathcal{Z}$ and $\mathcal{O}$ are smooth. In this instance, using the results of the previous section, one can easily translate all the results about strong rigidity of positive quaternion Kähler manifolds [**85, 88, 108**] (see the chapter in this volume on Quaternionic Kähler Manifolds by Salamon) to compact regular 3-Sasakian manifolds. In particular, we get

PROPOSITION 5.4.1. *Let $\mathcal{S}$ be a compact regular 3-Sasakian manifold of dimension $4n+3$. Then $\pi_1(\mathcal{S}) = 0$ unless $\mathcal{S} = \mathbb{RP}^{4n+3}$ and*

$$\pi_2(\mathcal{S}) = \begin{cases} \mathbb{Z} & \textit{iff } \mathcal{S} = SU(n+2)/S(U(n) \times U(1)), \\ \textit{finite} & \textit{otherwise.} \end{cases}$$

Furthermore, up to isometries, for each $n \geq 1$ there are only finitely many regular 3-Sasakian manifolds $\mathcal{S}$.

PROOF. Using the long exact homotopy sequence for the vertical map in diagram 5.1, this follows from the strong rigidity theorem of LeBrun and Salamon [**88, 85**] for positive quaternionic Kähler manifolds, and Salamon's theorem that a positive quaternionic Kähler manifold with vanishing Marchiafava-Romani class must be $\mathbb{HP}^n$. □

n	relation on Betti numbers or coefficients thereof
2	$b_2 = b_4$
3	$b_2 = b_6$
4	$2b_2 + b_4 = b_6 + 2b_8$
5	$5b_2 + 4b_4 = 4b_8 + 5b_{10}$
6	$5b_2 + 5b_4 + 2b_6 = 2b_8 + 5b_{10} + 5b_{12}$
7	$7b_2 + 8b_4 + 5b_6 = 5b_{10} + 8b_{12} + 7b_{14}$
8	$28b_2 + 35b_4 + 27b_6 + 10b_8 = 10b_{10} + 27b_{12} + 35b_{14} + 28b_{16}$
9	$12b_2 + 16b_4 + 14b_6 + 8b_8 = 8b_{12} + 14b_{14} + 16b_{16} + 12b_{18}$
10	$15, 21, 20, 14, 5$
16	$40, 65, 77, 78, 70, 55, 35, 12$
28	$126, 225, 299, 350, 380, 391, 385, 364, 330, 285, 231, 170, 104, 35$

Table 1

PROPOSITION 5.4.2. *The Betti numbers of a regular compact 3-Sasakian manifold $\mathcal{S}$ of dimension $4n+3$ satisfy*
(i) $b_2 \leq 1$, with equality iff $\mathcal{S} = SU(l+2)/S(U(l) \times U(1))$,

(ii) $\sum_{k=1}^{n} k(n+1-k)(n+1-2k)b_{2k} = 0.$

PROOF. (i) follows from Proposition 5.4.1 and (ii) for Salamon's relation on Betti numbers of $\mathcal{O}$ via Theorem 5.3.2. □

The following is a 3-Sasakian version of a theorem of Salamon [**54**]:

PROPOSITION 5.4.3. *Let $\mathcal{S}$ be a regular compact 3-Sasakian manifold of dimension $4n+3$. If $n = 3, 4$ and $b_4 = 0$, then $\mathcal{S}$ is either a sphere S^{4n+3} or a real projective space $\mathbb{RP}^{4n+3}$.*

The linear Betti number relations in Proposition 5.4.2(ii) exhibit an interesting symmetry of the coefficients which, for lower values of n, are listed in Table 1.

One can compute the Poincaré polynomials of all known regular 3-Sasakian manifolds, that is 3-Sasakian homogeneous space of Theorem 4.2.6. We get [**54**]

PROPOSITION 5.4.4. *The Poincaré polynomials of the homogeneous 3-Sasakian manifolds are as given in Table 2.*

G	H	$P(G/H,t)$
$SU(n+2)$	$SU(n)\times_{\mathbb{Z}_n} T^1$	$\sum_{i=0}^{n}(t^{2i}+t^{4n+3-2i})$
$SO(2k+3)$	$SO(2k-1)\times SU(2)$	$\sum_{i=0}^{k-1}(t^{4i}+t^{8k-1-4i})$
$Sp(n+1)$	$Sp(n)$	$1+t^{4n+3}$
$SO(2l+4)$	$SO(2l)\times SU(2)$	$t^{2l}+t^{6l+3}+\sum_{i=0}^{l}(t^{4i}+t^{8l+3-4i})$
E_6	$SU(6)$	$1+t^6+t^8+t^{12}+t^{14}+t^{20}+\cdots$
E_7	$\mathrm{Spin}(12)$	$1+t^8+t^{12}+t^{16}+t^{20}+t^{24}+t^{32}+\cdots$
E_8	E_7	$1+t^{12}+t^{20}+t^{24}+t^{32}+t^{36}+t^{44}+t^{56}+\cdots$
F_4	$Sp(3)$	$1+t^8+t^{23}+t^{31}$
G_2	$SU(2)$	$1+t^{11}$

Table 2

We conclude this section with a translation of two well-known classification results for positive quaternionic Kähler manifolds.

THEOREM 5.4.5. *Let $(\mathcal{S}, g)$ be a compact regular 3-Sasakian manifold of dimension $4n+3$. If $n < 3$ then then $\mathcal{S} = G/H$ is homogeneous, and hence one of the spaces listed in Theorem 4.2.6.*

The $n = 0$ case is trivial and it was an observation made by Tanno [**119**]. The $n = 1$ case is based on [**63, 49**] and it was first observed in [**47, 11**]. The $n = 2$ case is based on [**104**] and was stated in [**25**].

6. Killing Spinors and G_2-Structures

In this section we discuss some additional properties of Sasakian and Sasakian-Einstein manifolds which are connected with spin structure and eigenvalues of the Dirac operator.

6.1. Killing Spinors.

DEFINITION 6.1.1. *Let (M, g) be a complete n-dimensional Riemannian spin manifold, and let $S(M)$ be the spin bundle of M and ψ a smooth section of $S(M)$. We say that ψ is a Killing spinor if*

$$\nabla_X \psi = \alpha \cdot X \cdot \psi, \qquad \forall X \in \Gamma(TM),$$

where ∇ is the Levi-Civita connection of g and $X \cdot \psi$ denotes the Clifford product of X and ψ. We say that ψ is imaginary when $\alpha \in \mathrm{Im}(\mathbb{C}^*)$, ψ is parallel if $\alpha = 0$ and ψ is real if $\alpha \in \mathrm{Re}(\mathbb{C}^*)$.

From the point of view of Einstein geometry the importance of Killing spinors is an immediate consequence of the following theorem of Friedrich [**45**]:

THEOREM 6.1.2. *Let (M, g) be an n-dimensional; complete Riemannian spin manifold with a Killing spinor. Then M is Einstein with Einstein constant $\lambda = 4(n-1)\alpha^2$. In particular, when $\alpha \in \mathrm{Re}(\mathbb{C}^*)$, M is compact of positive scalar curvature.*

On the other hand, Friedrich showed that if M is a compact spin manifold of positive scalar curvature and R_0 is the minimum of the scalar curvature, then for all eigenvalues β of the Dirac operator D one has $\beta^2 \geq \frac{1}{4}\frac{nR_0}{n-1}$ [**45**]. If the equality holds than it follows that the corresponding eigenspinor must be a Killing spinor with $\alpha = \pm\frac{1}{2}[\frac{R_0}{n(n-1)}]^{1/2}$. We have the following important property of manifolds with Killing spinors [**11**]:

THEOREM 6.1.3. *Let (M^n, g) be a connected Riemannian spin manifold admitting a non-trivial Killing spinor with $\alpha \neq 0$. Then (M, g) is locally irreducible. Furthermore, if M is locally symmetric, or $n \leq 4$, then M is a space of constant sectional curvature equal $4\alpha^2$.*

From now on we will be interested only in the case of real Killing spinors. It was Friedrich and Kath [**46, 47**] who first noticed that in some low odd dimensions the existence of real Killing spinors leads naturally to the existence of Sasakian-Einstein or 3-Sasakian structures. Later, the problem found a simple classification in terms of the holonomy of the associated metric cone $C(M)$ [**10**]. First, we have the following definition:

DEFINITION 6.1.4. *We say that M is of type (p, q) if it carries exactly p linearly independent real Killing spinors with $\alpha > 0$ and exactly q linearly independent real Killing spinors with $\alpha < 0$, or vice versa.*

For, example, the standard sphere S^n is of type $(2^{[n/2]}, 2^{[n/2]})$. Bär shows that when M admits a real Killing spinor then the cone $(C(M), \bar{g})$ has a parallel spinor. In particular, $C(M)$ is always Ricci-flat and, when M is simply-connected, then only a few holonomy groups $\mathrm{Hol}(\bar{g})$ are possible [**129**]:

THEOREM 6.1.5. [**10**] *Let (M, g) be a simply-connected Riemannian spin manifold admitting a non-trivial Killing spinor and let $\mathrm{Hol}(\bar{g})$ be the holonomy group of the metric cone $(C(M), \bar{g})$. Then there are only the following 6 possibilities for the triple $(\dim M, \mathrm{Hol}(\bar{g}), (p, q))$:*

dim M	$\mathrm{Hol}(\bar{g})$	(p,q)
n	id	$(2^{[n/2]}, 2^{[n/2]})$
$4m+1$	$SU(2m+1)$	$(1,1)$
$4m+3$	$SU(2m+2)$	$(2,0)$
$4m+3$	$Sp(m+1)$	$(m+2,0)$
7	Spin(7)	$(1,0)$
6	G_2	$(1,1)$

Here $m \geq 1$, and $n > 1$.

The first case is special as M is the n-dimensional round sphere. Since M is assumed to be simply-connected, in the next two cases, by Proposition 2.1.4, M must be Sasakian-Einstein. In the fourth case above it follows from Definition 2.2.1 that M is 3-Sasakian. Specifically, we get the following theorem [**10**]:

THEOREM 6.1.6. [**10**] *Let (M^n, g) be a complete simply-connected Riemannian spin manifold admitting a non-trivial Killing spinor with $\alpha > 0$ or $\alpha < 0$. If $n = 4m+1$ with $m \geq 1$, then there are two possibilities:*
(i) $(M,g) = (S^n, g_{can})$,
(ii) (M,g) is of type $(1,1)$ and it is a Sasakian-Einstein manifold.

Conversely, if (M,g) is a complete simply-connected Sasakian-Einstein manifold of dimension $4m+1$, then M carries Killing spinors with $\alpha > 0$ and $\alpha < 0$.

REMARK 6.1. Note that in the converse statement we do not need to assume that M is spin. When $\pi_1(M) = 0$ this is automatic by Corollary 2.1.6. When $\pi_1(M) \neq 0$ then the 'if' part of Theorem 6.1.6 can be generalized and we still get two possibilities: (i) either M is a spin spherical space form, or (ii) it is of type $(1,1)$ with a Sasakian-Einstein structure and $\mathrm{Hol}(\bar{g}) = SU(m+1)$ [**129**].

THEOREM 6.1.7. [**10**] *Let (M^n, g) be a complete simply-connected Riemannian spin manifold admitting a non-trivial Killing spinor with $\alpha > 0$ or $\alpha < 0$. If $n = 4m+3$, $m \geq 2$, then there are three possibilities:*
(i) $(M,g) = (S^n, g_{can})$,
(ii) (M,g) is a Sasakian-Einstein manifolds of type $(2,0)$, but (M,g) is not 3-Sasakian,
(iii) (M,g) is of type $(m+2,0)$ and it is 3-Sasakian.

Conversely, if (M,g) is a complete simply-connected 3-Sasakian manifold, of dimension $4m+3$ which is not of constant curvature, then M carries $(m+2)$ linearly independent Killing spinors with $\alpha > 0$. If (M,g) is a complete simply-connected Sasakian-Einstein manifold of dimension $4m+3$ which is not 3-Sasakian then M carries 2 linearly independent Killing spinors with $\alpha > 0$.

REMARK 6.2. Note that in Theorem 6.1.7(ii) we are not excluding the possibility of M having another 3-Sasakian structure with a different metric g'. We are only saying that the holonomy group $\mathrm{Hol}(\bar{g}) = SU(2m+2)$ rather than $Sp(m+1) \subset SU(2m+2)$, which, by definition, means that g cannot be 3-Sasakian. However, we are not aware of any such example. We have excluded $\dim(M) = 7$ because in this case we have one more possibility due to Theorem 6.1.5 and we want to discuss the associated geometry in more detail later. Again, one can generalize Theorem 6.1.7 to $\pi_1(M) \neq 0$. For the full list of possible holonomy groups $\mathrm{Hol}(\bar{g})$ see [**128**]. The coresponding M are then only locally Sasakian-Einstein or locally 3-Sasakian

[**100, 103**]. The problem of the existence of Killing spinors on a Sasakian-Einstein or 3-Sasakian manifold with $\pi_1(M) \neq 0$ is, however, more subtle.

COROLLARY 6.1.8. *Let $(\mathcal{S}, g)$ be a compact Sasakian-Einstein manifold of dimension $2m+1$. Then $\mathcal{S}$ is locally symmetric if and only if $\mathcal{S}$ is of constant curvature. Moreover, $(\mathcal{S}, g)$ is locally irreducible as a Riemannian manifold.*

PROOF. If necessary, go to the universal cover $\tilde{\mathcal{S}}$. This is a compact simply-connected Sasakian-Einstein manifold; hence, it admits a non-trivial Killing spinor by Theorems 6.1.6 and 6.1.7. The statement then follows from the Theorem 6.1.3. □

COROLLARY 6.1.9. *Let $(\mathcal{S}, g)$ be a compact Sasakian-Einstein manifold of dimension $2m+1$. Then* $\mathrm{Hol}(g) = SO(2m+1)$.

PROOF. Let us consider universal cover $\tilde{\mathcal{S}}$. This is a compact simply-connected Sasakian-Einstein manifold; hence, it admits a non-trivial Killing spinor. By the previous corollary, it can be symmetric if only if it is isomorphic to a space of constant curvature, that is, a sphere. Then $\mathcal{S}$ is a spherical space form and $\mathrm{Hol}(g) = SO(2m+1)$. Assume $\mathcal{S}$ is not locally symmetric. By Corollary 6.1.8 $\mathcal{S}$ is locally Riemannian irreducible, so for dimensional reasons and Berger's famous classification theorem [**13**], the only possibilities for the restricted holonomy group $\mathrm{Hol}^0(g)$ are $SO(2m+1)$ and G_2 in dimension 7. But G_2 holonomy implies Ricci-flat and, hence, not Sasakian-Einstein. Hence, the restricted holonomy group $\mathrm{Hol}^0(g) = SO(2m+1)$. Since $\mathcal{S}$ is orientable this coincides with the holonomy group $\mathrm{Hol}(g)$. □

6.2. G_2-Structures. Recall, that geometrically G_2 is defined to be the Lie group acting on $\mathbb{R}^7$ and preserving the 3-form

$$
\begin{aligned}
\varphi = {} & \alpha_1 \wedge \alpha_2 \wedge \alpha_3 + \alpha_1 \wedge (\alpha_4 \wedge \alpha_5 - \alpha_6 \wedge \alpha_7) \\
& + \alpha_2 \wedge (\alpha_4 \wedge \alpha_6 - \alpha_7 \wedge \alpha_5) + \alpha_3 \wedge (\alpha_4 \wedge \alpha_7 - \alpha_5 \wedge \alpha_6),
\end{aligned} \tag{6.1}
$$

where $\{\alpha_i\}_{i=1}^7$ is a fixed orthonormal basis of the dual of $\mathbb{R}^7$. A G_2 structure on a 7-manifold M is, by definition, a reduction of the structure group of the tangent bundle to G_2. This is equivalent to the existence of a global 3-form $\varphi \in \Omega^3(M)$ which may be written locally as 6.1. Such a 3-form defines an associated Riemannian metric, an orientation class, and a spinor field of constant length. The following terminology is due to Gray [**59**]:

DEFINITION 6.2.1. *Let (M, g) be a complete 7-dimensional Riemannian manifold. We say that that (M, g) has weak holonomy G_2 if there exist a global 3-form $\varphi \in \Omega^3(M)$ which locally can be written in terms of a local orthonormal basis as in 6.1, and $d\varphi = c \star \varphi$, where $\star$ is the Hodge star operator associated to g and c is a constant whose sign is fixed by an orientation convention.*

The equation $d\varphi = c \star \varphi$ implies that φ is 'nearly parallel' in the sense that only a 1-dimensional component of $\nabla\varphi$ is different from zero [**42**]. Thus, a weak holonomy G_2 structure is sometimes called a *nearly parallel G_2 structure*. The case of $c = 0$ is somewhat special. In particular, it is known [**109**] that the condition $d\varphi = 0 = d\star\varphi$ is equivalent to the condition that φ be parallel, i.e., $\nabla\varphi = 0$ which is equivalent to the condition that the metric g has holonomy group contained in

G_2. For a discussion of this very interesting and very difficult case, see the article by D. Joyce in this volume. The following theorem provides the connection with the previous discussion on Killing spinors [**10**]

THEOREM 6.2.2. *Let (M, g) be a complete 7-dimensional Riemannian manifold with weak holonomy G_2. Then the holonomy group* $\mathrm{Hol}(\bar{g})$ *of the metric cone $(C(M), \bar{g})$ is contained in* $\mathrm{Spin}(7)$. *In particular, $C(M)$ is Ricci-flat and M is Einstein with positive Einstein constant $\lambda = 6$.*

REMARK 6.3. The sphere S^7 with its constant curvature metric is isometric to the isotropy irreducible space $Spin(7)/G_2$. The fact that G_2 leaves invariant (up to constants) a unique 3-form and a unique 4-form on $\mathbb{R}^7$ implies immediately that this space has weak holonomy G_2.

DEFINITION 6.2.3. *Let (M, g) be a complete 7-dimensional Riemannian manifold. We say that g is a proper G_2-metric if* $\mathrm{Hol}(\bar{g}) = \mathrm{Spin}(7)$.

THEOREM 6.2.4. *Let (M^7, g) be a complete simply-connected Riemannian spin manifold of dimension 7 admitting a non-trivial Killing spinor with $\alpha > 0$ or $\alpha < 0$. Then there are four possibilities:*
(i) (M, g) is of type $(1,0)$ *and it is a proper G_2-manifold,*
(ii) (M, g) is of type $(2,0)$ *and it is a Sasakian-Einstein manifold, but (M, g) is not 3-Sasakian,*
(iii) (M, g) is of type $(3,0)$ *and it is 3-Sasakian,*
(iv) $(M, g) = (S^7, g_{can})$ and is of type $(8,8)$.

Conversly, if (M, g) is a compact simply-connected proper G_2-manifold then it carries a Killing spinor with $\alpha > 0$. If (M, g) is a compact simply-connected Sasakian-Einstein 7-manifold which is not 3-Sasakian then M carries 2 *linearly independent Killing spinors with $\alpha > 0$. Finally, if (M, g) is a 3-Sasakian 7-manifold, which is not of constant curvature, then M carries* 3 *linearly independent Killing spinors with $\alpha > 0$.*

REMARK 6.4. The four possibilities of the Theorem 6.2.4 correspond to the sequence of inclusions

$$\mathrm{Spin}(7) \supset SU(4) \supset Sp(2) \supset \{\mathrm{id}\}.$$

All of the corresponding cases are examples of weak holonomy G_2 metrics. If we exclude the trivial case when the associated cone is flat, we have three types of the weak holonomy G_2 geometries. Following [**48**] we use the number of linearly independent Killing spinors to classify the types of weak holonomy G_2 geometries. We call these type I, II, and III corresponding to cases (i), (ii), and (iii) of Theorem 6.2.4, respectively.

REMARK 6.5. In the case $\pi_1(M) \neq 0$, then M is either a spin spherical space form or $\mathrm{Hol}(\bar{g})$ equals to $SU(4)$, $SU(4) \rtimes \mathbb{Z}_2$, $Sp(2)$, or $\mathrm{Spin}(7)$ of type $(2,0), (1,0), (3,0), (1,0)$, respectively. Hence, we have just one more possible geometry for M [**129**]. Note that in the case $\mathrm{Hol}(\bar{g}) = SU(4) \rtimes \mathbb{Z}_2$, the cone $C(M)$ is not Kähler so that M cannot be Sasakian, but it is locally so.

Recall that S^7, regarded as the space $Sp(2)/Sp(1)$ and fibering over S^4, admits a 'squashed' Einstein metric which does not have constant curvature. This metric also has weak holonomy G_2 since the associated cone metric has holonomy equal to $Spin(7)$ and therefore S^7 with this metric is a proper G_2-manifold. We can generalize this example to get [**54, 48**]:

THEOREM 6.2.5. *Let $(\mathcal{S}, g)$ be a 7-dimensional 3-Sasakian manifold. Then the metric g has weak holonomy G_2. Moreover, the second Einstein metric g' given by Theorem 3.4.1 has weak holonomy G_2. In fact g' is a proper G_2 metric.*

PROOF. For the second Einstein metric g' we have three mutually orthonormal 1-forms $\alpha^1 = \sqrt{t}\eta^1, \quad \alpha^2 = \sqrt{t}\eta^2, \quad \alpha^3 = \sqrt{t}\eta^3$, where t is the parameter of the canonical variation discussed in section 3.4. Let $\{\alpha^4, \alpha^5, \alpha^6, \alpha^7\}$ be local 1-forms spanning the annihilator of $\mathcal{V}_3$ in $T^*\mathcal{S}$ such that

$$\begin{aligned}\bar{\Phi}^1 &= 2(\alpha^4 \wedge \alpha^5 - \alpha^6 \wedge \alpha^7),\\ \bar{\Phi}^2 &= 2(\alpha^4 \wedge \alpha^6 - \alpha^7 \wedge \alpha^5),\\ \bar{\Phi}^3 &= 2(\alpha^4 \wedge \alpha^7 - \alpha^5 \wedge \alpha^6).\end{aligned}$$

Then the set $\{\alpha^1, \cdots \alpha^7\}$ forms a local orthonormal coframe for the metric g'. In terms of the 3-forms Υ and Θ of 5.1 we have $\varphi = \frac{1}{2}\sqrt{t}\Theta + \sqrt{t}^3\Upsilon$. One easily sees that this is of the type of Equation 6.1 and, therefore, defines a compatible G_2-structure. Moreover, a straightforward computation gives

$$d\varphi = \frac{1}{2}\sqrt{t}\Omega + \sqrt{t}(t+1)d\Upsilon, \qquad \star\varphi = -\frac{1}{2}td\Upsilon - \frac{1}{24}\Omega.$$

Thus, $d\varphi = c \star \varphi$ is solved with $\sqrt{t} = 1/\sqrt{5}$, and $c = -12/\sqrt{5}$. So g' has weak holonomy G_2. That g' is a proper G_2 metric is due to [**48**]. The idea is to use Theorem 6.2.4. Looking at the four possibilities given in that theorem, we see that it suffices to show that g' is not Sasakian-Einstein. The details are in [**48**]. □

EXAMPLE 6.1. 3-Sasakian 7-manifolds are plentiful and examples will be discussed in next section. These give, by Theorem 6.2.5, many examples of type I and type III geometries. Examples of simply-connected type I geometries that do not arise via Theorem 6.2.5 are the homogeneous Aloff-Wallach spaces $N_{k,l} = SU(3)/T^1_{k,l}$, with $\gcd(k,l) = 1$ and $(k,l) \neq (1,1)$ [**37, 11**] together with the homogeneous real Stiefel manifold $SO(5)/SO(3)$ [**36**]. All the known type II geometries are the 3 homogeneous examples from the list of Corollary 4.1.3(iii) (not 3-Sasakian) and the inhomogeneous simply-connected circle bundles over $P_k \times \mathbb{CP}^1$, where P_k is the del Pezzo surface with $2 < k < 9$ [**48**]. Actually, $N_{1,1}$ has three Einstein metrics. One is 3-Sasakian and is denoted by $\mathcal{S}(1,1,1)$ in section 8.4 below. The second is the proper G_2 metric of Theorem 6.2.5, while the third Einstein metric also has weak holonomy G_2 most likely of type I but we could not positively exclude type II as a possibility [**37**].

7. The Quotient Construction

In this section we give a general 3-Sasakian reduction procedure which constructs new 3-Sasakian manifolds from a given 3-Sasakian manifold $\mathcal{S}$ with a non-trivial 3-Sasakian isometry group $I_0(\mathcal{S}, g)$ [**26**]. Actually, this is a reduction that is associated with a quadruple of spaces of the fundamental diagram $\diamondsuit(\mathcal{S})$. At the level of the hyperkähler cone $C(\mathcal{S})$ the reduction was discovered by Lindström and Roček [**89**] in the context of supersymmetric σ-model and later rigorously described by Hitchin et al. in [**65**]. In the case of the quaternionic Kähler base $\mathcal{O}$ the reduction was discovered by the second author and Lawson [**51, 52**]. The lift of the quaternionic Kähler quotient to the twistor space $\mathcal{Z}$ was described by Hitchin [**64**].

In this section we restrict ourselves to describing the general procedure of reduction together with the homogeneous case arising from reduction by a circle group, as well as a brief discription of the singular case. The large class of 3-Sasakian toric manifolds obtained by reduction is relagated to a separate section, namely Section 8. It should also be understood that every 3-Sasakian reduction gives as well a reduction procedure for each of the spaces of the fundamental diagram $\diamondsuit(\mathcal{S})$.

7.1. The 3-Sasakian Moment Map. Let $(\mathcal{S},g)$ be a 3-Sasakian manifold with a nontrivial group $I_0(\mathcal{S},g)$ of 3-Sasakian isometries. By the Definition 2.2.1, $C(\mathcal{S})=\mathcal{S}\times\mathbb{R}^+$ is a hyperkähler manifold with respect to the cone metric $\bar{g}$. The isometry group $I_0(\mathcal{S},g)$ extends to a group $I_0(C(\mathcal{S}),\bar{g})\cong I_0(\mathcal{S},g)$ of isometries on $C(\mathcal{S})$ by defining each element to act trivially on $\mathbb{R}^+$. Furthermore, it follows easily from the definition of the complex structures I^a that these isometries $I_0(C(\mathcal{S}),\bar{g})$ are hyperkähler; that is, they preserve the hyperkähler structure on $C(\mathcal{S})$. Recall [**65**] shows that any subgroup $G\subset I_0(M,\bar{g})$ gives rise to a hyperkähler moment map $\mu: M\rightarrow\mathfrak{g}^*\otimes\mathbb{R}^3$, where $\mathfrak{g}$ denotes the Lie algebra of G and $\mathfrak{g}^*$ is its dual. Thus, we can define a 3-Sasakian moment map

$$\mu_{\mathcal{S}}:\mathcal{S}\longrightarrow\mathfrak{g}^*\otimes\mathbb{R}^3 \tag{7.1}$$

by restriction $\mu_{\mathcal{S}}=\mu\mid\mathcal{S}$. We denote the components of $\mu_{\mathcal{S}}$ with respect to the standard basis of $\mathbb{R}^3$, which we have identified with the imaginary quaternions, by $\mu^a_{\mathcal{S}}$. Recall that ordinarily moment maps determined by Abelian group actions (in particular, those associated to 1-parameter groups) are only specified up to an arbitrary constant. This is not the case for 3-Sasakian moment maps since we require that the group $Sp(1)$ generated by the Sasakian vector fields ξ^a acts on the level sets of $\mu_{\mathcal{S}}$. However, we shall see that 3-Sasakian moment maps are given by a particularly simple expression.

PROPOSITION 7.1.1. *Let $(\mathcal{S},g)$ be a 3-Sasakian manifold with a connected compact Lie group G acting on $\mathcal{S}$ by 3-Sasakian isometries. Let τ be an element of the Lie algebra $\mathfrak{g}$ of G and let X^τ denote the corresponding infinitesimal isometry. Then there is a unique 3-Sasakian moment map $\mu_{\mathcal{S}}$ such that the zero set $\mu_{\mathcal{S}}^{-1}(0)$ is invariant under the group $Sp(1)$ generated by the vector fields ξ^a. This moment map is given by*

$$<\mu^a_{\mathcal{S}},\tau>=\frac{1}{2}\eta^a(X^\tau).$$

Furthermore, the zero set $\mu_{\mathcal{S}}^{-1}(0)$ is G invariant.

PROOF. Using the Definition 2.2.1 we can define the 2-forms $\omega^a_{\mathcal{S}}$ on $\mathcal{S}$ as the restriction of the hyperkähler 2-forms ω^a. Then any 3-Sasakian moment map $\mu^a_{\mathcal{S}}(\tau)$ determined by $\tau\in\mathfrak{g}$ satisfies $2d\mu^a_{\mathcal{S}}(\tau)=2X^\tau\rfloor\omega^a_{\mathcal{S}}=-X^\tau\rfloor d\eta^a$. As X^τ is a 3-Sasakian infinitesimal isometry, Lemma 4.2.2 implies that $2<\mu^a_{\mathcal{S}},\tau>$ differs from $\eta^a(X^\tau)$ by a constant depending on a and τ. One then uses the invariance of the zero set $\mu_{\mathcal{S}}^{-1}(0)$ to show that these constants must vanish. See [**26**] for details. □

Henceforth by the 3-Sasakian moment map, we shall mean the moment map $\mu_{\mathcal{S}}$ determined in Proposition 7.1.1. Hence, the Definition 2.2.1 and Proposition 7.1.1 imply

THEOREM 7.1.2. *Let $(\mathcal{S},g)$ be a 3-Sasakian manifold with a connected compact Lie group G acting on $\mathcal{S}$ smoothly and properly by 3-Sasakian isometries. Let $\mu_{\mathcal{S}}$*

be the corresponding 3-Sasakian moment map and assume both that 0 is a regular value of $\mu_{\mathcal{S}}$ and that G acts freely on the submanifold $\mu_{\mathcal{S}}^{-1}(0)$. Furthermore, let $\iota : \mu_{\mathcal{S}}^{-1}(0) \longrightarrow \mathcal{S}$ and $\pi : \mu_{\mathcal{S}}^{-1}(0) \longrightarrow \mu_{\mathcal{S}}^{-1}(0)/G$ denote the corresponding embedding and submersion. Then $(\mathcal{S}/\!/\!/G = \mu_{\mathcal{S}}^{-1}(0)/G, \check{g})$ is a smooth 3-Sasakian manifold of dimension $4(n - \dim \mathfrak{g}) + 3$ with metric $\check{g}$ and characteristic vector fields $\check{\xi}^a$ determined uniquely by the two conditions $\iota^* g = \pi^* \check{g}$ and $\pi_*(\xi^a \,|\, \mu_{\mathcal{S}}^{-1}(0)) = \check{\xi}^a$.

We conclude this part with the following fact concerning 3-Sasakian isometries whose proof can be found in [**26**].

PROPOSITION 7.1.3. *Assume that the hypothesis of Theorem 7.1.2 holds. In addition assume that $(\mathcal{S}, g)$ is complete and hence compact. Let $C(G) \subset I_0(\mathcal{S}, g)$ denote the centralizer of G in $I_0(\mathcal{S}, g)$ and let $C_0(G)$ denote the subgroup of $C(G)$ given by the connected component of the identity. Then $C_0(G)$ acts on the submanifold $\mu_{\mathcal{S}}^{-1}(0)$ as isometries with respect to the restricted metric $\iota^* g$ and the 3-Sasakian isometry group $I_0(\mathcal{S}/\!/\!/G, \check{g})$ of the quotient $(\mathcal{S}/\!/\!/G, \check{g})$ determined in Theorem 7.1.2 contains an isomorphic copy of $C_0(G)$. Furthermore, if $C_0(G)$ acts transitively on $\mathcal{S}/\!/\!/G$, then $\mathcal{S}/\!/\!/G$ is a 3-Sasakian homogeneous space.*

It should be mentioned that it is not required that the isometry group $I_0(\mathcal{S}/\!/\!/G, \check{g})$ acts effectively.

7.2. Regular Quotients And Classical Homogeneous Metrics. We now apply the reduction procedure given in Theorem 7.1.2 to the round unit sphere S^{4n+3} to explicitly construct the Riemannian metrics for the 3-Sasakian homogeneous manifolds arising from the simple classical Lie algebras. These metrics are precisely the ones associated to the three infinite families appearing in Theorem 4.2.6. The quotient construction applied to $\Diamond(\mathcal{S}^{4n+3})$ explicitly describes all metrics in the fundamental diagrams $\Diamond(G/H)$, where G is either the special unitary $SU(n+1)$ or the orthogonal group $SO(n+1)$. To carry out this reduction we must set some conventions. We describe the unit sphere S^{4n+3} by its embedding in flat space and we represent an element $\mathbf{u} = (u_1, \cdots, u_{n+1}) \in \mathbb{H}^{n+1}$ as a column vector. The quaternionic components of this vector are denoted by $\mathbf{u}^0$ for the real component and by $\mathbf{u}^a$ for the three imaginary components so that we can write $\mathbf{u} = \mathbf{u}^0 + i\mathbf{u}^1 + j\mathbf{u}^2 + k\mathbf{u}^3$ using the quaternionic units $\{i, j, k\}$. We also define quaternionic conjugate $\bar{\mathbf{u}} = \mathbf{u}^0 - i\mathbf{u}^1 - j\mathbf{u}^2 - k\mathbf{u}^3$.

Now, the infinitesimal generators of the subgroup $Sp(1) \subset \mathbb{H}^*$ acting by the right multiplication on $\mathbf{u}$ are the defining vector fields ξ^a for the Sasakian 3-structure. These vector fields are given by

$$\xi^a_r \;=\; \mathbf{u}^0 \cdot \frac{\partial}{\partial \mathbf{u}^a} - \mathbf{u}^a \cdot \frac{\partial}{\partial \mathbf{u}^0} - \epsilon^{abc}\mathbf{u}^b \cdot \frac{\partial}{\partial \mathbf{u}^c},$$

where the dot indicates sum over the vector components u_i and the subscript r means that these vector fields are the generators of the right action.

We will first consider $G = U(1)$ acting on the sphere S^{4n+3} as follows

$$\varphi_t(\mathbf{u}) = \tau\mathbf{u}, \qquad \tau = e^{2\pi i t}, \qquad \mathbf{u} \in S^{4n+3}.$$

Note that this action is actually free on S^{4n+3} and hence it will be automatically free on the level set of the moment map. To compute the moment map we identify the imaginary quaternions $\mathbb{R}^3$ with the Lie algebra $\mathfrak{sp}(1)$ in equation 7.1 and the

Lie algebra of $U(1)$ with $\mathbb{R}$, so the moment map is $\mu_{\mathcal{S}} : S^{4n+3} \longrightarrow \mathbb{R} \otimes \mathfrak{sp}(1)$ and it can easily be computed

$$\mu_{\mathcal{S}}(\mathbf{u}) = \sum_{\alpha=1}^{n+1} \bar{u}_\alpha i u_\alpha.$$

One can easily identify the zero-level set of the moment map with the Stiefel manifold of complex 2-frames in $\mathbb{C}^{n+1}$, and the following proposition is then an immediate consequence of Theorem 7.1.2.

PROPOSITION 7.2.1. *Let $N = \mu_{\mathcal{S}}^{-1}(0)$ and $\iota : N \hookrightarrow S^{4n+3}$ be the inclusion. Then ι is an embedding and $(N, \iota^* g_{can})$ is the complex homogeneous Stiefel manifold $V^{\mathbb{C}}_{2,n+1} = SU(n+1)/SU(n-1)$ of 2-frames in $\mathbb{C}^{n+1}$. Hence, the 3-Sasakian quotient $S^{4n+3} /\!/\!/ U(1) = V^{\mathbb{C}}_{2,n+1}/U(1) = SU(n+1)/S(U(n-1) \times U(1))$ with the 3-Sasakian metric $\breve{g}$ given by inclusion ι and submersion $\pi : N \longrightarrow N/U(1)$, i.e., $\iota^* g_{can} = \pi^* \breve{g}$.*

REMARK 7.1. A similar construction can be carried out for the $Sp(1)$-action on S^{4n+3} defined by the left multiplication of $\mathbf{u}$ by a unit quaternion σ, i.e.,

$$\varphi_\sigma(\mathbf{u}) = \sigma\mathbf{u}, \qquad \sigma\bar{\sigma} = 1, \qquad \mathbf{u} \in S^{4n+3}.$$

This action is free on S^{4n+3} and the zero-level set of the corresponding moment map can be identified with the real Stiefel manifold $V^{\mathbb{R}}_{4,n+1} \simeq SO(n+1)/SO(n-3)$ of 4-frames in $\mathbb{R}^{n+1}$ with $n \geq 4$. Hence, the reduced space $S^{4n+3} /\!/\!/ Sp(1) = \frac{SO(n+1)}{SO(n-3) \times Sp(1)}$. For the more detailed and uniform description of the geometry of these two quotients see [**26**].

7.3. The Structure of Singular Quotients. In this section we will describe a more general situation, when the zero-level set of the 3-Sasakian moment map 7.1 is not necessarily smooth and the group action on the level set is not necessarily locally free.

Let G be a Lie group acting smoothly and properly on a manifold $\mathcal{S}$ and let $H \subset G$ be a subgroup. Using standard notation we will denote by $\mathcal{S}_H \subset \mathcal{S}$ the set of points in $\mathcal{S}$ where the stability group is exactly equal to H and by $\mathcal{S}_{(H)} \subset \mathcal{S}$ the set of points with stabilizer conjugate to H in G. It follows than that the normalizer $N(H)$ of H in G acts freely on $\mathcal{S}_{(H)}$. Then we have the following theorem due to Dancer and Swann [**39**].

THEOREM 7.3.1. *Let $(\mathcal{S}, g)$ be a 3-Sasakian manifold with a connected compact Lie group G acting on $\mathcal{S}$ smoothly and properly by 3-Sasakian isometries. Let $\mu_{\mathcal{S}}$ be the corresponding 3-Sasakian moment map. Then the quotient $\mu_{\mathcal{S}}^{-1}(0)/G$ is a union of the smooth, 3-Sasakian manifolds $(\mathcal{S}_{(H)} \cap \mu_{\mathcal{S}}^{-1}(0))/G$, where (H) runs over the conjugacy classes of stabilizers of points in $\mathcal{S}$.*

Quite often $\mathcal{S}_{(H)}$ does not meet the zero locus of the moment map. Then the stratum $(\mathcal{S}_{(H)} \cap \mu_{\mathcal{S}}^{-1}(0))/G$ is empty.

EXAMPLE 7.1. We start with the 3-Sasakian sphere S^{4n+3} in the notation of the previous section. But now we consider a different circle action $U(1)$, namely

$$\varphi_t^{p,q;m}(\mathbf{u}) = (\tau^p u_1, \ldots, \tau^p u_m, \tau^q u_{m+1}, \ldots, \tau^q u_{n+1}), \qquad \tau = e^{2\pi i t}, \qquad \mathbf{u} \in S^{4n+3},$$

where $0 \leq m \leq n+1$ and $p, q \in \mathbb{Z}$. Let $\mathcal{S}(p,q;m) = S^{4n+3} /\!/\!/ U(1)$ be the quotient.

(i) First, let p, q be relatively prime positive integers bigger than 1 and $2 \leq m \leq n-1$. Then, the stratified manifold $\mathcal{S}(p,q;m)$ consists of 3 strata. The stratum of the highest dimension corresponding to $H = \{\mathrm{id}\}$ is an open incomplete 3-Sasakian manifold. The two strata of lower dimension are easily seen to be the homogeneous spaces: one with $H = \mathbb{Z}_q$ is the homogeneous 3-Sasakian space of $SU(n+1-m)$ and the one with $H = \mathbb{Z}_p$ is the homogeneous 3-Sasakian space of $SU(m)$. In this case, $\mathcal{S}(p,q;m)$ is actually a compact 3-Sasakian orbifold and the stratification of Theorem 7.3.1 coincides with the orbifold stratification.
(ii) Consider $\mathcal{S}(0,p;m)$, where $p > 1$ and $2 \leq m \leq n-1$. There are two strata now: the stratum of the highest dimension corresponds to $H = \mathbb{Z}_p$ and the second stratum is just the sphere S^{4m-1} with $H = U(1)$. The space $\mathcal{S}(0,p;m)$ is not an orbifold but, as pointed out in [**39**], it does have a length space structure.
(iii) Consider $\mathcal{S}(0,1;n)$. Here H is either $U(1)$ or trivial but the set S^{4n+3}_{id} does not meet the zero locus of the moment map. Hence, there is only one stratum and $\mathcal{S}(0,1;n) = S^{4n-1}$.
(iv) Finally, consider $\mathcal{S}(0,1;n-1)$. The stability group H is either $U(1)$ or trivial. The stratum corresponding to $H = U(1)$ is the sphere S^{4n-5}. We leave it as an exercise to the reader to show that $\mathcal{S}(0,1;n-1)$ is an orbifold and that it can be identified with $S^{4n-1}/\mathbb{Z}_2$, where $(w_1, \ldots., w_n) \in S^{4n-1}$, where $\mathbb{Z}_2$ acts on the last quaternionic coordinate by multiplication by ± 1.

8. Toric 3-Sasakian Manifolds

In this section we shall describe the quotient construction of large families of 3-Sasakian manifolds $\mathcal{S}(\Omega)$. They all have the property that $I_0(\mathcal{S}(\Omega), g(\Omega)) \supset T^m$, where $\dim(\mathcal{S}(\Omega)) = 4m-1$, and following the ideas of [**18**] we shall call such 3-Sasakian manifolds *toric* (See 8.6.1 for a precise definition). We also describe some interesting geometric and topological properties of such spaces. Up until now all known examples of 3-Sasakian manifolds are either homogeneous or toric or discrete quotients of them.

8.1. Toral Reductions of Spheres. Using the notation of the previous section we start with the unit $(4n+3)$-dimensional sphere embedded in the quaternionic vector space $\mathbb{H}^{n+1}$. The subgroup of the full isometry group $O(4n+3)$ that preserves the quaternionic structure is $Sp(n+1)\cdot Sp(1)$ acting by

$$\varphi_{\mathbb{A},\sigma}(\mathbf{u}) = \mathbb{A}\mathbf{u}\sigma^{-1},$$

where $\mathbb{A} \in Sp(n+1)$ is the quaternionic $(n+1)\times(n+1)$ matrix of the quaternionic representation of $Sp(n+1)$, and $\sigma \in Sp(1)$ is a unit quaternion. As the diagonal $\mathbb{Z}_2$ acts trivially this is indeed an $Sp(n+1)\cdot Sp(1)$ action. The group $Sp(n+1)$ is the subgroup of $Sp(n+1)\cdot Sp(1)$ which preserves the 3-Sasakian structure on S^{4n+3}, so we have $I_0(S^{4n+3}, g_{can}) = Sp(n+1)$. We shall consider the maximal torus $T^{n+1} \subset Sp(n+1)$ and its subgroups. Every quaternionic representation of a k-torus T^k on $\mathbb{H}^{n+1}$ can be described by an exact sequence $0 \longrightarrow T^k \xrightarrow{f_\Omega} T^{n+1} \longrightarrow T^{n+1-k} \longrightarrow 0$. The monomorphism f_Ω can be represented by the matrix

$$f_\Omega(\tau_1,\ldots,\tau_k) = \begin{pmatrix} \prod_{i=1}^{k} \tau_i^{a_1^i} & \cdots & 0 \\ \vdots & \ddots & \vdots \\ 0 & \cdots & \prod_{i=1}^{k} \tau_i^{a_{n+1}^i} \end{pmatrix},$$

where $(\tau_1,..,\tau_k) \in S^1\times\cdots\times S^1 = T^k$ are the complex coordinates on T^k, and $a_j^i \in \mathbb{Z}$ are the coefficients of a $k\times(n+1)$ integral *weight* matrix $\Omega = (a_\alpha^l)_{\alpha=1,\ldots,n+1}^{l=1,\ldots,k} \in \mathcal{M}_{k,n+1}(\mathbb{Z})$.

Let $\{e_l\}_{l=1}^k$ denote the standard basis for $\mathfrak{t}_k^* \simeq \mathbb{R}^k$. Then the 3-Sasakian moment map $\mu_\Omega : S^{4n+3}\longrightarrow \mathfrak{t}_k^*\otimes\mathbb{R}^3$ of the k-torus action defined by $\varphi_{(\tau_1,\ldots,\tau_k)}(\mathbf{u}) = f_\Omega(\tau_1,\ldots,\tau_k)\mathbf{u}$, is given by $\mu_\Omega = \sum_l \mu_\Omega^l e_l$ with

$$\mu_\Omega^l(\mathbf{u}) = \sum_\alpha \bar{u}_\alpha i a_\alpha^l u_\alpha.$$

Let us further denote the triple $(T^k, f_\Omega, \varphi_{(\tau_1,\ldots,\tau_k)})$ by $T^k(\Omega)$.

DEFINITION 8.1.1. $N(\Omega) = \mu_\Omega^{-1}(0)$ *and* $\mathcal{S}(\Omega) = S^{4n+3}/\!/\!/T^k(\Omega) = N(\Omega)/T^k(\Omega)$.

Let S_H^{4n+3} denote all the points on the sphere where the stability group $H\subset T^k$ is exactly H. Because T^k is Abelian $S_H^{4n+3} = S_{(H)}^{4n+3}$. Furthermore, let $K_H = T^k/H$ and denote by $\mathcal{S}(\Omega;H) = S_H^{4n+3}\cap N(\Omega)/K_H$. Following Theorem 7.3.1 we have

PROPOSITION 8.1.2. *The quotient* $\mathcal{S}(\Omega) = \bigcup_H \mathcal{S}(\Omega;H)$ *is a disjoint union of 3-Sasakian manifolds, where each stratum* $\mathcal{S}(\Omega;H)$ *is smooth.*

We will be interested in the case when $\mathcal{S}(\Omega)$ is a compact orbifold (all stability groups H for which $S_H^{4n+3}\cap N(\Omega)$ are non-empty are discrete) or a compact smooth manifold (there is only one stratum). Necessary and sufficient conditions for this to happen can be expressed in terms of properties of the matrix Ω. First observe that, without loss of generality, we can assume that the rank of Ω equals k. Otherwise, one simply has an action of a torus of lower dimension and the whole problem reduces to considering another weight matrix Ω with fewer rows.

We introduce the following terminology: Consider the $\binom{n}{k}$ minor determinants

$$\Delta_{\alpha_1\ldots\alpha_k} = \det\begin{pmatrix} a_{\alpha_1}^1 & \cdots & a_{\alpha_k}^1 \\ \vdots & & \vdots \\ a_{\alpha_1}^k & \cdots & a_{\alpha_k}^k \end{pmatrix} \tag{8.1}$$

obtained by deleting $n+1-k$ columns of Ω.

DEFINITION 8.1.3. *Let* $\Omega\in\mathcal{M}_{k,n+1}(\mathbb{Z})$ *be the weight matrix.*
(i) If $\Delta_{\alpha_1\cdots\cdots\alpha_k}\neq 0, \forall \quad 1\le\alpha_1<\cdots<\alpha_k\le n+1$, *then we say that* Ω *is non-degenerate.*

Suppose Ω *is non-degenerate and let* g *be the* k*th determinantal divisor, i.e., the gcd of all the* k *by* k *minor determinants* $\Delta_{\alpha_1\cdots\cdots\alpha_k}$. *Then* Ω *is said to be admissible if in addition we have*
(ii) $\gcd(\Delta_{\alpha_2\cdots\alpha_{k+1}},\ldots,\Delta_{\alpha_1\cdots\hat{\alpha}_s\cdots\alpha_{k+1}},\ldots,\Delta_{\alpha_1\cdots\alpha_k}) = g$ *for all sequences of length* $(k{+}1)$ *such that* $1\le\alpha_1<\cdots<\alpha_s<\cdots<\alpha_{k+1}\le n+1$.

8.2. Equivalence Problem and Admissibility. Before we show how these properties of the matrix Ω impact on the geometry of the quotient $\mathcal{S}(\Omega)$ we need to discuss the notion of the equivalence of T^k-actions on S^{4n+3} and obtain a normal form for admissible weight matrices. We are free to change bases of the Lie algebra $\mathfrak{t}_k$. This can be done by the group of unimodular matrices $GL(k,\mathbb{Z})$. Moreover, if we fix a maximal torus T^{n+1} of $Sp(n+1)$, its normalizer, the Weyl group $\mathcal{W}(Sp(n+1)) \simeq \Sigma_{n+1} \bowtie (\mathbb{Z}_2)^{n+1}$, preserves the 3-Sasakian structure on S^{4n+3} and intertwines the T^k actions. Thus, there is an induced action of $GL(k,\mathbb{Z}) \times \mathcal{W}(Sp(n+1))$ on the set of weight matrices $\mathcal{M}_{k,n+1}(\mathbb{Z})$. The group $GL(k,\mathbb{Z})$ acts on $\mathcal{M}_{k,n+1}(\mathbb{Z})$ by matrix multiplication from the left, and the Weyl group $\mathcal{W}(Sp(n+1))$ acts by permutation and overall sign changes of the columns. Actually we want a slightly stronger notion of equivalence than that described above. If the ith row of Ω has a gcd d_i greater than one, then by reparameterizing the one-parameter subgroup $\tau_i' = \tau_i^{d_i}$ we obtain $\tau_i^{a_\alpha^j} = (\tau_i')^{b_\alpha^i}$, where $\gcd\{b_\alpha^i\}_\alpha = 1$. So the action obtained by using the matrix whose ith row is divided by its gcd d_i is the same as the original action. The integers d_i all divide the kth determinantal divisor g. We say that a non-degenerate matrix Ω is in *reduced* form (or simply *reduced*) if $g = 1$. The following easy lemma says that among non-degenerate matrices it is sufficient to consider matrices in a reduced form.

LEMMA 8.2.1. *Every non-degenerate weight matrix Ω is equivalent to a matrix in a reduced form.*

Henceforth, we shall only consider matrices in a reduced form.

DEFINITION 8.2.2. *Let $\mathcal{A}_{k,n+1}(\mathbb{Z}) \subset \mathcal{M}_{k,n+1}(\mathbb{Z})$ denote the subset of reduced admissible matrices. This subset is invariant under the action of $GL(k,\mathbb{Z}) \times \mathcal{W}(Sp(n+1))$, so the set $\mathcal{A}_{k,n+1}(\mathbb{Z})/GL(k,\mathbb{Z}) \times \mathcal{W}(Sp(n+1))$ of equivalence classes $[\Omega]$ is well defined. We let $\mathcal{E}_{k,n+1}(\mathbb{Z}) \subset \mathcal{A}_{k,n+1}(\mathbb{Z})$ denote a fundamental domain for the action.*

Our interest in $\mathcal{A}_{k,n+1}(\mathbb{Z})$ is the following:

THEOREM 8.2.3. *Let $\mathcal{S}(\Omega)$ be the quotient space of definition 8.1. Then*
(i) if Ω is non-degenerate, $\mathcal{S}(\Omega)$ is an orbifold.
(ii) If Ω is degenerate, then either $\mathcal{S}(\Omega)$ is a singular stratified space which is not an orbifold or it is an orbifold obtained by reduction of a lower dimensional sphere $S^{4n-4r-1}$ by a torus $T^{k-r}(\Omega')$ or a finite quotient of such, where $1 \leq r \leq k$ and Ω' is non-degenerate. (When $r = k$ the quotient is the sphere $S^{4n-4k-1}$).
(iii) Assuming that Ω is non-degenerate $\mathcal{S}(\Omega)$ is a smooth manifold if and only if Ω is admissible.

One can easily see that the non-degeneracy of Ω is not necessary for the quotient space $\mathcal{S}(\Omega)$ to be smooth or a compact orbifold (see Example 7.1(iv)). However, Theorem 8.2.3(ii) shows that then we can reformulate the whole problem in terms of another quotient and a new non-degenerate weight matrix Ω' and can be found in [**31**]. Theorem 8.2.3(iii) shows then the importance of admissible matrices in the construction and it easily follows from the fact that non-degeneracy implies that at most $n-k$ quaternionic coordinates u_j can simultaneously vanish on $N(\Omega)$ [**33**].

REMARK 8.1. Our discussion shows clearly that, if $\Omega, \Omega' \in \mathcal{A}_{k,n+1}(\mathbb{Z})$ such that $[\Omega] = [\Omega']$ then the quotients $\mathcal{S}(\Omega) \simeq \mathcal{S}(\Omega')$ are equivalent as 3-Sasakian manifolds.

We believe that the converse of this is also true, though we will establish it later only in certain cases.

8.3. Combinatorics and Admissibility. In general Theorem 8.2.3 is not yet an existence theorem, since $\mathcal{A}_{k,n+1}(\mathbb{Z})$ could be empty. Indeed, for many pairs (k,n) this is the case and we shall demonstrate this next.

Let $\Omega\in\mathcal{A}_{k,n+1}(\mathbb{Z})$. Since Ω is reduced there is a k by k minor determinant that is odd. By permuting columns if necessary this minor can be taken to be the first k columns. Now consider the mod 2 reduction $\mathcal{M}_{k,n+1}(\mathbb{Z})\longrightarrow\mathcal{M}_{k,n+1}(\mathbb{Z}_2)$. We have the following commutative diagram

$$\begin{array}{ccc} GL(k,\mathbb{Z})\times\mathcal{M}_{k,n+1}(\mathbb{Z}) & \longrightarrow & \mathcal{M}_{k,n+1}(\mathbb{Z}) \\ \downarrow & & \downarrow \\ GL(k,\mathbb{Z}_2)\times\mathcal{M}_{k,n+1}(\mathbb{Z}_2) & \longrightarrow & \mathcal{M}_{k,n+1}(\mathbb{Z}_2). \end{array} \tag{8.2}$$

Let $\tilde{\Omega}\in\mathcal{A}_{k,n+1}(\mathbb{Z}_2)$ denote the mod 2 reduction of $\Omega\in\mathcal{A}_{k,n+1}(\mathbb{Z})$. Since the first k by k minor determinant of Ω is odd, the mod 2 reduction of this minor in $\tilde{\Omega}$ is invertible. Thus, we can use the $GL(k,\mathbb{Z}_2)$ action to put $\tilde{\Omega}$ in the form

$$\tilde{\Omega}=\begin{pmatrix} 1 & 0 & \cdots & 0 & a^1_{k+1} & \cdots & a^1_{n+1} \\ 0 & 1 & \cdots & 0 & a^2_{k+1} & \cdots & a^2_{n+1} \\ \vdots & \vdots & \ddots & \vdots & \vdots & \cdots & \vdots \\ 0 & 0 & \cdots & 1 & a^k_{k+1} & \cdots & a^k_{n+1} \end{pmatrix} \tag{8.3}$$

with $a^i_j\in\mathbb{Z}_2$.

Lemma 8.3.1. *The set $\mathcal{A}_{k,n+1}(\mathbb{Z})$ is empty for $n>k+1$ and $k>4$.*

Proof. The second admissibility condition is equivalent to the condition that every k by $k+1$ submatrix of $\tilde{\Omega}$ has rank k. By considering $k-1$ of the first k columns and 2 of last $n+1-k$ columns, this condition implies $(a^j_l,a^j_m)\neq(0,0)$ for all $j=1,\cdots,k$, and $k+1\leq l<m\leq n+1$. Similarly, by considering $k-2$ of the first k columns and 3 of last $n+1-k$ columns 8.3 implies

$$\begin{pmatrix} a^i_l & a^i_m & a^i_r \\ a^j_l & a^j_m & a^j_r \end{pmatrix}\neq\begin{pmatrix} 1 & 1 & 1 \\ 1 & 1 & 1 \end{pmatrix},\qquad \begin{pmatrix} a^i_l & a^i_m & a^i_r \\ a^j_l & a^j_m & a^j_r \end{pmatrix}\neq\begin{pmatrix} 0 & 1 & 1 \\ 0 & 1 & 1 \end{pmatrix}, \tag{8.4}$$

where the last inequality is understood to be up to column permutation. Hence, it follows that, up to column and row permutations, that any four triples of the last $n-k$ columns of an admissible $\tilde{\Omega}$ must have the form

$$\begin{pmatrix} 1 & 1 & 1 \\ 0 & 1 & 1 \\ 1 & 0 & 1 \\ 1 & 1 & 0 \end{pmatrix}. \tag{8.5}$$

So we see that we cannot add another row without violating the above conditions. It follows that $k\leq 4$. □

Similar analysis shows that

Lemma 8.3.2. *The set $\mathcal{A}_{k,n+1}(\mathbb{Z})$ is empty if $k>1$ and $n-k\geq 4$.*

REMARK 8.2. In view of the above lemmas and the fact that in the remainder of this section we will be interested only in the smooth and compact quotients we are left with the following possibilities:
(i) Trivial case of $n = k$. Then there are many admissible matrices Ω but $\dim(\mathcal{S}(\Omega)) = 3$ and it follows that $\mathcal{S}(\Omega) = S^3/\mathbb{Z}_p$, where $p = p(\Omega)$ depends on Ω. This case is of little interest.
(ii) Bi-quotient geometry with $k = 1$ and $n > 1$ arbitrary. Here Ω is just a row vector $\mathbf{p}$. The admissibility condition means that the entries are non-zero and pairwise relatively prime. The quotient $\mathcal{S}(\mathbf{p})$ turns out to be a bi-quotient of the unitary group $U(n+1)$ and we shall discuss its geometry and topology in the next subsection.
(iii) The most interesting, 7-dimensional case of $k = n-1$. Here one easily sees that there are many admissible matrices and we analyze the geometry and topology of the quotients in a separate subsection.
(iv) "Special" quotients: $(k,n) = \{(2,4),(2,5),(3,5),(3,6),(4,6),(4,7)\}$. These quotients are 11- or 15-dimensional and we give examples of admissible weight matrices in each case. We shall show also that they provide counterexamples to certain Betti number relations that are satisfied in the regular case [**54**].

8.4. 3-Sasakian Structures on Bi-Quotients. When $k = 1$ we have $\Omega = \mathbf{p} = (p_1, \dots, p_{n+1})$ and we shall write $\mathcal{S}(\Omega) = \mathcal{S}(\mathbf{p})$, $N(\Omega) = N(\mathbf{p})$, $f_\Omega = f_{\mathbf{p}}$, and $\tau_1 = \tau$. The quotients $\mathcal{S}(\mathbf{p})$ are generalizations of the homogeneous examples discussed in Section 7.2. We get

$$\mathcal{A}_{1,n+1}(\mathbb{Z}) = \{\mathbf{p} \in (\mathbb{Z})^{n+1} \mid p_i \neq 0\ \forall i = 1, \dots, n+1 \text{ and } \gcd(p_i,p_j) = 1 \qquad \forall i \neq j\},$$

$$\mathcal{E}_{1,n+1}(\mathbb{Z}) = \{\mathbf{p} \in \mathbb{Z}^{n+1} \mid 0 < p_1 \leq \cdots \leq p_{n+1} \text{ and } \gcd(p_i,p_j) = 1 \qquad \forall i \neq j\}.$$

Note that $\mathcal{E}_{1,n+1}(\mathbb{Z})$ can be identified with a certain integral lattice in the positive Weyl chamber in $\mathfrak{t}^*_{n+1}$.

First, by studying the geometry of the foliations in the diagram $\diamondsuit(\mathcal{S}(\mathbf{p}))$ [**30**] one can solve the equivalence problem in this case. We get [**27**]:

PROPOSITION 8.4.1. *Let $n \geq 2$ and $\mathbf{p}, \mathbf{q} \in \mathcal{A}_{1,n+1}(\mathbb{Z})$ so the quotients $\mathcal{S}(\mathbf{p})$ and $\mathcal{S}(\mathbf{q})$ are smooth manifolds. Then $\mathcal{S}(\mathbf{p}) \simeq \mathcal{S}(\mathbf{q})$ are 3-Sasakian equivalent if and only if $[\mathbf{p}] = [\mathbf{q}]$.*

It is easy to see that for $\mathbf{p} \in \mathcal{A}_{1,n+1}(\mathbb{Z})$ the zero locus of the moment map $N(\mathbf{p})$ is always diffeomorphic to the Stiefel manifold $V^{\mathbb{C}}_{2,n+1}$ of complex 2-frames in $\mathbb{C}^{n+1}$. Hence, the quotient $\mathcal{S}(\mathbf{p}) = V^{\mathbb{C}}_{2,n+1}/S^1$. We first observe that one can identify $V^{\mathbb{C}}_{2,n+1}$ with the homogeneous space $U(n+1)/U(n-1)$. Using this identification we have

PROPOSITION 8.4.2. *For each $\mathbf{p} \in \mathcal{E}_{1,n+1}(\mathbb{Z})$, there is an equivalence $\mathcal{S}(\mathbf{p}) \simeq U(1)_{\mathbf{p}}\backslash U(n+1)/U(n-1)$ as smooth $U(1)_{\mathbf{p}} \times U(n-1)$-spaces, where the action of $U(1)_{\mathbf{p}} \times U(n-1) \subset U(n+1)_L \times U(n+1)_R$ is given by the formula*

$$\varphi^{\mathbf{p}}_{\tau,\mathbb{B}}(\mathbb{W}) = f_{\mathbf{p}}(\tau)\mathbb{W}\begin{pmatrix} \mathbb{I}_2 & \mathbb{O} \\ \mathbb{O} & \mathbb{B} \end{pmatrix}.$$

Here $\mathbb{W} \in U(n+1)$ and $(\tau, \mathbb{B}) \in S^1 \times U(n-1)$.

Note that the identification $\mathcal{S}(\mathbf{p}) \simeq U(1)_{\mathbf{p}}\backslash U(n+1)/U(n-1)$ is only true after assuming that all the weights are positive, as the right-hand side is not invariant under such sign changes. Proposition 8.4.2 shows that, in a way, the quotients $\mathcal{S}(\mathbf{p})$ can be though of as a discrete "bi-quotient deformation" of the homogeneous model $\mathcal{S}(\mathbf{1})$. Now let $\iota_{\mathbf{p}} : N(\mathbf{p}) \hookrightarrow S^{4n+3}$ be the inclusion and $\pi_{\mathbf{p}} : N(\mathbf{p}) \to \mathcal{S}(\mathbf{p})$ be the Riemannian submersion of the moment map. Then the metric $g(\mathbf{p})$ is the unique metric on $\mathcal{S}(\mathbf{p})$ that satisfies $\iota_{\mathbf{p}}^* g_{can} = \pi_{\mathbf{p}}^* g(\mathbf{p})$. Using the geometry of the inclusion $\iota_{\mathbf{p}}$ one can show the following [**27, 30**]

THEOREM 8.4.3. *Let $I_0(\mathcal{S}(\mathbf{p}), g(\mathbf{p}))$ be the group of 3-Sasakian isometries of $(\mathcal{S}(\mathbf{p}), g(\mathbf{p}))$ and let k be the number of 1's in $\mathbf{p}$. Then the connected component of I_0 is $S(U(k) \times U(1)^{n+1-k})$, where we define $U(0) = \{e\}$. Thus, the connected component of the isometry group is the product $S(U(k) \times U(1)^{n+1-k}) \times SO(3)$ if the sums $p_i + p_j$ are even for all $1 \leq i, j \leq n+1$, and $S(U(k) \times U(1)^{n+1-k}) \times Sp(1)$ otherwise.*

In the case that $\mathbf{p}$ has no repeated 1's, the cohomogeneity can easily be determined, viz. [**27**]

COROLLARY 8.4.4. *If the number of 1's in $\mathbf{p}$ is 0 or 1 then the dimension of the principal orbit in $\mathcal{S}(\mathbf{p})$ equals $n+3$ and the cohomogeneity of $g(\mathbf{p})$ is $3n-4$. In particular, the 7-dimensional $\mathcal{S}(\mathbf{p})$ the family $(\mathcal{S}(\mathbf{p}), g(\mathbf{p}))$ contains metrics of cohomogeneity 0,1, and 2.*

Combining Proposition 8.4.2 with techniques developed by Eschenburg [**40, 41**] in the study of certain 7-dimensional bi-quotients of $SU(3)$ one can compute the integral cohomology ring of $\mathcal{S}(\mathbf{p})$ [**26**]:

THEOREM 8.4.5. *Let $n \geq 2$ $\mathbf{p} \in \mathcal{E}_{1,n+1}(\mathbb{Z})$. Then, as rings,*

$$H^*(\mathcal{S}(\mathbf{p}), \mathbb{Z}) \cong \left(\frac{\mathbb{Z}[b_2]}{[b_2^{n+1} = 0]} \otimes E[f_{2n+1}] \right) / \mathcal{R}(\mathbf{p}).$$

Here the subscripts on b_2 and f_{2n+1} denote the cohomological dimension of each generator. Furthermore, the relations $\mathcal{R}(\mathbf{p})$ are generated by $\sigma_n(\mathbf{p}) b_2^n = 0$ and $f_{2n+1} b_2^n = 0$, where $\sigma_n(\mathbf{p}) = \sum_{j=1}^{n+1} p_1 \cdots \hat{p}_j \cdots p_{n+1}$ is the nth elementary symmetric polynomial in the entries of $\mathbf{p}$.

Notice that Theorem 8.4.5 shows that $H^{2n}(\mathcal{S}(\mathbf{p}); \mathbb{Z}) = \mathbb{Z}_{\sigma_n(\mathbf{p})}$ and hence has the following corollary.

COROLLARY 8.4.6. *The quotients $(\mathcal{S}(\mathbf{p}), g(\mathbf{p}))$ give infinitely many homotopy inequivalent simply-connected compact inhomogeneous 3-Sasakian manifolds in dimension $4n-1$ for every $n \geq 2$. In fact, there are infinite families that are not homotopy equivalent to any homogeneous space.*

REMARK 8.3. Corollary 8.4.6 shows that the finiteness results for regular 3-Sasakian manifolds discussed in Section 5 fail for non-regular 3-Sasakian manifolds. Moreover, combining our results with a well-known finiteness theorem of Anderson [**6**] we have

COROLLARY 8.4.7. *For each $n \geq 2$ there are infinitely many 3-Sasakian $4n-1$-manifolds with arbitrarily small injectivity radii.*

When $n = 2$ the spaces $\mathcal{S}(\mathbf{p}) = \mathcal{S}(p_1, p_2, p_3)$ give a subfamily of the more general bi-quotients of $U(3)$ studied by Eschenburg [**40, 41**]. This large collection of spaces contains not only our 3-Sasakian subfamily, but also the well-known Aloff-Wallach spaces [**5**] which are of much interest since they admit Einstein metrics of positive sectional curvature [**127**]. These two subfamilies intersect at the homogeneous 3-Sasakian manifold $\mathcal{S}(1,1,1)$, that is $\mathcal{S}(1,1,1)$ is diffeomorphic to the Aloff-Wallach space $N_{1,1}$ mentioned in 6.1. Then following Eschenburg [**40**] we can make use of the Cheeger ρ^*-topology on the space of Riemannian manifolds to show the existence of an infinite number of 3-Sasakian manifolds that admit metrics of positive sectional curvature. More precisely [**26**],

COROLLARY 8.4.8. *For all sufficiently large odd positive integers c, the 3-Sasakian manifolds $\mathcal{S}(c, c+1, c+2)$ admits a metric of positive sectional curvature.*

In the next subsection we give a result in the opposite direction. We shall exhibit an infinite family of 3-Sasakian manifolds that cannot admit any metric whose sectional curvature is bounded from below by a fixed arbitrary negative number.

We end this subsection with a discussion of topological and differential invariants of the 7-manifolds $\mathcal{S}(p_1, p_2, p_3)$. Homotopy invariants for Eschenburg space have been worked out independently by Kruggel [**80, 81**] and Milgram [**92**]. The homeomorphism and diffeomorphism classification was first done for a certain subclass of Eschenburg spaces which include some of the $\mathcal{S}(p_1, p_2, p_3)$ by Astey, Micha, and Pastor [**7**]. Later Kruggel [**82**] obtained the diffeomorphism and homeomorphism classification of all Eschenburg's bi-quotients by computing the Kreck-Stolz invariants [**78**]. This, in principle, gives a complete differential topological description of the 7-dimensional family $\mathcal{S}(p_1, p_2, p_3)$. Using this classification together with the help of a computer program, one would expect to find examples $\mathcal{S}(\mathbf{p})$ and $\mathcal{S}(\mathbf{q})$ with $[\mathbf{p}] \neq [\mathbf{q}]$ such that the quotients are homeomorphic, but not diffeomorphic, as well as examples that are diffeomorphic, but not 3-Sasakian equivalent. The later would show that a smooth 7-manifold can admit more than one inequivalent 3-Sasakian structure. In the case of the former, such exotic structures are known to exist for the family of Aloff-Wallach spaces [**79**], but the examples involve large integers and were obtained with help of a computer program. The analysis of the above mentioned invariants for our family $\mathcal{S}(p_1, p_2, p_3)$ proves even harder due to the positivity of the weights. For a fixed $\sigma_2 = p_1p_2 + p_2p_3 + p_3p_1$ there are only finitely many positive integer solution $\mathbf{p} \in \mathcal{E}_{1,3}(\mathbb{Z})$.

8.5. 7-dimensional Toric 3-Sasakian Manifolds. In this case we can easily see that there are many examples of admissible weight matrices Ω. The simplest family of examples is given by matrices of the form

$$\Omega = \begin{pmatrix} 1 & 0 & \dots & 0 & a_1 & b_1 \\ 0 & 1 & \dots & 0 & a_2 & b_2 \\ \vdots & \vdots & \ddots & \vdots & \vdots & \vdots \\ 0 & 0 & \cdots & 1 & a_k & b_k \end{pmatrix},$$

for which we have

PROPOSITION 8.5.1. *Let k be a positive integer, and let $\Omega \in \mathcal{M}_{k,k+2}(\mathbb{Z})$ be as in 8.5. Then $\Omega \in \mathcal{A}_{k,k+2}(\mathbb{Z})$ if and only if $(\mathbf{a}, \mathbf{b}) \in (\mathbb{Z}^*)^k \oplus (\mathbb{Z}^*)^k$ and the components*

(a^i, b^i) are pairs of relatively prime integers for $i = 1, \cdots, k$ such that if for some pair i, j $a^i = \pm a^j$ or $b^i = \pm b^j$ then we must have $b^i \neq \pm b^j$ or $a^i \neq \pm a^j$, respectively.

Proposition 8.5.1 shows that $\mathcal{A}_{k,k+2}(\mathbb{Z})$ is never empty and we have many examples of compact smooth 7-dimensional quotients $\mathcal{S}(\Omega)$ for arbitrary $k > 1$. Some of these examples were first mentioned in [**25**] and the idea of the quotient is based on the result of [**53**]. As we shall not present here the complete solution to the equivalence problem, we shall further assume that $\Omega \in \mathcal{A}_{k,k+2}(\mathbb{Z})$ is arbitrary and shall determine some important topological properties of the quotients $\mathcal{S}(\Omega)$. More explicitly,

THEOREM 8.5.2. *Let $\Omega \in \mathcal{A}_{k,k+2}(\mathbb{Z})$. Then $\pi_1(\mathcal{S}(\Omega)) = 0$ and $\pi_2(\mathcal{S}(\Omega)) = \mathbb{Z}^k$.*

Because of Corollary 5.2.3 and Poincare duality, Theorem 8.5.2 completely determines the rational homology of the 3-Sasakian 7-manifolds $\mathcal{S}(\Omega)$. The proof given below is a compilation with some simplifications of the proofs in [**33, 32**], while some of the more tedious details are left to those references.

PROOF. First note that the groups $T^{k+2} \times Sp(1)$ and $T^2 \times Sp(1)$ act as isometry groups on $N(\Omega)$ and $\mathcal{S}(\Omega)$, respectively. Let us define the following quotient spaces:

$$Q(\Omega) = N(\Omega)/T^{k+2} \times Sp(1), \qquad B(\Omega) = N(\Omega)/Sp(1).$$

We have the following commutative diagram

$$\begin{array}{ccc} N(\Omega) & \longrightarrow & B(\Omega) \\ \downarrow & & \downarrow \\ \mathcal{S}(\Omega) & \longrightarrow & Q(\Omega). \end{array} \tag{8.6}$$

The top horizontal arrow and the left vertical arrow are principal bundles with fibers $Sp(1)$ and T^k, respectively. The remaining arrows are not fibrations. The right vertical arrow has generic fibers T^{k+2}, while the lower horizontal arrow has generic fibers $T^2 \cdot Sp(1)$ homeomorphic either to $T^2 \times \mathbb{RP}^3$ or $T^2 \times S^3$ depending on Ω. The dimension of the orbit space $Q(\Omega)$ is 2.

The difficulty is in proving that both $N(\Omega)$ and $B(\Omega)$ are 2-connected. Once this is accomplished the result follows by applying the long exact homotopy sequence to the left vertical arrow in diagram 8.6.

LEMMA 8.5.3. *Both $N(\Omega)$ and $B(\Omega)$ are 2-connected.*

To prove this lemma we construct a stratification giving a Leray spectral sequence whose differentials can be analyzed. Let us define the following subsets of $N(\Omega)$: (Recall that, in this case, at most one quaternionic coordinate can vanish.)

$$N_0(\Omega) = \{\mathbf{u} \in N(\Omega) |\ u_\alpha = 0 \text{ for some } \alpha = 1, \cdots, k+2\},$$
$$N_1(\Omega) = \{\mathbf{u} \in N(\Omega) |\text{ for all } \alpha = 1, \cdots, k+2,\ u_\alpha \neq 0 \text{ and there is a pair } (u_\alpha, u_\beta) \text{ that lies on the same complex line in } \mathbb{H}\},$$
$$N_2(\Omega) = \{\mathbf{u} \in N(\Omega) |\text{ for all } \alpha = 1, \cdots, k+2,\ u_\alpha \neq 0 \text{ and no pair } (u_\alpha, u_\beta) \text{ lies on the same complex line in } \mathbb{H}\}.$$

Clearly, $N(\Omega) = N_0(\Omega) \sqcup N_1(\Omega) \sqcup N_2(\Omega)$ and $N_2(\Omega)$ is a dense open submanifold of $N(\Omega)$. This stratification is compatible with the diagram 8.6 and induces corresponding stratifications

$$(8.7) \qquad B(\Omega) = B_0(\Omega) \sqcup B_1(\Omega) \sqcup B_2(\Omega), \quad Q(\Omega) = Q_0(\Omega) \sqcup Q_1(\Omega) \sqcup Q_2(\Omega).$$

The $B_i(\Omega)$ fiber over the $Q_i(\Omega)$ whose fibers are tori T^{k+i}. The strata are labeled by the dimension of the cells in the resulting CW decomposition of $Q(\Omega)$. Using known results about cohomogeneity 2 actions [**35**] one can easily prove:

LEMMA 8.5.4.
(i) The orbit space $Q(\Omega)$ is homeomorphic to the closed disc $\bar{D}^2$, and the subset of singular orbits $Q_1(\Omega) \sqcup Q_0(\Omega)$ is homeomorphic to the boundary $\partial\bar{D}^2 \simeq S^1$.
(ii) $Q_2(\Omega)$ is homeomorphic to the open disc D^2.
(iii) $Q_1(\Omega)$ is homeomorphic to the disjoint union of $k+2$ copies of the open unit interval.
(iv) $Q_0(\Omega)$ is a set of $k+2$ points.

Next one can easily show that $\pi_1(B(\Omega))$ is Abelian; hence, $\pi_1(B(\Omega)) = H_1(B(\Omega))$. Now we claim that $H_1(B(\Omega)) = H_2(B(\Omega)) = 0$, and since the fibers of the top horizontal arrow of 8.6 are S^3's this together with Hurewicz will prove Lemma 8.5.3. To prove this claim we define $Y_0 = Q_0(\Omega)$, $Y_1 = Q_0(\Omega) \cup Q_1(\Omega)$, and $Y_2 = Q(\Omega)$. Then, we filter $B(\Omega)$ by $X_i = \pi^{-1}(Y_i)$ to obtain the increasing filtration

$$X_0 = B_0(\Omega), \; X_1 = B_0(\Omega) \cup B_1(\Omega), \text{ and } X_2 = B(\Omega).$$

The Leray spectral sequence associated to this filtration has E^1 term given by

$$E^1_{s,t} \;\cong\; H_{s+t}(X_t, X_{t-1}; \mathbb{Z})$$

with differential $d_1 : H_{s+t}(X_t, X_{t-1}; \mathbb{Z}) \longrightarrow H_{s+t-1}(X_{t-1}, X_{t-2}; \mathbb{Z})$, where we use the convention that $X_{-1} = \emptyset$.

To compute these E^1 terms notice that all the pairs (X_t, X_{t-1}) are relative manifolds so that one can apply the Alexander-Poincaré duality theorem. Hence, by 8.7

$$\begin{aligned} &(8.8) & H_s(X_0; \mathbb{Z}) &\cong H_s(\sqcup_{k+2} T^k; \mathbb{Z}); \\ &(8.9) & H_s(X_1, X_0; \mathbb{Z}) &\cong H^{k+2-s}(\sqcup_{k+2} T^{k+1}; \mathbb{Z}); \\ &(8.10) & H_s(X_2, X_1; \mathbb{Z}) &\cong H^{k+4-s}(T^{k+2}; \mathbb{Z}), \end{aligned}$$

where $\sqcup_j T^l$ means the disjoint union of j copies of T^l. Hence, the $E^1_{s,t}$ term of the spectral sequence is described by the diagram

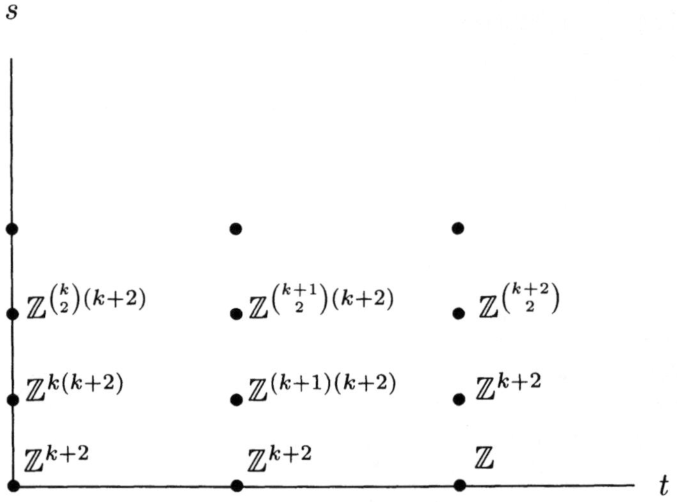

$$E^1_{s,t}$$

The computation of the differentials is fairly tedious and we refer the reader to [**32**] for details. Suffice it to say here that after making certain choices Lemma 8.5.4 can be used to represent $Q(\Omega)$ topologically as a polygon

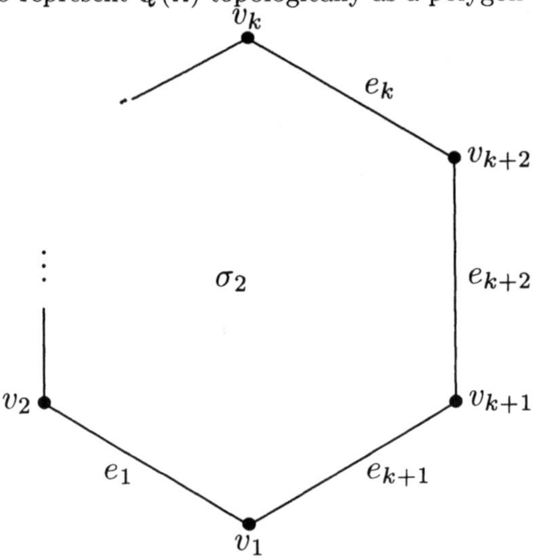

Diagram 8.5.12

The d_1 differential can then be computed and the result is that the $E^2_{s,t}$ term has zeros for $(s,t) = (1,0), (2,0), (1,1), (1,2), (0,1), (0,2)$. Then $E^2_{s,t} = E^\infty_{s,t}$ which converges to $H_{s+t}(B(\Omega), \mathbb{Z})$, so this proves Lemma 8.5.3 and hence, Theorem 8.5.2. $\square$

By further analysis of the differentials it should be possible to determine the torsion in $H_3(\mathcal{S}(\Omega), \mathbb{Z})$. This should be given in terms of symmetric functions of the invariants $|\Delta_{\alpha_1,\cdots,\alpha_k}|$.

REMARK 8.4. It was pointed out to the authors by Karsten Grove that if one takes the metric geometry into account, the internal angles in Diagram 8.5.12 are all less than 90 degrees. This indicates the presence of hyperbolic geometry.

Now, using Theorem 8.5.2 with Propositions 5.3.1, 5.3.2 gives:

PROPOSITION 8.5.5. *Let* $\Omega \in \mathcal{A}_{k,k+2}(\mathbb{Z})$ *so that* $\mathcal{S}(\Omega)$ *is a smooth manifold. Let* $\mathcal{Z}(\Omega)$ *and* $\mathcal{O}(\Omega)$ *be the associated twistor space and quaternionic Kähler orbifold, respectively. Then we have*

$$b_2(\mathcal{S}(\Omega)) = b_2(\mathcal{O}(\Omega)) = b_2(\mathcal{Z}(\Omega)) - 1 = k.$$

This shows that inequality $b_2 \leq 1$ in Proposition 5.4.2 does not hold for non-regular 3-Sasakian manifolds. Finally we give several interesting corollaries of our work.

COROLLARY 8.5.6. *There exists a simply-connected 3-Sasakian 7-manifold for every rational homology type allowed by Corollary 5.2.3.*

Our next corollary follows from the results of this section and remarkable theorem of Gromov [**60**]:

COROLLARY 8.5.7. *For any non-positive real number* κ *there are infinitely many 3-Sasakian 7-manifolds which do not admit metrics whose sectional curvatures are all greater than or equal to* κ.

For such an infinite family of 3-Sasakian 7-manifolds, the appearance of negative curvature is foretold by Remark 8.4. Corollary 8.5.7 can be contrasted with Corollary 8.4.8.

COROLLARY 8.5.8. *There exist 7-manifolds with arbitrary second Betti number having metrics of weak holonomy* G_2.

Of course, Corollary 8.5.7 also applies to these weak holonomy G_2 metrics.

COROLLARY 8.5.9. *There exist* $\mathbb{Q}$*-factorial contact Fano 3-folds* X *with* $b_2(X) = l$ *for any positive integer* l.

This corollary should be contrasted to the smooth case, where Mori and Mukai have proven that $b_2 \leq 10$ [**96**]:

COROLLARY 8.5.10. *If the second Betti number* $b_2(\mathcal{S}(\Omega)) = k > 3$, *the 3-Sasakian manifolds* $\mathcal{S}(\Omega)$ *are not homotopy equivalent to any homogeneous space.*

This corollary can be compared to Corollary 8.4.6. Finally we have

COROLLARY 8.5.11. *There exist compact,* T^2*-symmetric, self dual Einstein orbifolds of positive scalar curvature with arbitrary second Betti number.*

Again this should be contrasted to the smooth case where we must have $b_2 \leq 1$.

8.6. Higher Dimensional Toric 3-Sasakian Manifolds. We begin with the definition of a toric 3-Sasakian manifolds which is motivated by the hyperkähler case [**18**].

DEFINITION 8.6.1. *A 3-Sasakian manifold (orbifold) of dimension* $4m-1$ *is said to be a toric 3-Sasakian manifold (orbifold) if it admits an effective action of a* m*-torus* T^m *that preserves the 3-Sasakian structure.*

The importance of toric 3-Sasakian manifolds is underlined by the following recent Delzant-type theorem of Bielawski:

THEOREM 8.6.2. *Let $\mathcal{S}$ be a toric 3-Sasakian manifold of dimension $4n-1$. Then $\mathcal{S}$ is isomorphic as a 3-Sasakian T^n-manifold to a 3-Sasakian quotient of a sphere by a torus, that is to a $\mathcal{S}(\Omega)$ for some Ω.*

This theorem includes the degenerate case when the quotient is a sphere or a discrete quotient of such. The Betti numbers of a 3-Sasakian orbifold obtained by a toral quotient of a sphere were computed by Bielawski [**16**] using different techniques than the ones employed in Section 8.5:

THEOREM 8.6.3. *Let $\Omega \in \mathcal{M}_{k,n+1}(\mathbb{Z})$ be non-degenerate so that $\mathcal{S}(\Omega)$ is a compact 3-Sasakian orbifold of dimension $4(n-k)+3$. Then we have*

$$b_{2i} = \binom{k+i-1}{k}, \qquad i < n+1-k. \tag{8.11}$$

Furthermore, the Betti number constraints of Proposition 5.4.2(ii) can hold for $\mathcal{S}(\Omega)$ if and only if $k=1$.

Combining Theorems 8.6.2 and 8.6.3 with Lemmas 8.3.1 and 8.3.2 which give obstructions to smoothness gives the somewhat surprising result [**31**],

THEOREM 8.6.4. *Let $\mathcal{S}$ be a toric 3-Sasakian manifold.*
(i) If the dimension of $\mathcal{S}$ is 19 or greater, then $b_2(\mathcal{S}) \leq 1$.
(ii) If the dimension of $\mathcal{S}$ is 11 or 15, then $b_2(\mathcal{S}) \leq 4$.
(iii) If $b_2(\mathcal{S}) > 4$, then the dimension of $\mathcal{S}$ is 7.

A corollary due to Bielawski [**17**] is:

COROLLARY 8.6.5. *Let $\mathcal{S}$ be a regular toric 3-Sasakian manifold. Then $\mathcal{S}$ is one of the 3-Sasakian homogeneous spaces $S^{4n-1}, \mathbb{RP}^{4n-1}$ or $\frac{SU(n)}{S\left(U(n-2)\times U(1)\right)}$.*

Next we give an explicit construction of toric 3-Sasakian manifolds not eliminated by Theorem 8.6.4. It is enough to show that $\mathcal{A}_{4,8}$ and $\mathcal{A}_{4,7}$ are not empty as the rest follow by deletion of rows of the corresponding $\Omega \in \mathcal{A}_{4,*}$. We shall present two three parameter families of solutions, namely

$$A_1 = \begin{pmatrix} 1 & 1 & 1 \\ 2 & 1 & 1+2^l \\ 1 & 16 & 1+2^m \\ -1 & 3 & 2c \end{pmatrix}, \qquad A_2 = \begin{pmatrix} 1 & 1 & 1 & 2 \\ 2 & 1 & 1+2^{l'} & -1 \\ 1 & 16 & 1+2^{2n} & 3 \\ -1 & 3 & 2c' & -1 \end{pmatrix},$$

where $l, l', m, n \in \mathbb{Z}^+$, and $c, c' \in \mathbb{Z}$.

With the aid of MAPLE symbolic manipulation program, we find

LEMMA 8.6.6. [**31**] *Let $\Delta = 2(31c + 6 + 19 \cdot 2^{l-1} - 7 \cdot 2^{m-1})$.*
(i) $\Omega_1 = \begin{pmatrix}\mathbb{I}_4 & A_1\end{pmatrix}$ is admissible if and only if $c \neq 0$ and is not divisible by 3, and $\Delta \neq 0$ and is not divisible by 7, 19 nor 31.
(ii) $\Omega_2 = \begin{pmatrix}\mathbb{I}_4 & A_2\end{pmatrix}$ is admissible if and only if c and all minor determinants of A_2 are non-vanishing, and $c' \not\equiv 0 \pmod 3$, $l' \not\equiv 0 \pmod 4$, $c' \not\equiv 5 \pmod 7$, and $11, 19, 37, 71$ do not divide $\det A = 19 \cdot 2^{2n} - 63 - 148c' - 11 \cdot 2^{l'}$, and the following

conditions hold:

$$
(8.12)\qquad
\begin{aligned}
\gcd(3, 4c' + 2^{l'} + 1, 2c' - 2^{l'} - 1) &= 1 \\
\gcd(7, 2^{2n+1} - 2^{l'} + 1, 3 \cdot 2^{l'} + 2^{2n} + 4) &= 1 \\
\gcd(19, 2^{2n} - 2^{l'+4} - 15, 3 \cdot 2^{l'} + 2^{2n} + 4) &= 1 \\
\gcd(25, 32c' - 3 \cdot 2^{2n} - 3, 6c' + 2^{2n} + 1) &= 1.
\end{aligned}
$$

The conditions in this proposition guarantee that the quotient spaces denoted by $\mathcal{S}(c, l, m)$ and $\mathcal{S}(c', l', n)$ are smooth manifolds of dimension 11 and 15, respectively.

It is routine to verify that the three parameter infinite family given by

$$c \equiv 14 \pmod{21}, \qquad l \not\equiv 1 \pmod{5}, \qquad m \not\equiv \alpha(c) \pmod{18},$$

where $2^{\alpha(c)} = 22(31c + 6) \pmod{18}$ satisfies the conditions in (i) of Lemma 8.6.6. This gives examples in dimension 11. (Notice that as 2 is a primitive root of 19 the equation defining $\alpha(c)$ has a unique solution $\pmod{18}$ for each value of c.) Similarly, it is straightforward to verify that the infinite family given by

$$c' = 2, \qquad l' = 1, \qquad n \equiv 21 \pmod{90},$$

satisfies the conditions (ii) of Lemma 8.6.6. We have arrived at:

THEOREM 8.6.7. [**31**] *There exist toric 3-Sasakian manifolds $\mathcal{S}$ of dimensions* 11 *and* 15 *with $b_2(\mathcal{S}) = 2, 3, 4$. Consequently, the Betti number relations of Proposition 5.4.2 do not hold generally. More explicitly there are compact 11-dimensional 3-Sasakian manifolds for which $b_2 \neq b_4$, and compact 15-dimensional 3-Sasakian manifolds for which $b_2 \neq b_6$.*

9. Open Problems and Questions

We conclude this section with a short list of interesting problems. Some minor questions do appear in the text but these are usually of more technical ones. There, quite often, we simply could not provide a complete answer only because of the time constraint imposed by the fact that this chapter is a part of a collection of articles. Here we try to concentrate on, what we believe, are more fundamental questions.

PROBLEM 9.1. Classify all compact simply-connected Sasakian-Einstein manifolds in dimension 5.

All the known examples are regular and regular spaces were classified in [**11**]. Can one find irregular examples? Consider the connected sum $\mathcal{S}_k = S^5 \#^k (S^2 \times S^3)$. Now, $\mathcal{S}_k$ admits a Sasakian-Einstein metric for $k = 0, 1, 3, 4, 5, 6, 7, 8$. How about $k = 2$ and $k > 8$? In this case $\mathcal{S}_k$ would necessarily have to be a Seifert fibered space with the space of leaves a positive scalar curvature Kähler-Einstein orbifold (and not a smooth manifold) X_k with $b_2(X_k) = k$. If one could construct such structures for each remaining k, by the result of Smale, every compact simply-connected 5-manifold with no 2-torsion would admit an Einstein metric of positive scalar curvature. The same problem in any dimension $2m + 1$, $m > 2$ appears to be much more involved as it would necessarily have to include the classification of 3-Sasakian 7-manifolds [**23**].

PROBLEM 9.2. Classify all compact simply-connected 3-Sasakian manifolds in dimension 7.

Again, regular examples were classified in [**25, 47**]. This appears to be a difficult problem. Its solution would amount to a classification of good self-dual and Einstein orbifold of positive scalar curvature which, in smooth case, was done in [**63, 49**]. Certainly, a more modest, partial classification could be in reach. In particular, in terms of Definition 3.5.3 one easily sees that all toric 3-Sasakian manifolds are regular or of cyclic type. Is the converse true? That is:

QUESTION 9.3. Is every 3-Sasakian 7-manifold of cyclic type toric (which includes discrete quotients of a spheres as a degenerate case)?

In terms of the classification by symmetries one can ask:

QUESTION 9.4. Is every compact 3-Sasakian 7-manifold of cohomogeneity ≤ 2 toric?

QUESTION 9.5. Let $(\mathcal{S}, g)$ be a simply-connected 3-Sasakian 7-manifold. Can g be of maximal cohomogeneity 4?

We are not aware of any such examples. All toric examples are of cohomogeneity 0,1,2 and some new construction of [**23**] gives 3-Sasakian 7-manifolds of cohomogeneity 3.

Concerning topology and Problem 9.2, we can ask the following questions:

QUESTION 9.6. Let $(\mathcal{S}, g)$ be a compact simply-connected 3-Sasakian 7-manifold. Can $\mathcal{S}$ be topologically a product?

If so then $\mathcal{S}$ must be $S^2 \times S^5$. In the Sasakian-Einstein case it is known that such a splitting can occur. The simplest example is $S^2 \times S^3$ which has a Sasakian-Einstein structure [**121**]. Of course, the above problem and questions have versions in higher dimension. More generally,

QUESTION 9.7. Other than the vanishing of the odd Betti numbers up to the middle dimension and the finiteness of the fundamental group, what more can be said about the topology of a compact 3-Sasakian manifold? For instance, is $H_2(\mathcal{S}, \mathbb{Z})$ always torsion free? Are there further restrictions on the fundamental group?

Specifically in higher dimensions we ask:

QUESTION 9.8. Are there 3-Sasakian manifolds $\mathcal{S}$ of dimension 19 or greater with $b_2(\mathcal{S}) > 1$?

From a differentiable topological viewpoint we can ask:

QUESTION 9.9. Let $(\mathcal{S}, g)$, $(\mathcal{S}', g')$ be two compact simply-connected 3-Sasakian 7-mani- folds which are not 3-Sasakian equivalent. Can $\mathcal{S}$ be diffeomorphic (homeomorphic) to $\mathcal{S}'$? In particular, is there a non-standard 3-Sasakian structure on S^7? Can one have 3-Sasakian structures on exotic 7-spheres?

As pointed out at the end of section 8.4, we expect the positive answer to the first question. But the problem of existence of other 3-Sasakian structures on S^7 or exotic spheres lacks even the slightest hint, one way or the other. In general, due to the local rigidity, the moduli space of inequivalent 3-Sasakian structures must be discrete. So we have

QUESTION 9.10. Is the moduli space always finite, or can a 3-Sasakian manifold admit infinitely many inequivalent 3-Sasakian structures?

Concerning related geometries, we have

PROBLEM 9.11. Classify all compact simply-connected proper G_2-manifolds.

This appears to be more involved than Problem 9.2 because of Theorem 6.2.5. On the other hand, perhaps the G_2-structure can be investigated without reference to the 3-Sasakian geometry. It could happen that Problem 9.10 might admit a simpler solution and become the right approach to Problem 9.2. Maybe even one could try to classify all weak holonomy G_2-manifolds. At the moment we do not know if the converse of the Theorem 6.2.5 is true, that is if a proper G_2-manifold always admits a metric which is 3-Sasakian. This is unlikely though and one could start by looking for possible proper G_2-manifolds with $b_3 \neq 0$.

Next we turn to the regular case. All regular 3-Sasakian manifolds in dimension 7 and 11 are known as explained in Section 5.5. Any classification in higher dimensions would translate into the classification of positive quaternionic Kähler manifolds. Below we give the 3-Sasakian version of the conjecture that all compact positive quaternionic Kähler spaces are symmetric:

CONJECTURE 9.12. Let $(\mathcal{S}, g)$ be a compact regular 3-Sasakian manifold of dimension $4n+3$. Then $\mathcal{S}$ is homogeneous.

This is simply theorem 5.4.5 without $n < 3$ in the hypothesis. One might hope that 3-Sasakian geometry would provide some new input in the regular case. So far we have mostly used results about positive quaternionic Kähler manifolds to describe the properties of regular 3-Sasakian manifolds, but Section 5 does give some indication that 3-Sasakian geometry could be used to give new proofs of known theorems.

REMARK 9.1. Sasakian-Einstein, 3-Sasakian, and proper G_2-manifolds in the AdS/CFT Correspondence. Very recently Sasakian-Einstein geometry has emerged quite naturally in conformal field theory and string theory. In particular, Klebanov and Witten [**73**] cnosidered $\mathcal{S} = S^3 \times S^2$ in the context of superconformal field theory dual to the string theory on $AdS_5 \times \mathcal{S}$. Their article originates in a conjecture of Maldacena [**90**] who noticed that large N limit of certain conformal field theories in d dimensions can be described in terms of supergravity (and string theory) on a product of $(d+1)$-dimensional anti-de-Sitter AdS_{d+1} space with a compact manifold M. The idea was later examined by Witten who proposed a precise correspondence between conformal field theory observables and those of supergravity [**131**]. It turns out, and this observation has recently been made by Figueroa [**43**], that M necessarily has real Killing spinors and the number of them determines the number of supersymmetries preserved. Depending on the dimension and the amount of supersymmetry, the following geometries are possible: spherical in any dimension, Sasakian-Einstein in dimension $2k+1$, 3-Sasakian in dimension $4k+3$, 7-manifolds with weak G_2-holonomy, and 6-dimensional nearly Kähler manifolds [**1**]. The case when $\dim(M) = 5, 7$ seems to be of particular interest. For other results concerning Sasakian and 3-Sasakian manifolds in supersymmetric field theories see [**51, 56, 38**].

Appendix A. Fundamentals of Orbifolds

A.1. Orbifold and V-bundles. The notion of orbifold was introduced under the name V-manifold by Satake [**112**] in 1956, and subsequently he developed

Riemannian geometry on V-manifolds [**113**] ending with a proof of the Gauss-Bonnet theorem for V-manifolds. Contemporaneously, Baily introduced complex V-manifolds and generalized both the Hodge decomposition theorem [**8**], and Kodaira's projective embedding theorem [**9**] to V-manifolds. Somewhat later in the late 1970's and early 1980's Kawasaki generalized various index theorems [**70, 71, 72**] to the category of V-manifolds. It was about this time that Thurston [**123**] rediscovered the concept of V-manifold, under the name of orbifold, in his study of the geometry of 3-manifolds, and defined the orbifold fundamental group π_1^{orb}. By now orbifold has become the accepted term for these objects and we shall follow suit. However, we do use the name V-bundle for fibre bundles in this category.

Orbifolds arise naturally as spaces of leaves of Riemannian foliations with compact leaves, and we are particularly interested in this point of view. Conversely, every orbifold can be realized in this way. In fact, given an orbifold $\mathcal{O}$, we can construct on it the V-bundle of orthonormal frames whose total space P is a smooth manifold with a locally free action of the orthogonal group $O(n)$ such that $\mathcal{O} = P/O(n)$. Thus, every orbifold can be realized as the quotient space by a locally free action of a Lie group. We are not certain of the history of this connection, but it was surely well understood by Haefliger [**61**] in 1982 who developed the basic techniques for studying the topology of orbifolds.

DEFINITION A.1.1. *A smooth orbifold (or V-manifold) is a second countable Hausdorff space X together with a family $\{U_i\}_{i\in I}$ of open sets that satisfy:*
i) $\{U_i\}_{i\in I}$ is an open cover of X that is closed under finite intersections.
ii) For each $i \in I$ a local uniformizing system consisting of a triple $\{\tilde{U}_i, \Gamma_i, \varphi_i\}$, where $\tilde{U}_i$ is connected open subset of $\mathbb{R}^n$ containing the origin, Γ_i is a finite group of diffeomorphisms acting effectively and properly on $\tilde{U}_i$, and $\varphi_i : \tilde{U}_i \longrightarrow U_i$ is a continuous map onto U_i such that $\varphi_i \circ \gamma = \varphi_i$ for all $\gamma \in \Gamma_i$ and the induced natural map of $\tilde{U}_i/\Gamma_i$ onto U_i is a homeomorphism. The finite group Γ_i is called a local uniformizing group.
iii) Given $\tilde{x}_i \in \tilde{U}_i$ and $\tilde{x}_j \in \tilde{U}_j$ such that $\varphi_i(\tilde{x}_i) = \varphi_j(\tilde{x}_j)$, there is a diffeomorphism $g_{ji} : \tilde{V}_i \longrightarrow \tilde{V}_j$ from a neighborhood $\tilde{V}_i \subset \tilde{U}_i$ of $\tilde{x}_i$ onto a neighborhood $\tilde{V}_j \subset \tilde{U}_j$ of $\tilde{x}_j$ such that $\varphi_i = \varphi_j \circ g_{ji}$.

REMARK A.1. We can always take the finite subgroups Γ_i to be subgroups of the orthogonal group $O(n)$ and in the orientable case $SO(n)$. 2) Condition iii) implies that for each $\gamma_i \in \Gamma_i$ there exists a unique $\gamma_j \in \Gamma_j$ such that $g_{ji} \circ \gamma_i = \gamma_j \circ g_{ji}$. 3) One can define the notion of equivalence of families of open sets, any such family of open sets is contained in a unique maximal family satisfying the required properties. 4) The standard notions of smooth maps between orbifolds, and isomorphism classes of orbifolds, etc. can then be given in an analogous manner to manifolds (see [**112, 113, 8, 9**]). We leave this to the reader to fill in. Notice that in particular a diffeomorphism between orbifolds gives a homeomorphism of the underlying topological spaces. Similarly, a *complex orbifold* can be defined by making the obvious changes.

An alternative definition of orbifold given by Haefliger [**61**] can be obtained as follows: Let $G_{\Gamma,g}$ denote the groupoid of germs of diffeomorphisms generated by the germs of elements in Γ_i and the germs of the diffeomorphisms g_{ji} described above. Let $\tilde{U} = \sqcup_i \tilde{U}_i$ denote the disjoint union of the $\tilde{U}_i$. Then $x, y \in \tilde{U}$ are equivalent if there is a germ $\gamma \in G_{\Gamma,g}$ such that $y = \gamma(x)$. The quotient space $X = \tilde{U}/G_{\Gamma,g}$

defines an orbifold (actually an isomorphism class of orbifolds). In the case that an orbifold X is given as the space of leaves of a foliation $\mathcal{F}$ on a smooth manifold, the groupoid $G_{\Gamma,g}$ is the transverse holonomy groupoid of $\mathcal{F}$.

The following result relating to foliations, which is given in Molino, is fundamental to our work:

THEOREM A.1.2. *([**94**] Proposition 3.7) Let $(M,\mathcal{F},g)$ be a Riemannian foliation of codimension q with compact leaves and bundle-like metric g. Then the space of leaves $M/\mathcal{F}$ admits the structure of a q-dimensional orbifold such that the natural projection $\pi : M \longrightarrow M/\mathcal{F}$ is an orbifold submersion.*

Let X be an orbifold and choose a local uniformizing system $\{U,\Gamma,\varphi\}$. Let $x \in X$ be any point, and let $p \in \varphi^{-1}(x)$, then up to conjugacy the isotropy subgroup $\Gamma_p \subset \Gamma$ depends only on x, and accordingly we shall denote this isotropy subgroup by Γ_x. A point of X whose isotropy subgroups $\Gamma_x \neq \mathrm{id}$ is called a *singular* point. Those points with $\Gamma_x = \mathrm{id}$ are called *regular* points. The subset of regular points is an open dense subset of X. The isotropy groups give a natural stratification of X by saying that two points lie in the same stratum if their isotropy subgroups are conjugate. Thus, the dense open subset of regular points forms the principal stratum. In the case that X is the space of leaves of a foliation, the isotropy subgroup Γ_x is precisely the leaf holonomy group of the leaf x. An orbifold X is a smooth manifold or in the complex analytic category a complex manifold if and only if $\Gamma_x = \mathrm{id}$ for all $x \in X$. In this case we can take $\Gamma = \mathrm{id}$ and $\varphi = \mathrm{id}$, and the definition of an orbifold reduces to the usual definition of a smooth manifold.

Many of the usual differential geometric concepts that hold for smooth or complex analytic manifolds also hold in the orbifold category, in particular the important notion of a fiber bundle.

DEFINITION A.1.3. *A V bundle over an orbifold X consists of a bundle $B_{\tilde{U}}$ over $\tilde{U}$ for each local uniformizing system $\{\tilde{U}_i,\Gamma_i,\varphi_i\}$ with Lie group G and fiber F (independent of $\tilde{U}_i$) together with a homomorphism $h_{\tilde{U}_i} : \Gamma_i \longrightarrow G$ satisfying:*
i) If b lies in the fiber over $\tilde{x}_i \in \tilde{U}_i$ then for each $\gamma \in \Gamma_i$, $bh_{\tilde{U}_i}(\gamma)$ lies in the fiber over $\gamma^{-1}\tilde{x}_i$.
ii) If $g_{ji} : \tilde{U}_i \longrightarrow \tilde{U}_j$ is a diffeomorphism onto an open set, then there is a bundle map $g^*_{ij} : B_{U_j}|g_{ji}(\tilde{U}_i) \longrightarrow B_{\tilde{U}_i}$ satisfying the condition that if $\gamma \in \Gamma_i$, and $\gamma' \in \Gamma_j$ is the unique element such that $g_{ji} \circ \gamma = \gamma' \circ g_{ji}$, then $h_{\tilde{U}_i}(\gamma) \circ g^*_{ji} = g^*_{ji} \circ h_{\tilde{U}_j}(\gamma')$, and if $g_{kj} : \tilde{U}_j \longrightarrow \tilde{U}_k$ is another such diffeomorphism then $(g_{kj} \circ g_{ji})^* = g^*_{ji} \circ g^*_{kj}$.
If the fiber F is a vector space and G acts on F as linear transformations of F, then the V-bundle is called a vector V-bundle. Similarly, if F is the Lie group G with its right action, then the V-bundle is called a principal V-bundle.

The *total space* of a V-bundle over X is an orbifold E with local uniformizing systems $\{B_{\tilde{U}_i},\Gamma^*_i,\varphi^*_i\}$. By choosing the local uniformizing neighborhoods of X small enough, we can always take $B_{\tilde{U}_i}$ to be the product $\tilde{U}_i \times F$ which we shall heretofore assume. There is an action of the local uniformizing group Γ_i on $\tilde{U}_i \times F$ given by sending $(\tilde{x}_i, b) \in \tilde{U}_i \times F$ to $(\gamma^{-1}\tilde{x}_i, bh_{\tilde{U}_i}(\gamma))$, so the local uniformizing groups Γ^*_i can be taken to be subgroups of Γ_i. We are particularly interested in the case of a principal bundle. In the case the fibre is the Lie group G, so the image $h_{\tilde{U}_i}(\Gamma^*_i)$ acts freely on F. Thus the total space P of a principal V-bundle will be smooth if and only if $h_{\tilde{U}_i}$ is injective for all i.

REMARK A.2. We shall often denote a V-bundle by the standard notation $\pi : P \longrightarrow X$ and think of this as an "orbifold fibration". It must be understood, however, that an orbifold fibration is not a fibration in the usual sense. Shortly, we shall show that it is a fibration rationally. Again the standard notions of smooth maps between V-bundles, and isomorphism classes of V-bundles can be given in the usual manner. We let this description to the reader. An *absolute* V-bundle resembles a bundle in the ordinary sense, and corresponds to being able to take $h_{\tilde{U}} = \mathrm{id}$, for all local uniformizing neighborhoods $\tilde{U}$. In particular, the trivial V-bundle $X \times F$ is absolute. Another important notion introduced by Kawasaki [**71**] is that of proper. A V-bundle E is said to be *proper* if the local uniformizing groups Γ_i^* of E act effectively on X when viewed as subgroups of the local uniformizing groups Γ_i on X. Any V-bundle with smooth total space is clearly proper. The Kawasaki index theorems such as his Riemann-Roch Theorem used in section 3.2 require the V-bundles to be proper.

A.2. Orbifold Homology, Cohomology, and Homotopy Groups. Since an orbifold fibration is not a fibration in the usual sense, the usual techniques in topology for fibrations do not apply directly. However, Haefliger [**61**] has defined orbifold homology, cohomology, and homotopy groups which do have an analogue in the standard theory. Let X be an orbifold of dimension n and let P denote the bundle of orthonormal frames on X. It is a smooth manifold on which the orthogonal group $O(n)$ acts locally freely with the quotient X. Let $EO(n) \longrightarrow BO(n)$ denote the universal $O(n)$ bundle. Consider the diagonal action of $O(n)$ on $EO(n) \times P$ and denote the quotient by BX. Now there is a natural projection $p : BX \longrightarrow X$ with generic fiber the contractible space $EO(n)$, and Haefliger defines the orbifold cohomology, homology, and homotopy groups by

$$H^i_{orb}(X,\mathbb{Z}) = H^i(BX,\mathbb{Z}), \quad H_i^{orb}(X,\mathbb{Z}) = H_i(BX,\mathbb{Z}), \quad \pi_i^{orb}(X) = \pi_i(BX).$$

This definition of π_1^{orb} is equivalent to Thurston's better known definition [**123**] in terms of orbifold deck transformations, and when X is a smooth manifold these orbifold groups coincide with the usual groups. It should be noted that generally these groups are not topological invariants, but invariants of the orbifold structure only. Rationally, however, the orbifold groups coincide with the usual groups, and thus are topological invariants. Indeed from the Leray spectral sequence for the map p we have

PROPOSITION A.2.1. [**61**] *The map $p : BX \longrightarrow X$ induces an isomorphism*

$$H^i_{orb}(\mathcal{S},\mathbb{Z}) \otimes \mathbb{Q} \simeq H^i(\mathcal{S},\mathbb{Z}) \otimes \mathbb{Q}.$$

Now with this in hand for the orbifold category, the circle V-bundles over $\mathcal{S}$ are classified [**62**] by $H^2_{orb}(\mathcal{S},\mathbb{Z})$. Of course, rationally there is no difference by Proposition A.2.1. The rational Gysin sequence for orbifold sphere bundles whose generic fibres are spheres also holds. Haefliger's theory also applies to the following situation. Let G be a compact Lie group acting locally freely on an orbifold Y with quotient X. This gives rise to a fibration $EO(n) \times G \longrightarrow BY \longrightarrow BX$, which induces the long exact homotopy sequence

$$\cdots \longrightarrow \pi_i(G) \longrightarrow \pi_i^{orb}(Y) \longrightarrow \pi_i^{orb}(X) \longrightarrow \pi_{i-1}(G) \longrightarrow \cdots .$$

This was used by Haefliger and Salem [**62**] in their study of torus actions on orbifolds.

We are particularly interested in the case of circle V-bundles. Using the exponential exact sequence one sees as the usual case that $H^2_{orb}(X,\mathbb{Z})$ classifies equivalence classes of circle V-bundles over an orbifold X. Furthermore, in **[62]** it is shown that $H^2(X,\mathbb{Z})$ classifies circle V-bundles up to local equivalence. This gives a monomorphism $H^2(X,\mathbb{Z})\longrightarrow H^2_{orb}(X,\mathbb{Z})$ which is an isomorphism rationally.

In **[22]** we introduced the set $\mathrm{Pic}^{orb}(X)$ of equivalence classes holomorphic line V-bundles over a complex orbifold X and one easily sees **[22]**:

LEMMA A.2.2. *$Pic^{orb}(X)$ forms an Abelian group. Furthermore, there is a monomorphism $Pic(X)\longrightarrow Pic^{orb}(X)$ which is an isomorphism rationally.*

The notion of sections of bundles works just as well in the orbifold category.

DEFINITION A.2.3. *Let E be a V-bundle over an orbifold X. Then a section σ of E over the open set $V\subset X$ is a section σ_U of the bundle B_U for each local uniformizing system $\{U,\Gamma,\varphi\}\in\mathcal{F}_V$ such that for any $x\in U$ we have*
(i) For each $\gamma\in\Gamma$ $\sigma_U(\gamma^{-1}x)=h_U(\gamma)\sigma_U(x)$.
(ii) If $\lambda:\{U,\Gamma,\varphi\}\longrightarrow\{U',\Gamma',\varphi'\}$ is an injection, then $\lambda^*\sigma_{U'}(\lambda(x))=\sigma_U(x)$.

If each of the local sections σ_U is continuous, smooth, holomorphic, etc., we say that σ is continuous, smooth, holomorphic, etc., respectively. Given local sections σ_U of a vector V-bundle we can always construct Γ-invariant local sections by "averaging over the group", i.e., we define $\sigma^I_U=\frac{1}{|\Gamma|}\sum_{\gamma\in\Gamma}\sigma_U\circ\gamma$. A similar procedure holds for product structures. For example, if L is a holomorphic line V-bundle on X, and if σ is a holomorphic section, we can construct local invariant sections σ^I_U of $L^{|\Gamma|}$ by taking products, viz., $\sigma^I_U=\frac{1}{|\Gamma|}\prod_{\gamma\in\Gamma}\sigma_U\circ\gamma$.

The standard notions of tangent bundle, cotangent bundle, and all the associated tensor bundles all have V-bundle analogues **[8, 112, 113]**. In particular, if V is an open subset of $\varphi(U)$ then the integral of an n-form (measurable) σ is defined by $\int_V\sigma=\frac{1}{|\Gamma|}\int_{\varphi^{-1}(V)}\sigma_U$. All of the standard integration techniques, such as Stokes' theorem, hold on V-manifolds.

Riemannian metrics also exist by the standard partition of unity argument, and we shall always work with Γ-invariant metrics. Moreover, all the standard differential geometric objects involving curvature and metric concepts, such as the Ricci tensor, Hodge star operator, etc., hold equally well. On a complex orbifold there is a Γ-invariant tensor field J of type $(1,1)$ which describes the complex structure on the tangent V-bundle TX. The almost complex structure J gives rise in the usual way to the V-bundles $\Lambda^{r,s}$ of differential forms of type (r,s). The standard concepts of Hermitian and Kähler metrics hold equally well on V-manifolds, and all the special identities involving Kähler, Einstein, or Kähler-Einstein geometry hold. In particular, the standard Weizenböck formulas hold.

Finally, there is associated to every compact orbifold X an integer m_0 called the *order* of X and defined to be the least common multiple of the orders of the local uniformizing groups.

References

[1] B. S. Acharya, J. M. Figueroa-O'Farrill, C. M. Hull, and B. Spence, *Branes at Conical Singularities and Holography*, Adv. Theor. Math. Phys. 2 (1998) 1249-1286.

[2] D. N. Akhiezer, *Homogeneous complex manifolds*, in Enc. Math. Sci. vol 10, *Several Complex Variables IV*, S. G. Gindikin and G. M. Khenkin (Eds)., Springer-Verlag, New York, 1990.

[3] D. V. Alekseevski, *Riemannian manifolds with exceptional holonomy groups*, Functional Anal. Appl. 2 (1968), 106-114.

[4] D. V. Alekseevski, *Classification of quaternionic spaces with solvable group of motions*, Math. USSR-Izv. 9 (1975), 297-339.

[5] S. Aloff and N. Wallach, *An infinite family of distinct 7-manifolds admitting positively curved Riemannian structures*, Bull. Amer. Math. Soc. 81 (1975), 93-97.

[6] M. Anderson, *Convergence and rigidity of manifolds under Ricci curvature bounds*, Invent. Math. 102 (1990), 429-445.

[7] L. Astey, E. Micha, and G. Pastor, *Homeomorphism and diffeomorphism types of Eschenberg spaces*, Differential Geom. Appl. 7 (1997), 41-50.

[8] W. L. Baily, *The decomposition theorem for V-manifolds*, Amer. J. Math. 78 (1956), 862-888.

[9] W. L. Baily, *On the imbedding of V-manifolds in projective space*, Amer. J. Math. 79 (1957), 403-430.

[10] C. Bär, *Real Killing spinors and holonomy*, Comm. Math. Phys. 154 (1993), 509-521.

[11] H. Baum, T. Friedrich, R. Grunewald, and I. Kath, *Twistors and Killing Spinors on Riemannian Manifolds*, Teubner-Texte für Mathematik, vol. 124, Teubner, Stuttgart, Leipzig, 1991.

[12] L. Berard Bèrgery, *Variétés quaternionnienns*, Notes d'unde confŕence à la table ronde "Varieétés d'Einstein", Espalion (1997); (unpublished).

[13] M. Berger, *Sur les groupes d'holonomie des variétés à connexion et des variétés riemanniennes*, Bull. Soc. Math. France 83 (1955), 279-330.

[14] A. L. Besse, *Einstein Manifolds*, Springer-Verlag, New York (1987).

[15] R. Bielawski, *On the hyperkähler metrics associated to singularities of nilpotent varieties*, Ann. Global Anal. Geom. 14 (1996), 177-191.

[16] R. Bielawski, *Betti numbers of 3-Sasakian quotients of spheres by tori*, Bull. London Math. Soc. 29 (1997), 731-736.

[17] R. Bielawski, *Complete T^{n}-invariant hyperkähler $4n$-manifolds*, preprint MPI (1998). To appear in Math. Ann.

[18] R. Bielawski and A. Dancer, *The geometry and topology of toric hyperkähler manifolds*, to appear in Comm. Anal. Geom. (1998).

[19] D. E. Blair, *Contact Manifolds in Riemannian Geometry*, Lecture Notes in Mathematics 509, Springer-Verlag, New Yrok 1976.

[20] D. E. Blair and S. I. Goldberg, *Topology of almost contact manifolds*, J. Differential Geom. 1 (1967), 347-354.

[21] E. Bonan, *Sur les G-structures de type quaternionien*, Cah. Top. Geom. Differ. 9 (1967), 389-461.

[22] C. P. Boyer and K. Galicki, *The twistor space of a 3-Sasakian manifold*, Int. J. Math. 8 (1997), 31-60.

[23] C. P. Boyer and K. Galicki, *Sasakian-Einstein manifolds*, preprint DG/9811098, October 1998.

[24] C. P. Boyer and K. Galicki, *Polygons, gravitons, and Einstein manifolds*, in preparation.

[25] C. P. Boyer, K. Galicki, and B. M. Mann, *Quaternionic reduction and Einstein manifolds*, Comm. Anal. Geom. 1 (1993), 1-51.

[26] C. P. Boyer, K. Galicki, and B. M. Mann, *The geometry and topology of 3-Sasakian manifolds*, J. reine angew. Math. 455 (1994), 183-220.

[27] C. P. Boyer, K. Galicki, and B. M. Mann, *On strongly inhomogeneous Einstein manifolds*, Bull. London Math. Soc. 28 (1996), 401-408.

[28] C. P. Boyer, K. Galicki, and B. M. Mann, *New examples of inhomogeneous Einstein manifolds of positive scalar curvature*, Math. Res. Lett. 1 (1994), 115-121.

[29] C. P. Boyer, K. Galicki, and B. M. Mann, *3-Sasakian manifolds*, Proc. Japan Acad. vol. 69, Ser. A (1993), 335-340.

[30] C. P. Boyer, K. Galicki, and B. M. Mann, *Hypercomplex structures on Stiefel manifolds*, Ann. Global Anal. Geom. 14 (1996), 81-105.

[31] C. P. Boyer, K. Galicki, and B. M. Mann, *A note on smooth toral reductions of spheres*, Manuscripta Math. 95 (1998), 321-344.

[32] C. P. Boyer, K. Galicki, and B. M. Mann, *Hypercomplex structures from 3-Sasakian structures*, J. reine angew. Math., 501 (1998), 115-141.

[33] C. P. Boyer, K. Galicki, B. M. Mann, and E. Rees, *Compact 3-Sasakian 7-manifolds with arbitrary second Betti number*, Invent. Math. 131 (1998), 321-344.

[34] C. P. Boyer, K. Galicki, B. M. Mann, and E. Rees, *Einstein manifolds of positive scalar curvature with arbitrary second Betti number*, Balkan J. Geom. Appl. 1 no. 2 (1996), 1-8.

[35] G.E. Bredon, *Introduction to Compact Transformation Groups*, Academic Press, New York (1972).

[36] R. Bryant, *Metrics with exceptional holonomy*, Ann. Math. 126 (1987), 525-576.

[37] F. M. Cabrera, M. D. Monar, and A. F. Swann, *Classification of G_2-structures*, J. London Math. Soc. 53 (1996), 407-416.

[38] L. Castellani, A. Ceresole, R. D'Auria, S. Ferrara, P. Fré, and M. Trigiante, *G/H M-branes and AdS_{p+2} Geometries*, preprint, March 1998; hep-th/9803039.

[39] A. Dancer and A. Swann, *The geometry of singular quaternionic Kähler quotients*, Int. J. Math. 8 (1997), 595-610.

[40] J. H. Eschenburg, *New examples of manifolds with strictly positive curvature*, Invent. Math. 66 (1982), 469-480.

[41] J. H. Eschenburg, *Cohomology of biquotients*, Manuscripta Math. 75 (1992), 151-166.

[42] M. Fernández and A. Gray, *Riemannian manifolds with structure group G_2*, Ann. Mat. Pura Appl. 32 (1982), 19-45.

[43] J. M. Figueroa-O'Farrill, *Near-Horizon Geometries of Supersymmetric Branes*, preprint, July 1998; hep-th/9807149.

[44] A. Franc, *Spin structures and Killing spinors on lens spaces*, J. Geom. Phys. 4 (3) (1987), 277-287.

[45] T. Friedrich, *der erste Eigenwert des Dirac-Operators einer kompakten Riemannschen Mannigfaltigkeiten nichtnegativer Skalarkrümmung*, Math. Nach. 97 (1980), 117-146.

[46] T. Friedrich and I. Kath, *Compact five-dimensional Riemannian manifolds with parallel spinors*, Math. Nachr. 147 (1990), 161-165.

[47] T. Friedrich and I. Kath, *Compact seven-dimensional manifolds with Killing spinors*, Comm. Math. Phys. 133 (1990), 543-561.

[48] T. Friedrich, I. Kath, A. Moroianu, and U. Semmelmann, *On nearly parallel G_2-structures*, J. Geom. Phys. 23 (1997), 259-286.

[49] T. Friedrich and Kurke, *Compact four-dimensional self-dual Einstein manifolds with positive scalar curvature*, Math. Nach. 106 (1982), 271-299.

[50] A. Fujiki, *On the de Rham cohomology group of a compact Kähler symplectic manifold*, in Algebraic Geometry, Sendai, 1985 (Advanced Studies in Pure Mathematics 10), ed. T. Oda, North Holland (1987).

[51] K. Galicki, *Geometry of the Scalar Coupling in $N = 2$ Supergravity Models*, Class. Quan. Grav. 9(1) (1992), 27-40.

[52] K. Galicki and B. H. Lawson, Jr., *Quaternionic reduction and quaternionic orbifolds*, Math. Ann. 282 (1988), 1-21.

[53] K. Galicki and T. Nitta, *Nonzero scalar curvature generalizations of the ALE instantons*, J. Math. Phys. 33 (1992), 1765-1771.

[54] K. Galicki and S. Salamon, *On Betti numbers of 3-Sasakian manifolds*, Geom. Ded. 63 (1996), 45-68.

[55] H. Geiges, *Normal contact structures on 3-manifolds*, Tôhoku Math. J. 49 (1997), 415-422.

[56] G. W. Gibbons and P. Rychenkova, *Cones, tri-Sasakian structures and superconformal invariance*, preprint, September 1998; hep-th/9809158.

[57] S. I. Goldberg, *Nonegatively curved contact manifolds*, Proc. Amer. Math. Soc. 96 (1986), 651-656.

[58] A. Gray, *A note on manifolds whose holonomy group is a subgroup of $Sp(n)\Delta Sp(1)$*, Mich. Math. J. 16 (1965), 125-128.

[59] A. Gray, *Weak holonomy groups*, Math. Z. 123 (1971), 290-300.

[60] M. Gromov, *Curvature, diameter and Betti numbers*, Comment. Math. Helvetici 56 (1981) 179-195.

[61] A. Haefliger, *Groupoides d'holonomie et classifiants*, Astérisque 116 (1984), 70-97.

[62] A. Haefliger and E. Salem, *Actions of tori on orbifolds*, Ann. Global Anal. Geom. 9 (1991), 37-59.

[63] N. J. Hitchin, *Kählerian twistor spaces*, Proc. Lond. Math. Soc. 43 (1981), 133-150.

[64] N. J. Hitchin, unpublished.

[65] N. J. Hitchin, A. Karlhede, U. Lindström and M. Roček, *H*yperkähler metrics and supersymmetry, Comm. Math. Phys. 108 (1987), 535-589.
[66] S. Ishihara, *Q*uaternion Kählerian manifolds, J. Differential Geom. 9 (1974), 483-500.
[67] S. Ishihara, *Q*uaternion Kählerian manifolds and fibered Riemannian spaces with Sasakian 3-structure, Kodai Math. Sem. Rep. 25 (1973), 321-329.
[68] S. Ishihara and M. Konishi, *F*ibered Riemannian spaces with Sasakian 3-structure, Differential Geometry, in honor of K. Yano, Kinokuniya, Tokyo (1972), 179-194.
[69] T. Kashiwada, *A* note on a Riemannian space with Sasakian 3-structure, Nat. Sci. Reps. Ochanomizu Univ. 22 (1971), 1-2.
[70] T. Kawasaki, *T*he Riemann-Roch theorem for complex V-manifolds, Osaka J. Math. 16 (1979), 151-159.
[71] T. Kawasaki, *T*he signature theorem for V-manifolds, Topology 17 (1978), 75-83.
[72] T. Kawasaki, *T*he index of elliptic operators over V-manifolds, Nagoya Math. J. 84 (1981), 135-157.
[73] I. R. Klebanov and E. Witten, *S*uperconformal Field Theory on Threebranes at a Calabi-Yau Singularity, preprint, July 1998; hep-th/9807080.
[74] S. Kobayashi, *O*n compact Kähler manifolds with positive Ricci tensor, Ann. Math. 74 (1961), 381-385.
[75] S. Kobayashi and T. Ochiai, *C*haracterizations of complex projective spaces and hyperquadrics, J. Math. Kyoto. Univ. 13 (1973), 31-47.
[76] M. Konishi, *O*n manifolds with Sasakian 3-structure over quaternion Kählerian manifolds, Kodai Math. Sem. Reps. 26 (1975), 194-200.
[77] V. Kraines, *T*opology of quaternionic manifolds, Trans. Amer. Math. Soc. 122 (1966), 357-367.
[78] M. Kreck and S. Stolz, *A* diffeomorphism classification of 7-dimensional homogeneous Einstein manifolds with $SU(3)\,\Theta SU(2)\,\Theta U(1)$ symmetry, Ann. Math. 127 (1988), 373-388.
[79] M. Kreck and S. Stolz, *S*ome nondiffeomorphic homeomorphic homogeneous 7-manifolds with positive sectional curvature, J. Differential Geom. 33 (1991), 465-486.
[80] B. Kruggel, *A* homotopy classification of certain 7-manifolds, Trans. Amer. Math. Soc. 349 (1997), 2827-2843.
[81] B. Kruggel, *K*reck-Stolz invariants, normal invariants and the homotopy classification of generalized Wallach spaces, Quart. J. Math. Oxford Ser (2) 49 (1998), 469-485.
[82] B. Kruggel, *D*iffeomorphism and homeomorphism classification of Eschenburg spaces, preprint.
[83] Y.-Y. Kuo, *O*n almost contact 3-structure, Tôhoku Math. J. 22 (1970), 325-332.
[84] Y.-Y. Kuo and S. Tachibana, *O*n the distribution appeared in contact 3-structure, Taita J. Math. 2 (1970), 17-24.
[85] C. LeBrun, *A* finiteness theorem for quaternionic-Kähler manifolds with positive scalar curvature, Contemp. Math. 154 (1994), 89-101.
[86] C. LeBrun, *A* rigidity theorem for quaternionic-Kähler manifolds, Proc. Amer. Math. Soc. 103 (1988), 1205-1208.
[87] C. LeBrun, *F*ano manifolds, contact structures, and quaternionic geometry, Int. J. Math. 6 (1995), 115-127.
[88] C. LeBrun and S. M. Salamon, *S*trong rigidity of positive quaternion-Kähler manifolds, Invent. Math. 118 (1994), 109-132.
[89] U. Lindström and M. Roček, *S*calar tensor duality and $N = 1,2$ non-linear σ-models, Nucl. Phys. B222 (1983), 285-308.
[90] J. Maldacena, *T*he large N limit of superconformal field theories and supergravity, preprint, 1997; hep-th/9711200.
[91] S. Marchiafava and G. Romani, *S*ui fibrati con struttura quaternioniale generalizzata, Ann. Mat. Pura Appl. 107 (1976), 131-157.
[92] R. J. Milgram, *T*he classification of Aloff-Wallach manifolds and their generalizations, to appear in the volume in honor of the 60th birthday of C. T. C. Wall.
[93] Y. Miyaoka and S. Mori, *A* numerical criterion for uniruledness, Ann. Math. 124 (1986), 65-69.
[94] P. Molino, *R*iemannian foliations, Progress in Mathematics 73, Birkhäuser, Boston, 1988.
[95] C.C. Moore and C. Schochet, *G*lobal Analysis on Foliated Spaces, Springer-Verlag, New York, 1988.

[96] S. Mori and S. Mukai, Classification of Fano 3-folds with $B_2 \geq 2$, Manuscripta Math. 36 (1981), 147-162.
[97] A. Moroianu, Parallel and Killing spinors on Spin^c-manifolds, Comm. Math. Phys. 187 (1997), 417-427.
[98] Y. Nagatomo, Rigidity of c_1-self-dual connections on quaternionic Kähler manifolds, J. Math. Phys. 33 (1992), 4020-4025.
[99] B. O'Neill, Semi-Riemannian Geometry, Pure and Applied Math. 103, Academic Press, New York 1983.
[100] L. Ornea and P. Piccinni, Locally conformally Kähler structures in quaternionic geometry, Trans. Am. Math. Soc. 349 (1997), 641-655.
[101] H. Pedersen and Y. S. Poon, Deformations of hypercomplex Structures, J. reine angew. Math. 499 (1998), 81-99.
[102] H. Pedersen and Y. S. Poon, A note on rigidity of 3-Sasakian manifolds, Proc. Am. Math. Soc., to appear (1998).
[103] P. Piccinni, The geometry of positive locally quaternion Kähler manifolds, Ann. Global Anal. Geom. 16 (1998), 255-272.
[104] Y. S. Poon and S. Salamon, Eight-dimensional quaternionic Kähler manifolds with positive scalar curvature, J. Differential Geom. 33 (1990), 363-378.
[105] B. L. Reinhart, Foliated manifolds with bundle-like metrics, Ann. Math. 69(2) (1959), 119-132.
[106] S. Salamon, Quaternionic Kähler manifolds, Invent. Math. 67 (1982), 143-171.
[107] S. Salamon, Differential geometry of quaternionic manifolds, Ann. Sci. Ec. Norm. Sup. Paris 19 (1986), 31-55.
[108] S. Salamon, The Dirac operator and quaternionic manifolds, Proc. International Conference on Differential Geometry and its Applications, Opava (1992).
[109] S. Salamon, Riemannian geometry and holonomy groups, Pitman Research Notes in Mathematics Series, Longman Scientific & Technical, Essex UK, 1989.
[110] S. Sasaki, On differentiable manifolds with certain structures which are closely related to almost contact structure, Tôhoku Math. J. 2 (1960), 459-476.
[111] S. Sasaki, Spherical space forms with normal contact metric 3-structure, J. Differential Geom. 6 (1972), 307-315.
[112] I. Satake, On a generalization of the notion of manifold, Proc. Nat. Acad. Sci. U.S.A 42 (1956), 359-363.
[113] I. Satake, The Gauss-Bonnet theorem for V-manifolds, J. Math. Soc. Japan V.9 No 4. (1957), 464-476.
[114] S. Smale, On the structure of 5-manifolds, Ann. Math. 75 (1962), 38-46.
[115] A. F. Swann, Hyperkähler and quaternionic Kähler geometry, Math. Ann. 289 (1991), 421-450.
[116] S. Tachibana, On harmonic tensors in compact Sasakian spaces, Tôhoku Math. J. 17 (1965), 271-284.
[117] S. Tachibana and W. N. Yu, On a Riemannian space admitting more than one Sasakian structure, Tôhoku Math. J. 22 (1970), 536-540.
[118] S. Tanno, On the isometry of Sasakian manifolds, J. Math. Soc. Japan 22 (1970), 579-590.
[119] S. Tanno, Killing vectors on contact Riemannian manifolds and fiberings related to the Hopf fibrations, Tôhoku Math. J. 23 (1971), 313-333.
[120] S. Tanno, Sasakian manifolds with constant Φ-holomorphic sectional curvature, Tôhoku Math. J. 23 (1969), 21-38.
[121] S. Tanno, Geodesic flows on CL_1-manifolds and Einstein metrics on $S^3 \Theta S^2$, in *Minimal submanifolds and geodesics (Proc. Japan-United States Sem., Tokyo, 1977)*, pp. 283-292, North Holland, Amsterdam-New York, 1979.
[122] C.B. Thomas, Almost regular contact manifolds, J. Differential Geom. 11 (1976), 521-533.
[123] W. Thurston, The Geometry and Topology of 3-Manifolds, Mimeographed Notes, Princeton Univ. Chapt. 13 (1979).
[124] G. Tian and S.-T. Yau, Kähler-Einstein metrics on complex surfaces with $c_1 > 0$, Comm. Math. Phys. 112 (1987), 175-203.
[125] C. Udrişte, Structures presque coquaternioniennes, Bull. Math. de la Soc. Sci. Math. de Roumanie 12 (1969), 487-507.

[126] H.C. Wang, *Closed manifolds with homogeneous complex structures*, Am. J. Math. 76 (1954), 1-32.

[127] M. Wang, *Some examples of homogeneous Einstein manifolds in dimension seven*, Duke Math. J. 49 (1982), 23-28.

[128] M. Wang, *Parallel Spinors and Parallel Forms*, Ann. Global Anal. Geom. 7 (1989), 59-68.

[129] M. Wang, *Parallel Spinors and Parallel Forms*, Ann. Global Anal. Geom. 13 (1995), 31-42.

[130] M. Wang, *Einstein metrics and quaternionic Kähler manifolds*, Math. Z. 210 (1992), 305-325.

[131] E. Witten, *Anti-de Sitter space and holography*, preprint, 1998; hep-th/9802150.

[132] J. A. Wolf, *Complex homogeneous contact manifolds and quaternionic symmetric spaces*, J. Math. Mech. 14 (1965), 1033-1047.

[133] K. Yano and S. Bochner, *Curvature and Betti numbers*, Annals of Math. Studies 32, Princeton University Press (1953).

[134] K. Yano and M. Kon, *Structures on manifolds*, Series in Pure Mathematics 3, World Scientific Pub. Co., Singapore, 1984.

Department of Mathematics and Statistics, University of New Mexico, Albuquerque, N.M. 87131,

E-mail address: cboyer@math.unm.edu, galicki@math.unm.edu

Part II: Towards a General Theory of Einstein Manifolds

Ricci Flow and Einstein Metrics in Low Dimensions

Bennett Chow

1. Introduction

The purpose of this essay is to give an expository account of Hamilton's work during the 1980's on the Ricci flow on surfaces [**21**], 3-manifolds [**19**], and 4-manifolds [**20**]. We have restricted our attention to these papers both because of our personal familiarity with them, and because they deal directly with constructing Einstein metrics — in fact, constant sectional curvature metrics.

The study of the Ricci flow began with Hamilton's seminal 1982 paper 'Three-manifolds with positive Ricci curvature.' In this paper he not only introduced the notion of the Ricci flow, but applied it to classify closed 3-manifolds with positive Ricci curvature. Later, in another very important 1986 paper 'Four-manifolds with positive curvature operator,' Hamilton extended his methods to show that closed 4-manifolds with positive curvature operator are topologically either S^4 or $\mathbb{RP}^4$. An important development in this paper is the use of the 'weak and strong maximum principles for systems,' which enabled Hamilton to also classify both 3-manifolds with nonnegative Ricci curvature and 4-manifolds with nonnegative curvature operator. Furthermore, Hamilton also greatly simplified the computations in his original paper. The last of Hamilton's papers in the 80's on the Ricci flow appeared in 1988 and was entitled 'The Ricci flow on surfaces.' Here Hamilton proved for any initial metric on a surface the convergence of the Ricci flow to a constant curvature metric (except for the case of a metric on the 2-sphere with variable signed curvature, a condition which was later removed by similar methods.) The real jewel of this paper is the techniques that Hamilton introduced in this paper. In particular, he obtained both a 'Harnack estimate' and an 'entropy estimate.' The Harnack estimate is especially important in the analysis of singularities (see [**27**].)

The rest of this paper is organized as follows. In section 2, we describe the basic facts about the Ricci flow. Then in section 3 we quickly review the maximum principle, which is the main tool used in the study of the Ricci flow. In sections 4 through 6, we devote one section each to Hamilton's results on surfaces, 3-manifolds, and 4-manifolds. Finally, in the last section we reference some of the other important works on the Ricci flow.

2. Basic facts about the Ricci flow

2.1. The equation. The Ricci flow is a nonlinear heat equation which deforms metrics in the direction of minus the Ricci tensor. Let M be a differentiable manifold. A family of Riemannian metrics $g(t)$, $t \in [0,T)$, where $T \in (0,\infty]$, is called a solution to the *Ricci flow* if

$$\frac{\partial g}{\partial t}(x,t) = -2\,Rc\,(x,t)\,,$$

at all points $x \in M$ and times $t \in [0,T)$. In other words, for any tangent vectors X and Y at x we have:

$$\frac{\partial g}{\partial t}(X,Y)(x,t) = -2\,Rc(X,Y)(x,t),$$

for all $x \in M$ and $t \in [0,T)$. Taking $X = \partial/\partial x^i$ and $Y = \partial/\partial x^j$, we obtain the component form of the Ricci flow equation:

$$\frac{\partial}{\partial t} g_{ij} = -2\,R_{ij}\,,$$

which is the usual way we shall write the equation.

2.2. Short-time existence. The first question is that of short-time existence. On a compact manifold a solution exists for short time for any smooth initial metric:

THEOREM 2.1. (Hamilton 1982, DeTurck 1982) *Given any smooth, compact Riemannian manifold (M, g_o), there exists a unique smooth solution $g(t)$ to the Ricci flow with initial condition $g(0) = g_o$ on some time interval* $[0,\epsilon)$.

The [**19**] original proof of this result uses the Nash-Moser implicit function theorem and is rather involved. We suggest that the reader consult [**18**] for a vastly simplified proof.

2.3. Fundamental evolution equations. Once we are given the equation for the time evolution of the metric, in order to understand how the geometry of the metric evolves, we need to first derive the equations for the geometric quantities associated to the metric, such as the Christoffel symbols and the Riemann curvature tensor. In particular, we have (see [**19**] for proofs of all of the formulas in this subsection:)

LEMMA 2.2. *Under the Ricci flow, the Christoffel symbols evolve by*

$$\frac{\partial}{\partial t}\Gamma_{ij}^k = -g^{kl}\left(\nabla_i R_{jl} + \nabla_j R_{il} - \nabla_l R_{ij}\right).$$

Using the definition of the Riemann curvature tensor in terms of the Christoffel symbols, and applications of the Bianchi identities, one derives:

LEMMA 2.3. *Under the Ricci flow, the Riemann curvature (4,0)-tensor satisfies the following reaction-diffusion equation*

$$\frac{\partial}{\partial t} R_{ijkl} = \Delta R_{ijkl} + 2(B_{ijkl} - B_{ijlk} + B_{ikjl} - B_{iljk})$$

where $B_{ijkl} = R_{ipjq} R_{kplq}$.

Although B is quadratic in Rm, it is not exactly the square of Rm. To define the square, consider Rm as a self-adjoint operator on 2-forms

$$Rm : \wedge^2 M \to \wedge^2 M$$

defined by

$$Rm\left(U\right)_{ij} = \sum_{p,q,r,s} g^{pr} g^{qs} R_{ijpq} U_{rs} = R_{ijpq} U_{pq}$$

By one of the symmetries of the Riemann curvature tensor, we have

$$\langle Rm\left(U\right), V\rangle = \langle U, Rm\left(V\right)\rangle ,$$

that is, Rm is self-adjoint. Here the inner product on 2-forms is defined by

$$\langle U, V\rangle = g^{ik} g^{jl} U_{ij} V_{kl}.$$

Now we can square Rm as an operator to obtain

$$Rm^2 : \wedge^2 TM^* \to \wedge^2 TM^*$$

which is given by

$$Rm^2\left(U\right)_{ij} = Rm\left[Rm\left(U\right)\right]_{ij} = R_{ijpq} R_{pqrs} U_{rs}.$$

Hence we write

$$\left(Rm^2\right)_{ijkl} = R_{ijpq} R_{pqkl}.$$

Although this is obviously the most natural definition of the square of Riemann curvature operator, there is another concept of square which will be useful. This definition applies whenever one has a self-adjoint operator on a Lie algebra. The reason this is relevant is that $\wedge^2 M$ has a Lie algebra structure which makes it isomorphic to $so\left(n\right)$. In particular, we define the Lie bracket of two 2-forms by

$$[U, V]_{ij} = U_{ip} g^{pq} V_{qj} - V_{ip} g^{pq} U_{qj}.$$

Noting that the matrix of components of a 2-form is antisymmetric and that in coordinates where $g^{ij} = \delta_{ij}$ we have

$$[U, V]_{ij} = U_{ip} V_{pj} - V_{ip} U_{pj} = \left(UV - VU\right)_{ij}.$$

This gives the isomorphism between $\left(\wedge^2 M_x, [\ ,\ ]\right)$ and $so\left(n\right)$ for any given point $x \in M$. Choose any basis $\{\phi^a\}$ of $\mathfrak{g}$ and let c_c^{ab} denote the structure constants:

$$[\phi^a, \phi^b] = \sum_c c_{abc} \phi^c.$$

Now we can define the square using the Lie bracket. Given a Lie algebra $\mathfrak{g}$ with a Lie bracket $[\ ,\ ]$ and an inner product $\langle\ ,\ \rangle$, the **Lie square** $L^\# : \mathfrak{g} \to \mathfrak{g}$ of an operator $L : \mathfrak{g} \to \mathfrak{g}$ is defined by

$$L^\#_{ab} = c_{ace} c_{bdf} L_{cd} L_{ef}.$$

Computations yield

$$\left(Rm^2\right)_{ijkl} = 2(B_{ijkl} - B_{ijlk})$$

and

$$\left(Rm^\#\right)_{ijkl} = 2\left(B_{ikjl} - B_{iljk}\right).$$

Hence

$$\frac{\partial}{\partial t}R_{ijkl} = \Delta R_{ijkl} + \left(Rm^2\right)_{ijkl} + \left(Rm^{\#}\right)_{ijkl}. \tag{1}$$

Now we go on to the evolution equations for the Ricci tensor and scalar curvature. Since the Ricci tensor is the trace of the Riemann curvature tensor, one easily obtains from Lemma 2.3:

COROLLARY 2.4. *Under the Ricci flow, the Ricci tensor satisfies*

$$\frac{\partial}{\partial t}R_{jk} = \Delta R_{jk} + 2g^{pr}g^{qs}R_{pjqk}R_{rs} - 2g^{pq}R_{pj}R_{qk}.$$

Taking a second trace, one has:

COROLLARY 2.5. *Under the Ricci flow, the scalar curvature function evolves by*

$$\frac{\partial}{\partial t}R = \Delta R + 2\left|Rc\right|^2.$$

In order to make use of these nice equations, one needs the maximum principle, which in the parabolic case yields bounds for solutions to reaction-diffusion equations such as the ones above.

3. Maximum principles

In this section we recall the various versions of the maximum principle that are required for the study of the Ricci flow. We start with the scalar maximum principle and work our way up to the maximum principle for systems where the solution is a section of a vector bundle. We consider both the weak and the strong maximum principles. An excellent basic reference is [**47**]. See [**28**], p.99, for the scalar heat equation on a manifold, and [**19**], [**20**] for the parabolic maximum principle for tensors and systems.

3.1. Weak maximum principle.

3.1.1. *Scalar equations.* The heat equation is the prototype for parabolic equations. One of the most important properties it satisfies is the maximum principle, which says that for any smooth solution to the heat equation, whatever pointwise bounds hold at $t = 0$ also hold for $t > 0$.

THEOREM 3.1. (Scalar Maximum Principle I: pointwise bounds are preserved) *Let $u : M^n \times [0,T) \to \mathbb{R}$ be a C^2 solution to the heat equation*

$$\frac{\partial u}{\partial t} = \Delta u$$

on a complete Riemannian manifold. If $C_1 \leq u(x,0) \leq C_2$ for all $x \in M$, for some constants $C_1, C_2 \in \mathbb{R}$, then $C_1 \leq u(x,t) \leq C_2$ for all $x \in M$ and $t \in [0,T)$.

More generally, one may allow the metric to depend on time and also add in gradient and reaction terms. Namely, consider the semi-linear heat equation

$$\frac{\partial u}{\partial t} = \Delta u + \langle X, \nabla u\rangle + F(u) \tag{2}$$

where $\Delta = \Delta_{g(t)}$ is the laplacian with respect to a time-dependent metric $g(t)$, $X = X(t)$ is a time-dependent vector field, and $F : \mathbb{R} \to \mathbb{R}$ is a smooth function.

PROPOSITION 3.2. (Scalar Maximum Principle II: ODE gives pointwise bounds for PDE) *Let* $u : M^n \times [0,T) \to \mathbb{R}$ *be a* C^2 *solution to (2). If* $C_1 \le u(x,0) \le C_2$ *for all* $x \in M$, *for some constants* $C_1, C_2 \in \mathbb{R}$, *then* $\phi_1(t) \le u(x,t) \le \phi_2(t)$ *for all* $x \in M$ *and* $t \in [0,T)$, *where* $\phi_i(t)$, $i = 1,2$, *are the solutions to the associated ordinary differential equation*

$$\frac{d\phi_i}{dt} = F(\phi_i)$$
$$\phi_i(0) = C_i.$$

3.1.2. *Systems.* The maximum principle is quite robust, it applies to general classes of second-order parabolic equations and even to some systems, such as the Ricci flow. A simple example of the maximum principle for systems is the following.

PROPOSITION 3.3. (preserving nonnegativity of a 2-tensor) *Let* $(M, g(t))$ *be a time-dependent Riemannian manifold, where the metric depends smoothly on time (e.g.,* $g(t)$ *is a solution to the Ricci flow,) and* $\alpha(t) \in \Gamma(TM^* \otimes_S TM^*)$ *a symmetric 2-tensor satisfying the semi-linear heat equation*

$$\frac{\partial \alpha}{\partial t} = \Delta_{g(t)}\alpha + \beta,$$

where $\beta(t) = f(\alpha, g(t))$ *is a symmetric 2-tensor, which is a smooth (Lipschitz should be enough) function of* α *and* $g(t)$, *satisfying the condition that*

$$\beta_{ij} V^i V^j (x,t) \ge 0$$

whenever $V(x,t)$ *is a null eigenvector of* $\alpha(t)$:

$$\alpha_{ij} V^i (x,t) = 0.$$

If $\alpha(0) \ge 0$, *then* $\alpha(t) \ge 0$ *for all* $t \ge 0$.

Idea of the proof. Suppose $\alpha > 0$ at $t = 0$ and (x_0, t_0) is a point and time such that

$$\alpha_{ij} V^i (x_0, t_0) = 0$$

for the first time for some tangent vector V at (x_0, t_0). Then $\alpha_{ij} W^i W^j (x,t) \ge 0$ for all $x \in M$, $t \in [0, t_0]$, and tangent vectors W. Extend V to a neighborhood of (x_0, t_0) in space and time so that

$$\frac{\partial V}{\partial t}(x_0,t_0) = 0$$
$$\nabla V(x_0,t_0) = 0 \tag{3}$$
$$\Delta V(x_0,t_0) = 0.$$

In particular, this may be accomplished by parallel translating V in space along geodesic rays emanating from x_0, and taking V to be independent of time. Then at any point in the neighborhood of (x_0, t_0),

$$\frac{\partial}{\partial t}\left(\alpha_{ij} V^i V^j\right) = \left(\frac{\partial}{\partial t}\alpha_{ij}\right) V^i V^j$$
$$= \left(\Delta \alpha_{ij} + \beta_{ij}\right) V^i V^j.$$

On the other hand, by (3), at (x_0, t_0) we have

$$\left(\Delta \alpha_{ij}\right) V^i V^j = \Delta\left(\alpha_{ij} V^i V^j\right) \ge 0.$$

Combining this with our assumption

$$\beta_{ij} V^i V^j (x_0, t_0) \geq 0,$$

we conclude

$$\frac{\partial}{\partial t}\left(\alpha_{ij} V^i V^j\right) = \Delta\left(\alpha_{ij} V^i V^j\right) + \beta_{ij} V^i V^j \geq 0$$

at (x_0, t_0). Hence, if $\alpha_{ij} V^i V^j$ becomes zero, it wants to increase. Note that in the heuristic arguments above, we have not assumed that β is a smooth function of α and $g(t)$.

A fancier version of the maximum principle for systems, which is used for the Ricci flow, is as follows. Let $V \stackrel{\pi}{\to} M$ be a vector bundle over a manifold M with a time-dependent Riemannian metric $g(t)$ on M, a fixed metric h on the fibers of V, and a time-dependent connection (covariant derivative) $\nabla(t)$ on V compatible with h. The time-dependent laplacian $\Delta(t)$ acting on sections of V is defined by

$$\Delta = \operatorname{trace}_g\left(\nabla^2\right) = g^{ij}\nabla_i\nabla_j$$

where $\nabla : \Gamma(V) \to \Gamma(V \otimes TM^*)$ is the connection on V and $\nabla : \Gamma(V \otimes TM^*) \to \Gamma(V \otimes TM^* \otimes TM^*)$ is defined using the connection on V and the Levi-Civita connection on TM^*. Suppose that a time-dependent section $\sigma(t) \in \Gamma(V)$ satisfies the parabolic equation

$$\frac{\partial \sigma}{\partial t} = \Delta\sigma + F(\sigma), \tag{4}$$

where $F : V \to V$ is fiber preserving, i.e., F is a vertical vector field on V. Analogous, to the case of the semi-linear heat equation, we consider the corresponding ODEs to (4) obtained by dropping the laplacian term

$$\frac{ds}{dt} = F(s)$$

which are ordinary differential equations on the fibers $V_x = \pi^{-1}(x)$ for each $x \in M$. The analogue for systems to the initial pointwise bounds $C_1 \leq u(x,0) \leq C_2$, which we assumed in the maximum principle for the scalar heat equation, is to assume that the initial data $\sigma(0)$ lies in a subset $\mathcal{K} \subset V$ which is invariant under parallel translation in V and fiberwise convex, i.e., $\mathcal{K}_x = \mathcal{K} \cap V_x$ is a convex subset of V_x for all $x \in M$. The invariance under parallel translation corresponds to the interval $[C_1, C_2]$ being independent of $x \in M$, and the fiberwise convexity corresponds to the interval $[C_1, C_2]$ being convex in $\mathbb{R}$ (here $V = M \times \mathbb{R}$.)

The maximum principle for systems says that the associated ODE can give bounds for the PDE in the following sense

THEOREM 3.4. (maximum principle for systems) *Let $\sigma(t) \in \Gamma(V)$ be a solution to*

$$\frac{\partial \sigma}{\partial t} = \Delta\sigma + F(\sigma).$$

Suppose that $\mathcal{K}(t) \subset V$ is a time-dependent subset invariant under parallel translation and fiberwise convex such that for any solution $s(t) \in V_x$ to the ODE

$$\frac{ds}{dt} = F(s)$$

with $s(0) \in \mathcal{K}(0)_x$ *stays in* $\mathcal{K}_x$, *i.e.,* $s(t) \in \mathcal{K}(t)_x$ *for all* $x \in M$ *and* $t \geq 0$. *Then any solution* $\sigma(t)$ *to the PDE with* $\sigma(0) \in \mathcal{K}(0)$ *stays in* $\mathcal{K}$, *i.e.,* $\sigma(t) \in \mathcal{K}(t)$ *for all* $t \geq 0$.

Proof in a special case. We first consider the case of a flat trivial bundle, that is, where the solution is a function on M with values in $\mathbb{R}^k$. To understand why we need the fibers $\mathcal{K}_x$ to be convex, it is enough to consider the case of vector valued functions on the unit interval $[0,1]$

$$u : [0,1] \to \mathbb{R}^k,$$

where the values at the endpoints are fixed

$$u(0) = \vec{a}$$
$$u(1) = \vec{b}.$$

The heat equation smooths out the function u (in infinite time) to the linear function $u_\infty(s) := (1-s)\vec{a} + s\vec{b}$. That is, for the theorem to be true, we need that if $\vec{a}, \vec{b} \in \mathcal{K}$, then $(1-s)\vec{a} + s\vec{b} \in \mathcal{K}$. That is, $\mathcal{K}$ is convex.

The statement of the theorem in the special case of a flat trivial bundle is:

PROPOSITION 3.5. *Let* $u : M \to \mathbb{R}^k$ *be a solution to*

$$\frac{\partial u}{\partial t} = \Delta u + F(u) \tag{5}$$

where $F : \mathbb{R}^k \to \mathbb{R}^k$. *Let* $\mathcal{K} \subset \mathbb{R}^k$ *be a closed convex set such that any solution* U *to the ODE corresponding to (5)*

$$\frac{dU}{dt} = F(U)$$

which starts in $\mathcal{K}$ *stays in* $\mathcal{K}$, *i.e., if* $U(0) \in \mathcal{K}$, *than* $U(t) \in \mathcal{K}$ *for all* $t \geq 0$. *Then any solution to the PDE (5) which starts in* $\mathcal{K}$ *also stays in* $\mathcal{K}$.

Proof. Given $x \in \mathbb{R}^k$, let $d(x, \mathcal{K})$ denote the distance from x to $\mathcal{K}$ (with $d(x,\mathcal{K}) = 0$ for $x \in \mathcal{K}$.) Associated to the solution $u(x,t)$ to (5), let

$$s(t) := \sup_{x \in M} d(u(x,t), \mathcal{K})$$

be the maximum distance of u from the set $\mathcal{K}$ at time t. We shall show that $s(t)$ grows at most exponentially:

$$\frac{ds}{dt} \leq Cs$$

for some constant $C < \infty$. Since $s(0) = 0$, we conclude that $s(t) = 0$ for all $t \geq 0$, from which the proposition follows. For computational purposes, we describe the distance function $d(x, \mathcal{K})$ in terms of support functions for the convex set $\mathcal{K}$.

DEFINITION 1. We say that a linear function $l : \mathbb{R}^k \to \mathbb{R}$ is a *support function* for $\mathcal{K}$ at a point $v \in \partial\mathcal{K}$ if

1. $l(w - v) \leq 0$ for all $w \in \mathcal{K}$
2. $|l| = 1$.

We write $l \in \mathrm{supp}_v \mathcal{K}$. The distance function may be rewritten as

$$d(x,\mathcal{K}) = \left(\sup_{\substack{l \in \mathrm{supp}_v \mathcal{K} \\ v \in \partial \mathcal{K}}} l(x-v) \right)^+ . \tag{6}$$

Now the maximum distance of u to $\mathcal{K}$ may be written as

$$s(t) = \sup_{x \in M} \left\{ \sup_{\substack{l \in \mathrm{supp}_v \mathcal{K} \\ v \in \partial \mathcal{K}}} l(u(x,t)-v) \right\}^+ = \sup_{\substack{l \in \mathrm{supp}_v \mathcal{K} \\ v \in \partial \mathcal{K}}} \left\{ \sup_{x \in M} l(u(x,t)-v) \right\}^+ .$$

Since M, $\mathrm{supp}_v \mathcal{K}$ and $\partial \mathcal{K}$ are compact sets (we may assume $\mathcal{K}$ is a compact set by using a cutoff function,) we have

$$\begin{aligned} \frac{d}{dt} s(t) &\leq \sup \left\{ \left[\frac{\partial}{\partial t} m(u-w) \right] (y,t) : \begin{array}{l} m \in \mathrm{supp}_w \mathcal{K},\ w \in \partial \mathcal{K},\ y \in M \\ m(u(y,t)-w) = \sup_{l,v,x} l(u(x,t)-v) \end{array} \right\} \\ &= \sup \left\{ [m(\Delta u) + m(F(u))](y,t) : \begin{array}{l} m \in \mathrm{supp}_w \mathcal{K},\ w \in \partial \mathcal{K},\ y \in M \\ m(u(y,t)-w) = \sup_{l,v,x} l(u(x,t)-v) \end{array} \right\}, \end{aligned}$$

where we used

$$\frac{\partial}{\partial t} m(u-w) = m\left(\frac{\partial u}{\partial t} \right) = m(\Delta u + F(u)) .$$

Since m is linear, we have

$$m(\Delta u)(y,t) = \Delta(m(u))(y,t) \leq 0$$

by our assumptions on m and y. Given m and y as before, the point $w \in \partial \mathcal{K}$ is the unique point in $\mathcal{K}$ closest to $u(y,t)$

$$d(w,u(y,t)) = d(\mathcal{K},u(y,t)) = s(t) .$$

We have

$$m(F(w)) \leq 0,$$

so that

$$\begin{aligned} m(F[u(y,t)]) &= m(F(w)) + m(F[u(y,t)] - F(w)) \\ &\leq |m| \, |F[u(y,t)] - F(w)| \\ &\leq C|u(y,t) - w| = Cs(t) . \end{aligned}$$

Hence we conclude

$$\frac{d}{dt} s(t) \leq Cs(t) . \text{ q.e.d.}$$

Proof of the general case. Since the proof is similar, we only highlight the differences. Define

$$\begin{aligned} s(t) &:= \sup_{x \in M} d(\sigma(x,t), \mathcal{K}_x) \\ &= \sup_{x \in M} \left\{ \sup_{\substack{l \in \mathrm{supp}_v \mathcal{K}_x \\ v \in \partial \mathcal{K}_x}} l(\sigma(x,t)-v) \right\}^+ . \end{aligned}$$

As before, the time-derivative of the maximal distance function satisfies

$$\frac{d}{dt}s(t) \leq \sup\left\{ [m(\Delta\sigma) + m(F(\sigma))](y,t) : \begin{array}{l} m \in \mathrm{supp}_w \mathcal{K}_y,\ w \in \partial\mathcal{K}_y,\ y \in M \\ m(\sigma(y,t) - w) = \sup_{l,v,x} l(\sigma(x,t) - v) \end{array} \right\}.$$

We have $m(F(\sigma))(y,t) \leq 0$ as before. To show that $m(\Delta\sigma)(y,t) \leq 0$, we extend $m \in V_y^*$ to a neighborhood U of y by parallel translation along geodesics emanating from y. That is $m \in \Gamma(V^*|_U)$ and

$$\nabla m(y) = 0$$
$$\Delta m(y) = 0.$$

CLAIM 1. $m(x)$ is a support function for $\mathcal{K}_x$ for x in a neighborhood of y.

The theorem now follows from

$$m(\Delta\sigma)(y,t) = \Delta(m(\sigma))(y,t) \leq 0. \text{ q.e.d.}$$

Remark (how the convexity of $\mathcal{K}$ is used in the proof.) Only for convex sets do we have the property that

$$l(x - v) \leq 0 \text{ for all } x \in \mathcal{K}$$

for any support function l at v. In particular, if $\mathcal{K}$ is not convex, then (6) does not hold.

3.2. Strong maximum principle.

3.2.1. *Scalar heat equation.* The strong maximum principle says that if a supersolution to the heat equation is bounded from below by a constant C and if the bound is preserved in time, then for positive time, either the solution is strictly greater than C or identically equal to C. In other words, if we have a supersolution which initially is nonnegative everywhere and positive at one point, then for positive time, the supersolution is positive everywhere.

PROPOSITION 3.6. (Strong Maximum Principle, I.: scalar equation) *Let (M, g) be a complete Riemannian manifold and $u : M \times [0,T) \to \mathbb{R}$ be a solution to the equation*

$$\frac{\partial u}{\partial t} \geq \Delta u + \langle X, \nabla u \rangle$$

with $u \geq 0$ everywhere. If $u(x_0, t_0) = 0$ at some point $x_0 \in M$ and time $t_0 \in (0,T)$, then $u \equiv 0$ on $M \times [0,T)$.

Proof. See Protter-Weinberger [**47**], Theorem 4 of Ch. 3, sect. 2.

3.2.2. *Systems.* To apply the strong maximum principle to the reaction-diffusion equation satisfied by the curvature tensor under the Ricci flow, we need a version for systems of functions (i.e., tensors or sections of vector bundles.) Let $(M^n, g(t))$ be a manifold with a time-dependent Riemannian metric (e.g., a solution to the Ricci flow.) Let $\pi : V \to M$ be a vector bundle with a fixed metric h on the fibers and a time-dependent connection $A(t)$ compatible with h. The maximum principle for systems is (for the proof of the results in this section, see section 8 of [**20**]:)

PROPOSITION 3.7. *Let $\sigma(t) \in \Gamma(V)$ be a time-dependent section of V satisfying*

(7) $$\frac{\partial \sigma}{\partial t} = \Delta\sigma + F(\sigma),$$

where $F : V \to V$ is a fiber preserving map. Suppose that

$$k : V \to \mathbb{R}$$

is a function such that

1. *it is invariant under parallel translation: for every path in M and horizontal lift α in V, we have k is constant along α,*
2. *$k|_{V_x}$ is convex for all $x \in M$.*
Then

A. (weak maximum principle) *If for all $x \in M$, the sets*

$$\{v \in V_x : k(v) \leq c\}$$

are preserved by the ODE

$$\frac{ds}{dt} = F(s)$$

for some $c \in \mathbb{R}$, then the inequality

$$k(\sigma) \leq c$$

is also preserved by the PDE (7).

B. (strong maximum principle) *If $k(\sigma(x,0)) > c$ for some $x \in M$, then*

$$k(\sigma(x,t)) > c, \text{ for all } x \in M \text{ and } t > 0.$$

This is the general vector bundle statement of the strong maximum principle for systems. Since we are interested in obtaining bounds for the Riemann curvature tensor under the Ricci flow, and the Riemann curvature tensor is a bilinear form, the version of the strong maximum principle we shall use is.

Corollary 3.8. (Strong Maximum Principle, II.: bundle formulation) *Let $\beta(t) \in \Gamma(V^* \otimes_S V^*)$ be a time-dependent symmetric bilinear form on V satisfying the semi-linear parabolic equation*

$$\frac{\partial \beta}{\partial t} = \Delta \beta + F(\beta),$$

where $F : V^* \otimes_S V^* \to V^* \otimes_S V^*$ is a fiber preserving map with $F(\beta) \geq 0$ whenever $\beta \geq 0$. If $\beta \geq 0$ at $t = 0$, then $\beta \geq 0$ for $t > 0$ (by the weak maximum principle) and there exists a $\delta > 0$ such that on the time interval $[0, \delta]$

Proposition 3.9.

1. *the rank of β is constant,*
2. *the null space*

$$null(\beta(t)) := \{v \in V : \beta(t)(v,w) = 0 \text{ for all } w \in V \text{ with } \pi(w) = \pi(v)\}$$

is independent of time and invariant under parallel translation in space (in other words, invariant under parallel translation in space and time,) and furthermore

3. $$null(\beta(t)) \subset null(F[\beta(t)]).$$

Applying the above result to the curvature operator $Rm : \wedge^2 M \to \wedge^2 M$ of a solution to the Ricci flow yields:

THEOREM 3.10. (Strong Maximum Principle, III.: application to curvature) *Let $(M, g(t))$ be a solution to the Ricci flow with nonnegative curvature operator $Rm \geq 0$ on $U \times [0, \varepsilon)$ where $U \subset M$ is an open set and $\varepsilon > 0$. Then either*

1. (positive curvature) *$Rm > 0$ on $U \times (0, \varepsilon)$, or*
2. (holonomy group reduces) *for $t > 0$, the image of the Riemann curvature operator is invariant under parallel translation in space and independent of time. Furthermore, for each point $x \in M$ and $t \in (0, \varepsilon)$, the image of $Rm(x, t)$ is a proper Lie subalgebra of $\wedge^2 M_x \cong so(n)$ which is isomorphic to the holonomy algebra of $g(t)$.*

4. Surfaces

In this section we describe Hamilton's results on the Ricci flow on surfaces. The first thing to notice about the Ricci flow on surfaces is that the equation simplifies. Since when $n = 2$ we have $R_{ij} = \frac{1}{2} R g_{ij}$, the Ricci flow equation becomes

$$\frac{\partial}{\partial t} g_{ij} = -R g_{ij}.$$

This is a conformal flow, that is, the solutions $g(t)$ satisfy $g(t) = e^{u(t)} g(0)$ for some time-dependent function u on M^2.

4.1. Uniformization theorem and main result. The classical uniformization theorem in complex analysis is equivalent to the following differential geometric statement:

PROPOSITION 4.1. *Given an oriented Riemannian surface (M, h), there exists a function $u : M \to \mathbb{R}$ such that the metric $g = e^u h$ has constant Gaussian curvature $K_g \equiv 1$, 0, or -1.*

The Ricci flow provides a constructive way of proving the above result. Namely, we have

THEOREM 4.2. *(Hamilton 1988) If M^2 is a closed surface, then for any initial metric g_0 on M, the solution to the area normalized Ricci flow*

$$\begin{aligned} \frac{\partial}{\partial t} g &= (r - R) g \\ g(0) &= g_0, \end{aligned}$$

(where $r = \int R dA / \int dA$ is the average scalar curvature) exists for all time and has constant area. Moreover,

1. *If the Euler characteristic of M is non-positive, then the solution metric $g(t)$ converges to a smooth constant curvature metric as $t \to \infty$.*
2. *If the scalar curvature R of the initial metric g_0 is positive, then the solution metric $g(t)$ converges to a smooth positive constant curvature metric as $t \to \infty$.*

Since given a surface with positive Euler characteristic, it is easy to find a metric with positive scalar curvature, the uniformization theorem follows from the above result. In [**14**] the condition that the initial metric have positive scalar curvature was removed. In [**3**] a new proof of this result was given using the Alcksandrov reflection method analogous to the work of R. Schoen on the Yamabe problem (which in dimension 2 is the elliptic version of the Ricci flow.) In [**22**], a proof via an isoperimetric estimate and a singularity analysis was given.

4.2. The energy. In dimension 2, given a fixed metric h in the conformal class, the normalized Ricci flow is actually the gradient flow for a *relative energy functional* E_h on the space Met of all metrics which are in the conformal class of h and have the same area as h. Let

$$Met = \left\{ g = e^u h : \int_M e^u dA_h = \int_M dA_h \right\}.$$

The relative energy functional E_h is given by:

$$E_h : Met \to \mathbb{R},$$

where

$$E_h(g) = \int_M \ln(g/h)\,(R_g dA_g + R_h dA_h)\,.$$

Clearly

$$E_h(g) = -E_g(h).$$

The reason that E_h is called a relative energy functional is:

LEMMA 4.3. *If g, h, and k are any 3 metrics in a conformal class, then*

$$E_h(g) = E_k(g) + E_h(k).$$

To define the gradient of the energy functional, we need to define a metric on the infinite dimensional space of metrics in a conformal class with fixed area. We consider the L^2-metric:

$$\langle\ ,\ \rangle_{L^2} : T\left(Met\right)_g \to T\left(Met\right)_g$$

defined by

$$\langle a \cdot g, b \cdot g \rangle_{L^2} = 2 \int_M ab \cdot dA_g,$$

where a and b are smooth functions on M with $\int_M a \cdot dA_g = 0$ and $\int_M b \cdot dA_g = 0$. We compute that

$$\nabla E_h(g) = (R - r) \cdot g.$$

Thus the normalized Ricci flow is the same as the gradient flow of the energy:

$$\frac{d}{dt} E_h(g) = -\nabla E_h(g).$$

The time-derivative of the energy is given by:

LEMMA 4.4. *Under the normalized Ricci flow, the time-derivative of the energy is*

$$\frac{d}{dt} E_h(g) = -2 \int_M (R - r)^2 dA \leq 0.$$

An important fact is that the energy functional is bounded from below:

PROPOSITION 4.5. (Onofri 1982) *If $M \cong S^2$, then the energy E_h is bounded below on Met and the minimum of E_h is obtained on the real 5-dimensional family of constant curvature metrics with fixed area in the conformal class of h.*

The reader may find it surprising that the energy functional does not play an important role in the study of the Ricci flow on surfaces. In fact, most proofs of convergence results do not use the energy functional. However, the definition of the energy functional does extend to the Ricci flow on Kähler manifolds (see [**42**] and [**52**].)

4.3. Evolution of curvature. Similar to Lemma 2.5, we have (note that when $n = 2$, $Rc = \frac{1}{2}Rg$)

LEMMA 4.6. *Under the normalized Ricci flow,*

$$\frac{\partial}{\partial t}R = \triangle R + R(R - r). \tag{8}$$

This type of evolution is known as a reaction-diffusion equation. The Laplacian term is causing the diffusion of R, whereas the quadratic terms in R represent the reaction terms. If the right-hand-side only contained the Laplacian term, then the equation would be the heat equation (albeit the Laplacian is respect to a time-dependent metric) and R would tend to a constant as t approaches ∞. On the other hand, if the right-hand-side only contained the $R(R - r)$ term, then the equation would be an ODE and the solution would blow up in finite time for any initial data satisfying $R(0) > \max\{r, 0\}$. The answer to the question of how the scalar curvature behaves under the normalized Ricci flow depends on whether the diffusion or the reaction term dominates. It turns out, as we shall show later, that it is the diffusion term which dominates. Dropping the Laplacian term yields the following ordinary differential equation:

$$\frac{d}{dt}s = s(s - r), \tag{9}$$

where the function $s = s(t)$ plays the role of R. The solution to the ODE above with initial data $s(0) = s_o$ is:

$$s(t) = \frac{r}{1 - \left(1 - \frac{r}{s_o}\right)e^{rt}},$$

when $r \neq 0$ and $s_o \neq 0$. When $r = 0$, we have:

$$s(t) = \frac{s_o}{1 - s_o t},$$

and when $s_o = 0$,

$$s(t) \equiv 0.$$

We conclude that for all values of r, the solution blows up in finite time when $s_o > \max\{r, 0\}$:

$$\lim_{t \to T} s(t) = \infty,$$

where if $r = 0$, then $T = 1/s_o$, and if $r \neq 0$, then $T = -\frac{1}{r}\ln\left(1 - \frac{r}{s_o}\right) \in (0, \infty)$. Hence we cannot obtain an upper bound for the curvature under the normalized Ricci flow by directly applying the maximum principle to equation (8). On the other hand, the ODE behaves much better when $s_o < \min\{r, 0\}$, in which case we have:

$$s(t) - r \geq s_o - r.$$

In view of the solution $s(t)$ of equation (9), we may apply the maximum principle to the equation for R to conclude:

LEMMA 4.7. *Let $\{g(t)\}$ be a solution to the normalized Ricci flow on a surface* M.

1. *If* $r < 0$, *then*

$$R - r \geq \frac{r}{1 - \left(1 - \frac{r}{R_{\min}(0)}\right) e^{rt}} - r \geq (R_{\min}(0) - r)\, e^{rt}.$$

Note that in this case, $\lim_{t\to\infty} e^{rt} = 0$.

2. *If* $r = 0$, *then*

$$R \geq \frac{R_{\min}(0)}{1 - R_{\min}(0)\, t} > -\frac{1}{t}.$$

3. *If* $r > 0$ *and* $R_{\min}(0) < 0$, *then*

$$R - r \geq \frac{r}{1 - \left(1 - \frac{r}{R_{\min}(0)}\right) e^{rt}} \geq R_{\min}(0)\, e^{-rt}.$$

Hence we have uniform lower bounds for the curvature under the normalized Ricci flow, whereas our upper bounds for the curvature blow up in finite time. In the next section we shall obtain a uniform upper bound for the curvature when $r \leq 0$ and an exponential upper bound for the curvature when $r > 0$.

4.4. Ricci solitons and more estimates for curvature. In this subsection we consider Ricci solitons, which are fixed points of the normalized Ricci flow in the space of metrics (in a conformal class with fixed area) modulo the action of the group of conformal diffeomorphisms. These special solutions to the Ricci flow motivate certain quantities we consider in estimating the curvature. It turns out that the good quantities to estimate are the ones which are constant on Ricci solitons. In particular, by estimating such a quantity, we shall obtain upper bounds for the scalar curvature. These upper bounds are uniform, except in the case where $r > 0$, in which case they are exponential. We shall also show that the only Ricci solitons on a closed surface are the constant curvature metrics.

DEFINITION 2. A solution $\{g(t)\}$ to the normalized Ricci flow is a *Ricci soliton* if there exists a one-parameter family of conformal diffeomorphisms $\{\varphi(t)\}$ such that

$$g(t) = \varphi(t)^* g(0).$$

Differentiating this equation with respect to time implies:

$$\frac{\partial}{\partial t} g = L_X g, \tag{10}$$

where $\{X(t)\}$ is the one-parameter family of vector fields generated by $\{\varphi(t)\}$. Substituting the normalized Ricci flow into (10) yields:

$$(r - R) g_{ij} = \nabla_i X_j + \nabla_j X_i.$$

If $X = -\nabla f$ is minus the gradient of some time-dependent function f, then we obtain the equation:

$$(R - r) g_{ij} = 2 \nabla_i \nabla_j f.$$

In this case we say that $\{g(t)\}$ is a *gradient Ricci soliton.* Tracing implies:

$$\triangle f = R - r. \tag{11}$$

This equation is solvable for f since $\int_M (R-r)\, dA = 0$. The solution f is referred to as the *potential of the curvature*, and it is unique up to an additive constant since harmonic functions are constants. The reader should note that the potential f can be defined by equation (11) for any metric g, not just for Ricci solitons. Defining

$$M_{ij} = \nabla_i \nabla_j f - \frac{1}{2} \triangle f \cdot g_{ij}$$

to be the trace-free part of the Hessian of f, we see that the gradient Ricci soliton equation is equivalent to:

$$M_{ij} = 0.$$

Taking the divergence of M, we obtain:

$$\begin{aligned} div(M)_i &= \nabla_j M_{ij} = \nabla_j \nabla_i \nabla_j f - \frac{1}{2} \nabla_i \triangle f \\ &= \frac{1}{2}(\nabla_i R + R \nabla_i f). \end{aligned}$$

Note that dividing by R, we find that the gradient Ricci soliton equation implies:

$$\nabla(\ln R + f) = 0,$$

that is,

$$\ln R + f = C,$$

where C is a (time-dependent) constant.

More importantly, the gradient Ricci soliton equation implies:

$$\begin{aligned} 0 &= \nabla_i R + R \nabla_i f = \nabla_i R + (R - r) \nabla_i f + r \nabla_i f \\ &= \nabla_i R + 2 \nabla_i \nabla_j f \cdot \nabla_j f + r \nabla_i f = \nabla_i \left(R + |\nabla f|^2 + r f \right). \end{aligned}$$

That is, on a gradient Ricci soliton,

$$R + |\nabla f|^2 + r f = C,$$

where C is a (time-dependent) constant. Since $R + |\nabla f|^2 + r f$ is constant on Ricci solitons, we expect that it will satisfy a nice evolution equation in general. The potential f itself satisfies a nice evolution equation:

LEMMA 4.8. *Under the normalized Ricci flow, the potential of the curvature satisfies (provided we suitably adjust the additive constant in the definition of f):*

$$\frac{\partial}{\partial t} f = \triangle f + r f. \tag{12}$$

Applying the maximum principle to equation (12) yields:

COROLLARY 4.9. *Under the normalized Ricci flow, there exists a constant C such that*

$$|f| \leq C e^{rt}.$$

A consequence of this estimate is that when $r \leq 0$, as long as the solution exists, the metrics $g(t)$ are uniformly equivalent.

PROPOSITION 4.10. *If $r \le 0$, then under the normalized Ricci flow, there exists a constant $C \ge 1$ depending only on the initial metric g_o such that:*

$$\frac{1}{C} g(0) \le g(t) \le C\, g(0).$$

Define $H = R - r + |\nabla f|^2$. Since $H + rf$ is constant on Ricci solitons and f satisfies a nice evolution equation in general, it is natural that H should satisfy a nice evolution equation.

LEMMA 4.11. *Under the normalized Ricci flow:*

$$\frac{\partial}{\partial t} H = \Delta H - 2\left|M_{ij}\right|^2 + rH. \tag{13}$$

Applying the maximum principle to equation (13) implies:

COROLLARY 4.12. *Under the normalized Ricci flow, there exists a constant C depending only on the initial metric g_o such that:*

$$H \le C\, e^{rt},$$

in particular,

$$R - r \le C\, e^{rt}.$$

Combining our previous estimates for R, we have:

PROPOSITION 4.13. *Under the normalized Ricci flow, there exists a constant C depending only on the initial metric g_o such that:*

1. (a) If $r < 0$, then

$$-C\, e^{rt} \le R - r \le C\, e^{rt}.$$

   (b) If $r = 0$, then

$$-\frac{C}{1 + Ct} \le R \le C.$$

   (c) If $r > 0$, then

$$-C\, e^{-rt} \le R - r \le C\, e^{rt}.$$

Before going to the long-time existence theorem in the next subsection, which follows from the above estimates for the curvature, we conclude this subsection with a proof of the fact that the only Ricci solitons on a closed surface are the constant curvature metrics.

PROPOSITION 4.14. *If $(M, g(t))$ is a family of Ricci solitons, then $g(t) = g(0)$ is a constant curvature metric.*

PROOF. Multiplying the Ricci soliton equation

$$(r - R) g_{ij} = \nabla_i X_j + \nabla_j X_i$$

by $R\, g_{ij}$ yields:

$$-\int_M (R - r)^2\, dA = \int_M (r - R) R\, dA = \int_M R \operatorname{div} X\, dA.$$

Since X is a conformal Killing vector field, integrating by parts and applying the Kazdan-Warner identity implies:

$$\int_M (R-r)^2\,dA = \int_M \nabla R \cdot X\,dA = 0.$$

Hence $R \equiv r$ and the proposition is proved.

4.5. Bernstein-Shi estimates and long-time existence. The long time existence of the solution follows from Proposition 4.13 and estimates for the higher derivatives of the curvature. For the Ricci flow, such higher derivative estimates first appeared in [**19**] using interpolation inequalities. However, new higher derivative estimates were obtained by W. Shi [**49**] using the method of Bernstein (see [**39**].) These estimates are have also proved useful for the analysis of singularities.

PROPOSITION 4.15. *Suppose that $(M, g(t))$ is a solution to the normalized Ricci flow. There exists a universal constant C (depending on r if $r < 0$) such that if $R \leq M$ at $t = 0$ for some constant $M > 0$, then*

$$|\nabla R|(x,t) \leq \frac{C \cdot M}{t^{1/2}},$$

for all $x \in M$ and $t \in \left[0, \frac{1}{C\cdot M}\right]$.

Similarly, one can prove estimates for the higher covariant derivatives of the scalar curvature.

PROPOSITION 4.16. *Suppose that $(M, g(t))$ is a solution to the normalized Ricci flow. There exists a constant C_m depending only on m (and r if $r < 0$) such that if $R \leq M$ at $t = 0$ for some constant $M > 0$, then*

$$|\nabla^m R|(x,t) \leq \frac{C_m \cdot M}{t^{m/2}},$$

for all $x \in M$ and $t \in \left[0, \frac{1}{C_m\cdot M}\right]$.

Based on the curvature and higher derivative estimates, we have the following long-time existence result

THEOREM 4.17. *If (M, g_o) is a closed Riemannian surface, then a solution $g(t)$ to the normalized Ricci flow exists for all time.*

Concerning the convergence, we have

THEOREM 4.18. *If $r \leq 0$, then the metrics $g(t)$ converge uniformly in any C^k-norm to a smooth metric g_∞ as $t \to \infty$, and the metric has g_∞ has constant curvature.*

In the following subsections we shall consider the more difficult case where $R > 0$ initially.

4.6. Entropy. In this subsection we assume $R > 0$ initially. Since we have not been able to apply the maximum principle to obtain a uniform upper bound for the curvature, we consider integral quantities. Perhaps the most important such quantity for the Ricci flow on surfaces is the *entropy* N defined for a metric with positive curvature by:

$$N(g) = \int_M \ln R \cdot R\,dA.$$

The only reason this quantity is called the entropy is because it resembles other quantities called entropy which are the integral of a positive function times its logarithm. Our sign convention is the opposite of the usual one and we shall show that the entropy is decreasing (instead of increasing) under the normalized Ricci flow. The time-derivative of the entropy is given by (see [**21**]:)

LEMMA 4.19. *If $R(g_o) > 0$, then under the normalized Ricci flow*

$$\frac{dN}{dt} = -\int_M \frac{|\nabla R|^2}{R}\, dA + \int_M (R-r)^2\, dA$$

Hamilton originally proved that the entropy is non-increasing: $dN/dt \leq 0$ by showing

$$\frac{d}{dt}\left(\frac{dN}{dt}\right) \geq c_1 \left(\frac{dN}{dt}\right)^2 + c_2 \frac{dN}{dt}$$

and concluding that if dN/dt were ever positive, then it would tend to infinity in finite time, contradicting the long-time existence of the solution established in the previous subsection. A direct proof of this fact was given in [**15**]:

PROPOSITION 4.20. *If $R(g_o) > 0$, then under the normalized Ricci flow, then*

$$\frac{dN}{dt} = -\int_M \frac{|\nabla R + R\nabla f|^2}{R}\, dA - 2\int_M |M_{ij}|^2\, dA \leq 0.$$

As a consequence, we have

COROLLARY 4.21. *If $R(g_o) > 0$, then under the normalized Ricci flow, then the entropy is a strictly decreasing function of time unless $R(g_o) \equiv r$ in which case it is constant in time.*

4.7. Harnack estimate. In this section we describe Hamilton's Harnack inequality for the scalar curvature function under the normalized Ricci flow, which is modelled on the Li-Yau Harnack inequality [**41**] for the heat equation on a Riemannian manifold. Recall that for a Ricci soliton,

$$\nabla R + R\nabla f = 0. \tag{14}$$

Let

$$L = \ln R.$$

Dividing equation (14) by R and taking the divergence implies:

$$Q := \Delta L + R - r = 0.$$

The quantity Q is known as the *Harnack quantity*. We shall obtain a lower bound for Q depending only on the initial metric g_o by applying the maximum principle to the evolution equation which it satisfies. To compute its evolution equation, we first need to compute the evolution equation for L.

LEMMA 4.22. *Under the normalized Ricci flow,*

$$\frac{\partial}{\partial t} L = \Delta L + |\nabla L|^2 + R - r.$$

From the lemma we may rewrite the Harnack quantity as:

$$Q = \frac{\partial}{\partial t} L - |\nabla L|^2 .$$

The evolution for Q is given by:

LEMMA 4.23. *Under the normalized Ricci flow,*

$$\frac{\partial}{\partial t} Q = \Delta Q + 2\langle \nabla L, \nabla Q \rangle + 2 \left| \nabla\nabla L + \frac{1}{2}(R - r) g \right|^2 + rQ.$$

Using the inequality $|a_{ij}|^2 \geq \frac{1}{2}(\text{ trace}_g a)^2$ for any symmetric 2-tensor on a surface, we have

COROLLARY 4.24.

$$\frac{\partial}{\partial t} Q \geq \Delta Q + 2\langle \nabla L, \nabla Q \rangle + Q^2 + rQ. \tag{15}$$

Applying the maximum principle to equation (15) implies:

PROPOSITION 4.25. *There exists a constant $C > 1$ depending only on g_o such that:*

$$Q = \frac{\partial}{\partial t} \ln R - |\nabla \ln R|^2 \geq -\frac{Cr\, e^{rt}}{C\, e^{rt} - 1}.$$

This estimate for Q is known as a *differential Harnack inequality.* Integrating it along paths in space and time yields a classical Harnack inequality which gives a lower bound for the curvature at some point and time in terms of the curvature at an earlier time and another point. In particular, let $x_1, x_2 \in M$ be any two points and $0 \leq t_1 < t_2$ be two times. Define

$$A = A(x_1, t_1, x_2, t_2) = \inf_{\gamma} \int_{t_1}^{t_2} \left| \frac{d\gamma}{dt} \right|^2 dt,$$

where the infimum is taken over all C^1-paths $\gamma : [t_1, t_2] \to M$ joining x_1 and x_2. Then we have

PROPOSITION 4.26. *Let $(M, g(t))$ be a solution to the normalized Ricci flow. If $x_1, x_2 \in M$ and $0 \leq t_1 < t_2$, then there exists a constant $C > 1$ depending only on g_0 such that:*

$$\frac{R(x_2, t_2)}{R(x_1, t_1)} \geq \frac{C\, e^{rt_1} - 1}{C\, e^{rt_2} - 1} \cdot e^{-\frac{1}{4}A} \geq e^{-C(t_2 - t_1)} \cdot e^{-\frac{1}{4}A}.$$

4.8. Uniform bounds on R. In this subsection we show how the entropy estimate and the Harnack inequality may be used to obtain uniform positive upper and lower bounds for the scalar curvature in the case where $R(g_o) > 0$. The upper bound may also be obtained using the gradient estimate for the scalar curvature of subsection 4.5.

PROPOSITION 4.27. *If the initial metric g_o has positive scalar curvature $R(g_o) > 0$, then there exists a constant $C \in [1, \infty)$ such that*

$$R(x, t) \leq C$$

for all $x \in M$ and $t \in [0, \infty)$.

We divide the outline of the proof into several steps below. Our goal is to show that for any t_o, $R_{\max}(t_o)$ is bounded above by some constant depending only on g_o. First we show that $R_{\max}$ at most doubles on a small time-interval.

LEMMA 4.28. *Given any $t_o \in [0,\infty)$, we have*

$$R(x,t) \leq 2R_{\max}(t_o)$$

for all $t \in [t_o, t_o + \frac{1}{2R_{\max}(t_o)}]$ and $x \in M$.

This implies that the metrics are uniformly equivalent in that same time-interval.

COROLLARY 4.29. *Given any $t_o \in [0,\infty)$, we have*

$$\frac{1}{e} g(x,t_o) \leq g(x,t) \leq \sqrt{e} \cdot g(x,t_o)$$

for all $t \in [t_o, t_o + \frac{1}{2R_{\max}(t_o)}]$ and $x \in M$.

Let $p \in M$ be a point such that

$$R(p,t_o) = R_{\max}(t_o).$$

We next show that at time $t_o + \frac{1}{2R_{\max}(t_o)}$ in a small ball about p, R is comparable to $R(p,t_o)$.

LEMMA 4.30. *Assume that $R_{\max}(t_o) \geq 1$. Given any constant C', there exists a constant C depending only on g_o and C' such that if*

$$\rho(x,p) \leq \frac{C'}{\sqrt{R_{\max}(t_o)}},$$

then

$$R(x, t_o + \frac{1}{2R_{\max}(t_o)}) \geq C \cdot R_{\max}(t_o).$$

Next we show that the diameter of (M,g) is uniformly bounded from above.

LEMMA 4.31. *There exists a constant $C > 0$ depending only on g_0 such that*

$$\operatorname{diam}(M, g(t)) \leq C.$$

The lemma above enables us to apply the Harnack inequality again to obtain a positive lower bound for R.

LEMMA 4.32. *There exists a constant $c > 0$ depending only on g_0 such that*

$$R(x,t) \geq c > 0,$$

for all $x \in M$ and $t \in [0,\infty)$.

4.9. Asymptotic approach to soliton. In this subsection we show that when $R > 0$ the metric approaches a Ricci soliton as $t \to \infty$ under the normalized Ricci flow. Recall from subsection 4.4 that on a Ricci soliton

$$M_{ij} = \nabla_i \nabla_j f - \frac{1}{2}(R - r) g_{ij} \equiv 0.$$

LEMMA 4.33. *Under the normalized Ricci flow,*

$$\frac{\partial}{\partial t} M_{ij} = \Delta M_{ij} - 2RM_{ij} + rM_{ij}.$$

Next we compute the evolution equation for the square of the norm of M_{ij} :

LEMMA 4.34.

$$\frac{\partial}{\partial t}|M_{ij}|^2 = \Delta |M_{ij}|^2 - 2|\nabla_k M_{ij}|^2 - 2R|M_{ij}|^2.$$

Using the estimate $R > C > 0$ and the maximum principle yields:

PROPOSITION 4.35. *Under the normalized Ricci flow, there exists a constants* $C_1, C_2 > 0$ *such that*

$$|M_{ij}|^2 \leq C_1 \cdot e^{-C_2 t}.$$

4.10. Convergence of the normalized flow. We now consider the modified flow

$$\frac{\partial}{\partial t}g_{ij} = -Rg_{ij} + (L_{\nabla f}g)_{ij} = 2M_{ij}.$$

The solution $\bar{g}(t)$ of this flow with $\bar{g}(0) = g_0$ is equivalent to the solution $g(t)$ of the original normalized flow with $g(0) = g_0$ in that there exists a one-parameter family of diffeomorphisms ϕ_t of M^2 such that

$$\bar{g}(t) = \phi_t^* g(t).$$

Hence for the modified flow, we also have the estimate

$$|\bar{M}_{ij}|^2 \leq C_1 \cdot e^{-C_2 t}.$$

This implies that the modified flow converges exponentially to a limit metric $\bar{g}_\infty$ (one can obtain the necessary higher derivative estimates.) The limit metric for the modified flow satisfies

$$(\bar{M}_\infty)_{ij} \equiv 0.$$

Furthermore, the curvature and its derivatives converge to their limit exponentially fast. Now by Proposition 4.14, we conclude that $\bar{g}_\infty$ has constant curvature:

$$\bar{R}_\infty \equiv r.$$

This implies that $\bar{R}(t)$ tends to a constant exponentially fast, which in turn implies that the solution $g(t)$ to the original normalized flow converges exponentially fast to a constant curvature metric g_∞. This completes the outline of the proof of part 2 of Theorem 4.2.

5. Three-manifolds

In this section we present Hamilton's seminal result concerning the Ricci flow on closed 3-manifolds with positive Ricci curvature. In this exposition we closely follow Hamilton's paper [**19**], while omitting most of the detailed computations. We suggest that the reader consult [**20**] for a proof of the curvature estimates which simplified his earlier computations, but which also requires more machinery.

Let (M^n, g) be a closed Riemannian n-manifold with positive Ricci curvature. By Myers' Theorem, the fundamental group of M is finite (one shows that a positive lower bound for the Ricci curvature gives an upper bound for the diameter of the manifold, and then applies this result to the universal cover $\tilde{M}$ with the lifted metric $\tilde{g}$ to conclude that $\tilde{M}$ is compact.) When $n = 3$, $\tilde{M}^3$ is compact and simply-connected, and hence a homotopy 3-sphere. We have the well-known

CONJECTURE 1. *(Poincaré)* Any closed simply-connected 3-manifold is diffeomorphic to S^3.

Hence, if the Poincaré Conjecture is true, one concludes that $\tilde{M}^3$ is diffeomorphic to S^3. Furthermore, we also have

CONJECTURE 2. *(Spherical Space Form)* Any discrete group of diffeomorphisms acting freely on S^3 is conjugate to a group of isometrics.

Hence, if the Spherical Space Form Conjecture is also true, we have M^3 is diffeomorphic to S^3/Γ, where Γ is a discrete subgroup of $O(4)$. In particular, M^3 admits a metric with constant positive sectional curvature. It is this last statement that Hamilton proved, the existence of a constant positive sectional curvature metric on a positively Ricci curved closed 3 -manifold - which would be a consequence of Myers' Theorem and the Poincaré and Spherical Space Form Conjectures.

THEOREM 5.1. (Hamilton 1982) *Given any smooth, compact 3-dimensional Riemannian manifold (M, g_o) with positive Ricci curvature, there exists a unique smooth solution $g(t)$ to the normalized Ricci flow*

$$\frac{\partial}{\partial t} g_{ij} = -2R_{ij} + \frac{2}{3} r \cdot g_{ij},$$

with initial condition $g(0) = g_o$ on the time interval $[0, \infty)$. Moreover, the solution $g(t)$ converges exponentially fast to a constant sectional curvature metric. In particular, M^3 is diffeomorphic to a spherical space form.

We outline the proof in the following subsections.

5.1. Positivity of the Ricci tensor is preserved. We shall first study the (unnormalized) Ricci flow

$$\frac{\partial}{\partial t} g_{ij} = -R g_{ij},$$

and prove estimates for the curvature and its derivatives. These estimates will imply that the solution to the normalized Ricci flow converges to a constant curvature metric. We first recall the evolution equation for the scalar curvature function under the Ricci flow (Corollary 2.5 :)

$$\frac{\partial}{\partial t} R = \triangle R + 2 |Rc|^2 .$$

By the maximum principle, if the initial metric g_o has positive scalar curvature: $R_o > 0$, then the solution $g(t)$ has positive scalar curvature: $R(t) > 0$ as long as it exists.

In dimension 3, the Riemann curvature tensor is completely determined by the Ricci tensor:

$$R_{ijkl} = R_{il} g_{jk} + R_{jk} g_{il} - R_{ik} g_{jl} - R_{jl} g_{ik} - \frac{1}{2} R \left(g_{il} g_{jk} - g_{ik} g_{jl} \right) . \tag{16}$$

Recall that Corollary 2.4 says that under the Ricci flow, the evolution equation for the Ricci tensor is:

$$\frac{\partial}{\partial t} R_{jk} = \Delta R_{jk} + 2 g^{pr} g^{qs} R_{pjqk} R_{rs} - 2 g^{pq} R_{pj} R_{qk}.$$

Applying (16) to this yields

LEMMA 5.2. *Under the Ricci flow, the Ricci tensor satisfies the following reaction-diffusion equation:*

$$\frac{\partial}{\partial t}R_{jk} = \Delta R_{jk} - 6g^{pq}R_{jp}R_{qk} + 3R\,R_{jk} + \left(2\,|Rc|^2 - R^2\right)g_{jk},$$

where $|Rc|^2 = g^{pq}g^{rs}R_{pr}R_{qs}$ *is the square of the norm of the Ricci tensor.*

Applying the maximum principle for tensors Proposition 3.3, we have

COROLLARY 5.3. *If the initial metric* g_o *has semi-positive Ricci curvature:* $Rc_o \geq 0$*, then as long as the solution* $g(t)$ *to the Ricci flow exists,* $g(t)$ *has semi-positive Ricci curvature:* $Rc(t) \geq 0$.

PROOF. Let

$$S_{jk} = -6g^{pq}R_{jp}R_{qk} + 3R\,R_{jk} + \left(2\,|Rc|^2 - R^2\right)g_{jk}.$$

We need to show that if V is a null-eigenvector of R_{jk}, i.e., $R_{jk}V^j = 0$, then

$$S_{jk}V^jV^k \geq 0.$$

Diagonalizing the Ricci tensor with respect to the metric

$$R_{jk} = \begin{pmatrix} \lambda & & \\ & \mu & \\ & & \nu \end{pmatrix}$$

we find that the tensor S_{jk} is also diagonal and is given by:

$$S_{jk} = \begin{pmatrix} \begin{array}{l}-2\lambda^2 + \mu^2 + \nu^2 \\ +\lambda\mu + \lambda\nu - 2\mu\nu\end{array} & & \\ & \begin{array}{l}-2\mu^2 + \lambda^2 + \nu^2 \\ +\mu\lambda + \mu\nu - 2\lambda\nu\end{array} & \\ & & \begin{array}{l}-2\nu^2 + \lambda^2 + \mu^2 \\ +\nu\lambda + \nu\mu - 2\lambda\mu\end{array} \end{pmatrix}.$$

If $\lambda = 0$ with corresponding (unit) null-eigenvector V, then

$$S_{jk}V^jV^k = \mu^2 + \nu^2 - 2\mu\nu = (\mu - \nu)^2 \geq 0.$$

hence the null-eigenvector condition is satisfied and the proposition follows.

5.2. Pinching of the Ricci tensor is preserved. Let $\lambda \leq \mu \leq \nu$ denote the eigenvalues (in increasing order) of the Ricci tensor with respect to the metric. The corollary says that if $\lambda \geq 0$ at $t = 0$, then $\lambda \geq 0$ for all $t \geq 0$. Next we show that any positive pinching of the Ricci tensor is preserved. That is, if there exists an $\epsilon > 0$ such that

$$\lambda \geq \epsilon\,(\lambda + \mu + \nu)$$

at $t = 0$, then $\lambda \geq \epsilon\,(\lambda + \mu + \nu)$ for all $t \geq 0$. A computation yields

LEMMA 5.4. *Under the Ricci flow,*

$$\begin{aligned}\frac{\partial}{\partial t}\left(\frac{R_{jk}}{R}\right) &= \Delta\left(\frac{R_{jk}}{R}\right) + \frac{2}{R}g^{pq}\nabla_pR\,\nabla_q\left(\frac{R_{jk}}{R}\right) \\ &\quad + \frac{-6g^{pq}R_{jp}R_{qk} + 3R\,R_{jk} + \left(2\,|Rc|^2 - R^2\right)g_{jk}}{R} - \frac{R_{jk}}{R^2}\cdot 2\,|Rc|^2\end{aligned}$$

Again, applying the maximum principle for tensors, we obtain

COROLLARY 5.5. *If the initial metric g_o has positive scalar curvature $R_o > 0$ and satisfies the pinching condition: $Rc_o \geq \epsilon R_o g_o$, for some $\epsilon > 0$, then as long as the solution $g(t)$ to the Ricci flow exists, $g(t)$ also has positive scalar curvature $R(t) > 0$ and satisfies the pinching condition: $Rc(t) \geq \epsilon R(t) g(t)$.*

PROOF. We compute that

$$\frac{\partial}{\partial t}\left(\frac{R_{jk}}{R} - \epsilon\, g_{jk}\right) = \Delta\left(\frac{R_{jk}}{R} - \epsilon\, g_{jk}\right) + \frac{2}{R} g^{pq} \nabla_p R \nabla_q \left(\frac{R_{jk}}{R}\right) + T_{jk},$$

where

$$\begin{aligned} T_{jk} \;=\; & \frac{-6 g^{pq} R_{jp} R_{qk} + 3R\, R_{jk} + \left(2\,|Rc|^2 - R^2\right) g_{jk}}{R} \\ & - \frac{R_{jk}}{R^2} \cdot 2\,|Rc|^2 + 2\epsilon\, R_{jk}. \end{aligned}$$

Since the positivity of the scalar curvature is preserved under the Ricci flow, it is sufficient to show that if V is a null-eigenvector of $\frac{R_{jk}}{R} - \epsilon\, g_{jk}$, i.e., $\left(\frac{R_{jk}}{R} - \epsilon\, g_{jk}\right) V^j = 0$, then

$$R^2 T_{jk} V^j V^k \geq 0.$$

Diagonalizing the Ricci tensor as before and assuming

$$\lambda - \epsilon\,(\lambda + \mu + \nu) = 0$$

with corresponding (unit) null-eigenvector V, we find that the tensor $R^2 T_{jk}$ is also diagonal and:

$$R^2 T_{jk} V^j V^k = \lambda^2 \left(-2\lambda + \mu + \nu\right) + \left(\mu + \nu\right)\left(\mu - \nu\right)^2.$$

Since $0 < \lambda \leq \mu \leq \nu$, we have

$$R^2 T_{jk} V^j V^k \geq 0,$$

and the null-eigenvector condition is satisfied.

5.3. Pinching improves. The corollary says that the pinching constant of the Ricci tensor is preserved. Now we will show that the pinching constant improves. We consider the scalar quantity

$$|Rc|^2 - \frac{1}{3} R^2 = \left| Rc - \frac{1}{3} R\, g \right|^2.$$

Since $n = 3$, a metric is Einstein: $Rc - \frac{1}{3} R\, g = 0$ if and only if it has constant sectional curvature. Thus this quantity measures the difference of the metric from having constant sectional curvature. For a quantity to have geometric meaning independent of the size of the metric, it is necessary for it to be scale-invariant, i.e., if the metric is multiplied by a constant, then the quantity remains unchanged. A scale-invariant quantity measuring the difference of the metric from having constant sectional curvature is:

$$\frac{\left|Rc - \frac{1}{3} R\, g\right|^2}{R^2} = \frac{|Rc|^2}{R^2} - \frac{1}{3}.$$

From Corollary 5.5, we expect that the maximum of this quantity decreases in time, which is actually the case. However, more is true; namely the maximum of the quantity

$$\frac{\left|Rc - \frac{1}{3}R\, g\right|^2}{R^{2-\delta}}$$

is decreasing in time provided $\delta > 0$ is sufficiently small. Since we are assuming the initial metric has positive Ricci curvature, which is preserved under the Ricci flow, the metric is always shrinking under the Ricci flow. Furthermore, from the evolution equation for R, the minimum of the scalar curvature is increasing in time under the Ricci flow. Thus one would hope that the minimum of the scalar curvature increases to infinity as t approaches the final time. If this is the case then we can conclude that the scale-invariant quantity measuring the difference of the metric from having constant sectional curvature decreases to zero since:

$$\frac{\left|Rc - \frac{1}{3}R\, g\right|^2}{R^2} \leq C\, R^{-\delta}.$$

One would then expect that under the normalized Ricci flow, the metric converges to a constant curvature metric. We now proceed to prove the estimate for:

$$f := \frac{|Rc|^2 - \frac{1}{3}R^2}{R^{2-\delta}}.$$

The evolution equation for f is given by

LEMMA 5.6. *For $\delta \in [0,1]$,*

$$\frac{\partial}{\partial t} f \leq \Delta f + \frac{2\,(1-\delta)}{R}\langle \nabla R, \nabla f\rangle + \frac{2}{R^{3-\delta}}\left(2U + \delta\,|Rc|^2\left(|Rc|^2 - \frac{1}{3}R^2\right)\right),$$

where

$$\begin{aligned} U &= -\,|Rc|^4 - 2R\cdot tr_g\left(Rc^3\right) + \frac{5}{2}R^2\,|Rc|^2 - \frac{1}{2}R^4 \\ &= -\lambda^2\,(\lambda-\mu)\,(\lambda-\nu) - \left[\nu^2\,(\nu-\lambda) - \mu^2\,(\mu-\lambda)\right](\nu-\mu) \leq 0 \end{aligned}$$

By applying the maximum principle to the lemma, we have

COROLLARY 5.7. *If $Rc(g_o) > 0$, then there exists a $\bar{\delta} > 0$ depending only on g_o such that if $\delta \in [0,\bar{\delta}]$ and $R^{\delta-2}\left(|Rc|^2 - \frac{1}{3}R^2\right) \leq C$ at $t = 0$, then*

$$\frac{|Rc|^2 - \frac{1}{3}R^2}{R^2} \leq C\cdot R^{-\delta}$$

as long as the solution exists.

PROOF. It suffices show that

$$2U + \delta\,|Rc|^2\left(|Rc|^2 - \frac{1}{3}R^2\right) \leq 0 \tag{17}$$

for $\delta > 0$ sufficiently small depending on g_o. We compute:

$$U \leq -\lambda^2\,(\lambda-\mu)^2 - \nu^2\,(\nu-\mu)^2.$$

Since $Rc(g_o) > 0$ and M is compact, there exists a constant $\epsilon > 0$ such that $Rc(g_o) \geq \epsilon R_o g_o$. By lemma 5.5 , we have $Rc \geq \epsilon R g$ as long as the solution exists. Hence

$$U \leq -\epsilon^2 R^2 \left[(\lambda - \mu)^2 + (\nu - \mu)^2 \right].$$

On the other hand,

$$|Rc|^2 - \frac{1}{3}R^2 = \frac{1}{3}\left[(\lambda-\mu)^2 + (\lambda-\nu)^2 + (\mu-\nu)^2 \right] \leq \left[(\lambda-\mu)^2 + (\nu-\mu)^2 \right].$$

Thus

$$U \leq -\epsilon^2 R^2 \left(|Rc|^2 - \frac{1}{3}R^2 \right),$$

and inequality (17) holds for all

$$\delta \leq 2\epsilon^2.$$

5.4. The gradient estimate for the scalar curvature. In this subsection we obtain a gradient estimate for the scalar curvature. This estimate is important because it enables us to compare curvatures at different points, whereas the pinching estimate of the previous section is a pointwise estimate for the curvatures. Note that the contracted second Bianchi identity implies that an Einstein metric (which is a solution to the degenerate elliptic equation $0 = -R_{ij} + \frac{1}{n} r g_{ij}$) in dimension at least 3 has constant scalar curvature, i.e., the gradient of the scalar curvature is zero. In the case of the normalized Ricci flow on a closed 3-manifold, which is the degenerate parabolic equation

$$\frac{\partial}{\partial t} g_{ij} = -2R_{ij} + \frac{2}{n} r g_{ij},$$

we have that the metric approaches an Einstein metric where the scalar curvature becomes large. In particular, we have the following estimate (where the left-hand-side is a scale invariant quantity measuring the difference of the metric from being Einstein:)

$$\frac{|Rc|^2 - \frac{1}{3}R^2}{R^2} \leq C \cdot R^{-\delta}.$$

Hence it is natural to expect that the gradient of the scalar curvature approaches zero in some sense. Moreover, one might also expect that in order to prove this, one needs to use the contracted second Bianchi identity. Both are in fact the case, using the contracted second Bianchi identity we shall show that given any β sufficiently small, there exists a constant C depending only on β and g_o such that

$$|\nabla R|^2 \leq \beta R^{3-\bar{\delta}/2} + C.$$

This estimate may written as

$$\frac{|\nabla R|^2}{R^3} \leq \beta R^{-\bar{\delta}/2} + C\, R^{-3},$$

where the left-hand-side is a scale invariant quantity and the right-hand-side is small for R large. To prove this estimate, we need to compute several evolution equations, the first of which is for the square of the norm of the gradient of the scalar curvature. A computation gives

LEMMA 5.8.

$$\frac{\partial}{\partial t}|\nabla R|^2 = \Delta |\nabla R|^2 - 2|\nabla\nabla R|^2 + 4\langle \nabla R, \nabla |Rc|^2\rangle.$$

Next we divide $|\nabla R|^2$ by R (the reader may wonder why we divide by R and not some other power of R, the computations bear this out) and compute its evolution equation.

LEMMA 5.9.

$$\begin{aligned}\frac{\partial}{\partial t}\left(\frac{|\nabla R|^2}{R}\right) &= \Delta\left(\frac{|\nabla R|^2}{R}\right) - 2R\left|\nabla\left(\frac{\nabla R}{R}\right)\right|^2 \\ &\quad + \frac{4}{R}\langle \nabla R, \nabla |Rc|^2\rangle - \frac{2|Rc|^2}{R^2}|\nabla R|^2.\end{aligned}$$

The evolution equation for $\frac{|\nabla R|^2}{R}$ has only one bad (positive) term on the right-hand-side, which is

$$\frac{4}{R}\langle \nabla R, \nabla |Rc|^2\rangle.$$

We remedy this by adding to $\frac{|\nabla R|^2}{R}$ the quantity $|Rc|^2 - \frac{1}{3}R^2$, which introduces a good (negative) term which cancels out the bad term. The evolution equation for $|Rc|^2 - \frac{1}{3}R^2$ is:

LEMMA 5.10.

$$\begin{aligned}\frac{\partial}{\partial t}\left(|Rc|^2 - \frac{1}{3}R^2\right) &= \Delta\left(|Rc|^2 - \frac{1}{3}R^2\right) - 2\left(|\nabla Rc|^2 - \frac{1}{3}|\nabla R|^2\right) \\ &\quad - 8\,tr_g(Rc^3) + \frac{26}{3}R\,|Rc|^2 - 2R^3.\end{aligned}$$

The good term in the evolution equation for $|Rc|^2 - \frac{1}{3}R^2$ is

$$-2\left(|\nabla Rc|^2 - \frac{1}{3}|\nabla R|^2\right).$$

That this term dominates the bad term $\frac{4}{R}\langle \nabla R, \nabla |Rc|^2\rangle$ follows from:

LEMMA 5.11.

$$|\nabla Rc|^2 - \frac{1}{3}|\nabla R|^2 \geq \frac{1}{37}|\nabla Rc|^2.$$

PROOF. Using the fact that for any 2-tensor a_{ij} (not necessarily symmetric:)

$$|a_{ij}|^2 \geq \frac{1}{3}\left(g^{ij}a_{ij}\right)^2,$$

we have by the second contracted Bianchi identity

$$\begin{aligned}|\nabla Rc|^2 - \frac{1}{3}|\nabla R|^2 &= \left|\nabla_i R_{jk} - \frac{1}{3}\nabla_i R\, g_{jk}\right|^2 \geq \frac{1}{3}\left|g^{ij}\left(\nabla_i R_{jk} - \frac{1}{3}\nabla_i R\, g_{jk}\right)\right|^2 \\ &= \frac{1}{3}\left(\frac{1}{2} - \frac{1}{3}\right)^2 |\nabla R|^2 = \frac{1}{108}|\nabla R|^2,\end{aligned}$$

and the lemma follows easily.

COROLLARY 5.12.

$$\frac{\partial}{\partial t}\left(|Rc|^2-\frac{1}{3}R^2\right) = \Delta\left(|Rc|^2-\frac{1}{3}R^2\right)-\frac{2}{37}|\nabla Rc|^2 -8tr_g(Rc^3)+\frac{26}{3}R|Rc|^2-2R^3.$$

Since we have (using the estimates $|Rc|\le R$ and $|\nabla R|\le\sqrt{3}|\nabla Rc|$ to obtain the last inequality:)

$$\frac{4}{R}\left|\langle\nabla R,\nabla|Rc|^2\rangle\right|\le\frac{8|Rc|}{R}|\nabla R|\cdot|\nabla Rc|\le 8\sqrt{3}|\nabla Rc|^2,$$

we consider the quantity

$$V=\frac{|\nabla R|^2}{R}+\frac{37}{2}\left(8\sqrt{3}+1\right)\left(|Rc|^2-\frac{1}{3}R^2\right)$$

which satisfies the evolution equation:

$$\frac{\partial}{\partial t}V = \Delta V-2R\left|\nabla\left(\frac{\nabla R}{R}\right)\right|^2-\frac{2|Rc|^2}{R^2}|\nabla R|^2-|\nabla Rc|^2 +\frac{37}{2}\left(8\sqrt{3}+1\right)\left(-8\mathrm{tr}_g(Rc^3)+\frac{26}{3}R|Rc|^2-2R^3\right).$$

The only bad term on the right-hand-side is

$$W=-8\mathrm{tr}_g(Rc^3)+\frac{26}{3}R|Rc|^2-2R^3.$$

Some algebra yields:

LEMMA 5.13.

$$W\le\frac{50}{3}R\left(|Rc|^2-\frac{1}{3}R^2\right).$$

PROOF. We may rewrite W as:

$$W=-8\left\langle R_{ij}-\frac{1}{3}Rg_{ij},g^{kl}R_{ik}R_{jl}\right\rangle+6R\left(|Rc|^2-\frac{1}{3}R^2\right).$$

The first term on the right-hand-side may be estimated as follows:

$$-8\left\langle R_{ij}-\frac{1}{3}Rg_{ij},g^{kl}R_{ik}R_{jl}\right\rangle\le\frac{32}{3}R\left|R_{ij}-\frac{1}{3}Rg_{ij}\right|^2$$

and the lemma follows.

Combining the lemmas above, we have

$$\begin{aligned}\frac{\partial}{\partial t}V &\le \Delta V-|\nabla Rc|^2+\frac{37}{2}\left(8\sqrt{3}+1\right)\frac{50}{3}R\left(|Rc|^2-\frac{1}{3}R^2\right)\\ &\le \Delta V-|\nabla Rc|^2+CR^{3-\bar{\delta}}.\end{aligned}$$

On the other hand, we compute

$$\frac{\partial}{\partial t}R^{2-\bar{\delta}/2}=\Delta\left(R^{2-\bar{\delta}/2}\right)-\left(2-\bar{\delta}/2\right)\left(1-\bar{\delta}/2\right)R^{-\bar{\delta}/2}|\nabla R|^2+2(2-\bar{\delta}/2)R^{1-\bar{\delta}/2}|Rc|^2.$$

If we choose $\bar{\beta}$ depending only on g_o such that

$$\bar{\beta}\cdot\left(2-\bar{\delta}/2\right)\left(1-\bar{\delta}/2\right)R_{\min}(0)^{-\bar{\delta}/2}\leq\frac{1}{3},$$

then using the estimate

$$\left|\nabla R\right|^2\leq 3\left|\nabla Rc\right|^2,$$

we have for all $\beta\in[0,\bar{\beta}]$:

$$\begin{aligned}\frac{\partial}{\partial t}\left[V-\beta R^{2-\bar{\delta}/2}\right] &\leq \Delta\left[V-\beta R^{2-\bar{\delta}/2}\right]+CR^{3-\bar{\delta}}-\beta 2(2-\bar{\delta}/2)R^{1-\bar{\delta}/2}\left|Rc\right|^2\\ &\leq \Delta\left[V-\beta R^{2-\bar{\delta}/2}\right]+C(\beta,g_o).\end{aligned}$$

By the maximum principle, we conclude that

$$V-\beta R^{2-\bar{\delta}/2}\leq C,$$

for some constant C depending only on β and g_o. Hence we have the following:

PROPOSITION 5.14. *There exists a constants $\bar{\delta}$, $\bar{\beta}$ and C depending only on g_o such that for all $\beta\in[0,\bar{\beta}]$:*

$$\frac{\left|\nabla R\right|^2}{R}\leq\beta R^{2-\bar{\delta}/2}+C(\beta,g_o).$$

5.5. Convergence. Let $[0,T)$ denote the maximum time interval of existence of the solution $g(t)$ to the Ricci flow. As a consequence of the pinching improves estimate (Corollary 5.7) and the gradient estimate (Proposition 5.14,) one can show that

LEMMA 5.15.

$$\lim_{t\to T}\frac{R_{\max}(t)}{R_{\min}(t)}=1.$$

We now consider the solution $\tilde{g}\left(\tilde{t}\right)$ to the volume normalized equation with the same initial data

$$\begin{aligned}\frac{\partial}{\partial t}\tilde{g}_{ij} &= \frac{2}{3}\tilde{r}\tilde{g}_{ij}-2\tilde{R}_{ij}\\ \tilde{g}(0) &= g_o,\end{aligned}$$

which exists on a maximal time interval $[0,\tilde{T})$. Now combining the pinching estimate (all scale-invariant estimates for solutions of the unnormalized flow also hold for solutions of the normalized flow)

$$\tilde{R}_{ij}\geq\varepsilon\tilde{R}\tilde{g}_{ij}$$

and Myers' Theorem with the fact that the volume is preserved under the flow, yields a uniform upper bound for the minimum scalar curvature

$$\tilde{R}_{\min}\left(\tilde{t}\right)\leq C<\infty$$

for all $\tilde{t}\in[0,\tilde{T})$. Thus, by the lemma above, we also have

$$\tilde{R}_{\max}\left(\tilde{t}\right)\leq C<\infty.$$

Using Corollary 5.7 and the above estimates for the scalar curvature, Hamilton [**19**], section 17, now proves that the solution exists for all time, exponentially converges to a limit metric which has constant positive sectional curvature.

6. Four-manifolds

In this section we outline the proof of Hamilton's classification 4 -manifolds with positive curvature operator.

THEOREM 6.1. (Hamilton 1986) *Given any smooth, compact 4-dimensional Riemannian manifold* (M, g_o) *with positive curvature operator, there exists a unique smooth solution* $g(t)$ *to the normalized Ricci flow*

$$\frac{\partial}{\partial t} g_{ij} = -2R_{ij} + \frac{1}{2} r \cdot g_{ij},$$

with initial condition $g(0) = g_o$ *on the time interval* $[0, \infty)$. *Moreover, the solution* $g(t)$ *converges to a constant sectional curvature metric. In particular,* M^4 *is diffeomorphic to either* S^4 *or* $\mathbb{RP}^4$.

The general strategy of the proof is much along the lines of Hamilton's 3 -manifold result, except that the analysis of the curvature operator is significantly more complicated. We start by recalling that

$$Rm : \wedge^2 M \to \wedge^2 M$$

is a self-adjoint linear map at each point in the manifold, and satisfies the equation (1)

$$\frac{\partial}{\partial t} R_{ijkl} = \Delta R_{ijkl} + \left(Rm^2\right)_{ijkl} + \left(Rm^{\#}\right)_{ijkl},$$

where $Rm^{\#}$ is the square using the Lie algebra structure constants of $\wedge^2 M$. Now since $n = 4$, we have the decomposition of 2-forms

$$\wedge^2 M = \wedge^2_+ M \oplus \wedge^2_- M,$$

where

$$\begin{aligned} \wedge^2_+ &= \left\{\alpha \in \wedge^2 : *\alpha = \alpha\right\} \\ \wedge^2_- &= \left\{\alpha \in \wedge^2 : *\alpha = -\alpha\right\}, \end{aligned}$$

into self-dual and anti-self-dual 2-forms corresponding to the isomorphism

$$so(4) \cong so(3) \oplus so(3).$$

The Lie algebra bracket restricted to each of the factors $\wedge^2_+$ and $\wedge^2_-$ is the cross-product (since they are isomorphic to $so(3)$.) Hence if we decompose the Riemann curvature operator as

$$Rm = \begin{pmatrix} A & B \\ B^t & C \end{pmatrix},$$

where $A : \wedge^2_+ \to \wedge^2_+$, $B : \wedge^2_- \to \wedge^2_+$, and $C : \wedge^2_- \to \wedge^2_-$, then

$$Rm^{\#} = \begin{pmatrix} A^{\#} & B^{\#} \\ \left(B^{\#}\right)^t & C^{\#} \end{pmatrix}.$$

The ODE corresponding the PDE (1)

$$\frac{dS}{dt} = S^2 + S^{\#}.$$

may be rewritten as

$$\begin{aligned} \frac{\partial}{\partial t}A &= \Delta A + A^2 + 2A^{\#} + B^t B \\ \frac{\partial}{\partial t}B &= \Delta B + AB + BC + 2B^{\#} \\ \frac{\partial}{\partial t}C &= \Delta C + C^2 + 2C^{\#} + {}^t BB. \end{aligned}$$

Let $a_1 \leq a_2 \leq a_3$, $b_1 \leq b_2 \leq b_3$, $c_1 \leq c_2 \leq c_3$ denote the ordered eigenvalues of the symmetric matrices A, $\sqrt{B^t B} = \sqrt{BB^t}$, C, respectively. Under the above system of ODEs, we find that the eigenvalues evolve by

$$\begin{aligned} \frac{d}{dt}a_1 &\geq a_1^2 + a_2 a_3 + b_1^2 \\ \frac{d}{dt}a_3 &\leq a_3^2 + a_1 a_2 + b_3^2 \\ \frac{d}{dt}c_1 &\geq c_1^2 + c_2 c_3 + b_1^2 \\ \frac{d}{dt}c_3 &\leq c_3^2 + c_1 c_2 + b_3^2 \\ \frac{d}{dt}(b_2 + b_3) &\leq a_2 b_2 + a_3 b_3 + b_2 c_2 + b_3 c_3 + 2b_1 b_2 + 2b_1 b_3. \end{aligned}$$

The maximum principle for systems (Theorem 3.4,) reduces the problem of obtaining bounds for Rm to finding suitable convex sets in $so(4) \otimes_S so(4)$ which are preserved by the system of ordinary differential equations above. Using this method and the above inequalities for the evolution of the eigenvalues, Hamilton proved the following:

PROPOSITION 6.2. *Depending on the initial metric g_o, there exist constants $C < \infty$ and $\varepsilon > 0$ such that the following inequalities remain true under the flow as long as the solution exists:*

1. $a_3 \leq C \cdot a_1$
2. $c_3 \leq C \cdot c_1$
3. $(b_2 + b_3)^2 \leq C \cdot a_1 c_1$
4. $(b_2 + b_3)^{2+\varepsilon} \leq C \cdot a_1 c_1 (a - 2b + c)^{\varepsilon}$
5. $(b_2 + b_3)^{2+\varepsilon} \leq C \cdot a_1 c_1$
6. $a_3 \leq a_1 + C \cdot a_1^{1-\varepsilon}$
7. $c_3 \leq c_1 + C \cdot c_1^{1-\varepsilon}$.

The above estimates should be viewed as a 4-dimensional analogue to Corollary 5.5 and Corollary 5.7 in the 3-dimensional case. In particular, these estimates enable one to prove the same gradient estimate for the scalar curvature as in the 3-dimensional case (Proposition 5.14,) which in turn leads to the long-time existence and convergence of the solution to a constant curvature metric.

7. Concluding remarks

We have not included in this account the important recent work of Hamilton on the analysis of singularities which develop under the flow and the possibility of performing geometric surgeries to avoid these singularities **[27]**, **[24]**, **[23]** (see **[12]** for a brief survey of this material,) nor the important works of others, such as (we refer to **[27]** for a more complete bibliography:)

1. Positive curvature pinched manifolds - H. Chen **[13]**, Gerhard Huisken **[31]**, Christophe Margerin **[43]**, Seiki Nishikawa **[46]**
2. Kähler manifolds - S. Bando **[1]**, Huai-Dong Cao **[7]**, **[8]**, **[9]**, **[10]**, **[11]**, Ngaiming Mok **[45]**, Wan-Xiong Shi **[49]**, **[50]**, **[51]**
3. Ricci solitons - Robert Bryant **[5]**, Tom Ivey **[33]**, **[35]**, **[36]**, Koiso **[38]**
4. Negative curvature pinched manifolds - Maung Min-Oo **[44]**, Rugang Ye **[55]**
5. Surfaces - Bartz-Struwe-Ye **[3]**, Chow **[14]**, Daskalopoulos-del Pino **[17]**, Hamilton **[22]**, Lang-Fang Wu **[53]**, **[54]**
6. Special metrics - Cafora-Isenberg-Jackson **[6]**, Hamilton-Isenberg **[30]**, Isenberg-Jackson **[32]**, Dan Knopf **[37]**, Leviton-Rubinstein **[40]**
7. Smoothing properties - S. Bando **[2]**, Bemelmans-Min-Oo-Ruh **[4]**
8. Harnack inequalities and applications - Hamilton **[25]**, **[26]**, Sun-Chin Chu **[16]**
9. Manifolds with boundary - Ying Shen **[48]**
10. Compactness - Hamilton **[29]**

References

[1] S. Bando, *On three-dimensional compact Kähler manifolds of nonnegative bisectional curvature*, J.D.G. **19** (1984) 283-297.

[2] S. Bando, *Real analyticity of solutions of Hamilton's equation*, Math. Zeit. **195** (1987) 93-97.

[3] J. Bartz, M. Struwe, and R. Ye, *A new approach to the Ricci flow on* S^2, Annali de Scuola Normale Superiore di Pisa **21** (1994) 475-482.

[4] J. Bemelmans, M. Min-Oo, E.A. Ruh, *Smoothing Riemannian metrics*, Math. Zeit. **188** (1984) 69-74.

[5] R. Bryant, unpublished.

[6] M. Carfora, J. Isenberg and M. Jackson, *Convergence of the Ricci flow for metrics with indefinite Ricci curvature*, J. Differential Geom. **31** (1990) 249-263.

[7] H.-D. Cao, *Deformation of Kähler metrics to Kähler-Einstein metrics on compact Kähler manifolds*, Invent. Math. **81** (1985) 359-372.

[8] H.-D. Cao, *On Harnack's inequalities for the Kähler-Ricci flow*, Invent. Math. **109** (1992) 247-263.

[9] H.-D. Cao, *Limits of solutions to the Kähler-Ricci flow*, J. Differential Geom. **45** (1997) 257-272.

[10] H.-D. Cao, *Existence of gradient Kähler-Ricci solitons*, Elliptic and parabolic methods in geometry, B. Chow, R. Gulliver, S. Levy, J. Sullivan ed., AK Peters (1996) 1-16.

[11] H.-D. Cao and B. Chow, *Compact Kähler manifolds with nonnegative curvature operator*, Invent. Math. **83** (1986) 553-556.

[12] H.-D. Cao and B. Chow, *Recent Developments on the Ricci Flow*, Bull. AMS **36** (1999) 59-74.

[13] H. Chen, *Pointwise quarter-pinched 4 manifolds*, Ann.. Global Anal. Geom. **9** (1991) 161-176.

[14] B. Chow, *The Ricci flow on the 2-sphere*, J. Differential Geom. **33** (1991) 325-334.

[15] B. Chow, *On the entropy estimate for the Ricci flow on compact 2-orbifolds*, J. Differential Geom. **33** (1991) 597-600.

[16] B. Chow and S.-C. Chu, *A geometric interpretation of Hamilton's Harnack inequality for the Ricci flow*, Math. Research Letters **2** (1995) 701-718.

[17] P. Daskalopoulos and M.A. del Pino, On a singular diffusion equation, Comm. Anal. Geom. **3** (1995) 523-542.

[18] D. DeTurck, *Deforming metrics in the direction of their Ricci tensors*, J. Differential Geom. **18** (1983) 157-162; ibid., *improved version*, to appear in Selected Papers on the Ricci Flow, ed. H.-D. Cao, B. Chow, S.-C. Chu, and S.-T. Yau, International Press.

[19] R. S. Hamilton, *Three-manifolds with positive Ricci curvature*, J. Differential Geom. **17** (1982) 255-306.

[20] R. S. Hamilton, *Four-manifolds with positive curvature operator*, J. Differential Geom. **24** (1986) 153-179.

[21] R. S. Hamilton, *The Ricci flow on surfaces*, Contemporary Mathematics **71** (1988), 237-261.

[22] R. S. Hamilton, *An isoperimetric estimate for the Ricci flow on surfaces*, in Modern Methods in Complex Analysis, The Princeton conference in honor of Gunning and Kohn, pp. 191-200, ed. T. Bloom, etal., Annals of Math. Studies **137**, Princeton Univ. Press (1996).

[23] R. S. Hamilton, *Non-singular solutions of the Ricci flow on three-manifolds*, Comm. Anal. Geom., to appear.

[24] R. S. Hamilton, *Four-manifolds with positive isotropic curvature*, Comm. Anal. Geom. **5** (1997) 1-92.

[25] R. S. Hamilton, *The Harnack estimate for the Ricci flow*, J. Differential Geom. **37** (1993) 225-243.

[26] R. S. Hamilton, *Eternal solutions to the Ricci flow*, J. Differential Geom. **38** (1993) 1-11.

[27] R. S. Hamilton, *Formation of singularities in the Ricci flow*, Surveys in Diff. Geom. **2** (1995) 7-136, International Press, Boston.

[28] R.S. Hamilton, *Harmonic maps of manifolds with boundary*, Lecture Notes in Math., Vol 471, Springer, Berlin, 1975, 168 pp.

[29] R. S. Hamilton, *A compactness property for solutions of the Ricci flow*, Amer. J. Math. **117** (1995) 545-572.

[30] R. S. Hamilton and J. Isenberg, *Quasi-convergence of Ricci flow for a class of metrics*, Comm. Anal. Geom. **1** (1993) 543-559.

[31] G. Huisken, *Ricci deformation of the metric on a Riemannian manifold*, J. Diff. Geom. **17** (1985) 47-62.

[32] J. Isenberg and M. Jackson, *The Ricci flow of locally homogeneous geometries on closed manifolds*, J. Diff. Geom. **35** (1992) 723-741.

[33] T. Ivey, *Ricci solitons on compact three-manifolds*, Diff. Geom. and its Appl. **3** (1993) 301-307.

[34] T. Ivey, *The Ricci flow on radially symmetric* $\mathbb{R}^3$, Comm. P.D.E. **19** (1994) 1481-1500.

[35] T. Ivey, *New examples of complete Ricci solitons*, Proc. Amer. Math. Soc. **122** (1994) 241-245.

[36] T. Ivey, *Ricci solitons on compact Kähler surfaces*, Proc. Amer. Math. Soc. **125** (1997) 1203-1208.

[37] D. Knopf, *Quasi-convergence of the Ricci flow*, Comm. Anal. Geom., to appear.

[38] N. Koiso, *On rotationally symmetric Hamilton's equation for Kähler-Einstein metrics*, Recent Topics in Diff. Anal. Geom., Adv. Studies Pure Math., **18-I** (1990) 327-337.

[39] O.A. Ladyzenskaya, V.A. Solonnikov, N.N. Ural'ceva, *Linear and quasi-linear equations of parabolic type*, Transl. Amer. Math. Soc. Vol. 23, Amer. Math. Soc., Providence, RI, 1968.

[40] P. R. A. Leviton and J. H. Rubinstein, *Deforming Riemannian metrics on the 2-sphere*, **10** (1985); *Deforming Riemannian metrics on complex projective spaces*, Centre for Math Analysis **12** (1987) 86-95.

[41] Li, P. and S.-T. Yau, *On the parabolic kernel of the Schrödinger operator*, Acta Math. **156** (1968) 153-201.

[42] T. Mabuchi, *K-energy maps integrating Futaki invariants*, Tohoku Math. J. **38** (1986) 245-257.

[43] C. Margerin, *A sharp theorem for weakly pinched 4-manifolds*, C.R. Acad. Sci. Paris Serie 1 **17** (1986) 303; *Pointwise pinched manifolds are space forms*, Geometric Measure Theory Conference at Arcata, Proc. Symp. Pure Math. **44** (1986).

[44] M. Min-Oo, *Almost Einstein manifolds of negative Ricci curvature*, J. Diff. Geom. **32** (1990) 457-472.

[45] N. Mok, *The uniformization theorem for compact Kähler manifolds of nonnegative holomorphic bisectional curvature*, J. Differential Geom. **27** (1988) 179-214.

[46] S. Nishikawa, *Deformation of Riemannian metrics and manifolds with bounded curvature ratios*, Geometric Measure Theory Conference at Arcata, Proc. Symp. Pure Math.

44 (1986) 343-352; *On deformation of Riemannian metrics and manifolds with positive curvature operator*, Lecture Notes in Math. **1201** (1986) 201-211.

[47] M. H. Protter and H. F. Weinberger, *Maximum principles in differential equations*, Springer-Verlag (1984).

[48] Y. Shen, *On Ricci deformation of a Riemannian metric on manifold with boundary*, Pac. J. Math. **173** (1996) 203-221.

[49] W. X. Shi, *Deforming the metric on complete Riemannian manifolds*, J. Differential Geom. **30** (1989) 223-301; *Ricci deformation of the metric on complete noncompact Riemannian manifolds*, J. Differential Geom. **30** (1989) 303-394.

[50] W. X. Shi, *Complete noncompact Kähler manifolds with positive holomorphic bisectional curvature*, Bull. Amer. Math. Soc. **23** (1990) 437-440.

[51] W. X. Shi, *Ricci flow and the uniformization on complete noncompact Kähler manifolds*, J. Differential Geom. **45** (1997) 94-220.

[52] G. Tian, *The K-energy on hypersurfaces and stability,* Comm. Anal. Geom. **2** (1994) 239-265.

[53] L.-F. Wu, *The Ricci flow on 2-orbifolds with positive curvature*, J. Differential Geom. **33** (1991) 575-596.

[54] L.-F. Wu, *The Ricci flow on complete* $\mathbb{R}^2$, Comm. Anal. Geom. **1** (1993) 439-472.

[55] R. Ye, *Ricci flow, Einstein metrics, and space forms*, Trans. Amer. Math. Soc. **338** (1993) 871-896.

School of Mathematics, University of Minnesota, Minneapolis, MN 55455

Rigidity and Compactness of Einstein Metrics

Peter Petersen

ABSTRACT. We survey some rigidity and compactness results for Einstein metrics, with special emphasis on the degeneration results of Michael Anderson.

1. Introduction

In this essay, we shall explain some rigidity and compactness phenomena for Einstein manifolds and, more generally, for manifolds with bounded Ricci curvature. Many of the compactness results are due to Anderson and Anderson-Cheeger. Our proofs are not always rigorous. However, we have tried not to cheat too much, and to at least present the most important ideas.

The introduction is followed by a short section on notation, just to nail down our conventions on the various curvature tensors and L^p norms. After that, we proceed to establish a very general Bochner-Lichnerowicz-Weitzenböck formula which contains almost all the formulae of this type (spinor fomulæ excepted). In section 3 we begin to examine the Einstein condition and prove some results. The main focus here is on how the Einstein condition helps us go from weak to strong pinching conditions. This leads us to several rigidity/gap results. In section 4 we discuss harmonic coordinates and how they can be used to introduce pseudo-norms of Riemannian manifolds. In section 5 we use all of the previous material to study compactness and degeneration phenomena for Einstein manifolds. Also we have two pinching theorems that use these compactness results. First we show how to prove the compactness theorem from [**1**], when one has a lower bound for the injectivity radius. Then we slowly work our way up to proving the orbifold degeneration result from [**1**] for manifolds with an $L^{n/2}$ bound for the curvature. This is used to get some nice corollaries for 4-manifolds. In the last section we given some examples of orbifold degeneration of Einstein metrics.

We shall not always be careful with giving the original references, instead will rely heavily on the two books [**5**] and [**16**] whenever it is convenient.

I would like to thank C. LeBrun for convincing me to write this article and for advising me on many of the issues brought up here.

1991 *Mathematics Subject Classification.* Primary 53C20.

Supported in part by a grant from the National Science Foundation.

2. Notation

On a Riemannian n-manifold (M,g), usually denoted M, we have the metric and torsion free connection ∇. This enables us to define covariant derivatives of all tensors, in particular we get the curvature tensor as

$$\begin{aligned} R(X,Y)Z &= \left(\nabla^2_{X,Y} - \nabla^2_{Y,X}\right)(Z) \\ &= \nabla_X \nabla_Y Z - \nabla_{\nabla_X Y} Z - \nabla_Y \nabla_X Z + \nabla_{\nabla_Y X} Z \\ &= \nabla_X \nabla_Y Z - \nabla_Y \nabla_X Z - \nabla_{[X,Y]} Z. \end{aligned}$$

This gives us the Ricci tensor Ric which we think of as a $(1,1)$ tensor

$$\mathrm{Ric}(X) = \sum R(X, E_i) E_i,$$

where E_i is an orthonormal frame. The $(0,2)$ version of the Ricci tensor is defined by

$$\mathrm{ric}(X,Y) = g(\mathrm{Ric}(X), Y).$$

Finally we have the curvature operator $\mathfrak{R} : \Lambda^2 TM \to \Lambda^2 TM$ defined implicitly by

$$\langle \mathfrak{R}(X \wedge Y), V \wedge W \rangle = g(R(X,Y)W, V)$$

For a function $u : \Omega \to \mathbb{R}$ on a bounded domain in a Riemannian manifold M we define the L^p norm as

$$\|u\|_{p,\Omega} = \left(\frac{1}{\mathrm{vol}\Omega} \int_\Omega |u|^p \, d\mathrm{vol} \right)^{1/p},$$

where $d$vol denotes the Riemannian volume element. In case M is compact we usually let $\Omega = M$ and omit it as a subscript.

3. Bochner-Lichnerowicz-Weitzenböck Formulae

First we explain a very general Weitzenböck formula due to Lichnerowicz. In subsequent sections we shall see how this general formula behaves in some specific situations that lead to some interesting rigidity and regularity results for Einstein metrics.

Consider a Riemannian n-manifold M and a vector bundle E over M endowed with a metric and compatible connection. We shall use the notation ∇ and $R(X,Y)$ for the connection and curvature on both E and TM.

For an (E,p) tensor T on M, i.e., $T(X_1,\ldots,X_p)$ is a section of E for vector fields $X_1,\ldots,X_p$ on M, we have the covariant derivative and its dual defined by

$$\begin{aligned} (\nabla T)(X_0,\ldots,X_p) &= (\nabla_{X_0} T)(X_1,\ldots,X_p) \\ &= \nabla_{X_0} T(X_1,\ldots,X_p) - \sum_{i=1}^{p} T(X_1,\ldots,\nabla_{X_0} X_i,\ldots,X_p), \\ (\nabla^* T)(X_2,\ldots,X_p) &= -\sum_{i=1}^{n} (\nabla_{E_i} T)(E_i, X_2,\ldots,X_p), \end{aligned}$$

where E_i is an orthonormal frame. Note that if S has $p+1$ variables and T has p variables then we can construct a 1-form

$$\begin{aligned} \omega(X) &= \langle S(X,\cdots), T(\cdots) \rangle \\ &= \sum \left\langle S\left(X, E_{i_1},\cdots,E_{i_p}\right), T\left(E_{i_1},\cdots,E_{i_p}\right) \right\rangle. \end{aligned}$$

The divergence or codifferential of this form is $\nabla^*\omega$ and by the divergence theorem we know that the integral of $\nabla^*\omega$ is zero. This implies that ∇ and ∇^* are adjoints in L^2 since we have the identity

$$\nabla^*\omega = -\langle S, \nabla T\rangle + \langle \nabla^* S, T\rangle .$$

Note that this identity also gives us the basic identity

$$\begin{aligned} \Delta \langle T_1, T_2\rangle &= \nabla^*\nabla \langle T_1, T_2\rangle \\ &= -\langle \nabla T_1, \nabla T_2\rangle + \langle \nabla^*\nabla T_1, T_2\rangle + \langle T_1, \nabla^*\nabla T_2\rangle . \end{aligned}$$

When we can express the connection Laplacian $\nabla^*\nabla$ in terms of curvature and a different Laplacian this gives us a Bochner formula.

The generalized Lichnerowicz Laplacian Δ_L on E-valued tensors is defined as

$$\begin{aligned} \Delta_L T &= \nabla^*\nabla T + \mathrm{Ric}T, \\ (\mathrm{Ric}T)(X_1, \dots, X_p) &= \sum_{j=1}^{p}\sum_{i=1}^{n} (R(E_i, X_j)T)(X_1, \dots, E_i, \dots, X_k), \end{aligned}$$

where E_i replaces X_j in the jth spot. Note that if the curvature of both M and E is bounded, then $\mathrm{Ric}T$ is also bounded in terms of T. Using the Lichnerowicz Laplacian we get the Bochner formula

$$\begin{aligned} \Delta \langle T_1, T_2\rangle &= -\langle \nabla T_1, \nabla T_2\rangle + \langle \nabla^*\nabla T_1, T_2\rangle + \langle T_1, \nabla^*\nabla T_2\rangle \\ &= \langle \Delta_L T_1, T_2\rangle + \langle T_1, \Delta_L T_2\rangle \\ &\quad - \langle \nabla T_1, \nabla T_2\rangle - \langle \mathrm{Ric}T_1, T_2\rangle - \langle T_1, \mathrm{Ric}T_2\rangle \end{aligned}$$

In case T is skew symmetric we also have the exterior covariant derivative defined by

$$(dT)(X_0, \dots, X_p) = \sum_{i=0}^{p} (-1)^i (\nabla_{X_i} T)\left(X_0, \dots, \hat{X}_i, \dots, X_p\right).$$

Note that ∇^* is dual to both d and ∇. In the case of the exterior covariant derivative d this dual is often also denoted as d^* or δ. Moreover, one easily checks that we have the Weitzenböck formula

$$\Delta_L = dd^* + d^*d.$$

This demonstrates that the Lichnerowicz Laplacian is nonnegative in these cases and also that it is the Hodge Laplacian on forms.

4. Rigidity

In this section we shall see how an Einstein metric gives us extra control over the curvature tensor and how this leads to some rigidity results.

Consider the special case where $E = \Lambda^2 TM$ and the tensor T is the Riemannian curvature operator $\mathfrak{R} : \Lambda^2 TM \to \Lambda^2 TM$. In fact we shall modify this tensor a little by subtracting a constant tensor so that we can study some pinching phenomena. Thus we let $T = \mathfrak{R} - \lambda I$. It follows from Bianchi's second identity that $d\mathfrak{R} = 0$, thus also $d(\mathfrak{R} - \lambda I) = 0$. Furthermore the fact that the metric is Einstein also tells us, again via Bianchi's second identity, that $d^*\mathfrak{R} = 0$. Therefore we get

$$\nabla^*\nabla T + \mathrm{Ric}T = 0.$$

This leads to a Bochner formula of the form

$$\begin{aligned}\Delta\frac{1}{2}|T|^2 &= \langle\nabla^*\nabla T,T\rangle-|\nabla T|^2\\ &= -\langle \mathrm{Ric}T,T\rangle-|\nabla T|^2\\ &\leq C_1(n)|\mathfrak{R}|\,|T|^2-|\nabla T|^2.\end{aligned}$$

Since we also have

$$\Delta\frac{1}{2}|T|^2=|T|\,\Delta|T|-|\nabla|T||^2$$

Kato's inequality then tells us that

$$\begin{aligned}\Delta|T| &\leq C_1(n)|\mathfrak{R}|\,|T|\\ &\leq C_1(n)|T|^2+C_2(n)\alpha.\end{aligned}$$

We can now multiply both sides by $|T|^{2p-1}$, use integration by parts, and the Sobolev inequality

$$\left\|u-\frac{1}{\mathrm{vol}M}\int u\right\|_{\frac{2n}{n-2}}\leq C_S\,\|\nabla u\|_2,$$

to obtain an inequality of the form

$$\|T\|_{\frac{2pn}{n-2}}\leq C_3(n,p,\lambda,C_S)\left(\|T\|_{2p+1}+\left(\|T\|_{2p-1}\right)^{\frac{2p-1}{2p}}\right).$$

This does not look very promising for iteration purposes as we might have $2p+1>\frac{2pn}{n-2}$. However given some sort of L^q bound for T we can make things look a little better. Namely we can use Hölder's inequality to obtain

$$\|T\|_{\frac{2pn}{n-2}}\leq C_3(n,p,\lambda,C_S)\left(\|T\|_q\,\|T\|_{\frac{2pq}{q-1}}+\left(\|T\|_{2p-1}\right)^{\frac{2p-1}{2p}}\right).$$

When $q<n/2$ this doesn't give anything useful. However, as long as $q>n/2$ this can easily be iterated to yield a bound of the type

$$|T|\leq C_4(n,p,\lambda,C_S)\,\|T\|_q\,.$$

When $q=n/2$ we can bring the first term on the right-hand side to the left-hand side to obtain

$$\left(1-C_3(n,p,\lambda,C_S)\|T\|_{\frac{n}{2}}\right)\|T\|_{\frac{2pn}{n-2}}\leq C_3(n,p,\lambda,C_S)\left(\|T\|_{2p-1}\right)^{\frac{2p-1}{2p}}.$$

Provided $1-C_3(n,p,\lambda,C_S)\|T\|_{n/2}>0$, or in other words that $\|T\|_{n/2}$ is sufficiently small, this can be iterated to yield

$$|T|\leq C_5(n,p,\lambda,C_S)\,\|T\|_{n/2}\,.$$

Thus an L^q, $q>n/2$ bound for the curvature tensor of an Einstein metric immediately leads to a C^0 bound on the curvature. This is one of the crucial ingredients in all of our compactness results for Einstein metrics. As for pinching we see that if the $L^{n/2}$ norm of T is small then T is itself small and hence the eigenvalues for the curvature operator are pinched to be near λ. In order for these estimates to be truly interesting we must also have bounds for the Sobolev constant. Thanks to Gromov and Gallot (see **[9]**, **[10]**) we now know that upper diameter bounds and lower Ricci curvature bounds suffice to give bounds for C_S. In fact, this was recently generalized to the case where one allows for the Ricci curvature lying below a certain constant to be small in L^p, $p > n/2$ (see [**18**]).

Our first rigidity/gap theorem is the following $L^{n/2}$ pinching result.

THEOREM 4.1. *Let $\lambda > 0$ be given, there is a constant $\varepsilon(n, \lambda) > 0$ so that any Einstein metric with $\|\mathfrak{R} - \lambda I\|_{n/2} \leq \varepsilon$ has constant curvature.*

PROOF. Given a bound for the Sobolev constant the above result tells us that the curvature operator has eigenvalues close to one, in particular they are positive. One knows from a result of Tachibana that any Einstein metric with positive curvature operator has constant curvature (see [**16**, Chapter 7]). To get a bound for the Sobolev constant we first need a bound for the Ricci curvature. However, the $L^{n/2}$ pinching for the curvature tells us that the Einstein constant must be bigger than $(n-1)\lambda - C(n)\varepsilon$. Thus for small ε we get a positive lower bound for the Ricci curvature. Then Myers' Theorem gives us a diameter bound as well. □

N.B. Our conventions on L^p norms, established in §2, involve a volume normalization which is by no means standard. Without this normalization, the above result, and many other results in this article, would be false in the stated form. Readers accustomed to other conventions should thus exercise appropriate care in interpreting the results herein.

In case $\lambda = 0$ we cannot expect such a nice result since any Einstein metric can be scaled so as to have small $L^{n/2}$ norm on curvature. However, if we bound the diameter as well we get

THEOREM 4.2. *There is a constant $\varepsilon(n, D) > 0$ so that any Ricci flat metric with $\|\mathfrak{R}\|_{n/2} \leq \varepsilon$ and* diam $\leq D$ *is flat.*

PROOF. Simply observe that we get pinched curvature as in the almost flat manifold theorem of Gromov. In particular, the manifold must be $K(\pi, 1)$ (see [**12**]). Then it follows from the Cheeger-Gromoll splitting theorem that the manifold is flat. □

We shall later obtain a similar gap theorem for complete Ricci flat metrics which give us some very interesting compactness results. There is also a gap theorem when $\lambda < 0$ which is proved below (see 5.4).

To prepare for the non-compact result let us see what the above iterations can do for us. Let M be a complete Ricci flat manifold with $\text{vol}B(p,r) \geq v \cdot r^n$ and $\int |R|^{n/2} \text{dvol} \leq Q$. Note that these three condition are scaling invariant, i.e., if we multiply the metric by a constant these conditions will still hold with the same v and Q. We need to get some sort of smallness for the $L^{n/2}$ norm of the curvature. This is achieved as follows. Absolute volume comparison tells us that annuli of the form $A(r) = B(p, 2r) - B(p, r)$ satisfy $\text{vol}A(r) \geq v' \cdot r^n$ for some $v'(n, v)$. Moreover since $\int |R|^{n/2} \text{dvol} \leq Q$ we must have that $\int_{A(r)} |R|^{n/2} \text{dvol} \to 0$ as $r \to \infty$. The volume estimate for the annuli then tells us that

$$\|R\|_{n/2, A(r)} \leq \varepsilon(r) \cdot r^{-2},$$

where $\varepsilon(r) \to 0$ as $r \to \infty$. We now claim that this, in analogy with the compact case, gives us an inequality of the form

$$\|R\|_{\infty, A(r)} \leq C(n, v) \cdot \varepsilon(r) \cdot r^{-2},$$

in other words the curvature decays faster that quadratically at infinity. In order to prove this we have to use a Sobolev inequality of the form

$$\|u\|_{\frac{2n}{n-2},\Omega} \leq C_S \|\nabla u\|_{2,\Omega},$$

where u has compact support in the bounded domain $\Omega \subset M$. Results of Croke (see [**7**]) tell us that the volume growth condition and nonnegative Ricci curvature yield a bound for this Sobolev constant. The next problem is to bump $|R|$ down so that it has compact support in $A(r)$. To this end one selects an appropriate bump function ϕ with compact support in $A(r)$ and then multiply the inequality $\Delta|R| \leq C_1(n)|R|^2$ by $\phi^{2p}|R|^{2p-1}$. After some calculations and an iteration as above one then gets an estimate of the form

$$\|\phi R\|_{\infty,A(r)} \leq C(n,v) \cdot \varepsilon(r) \cdot r^{-2}.$$

While this is not precisely the promised estimate is good enough to give us the desired curvature decay condition. In section 5 we shall see how this curvature decay condition is used to "classify" all of the manifolds with these conditions and also how the space is Euclidean space provided Q is sufficiently small.

5. Harmonic coordinates

Harmonic coordinates give us even better control over the metric than we had in the previous section. A more in-depth account can be found in [**16**, Chapter 10].

Suppose we have harmonic coordinates $x = (x^1, \dots, x^n)$ on some open set $U \subset M$, i.e., $\Delta x^i = 0$. The Weitzenböck formula for the gradient fields ∇x^i then tells us that $\nabla^*\nabla(\nabla x^i) = \mathrm{Ric}(\nabla x^i)$. From this we can derive a Bochner formula

$$\frac{1}{2}\Delta\langle\nabla x^k, \nabla x^l\rangle = -\langle\nabla^2 x^k, \nabla x^l\rangle - \langle\nabla x^k, \nabla^2 x^l\rangle - \mathrm{ric}(\nabla x^k, \nabla x^l).$$

If we write out the gradient in terms of the coordinate vector fields ∂_i and define $g_{ij} = \langle\partial_i, \partial_j\rangle$, then one obtains

$$\frac{1}{2}\Delta g_{ij} = Q(g, \partial g) - \mathrm{ric}(\partial_i, \partial_j)$$

for some universal function Q that depends on the metric coefficients and its derivatives. What is interesting about this equation is that if one had C^1 bounds for the metric coefficients and C^0 bounds for the Ricci curvature then standard elliptic estimates tell us that one in fact has $C^{1,\alpha}$ bounds for the metric for any $\alpha \in (0,1)$. Moreover, if the metric is Einstein then one gets $C^{k,\alpha}$ bounds for any k and $\alpha \in (0,1)$. This is similar in spirit to what we saw above for the curvature tensor.

In order to make these estimates a little more precise and useful we introduce some more notation. Let $O \subset M$ be a subset. We say that the $C^{k,\alpha}$ norm of $O \subset M$ on the *scale of* r is bounded by K, denoted $\|O \subset M\|_{C^{k,\alpha},r} \leq K$, if we can find a covering U_s of O by *harmonic* coordinate charts such that

1. $x_s : U_s \to B(0,r) \subset \mathbb{R}^n$ is a diffeomorphism,
2. for each $x \in O$ the ball $B(x, re^{-K})$ lies in some chart U_s,
3. $|Dx| \leq e^K$, $\left|(Dx)^{-1}\right| \leq e^K$, and
4. g_{ij} as functions on $B(0,r)$ satisfy $r^{k+\alpha}\left\|\sum_{|I|=k} \partial^I g_{ij}\right\|_{C^\alpha} \leq K$.

This norm has many important and interesting properties. First we mention what happens with the above (interior) elliptic estimates. If $|\mathrm{Ric}| \leq \Lambda$, then for any scale $\bar{r} < r$ we have

$$\|O \subset M\|_{C^{1,\alpha},\bar{r}} \leq C(n,\alpha,\Lambda,r,\bar{r})\,\|O \subset M\|_{C^{1},r}.$$

Moreover, if the metric is Einstein $\mathrm{Ric} = \lambda I$, then

$$\|O \subset M\|_{C^{k,\alpha},\bar{r}} \leq C(n,k,\alpha,\lambda,r,\bar{r})\,\|O \subset M\|_{C^{1},r}.$$

We mention some further important facts about the norm:

1. If we multiply the metric on M by λ^2 we get a new Riemannian manifold N whose norm satisfies
$$\|O \subset N\|_{C^{k,\alpha},\lambda r} = \|O \subset M\|_{C^{k,\alpha},r}.$$
2. If (M_i,p_i) converges to (M,p) in the pointed $C^{k,\alpha}$ topology, then for any set $O \subset M$ we can find sets $O_i \subset M_i$ such that
$$\|O_i \subset M_i\|_{C^{k,\alpha},r} \to \|O \subset M\|_{C^{k,\alpha},r}.$$
3. Given r the norm $\|O \subset M\|_{C^{k,\alpha},r}$ is realized at some $p \in O$ if O is compact, in other words
$$\|O \subset M\|_{C^{k,\alpha},r} = \|\{p\} \subset M\|_{C^{k,\alpha},r}.$$
For the latter norm it suffices to use one chart. In case O is open or unbounded we can at least find $p \in O$ such that
$$\frac{1}{2}\|O \subset M\|_{C^{k,\alpha},r} < \|\{p\} \subset M\|_{C^{k,\alpha},r} \leq \|O \subset M\|_{C^{k,\alpha},r}.$$
4. For a compact set O the norm satisfies
$$\|O \subset M\|_{C^{k,\alpha},r} \to 0 \text{ as } r \to 0.$$
5. Euclidean space is the only Riemannian manifold such that all of its norms are zero on all scales r. In fact
$$\|M\|_{C^{k,\alpha},r} \to \infty \text{ as } r \to \infty$$
unless M is Euclidean space.

6. Compactness

In this section we shall work our way towards understanding certain classes of manifolds with bounded Ricci curvature. The first part of the material is covered in [**16**, Chapter 10] for the rest be have supplied references to the appropriate research articles.

We begin by mentioning the following finiteness and compactness theorem essentially due to Cheeger.

THEOREM 6.1. *Given n,k,α and $r,K,D > 0$ we have that the class of Riemannian n-manifolds with $\|M\|_{C^{k,\alpha},r} \leq K$ and* diam $\leq D$ *contains only finitely many diffeomorphism types and is compact in the $C^{k,\beta}$ topology for any $\beta < \alpha$.*

In case we allow for complete manifolds and don't have a diameter bound we can no longer get finiteness for diffeomorphism types, but we can still get compactness in the pointed $C^{k,\beta}$ topology. More precisely this means that for each sequence M_i with $\|M_i\|_{C^{k,\alpha},r} \leq K$ and $p_i \in M_i$ we can find a subsequence (again indexed by i) and a limit manifold M with a point $p \in M$ such that for each $R > 0$ there are embeddings $\phi_i : B(p,R) \to M_i$ which contain $B(p_i, R)$ and such that the pull-back metrics $\phi_i^* g_i \to g$ in the $C^{k,\beta}$ topology on $B(p,R)$.

In order for all this to be useful it is of course necessary that we have some sort of method that allows us to get bounds for these norms. This is a achieved by a very interesting rescaling argument which was first explored in detail by Anderson. The simplest result along these lines is (see [**1**] and [**16**, Chapter 10])

THEOREM 6.2. *Given n, i_0, D, and Λ the class of Riemannian n-manifolds with*

$$\begin{array}{rcl} \text{inj} & \geq & i_0 \\ \text{diam} & \leq & D \\ |\text{Ric}| & \leq & \Lambda \end{array}$$

has the property that for every $K > 0$ we can find $r(n, i_0, D, \Lambda, K, \alpha) > 0$ such that any manifold in this class satisfies $\|M\|_{C^{1,\alpha},r} \leq K$. In particular, this class is compact in any $C^{1,\beta}$ topology.

PROOF. The proof goes by contradiction. Thus suppose that we have a sequence of manifolds M_i in this class such that $\|\{p_i\} \subset M_i\|_{C^{1,\alpha},r_i} = \|M_i\|_{C^{1,\alpha},r_i} = K$ for a sequence $r_i \to 0$. If we rescale these Riemannian manifolds by r_i^{-2}, the norms stay the same on the new scale of 1. Thus we have a new sequence of Riemannian manifolds N_i which satisfy

$$\begin{array}{rcl} \|\{p_i\} \subset N_i\|_{C^{1,\alpha},1} & = & K \\ \text{inj} N_i & \to & \infty \\ |\text{Ric} N_i| & \to & 0 \end{array}$$

From the above compactness theorem we can conclude that a subsequence (not renumbered) will converge in the pointed $C^{1,\beta}$ topology to some complete Riemannian manifold N. First we note that since all of the manifolds have bounded Ricci curvature we can in fact assume that their $C^{1,\gamma}$ norms are bounded for any $\gamma \in (0,1)$, but on some slightly smaller scale. Thus we can also assume that the manifolds converge in the $C^{1,\alpha}$ topology. Since the $C^{1,\alpha}$ norm is continuous with respect to the $C^{1,\alpha}$ topology we therefore get that $\|\{p\} \subset N\|_{C^{1,\alpha},1} = K$, in particular, the manifold cannot be Euclidean space. On the other hand if we look at the formula for the Ricci tensor in harmonic coordinates on N_i we see that the limit metric must be a weak solution to

$$\frac{1}{2}\Delta g = Q(g, \partial g),$$

since $|\text{Ric} N_i| \to 0$. Elliptic regularity theory then tells us that the metric on N is smooth of any order and Ricci flat. Now we come to the crucial point. Since $\text{inj} N_i \to \infty$ the limit manifold also has $\text{inj} N = \infty$, thus the Cheeger-Gromoll splitting theorem tells us that the manifold is the standard Euclidean space. We have therefore arrived at a contradiction. □

If we insist on only considering Einstein metrics the class becomes compact in the $C^{k,\alpha}$ topology for any k, α. In the Einstein case we can also get a similar result which lies closer to some of the stuff we are aiming for.

THEOREM 6.3. *Given $n, q > n/2, v_0, D, Q$ and λ the class of Einstein n-manifolds with*

$$\begin{array}{rcl} \mathrm{vol} & \geq & v_0 \\ \mathrm{diam} & \leq & D \\ \mathrm{Ric} & = & \lambda I \\ \|R\|_q & \leq & Q \end{array}$$

has the property that for every $K > 0$ we can find $r(n, q, v_0, D, Q, \Lambda, K, k, \alpha) > 0$ such that any manifold in this class satisfies $\|M\|_{C^{k,\alpha},r} \leq K$. In particular, this class is compact in the $C^{k,\beta}$ topology.

PROOF. We know from above that the curvature is bounded not just in L^q but in C^0. Cheeger's lemma then tells us that this class must have a lower bound for the injectivity radius (see [**16**, Chapter 10]). The above theorem then takes care of the rest. □

From the proof of this theorem we can also get the promised gap theorem for negative Einstein metrics.

THEOREM 6.4. *Given n, D and $\lambda < 0$ there is an $\varepsilon(n, D, \lambda) > 0$ such that any Einstein metric with* diam $\leq D$ *and $\|\mathfrak{R} - \lambda I\|_{n/2} \leq \varepsilon$ has constant curvature.*

PROOF. The iterations from the above section tell us that the metric satisfies $|\mathfrak{R} - \lambda I| \leq C \cdot \varepsilon$. In particular, the manifolds have pinched negative curvature. This together with the diameter bound tells us that the manifold has a lower volume bound (see [**11**]). Therefore, if the theorem were false we would have a sequence of Einstein metrics converging to a hyperbolic metric in any $C^{k,\alpha}$ topology. This however contradicts a rigidity result of Koiso (see [**5**, 12.F and 12.H]). □

It is interesting to see what happens if relax the Einstein condition so that we only have $|\mathrm{Ric}| \leq \Lambda$. In this case we still get compactness in the $C^{1,\alpha}$ topology. The argument goes by contradiction and uses rescaling as above. The way in which we get that the limit manifold is flat is to note that $\int |R_i|^q \to 0$ after the rescaling (a slightly stronger convergence coming from elliptic L^p estimates is needed here (see [**13**, p167-202])). Also, as we have a global diameter bound and a lower volume bound, relative volume comparison gives us that the limit manifold has a volume growth condition $\mathrm{vol}B(p, r) \geq v \cdot r^n$ for some $v(n, v_0, \Lambda)$. Now the only flat manifold with such volume growth is $\mathbb{R}^n$.

Finally we could try to examine the borderline case $q = n/2$. We already studied what happened when $\|R - \lambda I\|_{n/2}$ was small in the Einstein case. Furthermore we showed that Ricci flat manifolds with volume growth $\mathrm{vol}B(p, r) \geq v \cdot r^n$ and $\int |R|^{n/2} \leq Q$ have faster than quadratic curvature decay.

Suppose we have a sequence of manifolds with

$$\begin{array}{rcl} \mathrm{vol} & \geq & v_0, \\ \mathrm{diam} & \leq & D, \\ |\mathrm{Ric}| & \leq & \Lambda, \\ \int |R|^{n/2} & < & Q. \end{array}$$

If we blow up these metrics as above, then we have that the volume condition gives us a volume growth condition in the limit, the Ricci curvature makes the limit Ricci flat, and finally since $\int |R|^{n/2}$ is a scale invariant quantity the limit satisfies $\int |R|^{n/2} \leq Q$. The next result tells us what happens when Q is small.

THEOREM 6.5. *Given n, v there exists $\varepsilon(n,v) > 0$ such that any Ricci flat manifold with* $\text{vol}B(p,r) \geq v \cdot r^n$ *and* $\int |R|^{n/2} \leq \varepsilon$ *is Euclidean space.*

PROOF. First we exploit the fact that the metric has faster than quadratic curvature decay. Note that relative volume comparison implies that the volume growth condition is independent of the base point. Thus we have good lower volume bounds for all balls in M. If we take x so that $d(x,p) = r$, then the ball $B(x, r/4)$ will have volume $\geq v \cdot (r/4)^n$ and the curvature on this ball will be smaller than $\varepsilon(r) r^{-2}$. These two facts tell us that the injectivity radius at x must be larger than $c(n,v) \cdot r$ (this requires a slight improvement on Cheeger's original argument which can be found in [**6**]). In particular, any of the norms $\|M\|_{C^{k,\alpha},r}$ will be finite and bounded uniformly in terms of v and $\int |R|^{n/2}$.

To prove the theorem we now proceed by contradiction. Thus suppose we have a sequence of non-flat Ricci flat manifolds M_i satisfying $\text{vol}B(p_i, r) \geq v \cdot r^n$ and $\int |R|^{n/2} = \varepsilon_i \to 0$ as $i \to \infty$. Since any manifold which is not Euclidean space has the property that its norm goes to infinity as the scale goes to infinity, we can find r_i such that $\|M_i\|_{C^{k,\alpha},r_i} = 1$. Now rescale these metrics by r_i^{-2} so that we have new manifolds N_i with $\|N_i\|_{C^{k,\alpha},1} = 1$. Thus we can find $q_i \in N_i$ such that $\|\{q_i\} \subset N_i\|_{C^{k,\alpha},1} \geq 1/2$. Since all of the conditions for M_i are scale invariant we also have that $\|N_i\|_{C^{k,\beta},1} \leq K\left(n, v, \int |R|^{n/2}\right)$, where $\beta > \alpha$. This means that we can assume that N_i converges in the pointed $C^{k,\alpha}$ topology to a complete Ricci flat manifold with $\text{vol}B(q,r) \geq v \cdot r^n$. Moreover, since the norm is continuous in this topology we must also have $\|\{q\} \subset N\|_{C^{k,\alpha},1} \geq 1/2$. On the other hand $\int |R|^{n/2} = \varepsilon_i \to 0$, showing that the limit space is flat. This together with the volume growth condition tells us that the manifold is Euclidean space. Thus we have arrived at a contradiction. □

We can use this result to obtain a very general pinching theorem.

THEOREM 6.6. *Given $n, q > n/2, v_0, D$, and Λ, λ we can find $\varepsilon(n, q, v_0, D, \Lambda, \lambda) > 0$ such that any n-manifold with*

$$\begin{aligned} \text{vol} &\geq v_0, \\ \text{diam} &\leq D, \\ \|\text{Ric}\|_q &\leq \Lambda, \\ \|R - \lambda I\|_{n/2} &\leq \varepsilon \end{aligned}$$

is C^α, $\alpha < 2 - n/q$ close to a metric with constant curvature λ.

PROOF. For this to work we actually need to work with $L^{k,p}$ norms of manifolds as in [**13**, p167-202]. Also in order to get relative volume comparison as in [**17**] we need to know that the amount of Ricci curvature below a certain constant is small in $L^{q'}$ for some $q' > n/2$. However the fact that $\|R - \lambda I\|_{n/2} \leq \varepsilon, \|\text{Ric}\|_q \leq \Lambda$ implies that for any $q' \in [n/2, q)$ we have $\|\text{Ric} - (n-1)\lambda I\|_{q'} \leq C(n, q, \Lambda, \lambda, \varepsilon)$. As $\varepsilon \to 0$ also $C(n, q, \Lambda, \lambda, \varepsilon) \to 0$, so we get the desired pinching. This means we have relative volume comparison and hence that the above arguments kick in to finish the proof. □

We are now ready to study the more general class of manifolds which satisfy

$$\begin{aligned} \text{vol} &\geq v_0, \\ \text{diam} &\leq D, \\ |\text{Ric}| &\leq \Lambda, \\ \|R\|_{n/2} &\leq Q. \end{aligned}$$

When rescaling (blowing up) manifolds in this class we end up, as already pointed out, with complete Ricci flat metrics satisfying $\text{vol}B(p,r) \geq v \cdot r^n$ and $\int |R|^{n/2} \leq Q$. Such metrics are called ***almost locally Euclidean gravitational instantons***. If Q is not small these manifolds are not necessarily Euclidean space. The Eguchi-Hanson metric is an example of a non-flat limit space of this type. Still, one can say something intelligent about these spaces. We already know that they have faster than quadratic curvature decay. This implies that if we multiply the metric by a constant $\varepsilon \to 0$, then the curvature will converge to zero outside any compact set. Moreover, we also showed that the injectivity radius was large outside compact sets. Thus we get that the metric outside any compact set converges (in any topology) to a flat metric with Euclidean volume growth. This means that the limit as $\varepsilon \to 0$ must look like a cone over a space form of the type S^{n-1}/Γ, where Γ is a finite group of isometries acting freely on the sphere (see also [**4**]). The ALE gravitational instanton is then Euclidean space precisely when Γ is trivial. Note that when the dimension is odd, the space from must be even dimensional, hence it is either a sphere (total space is Euclidean) or a real projective space. The latter case, however, cannot occur as an even dimensional projective space cannot be the boundary of a compact manifold. As for the Eguchi-Hanson metric we know that it is a Ricci flat metric on TS^2, so when we scale the metric it will converge to the flat cone over $\mathbb{R}P^3$. The type of cone singularities that develop in this way are also known as orbifold point singularities.

From the above observations we obtain the following result of Anderson (see [**1**])

THEOREM 6.7. *A sequence of n-manifolds satisfying*

$$\begin{aligned} \text{vol} &\geq v_0, \\ \text{diam} &\leq D, \\ |\text{Ric}| &\leq \Lambda, \\ \|R\|_{n/2} &\leq Q. \end{aligned}$$

will subconverge to an orbifold with finitely many point singularities. Away from these points the convergence is in $C^{1,\alpha}$. In case n is odd the limit space is a manifold and the class is compact in the $C^{1,\alpha}$ topology.

It is important that one only gets finitely many singularities. The reason for this is that each singularity takes up a certain amount of $\int |R|^{n/2}$, otherwise the above gap theorem tells us that there isn't a singularity. Given that $\int |R|^{n/2}$ is bounded we can therefore only develop singularities at a given number of points.

Studying in detail this special type of possible degeneration Anderson and Cheeger (see [**3**]) were able to obtain a very attractive finiteness theorem for this class of manifolds.

THEOREM 6.8. *The class of n-manifolds satisfying*

$$\begin{aligned} \text{vol} &\geq v_0, \\ \text{diam} &\leq D, \\ |\text{Ric}| &\leq \Lambda, \\ \|R\|_{n/2} &\leq Q. \end{aligned}$$

contains only finitely many diffeomorphism types.

These two results take a particularly nice form on dimension 4. Namely, given the bound on the Ricci curvature one can obtain a bound on the $L^{n/2} = L^2$ norm of the curvature from the Euler characteristic. This is done using the Allendoerfer-Weil formula for the Gauss-Bonnet integrand in dimension 4

$$\begin{aligned} \chi(M) &= \frac{1}{8\pi^2}\int\left(|R|^2 - \left|\text{Ric} - \frac{\text{scal}}{4}I\right|^2\right) \\ &= \frac{1}{8\pi^2}\int |R|^2 - \frac{1}{8\pi^2}\int \left|\text{Ric} - \frac{\text{scal}}{4}I\right|^2. \end{aligned}$$

Thus

$$\int |R|^2 \leq C\left(|\chi(M)| + \Lambda\right).$$

This means that in these two theorems we can replace the $L^{n/2}$ bound on curvature by a bound on the Euler characteristic. Note that for an Einstein metric $\text{Ric} = \frac{\text{scal}}{4}I$, so in this case we don't need to know the Einstein constant in order to bound the $L^{n/2}$ norm of the curvature. Moreover, the Einstein constant is actually bounded by the Euler characteristic provided we have a lower volume bound.

From the discussion on Euler characteristic and L^2 norms on curvature we now obtain

COROLLARY 6.9. *Given D, v, C there are only finitely many diffeomorphism classes of Einstein 4-manifolds satisfying*

$$\begin{aligned} \text{vol} &\geq v, \\ \text{diam} &\leq D, \\ \chi &\leq C. \end{aligned}$$

Moreover, a sequence of such manifolds always has a subsequence which converges to an orbifold.

Having such a finiteness theorem means that we only need to study metrics on a fixed manifold. Thus we are left with studying the moduli of Einstein metrics on a fixed 4-manifold M. It is standard practice to fix something like the volume, but here it seems more natural to fix the Einstein constant. This leaves us with 3 different cases according to whether the Einstein constant is positive, zero or negative. Furthermore in the positive/negative case we can fix the constant to be ± 3. In the positive case the degeneration results then take the form

COROLLARY 6.10. *Let $v > 0$ be given. A sequence of 4-dimensional Einstein manifolds with Einstein constant 3 and volume $\geq v$ has a subsequence which converges to an orbifold with only point singularities.*

One can also establish a Gauss-Bonnet type formula for the ALE gravitational instantons. First we note that the volume growth together with the nonnegative Ricci curvature imply that the fundamental group is finite. From Poincaré duality it then follows that the Euler characteristic is $\chi = 1 + b_2 = 1 + \dim H^2(M, \mathbb{R})$. The Gauss-Bonnet formula now gives a formula for the modified Euler characteristic

$$\tilde{\chi}(M) = \left(1 - \frac{1}{|\Gamma|}\right) + b_2 = \frac{1}{8\pi^2}\int |R|^2,$$

where Γ is the finite fundamental group at infinity. The formula is obtained by applying the Gauss-Bonnet formula to a suitable sequence of sets that exhaust M. The $1/|\Gamma|$ term comes from the boundary terms as we pass to the limit. This means that the L^2 norm of the curvature is quantized, i.e., can only take certain values.

Moreover, if

$$\frac{1}{8\pi^2}\int |R|^2 < \frac{1}{2},$$

then the space must be Euclidean space. The Eguchi-Hanson metric in fact has a free $\mathbb{Z}_2$ action by isometries. If we pass to the quotient $TS^2/\mathbb{Z}_2$ we obtain a complete manifold where the Betti numbers satisfy $b_1 = b_2 = b_3 = b_4 = 0$. The topology of the example dictates that $\frac{1}{8\pi^2}\int |R|^2 = \frac{3}{4}$. It is not hard to show that this is the smallest possible nontrivial value for $\frac{1}{8\pi^2}\int |R|^2$.

7. Examples

It is worthwhile to study the orbifold degeneration in a little more detail. First we briefly mention some examples which show that such degeneration does occur.

The first examples of orbifold degeneration on compact manifolds come from [**14**]. Consider the standard flat 4-torus $T^4 = \mathbb{R}^4/\mathbb{Z}^4$. Let $\sigma : T^4 \to T^4$ be a Cartan involution, i.e., an involution in a point. Such an isometry has 16 fixed points. If we divide out by σ then we get a flat orbifold T^4/σ with 16 singularities each of which looks like a cone over $\mathbb{R}P^3$. We can blow each of these singularities up to get a K3 surface. On this K3 surface it is now possible to construct a sequence of Ricci flat metrics which converge in the above sense to the flat orbifold T^4/σ.

Tian in [**19**] shows that surfaces of the type $\mathbb{C}P^2 \sharp k \overline{\mathbb{C}P^2}$, where $3 \leq k \leq 8$, admit positive Kähler-Einstein metrics. Moreover, it is also established that when $5 \leq k \leq 8$ it is possible for these Kähler-Einstein metrics to degenerate to orbifolds. Tian even gives very explicit possibilities for the exact type of degenerations that might occur. We just mention a folklore example which looks similar to the above K3 surface example. Take X to be the orbifold obtained from $S^2 \times S^2$ by dividing out by the Cartan involution that rotates π on each sphere. One can blow up the 4 resulting singularities to obtain $M = \mathbb{C}P^2 \sharp 5 \overline{\mathbb{C}P^2}$, and on this manifold it is possible using Tian's work to construct Kähler-Einstein metrics which converge to X.

References

[1] M. T. Anderson, *Convergence and rigidity of manifolds under Ricci curvature bounds*, Inv. Math. 102 (1990), 429-445.

[2] M. T. Anderson, *The L^2 Structure of Moduli Spaces of Einstein Metrics on 4-Manifolds*, Geom. Func. An. 2 (1992) 29–89.

[3] M. T. Anderson and J. Cheeger, *Diffeomorphism finiteness for manifolds with Ricci curvature and $L^{n/2}$ curvature bounded*, GAFA 1 (1991), 231-251.

[4] S. Bando, A. Kasue and H. Nakajima, *On a construction of coordinates at infinity on manifolds with fast curvature decay and maximal volume growth*, Inv. Math. (1989) 313-349.

[5] A. Besse, *Einstein Manifolds*, Springer-Verlag, 1987.

[6] J. Cheeger, M. Gromov and M. Taylor, *Finite propagation speed, kernel estimates for functions of the Laplace operator, and the geometry of complete Riemannian manifolds*, J. Diff. Geom. 17 (1982) 15-54.

[7] C. B. Croke, *Some isoperimetric inequalities and eigenvalue estimates*, Ann. Scient. Ec. Norm. Sup. 13 (1980) 419-435.

[8] M. Freedman, *The topology of 4-dimensional manifolds*, J. Diff. Geom. 17 (1982) 357-453.

[9] S. Gallot, *Isoperimetric inequalities based on integral norms of Ricci curvature*, Astérisque No. 157-158 (1988), 191-216.

[10] S. Gallot, *Inégalités isopérimétriques et analytiques sur les variétés riemanniennes*, Astérisque No. 163-164 (1988), 31-91.

[11] M. Gromov, *Manifolds of negative curvature*, J. Diff. Geom 13 (1978) 223-230.

[12] M. Gromov, *Almost flat manifolds*, J. Diff. Geom 13 (1978) 231-241.

[13] K. Grove and P. Petersen, Eds, *Comparison Geometry*, MSRI publications vol. 30, published by Cambridge University Press.

[14] R. Kobayashi and A. Todorov, *Polarized period map for generalized K3 surfaces and the moduli of Einstein metrics*, Tohoku. Math J. (2) 39 (1987) 341–363.

[15] P.B. Kronheimer, *A Torelli-type theorem for gravitational instantons*, J. Diff. Geom. 29 (1989) 685-697.

[16] P. Petersen, *Riemannian geometry*, GTM 171, New York: Springer-Verlag, 1997.

[17] P. Petersen and G. Wei, *Relative volume comparison with integral curvature bounds*, GAFA 7 (1997) 1031-1045.

[18] P. Petersen and C. Sprouse, *Integral curvature bounds, distance estimates and applications*, preprint from UCLA.

[19] G. Tian, *On Calabi's conjecture for complex surfaces with positive first Chern class*, Inv. Math. 101 (1990) 101-172.

[20] J. A. Wolf, *Spaces of constant curvature*, 5th ed. Publish or Perish 1984.

Department of Mathematics, University of California, Los Angeles, CA 90095-1555
E-mail address: petersen@math.ucla.edu

Einstein Deformations of Hyperbolic Metrics

Olivier Biquard

Introduction

The simplest non-trivial examples of Einstein metrics are the rank-one symmetric spaces. In this essay, we will be interested in those of negative curvature — i.e. the hyperbolic spaces $\mathbb{K}\mathbf{H}^m$ ($m \geqslant 2$), where $\mathbb{K}$ is the field/algebra of real numbers ($\mathbb{R}$), complex numbers ($\mathbb{C}$), quaternions ($\mathbb{H}$) or octonions ($\mathbb{O}$); in the last case, only the Cayley hyperbolic plane $\mathbb{O}\mathbf{H}^2$ is well-defined. These spaces are the non-compact duals of the projective spaces $\mathbb{K}\mathbf{P}^m$. We normalize the metric so that the maximum of the sectional curvature is -1, and denote the real dimension of $\mathbb{K}$ by d ($= 1$, 2, 4 or 8), and the real dimension of $\mathbb{K}\mathbf{H}^m$ by $n = md$.

The boundary sphere $\mathbb{S}^{n-1}$ of a hyperbolic space carries a rich geometric structure, namely a conformal Carnot-Carathéodory metric. Let see this first in the real and complex examples.

The real hyperbolic space (with constant sectional curvature -1) is the unit ball $\mathbb{B}^n$ in $\mathbb{R}^n$, with the metric

$$g = 4\frac{\mathrm{euc}}{(1-\rho^2)^2},$$

where euc is the flat metric on $\mathbb{R}^n$ and ρ the radius. The metric g induces a metric on the boundary $\mathbb{S}^{n-1}$

(1) $$\gamma = \lim_{\rho\to 1}(1-\rho^2)^2 g_{\mathbb{S}_\rho};$$

the function $(1-\rho^2)$ is a *defining function* for the boundary, and the metric γ depends on the choice of the defining function only up to a conformal factor, so that the conformal class $[\gamma]$ is well defined. We shall say (following LeBrun's terminology) that $[\gamma]$ is the *conformal infinity* of g.

The complex hyperbolic space (with constant holomorphic sectional curvature -4) is the unit ball of $\mathbb{C}^m$ with the Bergman metric

$$g = \frac{\mathrm{euc}}{1-\rho^2} + \frac{\rho^2(d\rho^2 + (Id\rho)^2)}{(1-\rho^2)^2}.$$

Now equation (1) would lead to a very degenerate tensor on the boundary, so we consider instead

(2) $$\gamma = \lim_{\rho\to 1}(1-\rho^2) g_{\mathbb{S}_\rho};$$

this metric is infinite, except on the distribution $V = \ker \eta$, where $\eta = Id\rho$ is a connection 1-form for the $\mathbb{S}^1$-bundle $\mathbb{S}^{2m-1} \to \mathbb{CP}^{m-1}$. Such a metric defined on a contact distribution is called a Carnot-Carathéodory metric. Again, only the conformal class $[\gamma]$ is well defined and we extend the previous terminology to call it the *conformal infinity* of g.

These two examples fit in with the following more general picture. Fix a base point $*$ in the hyperbolic space and denote by r the distance to $*$ and by $\mathbb{S}_r$ the radius r sphere around $*$. The metric γ on the boundary sphere $\mathbb{S}$ of the hyperbolic space $\mathbb{KH}^m$ is defined as

$$\gamma = \lim_{r\to\infty} e^{-2r} g_{\mathbb{S}_r} . \tag{3}$$

This metric is infinite except on a distribution V of codimension 1 (complex case), 3 (quaternionic case) or 7 (octonionic case). In the real case it is finite and $V = T\mathbb{S}$. The brackets of vector fields in V generate the whole tangent bundle $T\mathbb{S}$, making γ into a Carnot-Carathéodory metric. Moreover, there is a contact form η with values in $\mathrm{Im}(\mathbb{K}) = \mathbb{R}$, $\mathbb{R}^3$ or $\mathbb{R}^7$, such that the metric is exactly

$$g = dr^2 + \sinh^2(2r)\eta^2 + \sinh^2(r)\gamma .$$

In the real case, the η^2 term does not appear. To give a sense to the formula in the three other cases, we have to choose a supplementary subspace to the distribution $V \subset T\mathbb{S}$. This is given here by the fibers of the fibration

$$\begin{array}{ccc} \mathbb{S}^{d-1} & \longrightarrow & \mathbb{S}^{n-1} \\ & & \downarrow \\ & & \mathbb{KP}^{m-1} \end{array}$$

Of course, all this depends on the choice of the base point $*$, but the conformal class $[\gamma]$ is well defined by (3) and will be called the *conformal infinity* of g.

The symmetric metrics are Einstein, $\mathrm{Ric}^g = -\lambda g$ with

$$\lambda = n-1,\ n+2,\ n+8,\ 36$$

in the real, complex, quaternionic and octonionic cases respectively. In this article, we will explain how all (in a sense to be precised) Einstein deformations of the hyperbolic metric are obtained as solutions of the following problem: given a Carnot-Carathéodory metric γ on the boundary, compatible in some sense with a contact structure, find a metric g in the interior such that

(i) $\mathrm{Ric}^g = -\lambda g$;
(ii) g has $[\gamma]$ as conformal infinity.

This problem has a long history. In the complex case, one can try to find Kähler-Einstein deformations. The problem is solved by the theorem of Cheng and Yau **[3]**: they prove in particular that any smooth strictly pseudoconvex domain in $\mathbb{C}^m$ admits a unique complete Kähler-Einstein metric, which is asymptotic to the CR-structure of the boundary as in equation (2). High order approximate formal solutions near the boundary had been constructed earlier by Fefferman **[5]** for this complex Monge-Ampère equation, and the regularity of the solution near the boundary has been determined by Lee and Melrose **[15]**.

In the real case, LeBrun **[12]** has solved, in dimension 4, a local problem near $\mathbb{S}^3$, or more generally near any real-analytic 3-manifold M. Since equation (i) with initial data (ii) is an underdetermined local Cauchy problem, he needs the following additional condition:

(iii) g is self-dual.

This means that the Weyl tensor W^g of the metric is a selfdual 2-form. Using Penrose twistor correspondence, he proves that for any real analytic γ on M, there is a unique solution g of equations (i), (ii) and (iii) in a neighborhood of M, up to diffeomorphisms. Moreover, if t is a defining function for the boundary M, the tensor $t^2 g$ is smooth up to the boundary, so that $[\gamma]$ is the conformal infinity of g in a strong sense. We have given earlier a weaker definition, since, for example in the complex case, such a regularity does not hold for the solutions produced by Cheng and Yau: the asymptotics provided by Lee and Melrose contain logarithmic terms.

Now, condition (iii) is special to dimension 4; in order to get a generalization to any dimension, Fefferman and Graham **[6]** have replaced condition (iii) by a less geometric condition: in a coordinate system such that

$$g = t^{-2}\left(dt^2 + \sum_{1}^{n-1} g_{ij}(x,t)dx^i dx^j\right),$$

where t is a defining function for the boundary M, they ask

(iii') $g_{ij}(x,t)$ is an even function of t.

This condition is independent of the coordinate system. They prove that, for n even, given a metric γ on M^{n-1}, the equations (i), (ii) and (iii') have a unique formal solution, which converges in a neighborhood of M if γ is real analytic (for n odd, there exist conformal structures $[\gamma]$ for which there is no formal solution). The aim of their study was to construct conformal invariants of $[\gamma]$ from Riemannian invariants of the canonical metric g. It seems that this theorem is a rediscovery of a theorem of Schouten and Haantjes **[22, 23]**.

Finally, in the case of quaternionic hyperbolic space, LeBrun **[13]** has constructed an infinite dimensional family of quaternionic Kähler deformations, using the twistor correspondence for quaternionic Kähler metrics.

Explicit solutions are known only in one case: SU_2-invariant solutions of (i), (ii) and (iii) in $\mathbb{B}^4$. Given any left-invariant metric or Carnot-Carathéodory metric γ on $\mathbb{S}^3$, Hitchin **[10]**, using twistor theory, has found formulas for the solutions in terms of elliptic functions. The case when γ is the metric of a Berger sphere had been established earlier by Pedersen **[21]**.

These explicit examples are very special, since for general boundary data, one can not hope to solve more than equations (i) and (ii) if one wants global solutions. These have been produced, when γ is close to the standard metric on the sphere at infinity, by Graham and Lee **[7]** in the real case, by the author **[2]** in the three other cases. In the rest of this survey, we explain this solution. In section 1, we define the Carnot-Carathéodory metrics needed at infinity, and then state the main theorem (theorem 3). In section 2, we give an idea of the proof, and in section 3, we ask some questions related to the problem.

1. Asymptotically Symmetric Metrics

Given a conformal Carnot-Carathéodory metric γ on $\mathbb{S}$, in order to solve problem (i)-(ii), we begin by producing a first order solution. This leads to the definition of special Carnot-Carathéodory metrics on the boundary, and then to the notion of "asymptotically symmetric metrics". We close the section by stating the theorem giving the solution to problem (i)-(ii).

1.1. Carnot-Carathéodory Metrics. If γ is defined on a distribution V, choose a supplementary V_2 in $T\mathbb{S}$, and a 1-form η with values in $\mathbb{R}^{d-1}$ and kernel V. We define

$$g = dr^2 + \sinh^2(2r)\eta^2 + \sinh^2(r)\gamma\,; \tag{4}$$

this metric is not smooth at $r = 0$, so we modify it for $r \leqslant 1$ so that it extends smoothly in the interior; we can do so with smooth dependence to the parameter γ. Choose a local basis of vector fields on $\mathbb{S}$ in the following way: $(X_1, \dots, X_{d-1})$ is a basis of V_2 such that $(\eta(X_i))$ is the standard basis of $\mathbb{R}^{d-1}$, and $(X_d, \dots, X_n)$ is a γ-orthonormal basis of V. Define now the orthonormal basis $\xi_0 = \partial_r$, $\xi_i = X_i/\sinh(2r)$ for $1 \leqslant i < d$, and $\xi_i = X_i/\sinh(r)$ for $i \geqslant d$. Using the form of the metric, we see easily that

$$\begin{aligned} [\xi_0, \xi_i] &= -2\xi_i + O(e^{-r}), && 1 \leqslant i < d, \\ [\xi_0, \xi_i] &= -\xi_i + O(e^{-r}), && i \geqslant d, \\ [\xi_i, \xi_j] &= O(e^{-r}), && i \geqslant 1,\ 1 \leqslant j < d, \\ [\xi_i, \xi_j] &= \textstyle\sum_1^d b_{ij}^k \xi_k + O(e^{-r}), && i, j \geqslant d, \end{aligned} \tag{5}$$

where the $b_{ij}^k = -d\eta_k(X_i, X_j)$ are the coefficients of the tensor induced by the bracket $V \otimes V \to V_2$. In particular, we see that at infinity, the bracket structure on an orthonormal basis depends only on this tensor. We shall require that this tensor is the same as in our model, that is the symmetric space.

This motivates the following definition. Let $H = U_{m-1}$, $Sp_{m-1}Sp_1$ or $Spin_7$ in the complex, quaternionic or octonionic cases. As is well known, rank one symmetric spaces are characterized by the fact that their spheres are homogeneous, and the group H is precisely the isotropy group of $\mathbb{S}^{n-1}$ when we represent it as a quotient under isometries of $\mathbb{KH}^m$, that is as a homogeneous sphere U_m/U_{m-1}, $Sp_mSp_1/Sp_{m-1}Sp_1$ or $Spin_9/Spin_7$.

DEFINITION 1. *A* Carnot-Carathéodory H-metric *on the sphere* $\mathbb{S}^{n-1}$ *is the data of a Carnot-Carathéodory metric* γ *on a distribution of codimension* $d-1$, *such that there exists a 1-form* η *with values in* $\operatorname{Im}(\mathbb{K}) = \mathbb{R}^{d-1}$ *and kernel* V, *satisfying*

- complex case: *the restriction to* V *of* $d\eta$ *is a symplectic form, compatible with* γ *(that is,* $d\eta(\cdot,\cdot) = \gamma(I\cdot,\cdot)$ *with* I *an almost complex structure on* V*);*
- quaternionic case: *the three 2-forms* $(d\eta_1, d\eta_2, d\eta_3)$ *on* V *give a quaternionic structure on* V, *compatible with* γ *(that is,* $d\eta_i(\cdot,\cdot) = \gamma(I_i\cdot,\cdot)$ *for almost complex structures* I_i *which satisfy the commutation relations of the quaternions);*
- octonionic case: *the seven 2-forms* $(d\eta_i)_{i=1,\dots,7}$ *give an octonionic structure on* V, *compatible with* γ *(that is,* $d\eta_i(\cdot,\cdot) = \gamma(I_i\cdot,\cdot)$ *for almost complex structures* I_i *which satisfy the commutation relations of the octonions).*

The meaning of this definition is that, in each case, we get a H-structure on V, with $H \subset SO(\gamma)$, compatible in some sense with the symplectic form $d\eta$. In the real case, the isotropy group is $H = SO(n-1)$, and we need nothing more than a metric γ on $\mathbb{S}^{n-1}$. In the complex case, there are lots of metrics compatible with a given contact form, but any deformation of a contact structure V is diffeomorphic to V: when we study this case, we may fix V and vary the almost complex structure on V. In the quaternionic and octonionic cases, the situation is completely different, because the metric is completely determined by the contact form η, and more precisely by the fundamental 4-form $\sum d\eta_i^2$, whose stabilizer is H (except for $m = 2$ in the quaternionic case). Finally, remark that given the H-structure on V, the contact form η with values in $\mathbb{R}^{d-1}$ is unique in the complex case, defined up to the action of the sections of the trivial SO_3-bundle (resp. SO_7-bundle) on $\mathbb{R}^3$ (resp. $\mathbb{R}^7$) in the quaternionic (resp. octonionic) case.

In all cases, the definition means that for a $\mathbb{K}$-basis (e_i) on V, we have

$$d\eta\left(\sum x_i e_i, \sum y_i e_i\right) = -\operatorname{Im} \sum \overline{x_i} y_i \,.$$

For the question of existence of such structures in the quaternionic and octonionic cases, see section 3.5.

1.2. Solution of (i)-(ii). We can now come back to our metric g defined from a Carnot-Carathéodory H-metric γ, and to formula (5). We see that the coefficients b_{ij}^k are the same for g and for the symmetric metric, and can be chosen constant in an adapted basis. From this follows easily that the curvature tensor of g at infinity is asymptotic to the curvature tensor of the symmetric metric, that is

$$|R^g - R^{sym}| = O(e^{-r})\,; \tag{6}$$

in particular,

$$|\mathrm{Ric}^g + \lambda g| = O(e^{-r})\,,$$

so we get the promised first order solution.

Because of the form (6) for the curvature, we shall say that g is asymptotically symmetric. We want to define more precisely this notion. For this, we introduce a little analysis: we need the usual Hölder spaces $C^{k,\alpha}$ for the metric g, and the weighted versions

$$C_\delta^{k,\alpha} = \cosh(-\delta r) C^{k,\alpha}\,.$$

This weighted space corresponds to functions decreasing as $e^{-\delta r}$, with their derivatives. Now we suppose that the Carnot-Carathéodory metric has regularity $C^{2,\alpha}$; then it is easy to see that $R^g - R^{sym}$ belongs to C_1^α. Therefore, the following definition is natural.

DEFINITION 2. *We say that a metric h is* asymptotically symmetric *if, for some g constructed as above, one has $h - g \in C_1^{2,\alpha}$. We say that the* conformal infinity *of h is $[\gamma]$.*

Of course, the order $\delta = 1$ is arbitrary, as is the number of derivatives that we require. We are now able to state the theorem giving the solution to the problem (i)-(ii) of the introduction.

THEOREM 3. *If a conformal Carnot-Carathéodory H-metric $[\gamma]$ is close enough (in $C^{2,\alpha}$ norm) to the symmetric conformal infinity, then there is an asymptotically symmetric metric h, solution to the problem (i)-(ii). This metric is locally unique, that is if we have another solution h' close enough to h, then there exists a diffeomorphism ϕ, equal to the identity at the boundary, such that $h' = \phi_* h$.*

In the real case, this is the theorem of Graham and Lee [**7**], with a slight technical difference in dimension 4, because they had to ask a regularity $C^{3,\alpha}$ on γ: this reflects the difference between their method and ours.

In the complex case, in dimension $2m - 1 \geqslant 5$, the integrability of the CR-structure (sufficiently close to the standard CR-structure) is equivalent to the solution h being Kähler-Einstein. Indeed, if a CR-deformation of $\mathbb{S}^{2m-1}$ is integrable, it can be realized as a strongly pseudoconvex hypersurface in $\mathbb{C}^m$ (see [**24**, theorem 9.4]); then the theorem of Cheng and Yau furnishes a Kähler-Einstein solution to the problem (i)-(ii). Conversely, if the CR-structure is not integrable, the solution h cannot be Kähler-Einstein.

Again in the complex case, recall that all nearby contact structures are diffeomorphic, so we may suppose that the distribution V is fixed. It follows that all the metrics constructed by the metric in the real or complex cases are mutually bounded; in fact, it will be shown that *these metrics exhaust all the bounded Einstein deformations of the symmetric metric in the real and complex cases* (with sufficient regularity at infinity). On the contrary, *there is no bounded Einstein deformation of the symmetric metric in the quaternionic and octonionic cases.* Finally, the theorem holds around other metrics than the symmetric one, provided that some L^2-obstruction space (see definition 4) vanishes. For all these complements to the theorem, see section 2.3.

2. Proof of the Theorem

Recall that we want to solve the problem (i)-(ii) of the introduction. The symmetric metric is a solution; given a perturbation of the data at the boundary, we have constructed a first order approximate solution g, already satisfying the boundary condition (ii).

2.1. A Gauge-Fixing Condition. We now want to solve the equation (i), that is $\mathrm{Ric}^h + \lambda h = 0$, for h close to g. As is well known, this is not an elliptic problem, because the equation is invariant under the action of the group of diffeomorphisms. This reflects also in the fact that the Ricci tensor always satisfies the Bianchi identity $\delta^h \mathrm{Ric}^h = -\frac{1}{2} ds^h$, where s^h is the scalar curvature. However, the method to overcome this difficulty is now well known: one can use harmonic coordinates to break the invariance, or more globally require that the identity map $(X,h) \to (X, \mathrm{ref})$ is a harmonic map, where ref if a reference metric. In the real case, Graham and Lee [**7**] choose for ref the first approximation g. We choose another condition, much in the spirit of the Coulomb gauge in gauge theory: it is a linear condition, essentially the infinitesimal version of the previous harmonicity condition:

$$B^g(h) := \delta^g h + \frac{1}{2} \mathrm{dtr}^g h = 0 . \tag{7}$$

There are two reasons why this is the correct choice. First we want to prove that this is a "gauge fixing condition" for the action of the diffeomorphism group, that is: given h close to g, prove that there is a unique diffeomorphism ϕ (equal to the identity on the boundary), such that $\phi_* h$ satisfies condition (7). It is sufficient to check this infinitesimally: the diffeomorphism group acts infinitesimally on g by taking the vector field X to the symmetrized covariant derivative $(\delta^g)^* X$, so the problem to solve is

$$B^g\left((\delta^g)^* X\right) = -B^g(h) .$$

But one has the formula

$$B^g(\delta^g)^* = \frac{1}{2}((D^g)^* D^g - \mathrm{Ric}^g) , \tag{8}$$

where D^g is the covariant derivative; as we consider essentially metrics g with negative curvature, the analysis explained later in section 2.3 proves easily that the operator $B^g(\delta^g)^*$ is an isomorphism.

The second reason why the gauge condition (7) is the right one is that a solution of equation (i) put in this gauge will satisfy

$$\Phi^g(h) := \mathrm{Ric}^h + \lambda h + (\delta^h)^* \left(\delta^g h + \frac{1}{2} d\mathrm{tr}^g h \right) = 0 . \tag{9}$$

The linearization of the nonlinear second order differential operator Φ^g at g is now very simple:

$$d_g \Phi^g(\dot{h}) = \frac{1}{2}(D^g)^* D^g \dot{h} + \frac{1}{2}(\mathrm{Ric}^g \circ \dot{h} + \dot{h} \circ \mathrm{Ric}^g + 2\lambda \dot{h}) - \overset{\circ}{R^g} \dot{h} .$$

Here, $(\overset{\circ}{R^g} k)_{X,Y} = \sum k(R^g_{e_i,X} Y, e_i)$. In particular the operator Φ^g becomes elliptic. Conversely, using the Bianchi identity $B^h(\mathrm{Ric}^h) = 0$, a solution h of (9) will satisfy

$$B^h(\delta^h)^* \left(\delta^g h + \frac{1}{2} d\mathrm{tr}^g h \right) = 0 .$$

Therefore, using equation (8), we see that h actually satisfies the gauge condition (7) and the initial equation (i).

We deduce from these considerations that the resolution of equation (i) modulo diffeomorphisms near g is completely reduced to problem (9).

2.2. An L^2 Obstruction. Let us now look at the linearization $d_g \Phi^g$ in the case where g is Einstein. The operator is then reduced to

$$d_g \Phi^g(\dot{h}) = \frac{1}{2}(D^g)^* D^g \dot{h} - \overset{\circ}{R^g} \dot{h} .$$

The kernel consists of the infinitesimal Einstein deformations of g.

DEFINITION 4. *If g is an Einstein asymptotically symmetric metric, we define the L^2-infinitesimal deformation space, $L^2\mathbf{H}^1(g)$, as the L^2-kernel of $d_g \Phi^g$.*

By a Weitzenböck formula (see [**1**, lemma 12.71]), it is easy to prove that

$$2 \int \langle d_g \Phi^g(\dot{h}), \dot{h} \rangle \geqslant (n-2)(-\sup K^g) \int |\dot{h}|^2 . \tag{10}$$

Therefore the operator $d_g \Phi^g$ is an isomorphism in L^2 if g has negative curvature, and in particular if g is symmetric. This is the argument used by Koiso [**11**] to prove that compact quotients of irreducible symmetric spaces of noncompact type and dimension greater than 2 do not admit Einstein deformations.

However, in the noncompact case, L^2-theory for the operator $d_g \Phi^g$ is not enough. Since this is a nonlinear problem, we need to work in Hölder spaces rather than L^2-spaces. More importantly, the L^2-condition gives a strong decay at infinity: functions like $\exp(-\delta r)$ are in L^2 if $\delta > \mathcal{H}$, where the critical exponent $\mathcal{H}$ is easily seen from (4) to be

$$\mathcal{H} = \frac{n-1}{2}, \frac{n}{2}, \frac{n}{2} + 4, 11 \tag{11}$$

in the real, complex, quaternionic and octonionic cases respectively. Therefore, in order to understand bounded deformations of g, we need to understand the behavior of the operator $d_g \Phi^g$ in the weighted Hölder spaces $C^{k,\alpha}_\delta$, for $\delta = 0$.

We have also another problem to solve: if g is now the first order approximate solution to the problem (i)-(ii) that we have constructed before for some $C^{2,\alpha}$ data γ on the boundary, then we have $\Phi^g(g) \in C_1^\alpha$ and we want to find an exact solution h of $\Phi^g(h) = 0$ with $h - g \in C_1^{2,\alpha}$. Basically, if g is a good enough approximate solution, this has a chance to be true if the differential $d_g\Phi^g$ is an isomorphism for the weight $\delta = 1$. The space $L^2\mathbf{H}^1(g)$ then appears as the obstruction for $d_g\Phi^g$ being an isomorphism in L^2. We shall see below that there is no other obstruction.

Before we proceed to the analysis, we need the following lemma on the eigenvalues of the curvature acting on symmetric 2-tensors. The only proof I know is by checking case by case.

LEMMA 5. *For the rank 1 symmetric metric g, the highest eigenvalue of $\overset{\circ}{R}{}^g$ is 4 (except in the real case, it is 1), and the other eigenvalues are negative.*

The value 1 for $\mathbb{R}\mathbf{H}^m$ if due to the choice of the sectional curvature equal to -1, instead of -4 for example for the holomorphic sectional curvature of $\mathbb{C}\mathbf{H}^m$.

2.3. Analysis and Resolution. For P a zero-th order homogeneous self-adjoint operator on hyperbolic space, analyzing the behavior of the operator $\Phi = D^*D + P$ is quite subtle, and we will confine ourselves here to a heuristic discussion of the problem. Suppose, for example, that we know that Φ is an isomorphism in L^2, that is essentially for the weight $\delta = \mathcal{H}$. The question is: for which range of weights (δ_0, δ_1) does the operator Φ remain an isomorphism?

There is an elementary (but non optimal) approach: using Kato's inequality $|Ds| \geqslant |d|s||$ and the maximum principle, one can prove that the interval (δ_0, δ_1) contains the interval (δ_0', δ_1') for the scalar operator $d^*d + \nu$, where ν is the smallest eigenvalue of P. Now one can see that for functions $f(r)$ depending only on r, one has

$$(d^*d + \nu)f = -\partial_r^2 f - 2\mathcal{H}\partial_r f + \nu f + O(e^{-r})(f, \partial_r f)\,. \tag{12}$$

The operator $-\partial_r^2 - 2\mathcal{H}\partial_r + \nu$ is called the *indicial operator.* It governs the behavior at infinity of the operator $d^*d + \nu$ because differentiating along the other directions always has a weight $\exp(-r)$ or $\exp(-2r)$. It is not difficult to see that the solutions $\exp(-\delta r)$ of the indicial operator give the values of δ_0' and δ_1', that is

$$\delta_0' = \mathcal{H} - \sqrt{\mathcal{H}^2 + \nu}\,, \tag{13}$$

$$\delta_1' = \mathcal{H} + \sqrt{\mathcal{H}^2 + \nu}\,. \tag{14}$$

Let us apply this result to $d_g\Phi^g$, using lemma 5. We have $\nu = -8$ (-2 in the real case) so we cannot catch the weight $\delta = 0$ but we can try to catch the weight $\delta = 1$. One can see easily, using (11), that $\delta_0' < 1$ if $n > 4$ in the real case, $n > 9$ in the complex case, and always in the quaternionic and octonionic cases. This is the analysis result used in the real case by Graham and Lee, and the technical restriction in dimension 4 we mentioned after theorem 3 comes from here: in this case, they have to find a higher order approximation before applying this analysis.

Now come back to our operator $\Phi = D^*D + P$. In order to get the optimal values of δ_0 and δ_1, one cannot use Kato's inequality, because this neglects zero order terms in D^*D. Actually, there is an indicial operator as in (12) for D^*D itself, given by the asymptotic behavior of D^*D:

$$-\partial_r^2 - 2\mathcal{H}\partial_r + C\,,$$

where $\mathcal{C}$ is now some zero order operator. The above discussion remains true when we replace the smallest eigenvalue ν of P by the smallest eigenvalue μ of $\mathcal{C} + P$, and we get

$$\delta_0 = \mathcal{H} - \sqrt{\mathcal{H}^2 + \mu}, \tag{15}$$

$$\delta_1 = \mathcal{H} + \sqrt{\mathcal{H}^2 + \mu}. \tag{16}$$

This is a considerably more difficult result, because one cannot use the maximum principle, which forgets the zero order term $\mathcal{C}$. These operators are probably a matter for the beautiful theory of *edge operators*, see for example [**17**] in the real case, [**4, 19**] in the complex case. In our symmetric case, there is an alternative approach using some elementary harmonic analysis [**2**]. Now apply this theory to the operator $d_g\Phi^g$. A calculation gives $\mu = 0$ in the real and complex cases, and $\mu > 0$ in the quaternionic and octonionic cases. In the last two cases, we deduce that $d_g\Phi^g$ is an isomorphism $C^{2,\alpha} \to C^\alpha$, so Φ^g is a local isomorphism, which means that there is no bounded Einstein deformation. On the contrary, in the real and complex cases, there are lots of bounded Einstein deformations, corresponding to sections on the sphere at infinity of the eigenbundle associated to the eigenvalue $\mu = 0$. This eigenbundle can be made explicit: in the real case, it is $\mathrm{Sym}_0^2 T\mathbb{S}^{n-1}$, so that Einstein infinitesimal deformations are given by conformal deformations of the boundary metric. In the complex case, recall that we have the contact distribution V with a symplectic form and a complex structure I, and the eigenbundle can be seen to be the subspace of $\mathrm{Sym}^2_{\mathbb{R}} V$ consisting of symmetric 2-tensors k on V such that $k(I\cdot, I\cdot) = -k(\cdot,\cdot)$. This is exactly the tangent space to metrics on V which remain compatible with the symplectic form. Thus we see that our theorem 3 gives all bounded Einstein deformations of the symmetric metric, such that the data on the boundary has regularity $C^{2,\alpha}$.

Now pass to the problem of actually producing the Einstein deformations. Denote the symmetric metric by g_0, and g the first approximation to the solution of problem (i)-(ii). In the complex case, one can fix the contact structure and deform only the almost complex structure. In the real and complex cases, the metrics remain mutually bounded, so that the weighted Hölder spaces for g and g_0 remain equal, and the problem is solved by applying the implicit function theorem to the equation $\Phi^g(h) = 0$ at $g = g_0$, using the analysis above for the weight $\delta = 1$. In the quaternionic and octonionic cases, this is not possible, because g and g_0 are no more close, but a more constructive method proves that if γ is close enough to the standard metric on the boundary, so that $h = g$ is a very good approximate solution, then one can deform g into a solution h of $\Phi^g(h) = 0$. This proves theorem 3.

The analysis above for the symmetric space has a counterpart for any asymptotically symmetric metric. Indeed, in the complex case, any contact structure is locally diffeomorphic to the standard contact structure, so that it is locally possible to approximate an asymptotically symmetric metric by a symmetric metric. In the real case, this is of course even simpler. In both cases, using this local approximation, one can graft the isomorphism obtained for the symmetric model to construct a parametrix for an operator $\Phi = D^*D + P$, and prove that Φ is Fredholm for $\delta \in (\delta_0, \delta_1)$, where δ_0 and δ_1 are given by formulas (15)-(16). In particular, if Φ is an isomorphism in L^2, it remains an isomorphism $C^{2,\alpha}_\delta \to C^\alpha_\delta$ for δ in this range.

In the quaternionic and octonionic cases, our special contact structures are not locally diffeomorphic, so that such an approximation by the symmetric model seems difficult; at least, one can use the first more elementary method above to prove a similar statement, but with the weight δ restricted to (δ_0', δ_1'), where δ_0' and δ_1' are given by formulas (13)-(14); as we have seen earlier, this is probably not the optimal interval, but it is sufficient for these two cases. The application of these considerations is that theorem 3 remains true around any asymptotically symmetric Einstein metric g, provided that the L^2-obstruction space $L^2\mathbf{H}^1(g)$ vanishes. In view of (10), this is true in particular if g has negative curvature.

3. Open Questions

3.1. Regularity. There are two questions on regularity. The first question is: suppose γ is smooth, what can be said on the regularity of the solution g up to the boundary? In the real case, Graham and Lee have constructed a high order approximate formal solution: the resolution stops at the critical weight $\delta_1 = 2\mathcal{H} = n-1$ in the notations of section 2.3; this enables them to prove that, if $n > 4$, the solution satisfies $t^2 g \in C^{n-2,\alpha}(\mathbb{B}^n)$, where t is some defining function of the boundary and the Hölder spaces are taken with respect to the flat metric on the ball. There is no doubt that such a high order approximate formal solution can be constructed in the other cases and that the resolution stops at the weight δ_1. Is it possible to construct an expansion in power series? this expansion should eventually contain logarithmic terms, as does the Lee-Melrose expansion for the Cheng-Yau metric.

The second question on regularity goes in the opposite direction: we have required a regularity $C^{2,\alpha}$ for the boundary metric γ, because the first order approximation can then be merely defined by (4), but it is obvious that this is not optimal. What is the lowest regularity on the Carnot-Carathéodory metric which enables to solve the problem?

3.2. Uniqueness. The Einstein metric with given conformal infinity is locally unique modulo the action of the group of diffeomorphisms, so an important question is to get a more global answer: namely, if g and h are two asymptotically symmetric Einstein metrics, with the same conformal infinity, does there exist a diffeomorphism ϕ, equal to the identity on the sphere at infinity, such that $h = \phi_* g$?

There are partial answers to this question, when g is the real or complex hyperbolic metric; these answers are rigidity results for the scalar curvature rather than for the Ricci tensor, and the idea comes from Witten's spinorial proof of the positive mass conjecture [**25**]. Namely, in [**20**], Min-Oo has proven that a metric h which is strongly asymptotic to the real hyperbolic metric g and such that $s^h \geqslant s^g$, is equal to g; here strongly asymptotic means that $h-g$ and one derivative are $O(e^{-(n+\varepsilon)r})$ (one can give weaker L^1 and L^2 bounds instead). For other rigidity results based on Min-Oo's theorem, see [**16**]. In [**9**], Herzlich has proven a similar statement in the case where g is the complex hyperbolic metric and h is a Kähler metric, strongly asymptotic to g, such that $s^h \geqslant s^g$; here, strongly asymptotic means that $h-g$ and one derivative are $O(e^{-(n+2+\varepsilon)r})$, and the same for the difference between the two complex structures (again one can use some integral bounds instead).

3.3. Global Existence. Related to the uniqueness statement is the very difficult question to know for which $[\gamma]$ on the boundary one can solve problem (i)-(ii).

The problem is completely open, and is probably related to a better understanding of the obstruction space $L^2\mathbf{H}^1(g)$.

3.4. Conformal Geometry. If one believes in the uniqueness of the solution (up to diffeomorphism), then it is clear that the conformal geometry of $[\gamma]$ must be related to the Riemannian geometry of g. An interesting step in this direction in the real case is the following result of Lee [**14**], relating the Yamabe invariant of $[\gamma]$ and the spectrum of g. The essential L^2 spectrum of an asymptotically real hyperbolic metric is $[(n-1)^2/4, \infty)$, without embedded eigenvalues [**18**]; moreover the real hyperbolic metric has purely continuous spectrum consisting of this ray. If g is an Einstein asymptotically hyperbolic metric with conformal infinity $[\gamma]$, such that the conformal class $[\gamma]$ on $\mathbb{S}^{n-1}$ contains a metric with positive scalar curvature, then the infimum of the L^2-spectrum of the scalar Laplacian of g is again $(n-1)^2/4$ (that is, there is no discrete eigenvalues below the continuous spectrum, as for the symmetric metric).

3.5. The Quaternionic and Octonionic Cases. In the quaternionic and octonionic cases, even the basis theorem 3 remains mysterious. In the real and complex cases, we know precisely that the Carnot-Carathéodory metrics are given by sections on some bundles. In the quaternionic and octonionic cases, existence of the corresponding Carnot-Carathéodory metrics is already a complicated differential system! it is an important question to understand these metrics. In the quaternionic case, examples are provided by LeBrun's twistorial construction in [**13**], but the metrics g share the same holonomy Sp_mSp_1 as the hyperbolic metric; are there $[\gamma]$ for which the metric g is not quaternionic Kähler? In the octonionic case, we have no example at all; note that in this case, it is impossible to deform the hyperbolic metric preserving the holonomy $Spin_9$, since any metric with holonomy $Spin_9$ must be locally symmetric.

Added in proof. Since this article was written, some new developments have occurred. In a new version of [**2**] (to appear in *Astérisque*), it is proven that in dimension $4m-1$ greater than 7, a Carnot-Carathéodory $Sp_{m-1}Sp_1$-metric is the conformal infinity of a unique quaternionic Kähler, asymptotically symmetric metric defined in a neighborhood. This gives a higher dimensional version of LeBrun's theorem (problem (i)-(ii)-(iii) in the introduction). In dimension 7, a similar statement is probably true with some added integrability hypothesis.

In a different direction, the correspondence between conformal metrics and Einstein metrics has recently led to exciting developments in physics concerning the relationship between conformal field theory and string theory; cf. [**26, 8**] and the references there.

References

[1] A. L. Besse, *Einstein manifolds*, vol. 10 of Ergebnisse der Mathematik und ihrer Grenzgebiete (3) [Results in Mathematics and Related Areas (3)], Springer-Verlag, Berlin, 1987.

[2] O. Biquard, *Métriques d'Einstein asymptotiquement symétriques*. Preprint, École Polytechnique, 1997.

[3] S. Y. Cheng and S. T. Yau, *On the existence of a complete Kähler metric on noncompact complex manifolds and the regularity of Fefferman's equation*, Comm. Pure Appl. Math., 33 (1980), pp. 507–544.

[4] C. L. Epstein, R. B. Melrose, and G. A. Mendoza, *Resolvent of the Laplacian on strictly pseudoconvex domains*, Acta Math., 167 (1991), pp. 1–106.

[5] C. L. Fefferman, *Monge-Ampère equations, the Bergman kernel, and geometry of pseudoconvex domains*, Ann. Math. (2), 103 (1976), pp. 395–416.

[6] C. L. Fefferman and C. R. Graham, *Conformal invariants*, Astérisque, hors série (1985), pp. 95–116. The mathematical heritage of Élie Cartan (Lyon, 1984).

[7] C. R. Graham and J. M. Lee, *Einstein metrics with prescribed conformal infinity on the ball*, Adv. Math., 87 (1991), pp. 186–225.

[8] C. R. Graham and E. Witten, *Conformal anomaly of submanifolds observables in ADS/CFT correspondance*. hep-th/9901021.

[9] M. Herzlich, *Scalar curvature and rigidity of odd-dimensional complex hyperbolic spaces*, Math. Ann., 312 (1998), pp. 641–657.

[10] N. J. Hitchin, *Twistor spaces, Einstein metrics and isomonodromic deformations*, J. Differential Geom., 42 (1995), pp. 30–112.

[11] N. Koiso, *Nondeformability of Einstein metrics*, Osaka J. Math., 15 (1978), pp. 419–433.

[12] C. R. LeBrun, *H-space with a cosmological constant*, Proc. Roy. Soc. London Ser. A, 380 (1982), pp. 171–185.

[13] ———, *On complete quaternionic-Kähler manifolds*, Duke Math. J., 63 (1991), pp. 723–743.

[14] J. M. Lee, *The spectrum of an asymptotically hyperbolic Einstein manifold*, Comm. Anal. Geom., 3 (1995), pp. 253–271.

[15] J. M. Lee and R. B. Melrose, *Boundary behaviour of the complex Monge-Ampère equation*, Acta Math., 148 (1982), pp. 159–192.

[16] M. C. Leung, *Pinching theorem on asymptotically hyperbolic spaces*, Internat. J. Math., 4 (1993), pp. 841–857.

[17] R. Mazzeo, *Elliptic theory of differential edge operators. I*, Comm. Partial Differential Equations, 16 (1991), pp. 1615–1664.

[18] ———, *Unique continuation at infinity and embedded eigenvalues for asymptotically hyperbolic manifolds*, Amer. J. Math., 113 (1991), pp. 25–45.

[19] R. B. Melrose, *Calculus of conormal distributions on manifolds with corners*, Internat. Math. Res. Notices, 3 (1992), pp. 51–61.

[20] M. Min-Oo, *Scalar curvature rigidity of asymptotically hyperbolic spin manifolds*, Math. Ann., 285 (1989), pp. 527–539.

[21] H. Pedersen, *Einstein metrics, spinning top motions and monopoles*, Math. Ann., 274 (1986), pp. 35–59.

[22] J. A. Schouten and J. Haantjes, *Beitraege zur allgemeinen (gekruemmten) konformen Differentialgeometrie*, Math. Ann., 112 (1936), pp. 594–629.

[23] ———, *Beitraege zur allgemeinen (gekruemmten) konformen Differentialgeometrie. II*, Math. Ann., 113 (1936), pp. 568–583.

[24] N. Tanaka, *A differential geometric study on strongly pseudo-convex manifolds*, Kinokuniya Book-Store Co. Ltd., Tokyo, 1975. Lectures in Mathematics, Department of Mathematics, Kyoto University, No. 9.

[25] E. Witten, *A new proof of the positive energy theorem*, Comm. Math. Phys., 80 (1981), pp. 381–402.

[26] ———, *Anti de Sitter space and holography*, Adv. Theor. Math. Phys., 2 (1998), pp. 253–291. hep-th/9802150.

CMAT, École Polytechnique, F-91128 Palaiseau Cedex, France,
E-mail address: biquard@math.polytechnique.fr

Four-Dimensional Einstein Manifolds, and Beyond

Claude LeBrun

1. Introduction

The aim of modern Riemannian geometry is to understand the relationship between topology and curvature. A case in point is that one would like to know when a given smooth compact n-manifold M admits an Einstein metric — that is, a Riemannian metric g such that

$$r = \lambda g,$$

where r is the Ricci tensor of g and λ is some real constant. When such a metric exists, moreover, it is natural to ask to what extent it is unique; in other words, one would like to understand the Einstein *moduli space* of M — i.e. the set of unit-volume Einstein metrics on M, modulo the action of the diffeomorphism group.

These existence and uniqueness questions are easily answered in dimensions 2 and 3, because a Riemannian manifold of dimension $n < 4$ is Einstein iff it has constant sectional curvature. In low dimensions, the sign of λ is therefore completely determined by the topology of M — indeed, by the size of $\pi_1(M)$. Moreover, the moduli space of Einstein metrics on a 2- or 3-manifold is always connected [**43**], so the value of λ, for *unit-volume* Einstein metrics g, is actually an invariant of M. In dimension 2, the moduli space is never empty, and has positive dimension if $\lambda \leq 0$. By contrast, the moduli space of a 3-manifold is [**43**] a single point if $\lambda < 0$. On the other hand, the Einstein moduli space is empty [**8, 61**] for any 3-manifolds with $\pi_2 \neq 0$; cf. [**3**].

In dimension $n \geq 4$, the curvature tensor of g is no longer determined by the Einstein condition in a point-wise manner, and Einstein metrics are no longer describable in terms of universal local models. While this, of course, is precisely what gives the subject its interest, the existence and uniqueness problems are commensurately harder when $n \geq 4$. Indeed, there are, to date, no non-existence or uniqueness results known when $n > 4$. Fortunately, however, a constellation of low-dimensional accidents makes the borderline case of $n = 4$ comparatively tractable. My aim here is to survey the current state of our knowledge regarding the existence and uniqueness of Einstein metrics on 4-manifolds, and point out some hints these give us regarding higher dimensions.

Supported in part by NSF grant DMS-9802722.

2. The Hitchin-Thorpe Inequality

Four-dimensional Riemannian geometry displays many features which have no adequate analogues in other dimensions. These are largely attributable to a single Lie-group-theoretic fluke: the rotation group $SO(4)$ isn't simple. Indeed,

$$\mathfrak{so}(4) \cong \mathfrak{so}(3) \oplus \mathfrak{so}(3),$$

so the adjoint action of $SO(4)$ on its Lie algebra preserves a decomposition into two 3-dimensional subspaces. Now $\mathfrak{so}(n)$ and $\Lambda^2(\mathbb{R}^n)$ are isomorphic as $SO(n)$-modules. Thus the rank-6 bundle of 2-forms on an oriented Riemannian 4-manifold decomposes invariantly into two rank-3 bundles:

$$\Lambda^2 = \Lambda^+ \oplus \Lambda^-. \tag{1}$$

Schur's lemma tells us that these bundles must coincide with the eigenspaces of the Hodge duality operator

$$\star : \Lambda^2 \to \Lambda^2.$$

With appropriately chosen conventions, sections of Λ^+ are thus characterized by $\star\varphi = \varphi$, and sections of Λ^- satisfy $\star\varphi = -\varphi$.

Definition 2.1. On any smooth oriented 4-manifold, sections of Λ^+ are called *self-dual 2-forms*, whereas sections of Λ^- are called *anti-self-dual 2-forms.*

Now let us suppose that (M, g) is a *compact* oriented Riemannian 4-manifold. The Hodge theorem then tells us that every de Rham class on M has a unique harmonic representative; in particular, there is a canonical identification

$$H^2(M, \mathbb{R}) = \{\varphi \in \Gamma(\Lambda^2) \mid d\varphi = 0,\ d\star\varphi = 0\}.$$

But the Hodge star operator $\star$ defines an involution of the right-hand side. We therefore have a direct sum decomposition

$$H^2(M, \mathbb{R}) = \mathcal{H}^+_g \oplus \mathcal{H}^-_g, \tag{2}$$

where

$$\mathcal{H}^\pm_g = \{\varphi \in \Gamma(\Lambda^\pm) \mid d\varphi = 0\}$$

are the spaces of self-dual and anti-self-dual harmonic forms. Notice that $\star$ is conformally invariant in the middle dimension, so the decomposition (2) remains unchanged if the metric g is multiplied by a smooth positive function.

The intersection form

$$\begin{aligned} \smile : H^2(M, \mathbb{R}) \times H^2(M, \mathbb{R}) &\longrightarrow \mathbb{R} \\ ([\varphi]\, ,\, [\psi]) &\mapsto \int_M \varphi \wedge \psi \end{aligned}$$

becomes positive-definite when restricted to $\mathcal{H}^+_g$, and negative-definite when restricted to $\mathcal{H}^-_g$; and the two are mutually orthogonal with respect to $\smile$. Thus, combining an L^2-orthonormal basis for $\mathcal{H}^+_g$ with an L^2-orthonormal basis for $\mathcal{H}^+_g$ gives us a basis for $H^2(\mathbb{R})$ in which the intersection form is represented by the diagonal matrix

$$\begin{bmatrix} \underbrace{\begin{matrix} 1 & & \\ & \ddots & \\ & & 1 \end{matrix}}_{b_+(M)} & \\ & b_-(M)\left\{ \begin{matrix} -1 & & \\ & \ddots & \\ & & -1 \end{matrix} \right. \end{bmatrix}.$$

The numbers $b_\pm(M) = \dim \mathcal{H}_g^\pm$ are therefore oriented homotopy invariants of M; namely, b_+ (respectively, b_-) is the dimension of any maximal linear subspace of $H^2(M, \mathbb{R})$ on which the restriction of $\smile$ is positive (respectively, negative) definite.

The intersection form described above is a bilinear form over $\mathbb{R}$. But of course, the cup product is also defined on integer cohomology, and one should therefore think of the intersection form over $\mathbb{R}$ as a mere shadow of a more fundamental object

$$\smile : H^2(M, \mathbb{Z}) \times H^2(M, \mathbb{Z}) \to \mathbb{Z},$$

concretely representable as a $b_2 \times b_2$ integer matrix of determinant ± 1. While such an integer quadratic form can of course be diagonalized over the reals, the analogous assertion fails over the integers. For example, the intersection form

$$\begin{bmatrix} 0 & 1 \\ 1 & 0 \end{bmatrix}$$

of $S^2 \times S^2$ is an *even* form, meaning that $\alpha \smile \alpha \equiv 0 \bmod 2$ for all $\alpha \in H^2(M, \mathbb{Z})$. By contrast, of course, the diagonal form

$$\begin{bmatrix} 1 & 0 \\ 0 & -1 \end{bmatrix}$$

is *odd* — which, by definition, just means that it is *not even*!

Manifolds with any specified values of $b_\pm$ can easily be constructed by the following operation:

DEFINITION 2.2. Let M_1 and M_2 be connected compact oriented 4-manifolds. Their connected sum $M_1 \# M_2$ is then the oriented 4-manifold obtained by deleting a small ball from each manifold and gluing together the resulting S^3 boundaries via a reflection.

For example, the 2×2 diagonal form considered above can be realized as the intersection form of $\mathbb{CP}_2 \# \overline{\mathbb{CP}}_2$, where $\mathbb{CP}_2$ is the complex projective plane with its standard orientation, and $\overline{\mathbb{CP}}_2$ is the same smooth 4-manifold with the *opposite* orientation. Similarly, the iterated connected sum

$$k\mathbb{CP}_2 \# \ell\overline{\mathbb{CP}}_2 = \underbrace{\mathbb{CP}_2 \# \cdots \# \mathbb{CP}_2}_{k} \# \underbrace{\overline{\mathbb{CP}}_2 \# \cdots \# \overline{\mathbb{CP}}_2}_{\ell}$$

has diagonal intersection form, with $b_+ = k$ and $b_- = \ell$. Notice that $n(S^2 \times S^2)$ and $n\mathbb{CP}_2 \# n\overline{\mathbb{CP}}_2$ are simply connected 4-manifolds with the same invariants $b_\pm$, but are not homotopy equivalent because one has even intersection form and one has odd intersection form. This distinction can be restated by saying that one is *spin* and the other is *non-spin*. An oriented manifold is called *spin* iff it satisfies

$w_2 = 0$, where $w_2 \in H^2(\mathbb{Z}_2)$ denotes the second Stiefel-Whitney class of the tangent bundle. In dimension 4, this is equivalent to the statement that every $\alpha \in H^2(\mathbb{Z}_2)$ satisfies $\alpha \smile \alpha = 0 \in \mathbb{Z}_2$, as a consequence of the Wu relation

$$w_2 \smile \alpha = \alpha \smile \alpha \in \mathbb{Z}_2.$$

In particular, a simply connected 4-manifold is spin iff its intersection form on $H^2(\mathbb{Z})$ is even.

Once this distinction between spin and non-spin 4-manifolds is understood, the topological classification of smooth simply connected 4-manifolds is easily stated.

THEOREM 2.1 (Freedman). *Two smooth simply connected oriented 4-manifolds are orientedly homeomorphic iff*

- *they have the same invariants b_+ and b_-; and*
- *both are spin, or both are non-spin.*

Freedman's result was originally stated [**19**] in terms of the equivalence of intersection forms; but Donaldson's celebrated theorem [**17**] on the diagonalizability of definite intersection forms and the Minkowski-Hasse classification of indefinite forms [**28**] allow one to make the simplified statement given here. On the other hand, the reader should immediately be warned that the classification of 4-manifolds up to *diffeomorphism*, while still poorly understood, is at least known to be much more complicated. In particular, the Seiberg-Witten invariants discussed in §4 allow one to show that some of the homeotypes treated by Theorem 2.1 can be realized by infinitely many distinct diffeotypes.

The difference $\tau(M) = b^+(M) - b^-(M)$ is called the *signature* of M. It is precisely the index of an elliptic operator

$$d^- d^* : \Gamma(\Lambda^+) \to \Gamma(\Lambda^-),$$

and the Atiyah-Singer index theorem therefore predicts [**4**] that it must be calculable by integrating an invariant polynomial in curvature; and indeed, this had been been discovered much earlier by Hirzebruch [**25**], using a less general argument. Of course, the same is also true of the Euler characteristic $\chi(M) = 2-2b_1(M)+b_2(M)$, which is the index of

$$d + d^* : \Gamma(\Lambda^{\text{even}}) \to \Gamma(\Lambda^{\text{odd}});$$

in this case, the corresponding Gauss-Bonnet formula was first proved by Allendoerfer and Weil [**1**]. In both cases, the integrand is quadratic in curvature, as is forced on one by invariance under rescalings $g \to cg$, where $c > 0$ is any real constant.

Now let g be an arbitrary Riemannian metric on an oriented 4-manifold M, and, by raising an index, identify its curvature tensor with the *curvature operator* $\mathcal{R} : \Lambda^2 \to \Lambda^2$. Decomposing the 2-forms as in (1), this linear endomorphism of Λ^2 can then be decomposed into primitive pieces

$$\mathcal{R} = \left(\begin{array}{c|c} W^+ + \frac{s}{12} & \overset{\circ}{r} \\ \hline \overset{\circ}{r} & W^- + \frac{s}{12} \end{array} \right). \tag{3}$$

Here $W_\pm$ are the trace-free pieces of the appropriate blocks, and are called the self-dual and anti-self-dual Weyl curvatures, respectively. The scalar curvature s is understood to act by scalar multiplication, whereas the trace-free Ricci curvature $\overset{\circ}{r} = r - \frac{s}{4}g$ acts on 2-forms by

$$\varphi_{ab} \mapsto \overset{\circ}{r}_{ac}\, \varphi^c{}_b - \overset{\circ}{r}_{bc}\, \varphi^c{}_a.$$

Each of these curvatures corresponds to a different irreducible representation of $SO(4)$, and so any invariant quadratic polynomial in curvature must be a linear decomposition of s^2, $|\overset{\circ}{r}|^2$, $|W^+|^2$ and $|W^-|^2$, and the signature and Euler characteristic are thus expressible as a linear combination of their integrals. The coefficients, of course, may then be deduced by inspecting a handful of well-chosen examples. Thus the 4-dimensional Gauss-Bonnet formula may explicitly be written as

$$\chi(M) = \frac{1}{8\pi^2}\int_M \left[|W^+|^2 + |W^-|^2 + \frac{s^2}{24} - \frac{|\overset{\circ}{r}|^2}{2}\right] d\mu,$$

whereas the Hirzebruch signature theorem takes the form

$$\tau(M) = \frac{1}{12\pi^2}\int_M \left[|W^+|^2 - |W^-|^2\right] d\mu.$$

Here the curvatures, norms $|\cdot|$, and volume form $d\mu$ are, of course, those of our chosen Riemannian metric g.

In particular, it follows that

$$(4) \qquad (2\chi \pm 3\tau)(M) = \frac{1}{4\pi^2}\int_M \left[2|W_\pm|^2 + \frac{s^2}{24} - \frac{|\overset{\circ}{r}|^2}{2}\right] d\mu.$$

Since the above integrand is non-negative for any Einstein metric, we therefore have the following celebrated result of Thorpe [**60**] and Hitchin [**26**]:

THEOREM 2.2 (Hitchin-Thorpe Inequality). *If the smooth compact oriented 4-manifold M admits an Einstein metric g, then*

$$2\chi(M) \geq 3|\tau(M)|,$$

with equality iff the g-induced connection on one of the bundles $\Lambda^\pm$ is flat.

The last statement follows from the observation [**53**] that the self-dual and anti-self-dual parts of the curvature of Λ^+ are precisely represented by the two left-hand blocks of (3), whereas the two right-hand blocks represent the self-dual and anti-self-dual parts of the curvature of Λ^-. An oriented Riemannian 4-manifold is called *locally hyper-Kähler* if Λ^+ is flat; and Λ^- is therefore flat iff the orientation-reverse of the manifold is locally hyper-Kähler. We will discuss the classification of locally hyper-Kähler manifolds in the next section. For now, suffice it observe that the bundle Λ^+ becomes trivial when pulled back to the universal cover of any locally hyper-Kähler manifold, so that the universal cover must, in particular, be *spin*.

EXAMPLE 2.1. The simply connected non-spin 4-manifold $k\mathbb{CP}_2\#\ell\overline{\mathbb{CP}}_2$ has $\chi = 2 + k + \ell$ and $\tau = k - \ell$, and so cannot admit an Einstein metric unless $4 + 5k > \ell > (k-4)/5$.

It is worth pointing out that the invariant $2\chi + 3\tau$ has an intrinsic importance: it is the first Pontrjagin number of the bundle Λ^+. Indeed, the above description of the curvature of Λ^+ tells us that our integral formula for $2\chi + 3\tau$ thus coincides with the usual integral formula for $p_1(\Lambda^+)$. Notice that the Riemannian connection on Λ^+ is self-dual iff g is Einstein, so the Hitchin-Thorpe inequality is a special case of the celebrated fact that a bundle with self-dual connection must have non-negative *instanton number* [**18**].

Note that the Hitchin-Thorpe inequality only involves homotopy invariants of the 4-manifold in question. Thus, for instance, we could have reached precisely the same conclusion in the above example if M were merely *homeomorphic* to one of the connected sums $k\mathbb{CP}_2\#\ell\overline{\mathbb{CP}}_2$ considered in the above example. On the other hand, the scalar and Weyl terms have effectively been treated as junk terms. My primary aim in this essay will be to describe some interesting new estimates on these terms which allow one to improve on the Hitchin-Thorpe result. At times, however, this will be done at the price of sacrificing the homotopy invariance of the obstruction.

Let me conclude this section by mentioning an amusing elementary interpretation [**53**] of the 4-dimensional Einstein equations. By (3), one sees that a 4-manifold is Einstein iff the curvature operator $\mathcal{R}$ commutes with the Hodge star operator $\star$. But this is clearly the same as asking that the sectional curvature assigned to any 2-plane be the same as that assigned to its orthogonal complement:

$$(M^4, g) \text{ Einstein } \Longleftrightarrow K(P) = K(P^\perp) \ \forall \text{ 2-plane } P \subset TM.$$

Of course, this can also be proved in a completely elementary manner. Indeed, the definition of the Ricci tensor and the symmetries of the Riemann tensor tell one that

$$(r_{11} + r_{22}) - (r_{33} + r_{44}) = 2(R_{1212} - R_{3434})$$

in any orthonormal frame on a 4-manifold. But the left-hand side obviously vanishes for every orthonormal frame iff the eigenvalues of r are all equal.

3. Complex and Almost-Complex Structures

In order to give our discussion some substance, we need to have some examples. The simplest examples of Einstein manifolds are of course the spaces of constant curvature. A much richer and more illuminating family of examples, however, is provided by the Kähler-Einstein manifolds. Let us begin our description of these by first recalling the notion of an almost-complex structure.

An almost-complex structure on a smooth n-manifold M is by definition an endomorphism $J : TM \to TM$ of the tangent bundle such that $J^2 = -1$. Such an object may be thought of as scalar multiplication by $\sqrt{-1}$, and so makes TM into a complex vector bundle, denoted by $T^{1,0}$; in particular, such a structure can exist only if M has even dimension $n = 2m$. Sections of the dual $\Lambda^{1,0}$ of $T^{1,0}$ may concretely be identified with those complex-valued 1-forms on M which convert J into multiplication by i:

$$\phi \in \Lambda^{1,0} \Longleftrightarrow \phi(Jv) = i\phi(v) \ \forall v \in TM.$$

The sections of the rank-m complex vector bundle $\Lambda^{1,0} \to M^{2m}$ are therefore called $(1,0)$-*forms*. More generally, a complex-valued $(p+q)$-form on M is called a (p,q)-form (with respect to J) if it is a section of

$$\Lambda^{p,q} = \wedge^p(\Lambda^{1,0}) \otimes \wedge^q(\overline{\Lambda^{1,0}}).$$

DEFINITION 3.1. Let (M, J) be an almost-complex manifold of real dimension $2m$. The rank 1 bundle

$$K = \Lambda^{m,0} \to M^{2m}$$

is called the *canonical line bundle* of (M, J). Its dual

$$K^{-1} = \wedge^m T^{1,0}$$

is called the *anti-canonical* line bundle.

Notice that we thus have a number of equivalent expressions for the first Chern class of (M, J):

$$c_1(M, J) := c_1(T^{1,0}) = c_1(K^{-1}) = -c_1(K) = -c_1(\Lambda^{1,0}).$$

A Riemannian metric g and an almost-complex structure J on M are said to be *compatible* iff J is an orthogonal transformation with respect to g:

$$g(\cdot, \cdot) = g(J\cdot, J\cdot).$$

This is the same as requiring that the tensor field

$$\omega(\cdot, \cdot) = g(J\cdot, \cdot)$$

be skew-symmetric. When this happens, ω will be called the *associated 2-form* of (g, J). Notice that ω is automatically J-invariant, in the sense that

$$\omega(J\cdot, J\cdot) = \omega(\cdot, \cdot),$$

which is to say that ω is a (real) $(1,1)$-form with respect to J. If J is an almost-complex structure, and if ω is a real $(1,1)$-form, then we may, conversely define a symmetric tensor field g by

$$g(\cdot, \cdot) = \omega(\cdot, J\cdot);$$

if g is positive-definite, it is then a J-compatible metric for which ω is the associated 2-form.

If g is any Riemannian metric on M, and if J is any almost complex structure, then we can produce a J-compatible metric h by setting $h = [g + g(J\cdot, J\cdot)]/2$. But any metric h on M may be uniquely written as $h = g(H\cdot, H\cdot)$, where the 'symmetric' endomorphism H of TM corresponds to yet another Riemannian metric $g(H\cdot, \cdot)$. Since the set of such H's is convex, this provides us with a deformation-retraction $J \mapsto HJH^{-1}$ of the space of almost-complex structures J onto the space of g-compatible almost-complex structures on M. In particular, M admits an almost-complex structure iff it admits some J compatible with any given metric g.

Now if (M, g) is an oriented Riemannian 4-manifold, and if J is a compatible almost-complex structure, then J has matrix

$$(5) \qquad (a) \begin{bmatrix} & -1 & & \\ 1 & & & \\ & & & -1 \\ & & 1 & \end{bmatrix} \quad \text{or} \quad (b) \begin{bmatrix} & -1 & & \\ 1 & & & \\ & & & 1 \\ & & -1 & \end{bmatrix}$$

in an appropriate oriented orthonormal frame $e_1, \ldots, e_4$. The associated 2-form

$$\omega = e^1 \wedge e^2 \pm e^3 \wedge e^4$$

is therefore always either self-dual or anti-self-dual, and has norm $\sqrt{2}$. The self-dual/anti-self-dual distinction amounts to whether or not J determines the given orientation on M. Conversely, every self-dual or anti-self-dual 2-form of norm $\sqrt{2}$

arises from a g-compatible J. Thus a smooth compact oriented 4-manifold M admits an orientation-compatible almost-complex structure iff Λ^+ admits a nowhere-zero section.

In fact, the specification of an almost-complex structure J compatible with g and the orientation gives us a concrete alternate description of Λ^+. Indeed, if $e_1, \dots, e_4$ is an oriented orthonormal frame in which J is given by (5a), then K is spanned by

$$(e^1 + ie^2) \wedge (e^3 + ie^4) = (e^1 \wedge e^3 - e^2 \wedge e^4) + i(e^1 \wedge e^4 + e^2 \wedge e^3),$$

the real and imaginary parts of which are self-dual 2-forms. Thus

$$\Lambda^+ = \mathbb{R}\omega \oplus \Re e K,$$

and

$$\Lambda^+ \otimes \mathbb{C} = \mathbb{C}\omega \oplus K \oplus \bar{K}.$$

In particular,

$$(2\chi + 3\tau)(M) = p_1(\Lambda^+) = -c_2(\mathbb{C} \oplus K \oplus K^{-1}) = [c_1(K^{-1})]^2 = c_1^2(M, J).$$

We also see that the first Chern class satisfies the constraint

$$w_2(M) = w_2(\Lambda^+) = w_2(\Re e K) \equiv c_1(K) \equiv c_1(M, J) \bmod 2.$$

Conversely, if $\alpha \in H^2(M, \mathbb{Z})$ is any element satisfying

$$\alpha^2 = 2\chi + 3\tau \tag{6}$$

$$\alpha \equiv w_2 \bmod 2 \tag{7}$$

we may take K to be a complex line bundle with $c_1(K) = -\alpha$, and notice that $\mathbb{R} \oplus \Re e K$ then has the same characteristic classes p_1 and w_2 as Λ^+. Since these characteristic classes completely classify $SO(3)$-bundles over any 4-manifold [**16**], it follows that Λ^+ has a non-zero section, and that M admits an orientation-compatible almost-complex structure, iff equations (6) and (7) have a solution $\alpha \in H^2(M, \mathbb{Z})$.

EXAMPLE 3.1. The 4-sphere S^4 does not admit an almost-complex structure, since $H^2(S^4) = 0$, whereas $(2\chi + 3\tau)(S^4) = 4 \neq 0$. Notice, by the way, that the rank-3 bundle $\Lambda^+ \to S^4$ therefore does not admit a nowhere-zero section, even though its Euler class $e(\Lambda^+) \in H^3(S^4)$ is of course zero.

An almost-complex structure J on a $2m$-manifold M is said to be *integrable* if there is an atlas of charts on M in which J becomes the standard, constant-coefficient almost-complex structure on $\mathbb{R}^{2m} = \mathbb{C}^m$. For such an atlas, the transition functions are biholomorphisms, and M acquires the structure of a *complex m-manifold.* In this case, we will therefore say that J is a *complex structure* on M. If ∇ is any torsion-free connection on TM, the Newlander-Nirenberg theorem asserts that that the obstruction to integrability is precisely the $(\Lambda^{2,0} \oplus \Lambda^{0,2}) \otimes TM$ component of ∇J; the latter is usually called the *Nijenhuis tensor* or the *Fröhlicher torsion.* An easy partition-of-unity argument therefore shows that J is integrable iff there is a torsion-free connection ∇ such that $\nabla J = 0$.

A Riemannian metric g is said to be Kähler with respect to a compatible almost-complex structure J iff $\nabla J = 0$, where ∇ is now the *Riemannian* (Levi-Civita) connection. When this happens, (M, J) is a complex manifold, per the above discussion. Moreover, the 2-form ω, which is now known as the *Kähler form,* satisfies $\nabla\omega = 0$, and so is closed. Conversely, g is Kähler with respect to J iff J is integrable and ω is closed. Since g is completely determined by J and ω, this allows one to construct all Kähler manifolds as complex manifolds equipped with closed, real, non-degenerate $(1,1)$-forms.

The Kähler concept may be further clarified by a discussion of holonomy. On any Riemannian manifold (M,g), parallel transport around a piece-wise smooth loop γ based at $x \in M$ gives rise to a so-called *holonomy transformation* $L_\gamma : T_xM \to T_xM$. Of course, L_γ is automatically an orthogonal transformation, since Riemannian parallel transport preserves g. The Kähler condition may now be restated as requiring that every L_γ be a *unitary* transformation. (When this happens, the relevant complex structure on T_xM can be declared to be $J|_x$, and this can be uniquely extended to an almost-complex structure J on M by Riemannian parallel transport.) Since curvature just represents parallel transport around infinitesimal loops, it follows the curvature tensor of a Kähler manifold is a 2-form with values in the *skew-Hermitian* endomorphisms of the tangent space. But index-lowering with g identifies the skew-Hermitian endomorphisms of TM with the bundle $\Lambda^{1,1}_{\mathbb{R}}$ of real $(1,1)$-forms. This tells us that the curvature operator $\mathcal{R}$ of a Kähler manifold is just an endomorphism of $\Lambda^{1,1}_{\mathbb{R}}$, since the first Bianchi identity always tells one that $\mathcal{R}$ is self-adjoint.

In particular, the 2-form $\rho = \mathcal{R}(\omega/2)$ is of type $(1,1)$ on any Kähler manifold. Now one can use the first Bianchi identity and the fact that $\nabla\nabla J = 0$ to show that

$$\rho(\cdot,\cdot) = r(J\cdot,\cdot),$$

and the $(1,1)$-form ρ is therefore called the *Ricci form*. On the other hand, ρ represents the half the real trace of the infinitesimal holonomy composed with J, and so is $-i$ times the curvature of the canonical line bundle K with its induced connection. The latter connection is called the *Chern connection*, and can be characterized by the fact that it preserves the induced inner product, and that its $(0,1)$ component is

$$\begin{array}{ccc} \Gamma(K) & \xrightarrow{\bar\partial} & \Gamma(\Lambda^{0,1}\otimes K) \\ \| & & \| \\ \Gamma(\Lambda^{m,0}) & \xrightarrow{d} & \Gamma(\Lambda^{m,1}). \end{array}$$

Because the Ricci tensor and Ricci form are related in exactly the same way as are the metric and Kähler form, a Kähler manifold is Einstein iff

$$\rho = \lambda\omega.$$

When this happens, g is called a compatible *Kähler-Einstein metric* on the complex manifold (M,J), and (M,g,J) is called a Kähler-Einstein manifold. If $\lambda < 0$, this says that K is a 'positive' holomorphic line bundle, and the Kodaira embedding theorem tells us that K is *ample*, meaning that there is a holomorphic embedding of (M,J) in complex projective space defined by the holomorphic sections of $K^{\otimes \ell}$ for any sufficiently large ℓ. If $\lambda = 0$, one instead concludes that $K^{\otimes \ell}$ is holomorphically trivial for some $\ell \neq 0$. Finally, $\lambda > 0$ would imply that K^{-1} is ample.

In the $\lambda > 0$ case, however, the ampleness of K^{-1} is not enough to guarantee the existence of a Kähler-Einstein metric. Indeed, if there *were* such a metric, it would follow [**41**] that the identity component of the biholomorphism group would be a complexification of the identity component of the isometry group. Since the latter group is *compact*, this constrains the Lie algebra of holomorphic vector fields to be a *reductive* Lie algebra. Thus we have one extra necessary condition for the existence of a Kähler-Einstein metric in the $\lambda > 0$ case. But amazingly, the necessary conditions we have described also turn out to be sufficient [**5, 68, 55, 63, 62**] in real dimension 4:

THEOREM 3.1 (Aubin/Yau). *A compact complex manifold (M, J) admits a compatible Kähler-Einstein metric with $\lambda < 0$ iff its canonical line bundle K is ample.*

THEOREM 3.2 (Yau). *A compact complex manifold (M, J) admits a compatible Kähler-Einstein metric with $\lambda = 0$ iff (M, J) admits a Kähler metric and $K^{\otimes \ell}$ is trivial for some positive integer ℓ.*

THEOREM 3.3 (Tian). *A compact complex surface (M^4, J) admits a compatible Kähler-Einstein metric with $\lambda > 0$ iff its Lie algebra of holomorphic vector fields is reductive and its anti-canonical line bundle K^{-1} is ample.*

For further discussion, see the essays by Tian and Yau in this volume.

EXAMPLE 3.2. Consider the Fermat hypersurface

$$\{[u : v : w : z] \in \mathbb{CP}_3 \mid u^k + v^k + w^k + z^k = 0\}$$

of degree k in complex projective 3-space. The canonical line bundle K of such a surface is the restriction of the hyperplane line bundle raised to the power $k - 4$. Moreover, the Lie algebra of holomorphic vector fields is trivial, except for $k = 1$, where it is the reductive Lie algebra $\mathfrak{sl}(3, \mathbb{C})$, and $k = 2$, where it is the reductive Lie algebra $\mathfrak{so}(4, \mathbb{C})$. Thus these complex algebraic surfaces all admit compatible Kähler-Einstein metrics.

Notice that λ has the same sign as $4 - k$. All of these surfaces are simply connected (by the Lefschetz theorem), so we see that knowing the fundamental group alone cannot allow one to predict the sign of the Einstein constant λ.

The first two of these surfaces are just $\mathbb{CP}_2$ and $\mathbb{CP}_1 \times \mathbb{CP}_1$, and their Kähler-Einstein metrics are just the obvious homogeneous ones. The cubic surface $k = 3$ is much more interesting; it is diffeomorphic to $\mathbb{CP}_2 \# 6\overline{\mathbb{CP}}_2$, and its $\lambda > 0$ Kähler-Einstein metric is not known explicitly.

The quartic ($k = 4$) surface has trivial canonical line bundle, and carries Ricci-flat Kähler metrics. Notice that this manifold has $2\chi + 3\tau = c_1^2 = 0$, and so exactly saturates the Hitchin-Thorpe inequality of Theorem 2.2. Generalizations of this quartic, called $K3$ surfaces, will be discussed at length below.

Finally, notice that most of the Einstein manifolds under consideration have $\lambda < 0$. As we let $k \to \infty$, we run through infinitely many different homeotypes. As we will see in a moment, these $k > 4$ surfaces are examples of *surfaces of general type.*

The quartic in $\mathbb{CP}_3$ provides us with the prototypical example of a $K3$ *surface.* By the usual definition [**7**], a compact complex surface is called a $K3$ iff it is simply connected with $c_1 = 0$. (As it turns out, however, a compact complex surface is a $K3$ iff it is diffeomorphic to our quartic prototype.) Every $K3$ admits Kähler metrics [**54**], and in light of Theorem 3.2, therefore admits Ricci-flat Kähler metrics. Now recall that $\Lambda^+ = \mathbb{R} \oplus K$ for a Kähler surface, and K is flat iff the Ricci curvature vanishes. Thus any Ricci-flat Kähler surface is locally hyper-Kähler in the sense of Theorem 2.2. In fact [**26**], this is essentially the general case.

PROPOSITION 3.4 (Hitchin). *Let (M, g) be a compact oriented Einstein 4-manifold with $2\chi + 3\tau = 0$. Then the pull-back of g to some finite cover of M*

is either a Ricci-flat Kähler metric on a $K3$ surface, or else a flat metric on a 4-torus.

PROOF. The proof of Theorem 2.2 tells us that g is Ricci-flat, and induces a flat connection on Λ^+. But the Cheeger-Gromoll splitting theorem asserts that any compact Ricci-flat manifold has universal cover equal to the Riemannian product of a compact, simply connected Ricci-flat manifold with a Euclidean space. Since any Ricci-flat manifold of dimension < 4 is necessarily flat, this tells us that the universal cover $\tilde{M}$ of our 4-manifold M must either be compact, or else is Euclidean. In the latter case, Bieberbach's theorem [**10, 67**] asserts that M is finitely covered by a flat torus.

We are left with the case in which $\tilde{M}$ is compact. But the pulled-back metric $\tilde{g}$ induces a flat connection on Λ^+, and the simple-connectivity of $\tilde{M}$ then guarantees that Λ^+ is then spanned by parallel 2-forms. An arbitrary such form ω of norm $\sqrt{2}$ corresponds to a parallel almost-complex structure J on $\tilde{M}$, and makes $(\tilde{M}, \tilde{g})$ into a Kähler manifold. We then have $\Lambda^+ = \mathbb{R}\omega \oplus K$, and since Λ^+ is flat and trivial, so is K. Thus $(\tilde{M}, J)$ is a $K3$ surface, and $\tilde{g}$ is a compatible Ricci-flat Kähler metric on this $K3$. □

Let us now consider how the Kähler-Einstein complex surfaces fit into Kodaira's general scheme of surface classification. The single most important invariant of a compact complex surface is its *Kodaira dimension.* Let (M^4, J) be a compact complex 2-manifold, and let $K = \Lambda^2(T^{1,0}M)^*$ be its canonical line bundle. For each positive integer ℓ, we have a tautological map $K^{-\ell} \to [\Gamma(M, \mathcal{O}(K^\ell))]^*$ defined by evaluation of a global holomorphic section of K^ℓ on an element of its dual line bundle. This map descends to a holomorphic map $M - B_\ell \to \mathbb{P}([\Gamma(M, \mathcal{O}(K^\ell))]^*)$ with values in a projective space, but at the price of throwing out the *base locus* B_ℓ where all the holomorphic sections of K^ℓ vanish. The Kodaira dimension is defined to be the maximal complex dimension of the image of $M - B_\ell$ as ℓ ranges over the positive integers. Here $\emptyset$ is assigned dimension $-\infty$, so the Kodaira dimension is an element of $\{-\infty, 0, 1, 2\}$. The classification of complex surfaces with Kodaira dimension < 2 and b_1 even is thoroughly understood. A complex surface is said to be of *general type* if its Kodaira dimension is 2.

A following procedure [**7**] provides a simple, beautiful way of modifying a complex surface without changing its Kodaira dimension.

DEFINITION 3.2. Let (M, J) be a compact complex surface, and let $x \in M$ be any point. The *blow-up* of M at x is the unique compact complex surface $(\hat{M}, \hat{J})$ obtained by replacing x with a complex projective line $\mathbb{CP}_1$.

The introduced $\mathbb{CP}_1$ has self-intersection -1, and so is called a (-1)-curve. The blow-up can be explicitly constructed by replacing a small ball around x with a tubular neighborhood of the zero section in the Chern class -1 line bundle over $\mathbb{CP}_1$. Since the one-point compactification of this line bundle is diffeomorphic to $\mathbb{CP}_2$ in an orientation-reversing manner, the blow-up $\hat{M}$ is diffeomorphic to the connected sum $M \# \overline{\mathbb{CP}}_2$. Notice that the blow-up procedure can be iterated as many times as we like, and so gives us complex structures on $M \# k\overline{\mathbb{CP}}_2$ for each positive integer k.

There is an inverse process, called *blowing down.* Indeed, if a complex surface $(\hat{M}, \hat{J})$ contains a $\mathbb{CP}_1$ of self-intersection -1, it is necessarily the blow-up of some other surface. Moreover, one can iterate this procedure until one finally produces a surface without (-1)-curves. (The process must terminate after a finite number of steps because each blow-down reduces b_2 by 1.) A complex surface X without (-1)-curves is called a *minimal surface*. If M is obtained from X by some sequence of blow-ups, we say that X is a *minimal model* for M. If M has Kodaira dimension ≥ 0, moreover, its minimal model is unique.

Using Nakai's criterion, the Kodaira-Enriques classification [**7**] and a result of Siu [**54**], the previous criteria for the existence of Kähler-Einstein metrics can be restated as follows:

COROLLARY 3.5. *Let (M, J) be a compact complex surface. Then the following are equivalent:*

- *(M, J) admits a compatible Kähler-Einstein metric with $\lambda < 0$;*
- *(M, J) has ample canonical line bundle;*
- *$(2\chi + 3\tau)(M) > 0$, and every $\mathbb{CP}_1 \subset (M, J)$ has self-intersection ≤ -3;*
- *(M, J) is minimal, of general type, and contains no (-2)-curves.*

Here a (-2)-curve means a $\mathbb{CP}_1$ of self-intersection -2. If a minimal complex surface of general type contains such curves, we can collapse them all to obtain a complex *orbifold* which has K ample in the orbifold sense. The Aubin/Yau proof then constructs [**29, 64**] a Kähler-Einstein orbifold metric on this so-called pluri-canonical model. This shows that a complex surface is of general type iff it can be obtained from a Kähler-Einstein orbifold with $\lambda < 0$ by resolving the singularities and blowing up.

COROLLARY 3.6. *Let (M, J) be a compact complex surface. Then the following are equivalent:*

- *(M, J) admits a compatible Kähler-Einstein metric with $\lambda = 0$;*
- *(M, J) is finitely covered by a K3 surface or complex torus;*
- *(M, J) is minimal, of Kodaira dimension 0, and has b_1 even.*

COROLLARY 3.7. *Let (M, J) be a compact complex surface. Then the following are equivalent:*

- *(M, J) admits a compatible Kähler-Einstein metric with $\lambda > 0$;*
- *(M, J) has ample anti-canonical line bundle and reductive automorphism algebra;*
- *(M, J) is $\mathbb{CP}_2$, $\mathbb{CP}_1 \times \mathbb{CP}_1$, or the blow-up of $\mathbb{CP}_2$ at k distinct points, $3 \leq k \leq 8$, with no three on a line and no six on a conic.*

While there is no Kähler-Einstein metric on the blow-up of $\mathbb{CP}_2$ at one or two points, there is [**8, 46**] an Einstein metric on the one-point blow-up which is *conformally* Kähler. There is reason to hope that this so-called Page metric on $\mathbb{CP}_2\#\overline{\mathbb{CP}}_2$ has a companion on the two-point blow-up $\mathbb{CP}_2\#2\overline{\mathbb{CP}}_2$. On the other hand, one can show [**37**] that the only compact complex surfaces which might admit Hermitian but non-Kähler Einstein metrics are the blow-ups of $\mathbb{CP}_2$ at one, two, or three points in general position.

4. Seiberg-Witten Estimates

The Hitchin-Thorpe argument treats the L^2 norms of s and W^+ as 'junk terms,' about which one knows nothing except that they are non-negative. Seiberg-Witten theory [**33, 66**], however, provides remarkable information about both these terms

[**36, 38**]. In this section, we will develop the rudiments of Seiberg-Witten theory, and explore some of its ramifications regarding the scalar curvature.

Let (M, g) be a compact oriented Riemannian 4-manifold, and suppose that M admits an almost-complex structure. As we saw in §3, we can then find almost complex structures J which are compatible with g in the sense that $J^*g = g$. Choose such a J, and consider the rank-2 complex vector bundles

$$V_+ = \Lambda^{0,0} \oplus \Lambda^{0,2} \tag{8}$$

$$V_- = \Lambda^{0,1}. \tag{9}$$

Notice that g induces canonical Hermitian inner products on these bundles.

As described, these bundles depend on the choice of a particular almost-complex structure, but they have a deeper meaning [**26**] that is invariant under deformations of J. Indeed, on any contractible open subset of M one can define Hermitian vector bundles

$$\begin{array}{ccc} \mathbb{C}^2 & \to & \mathbb{S}_\pm \\ & & \downarrow \\ & & M \end{array}$$

called spin bundles, characterized by the fact that their determinant line bundles $\wedge^2\mathbb{S}_\pm$ are canonically trivial and that their projectivizations

$$\begin{array}{ccc} \mathbb{CP}_1 & \to & \mathbb{P}(\mathbb{S}_\pm) \\ & & \downarrow \\ & & M \end{array}$$

are exactly the unit 2-sphere bundles $S(\Lambda^\pm)$. On the other hand, one cannot generally define the bundles $\mathbb{S}_\pm$ globally on M; manifolds on which this can be done are called *spin*, and are characterized by the vanishing of the Stiefel-Whitney class $w_2 = w_2(TM)$. However, our bundles $V_\pm$ still satisfy

$$\mathbb{P}(V_\pm) = S(\Lambda^\pm),$$

and we formally have

$$V_\pm = \mathbb{S}_\pm \otimes L^{1/2},$$

where the Hermitian complex line bundle $L = \wedge^2 V_\pm$ is just the anti-canonical line-bundle K^{-1} associated with J.

The isomorphism class $\mathfrak{c}$ of such a choice of $V_\pm$ is called a *spin^c structure* on M. The cohomology group $H^2(M, \mathbb{Z})$ acts freely and transitively on the spin^c structures by tensoring $V_\pm$ with complex line bundles. Each spin^c structure has a first Chern class $c_1 := c_1(L) = c_1(V_\pm) \in H^2(M, \mathbb{Z})$ such that

$$c_1 \equiv w_2 \mod 2, \tag{10}$$

and the previously mentioned $H^2(M, \mathbb{Z})$-action induces the action $c_1 \mapsto c_1 + 2\alpha$, $\alpha \in H^2(M, \mathbb{Z})$, on first Chern classes. Thus, if $H^2(M, \mathbb{Z})$ has trivial 2-torsion — as can always be arranged by replacing M with a finite cover — the spin^c structures are precisely in one-to-one correspondence with the set of cohomology classes $c_1 \in H^2(M, \mathbb{Z})$ satisfying (10). A spin^c structures $\mathfrak{c}$ arises from some almost-complex structure J iff its first Chern class satisfies the additional constraint

$$c_1^2 = 2\chi + 3\tau.$$

It is with these spin^c structures of *almost-complex type* we will concern ourselves here.

The Levi-Civita connection ∇ of g naturally induces Hermitian connections on the locally defined bundles $\mathbb{S}_{\pm}$. Given a spin^c structure $\mathfrak{c}$ and a Hermitian connection A on the anti-canonical line bundle L, we therefore have induced Hermitian connections ∇_A on $V_{\pm}$. On the other hand, there is a canonical isomorphism $\Lambda^1 \otimes \mathbb{C} = \text{Hom}\,(\mathbb{S}_+, \mathbb{S}_-)$, so that $\Lambda^1 \otimes \mathbb{C} \cong \text{Hom}\,(V_+, V_-)$ for any spin^c structure, and this induces a canonical homomorphism

$$\cdot\, : \Lambda^1 \otimes V_+ \to V_-$$

called *Clifford multiplication.* Composing these operations allows us to define a so-called *twisted Dirac operator*

$$D_A : \Gamma(V_+) \longrightarrow \Gamma(V_-)$$

by $D_A\Phi = \nabla_A \cdot \Phi$. This is an elliptic operator of index

$$\text{ind}_{\mathbb{C}} D_A = \dim_{\mathbb{C}} \ker D_A - \dim_{\mathbb{C}} \ker D_A^* = \frac{c_1^2 - \tau(M)}{8}.$$

If $\mathfrak{c}$ is of almost-complex type, this number becomes the *Todd genus* $(\chi + \tau)/4$ of the almost-complex manifold (M, J).

EXAMPLE 4.1. Let (M, g, J) be a Kähler manifold of complex dimension 2. Let $\mathfrak{c}$ be the spin^c structure induced by J, and let A be the usual (Chern) connection on the anti-canonical line bundle $L = K^{-1}$. Then

$$D_A = \sqrt{2}(\overline{\partial} \oplus \overline{\partial}^*) : \Gamma(\Lambda^{0,0} \oplus \Lambda^{0,2}) \to \Gamma(\Lambda^{0,1})$$

where $\overline{\partial}$ is the Dolbeault operator and $\overline{\partial}^*$ is its formal adjoint. In particular, the index of D_A is just the alternating sum of the dimensions of the Dolbeault cohomology groups $H^{0,q}(M)$, and our (Noether) formula for its index (the Todd genus) is an elementary consequence of the Hodge decomposition.

For any spin^c structure, we have already noted that there is a canonical diffeomorphism $\mathbb{P}(V_+) \xrightarrow{\cong} S(\Lambda^+)$. In polar coordinates, we now use this to define the angular part of a unique continuous map

$$\sigma : V_+ \to \Lambda^+$$

with

$$|\sigma(\Phi)| = \frac{1}{2\sqrt{2}}|\Phi|^2.$$

This map is actually real-quadratic on each fiber of V_+; indeed, assuming our spin^c structure is induced by a complex structure J, then, in terms of (8), σ is explicitly given by

$$\sigma(f, \phi) = (|f|^2 - |\phi|^2)\frac{\omega}{4} + \Im m(\bar{f}\phi),$$

where $f \in \Lambda^{0,0}$, $\phi \in \Lambda^{0,2}$, and where $\omega(\cdot,\cdot) = g(J\cdot,\cdot)$ is the associated 2-form of (M, g, J).

We are now in a position to introduce the Seiberg-Witten equations

$$D_A\Phi = 0 \tag{11}$$

$$F_A^+ = i\sigma(\Phi), \tag{12}$$

where the unknowns are a Hermitian connection A on L and a section Φ of V_+. Here F_A^+ is the self-dual part of the curvature of A, and so is a purely imaginary 2-form. A solution of (11–12) is said to be *reducible* if $\Phi \equiv 0$; otherwise, it is called *irreducible*.

EXAMPLE 4.2. Let (M, g, J) be a Kähler surface of scalar curvature $s \equiv -1$. Let $\Phi = (1,0) \in \Lambda^{0,0} \oplus \Lambda^{0,2}$, and let A be the Chern connection on $L = K^{-1}$. Since $F_A = -i\rho = -ir(J\cdot, \cdot)$, its self-dual part corresponds to the trace piece of the Ricci tensor, and so is given by $F_A^+ = -is\omega/4 = i\omega/4$. On the other hand, $D_A\Phi = \bar\partial(1) + \bar\partial^*(0) = 0$, and $\sigma(\Phi) = \sigma(1,0) = \omega/4$. Thus (Φ, A) is an irreducible solution of the Seiberg-Witten equations (11–12).

The geometric character of the Seiberg-Witten equations makes them invariant under automorphisms of L. Thus the so-called *gauge group* of smooth maps $f : M \to S^1$ acts on the space of smooth solutions of the Seiberg-Witten equations (11–12) by

$$(A, \Phi) \mapsto (A - 2f^{-1}df, f\Phi).$$

Notice that this action is free on the set of irreducible solutions, whereas the stabilizer is precisely S^1 if $\Phi \equiv 0$. In particular, the solution space is always either infinite-dimensional or empty. However, one can compensate for the action of the gauge group by choosing some back-ground connection A_0 and then imposing the *gauge-fixing condition*

$$d^*(A - A_0) = 0. \tag{13}$$

The system (11–13) is then elliptic, and the solution space is finite dimensional. There is still a residual part of the action of the gauge group; the constant S^1-valued functions still act, and after modding out by these there is still an action of the discrete group $H^1(M, \mathbb{Z})$ of homotopy classes of maps $M \to S^1$. After dividing out by these, however, we obtain the *moduli space* $\mathfrak{M}_{c,g}$, which is by definition the set solutions of (11–12) modulo the action of the gauge group.

Not only is this moduli space finite-dimensional — it is also compact. This is because (11–12) imply the *Weitzenböck formula*

$$0 = 4\nabla_A^*\nabla_A\Phi + s\Phi + |\Phi|^2\Phi. \tag{14}$$

Taking the inner product with Φ, we have

$$0 = 2\Delta|\Phi|^2 + 4|\nabla_A\Phi|^2 + s|\Phi|^2 + |\Phi|^4,$$

and at the *maximum* of $|\Phi|^2$ we therefore have

$$0 \geq 4|\nabla_A\Phi|^2 + |\Phi|^2(s + |\Phi|^2),$$

so that any irreducible solution must satisfy the C^0 estimate

$$|\Phi|^2 \leq -\min s. \tag{15}$$

Moreover, equality can only occur at points where $\nabla_A\Phi = 0$. In particular, one has uniform L^p-bounds on Φ for all solutions, and compactness therefore follows **[33, 42]** via the L^p versions of the Gårding inequality for (11–13) and the Rellich lemma.

Now consider a perturbed versions of the Seiberg-Witten equations, obtained by replacing (12) with

$$iF_A^+ + \sigma(\Phi) = \varepsilon, \tag{16}$$

where ε is some self-dual 2-form. For generic ε, Smale's infinite-dimensional version of Sard's theorem implies that the corresponding 'perturbed' moduli space $\mathfrak{M}_{c,g,\varepsilon}$ is a smooth manifold whose dimension is given by the (real) index[1] of the linearization of (11–13), which is to say that

$$
\begin{aligned}
\dim \mathfrak{M}_{\mathfrak{c},g,\varepsilon} &= (b_1 - 1 - b^+) + 2\,\mathrm{ind}_{\mathbb{C}} D_A \\
&= -\frac{\chi+\tau}{2} + 2\left(\frac{c_1^2-\tau}{8}\right) \qquad (17)\\
&= \frac{c_1^2 - (2\chi+3\tau)}{4}.
\end{aligned}
$$

If our spin^c structure $\mathfrak{c}$ is of almost-complex type, the moduli space is therefore discrete. Moreover, a slight variation on the previous Weitzenböck argument shows that these moduli spaces are compact. Again assuming that our spin^c structure is of almost-complex type, the moduli spaces $\mathfrak{M}_{\mathfrak{c},g,\varepsilon}$ are thus finite for generic ε.

As we vary g and ε, the moduli spaces remain cobordant as long as one can avoid hitting reducible solutions. Now a reducible solution can only occur when the self-dual part $2\pi c_1^+$ of the harmonic representative of $2\pi c_1 = [iF_A]$ agrees with the harmonic part of ε. Since $(c_1^+)^2 \geq c_1^2$, it follows that we can avoid reducible solutions if we assume that

$$
2\chi + 3\tau = c_1^2 > 0, \qquad (18)
$$

and if, for each metric, we only consider ε with L^2 norm smaller than $2\pi\sqrt{c_1^2}$. Thus (18) is enough to guarantee that we have a cobordism class of $\mathfrak{M}_{\mathfrak{c},g,\varepsilon}$ determined by the smooth structure of M and the spin^c structure $\mathfrak{c}$ alone.

DEFINITION 4.1. Let $(M, \mathfrak{c})$ be a smooth compact 4-manifold, equipped with the spin^c structure and orientation determined by some almost-complex structure J. Assume that (18) holds. Then the (mod 2) Seiberg-Witten invariant $n_{\mathfrak{c}}(M) \in \mathbb{Z}_2$ is defined to be

$$
n_{\mathfrak{c}}(M) = \#\mathfrak{M}_{\mathfrak{c},g,\varepsilon} \bmod 2,
$$

where g is any Riemannian metric on M and ε is a generic self-dual form of small L^2 norm on (M, g).

Notice that (18) implies that $b_+(M) \geq 1$. On the other hand, if $b_+(M) \geq 2$, the set of ε for which there is a reducible solution has codimension ≥ 2; it is then easy to see that the generic moduli spaces $\mathfrak{M}_{\mathfrak{c},g,\varepsilon}$ are all cobordant, and one can thus define the Seiberg-Witten invariant even if (18) fails. However, the Hitchin-Thorpe inequality makes (18) a very natural hypothesis for investigations concerning Einstein manifolds, and adopting it here will enable us to treat the $b^+ = 1$ and $b^+ \geq 2$ cases simultaneously.

We have now defined an elegant invariant of a smooth compact 4-manifold by counting solutions of a non-linear system of partial differential equations. But have we merely given a complicated definition of zero? Fortunately not!

THEOREM 4.1 (Witten/Kronheimer). *Let (M, J) be a complex surface of general type for which (18) holds. Then $n_{\mathfrak{c}}(M) \neq 0$, where $\mathfrak{c}$ is the spin^c structure induced by J.*

---

[1]This dimension count actually involves a subtle cancellation which is often overlooked. Namely, the contribution due to the 1-dimensional cokernel of $d^* : \Gamma(\Lambda^1) \,!\, \Gamma(\Lambda^0)$ is canceled out by the action of the 1-dimensional group S^1 of constant gauge transformations.

For simplicity, let us just sketch a proof assuming that (M, J) satisfies any of the equivalent conditions catalogued by Corollary 3.5. There is then a J-compatible Kähler-Einstein metric g on M of scalar curvature $s \equiv -1$, and hence there is an irreducible solution of the Seiberg-Witten equations obtained by taking $\Phi = (1, 0) \in \Gamma(\Lambda^{0,0} \oplus \Lambda^{0,2})$ and letting A be the Chern connection on $L = K^{-1}$. It is not hard to see that the linearization of (11–12) is surjective at this solution, so it suffices to show that any other solution is gauge-equivalent to the constructed one. But the C^0 estimate (15) implies that

$$8 \int_M |F_A^+|^2 d\mu = \int_M |\Phi|^4 d\mu \leq \int_M s^2 d\mu = 32\pi^2 c_1^2(M),$$

and equality can only hold if $|\Phi|^2 \equiv -s = 1$ and $\nabla_A \Phi = 0$. But since A is a connection on $L = K^{-1}$, we have

$$4\pi^2 c_1^2 = \int_M \left[|F_A^+|^2 - |F_A^-|^2\right] d\mu,$$

and the reverse inequality also holds. Thus Φ has unit length, and is parallel with respect to ∇_A. It follows that (A, Φ) is gauge-equivalent to our explicit solution.

If (18) holds and the Seiberg-Witten invariant $n_{\mathfrak{c}}(M)$ is non-zero, it follows that the unperturbed equations (11–12) have an irreducible solution for each metric. Via the Weitzenböck formula (14), this severely constrains the geometry of Riemannian metrics on the given manifold:

THEOREM 4.2. *Let $(M, \mathfrak{c})$ be a smooth compact oriented 4-manifold with a spin^c structure of almost-complex type. Assume that (18) holds, and that $n_{\mathfrak{c}}(M) \neq 0$. Let g be any Riemannian metric on M. Then the scalar curvature s_g of g is negative somewhere, and*

$$\int_M s_g^2 d\mu_g \geq 32\pi^2 (c_1^+)^2,$$

with equality only if g has constant negative scalar curvature, and is Kähler with respect to a $\mathfrak{c}$-compatible complex structure.

PROOF. The C^0 estimate (15) for any irreducible solution of (11–12) forces any metric to satisfy $\min s < 0$.

Now any conformal class of metrics $[g] = \{u^2 g | u : M \to \mathbb{R}^+\}$ contains a metric of constant scalar curvature [**40, 51**], and when the constant is negative, such a metric is unique up to scale. Moreover, such a metric minimizes the functional $\int s^2 d\mu$ within its conformal class. It thus suffices to prove the stated lower bound for $\int s^2 d\mu$ assuming that g has constant scalar curvature.

Now when s is constant, the C^0 estimate (15) takes the form

$$|\Phi|^2 \leq -s,$$

and if equality holds identically we have $\nabla_A \Phi \equiv 0$. Thus

$$\int s^2 d\mu \geq \int |\Phi|^2 d\mu = 8 \int |F_A^+|^2 d\mu \geq 32\pi^2 (c_1^+)^2.$$

If equality holds, $\nabla \sigma(\Phi) \equiv 0$, and the metric is Kähler with respect to a complex structure J for which $\sigma(\Phi)$ is a constant positive multiple of the Kähler form ω. Now

$$V_+ = (\Lambda^{0,0} \oplus \Lambda^{0,2}) \otimes E$$

for some Hermitian line bundle E with connection, and $\Phi \neq 0$ is a parallel section of $\Lambda^{0,0} \otimes E \subset V_+$. Thus E is trivial, and $\mathfrak{c}$ is exactly the spin^c structure determined by J. □

This has an immediate application to the sign-of-the-Einstein-constant problem. Recall that the Hitchin-Thorpe inequality implies that if a 4-manifold admits a Kähler-Einstein metric with $\lambda = 0$, any other Einstein metric on M also has $\lambda = 0$. Seiberg-Witten theory implies the analogous conclusion in the negative case:

COROLLARY 4.3. *Let M be a smooth compact 4-manifold which admits a* Kähler-*Einstein metric g with $\lambda < 0$. Then any other Einstein metric $\tilde{g}$ on M also has $\lambda < 0$.*

PROOF. Give M the orientation and spin^c structure $\mathfrak{c}$ induced by the complex structure. Then (18) holds by the Hitchin-Thorpe inequality, and has $n_{\mathfrak{c}}(M) \neq 0$ by Theorem 4.1. Theorem 4.2 thus tells us that $\tilde{g}$ cannot have $s \geq 0$, and so must have $\lambda < 0$. □

But we also have the following remarkable estimate:

COROLLARY 4.4. *Let M be an oriented smooth compact 4-manifold equipped with a spin^c structure $\mathfrak{c}$ of almost-complex type. Assume that (18) holds, and that $n_{\mathfrak{c}}(M) \neq 0$. Then any Riemannian metric g on M satisfies*

$$\frac{1}{32\pi^2}\int_M s_g^2 d\mu_g \geq (2\chi + 3\tau)(M),$$

with equality iff g is Kähler-Einstein.

PROOF. One has

$$(c_1^+)^2 \geq c_1^2 = 2\chi + 3\tau,$$

with equality iff the harmonic representative of c_1 is self-dual. But for a Kähler metric of constant scalar curvature, the Ricci form ρ is the harmonic representative of $2\pi c_1$, and ρ is self-dual iff the metric is Einstein. The claim therefore follows from Theorem 4.2. □

This implies a Riemannian version [**35**] of the so-called Miyaoka-Yau [**7, 68**] inequality:

THEOREM 4.5 (LeBrun). *Let (M, g) be a non-flat compact Einstein 4-manifold which admits an almost-complex structure. Give M the induced orientation and spin^c structure $\mathfrak{c}$, and assume that $n_{\mathfrak{c}}(M) \neq 0$. Then Euler characteristic χ and signature τ of M satisfy*

$$\chi \geq 3\tau,$$

with equality only if the universal cover of (M, g) is complex-hyperbolic 2-space $\mathbb{C}\mathcal{H}_2 := SU(2,1)/U(2)$, with a constant multiple of its standard metric.

PROOF. By Corollary 4.4 and the Gauss-Bonnet formula, we have

$$\begin{aligned} 3(2\chi - 3\tau)(M) &= \frac{3}{4\pi^2}\int_M \left(2|W^-|^2 + \frac{s^2}{24}\right) d\mu \\ &\geq \frac{1}{32\pi^2}\int_M s^2 d\mu \\ &\geq (2\chi + 3\tau)(M), \end{aligned}$$

with equality only if the metric is Kähler-Einstein and $W^- \equiv 0$.

If the latter happens, $\Re e K$, $\mathbb{R}\omega$, and Λ^- are eigenspaces of the curvature operator $\mathcal{R}$, with respective eigenvalues 0, $s/4$, and $s/12$. Since s is constant and ω is parallel, this implies that $\nabla\mathcal{R} \equiv 0$, and g is therefore locally symmetric. Since $2\chi + 3\tau > 0$ by the Hitchin-Thorpe inequality, the assumption that $n_{\mathfrak{c}} \neq 0$ forces s to be negative, and the point-wise form of $\mathcal{R}$ then tells us that the universal cover is isometric to a rescaled version of the symmetric space $\mathbb{C}\mathcal{H}_2$. □

In particular, we get a uniqueness result [**35**]:

COROLLARY 4.6 (LeBrun). *Let $M = \mathbb{C}\mathcal{H}_2/\Gamma$ be a compact complex-hyperbolic 4-manifold, and let g_0 be its tautological metric. Then every Einstein metric g on M is of the form $g = \varphi^* c g_0$, where $\varphi : M \to M$ is a diffeomorphism and $c > 0$ is a constant.*

PROOF. Because M carries a tautological Kähler-Einstein metric with $\lambda < 0$, Theorem 4.1 guarantees that M has a non-trivial Seiberg-Witten invariant $n_{\mathfrak{c}}$. Up to rescaling, any Einstein metric g on M is therefore complex-hyperbolic by Theorem 4.5. But Mostow rigidity [**43**] tells us that the fundamental group of a complex hyperbolic 4-manifold determines the manifold up to isometry, so the result follows. □

For constructions of compact complex-hyperbolic hyperbolic 4-manifolds, see [**13, 43**]. For a generalization of the above result to the non-compact, finite-volume setting, see [**11**].

The analogous result for real-hyperbolic 4-manifolds is also true, and indeed was proved by Besson-Courtois-Gallot [**9**] several months before the discovery of the above result [**35**]. However, as will explained in §9, the proof of the real-hyperbolic result proceeds upon completely different lines.

Seiberg-Witten theory also allows one to prove non-existence results for Einstein metrics [**36**]. To see this, first notice that Theorem 4.4 can be sharpened in the following direction:

THEOREM 4.7. *Let M be a 4-manifold which has a smooth connected sum decomposition*

$$M = X \# \underbrace{\overline{\mathbb{CP}}_2 \# \cdots \# \overline{\mathbb{CP}}_2}_{k}$$

for some positive integer k. Assume that (18) holds, and suppose that M has a spin^c structure of almost-complex type for which $n_{\mathfrak{c}}(M) \neq 0$. Then every Riemannian metric g on M satisfies

$$\frac{1}{32\pi^2}\int_M s_g^2 d\mu_g > (2\chi + 3\tau)(X) = (2\chi + 3\tau)(M) + k.$$

PROOF. The first Chern class c_1 of $\mathfrak{c}$ can be uniquely expressed as

$$c_1 = c_1(X) + \sum_{j=1}^{k} a_j E_j,$$

where $c_1(X) \in H^2(X, \mathbb{Z})$ and the E_j are generators for the k relevant copies of $H^2(\overline{\mathbb{CP}}_2, \mathbb{Z})$. Since $c_1 \equiv w_2 \bmod 2$ and $E_j \cdot E_j = -1$, the integers a_j are all necessarily odd. Let g be the arbitrary metric on M which we wish to examine, and let us fix our choice of generators E_j so that $c_1(X)^+ \cdot E_j \geq 0$ for all j.

Now notice that complex conjugation $[z^1 : z^2 : z^3] \mapsto [\bar{z}^1 : \bar{z}^2 : \bar{z}^3]$ defines a self-diffeomorphism of $\mathbb{CP}_2$ with non-empty fixed-point set which acts by -1 on H^2. Using this as a model, one can construct self-diffeomorphisms of M which act trivially on $H^2(X)$ and reverse the sign of exactly one E_j. Moving $\mathfrak{c}$ by a sequence of such diffeomorphisms, one can thus obtain a spin^c structure $\hat{\mathfrak{c}}$ with $n_{\hat{\mathfrak{c}}} \neq 0$ and

$$\widehat{c_1} = c_1(X) + \sum_{j=1}^{k} |a_j| E_j.$$

Thus

$$\begin{aligned}
(\widehat{c_1}^+)^2 &= [c_1(X)^+]^2 + (\sum |a_j| E_j^+)^2 + 2\sum |a_j| (c_1(X)^+ \cdot E_j) \\
&\geq [c_1(X)^+]^2 \\
&\geq [c_1(X)]^2 \\
&= \widehat{c_1}^2 + \sum a_j^2 \\
&\geq (2\chi + 3\tau)(M) + k.
\end{aligned}$$

Theorem 4.2 therefore tells us that

$$\frac{1}{32\pi^2}\int_M s^2 d\mu \geq (2\chi + 3\tau)(M) + k = (2\chi + 3\tau)(X).$$

If equality held, g would be necessarily be Kähler with respect to a $\hat{\mathfrak{c}}$-compatible complex structure. Also, our specification of $\hat{\mathfrak{c}}$ would have not been unique, since we would also have $[c_1(X)^+] \cdot E_j = 0$ for all j. The same reasoning could thus be applied to 2^k different spin^c structures to give us 2^{k+1} different parallel complex structures, and Λ^+ would therefore have to be flat and trivial. But any two parallel sections would then by deformation-equivalent, and hence determine the same spin^c structure, contradicting the construction. Hence the inequality is necessarily strict. □

In particular, we get the following non-existence result:

THEOREM 4.8 (LeBrun). *Let X be a minimal complex algebraic surface of general type, and let $M = X\#k\overline{\mathbb{CP}}_2$ be obtained from X by blowing up $k > 0$ points. If $k \geq \frac{2}{3}(2\chi + 3\tau)(X)$, then M does not admit Einstein metrics.*

PROOF. If M admits an Einstein metric g, it satisfies (18) by the Hitchin-Thorpe inequality, and has a non-trivial Seiberg-Witten invariant by Theorem 4.1. Using the scalar curvature estimate of Theorem 4.7 and the Gauss-Bonnet formula (4), we have

$$\begin{aligned}
(2\chi + 3\tau)(X) - k &= (2\chi + 3\tau)(M) \\
&= \frac{1}{4\pi^2}\int_M \left(2|W^+|^2 + \frac{s^2}{24}\right) d\mu \\
&\geq \frac{1}{3\cdot 32\pi^2}\int_M s^2 d\mu \\
&> \frac{1}{3}(2\chi + 3\tau)(X),
\end{aligned}$$

so that

$$\frac{2}{3}(2\chi + 3\tau)(X) > k.$$

Hence M cannot admit an Einstein metric if $k \geq \frac{2}{3}(2\chi + 3\tau)(X)$. □

COROLLARY 4.9 (LeBrun). *Even up to homeomorphism, there are infinitely many smooth compact* simply connected *4-manifolds which satisfy the strict Hitchin-Thorpe inequality* $2\chi > 3|\tau|$, *but nevertheless do not admit Einstein metrics.*

PROOF. If X is any minimal complex surface of general type with $2\chi + 3\tau \geq 3$, there is then at least one integer k satisfying $(2\chi + 3\tau)(X) > k \geq \frac{2}{3}(2\chi + 3\tau)(X)$. The complex surface $M = X \# k\overline{\mathbb{CP}}_2$ then satisfies $2\chi > 3|\tau|$, but does not admit Einstein metrics by the above result. Theorem 4.9 therefore follows by considering the sequence of X's given by the hypersurfaces of degree > 4 in $\mathbb{CP}_3$. □

It should be pointed out that, even when $b^+ = 1$, Seiberg-Witten theory can be used to prove results along the lines of Theorem 4.7 without assuming (18). However, the proofs are complicated by the metric-dependence of the moduli spaces, and one is saved only by considering different spin^c structures for different metrics. For details, see [**20, 36**].

It should also be observed that the results of this section really depend only on the existence of solutions of the Seiberg-Witten equations for *each metric* on the given manifold. This may occur even when $n_{\mathfrak{c}} \in \mathbb{Z}_2$ vanishes. In particular, it turns out that the moduli spaces $\mathfrak{M}_{\mathfrak{c},g}$ can be oriented in a natural way, and this gives rise to an invariant $SW_{\mathfrak{c}} \in \mathbb{Z}$ whose mod 2 reduction is $n_{\mathfrak{c}}$. One can also generalize the definition of $SW_{\mathfrak{c}}$ so as to allow [**59**] for spin^c structures which are not of almost-complex type. Some of these invariants turn out, moreover, to be non-trivial [**58**] on any *symplectic* 4-manifold with $b_+ \geq 2$.

On the other hand, Kronheimer [**34**] recently showed that certain 4-manifolds with $SW \equiv 0$ nonetheless admit solutions of the Seiberg-Witten equations for each and every metric. A different construction of such examples, with direct relevance to the theory of Einstein manifolds, is described in §5 below. In any case, it would seem that the Seiberg-Witten equations have ramifications for the theory of Einstein manifolds which in contexts beyond the scope of the invariants which have been explored to date.

5. Surgery and Scalar Curvature

We have already observed that lower bounds for the L^2 norm of the scalar curvature have natural applications to the theory of Einstein metrics on 4-manifolds. Let us consider such bounds in a broader context. If M is a smooth compact n-manifold, consider the diffeomorphism invariant

$$\mathcal{I}(M^n) = \inf_g \int_M |s_g|^{n/2} d\mu_g,$$

where the infimum is taken over all metrics on M. Notice that choice of the power $n/2$ is dictated by scale invariance; for any other power, the analogous infimum would perforce be zero.

The invariant $\mathcal{I}$ is well behaved with respect to under surgeries in high codimension [**49**]; this fact is essentially a quantitative refinement of results of Gromov-Lawson [**22**] and Schoen-Yau [**52**] concerning metrics of positive scalar curvature. Recall that if M is any smooth compact n-manifold, and if $S^q \subset M$ is a smoothly embedded q-sphere with trivial normal bundle, we may construct a new n-manifold

$\widehat{M}$ by replacing a tubular neighborhood $S^q \times \mathbb{R}^{n-q}$ of S^q with $S^{n-q-1} \times \mathbb{R}^{q+1}$. One then says $\widehat{M}$ is obtained from M by performing a *surgery* in codimension $n-q$ (or dimension q). This operation precisely describes the way that level sets of a Morse function change as one passes a critical point of index $q+1$, and two manifolds are therefore cobordant iff one can be obtained from the other by such a sequence of surgeries.

PROPOSITION 5.1 (Petean-Yun). *Let M be any smooth compact n-manifold, and let $\widehat{M}$ be obtained from M by performing a surgery in codimension* ≥ 3. *Then*

$$\mathcal{I}(\widehat{M}) \leq \mathcal{I}(M).$$

PROOF. We may assume that $n \geq 3$, as otherwise there is nothing to prove. But with this assumption, $\mathcal{I}$ can be rewritten as

$$\mathcal{I}(M) = \inf_g \int_M |s_{-g}|^{n/2} d\mu_g,$$

where

$$s_- = \begin{cases} 0 & s \geq 0 \\ s & s \leq 0. \end{cases}$$

Indeed, if M admits a metric of positive scalar curvature, it also [**5**] admits a metric with $s \equiv 0$, so both infima vanish. If, on the other hand, M does *not* admit a metric of positive scalar curvature, both functionals [**9**] are minimized in each conformal class by a metric of constant scalar curvature ≤ 0, and the claim is then an immediate consequence. .

Now let g be a metric on M such that

$$\int_M |s_{-g}|^{n/2} d\mu_g < \mathcal{I}(M) + \frac{\epsilon}{2},$$

and suppose that $S^q \subset M$ is any embedded sphere of codimension $n-q \geq 3$. By making a conformal change which is trivial outside a small tubular neighborhood of the sphere, one may produce a conformally related metric $\tilde{g} = ug$ which has positive scalar curvature along S^q, but still satisfies

$$\int_M |s_{-\tilde{g}}|^{n/2} d\mu_{\tilde{g}} < \mathcal{I}(M) + \epsilon.$$

But on the manifold $\widehat{M}$ obtained by surgery on S^q, a celebrated local construction of Gromov-Lawson [**22**] then gives us a metric $\hat{g}$ which has positive scalar curvature in the surgered region, and agrees with $\tilde{g}$ on the set where $s_{\tilde{g}} \leq 0$. Thus

$$\int_{\widehat{M}} |s_{-\hat{g}}|^{n/2} d\mu_{\hat{g}} = \int_M |s_{-\tilde{g}}|^{n/2} d\mu_{\tilde{g}} < \mathcal{I}(M) + \epsilon,$$

so that

$$\mathcal{I}(\widehat{M}) = \inf_{\hat{g}} \int_{\widehat{M}} |s_{-\hat{g}}|^{n/2} d\mu_{\hat{g}} \leq \mathcal{I}(M),$$

as claimed. □

Because any surgery can be undone by another surgery, this implies [**47**]

COROLLARY 5.2 (Petean). *If M is any smooth compact 4-manifold, then*

$$\mathcal{I}(M \# [S^1 \times S^3]) = \mathcal{I}(M).$$

PROOF. One may obtain $S^1 \times S^3$ from S^4 by a surgery in codimension 4, and S^4 may be obtained from $S^1 \times S^3$ by a surgery in codimension 3. Taking connected sums with M, we see that $M = M\#S^4$ and $M\#[S^1 \times S^3]$ can each be obtained from the other by a surgery in codimension ≥ 3. By Theorem 5.1, it follows that

$$\mathcal{I}(M) \leq \mathcal{I}(M\#[S^1 \times S^3]) \leq \mathcal{I}(M),$$

and the result follows. □

This and Theorem 4.7 now imply [**47**]

THEOREM 5.3. *Let $X = \mathbb{C}\mathcal{H}_2/\Gamma$ be any compact complex-hyperbolic 4-manifold. Then $X\#\ell(S^1 \times S^3)$ does not admit Einstein metrics for any $\ell > 0$. Moreover, $X\#k\overline{\mathbb{CP}}_2\#\ell(S^1 \times S^3)$ does not admit Einstein metrics for any $\ell \geq \frac{5}{4}k > 0$.*

PROOF. Let $M = X\#k\overline{\mathbb{CP}}_2$ and $\widehat{M} = M\#\ell(S^1 \times S^3) = X\#k\overline{\mathbb{CP}}_2\#\ell(S^1 \times S^3)$ for some $k \geq 0$ and $\ell > 0$. If $\widehat{M}$ admitted an Einstein metric, then the Hitchin-Thorpe inequality would then tell us that

$$(2\chi + 3\tau)(M) = (2\chi + 3\tau)(\widehat{M}) + 4\ell > 0,$$

and M would thus satisfy (18). But M is the underlying 4-manifold of a complex surface of general type, and hence has a non-zero Seiberg-Witten invariant by Theorem 4.1. Thus Theorem 4.2 tells us that

$$\mathcal{I}(M) = \inf_g \int_M s^2 d\mu \geq 32\pi^2(2\chi + 3\tau)(X).$$

Hence

$$\mathcal{I}(\widehat{M}) = \mathcal{I}(M) \geq 32\pi^2(2\chi + 3\tau)(X)$$

by Theorem 5.1. Now suppose that $\hat{g}$ is an Einstein metric on $\widehat{M}$. Then the Gauss-Bonnet formula (4) tells us that

$$\begin{aligned}(2\chi - 3\tau)(\widehat{M}) &= \frac{1}{4\pi^2}\int_{\widehat{M}} \left(2|W^-|^2 + \frac{s^2}{24}\right) d\mu_{\hat{g}} \\ &\geq \frac{1}{3 \cdot 32\pi^2}\mathcal{I}(\widehat{M}) \\ &\geq \frac{1}{3}(2\chi + 3\tau)(X) \\ &= (2\chi - 3\tau)(X),\end{aligned}$$

where in the last step we have used the fact that $\chi(X) = 3\tau(X)$ for any complex-hyperbolic 4-manifold X. But since

$$(2\chi - 3\tau)(\widehat{M}) = (2\chi - 3\tau)(X) + 5k - 4\ell,$$

this tells us that

$$5k - 4\ell \geq 0$$

if $\widehat{M} = X\#k\overline{\mathbb{CP}}_2\#\ell(S^1 \times S^3)$ admits an Einstein metric. The claim now follows by contraposition. □

In particular, this yields a new proof of a beautiful result of Sambusetti [**50**], whose own proof will be discussed in §9 below.

COROLLARY 5.4 (Sambusetti). *Let (a, b) be any pair of integers with $a \equiv b$ mod 2. Then there is a smooth compact oriented 4-manifold M which does not admit Einstein metrics, such that $\chi(M) = a$, $\tau(M) = b$.*

PROOF. Let X_0 be any compact complex-hyperbolic 4-manifold $\mathbb{C}\mathcal{H}_2/\Gamma$. Since $\Gamma \subset SU(1,2)$ is a finitely generated matrix group, Mal'tsev's theorem [**69**, p. 151] asserts that it is *residually finite*, and in particular has a non-trivial homomorphism to a finite group. The kernel Γ_1 of such a homomorphism then defines a finite cover $X_1 = \mathbb{C}\mathcal{H}_2/\Gamma_1$ of X_0. Iterating this procedure, we obtain an infinite tower

$$\cdots \to X_i \to \cdots \to X_1 \to X_0$$

of finite covers of X_0. Thus there is a sequence of integers $j_i \to \infty$ such that each $(3j_i, j_i) = (\chi(X_i), \tau(X_i))$ for some complex-hyperbolic 4-manifold X_i. Now choose i to be large enough so that

$$j_i > \max(b, \frac{a+b}{4}, \frac{2a}{3} - b).$$

The positive integers

$$\begin{aligned} k &= j_i - b \\ \ell &= 2j_i - \frac{a+b}{2} \end{aligned}$$

then satisfy $\ell > \frac{5}{4}k$. Thus $M = X_i \# k\overline{\mathbb{CP}}_2 \# \ell(S^1 \times S^3)$ has

$$\begin{aligned} \chi(M) &= 3j_i + k - 2\ell = a \\ \tau(M) &= j_i - k = b, \end{aligned}$$

and yet, by Theorem 5.3, does not admit Einstein metrics. □

The reader should note that the χ and τ are necessarily congruent mod 2, so that the above result is optimal.

Probing the scalar curvature estimates provided by Seiberg-Witten theory with concrete sequences of test metrics actually allows one to compute $\mathcal{I}$ for any complex surface of general type [**36**]. Applying Corollary 5.2 then yields

PROPOSITION 5.5. *If X is any complex surface of general type, then*

$$\mathcal{I}(X \# k\overline{\mathbb{CP}}_2 \# \ell[S^1 \times S^3]) = 32\pi^2(2\chi + 3\tau)(X).$$

This of course immediately implies non-existence results for Einstein metrics for certain values of k and ℓ. The curious thing about this argument, however, is that it ultimately exploits the existence of solutions of the Seiberg-Witten equations (11–12) on one manifold to obtain scalar curvature estimates on another! Might it not be more satisfying to show that there are actually solutions of the Seiberg-Witten equations on the manifold in question? Fortunately, as was recently proved by Ozsváth and Szabó [**45**], and noticed independently by the present author and various others, such Seiberg-Witten solutions do in fact exist.

To this end, suppose that we have a spin^c structure $\mathfrak{c}$ on a 4-manifold M for which (17) predicts that the moduli space $\mathfrak{M}_{\mathfrak{c},g,\varepsilon}$ generically has dimension $\ell > 0$. Fix ℓ loops $\gamma_1, \ldots, \gamma_\ell$ in M, and define a map $\mathfrak{M}_{\mathfrak{c},g,\varepsilon} \to T^\ell$ by sending (Φ, A) to the holonomies of A around the ℓ given loops. For a fixed $(\mathfrak{c}, g, \varepsilon)$, the homotopy class of this map only depends on the homology classes $[\gamma_i] \in H_1(M, \mathbb{Z})$, and we may therefore define $n_{\mathfrak{c}}(M, [\gamma_1], \cdots, [\gamma_\ell]) \in \mathbb{Z}_2$ to be the degree mod 2 of this map. If this invariant is non-zero for some choice of $[\gamma_i]$, it of course follows in particular that there must be a Seiberg-Witten solution for every metric g on M.

THEOREM 5.6. *Let N be a complex surface of general type, let $M = N\#\ell[S^1 \times S^3]$, and let $\gamma_1, \dots, \gamma_\ell$ be S^1 factors of the ℓ relevant copies of $S^1 \times S^3$. Assume, for simplicity, that M satisfies (18), and let $\mathfrak{c}'$ be the spin^c structure on M obtained by pulling back the canonical spin^c structure $\mathfrak{c}$ from the complex surface N. Then* $n_{\mathfrak{c}}(M, [\gamma_1], \cdots, [\gamma_\ell]) \neq 0$.

One way of proving this is to consider metrics on M which approximate standard product metrics on each $S^1 \times S^3$, where the S^1 factor is taken to be extremely long. By cutting out an S^3 and capping off, each such metric can be approximated by a metric on N containing two long cylinders $[0, b] \times S^3$ for each $S^1 \times S^3$. On the other hand, the Weitzenböck formula (14) forces $|\Phi|^2$ to fall off exponentially along such a cylinder because of the positivity of the scalar curvature. Hence one can use a cut-off function to pass from a solution of any small perturbation of the Seiberg-Witten equations on N to a solution of a small perturbation of the Seiberg-Witten equations on M which has any specified holonomy around the γ_i; conversely, solutions on M can be pasted back onto N. This allows one to conclude that $n_{\mathfrak{c}'}(M, [\gamma_1], \dots, [\gamma_\ell]) = n_{\mathfrak{c}}(N) = 1$. For a different argument, see [**45**].

Thus we see that the 4-dimensional scalar curvature estimates obtainable by Theorem 5.1 can, in practice, actually be deduced directly from the theory of the Seiberg-Witten equations. However, the most striking consequence of Theorem 5.1 is to be found in dimensions *bigger* than four. Indeed, this surgical argument implies [**48**]

THEOREM 5.7 (Petean). *Let M^n be any simply connected smooth compact n-manifold, where $n \geq 5$. Then $\mathcal{I}(M) = 0$.*

The proof builds on a circle of ideas due to Gromov and Lawson [**22**], using Theorem 5.1 to reduce the problem to that of finding a suitable system of generators for the spin-cobordism ring.

It follows that any simply connected n-manifold, $n \geq 5$, has unit volume metrics of scalar curvature $-\varepsilon$, for any $\varepsilon > 0$. If the manifold is also non-spin, one can even find unit-volume metrics of constant scalar curvature > 0 by the earlier result of Gromov-Lawson [**22**]. Thus, while Seiberg-Witten theory tells us that a Kähler-Einstein metric with $\lambda < 0$ maximizes the scalar curvature among constant-scalar-curvature metrics of fixed volume, the analogous assertion is dramatically false on simply connected manifolds of higher dimension. Thus, one might suspect that the sign of the Einstein constant is not determined by the smooth topology in high dimensions. In the next section, we shall see that the facts show that this suspicion is completely justified.

6. The Sign of the Einstein Constant

We have already seen that the fundamental group alone does not contain enough information to determine the sign of the Einstein constant. However, one might still hope [**8**] that the sign of λ is somehow determined by the topology of M. Indeed, Corollary 4.3 seems to support such a hope in dimension 4. In higher dimensions, however, the theory of Kähler-Einstein manifolds allow one to actually construct counter-examples to such a conjecture [**14, 31**]. The first step is to observe that Corollary 4.3 becomes false if the rules are altered so as to allow one to change not only the metric, but also the *differentiable structure*, on a fixed topological 4-manifold.

THEOREM 6.1. *There is a homeomorphic pair of 4-manifolds (M_1, M_2) such that M_1 admits a Kähler-Einstein metric g_1 with $\lambda < 0$, and such that M_2 admits a Kähler-Einstein metric g_2 with $\lambda > 0$.*

In higher dimensions, it therefore turns out that the sign of λ cannot be deduced from the smooth topology.

THEOREM 6.2 (Catanese-LeBrun). *There is a smooth 8-manifold M which admits a pair of Einstein metrics for which the Einstein constants λ have opposite signs. Moreover, one may arrange for both of these Einstein metrics to be Kähler, albeit with respect to wildly unrelated complex structures.*

Indeed, one may take the 4-manifold M_2 to be $\mathbb{CP}_2\#8\overline{\mathbb{CP}}_2$, which, as we saw in Theorem 3.3, admits Kähler-Einstein metrics with $\lambda > 0$. On the other hand, M_1 may be taken to be the underlying smooth 4-manifold of the *Barlow surface.* The Barlow surface [**6**] is a simply connected minimal complex surface of general type with the same $b_\pm$ as $\mathbb{CP}_2\#8\overline{\mathbb{CP}}_2$. With Barlow's complex structure, M_1 contains four (-2)-curves, and so does not have K ample, but one can deform this complex structure [**14**] so as to destroy these (-2)-curves. Thus M_1 admits other complex structures for which K is ample, and so admits Kähler-Einstein metrics with $\lambda < 0$ by Theorem 3.1. In particular, by taking the product metrics, it follows that $M_1 \times M_1$ and $M_2 \times M_2$ admit Kähler-Einstein metrics with Einstein constants λ of opposite signs.

On the other hand, the intersection forms

$$\smile: H^2(\mathbb{Z}) \times H^2(\mathbb{Z}) \to \mathbb{Z}$$

of M_1 and M_2 are isomorphic because the Minkowski-Hasse classification [**28**] asserts there is only one isomorphism class when b_+ and b_- are both non-zero and $\tau = b_+ - b_- \not\equiv 0 \bmod 8$. A theorem of Wall [**65**] therefore shows that M_1 and M_2 are h-cobordant; that is, there is a 5-manifold-with-boundary V with $\partial V = M_1 \coprod M_2$, such that the inclusions $M_1, M_2 \hookrightarrow W$ are both homotopy equivalences. Hence $M_1 \times M_1$ is h-cobordant to $M_2 \times M_2$, via $(M_1 \times W) \cup (W \times M_2)$. But Smale's h-cobordism theorem [**56**] asserts that simply connected h-cobordant smooth manifolds of dimension ≥ 5 are necessarily diffeomorphic. Thus $M_1 \times M_1$ is diffeomorphic to $M_2 \times M_2$, and the Kähler-Einstein metrics under discussion may therefore be considered to live on the same manifold $M = M_1 \times M_1$.

On the other hand, Corollary 4.3 makes it painfully obvious that M_1 and M_2 are not diffeomorphic — a fact which was first proved [**30**] using Donaldson theory [**18**]; cf. [**44**]. In other words, the h-cobordism theorem breaks down in dimension 4. However, Freedman did manage to salvage the *topological* part of Smale's proof in dimension 4, and Theorem 2.1 thus allows one to still conclude that M_1 and M_2 are *homeomorphic.*

7. Weyl Estimates

So far, we have seen that the Seiberg-Witten equations give rise to scalar-curvature estimates on 4-manifolds. We will now see that that also give rise [**38**] to estimates concerning the Weyl curvature.

LEMMA 7.1. *Let (M, g) be an oriented Riemannian 4-manifold, and let $\mathfrak{c}$ be a spin^c structure on M. Let $\tilde{g}$ be a Yamabe metric conformal to g. If there is an irreducible solution (Φ, A) of the Seiberg-Witten equations (11–12) on $(M, \tilde{g}, \mathfrak{c})$,* *then the L^2-norms of the self-dual Weyl curvature and scalar curvature of g must satisfy*

$$\frac{1}{\sqrt{3}}\|W^+\|_2 + \frac{1}{2\sqrt{2}}\|s\|_2 \geq \frac{8\pi}{3}|c_1^+|.$$

Moreover, equality occurs iff g is Yamabe and also Kähler, with respect to some $\mathfrak{c}$-compatible complex structure.

PROOF. By conformal rescaling, we may assume that the scalar curvature s is a negative constant. Now consider the Weitzenböck formula

$$(d+d^*)^2\phi = \nabla^*\nabla\phi - 2W^+(\phi,\cdot) + \frac{s}{3}\phi,$$

which holds for any self-dual 2-form ϕ. Assuming that $\phi \not\equiv 0$,this formula implies that

$$\|W^+\|_2 \geq \frac{|s|\ \|\phi\|_2}{2\sqrt{2}\|\phi\|_4^2}\left[\frac{1}{\sqrt{3}}\|\phi\|_2 - \frac{\sqrt{3}}{|s|}\frac{\|\nabla\phi\|_2^2}{\|\phi\|_2}\right], \tag{19}$$

where again we have assumed that the scalar curvature s is a negative constant. We now apply this to the particular 2-form $\phi = \sigma(\Phi) = -iF_A^+$ associated with a solution of the Seiberg-Witten equations. To do so, first notice that (14) and the Cauchy-Schwarz inequality tell us that

$$\left(\|s\|_2 - \sqrt{8}\|\phi\|_2\right)\sqrt{8}\|\phi\|_2 \geq \int\left[(-s)|\Phi|^2 - |\Phi|^4\right]d\mu = \int 4|\nabla_A\Phi|^2 d\mu,$$

since $|\Phi|^4 = 8|\phi|^2$. On the other hand, $|\nabla\phi|^2 \leq \frac{1}{2}|\Phi|^2|\nabla_A\Phi|^2$, and (15) tells us that $|\Phi|^2 \leq |s|$. Since harmonic theory tells us that

$$\|\phi\|_2 \geq 2\pi|c_1^+|, \tag{20}$$

we therefore have

$$\frac{\|\nabla\phi\|_2^2}{|s|\ \|\phi\|_2} \leq \frac{\|s\|_2}{2\sqrt{2}} - 2\pi|c_1^+|. \tag{21}$$

Finally, another application of (15) gives us

$$\frac{|s|\ \|\phi\|_2}{2\sqrt{2}\|\phi\|_4^2} \geq 1. \tag{22}$$

Plugging (20–22) into (19) then proves the lemma. □

This lemma can be usefully exploited by interpreting the left-hand side as a dot product in $\mathbb{R}^2$:

$$\frac{1}{\sqrt{3}}\|W^+\|_2 + \frac{1}{2\sqrt{2}}\|s\|_2 = \left(\frac{1}{\sqrt{6}}, \sqrt{3}\right)\cdot\left(\sqrt{2}\|W^+\|_2, \frac{\|s\|_2}{2\sqrt{6}}\right).$$

The Cauchy-Schwarz inequality therefore tells us that

$$\left(\frac{1}{6}+3\right)\left(2\|W^+\|_2^2 + \frac{\|s\|_2^2}{24}\right) \geq \left(\frac{1}{\sqrt{3}}\|W^+\|_2 + \frac{1}{2\sqrt{2}}\|s\|_2\right)^2 \geq \frac{64\pi^2}{9}(c_1^+)^2,$$

or in other words that

$$\frac{1}{4\pi^2}\int_M\left(2|W^+|_g^2 + \frac{s_g^2}{24}\right)d\mu_g \geq \frac{32}{57}(c_1^+)^2.$$

Now there is no reason to believe that the constant $\frac{32}{57}$ is sharp, so there is little to lose if we replace it here with $\frac{5}{9}$, which is only 1% smaller, and much more easily remembered. Doing so yields

THEOREM 7.2. *Let (M, g) be a compact oriented Riemannian 4-manifold with a non-trivial Seiberg-Witten invariant. Let $c_1(L) \in H^2(M, \mathbb{R})$ be the first Chern class of the corresponding spin^c structure on M, and let $c_1^+ \neq 0$ denote its projection into the space of g-self-dual harmonic 2-forms. Then*

$$\frac{1}{4\pi^2}\int_M \left(2|W^+|_g^2 + \frac{s_g^2}{24}\right) d\mu_g > \frac{5}{9}(c_1^+)^2.$$

This leads to yet more obstructions to the existence of Einstein metrics. Indeed, the Gauss-Bonnet formula (4) tells us that the left-hand side of the inequality in Theorem 7.2 is just $(2\chi + 3\tau)(M)$ if g is Einstein. This then gives us the following improvement of Theorem 4.8:

THEOREM 7.3 (LeBrun). *Let X be a minimal complex algebraic surface of general type, and let $M = X\#k\overline{\mathbb{CP}}_2$ be obtained from X by blowing up $k > 0$ points. If $k \geq \frac{4}{9}(2\chi + 3\tau)(X)$, then M does not admit Einstein metrics.*

Again, the constant of $\frac{4}{9}$ is not sharp, but suffices for our present purposes.

EXAMPLE 7.1. Let X_ℓ be the Fermat surface of degree $\ell \geq 8$ in $\mathbb{CP}_3$, and let $M_\ell = X_\ell\#k\overline{\mathbb{CP}}_2$ be obtained from X_ℓ by blowing up $k = \ell(\ell-4)^2 - 2\binom{\ell-1}{3} + 4$ points. Since $c_1^2(X) = \ell(\ell-4)^2 \sim \ell^3$, whereas $k \sim \frac{2}{3}\ell^3$, we must have $k > \frac{4}{9}c_1^2(X)$ for sufficiently large ℓ; and indeed, closer inspection shows that this actually happens for all $\ell \geq 8$. Thus Theorem 7.3 implies that none of these 4-manifolds M_ℓ admits an Einstein metric.

Now assume, for simplicity, that ℓ is odd, so that $\binom{\ell-1}{3} \equiv 0 \bmod 4$, and notice that M_ℓ has

$$\begin{aligned} b_+ &= 2\binom{\ell-1}{3} + 1, \\ b_- &= 8\binom{\ell-1}{3} + 13, \end{aligned}$$

exactly like the surface N_ℓ gotten by taking the the double branched cover of $\mathbb{CP}_1 \times \mathbb{CP}_1$ ramified over a smooth holomorphic curve of bidegree $(6, \binom{\ell-1}{3} + 2)$. (The latter is an example of a *Horikawa surface* [**27**].) Since the simply connected complex surfaces M_ℓ and N_ℓ both have $\tau = -6[\binom{\ell-1}{3} + 2] \equiv 4 \bmod 8$, both are non-spin, so Theorem 2.1 tells us that M_ℓ and N_ℓ are homeomorphic. But N_ℓ is a minimal surface of general type, and contains no (-2)-curves. Corollary 3.5 therefore tells us that N_ℓ carries an Einstein metric, even though it is homeomorphic to M_ℓ, which does not.

Thus Theorem 7.3 gives us a simple proof of a result originally deduced by Kotschick [**31**], who instead applied Theorem 4.8 to some rather more exotic algebraic-geometric examples.

THEOREM 7.4 (Kotschick). *For infinitely many homeotypes of compact simply connected non-spin 4-manifolds, there are some choices of smooth structure which admit Einstein metrics, and others which do not.*

Presumably this also occurs in the spin case. However, $K3$ provides the only spin homeotype where this phenomenon has been observed to date.

Notice that the holonomy-modified Seiberg-Witten invariants of Theorem 5.6 also give rise to obstructions to the existence of Kähler-Einstein metrics. For example, one has

THEOREM 7.5. *Let X be a minimal surface of general type. Then $M = X\#k\overline{\mathbb{CP}}_2\#\ell[S^1 \times S^3]$ does not admit Einstein metrics if $k + 4\ell \geq \frac{4}{9}(2\chi + 3\tau)(X)$.*

The proof imitates that of Theorem 4.7, but uses Theorem 7.2 in place of Theorem 4.2. Details are left to the reader.

8. Minimal Volumes

If M is a compact n-manifold, multiplying any given metric on M by a large enough positive constant will yield a new metric on M of sectional curvature > -1. This rescaling process, however, will also typically make the volume of M enormous. Gromov [**21**] thus realized that it is natural to define a a diffeomorphism invariant, called the *minimal volume*, by setting

$$\mathrm{Vol}_K(M) = \inf\{\mathrm{Vol}(M, g) \mid g \text{ has } K \geq -1\}.$$

But it is equally natural to consider minimal volumes with respect to lower bounds on the Ricci or scalar curvatures:

$$\begin{aligned}\mathrm{Vol}_r(M^n) &= \inf\{\mathrm{Vol}(M, g) \mid g \text{ satisfies } r \geq -(n-1)g\}\\ \mathrm{Vol}_s(M^n) &= \inf\{\mathrm{Vol}(M, g) \mid g \text{ has } s \geq -n(n-1)\}.\end{aligned}$$

Notice that our conventions have been chosen so that

$$\mathrm{Vol}_K(M) \geq \mathrm{Vol}_r(M^n) \geq \mathrm{Vol}_s(M^n)$$

tautologically.

For any manifold of dimension $n \geq 3$, one can show, by first considering one conformal class at a time, that the minimal volume for s is given by

$$\mathrm{Vol}_s(M^n) = \frac{\mathcal{I}(M)}{n^2(n-1)^2},$$

where the invariant $\mathcal{I}$ was defined in §5. Inspection of the Gauss-Bonnet formula (4) therefore shows that an oriented 4-manifold M can admit an Einstein metric g only if

$$2\chi(M) - 3|\tau(M)| \geq \frac{1}{96\pi^2}\mathcal{I}(M) = \frac{3}{2\pi^2}\mathrm{Vol}_s(M),$$

with equality iff g is half-conformally flat and $\mathrm{Vol}_s(M)$ is realized by a suitable rescaling of g. Much of what we have done so far simply consists of making this inequality effective by introducing non-trivial estimates for $\mathrm{Vol}_s(M)$.

In a sense, however, this inequality is quite wasteful; after all, if g is an Einstein metric, its Ricci curvature is determined by its scalar curvature. Thus, the same argument actually proves the following:

LEMMA 8.1. *Let (M, g) be a 4-dimensional Einstein manifold. Then*

$$2\chi(M) \geq 3|\tau(M)| + \frac{3}{2\pi^2}\mathrm{Vol}_r(M),$$

with equality iff g is half-conformally flat and can be rescaled so as to realize the minimal Ricci volume

Of course, such an inequality only acquires content in conjunction with an effective method for estimating the invariant $\mathrm{Vol}_r(M)$. The first result in this direction was discovered by Gromov [**21**], and involves an invariant $||M||$ of a compact topological n-manifold known as its *simplicial volume*, and defined as the infimum of expressions of the form $\sum |c_j|$, where $\sum c_j\sigma_j$ is any singular homology cycle with *real coefficients* c_j representing the fundamental cycle $[M] \in H_n(M, \mathbb{R})$.

PROPOSITION 8.2 (Gromov). *For every smooth compact n-manifold M,*

$$\mathrm{Vol}_r(M) > \frac{1}{(n-1)^n n!}||M||.$$

We will say a bit about the proof of this result in the next section. For the moment, let us merely notice that, with Lemma 8.1, it immediately implies

THEOREM 8.3 (Gromov/Kotschick). *Let (M, g) be a 4-dimensional Einstein manifold. Then*

$$2\chi(M) \geq 3|\tau(M)| + \frac{||M||}{1296\pi^2}.$$

Curiously, Gromov only derived the weaker inequality obtained from this by dropping the τ term. The fact that Gromov's results actually imply an improved version of the Hitchin-Thorpe inequality was only recently brought to light by Kotschick [**32**]. Notice that, in contrast to results derived by Seiberg-Witten methods, the Gromov/Kotschick inequality only involves terms depending on the homotopy type of M. However, the simplicial volume $||M||$ turns out to vanish for any simply connected manifold, so the inequality only improves upon the Hitchin-Thorpe inequality in cases where the fundamental group is infinite.

On the other hand, Theorem 8.3 does represent an honest improvement over the Hitchin-Thorpe inequality. For example, let X is a hyperbolic 4-manifold, and recall that Mal'tsev's Theorem [**69**] predicts that there are ℓ-fold covers X_ℓ of X for arbitrarily large ℓ. If $M = X_\ell \# m\overline{\mathbb{CP}}_2$, then $||M|| \geq \ell ||X||$, and $||X||$ in turn is positive — in fact, $||X|| = \frac{4\pi^2}{3v_4}\chi(X)$, where v_4 is the volume of a regular ideal hyperbolic 4-simplex. Since the Gromov-Kotschick inequality requires that

$$\left(2 - \frac{1}{972v_4}\right)\chi(X) > \frac{m}{\ell},$$

whereas the Hitchin-Thorpe inequality would merely stipulate that

$$2\chi(X) > \frac{m}{\ell},$$

Theorem 8.3 actually predicts non-existence in a certain range of m missed by Hitchin-Thorpe, provided that ℓ is sufficiently large. But in the next section, we will see that one can do a great deal better: *none* of these manifolds admits an Einstein metric!

Even without the signature term, Gromov was able to predict non-existence in cases missed by the Hitchin-Thorpe inequality by considering examples of the form $M = 2(\Sigma \times \Sigma)\#k[S^1 \times S^3]$, where Σ is a Riemann surface of large genus. It is for this reason that simple connectivity was emphasized in Corollary 4.9.

9. Entropy and Ricci Curvature

Let (M, g) be a compact Riemannian manifold, and let $(\tilde{M}, g)$ be its universal cover. Let $x \in \tilde{M}$, and let $B_\varrho(x) \subset \tilde{M}$ denote the open distance ball, consisting of of points of distance $< \varrho$ from x; let $\mathrm{Vol}(B_\varrho(x))$ denote the Riemannian volume of this distance-ball. Then the *volume entropy* of (M, g) is defined to be

$$h_{vol}(M,g) = \lim_{\varrho\to\infty} \frac{\log \mathrm{Vol}(B_\varrho(x))}{\varrho}.$$

This is independent of the base-point x, but of course can be non-zero only if the fundamental group of M is infinite.

An easy calculation shows that an n-manifold of constant sectional curvature $K \leq 0$ has entropy $h_{vol} = (n-1)\sqrt{|K|}$. After a bit of of pure thought, we therefore get the the following:

LEMMA 9.1. *Any compact Riemannian manifold (M, g) with $r \geq -(n-1)g$ has volume entropy*

$$h_{vol}(M,g) \leq n-1.$$

Indeed, this is an immediate consequence of *Bishop's inequality* [**12, 8**], which, in light of our assumption that the Ricci curvature of $(\tilde{M}, g)$ is no smaller than that of hyperbolic space $\mathcal{H}^n$, says a ball of radius ϱ in $(\tilde{M}, g)$ must have volume no bigger than that of the corresponding ball in $\mathcal{H}^n$.

Of course, the entropy $h_{vol}(M, g)$ is not invariant under rescalings, and indeed it is easy to show that

$$h_{vol}(M, cg) = c^{-1/2} h_{vol}(M, g).$$

Fortunately, this is easily remedied by instead considering the scale-invariant quantity

$$\mathcal{E}(M^n, g) = [h_{vol}(M,g)]^n \mathrm{Vol}(M, g).$$

This invariant was already considered by Gromov [**21**], who showed that any metric on any compact n-manifold M satisfies

$$\mathcal{E}(M,g) > \frac{1}{n!}||M||.$$

With Lemma 9.1, this then implies Proposition 8.2.

While Gromov's lower bound on $\mathcal{E}(M, g)$ opened up several new frontiers of mathematical research, it is, in practice, far from sharp. It was therefore a development of the greatest significance when Besson, Courtois, and Gallot [**9**] were able to prove that locally symmetric metrics of strictly negative curvature actually *minimize* this functional:

THEOREM 9.2 (Besson-Courtois-Gallot). *Let M be any compact quotient of a real, complex, quaternionic, or octonionic hyperbolic space, and let g_0 be the standard metric on M. Then any other metric g on M satisfies*

$$\mathcal{E}(M,g) \geq \mathcal{E}(M, g_0),$$

with equality iff g is locally symmetric.

SKETCH OF PROOF. Let S^∞ denote the unit sphere in the real Hilbert space $L^2(\partial\tilde{M})$ of square-integrable half-densities on the sphere-at-infinity of $\tilde{M}$, and let $S^\infty_+ \subset S^\infty$ denote its intersection with the open cone of *positive* half-densities. We will consider smooth $\pi_1(M)$-equivariant maps $\Phi : \tilde{M} \to S^\infty_+$. Each such map induces a (possibly degenerate) metric g_Φ on $\tilde{M}$ which is $\pi_1(M)$-invariant, and so descends to M. The volume $\mathrm{Vol}(M, g_\Phi)$ may then be viewed as a sort of equivariant volume of the image of Φ, and it is natural to ask whether some choice of Φ minimizes this volume.

Such a minimizer actually does exist. Indeed, for each $x \in \tilde{M}$, consider the map from the unit sphere in $T_x\tilde{M}$ to $\partial\tilde{M}$ obtained by following the geodesics of g_0 all the way to infinity. The *Poisson measure* $d\wp_x$ on $\partial\tilde{M}$ is the push-forward of the usual probability measure on the unit sphere in $T_x\tilde{M}$ via this map. The promised minimizer is then given by $\Phi_0(x) = \sqrt{\wp_x}$. The fact that this *is* a minimizer is proved by a calibrated geometry argument. Namely, there is a canonical *baricenter map* $\varpi : S^\infty_+ \to \tilde{M}$, and one may therefore define a closed n-form ω on S^∞_+ by

$$\omega = \left(\frac{h_{vol}(g_0)}{2\sqrt{n}}\right)^n \varpi^* d\mu_{g_0},$$

where $d\mu_{g_0}$ is (traditional bad notation for) the volume n-form of g_0. The integral of ω on any n-manifold is then less than or equal to the manifold's volume, and equality holds for the image of Φ_0. Since any other Φ is smoothly, equivariantly homotopic to Φ_0, Stokes' theorem tells us that

$$\mathrm{Vol}(M, g_\Phi) \geq \int_M \Phi^*\omega = \int_M \Phi_0^*\omega = \mathrm{Vol}(M, g_{\Phi_0}) = (4n)^{-n/2}\mathcal{E}(M, g_0).$$

On the other hand, given any metric g on M and any constant $c > h_{vol}(g)$, one may define a smooth equivariant map by

$$\Phi_{g,c}(x) = \left[\frac{\int_{\tilde{M}} e^{-c\delta_g(x,y)} d\wp_y d\mu_{y,g}}{\int_{\partial\tilde{M}}\int_{\tilde{M}} e^{-c\delta_g(x,y)} d\wp_y d\mu_{y,g}}\right]^{1/2}$$

where δ_g is the Riemannian distance. One is then able to show that

$$\frac{c^2}{4n} g \geq g_{\Phi_{g,c}},$$

so that

$$\mathcal{E}(M, g) \geq (4n)^{n/2} \inf_c \mathrm{Vol}(M, g_{\Phi_{g,c}}).$$

Thus $\mathcal{E}(M, g) \geq \mathcal{E}(M, g_0)$, as claimed. □

In light of Lemma 9.1, Theorem 9.2 implies

COROLLARY 9.3. *Let M be any compact hyperbolic 4-manifold. Then*

$$\mathrm{Vol}_r(M) = \frac{4\pi^2}{3}\chi(M).$$

Of course, we also get similar results in other dimensions; for example, if M^{2m} is any even-dimensional hyperbolic manifold, $\mathrm{Vol}_r(M) = (-4\pi)^m \frac{m!}{(2m)!}\chi(M)$. In high dimensions, alas, this tells us essentially nothing about Einstein metrics. But in dimension 4, we find that any Einstein metric on a hyperbolic 4-manifold saturates the inequality

$$2\chi \geq 3|\tau| + \frac{3}{2\pi^2}\mathrm{Vol}_r$$

of Lemma 8.1, is therefore conformally flat, and hence has constant curvature. With Mostow rigidity, this yields:

THEOREM 9.4 (Besson-Courtois-Gallot). *Let M^4 be a smooth compact quotient of hyperbolic 4-space $\mathcal{H}^4 = SO(4,1)/SO(4)$, and let g_0 be its standard metric of constant sectional curvature. Then every Einstein metric g on M is of the form $g = \lambda\varphi^* g_0$, where $\varphi : M \to M$ is a diffeomorphism and $\lambda > 0$ is a constant.*

Notice that Theorem 9.2 also makes an assertion about complex-hyperbolic 4-manifolds, so one might expect to also be able to prove Theorem 4.6 by this method. Unfortunately, however, because Lemma 9.1 uses the Bishop comparison theorem to compare metrics with *real* hyperbolic space, the associated lower bound for Vol_r turns out to be too small by a factor of $\frac{64}{81}$. It would of course be of fundamental interest to find some over-arching point of view which could explain both results simultaneously.

By imitating the proof of Theorem 9.2, one can also prove the following:

THEOREM 9.5 (Besson-Courtois-Gallot). *Let (X, g_0) be a compact oriented locally symmetric space of negative curvature, and let M be a compact manifold of the same dimension. Let $f : M \to X$ be any smooth map. Then any metric g on M satisfies*

$$\mathcal{E}(M,g) \geq |\deg(f)|\mathcal{E}(X,g_0),$$

where $\deg(f)$ *denotes the degree of f.*

COROLLARY 9.6. *Let X be a compact oriented hyperbolic-4-manifold, and suppose M is a compact oriented 4-manifold which admits a map $f : M \to X$ of degree q. Then*

$$\mathrm{Vol}_r(M) \geq q\frac{4\pi^2}{3}\chi(X).$$

EXAMPLE 9.1. Let X be a compact oriented hyperbolic 4-manifold, and let

$$M = X\#\ell(S^1 \times S^3)\#m\mathbb{CP}_2.$$

Then M admits a degree-1 map to X, and hence

$$\mathrm{Vol}_r(M) \geq \frac{4\pi^2}{3}\chi(X).$$

On the other hand, we have

$$\begin{aligned} \chi(M) &= \chi(X) - 2\ell + m, \\ \tau(M) &= m. \end{aligned}$$

Plugging these numbers into the inequality

$$2\chi \geq 3\tau + \frac{3}{2\pi^2}\mathrm{Vol}_r$$

of Lemma 8.1, we conclude that M can admit an Einstein metric only if

$$2\chi(X) - 4\ell + 2m \geq 3m + 2\chi(X).$$

If either ℓ or m is positive, such a manifold therefore never admits an Einstein metric. Moreover, since our estimate of Vol_r only depends on the existence of a map of a degree-1 map to X, the same conclusion applies to any 4-manifold which is even *homotopy equivalent* to one of these examples.

Now for any compact oriented hyperbolic 4-manifold X, the Euler characteristic $\chi(X)$ is even and positive, and we may arrange for it to be as large as we like by passing to finite covers — which exist in abundance by Mal'tsev's theorem [**69**]. By choosing X and ℓ appropriately, we may, for any $m \geq 0$, therefore construct manifolds M as above such that $\chi(M)$ is any given integer $\equiv m \bmod 2$. Since one also has $\tau(M) = m$, this family of manifolds M, together with their orientation reversed versions $\overline{M}$, suffices to prove a stronger version of Corollary 5.4:

THEOREM 9.7 (Sambusetti). *Any integer pair (χ, τ) with $\tau \equiv \chi \bmod 2$ can be realized as the Euler characteristic and signature of a smooth compact oriented 4-manifold M (with infinite fundamental group) which is not homotopy equivalent to any 4-dimensional Einstein manifold.*

Once again, notice how these entropy arguments involve the gigantic size of the fundamental group in an essential way. These beautiful results can therefore shed no light at all on the simply connected case.

10. The Positive Case

This essay has focused almost exclusively on recent results concerning 4-dimensional Einstein manifolds with $\lambda < 0$. However, recent years have also witnessed remarkable progress in our knowledge of the $\lambda > 0$ case. The most striking result in this direction is the weak compactness theorem of Anderson [**2**], which shows that, while the moduli space of $\lambda > 0$ Einstein metrics on a 4-manifold M is generally non-compact, it can always be compactified by adding points representing orbifold Einstein metrics on spaces obtained from M by collapsing chains of 2-spheres. For a description of results in this direction, see the essay by Petersen in this volume.

In the negative case, we have seen that the presence of a non-vanishing Seiberg-Witten invariant is enough to guarantee that any Riemannian metric on a manifold satisfies $\int s^2 d\mu \geq 32\pi^2(2\chi + 2\tau)$. For an Einstein metric, one may use (4) to rewrite this in the form

$$\int_M \frac{s^2}{24} d\mu \geq \int_M |W^+|^2 d\mu.$$

It may therefore come as something of a surprise to learn that Gursky [**23, 24**] has proved that any Einstein 4-manifold with $\lambda > 0$ satisfies exactly the *opposite* inequality, unless it is anti-self-dual:

THEOREM 10.1 (Gursky). *Let (M, g) be a compact oriented Einstein 4-manifold with $s > 0$ and $W^+ \not\equiv 0$. Then*

$$\int_M |W_g^+|^2 d\mu_g \geq \int_M \frac{s_g^2}{24} d\mu_g,$$

with equality iff $\nabla W^+ \equiv 0$.

SKETCH OF PROOF. The Einstein equations and the second Bianchi identity imply that W^+ is divergence-free. This implies that

$$0 \geq \Diamond |W^+|^{1/3},$$

where $\Diamond = 6\Delta + s - 2\sqrt{6}|W^+|$. However, the quantity $\mathfrak{S} = s - 2\sqrt{6}|W^+|$ transforms under conformal transformations according to the rule

$$\mathfrak{S}_{u^2 g} = u^{-3} \Diamond u.$$

Thus, assuming that $W^+ \not\equiv 0$, we may take u to be a smooth positive approximation of $|W^+|^{1/3}$, and thereby construct a conformally rescaled metric $\hat{g} = u^2 g$ such that

$$\int_M \left[s_{\hat{g}} - 2\sqrt{6}|W^+_{\hat{g}}| \right] d\mu_{\hat{g}} \leq 0.$$

Now g minimizes $\int s \, d\mu$ among all metrics of fixed volume in its conformal class, so that

$$\begin{aligned}
\left(\int_M s_g^2 d\mu_g \right)^{1/2} &= \frac{\int_M s_g d\mu_g}{\sqrt{\int_M d\mu_g}} \\
&\leq \frac{\int_M s_{\hat{g}} d\mu_{\hat{g}}}{\sqrt{\int_M d\mu_{\hat{g}}}} \\
&\leq \frac{\int_M 2\sqrt{6}|W^+_{\hat{g}}| d\mu_{\hat{g}}}{\sqrt{\int_M d\mu_{\hat{g}}}} \\
&\leq \left(24 \int |W^+_{\hat{g}}|^2 d\mu_{\hat{g}} \right)^{1/2} \\
&= \left(24 \int |W^+_g|^2 d\mu_g \right)^{1/2}
\end{aligned}$$

by the positivity of s_g and the conformal invariance of $|W^+|^2 d\mu$. □

The resulting estimate $\int s^2 d\mu \leq 32\pi^2(2\chi + 3\tau)$ is certainly interesting, but not particularly powerful in the absence of other geometric assumptions. If one assumes, however, that the *sectional curvature* of g is non-negative, one also has the point-wise estimate

$$\frac{s}{\sqrt{6}} \geq |W^+| + |W^-|,$$

and the Gauss-Bonnet formula therefore tells one that $\chi < \frac{5}{8\pi^2} \int_M \frac{s^2}{24} d\mu$. Putting these two inequalities together, reversing the orientation if necessary, and using Bishop's inequality, one thus obtains

PROPOSITION 10.2. *Let (M, g) be a smooth compact oriented Einstein 4-manifold with non-negative sectional curvature. Assume, moreover, that g is neither self-dual nor anti-self-dual. Then the Euler characteristic χ and the signature τ of M satisfy*

$$9 \geq \chi > \frac{15}{4}|\tau|.$$

In particular, if a 4-manifold M has $b_- = 0$ and $b_+ \neq 0$, any Einstein metric of non-negative sectional curvature on M must be self-dual. Since Hitchin has proved [**8**] that the only self-dual Einstein manifolds with positive scalar curvature are the symmetric spaces S^4 and $\mathbb{CP}_2$, this gives us a clean characterization of the Fubini-Study metric [**24**]:

THEOREM 10.3 (Gursky-LeBrun). *Let M be a smooth compact oriented 4-manifold with strictly positive intersection form. Suppose that g is an Einstein metric on M which has non-negative sectional curvature. Then (M, g) is isometric to $\mathbb{CP}_2$, equipped with a constant multiple of its standard Fubini-Study metric.*

One might thus hope that any Einstein 4-manifold with non-negative sectional curvature is actually locally symmetric. One piece of evidence in favor of such a conjecture is the fact [**57**] that any Einstein manifold of positive *curvature operator* $\mathcal{R} : \Lambda^2 \to \Lambda^2$ is locally symmetric. For a further discussion of the (extremely strong) condition $\mathcal{R} \geq 0$, see the essay by Chow in this volume.

In any case, there are, up to diffeomorphism, only finitely many compact 4-manifolds with Einstein metrics of non-negative sectional curvature. The flat 4-manifolds, of course, nominally form a subclass of the the manifolds under discussion, but Bieberbach's theorem [**10**] in any case tells us that there are finitely many diffeomorphism types of these. For the others, the Ricci curvature must be positive, and we may thus rescale the metric so that, for example, $r = 3g$. The definition of the Ricci curvature then tells us that the sectional curvatures all satisfy $0 \leq K(P) \leq 3$. Gauss-Bonnet therefore tells us that the volume is $\geq 8\pi^2/15$. On the other hand, Myers' theorem predicts that the diameter is $\leq \pi$. Given such bounds, Cheeger's finiteness theorem [**15**] then predicts that there are only finitely many diffeomorphism types — although, of course, the actual number could still be astronomical. By contrast, Proposition 10.2 and Freedman's classification [**19**] tell us that there are at most twelve *homeotypes* of simply connected compact Einstein 4-manifolds with non-negative sectional curvature.

REMARK 10.1. If an Einstein manifold instead has *non-positive* sectional curvature, one still has the inequality

$$\frac{|s|}{\sqrt{6}} \geq |W^+| + |W^-|,$$

and it is straightforward to show that consequently

$$\chi > \frac{15}{8}|\tau|.$$

This is actually a minor improvement on a result of Hitchin [**26**], who observed that such an inequality holds for the somewhat smaller coefficient of $(\frac{3}{2})^{3/2}$. In all likelihood, however, the present constant of $\frac{15}{8}$ isn't sharp, either. In any case, it would be extremely interesting to construct some non-locally-symmetric examples, and give this discussion some substance!

11. Concluding Remarks

In this essay, we have explored several recent streams of thought which bear upon the existence and uniqueness of Einstein metrics on 4-dimensional manifolds. For example, we have seen that Seiberg-Witten theory gives one control of the L^2-norms of scalar and Weyl curvature when certain diffeomorphism invariants are non-zero. Entropy estimates instead allow one to control the Ricci curvature under certain homotopy-theoretic assumptions. The mystery is that, while these techniques sometimes lead to analogous results, they seem completely unrelated. One might hope for a deeper, unified explanation of these results involving principles which remain to be discovered.

On the other hand, it could be that the striking parallels between these two sets of results are merely ephemeral. For example, the parallel formulations of Theorem 9.4 and Corollary 4.6 hide an important technical distinction. The proof of Theorem 9.4 actually shows that any Einstein 4-manifold which is homotopy equivalent to a hyperbolic manifold must itself be hyperbolic, whereas the proof of

Corollary 4.6 yields no such conclusion in the complex-hyperbolic case. Does this merely illustrate a limitation of the methods of proof, or does it capture a factual difference between the real- and complex-hyperbolic cases?

In the same vein, it is interesting to compare the information that these very different sets of techniques provide concerning blow-ups of complex-hyperbolic manifolds. Theorem 7.3 tells us that blowing up such a space at, say, 4τ points will result in a smooth manifold without Einstein metrics. Theorem 9.5 is less efficient, but it does reach a similar conclusion **[50]** if something over 6τ points are blown up. However, the entropy argument yields non-existence for every smooth structure on the manifold.

While we have described a number of techniques for showing that Einstein metrics do not exist on certain 4-manifolds, a direct variational approach to the existence problem **[3]** might suggest that one should instead try to construct sequences of metrics on a given 4-manifold which geometrically converge to a disjoint union of Einstein pieces. For example **[36]**, while complex surfaces of general type do not generally admit Einstein metrics, they do always admit minimizing sequences for the functional $\int s^2 d\mu$ which converge to orbifold Einstein metrics on their pluricanonical models, at the price of 'bubbling off' some topology. However, there are circumstances **[3, 39]** in which such minimizing sequences instead 'collapse' to a lower-dimensional object. At any rate, while most 4-manifolds do not admit Einstein metrics, one might still hope that unions of special Einstein manifolds will eventually play a rôle in 4-dimensional smooth topology similar to that played by minimal models in complex surface theory.

In a different direction, we have seen that the sign of the Einstein constant is definitely not a diffeomorphism invariant in high dimensions. On the other hand, we have seen some weak indications that just the opposite may hold in dimension 4. Further exploration of this issue would seem to be one of the most compelling potential directions for future research.

Finally, it is worth comparing the general state of our knowledge concerning the positive and negative cases. For example, we now know that there aren't any non-standard Einstein metrics on compact quotients of $\mathcal{H}^4$ or $\mathbb{C}\mathcal{H}_2$. What about non-standard Einstein metrics on S^4 or $\mathbb{CP}_2$? The question seems fair enough. Yet the only results currently available in this direction pertain to metrics of positive sectional curvature. The need for such an extraneous hypothesis should serve as a clear indication of the depth of our present ignorance.

References

[1] C. Allendoerfer and A. Weil, *The Gauss-Bonnet theorem for Riemannian polyhedra*, Trans. Am. Math. Soc., 53 (1943), pp. 101–129.

[2] M. Anderson, *The L^2 structure of moduli spaces of Einstein metrics on 4-manifolds*, Geom. Func. An., 2 (1992), pp. 29–89.

[3] ———, *Extrema of curvature functionals on the space of metrics on 3-manifolds*, Calc. Var. Partial Differ. Eq., 5 (1997), pp. 199–269.

[4] M. Atiyah, R. Bott, and V. Patodi, *On the heat equation and index theorem*, Inventiones Mathematicae, 19 (1973), pp. 279–330.

[5] T. Aubin, *Equations du type Monge-Ampère sur les variétés Kählériennes compactes*, C. R. Acad. Sci. Paris, 283A (1976), pp. 119–121.

[6] R. Barlow, *A simply connected surface of general type with* pg $= 0$, Inventiones Mathematicae, 79 (1985), pp. 293–301.

[7] W. Barth, C. Peters, and A. V. de Ven, *Compact Complex Surfaces*, Springer-Verlag, 1984.

[8] A. BESSE, *Einstein Manifolds*, Springer-Verlag, 1987.

[9] G. BESSON, G. COURTOIS, AND S. GALLOT, *Entropies et rigidités des espaces localement symétriques de courbure strictement négative*, Geom. and Func. An., 5 (1995), pp. 731–799.

[10] L. BIEBERBACH, *Über die Bewegungsgruppen des Euklidischen Räume I*, Mathematische Annalen, 70 (1911), pp. 297–336.

[11] O. BIQUARD, *Métriques d'Einstein à cusps et Équations de Seiberg-Witten*, J. Reine Angew. Math., 490 (1997), pp. 129–154.

[12] R. BISHOP, *A relationship between volume, mean curvature, and diameter*, Notices Am. Math. Soc., 10 (1963), p. 364.

[13] A. BOREL, *Compact Clifford-Klein forms of symmetric spaces*, Topology, 2 (1963), pp. 111–222.

[14] F. CATANESE AND C. LEBRUN, *On the scalar curvature of Einstein manifolds*, Mathematics Research Letters, 4 (1997), pp. 843–854.

[15] J. CHEEGER, *Finiteness theorems for Riemannian manifolds*, Amer. J. Math., 92 (1970), pp. 61–74.

[16] A. DOLD AND H. WHITNEY, *Classification of oriented sphere bundles over a 4-complex*, Annals of Mathematics, 69 (1959), pp. 667–677.

[17] S. DONALDSON, *An application of gauge theory to four-dimensional topology*, Journal of Differential Geometry, 18 (1983), pp. 279–315.

[18] S. DONALDSON AND P. KRONHEIMER, *The Geometry of Four-Manifolds*, Oxford University Press, Oxford, 1990.

[19] M. FREEDMAN, *On the topology of 4-manifolds*, Journal of Differential Geometry, 17 (1982), pp. 357–454.

[20] R. FRIEDMAN AND J. MORGAN, *Algebraic surfaces and Seiberg-Witten invariants*, J. Alg. Geom., 6 (1997), pp. 445–479.

[21] M. GROMOV, *Volume and bounded cohomology*, Publ. Math. IHES, 56 (1982), pp. 5–99.

[22] M. GROMOV AND J. H.B. LAWSON, *The classification of simply connected manifolds of positive scalar curvature*, Annals of Mathematics, 111 (1980), pp. 423–434.

[23] M. GURSKY, *Four-manifolds with $\delta W^+ = 0$ and Einstein constants on the sphere*. preprint, 1997.

[24] M. GURSKY AND C. LEBRUN, *On Einstein manifolds of positive sectional curvature*, Ann. Glob. An. Geom.

[25] F. HIRZEBRUCH, *Topological Methods in Algebraic Geometry*, Springer-Verlag, 1966.

[26] N. HITCHIN, *On compact four-dimensional Einstein manifolds*, Journal of Differential Geometry, 9 (1974), pp. 435–442.

[27] E. HORIKAWA, *Algebraic surfaces of general type with small c_1^2, I*, Annals of Mathematics, 104 (1976), pp. 357–387.

[28] D. HUSEMOLLER AND J. MILNOR, *Symmetric Quadratic Forms*, Springer-Verlag, 1973.

[29] R. KOBAYASHI, *Einstein-Kähler V-metrics on open satake v-surfaces with isolated quotient singularities*, Mathematische Annalen, 272 (1985), pp. 385–398.

[30] D. KOTSCHICK, *On manifolds homeomorphic to $\mathbb{CP}_2\#8\overline{\mathbb{CP}}_2$*, Inventiones Mathematicae, 95 (1989), pp. 591–600.

[31] ———, *Einstein metrics and smooth structures*, Geom. Topol., 2 (1998), pp. 1–10.

[32] ———, *On the Gromov-Hitchin-Thorpe inequality*, C. R. Acad. Sci. Paris, 326 (1998), pp. 727–731.

[33] P. KRONHEIMER AND T. MROWKA, *The genus of embedded surfaces in the complex projective plane*, Mathematics Research Letters, 1 (1994), pp. 797–808.

[34] P. B. KRONHEIMER, *Minimal genus in $S^1 \Theta M^3$*, Invent. Math., 135 (1999), pp. 45–61.

[35] C. LEBRUN, *Einstein metrics and Mostow rigidity*, Mathematics Research Letters, 2 (1995), pp. 1–8.

[36] ———, *Four-manifolds without Einstein metrics*, Mathematics Research Letters, 3 (1996), pp. 133–147.

[37] C. LEBRUN, *Einstein metrics on complex surfaces*, in Geometry and physics (Aarhus, 1995), Dekker, New York, 1997, pp. 167–176.

[38] ———, *Weyl curvature, Einstein metrics, and Seiberg-Witten theory*, Math. Res. Lett., 5 (1998), pp. 423–438.

[39] ———, *Kodaira dimension and the Yamabe problem*, Comm. Anal. Geom., 7 (1999), pp. 133–156.

[40] J. Lee and T. Parker, *The Yamabe problem*, Bull. Am. Math. Soc., 17 (1987), pp. 37–91.
[41] Y. Matsushima, *Sur la structure du groupe d'homéomorphismes d'une certaine varieté Kählérienne*, Nagoya Math. J., 11 (1957), pp. 145–150.
[42] J. Morgan, *The Seiberg-Witten Equations and Applications to the Topology of Smooth Four-Manifolds*, vol. 44 of Mathematical Notes, Princeton University Press, 1996.
[43] G. Mostow, *Quasiconformal mappings in n-space and the strong rigidity of hyperbolic space forms*, Publ. Math. IHES, 34 (1968), pp. 53–104.
[44] C. Okonek and A. V. de Ven, *Γ-type-invariants associated to* pu(2)*-bundles and the differentiable structure of barlow's surface*, Inventiones Mathematicae, 95 (1989), pp. 601–614.
[45] P. Ozsváth and Z. Szabó, *Higher type adjunction inequalities in Seiberg-Witten theory.* Princeton University preprint, 1998.
[46] D. Page, *A compact rotating gravitational instanton*, Phys. Lett., 79B (1979), pp. 235–238.
[47] J. Petean, *Computations of the Yamabe invariant*, Math. Res. Lett., 5 (1998), pp. 703–709.
[48] J. Petean, *Yamabe invariants of simply connected manifolds.* e-print # math.DG/9808062, 1998.
[49] J. Petean and G. Yun, *Surgery and the Yamabe invariant*, Geom. Func. An.
[50] A. Sambusetti, *An obstruction to the existence of Einstein metrics on 4-manifolds*, C. R. Acad. Sci. Paris, 322 (1996), pp. 1213–1218.
[51] R. Schoen, *Conformal deformation of a Riemannian metric to constant scalar curvature*, Journal of Differential Geometry, 20 (1984), pp. 478–495.
[52] R. Schoen and S. T. Yau, *On the structure of manifolds with positive scalar curvature*, Manuscripta Math., 28 (1979), pp. 159–183.
[53] I. M. Singer and J. A. Thorpe, *The curvature of* 4*-dimensional Einstein spaces*, in Global Analysis (Papers in Honor of K. Kodaira), Univ. Tokyo Press, 1969, pp. 355–365.
[54] Y. Siu, *Every K3 surface is Kähler*, Inventiones Mathematicae, 73 (1983), pp. 139–150.
[55] ———, *The existence of Kähler-Einstein metrics on manifolds with positive anti-canonical line bundle and suitable finite symmetry group*, Annals of Mathematics, 127 (1988), pp. 585–627.
[56] S. Smale, *On the structure of manifolds*, Annals of Mathematics, 87 (1964), pp. 387–399.
[57] S. Tachibana, *A theorem on Riemannian manifolds with positive curvature operator*, Proc. Japan Acad., 50 (1974), pp. 301–302.
[58] C. Taubes, *The Seiberg-Witten invariants and symplectic forms*, Mathematics Research Letters, 1 (1994), pp. 809–822.
[59] ———, *The Seiberg-Witten and Gromov invariants*, Mathematics Research Letters, 2 (1995), pp. 221–238.
[60] J. Thorpe, *Some remarks on the Gauss-Bonnet formula*, J. Math. Mech., 18 (1969), pp. 779–786.
[61] W. Thurston, *Three-dimensional manifolds, Kleinian groups and hyperbolic geometry*, Bull. Am. Math. Soc., 6 (1982), pp. 357–381.
[62] G. Tian, *On Calabi's conjecture for complex surfaces with positive first chern class*, Inventiones Mathematicae, 101 (1990), pp. 101–172.
[63] G. Tian and S. Yau, *Kähler-Einstein metrics on complex surfaces with* $\mathbf{c}_1 > 0$, Communications in Mathematical Physics, 112 (1987), pp. 175–203.
[64] H. Tsuji, *Existence and degeneration of Kähler-Einstein metrics on minimal algebraic varieties of general type*, Mathematische Annalen, 281 (1988), pp. 123–133.
[65] C. Wall, *On simply connected 4-manifolds*, J. Lond. Math. Soc., 39 (1964), pp. 141–149.
[66] E. Witten, *Monopoles and four-manifolds*, Mathematics Research Letters, 1 (1994), pp. 809–822.
[67] J. Wolf, *Spaces of Constant Curvature*, McGraw-Hill, 1967.
[68] S.-T. Yau, *Calabi's conjecture and some new results in algebraic geometry*, Proc. Nat. Acad. USA, 74 (1977), pp. 1789–1799.
[69] A. Zalesskij, *Linear groups*, in Algebra. IV, Kostrikin and Shafarevich, eds., vol. 37 of Encyclopædia of Mathematical Sciences, Berlin, 1993, Springer-Verlag.

Department of Mathematics, SUNY at Stony Brook, Stony Brook, NY 11794-3651
E-mail address: claude@math.sunysb.edu

Einstein Metrics from Symmetry and Bundle Constructions

McKenzie Y. Wang

Introduction.

In this article we will primarily discuss the construction of Einstein metrics whose holonomy group is generic, i.e., the restricted holonomy is $SO(n)$, where n is the dimension of the manifold. Unfortunately, such Einstein metrics are not at all well-understood. There is no known obstruction for Einstein metrics in dimensions greater than 4, nor is there a general existence theorem for Einstein metrics with generic holonomy. For a discussion of obstructions in dimension 4, see the essay by LeBrun in this volume.

Recall that the Einstein equation $Ric(g) = \Lambda g$ is a non-linear second order system of partial differential equations which is invariant under the action of the diffeomorphism group of the manifold. (We will call the constant Λ the *Einstein constant*, while physicists call it the *cosmological constant.*) In the absence of any general understanding of the solutions of this system, the current strategy for constructing examples is to employ either symmetry or bundle structures to reduce the Einstein equation to more manageable systems of equations.

By the use of symmetry we mean constructing Einstein metrics having a finite-dimensional Lie group of isometries. Generally speaking, progress has been made only when the Lie group acts transitively on the manifold or acts with hypersurface principal orbits. Under these assumptions, the Einstein equation becomes respectively a system of algebraic or ordinary differential equations.

By the use of bundle structures we mean constructing Einstein metrics on the total spaces of bundles which are put together from special families of metrics on the fibres and base using suitable connections. In this situation the Einstein condition translates into a coupled system of equations involving the Ricci curvatures of the fibres and base, as well as the curvature of the connection. Since bundles have structural groups which play a role in the construction, we may, in the spirit of physicists, regard bundle constructions as exploiting the "internal" symmetry of the manifolds. Indeed, these bundle constructions originated from Kaluza-Klein theories of supergravity.

Where the methods surveyed here also produce Einstein metrics with special holonomy, a brief account of the results will be given. The reader is referred to the relevant chapters in this volume for further information.

We do not claim to give a complete survey of all work done in the above topics. Rather, this article only surveys those developments which the author knows how to link into a coherent whole. Also, in view of the excellent book of A. Besse [**17**], we will concentrate only on developments in the last decade.

Acknowledgements: I would like to thank Christoph Böhm, Andrew Dancer, and Wolfgang Ziller for their careful reading of earlier versions of this article and for their many helpful suggestions and corrections. Thanks also go to the taxpayers of Canada for their partial support through NSERC operating grant no. OPG0009421.

1. Kaluza-Klein Constructions on Principal and Fibre Bundles.

Let $\pi : P \to M$ be a smooth principal G-bundle, where G is a compact Lie group, and n and d denote respectively the dimensions of P and M. Let ϕ be a connection on P with curvature form $\Omega = d\phi + [\phi, \phi]$. Given a left-invariant metric $\langle\ ,\ \rangle$ on G and a metric g^* on M, we may use ϕ to construct a metric g on P given by the formula

$$g(X,Y) = g^*(\pi_*(X), \pi_*(Y)) + \langle \phi(X), \phi(Y)\rangle. \tag{1.1}$$

Then $\pi : (P, g) \to (M, g^*)$ becomes a Riemannian submersion with totally geodesic fibres. We refer readers to Chapter 9 of [**17**] for the basic theory of Riemannian submersions. The connection ϕ is said to be *Yang-Mills* if Ω is coclosed as an $ad(\mathfrak{g})$-valued 2-form on M. Clearly, this notion depends on the choice of g^*. The *Kaluza-Klein ansatz* is the construction of Einstein metrics g on P of the type described. In physical theories, (M, g^*) is space-time and matter fields are sections of vector bundles associated to the principal G-bundles P. A good reference for Kaluza-Klein theory from the physical viewpoint is [**50**]. For a good mathematical account, see [**23**].

The Einstein condition for g is equivalent to the system

(1.2)

$$Ric_G(\phi(U), \phi(V)) + \frac{1}{4}\sum_{i,j}\langle \Omega(e_i, e_j), \phi(U)\rangle\langle \Omega(e_i, e_j), \phi(V)\rangle = \Lambda\langle \phi(U), \phi(V)\rangle,$$

$$Ric(g^*)(\pi_*(X), \pi_*(Y)) - \frac{1}{2}\sum_i \langle \Omega(X, e_i), \Omega(Y, e_i)\rangle = \Lambda g^*(\pi_*(X), \pi_*(Y), \tag{1.3}$$

together with the Yang-Mills condition for ϕ. (In the above, Λ is the Einstein constant, U, V are vertical tangent vectors, X, Y are horizontal tangent vectors, and $\{e_1, \cdots, e_d\}$ is a g-orthonormal basis of horizontal tangent vectors.) This Einstein condition follows immediately from (9.61) in [**17**], and the "unknowns" in the above system are ϕ, $\langle\ ,\ \rangle$, and g^*. The Yang-Mills condition is equivalent to the condition that the horizontal and vertical distributions of ϕ are orthogonal with respect to $Ric(g)$.

As was pointed out in [**17**, (9.62)], the Einstein condition implies that g^* has constant scalar curvature and the pointwise norm of Ω must be constant. The latter condition obviously holds if Ω is parallel or if M is homogeneous and ϕ is an invariant connection. However, it is not clear to the author how strong this condition is, especially when G is non-abelian.

The equations (1.2) and (1.3) are in general coupled equations on P. If $\langle\ ,\ \rangle$ is a bi-invariant metric, then (1.3) becomes an equation on M and (1.2) is invariant under the right action of G. However, it is a non-vacuous condition for the second term of the left-hand side of (1.2) to be invariant under the left action of G, contrary to the claims in (9.63) of [**17**] and the ensuing corollary.

As in the situation of Kaluza-Klein theory, the case of an abelian G is more approachable and we will discuss this case first.

Let G be an r-torus T^r. Notice that all left-invariant metrics on T^r are bi-invariant. We will think of T^r as an r-fold product of circles $S^1 = \sqrt{-1}\mathbb{R}/2\pi\sqrt{-1}\mathbb{Z}$. Then a principal torus bundle P is classified by r cohomology classes $\chi_1, \cdots, \chi_r$ in $H^2(M;\mathbb{Z})$, which can be thought of as the Euler classes of the circle bundles P/T^{r-1}, where T^{r-1} ranges over the r codimension 1 subtori obtained by omitting one of the circle factors. Given a connection ϕ on P, the $\mathbb{R}^r$-valued 2-form $\frac{\sqrt{-1}}{2\pi}\Omega$ is the pull-back of an $\mathbb{R}^r$-valued 2-form $\eta = \eta_1 + \cdots + \eta_r$ on M whose components η_i represent χ_i. If a metric g^* is chosen on M, then there is a connection on P such that the corresponding 2-forms η_i are harmonic. If in addition $H^1(M;\mathbb{R}) = 0$, then the choice of ϕ is unique up to gauge equivalence. Thus when G is abelian, the Yang-Mills condition is easily satisfied. Recall, however, that the pointwise norm of η_i must also be constant.

In order to increase the chances of solving (1.2) and (1.3), we need to be able to vary g^* in a family of metrics whose Ricci tensors are simple and whose scalar curvature functions are constant. In general, the harmonic forms η_i will vary with g^*, so at least some information about this variation is required in solving the Einstein equation.

With these considerations in mind, let (M_j, J_j^*), $j = 1, \cdots, m$, be Fano manifolds, i.e., Kähler manifolds with positive first Chern class. By [**129**] they admit a Kähler metric with positive definite Ricci tensor, so by [**73**] they are simply connected. The cohomology group $H^2(M_j;\mathbb{Z})$ is torsion free and so the first Chern class $c_1(M_j)$ can be written as $p_j\alpha_j$ where p_j is a positive integer and α_j is an indivisible class in $H^2(M_j;\mathbb{Z})$.

We assume further that these Fano manifolds are equipped with a Kähler-Einstein metric g_j^* normalized so that $Ric(g_j^*) = p_j g_j^*$. This assumption is non-trivial and we refer the reader to Tian's article in this volume for up-to-date information. We will denote the Kähler form of g_j^* by ω_j^* and its Ricci form by ρ_j^*.

Now let $M = M_1 \times \cdots \times M_m$ and π_j be the projection map onto M_j. We will consider principal T^r bundles P_χ over M which are classified by cohomology classes χ_i of the form

$$\chi_i = \sum_{j=1}^{m} b_{ij}\pi_j^*\alpha_j, \qquad 1 \leq i \leq r,$$

where b_{ij} are integers. On M we let g^* denote a general product metric of the form $x_1 g_1^* + \cdots + x_m g_m^*$ with $x_j > 0$. Every such metric is Kähler with respect to the product complex structure on M. Furthermore, the 2-forms $\eta_i = \frac{1}{2\pi}\sum_j b_{ij}\omega_j$ are harmonic with respect to any of the product metrics g^*. We equip P_χ with a connection ϕ such that $\frac{\sqrt{-1}}{2\pi}d\phi = \pi^*(\eta_1 + \cdots + \eta_r)$.

THEOREM 1.1. [**124**] *Let $\pi : P_\chi \longrightarrow M$ be a principal r-torus bundle with characteristic classes $\chi = (\chi_1, \cdots, \chi_r)$ as described above. If the matrix $B = (b_{ij})$ has maximal rank, then there is an Einstein metric g with positive scalar curvature* *on P_χ of the form (1.1) where $\langle \ , \ \rangle$ is a certain left-invariant metric on T^r and g^* is a certain product metric.*

Because the connection form has been fixed and the metrics g_i^* are Einstein, the Einstein condition in the situation of the theorem becomes a system of algebraic equations in the scaling parameters $x_1, \cdots, x_m$ and in the components of the left-invariant metric $\langle \ , \ \rangle$. It turns out that the latter are determined by the former, and so we are reduced to a system involving only the x_j. This is then solved by a degree argument. Notice that the rank assumption on B is necessary in view of Bonnet-Myers, as the fundamental group of P is finite iff the rank of B is maximal. Note that the submersed product metric g^* is generally not Einstein.

When $r = 1$ and $m = 1$ in the above theorem, we recover the well-known theorem of S. Kobayashi [**74**]. The Einstein metrics on circle bundles over $\mathbb{CP}^1 \times \mathbb{CP}^2$ and $\mathbb{CP}^1 \times \mathbb{CP}^1 \times \mathbb{CP}^1$ were independently found by the physicists D'Auria, Castellani, Fré, and van Nieuwenhuizen [**35**], [**46**] in their quest for 11-dimensional supergravity theories. Circle bundles over an m-fold product of $\mathbb{CP}^1$ was studied by Rodionov [**103**] in the context of homogeneous Einstein metrics.

The Einstein manifolds constructed in Theorem 1.1 display many interesting geometrical and topological properties. Especially noteworthy are the following, whose details can be found in [**124**].

1. There are compact simply connected manifolds in all odd dimensions greater than 4 which admit infinitely many pairwise non-isometric Einstein metrics (with positive scalar curvature) belonging to different path components of the moduli space of Einstein structures. If the volumes of these Einstein metrics are normalized to be 1, then the Einstein constants have 0 as an accumulation point. For example, for each $k \geq 1, S^2 \times S^{2k+1}$ and certain non-trivial $\mathbb{RP}^{2k+1}$ or S^{4k+1} bundles over S^2 exhibit this property. Furthermore, the infinitely many Einstein metrics on any of these manifolds all have isomorphic transitive isometry groups which are not conjugate in the diffeomorphism group, and hence represent inequivalent actions by the same abstract group.

   In §2D we will describe some recent examples of C. Böhm [**19**] which include even-dimensional manifolds, e.g., S^6, S^8, admitting infinitely many inhomogeneous Einstein metrics of volume 1 such that the sequence of Einstein constants converge to a *positive* value.
2. In dimension 7, among the circle bundles over $\mathbb{CP}^1 \times \mathbb{CP}^2$, there are certain bundles P_χ such that for each homotopy 7-sphere Σ, the manifold $P_\chi \sharp \Sigma$ (connected sum) exhibits the phenomena described in (1) above. For different homotopy spheres, the spaces are homeomorphic but not diffeomorphic. These results follow from Theorem 1.1 and the classification theorem of Kreck and Stolz [**79**]. Thus it would appear that Einstein metrics with positive scalar curvature do not always show a preference for one differential structure over another.
3. Condition C of Palais-Smale consequently fails in general for the total scalar curvature functional on the space of Riemannian structures of volume 1.
4. There are Einstein metrics of positive scalar curvature (in odd dimensions) whose (connected) isometry group acts with arbitrarily large cohomogeneity. (Recall that the *cohomogeneity* of a compact Lie group action is the codimension of any principal (generic) orbit.) Indeed, provided that the

characteristic class χ is complicated enough in a suitable sense, the cohomogeneity of P_χ is the sum of the cohomogeneities of the factors of M. Hence the above fact follows from the existence of cohomogeneity 1 Kähler-Einstein Fano manifolds [**75**]. (In [**76**] Kähler-Einstein Fano manifolds of arbitrary large cohomogeneity are constructed by a blowing-down process.)

5. There are odd-dimensional Einstein manifolds with positive scalar curvature which have Einstein moduli spaces of positive dimension. Indeed these are circle bundles of sufficiently complicated topology over Kähler-Einstein Fano manifolds with positive-dimensional Kähler-Einstein moduli.

One may be tempted to extend Theorem 1.1 by letting the base (M, g^*) vary over all Kähler manifolds with constant scalar curvature, or by choosing more general elements of $H^2(M; \mathbb{Z})$ to be the characteristic classes of P_χ. However, at least in the case of circle bundles, we have the following converse.

THEOREM 1.2. [**116**] *Let $\pi : (P, g) \to (M, g^*)$ be a principal circle bundle such that g is an Einstein metric making π into a Riemannian submersion with totally geodesic fibres onto a compact Kähler manifold. Suppose further that the Euler class of P is a cohomology class of type $(1,1)$ with respect to the complex structure of M. Then (M, g^*) is isometric to a Kählerian product $\prod_j (M_j, g_j^*)$ where g_j^* is a Kähler-Einstein metric on a Fano manifold M_j and the Euler class of P is a linear combination of the first Chern classes of M_j.*

We shall give the principal ideas in the proof of the above theorem. First, since the scalar curvature of g^* must be constant, the contracted second Bianchi identity implies that the Ricci form of g^* is harmonic. Therefore, the 2-form corresponding to the second term of the left-hand side of (1.3) is also harmonic. Using these facts, one shows that the eigenvalues of the symmetric operator S given by $g^*(S(X), Y) = -\Omega(J^*(X), Y)$ are constant over M and the eigenspaces corresponding to distinct eigenvalues have constant dimension. The eigenbundles $\mathcal{E}_j$ are therefore well-defined. They are actually J^*-invariant and satisfy a strong integrability condition: $[\mathcal{E}_i, \mathcal{E}_i] \subset \mathcal{E}_i$ for all i and $[\mathcal{E}_i \oplus \mathcal{E}_j, \mathcal{E}_i \oplus \mathcal{E}_j] \subset \mathcal{E}_i \oplus \mathcal{E}_j$ for all $i \neq j$. We consider next the leaves of eigenspace foliation $\mathcal{E}_j$, which are complex submanifolds. Using the Riemannian submersion structure, one checks that in the induced metric the leaves all have Ricci curvature bounded below by that of g^*. The compactness of M then implies that all the leaves are compact simply connected regular submanifolds of M. Finally, using a Bochner argument, one shows that all the leaves are totally geodesic and give a de Rham decomposition of (M, g^*). Equation (1.3) then implies that each de Rham factor is Kähler-Einstein and that the curvature form of the circle bundle is a linear combination of the Kähler classes of the factors.

There are, however, Einstein metrics of type (1.1) on circle bundles over Kähler base manifolds. Of course, the submersed metric on the base is not Kähler.

THEOREM 1.3. *There are Einstein metrics on the total spaces of the following principal S^1 bundles over the specified coadjoint orbits:*

(i) [**119, 36, 92, 56, 77, 26**] *any non-trivial S^1 bundle over $SU(3)/T^2$,*
(ii) [**110**] *any non-trivial S^1 bundle over $SU(p+q+r)/S(U(p)U(q)U(r))$ and $SO(2n)/U(n-1)U(1), n \geq 3$,*
(iii) [**111**] *Let G/L be a coadjoint orbit where G is semisimple and the Lie algebra of L is obtained by deleting a simple root from the Dynkin diagram of $\mathfrak{g}$ which has a coefficient of 2 in the expression of the maximal root of G as a*

linear combination of the simple roots. Then on the principal circle bundle corresponding to the $U(1)$ factor in L there is an Einstein metric other than that from Kobayashi's theorem [**74**].

In the above theorem, note that since all base manifolds are coadjoint orbits, for any homogeneous complex structure one chooses, the first Chern class is positive. Thus all 2-forms are of type $(1,1)$. In (i) the principal circle bundles P_χ can be indexed by 2 integers k, l, where in order to eliminate covering manifolds one assumes that they are relatively prime. Furthermore, notice that the Weyl group $N(T)/T$ acts on the right on $SU(3)/T$, and so there are some obvious diffeomorphisms among the bundles, e.g., $P_{1,1} \cong P_{0,1}$ and $P_{1,2} \cong P_{1,-1}$. (We caution the reader that our notation is such that $P_{2,2}$ corresponds to the first Chern class of $SU(3)/T$.) Similar remarks apply to the other cases where the subgroup has a non-trivial normalizer.

The existence of an Einstein metric in (i) was first obtained in [**119**] in order to show that in a fixed dimension there can *already* be infinitely many homotopy types among *homogeneous* Einstein manifolds. These metrics were rediscovered in [**36**] in a more explicit form. A second Einstein metric was constructed by Page and Pope [**92**]. These physicists also showed that the Einstein metrics have Killing spinors, a fact later rediscovered by Friedrich and Kath [**56**]. Finally, Kowalski and Vlásek [**77**], in a very careful study of these examples, discovered that for large k, one of the Einstein metrics on $P_{-k-1,k}$ also has positive sectional curvature. A third Einstein metric was discovered on $P_{-1,1}$ in [**26**]. All the Einstein metrics in (i) are also related to G_2 structures, as was discovered in [**30**].

If we put $p = q = r = 1$ in (ii), we recover (i). The second Einstein metric in (iii) lies in the canonical variation [**17**, 9.70] of the Kobayashi metric.

As for examples with non-abelian G, the following framework unifies many known Einstein metrics. Suppose that M is an irreducible Riemannian manifold such that the structural group G of its holonomy bundle $\tilde{P}$ is non-simple. Let $G = H \cdot K$ where $\cdot$ means the quotient of the product by a finite normal subgroup. Then $P = \tilde{P}/H$ is a principal $\overline{K}$ bundle over M for a certain quotient $\overline{K}$ of K. We can ask for an Einstein metric of type (1.1) on P.

EXAMPLE 1.1. If (M, g^*) is Kähler-Einstein Fano, then $G = U(n)$ and we can let $H = SU(n)$. One is then precisely in the situation of Kobayashi's theorem [**74**].

EXAMPLE 1.2. If (M, g^*) is quaternionic-Kähler with positive scalar curvature, then either $G = Sp(n) \cdot Sp(1)$ and we can let $H = Sp(n)$, or M is quaternionic symmetric and $G = H \cdot Sp(1)$ with $H \subset Sp(n)$. Then $\overline{K} = SO(3)$ unless $M = \mathbb{HP}^n$, in which case $K = \overline{K} = Sp(1)$. The connection on P induced by the Levi-Civita connection is Yang-Mills with constant norm, as was observed by independently in [**32**] and [**90**]. Using this connection, one can construct two non-isometric Einstein metrics of type (1.1) on P [**17**, 14.85]. When $M = \mathbb{HP}^n$, $P = S^{4n+3}$, and the two Einstein metrics are the constant curvature metric and the Jensen metric [**68**]. Alternatively, since P is a principal circle bundle over the quaternionic-Kähler twistor space of M, we can also appeal to Kobayashi's theorem and the canonical variation [**17**, 9.70] to obtain the two Einstein metrics. These metrics also occur among those in Theorem 1.3(iii). Still another viewpoint is that the Einstein metrics come from 3-Sasakian structures [**25, 26**].

EXAMPLE 1.3. If M is a compact irreducible hermitian symmetric space, $G = K \cdot U(1)$. Then if we let $H = U(1)$, we obtain two non-isometric Einstein metrics on P. Except in the case $K = SU(p)SU(q), p \neq q$, one of these Einstein metrics was found by Jensen [**68**]. For the remaining case and the second Einstein metric (which comes from the canonical variation), see [**120**, Theorem 4].

EXAMPLE 1.4. If M is a compact quaternionic symmetric space, $G = K \cdot Sp(1)$, and if we let $H = Sp(1)$, there are again two non-isometric Einstein metrics on P. When K is simple, one of the Einstein metrics was again found in [**68**]. For the rest, see [**120**, Theorem 2].

EXAMPLE 1.5. If M is a compact irreducible symmetric space whose isotropy group is non-simple, and H is not one of the choices already discussed, then Einstein metrics on the bundle P for such a choice of H were again obtained in [**68**].

Instead of principal bundles, we can also consider Kaluza-Klein constructions on associated fibre bundles of principal bundles. As before, let $\pi : P \to M$ be a principal G-bundle with connection ϕ whose curvature form is Ω. Let G act almost effectively on a manifold F and let $W = P \times_G F$. If g^* is a metric on M and $\langle\ ,\ \rangle$ now denotes a G-invariant metric on F, then (1.1) defines a metric g so that the projection $\pi : (W, g) \to (M, g^*)$ is a Riemannian submersion with totally geodesic fibres. The Einstein condition for g is again equivalent to the Yang-Mills condition on ϕ and equations similar to (1.2) and (1.3).

In order to describe these equations precisely, recall that a point in W is an equivalent class $[p, x]$ where $p \in P, x \in F$ and $(p, x) \sim (pg, g^{-1}x)$. Having chosen a representative (p, x), there is an inclusion $i_p : F \to W$ given by $i_p(x) = [p, x]$. Because $i_{pg} = i_p \circ g$, i_p is an isometry between $(F, \langle\ ,\ \rangle)$ and the fibre through $[p, x]$ with the metric induced from g. To take care of horizontal directions, we make use of $j_x : P \to W$ given by $j_x(p) = [p, x]$, which satisfies $j_{gx} = j_x \circ R_g$. Then the equations analogous to (1.2) and (1.3) are respectively

$$(1.4) \quad Ric_F(i_{p*}^{-1}(U), i_{p*}^{-1}(V)) + \frac{1}{4}\sum_{i,j} \langle \overline{\Omega(\tilde{e}_i, \tilde{e}_j)}_x, i_{p*}^{-1}(U)\rangle \langle \overline{\Omega(\tilde{e}_i, \tilde{e}_j)}_x, i_{p*}^{-1}(V)\rangle = \Lambda\langle i_{p*}^{-1}(U), i_{p*}^{-1}(V)\rangle,$$

$$(1.5) \quad Ric(g^*)(\pi_*(X), \pi_*(Y)) - \frac{1}{2}\sum_i \langle \overline{\Omega(\tilde{X}, \tilde{e}_i)}_x, \overline{\Omega(\tilde{Y}, \tilde{e}_i)}_x\rangle = \Lambda g^*(\pi_*(X), \pi_*(Y)),$$

where $\sim$ denotes horizontal lifts and for $Z \in \mathfrak{g}$, $\overline{Z}_x$ denotes the value of the Killing field induced by Z on F at x. Unlike the principal bundle case, it is possible for $\overline{Z}$ to vanish at some points.

We now describe some Einstein metrics on bundles for which G acts transitively on F. The first family gives quaternionic analogues of Einstein metrics given by Theorem 1.1.

THEOREM 1.4. [**120**] *Let $(M_j, g_j^*), 1 \leq j \leq m$, be quaternionic Kähler manifolds with positive scalar curvature and P_j be the canonical $SO(3)$-bundle over M_j associated with the quaternionic-Kähler structure. Let $P = P_1 \times \cdots \times P_m$, $G = SO(3) \times \cdots \times SO(3)$, (m factors), and $F = G/\Delta SO(3)$ where $\Delta SO(3)$ denotes the diagonally embedded subgroup. Then $W = P \times_G F$ admits an Einstein* *metric with positive scalar curvature of type (1.1) submersing onto a product of the metrics g_j^* and having a normal homogeneous fibre metric.*

(A *normal homogeneous* metric on G/K is a G-invariant Riemannian metric induced by some bi-invariant metric on G, not necessarily positive definite.)

This theorem is proved in a similar way as Theorem 1.1. On the other hand, the Einstein metric can also be deduced as a special case of 3-Sasakian reduction discovered by Boyer, Galicki and Mann [**25, 26**]. See the article by the first two authors in this volume for details and up-to-date information.

Constructions similar to those in Theorem 1.4 can be performed with the bundles P in Examples 1.3 and 1.4. Namely, let $M_1 \times \cdots \times M_m$ be the m-fold product of the same compact quaternionic (resp. irreducible hermitian) symmetric space M, and let $\tilde{P}_j$ be the holonomy bundle of M_j with group $G = H \cdot K$ where $H = Sp(1)$ (resp. $U(1)$). Denote by P_j the quotient $\tilde{P}_j/H$, which is a principal $\overline{K}$-bundle. Then under certain conditions there are Einstein metrics of type (1.1) on $W = (P_1 \times \cdots \times P_m)/\Delta\overline{K}$. We refer the reader to Theorems 3 and 5 in [**120**] for details. Here we only mention two examples to indicate the possibilities.

EXAMPLE 1.6. For $M = \mathbb{HP}^n, n \geq 1$, there is an Einstein metric of type (1.1) on W if the number of factors m satisfies

$$2n^2(m-2) \leq n(3m^2 - 7m + 6) + 5m^2 - 5m + 2.$$

EXAMPLE 1.7. For $M = \mathbb{CP}^n$, there is an Einstein metric on W of type (1.1) provided

$$n^2(2-m) + n(m^2 - 2m) + 2m^2 + m - 2 \geq 0.$$

2. Einstein Metrics of Cohomogeneity One.

A. Generalities. Let G be a compact Lie group. A connected G-manifold is said to be of *cohomogeneity* 1 if the principal orbits are hypersurfaces. In this section we will be concerned with G-invariant Einstein metrics on such manifolds whose full isometry groups do not act transitively. The orbit space of a cohomogeneity 1 manifold is either an interval $\hat{I}$ whose boundary points represent singular orbits, or it is a circle. We will only concern ourselves with the former situation. For cohomogeneity 1 metrics, the Einstein condition reduces to a system of nonlinear ordinary differential equations on $\hat{I}$ together with appropriate boundary conditions to ensure that we have a smooth metric. The first systematic study of cohomogeneity 1 Einstein metrics was carried out in [**16**]. Some recent works about manifolds of cohomogeneity 1 which contain useful information include [**1, 8, 87, 97, 113**].

We will give first a geometric description of the Einstein condition for a cohomogeneity 1 metric following [**55**].

Let $(\hat{M}, \hat{g})$ be a cohomogeneity 1 G-manifold of dimension $n+1$ with a G-invariant metric. Let $P = G/K$ be the principal orbit type and $Q_i = G/H_i$ be the singular orbit types. There are at most 2 singular orbits, and when we are concentrating on one of them, we will use Q and H respectively to denote the orbit and its corresponding isotropy group. We can easily arrange for $K \subset H_i$. For example, we can choose a unit speed geodesic that starts from a singular orbit and intersects each principal orbit orthogonally. Then the points in the geodesic belonging to principal orbits all have the same isotropy group K, which then lies in the isotropy groups of the points on the geodesic belonging to the singular orbits. It follows from the cohomogeneity 1 condition that H_i must act transitively on the unit sphere in the normal slice to Q_i. So $H_i/K \cong S^{k_i}$, and P may be viewed as the unit sphere bundle of the normal bundle of Q_i in $\hat{M}$, which has the form $\nu(Q_i) = G \times_{H_i} V_i$, where H_i acts orthogonally on the slice representation $V_i \cong \mathbb{R}^{k_i+1}$. (This last identification is given by the normal exponential map.)

Let $\hat{M}_0$ denote the union of the principal orbits in $\hat{M}$. The geodesic chosen above gives a diffeomorphism $\hat{M}_0 \cong I \times P$, where $I = int(\hat{I})$. The pull-back of $\hat{g}$ via this diffeomorphism takes the form

$$dt^2 + g_t, \quad t \in I,$$

where g_t is a 1-parameter family of G-invariant metrics on P. It is occasionally useful to fix a background metric g_b on P of type (1.1) where g^* is a G-invariant metric on Q, ϕ is a connection for the principal bundle $H \to G \to G/H$, and $\langle\ ,\ \rangle$ is the constant curvature 1 metric on $H/K \cong S^k$. In terms of g_b, we can think of g_t as a g_b-symmetric endomorphism of TP. The Ricci tensor of g_t can be thought of as an endomorphism r_t of TP, symmetric with respect to g_t but not in general so with respect to g_b.

If we can construct a *smooth* metric $\hat{g}$ on $\hat{M}$ such that on $\hat{M}_0$ the Einstein equation is satisfied, then by continuity we have an Einstein metric on $\hat{M}$. In order to write down the Einstein equation on $\hat{M}_0$, we introduce the shape operator $\mathcal{L}_t$ of the principal orbits $\{t\} \times P$. This is the endomorphism of TP given by $\mathcal{L}_t(X) = \hat{\nabla}_X N$, where N is the unit vector field $\partial/\partial t$.

By using the Gauss and Codazzi equations, we easily obtain the Einstein equation for $\hat{g}$ on $\hat{M}_0$ as a system on P. This is the system below corresponding to the choice $\epsilon = 1$.

$$g' = 2g\mathcal{L}, \tag{2.1}$$

$$\mathcal{L}' + tr(\mathcal{L})\mathcal{L} - \epsilon r_t = -\epsilon\Lambda \cdot \mathbf{I}, \tag{2.2}$$

$$tr(\mathcal{L}') + tr(\mathcal{L}^2) = -\epsilon\Lambda, \tag{2.3}$$

$$tr(X \neg\, d^{\nabla}\mathcal{L}) = 0, \tag{2.4}$$

for all $X \in TP$, where Λ is the Einstein constant, $\neg$ denotes interior multiplication, and d^{∇} is the exterior covariant derivative $T^*P \otimes TP \to \Lambda^2(T^*P) \otimes TP$ formed using the Levi Civita connection ∇^t of g_t. If we take $\epsilon = -1$ instead, we obtain the Einstein condition for the Lorentz metric $-dt^2 + g_t$.

Note that (2.1) is essentially the definition of $\mathcal{L}_t$, which must also be symmetric with respect to g_t. Equation (2.4) is just $\widehat{Ric}(X, N) = 0$, and equations (2.2–2.3) represent the Einstein condition in the direction of the principal orbit and N respectively. Let s_t denote the scalar curvature of r_t. Then if we take the trace of (2.2) and use (2.3), we immediately obtain the equation

$$\epsilon s - (tr(\mathcal{L}))^2 + tr(\mathcal{L}^2) = (n-1)\epsilon\Lambda. \tag{2.5}$$

It is possible to interpret this equation as a first integral of a suitable Hamiltonian system.

By using the contracted second Bianchi identity, A. Back has deduced the following useful lemma [**13**].

LEMMA 2.1. *Let $\hat{g} = dt^2 + g_t$ be an equidistant family of hypersurfaces $I \times P$ satisfying (2.1) and (2.2) for some constant Λ. Let the scalar curvature s_t of g_t be constant for each $t \in I$ and v_t be the volume distortion of g_t with respect to some background metric on P. Then $\widehat{Ric}(X, N)v$ is constant in t for any $X \in TP$. Furthermore, if (2.4) is also satisfied, then $(\widehat{Ric}(N, N) - \Lambda)v^2$ is constant in t.*

Applying this lemma together with Theorem 5.2 in [**48**] gives

PROPOSITION 2.2. *Let $\hat{M}$ be a cohomogeneity 1 G-manifold with at least one singular orbit of dimension strictly smaller than that of the principal orbits. If $\hat{g}$ is a G-invariant metric of class C^3 such that (2.1) and (2.2) are satisfied on $I \times P$, then $\hat{g}$ is actually a smooth Einstein metric and hence real analytic.*

Proofs of the above statements can be found in [**55**]. Proposition 2.2 implies that we can focus on equation (2.2), provided we can ensure that the solution represents a smooth enough metric. Here, a C^3 metric is needed because the contracted second Bianchi identity is used in the proof. In special cases, the smoothness requirement can sometimes be weakened. We describe now a practical criterion for smoothness for the metrics $\hat{g}$, following [**55**], and then give an example illustrating how one applies this criterion in practice.

Let $\mathfrak{p}_+$ (resp. $\mathfrak{p}_-$) denote the subspace of the tangent space of G/K at the coset (K) corresponding to H/K (resp. G/H). For example, we could choose an $Ad(K)$-invariant decomposition

$$\mathfrak{g} = \mathfrak{k} \oplus \mathfrak{p}_+ \oplus \mathfrak{p}_-$$

such that $\mathfrak{h} = \mathfrak{k} \oplus \mathfrak{p}_+$ and $\mathfrak{p}_-$ are $Ad(H)$-invariant. A smooth G-invariant metric $\hat{g}$ on $G \times_H V$ is equivalent to an H-equivariant smooth map

$$\psi : V \longrightarrow S^2(V \oplus \mathfrak{p}_-),$$

where H acts on V via the slice representation and on $\mathfrak{p}_-$ by the isotropy representation of G/H. We can approximate ψ near the origin by Taylor polynomials whose homogeneous parts are H-equivariant polynomials of degree p on V with coefficients in $A := S^2(V \oplus \mathfrak{p}_-)$, i.e., elements of $Hom_H(S^p(V), A)$.

On the other hand, in writing $\hat{g}$ in the form $dt^2 + g_t$, we are really restricting ψ to a ray in V emanating from the origin. We then obtain a smooth curve $a(t)$ in A^K, the K-invariant elements in A. Conversely, given such a smooth curve $a : \mathbb{R}_+ \to A^K$, we obtain a smooth map $V \setminus 0 \to A$ by using the H-action. The smoothness question is when such a map extends smoothly to an H-equivariant map $\psi : V \to A$.

LEMMA 2.3. [**55**] *A smooth map $a : \mathbb{R}_+ \to A^K$ extends to a smooth map $\psi : V \to A$ as above iff each Taylor coefficient a_p of $a(t)$ is the restriction of an element of $Hom_H(S^p(V), A)$ to the unit sphere $S^k \subset V$.*

Clearly, entirely analogous criteria exist for smoothness of G-invariant tensors of other types on $G \times_H V$. One just has to replace A above by the relevant H-representation.

We now make some observations regarding the lowest degree Taylor coefficients. First, note that smoothness implies that $a_0 \in A^H$. Now

$$A^H = S^2(V \oplus \mathfrak{p}_-)^H = S^2(V)^H \oplus (V \otimes \mathfrak{p}_-)^H \oplus S^2(\mathfrak{p}_-)^H,$$

and V is an irreducible H-representation since H acts transitively on the unit sphere in V. The component of a_0 in $S^2(V)^H = \mathbf{1}$ is the Euclidean metric because in the exponential coordinate system, spheres with decreasing radii must become round to first order. The component of a_0 in $S^2(\mathfrak{p}_-)^H$ is just the G-invariant metric on Q induced by $\hat{g}$. a_0 has no component in $(V \otimes \mathfrak{p}_-)^H$ because V is the normal slice to Q at the coset $(H) \in G/H$. Thus a_0 is just the identity map relative to a suitable background metric.

Next we consider the first order Taylor coefficient a_1. Smoothness implies that it is an H-equivariant linear map $V \to A$. It is not difficult to see that there are no non-zero H-equivariant linear maps $V \to S^2(V)$. Hence, $tr(a_1)$ comes only from $V \to S^2(\mathfrak{p}_-)$. This part of a_1 is just the shape operator of Q by (2.1). Since $tr(a_1)$ is an H-invariant linear function on V, it must be zero. Hence we have deduced the following corollary using only local smoothness considerations.

COROLLARY 2.4. [**65**] *If $(\hat{M}, \hat{g})$ is a smooth Riemannian manifold of cohomogeneity* 1 *with a singular orbit Q, then Q is a minimal submanifold.*

In [**65**], the above corollary followed from an equivariant variational principle.

EXAMPLE 2.1. Let $\hat{M} = S^4$ be the unit sphere in $\mathbb{R}^5$, viewed as the space of 3×3 symmetric matrices with real entries and trace 0. Let $G = O(3)$ act by conjugation on these symmetric matrices. Then the principal orbits consist of matrices in S^4 with distinct eigenvalues and the principal isotropy group is $K = O(1)^3$. The two singular orbits comprise matrices in S^4 with 2 distinct eigenvalues. The isotropy group H is, up to conjugation, $O(2)\times O(1)$, and Q is the projective plane, minimally embedded as the Veronese surface. The isotropy representation of G/K is

$$(-1 \otimes -1 \otimes 1) \oplus (-1 \otimes 1 \otimes -1) \oplus (1 \otimes -1 \otimes -1),$$

where ± 1 denote respectively the trivial/non-trivial representation of $O(1) \approx \mathbb{Z}/2$. With the above choice of H, $\mathfrak{p}_+ = -1 \otimes -1 \otimes 1$. The slice representation at the singular orbit Q is $\rho^2 \otimes \mathbf{1}$, where ρ^m is the irreducible 2-dimensional representation of $O(2)$ lying in the mth symmetric power of the usual representation ρ^1 which does not already lie in the $(m-2)$nd symmetric power. Then we have H-module decompositions

$$S^2(\mathfrak{p}_-) = (\rho^2 \otimes \mathbf{1}) \oplus (\mathbf{1} \otimes \mathbf{1}),$$

$$S^m(V) = (\rho^{2m} \otimes \mathbf{1}) \oplus S^{m-2}(V).$$

Hence for $m \geq 1$, $Hom_H(S^{2m}(V), S^2(\mathfrak{p}_-)) \approx Hom_H(S^2(V), S^2(\mathfrak{p}_-))$, which is 1-dimensional and is generated by t^2 times the identity matrix. Likewise, we have $Hom_H(S^{2m-1}(V), S^2(\mathfrak{p}_-)) \approx Hom_H(V, S^2(\mathfrak{p}_-))$, which is again 1-dimensional, generated by

$$\begin{pmatrix} t_1 & t_2 \\ t_2 & -t_1 \end{pmatrix}$$

where (t_1, t_2) are Euclidean coordinates in $V \approx \mathbb{R}^2$ and $t^2 = t_1^2 + t_2^2$. Up to a constant, this is the shape operator of the Veronese surface in S^4. On the other hand, it is a general fact (see [**55**, §1, Lemma 2]) that for a compact linear group H acting transitively on the unit sphere in V, one has $Hom_H(S^{2m-1}(V), S^2(V)) = 0$ and $Hom_H(S^{2m}(V), S^2(V)) \approx Hom_H(S^2(V), S^2(V))$. In the present example, this last space has dimension 2 and for the generators one can take t^2 times the identity matrix and

$$\begin{pmatrix} t_2^2 & -t_1t_2 \\ -t_1t_2 & t_1^2 \end{pmatrix}.$$

However, only multiples of the second generator are candidates for the second order Taylor coefficient of a smooth metric.

Let B denote the bi-invariant metric on $O(3)$ given by $-tr(XY)$. We express g_t as

$$h(t)^2\ 2B|_{\mathfrak{p}_+} \oplus f_1(t)^2\ \frac{3}{2}B|_{\mathfrak{p}'_-} \oplus f_2(t)^2\ \frac{3}{2}B|_{\mathfrak{p}''_-}.$$

(The coefficients in front of B are chosen so that $dt^2 + 2B|_{\mathfrak{p}_+}$ is the Euclidean metric on $V = \mathbb{R}^2$.) It follows that smoothness of $\hat{g}$ means that

$$\begin{pmatrix} f_1(t)^2 \\ f_2(t)^2 \end{pmatrix} = \sum_{j=0}^{\infty} \left\{ a_{2j+1} \begin{pmatrix} 1 \\ -1 \end{pmatrix} t^{2j+1} + a_{2j} \begin{pmatrix} 1 \\ 1 \end{pmatrix} t^{2j} \right\}.$$

As for $h(t)$, smoothness is equivalent to it being odd with $h'(0) = 1$. Note that for the usual metric on S^4, in terms of B above, $h(t) = \sin t$, $f_1(t) = \cos t - \frac{1}{\sqrt{3}} \sin t$, and $f_2(t) = \cos t + \frac{1}{\sqrt{3}} \sin t$ over the interval $[0, \frac{\pi}{3}]$.

B. Initial Value Problem. A basic analytical question about the Einstein system (2.1–2.4) is the initial value problem. The easier case is the initial value problem at a principal orbit. Considerably subtler is the initial value problem at a singular orbit. We begin with the easier case.

THEOREM 2.5. [**55**] *Let G be a compact Lie group and K be a closed subgroup such that G/K is connected. Let h be a given G-invariant metric on G/K and $\mathcal{L}_0$ be an h-symmetric endomorphism of $T(G/K)$ such that for all $X \in T(G/K)$ we have $tr(X \neg d^{\nabla^h} \mathcal{L}_0) = 0$. Then there is a unique Einstein metric $\hat{g} = dt^2 + g_t$ defined on $(-\epsilon, \epsilon) \times G/K$, for some $\epsilon > 0$, with $g_0 = h$ and $\mathcal{L}_0$ equal to the shape operator of $\{0\} \times G/K$. Furthermore, $\hat{g}$ depends continuously on the initial values h and $\mathcal{L}_0$.*

Let us now assume that there is a singular orbit of strictly smaller dimension than the principal orbits. By Proposition 2.2, for the initial value problem, we need only consider the equations (2.1) and (2.2). In a neighbourhood around Q, the term $\mathcal{L}_t$ has t^{-1} dependence while r_t has t^{-2} dependence. So the differential equations have a singularity at $t = 0$. Of course, these equations are very nonlinear, especially because of the Ricci term, whose dependence on the metric g_t cannot be very explicitly written down if we want to leave G/K general. (r_t is a rational function of the components of g_t, but the constants in the expression depend on the specific G/K.) The linearization of (2.2) has the form $z' = t^{-2}\mathcal{A}(t)z$ where $\mathcal{A}(0)$ is a lower triangular matrix. The initial value problem for the linear case, though well-understood, is not completely trivial. In particular, a formal power series solution cannot be expected in all cases.

The singular initial value problem has been solved under an additional assumption.

THEOREM 2.6. [**55**] *Assume that as K-representations, V and $\mathfrak{p}_-$ have no irreducible sub-representations in common. Then, given any G-invariant metric g^* on Q and any G-equivariant homomorphism $\mathcal{L}_1 : \nu(Q) \longrightarrow S^2(T^*Q)$, there exists* *a smooth G-invariant Einstein metric on some open disk bundle of $\nu(Q)$ with any prescribed sign (positive, zero, or negative) of the Einstein constant Λ and having g^* and $\mathcal{L}_1$ as initial metric and shape operator on Q.*

The theorem is proved by the classical method of asymptotic series. The key step is to show that there is a formal power series solution any finite truncation of which defines a smooth metric on $\nu(Q)$. This involves input from geometry and representation theory since from a purely analytic point of view there is no reason to expect *power series* solutions at all. (From the smoothness discussion above, asymptotic series which are not power series do not give rise to a smooth metric.) One then applies a Picard iteration scheme to sufficiently high order truncations of the formal power series solution to get a smooth metric defined in a tube around Q. Alternatively, for this last step, one may quote a theorem of Malgrange [**83**].

Uniqueness is not true for the above singular initial value problem. It turns out that in general one needs to prescribe a finite number of additional Taylor coefficients in order to obtain a unique solution. These parameters can be calculated explicitly using representation theory once the triple $K \subset H \subset G$ is given. Non-uniqueness can be explained as follows. In constructing the formal power series solution, as is customary, one has to solve for Taylor coefficients recursively in terms of Taylor coefficients of lower degrees. The linear operators involved in this process are only injective above a certain critical degree which varies from situation to situation. Non-uniqueness comes from the kernels of these operators in lower degrees. In fact, there are sequences of examples for which the critical degrees tend to infinity (see example 3, §5 of [**55**]).

When the assumption on V and $\mathfrak{p}_-$ as K-representations does not hold, the initial value problem has been solved in the special case of the Kervaire spheres in [**13**]. The statement of the result is the same as in Theorem 2.6. It is conceivable that Theorem 2.6 holds without the technical assumption on V and $\mathfrak{p}_-$.

C. Examples With Special Holonomy. Under the further assumption of special holonomy, classification theorems are often available in addition to the construction of examples. We shall begin with cohomogeneity 1 hyperkähler metrics, which are metrics on $4n$-dimensional manifolds whose holonomy lies in $Sp(n)$. Alternatively, these are Riemannian manifolds which are Kähler with respect to 3 complex structures satisfying the multiplicative relations between the quaternions i, j, and k. See the article by A. Dancer in this volume for further information.

If we assume that the hyperkähler metric is irreducible, then since the Ricci tensor is zero, a cohomogeneity 1 metric exists only on a noncompact manifold. Calabi constructed [**31**] a complete hyperkähler metric on $T^*\mathbb{CP}^n$ of cohomogeneity 1 under $PSU(n+1)$. When $n = 1$, this metric was discovered earlier by Eguchi-Hanson [**54**]. In dimensions greater than 4, one has the following classification theorem.

THEOREM 2.7. [**43**] *Let $(\hat{M}, \hat{g})$ be an irreducible hyperkähler manifold of dimension greater than 4 which is of cohomogeneity 1 with respect to a compact simple Lie group G. Then, up to coverings, $\hat{M}$ is an open subset of either $T^*\mathbb{CP}^n$ with the Calabi metric or the $\mathbb{H}^*$ or $\mathbb{H}^*/\mathbb{Z}_2$ bundle over a quaternionic symmetric space of compact type with the Swann metric. If g is in addition complete, then it is isometric to the Calabi metric.*

R. Bielawski [**18**] independently obtained the classification theorem under the additional assumption of completeness.

To describe the Swann metric, recall from Example 1.2 that every quaternionic Kähler manifold has a canonical $SO(3)$ bundle over it. Therefore there is an associated $\mathbb{H}^*/\mathbb{Z}_2$ bundle, which is an $\mathbb{H}^*$ bundle in the case of the quaternionic projective space. A. Swann constructed an incomplete hyperkähler metric on this bundle in [**114**].

The above classification is also valid for a compact semisimple cohomogeneity one group action provided that any $\mathfrak{su}(2)$ ideal in $\mathfrak{g}$ acts trivially on the three complex structures on $\hat{M}$.

THEOREM 2.8. *A non-flat hyperkähler* 4*-manifold of cohomogeneity* 1 *with respect to a compact connected simple group is one of the following.*

(i) [**15**] *a member of a* 2*-parameter family of* $SU(2)$*-invariant incomplete examples or the Eguchi-Hanson metric on* $T^*\mathbb{CP}^1$,
(ii) [**60**] *the* $U(2)$*-invariant Taub-NUT metric on* $\mathbb{R}^4$,
(iii) [**11**] *up to a double covering, the* 2*-monopole space* M_2^0, *which is the unique complete hyperkähler* 4*-manifold with cohomogeneity* 1 *under* $G = SO(3)$ *and such that* G *rotates the complex structures,*
(iv) [**58**] *a member of a family of incomplete examples with* $G = SU(2)$, *which also acts transitively on the complex structures.*

Cohomogeneity 1 Kähler-Einstein metrics of non-positive scalar curvature on holomorphic line bundles over Kähler manifolds can be found among the bundle constructions of Calabi [**31**], Bérard Bergery [**16**], Page and Pope [**93**]. For these authors, the Euler class of the line bundle is proportional to the first Chern class of the base. Theorem 3.2 generalizes these examples in the bundle context to line bundles over a product of Fano manifolds such that the Euler class is a linear combination of the first Chern classes of the de Rham factors of the base. Furthermore, certain blow-downs of the zero section are also allowed, as was anticipated by Calabi [**31**, p. 277]. In the cohomogeneity 1 context, the choices for the Euler class of the line bundles are even more numerous. We have the following classification/existence theorem.

THEOREM 2.9. [**45**] *Let* G *be a compact connected semisimple Lie group acting with cohomogeneity* 1 *via isometries on a Kähler-Einstein manifold* $(\hat{M}, \hat{g})$ *which is irreducible and not hyperkähler. Suppose further that the isotropy representation of the principal orbit* G/K *splits into pairwise inequivalent irreducible subrepresentations.*

(i) *There is a coadjoint orbit* G/L *with a fixed invariant complex structure* J^* *so that* $K \subset L, L/K \cong S^1$ *and the induced metric on each principal orbit gives* $G/K \to G/L$ *the structure of a Riemannian submersion with totally geodesic fibres onto an invariant Kähler metric on* G/L.
(ii) *The complex structure on* $\hat{M}$ *is induced by* J^*, *the underlying connection of the Riemannian submersions, and the metric on the fibres. On* $\hat{M}_0$, *the union of all the principal orbits, the Kähler-Einstein metric can be expressed explicitly in terms of rational functions which depend on* $\dim H^2(G/L;\mathbb{R})$ *continuous parameters in the Ricci flat case and on a single constant of integration otherwise.*

(iii) *When there is a singular orbit G/H, then it is also a coadjoint orbit with an invariant complex structure induced from J^*. Moreover, it is a totally geodesic Kähler submanifold of $\hat{M}$ and H/L is analytically isomorphic to a complex projective space $\mathbb{CP}^{l-1}$.*

(iv) *Let χ denote the Euler class of the circle bundle $L/K \to G/K \to G/L$. Then the cohomology class*

$$c_1(G/L, J^*) + l\chi$$

is 0 when restricted to H/L, and, as an element of $H^2(G/H;\mathbb{R})$, is positive, zero, or negative depending on the sign of the Einstein constant.

(v) *The geometric data in (i), (iii), (iv) are sufficient for the construction of a smooth G-invariant Kähler-Einstein metric on a neighborhood of the zero section of the bundle $G \times_H \mathbb{C}^l$, and this metric extends to a complete metric on the underlying smooth vector bundle when the Einstein constant is non-positive.*

(vi) *If $(\hat{M}, \hat{g})$ is complete, then either there is a singular orbit G/H as above and $\hat{M} \cong G \times_H \mathbb{C}^l$, or else M is compact and the Einstein constant is positive.*

Of course, the condition on the isotropy representation of G/K is not always satisfied, but since it is satisfied for all coadjoint orbits G/L (L has maximal rank in G) a generic choice of K with $L/K \cong S^1$ will result in a G/K with the same property. In any event, the existence part of the theorem (i.e., part (v)) remains valid without this condition on the isotropy representation.

For the above theorem, the semisimplicity of G provides us with a moment map which takes orbits in $\hat{M}$ to coadjoint orbits in $\mathfrak{g}^*$. Under the assumption on the isotropy representation of G/K we obtain (i). The Einstein condition is then seen to be the same as (3.2-3.4) in the bundle situation discussed in the next section. One therefore gets explicit local solutions in the same manner. Note that the analysis of the singular orbits shows that the admissible quadruples (G, H, L, K) can be enumerated in terms of combinatorial data. Also, moduli of the Ricci-flat Kähler metrics come from the choice of an invariant Kähler metric on G/H. When the Einstein constant Λ is non-zero, the cohomology class in (iv) is really Λ times the Kähler class of the metric on G/H.

While the condition on the isotropy representation of G/K is generically satisfied, interesting Kähler-Einstein metrics nevertheless exist in situations where the condition does not hold. The Calabi metric on $T^*\mathbb{CP}^n$ is one example. We also have

THEOREM 2.10. [**112**] *There exists a complete Ricci-flat Kähler metric of cohomogeneity* 1 *on the cotangent bundle of a compact symmetric space of rank* 1.

The complex structures on the above spaces are special cases of adapted complex structures on tubes of zero sections of tangent bundles of real analytic manifolds constructed by Lempert, Szöke, [**82, 115**] and Guillemin and Stenzel [**59**].

Cohomogeneity 1 Kähler-Einstein metrics of positive scalar curvature were first constructed by Sakane [**110**] on certain $\mathbb{CP}^1$ bundles over a product of two compact Hermitian symmetric spaces. Later, Koiso and Sakane [**75, 76**] generalized this construction to the bundle (rather than the strictly cohomogeneity 1) situation and discovered the *sufficiency* of the vanishing of the Futaki invariant for existence in this set-up. (This is not true for the general existence problem in the Fano case, cf Tian's article.) These constructions will be discussed further in §3 below. As in the non-positive case there is the following classification/existence theorem.

THEOREM 2.11. [**75, 76, 45, 102**] *Let G and $(\hat{M}, \hat{g})$ be as in Theorem 2.9 and suppose that the Einstein constant is positive. In addition to (i) and (ii), we have the following analogues of (iii) and (iv):*

(iii)* *Each singular orbit G/H_i, $i = 1, 2$, is a coadjoint orbit with an invariant complex structure induced from J^*. They are totally geodesic Kähler submanifolds of $\hat{M}$. Furthermore, $H_i/L \cong \mathbb{CP}^{l_i - 1}$ and their isotropy representations have no common root spaces.*

(iv)* *Let χ be as in Theorem 2.9. Then for $i = 1, 2$, the class*

$$c_1(G/L, J^*) + (-1)^{i+1} l_i \chi$$

restricts to 0 on H_i/L and lies in the Kähler cone in $H^2(G/H_i; \mathbb{R})$.

The geometric data in (i), (iii) and (iv)* together with the vanishing of the Futaki integral*

$$\int_{-l_1}^{l_2} \prod_j (\lambda_j x - 1)^{d_j/2} x dx$$

are sufficient for the existence of a G-invariant Kähler-Einstein metric with positive constant on $\hat{M}$ having the stated orbit types.

In the above, d_j is the (real) dimension of the jth irreducible summand in the isotropy representation of G/L and λ_j is the corresponding eigenvalue of the curvature form of the circle bundle $L/K \to G/K \to G/L$, which can be expressed in terms of the first Chern class of G/L and the Euler class of the circle bundle. As in Theorem 2.9 the existence part does not require the condition on the isotropy representation of the principal orbit.

The special case of 4-dimensional Kähler-Einstein manifolds with cohomogeneity 1 has also been analysed. Here, the Ricci flat case is precisely the hyperkähler case, which has already been mentioned.

When $G = SU(2)$, Dancer and Strachan [**42**] proved that the complete cohomogeneity 1 Kähler-Einstein metrics with negative Einstein constant form two families. One of the families consists of $U(2)$-invariant metrics on complex line bundles over $\mathbb{CP}^1$ with Chern class < -2. These are just the noncompact Kähler examples discovered independently in [**16**], [**31**], and [**58**], and can be viewed as special cases of Theorem 3.2(ii) below. The second family consists of triaxial metrics, i.e., the metric components in the 3 independent directions in the principal orbits (S^3) are unequal. On the other hand, compact solutions must be the canonical Einstein metrics on $\mathbb{CP}^2$ and $\mathbb{CP}^1 \times \mathbb{CP}^1$.

Quaternionic-Kähler manifolds with positive scalar curvature and of cohomogeneity 1 with respect to a compact connected isometry group have been investigated in [**8**], resulting in a partial classification. Recently, Dancer and Swann [**44**] proved that a complete quaternionic-Kähler manifold with positive scalar curvature which has a semisimple compact group of isometries with cohomogeneity 1 must be quaternionic symmetric. The methods in [**44**] involve the associated twistor space of the quaternionic-Kähler manifold and its complex contact geometry. As a result, they also obtain information in the incomplete as well as non-compact cases.

For metrics of cohomogeneity 1 with holonomy G_2 or $Spin(7)$, see Theorem 3.7.

In closing this subsection on cohomogeneity one Einstein metrics with special holonomy, we would like to mention Hitchin's classification [**64**] of the cohomogeneity one $SU(2)$-invariant anti-self-dual Einstein metrics on 4-manifolds. Recall (cf LeBrun's article) that an oriented Riemannian 4-manifold is anti-self-dual (ASD) if the self-dual part of its Weyl tensor vanishes identically. While ASD Einstein metrics do not have special holonomy in general, the anti-self-duality gives an extra structure which can be used to analyse the Einstein condition via twistor theory. Furthermore, an ASD Einstein metric with zero scalar curvature is locally hyperkählerian, so Hitchin's classification includes the Einstein manifolds in Theorem 2.8.

THEOREM 2.12. [**64**] *Suppose that $(\hat{M}, \hat{g})$ is a complete ASD Einstein manifold with an isometric $SU(2)$ action with cohomogeneity* 1.

(i) *If the scalar curvature is positive, M is either S^4 or $\mathbb{CP}^2$ with the canonical metric.*
(ii) *If the scalar curvature is zero, then M is isometric to flat $\mathbb{R}^4$, $\mathbb{R}^4$ with the Taub-NUT metric, T^*S^2 with the Eguchi-Hanson metric, or the Atiyah-Hitchin* 2*-monopole space.*
(iii) *If the scalar curvature is negative, M is either the unit* 4*-ball with the flat metric, the Bergmann metric, Pedersen's metric* [**99**], *or a member of a family of metrics arising from solutions of Painlevé VI, or else M is the complex line bundle over S^2 with Euler class < -2 equipped with the Bérard Bergery metric.*

Part(i) of the above result recovers a well-known earlier theorem of Hitchin [**63**]. The conformal structure of the Bérard Bergery metric in (iii) was studied by Pedersen [**99**] and LeBrun [**81**]. In [**64**], Hitchin actually gives a local classification, from which the above global classification follows by examining completeness issues. The proof of the local classification is twistorial in nature. The $SU(2)$ action can be lifted to a Lie algebra of holomorphic vector fields on the twistor space Z. Generically, one obtains from this a section of the anti-canonical line bundle over Z and a flat connection on the trivial $SU(2)^{\mathbb{C}}$ bundle over the complement of the zero set of the above-mentioned section. Restricting the connection to a connected family of twistor lines intersecting the zero set transversally, one obtains an isomonodromic deformation of connections over $\mathbb{CP}^1$, whose residues can be associated to a solution of Painlevé's sixth. The Einstein condition then gives strong restrictions on the above data, and the local classification results from a detailed analysis of the possibilities. The non-generic situation corresponds to the locally hypercomplex case.

D. Examples with Generic Holonomy. Solutions of (2.1–2.4) with generic holonomy include some of the very first examples of cohomogeneity 1 Einstein metrics, e.g., the Page metric on $\mathbb{CP}^2\sharp(-\mathbb{CP}^2)$ [**91**] and its generalizations [**16, 93**]. These however will be dealt with in the broader context of §3, where the bundle structure plays a more important role and allows examples with little or no symmetry to be constructed. We would like to mention, however, that 4-dimensional Einstein orbifolds with $U(2)$-actions of cohomogeneity 1 have been studied in detail in [**100**]. Both Kähler and non-Kähler Einstein orbifolds with positive Einstein constants were found, but not with zero or negative constants. It is interesting to compare this study with Theorems 2.9, 2.11, and Theorems 3.1–3.5 below since these results show that blow-downs of the singular orbits can be realized on manifolds when the base of the bundle is more complicated. On the other hand, when there are no manifold solutions in these situations, it might in turn be possible to find many orbifold solutions as in [**100**].

We turn now to the work of C. Böhm, who studied the cohomogeneity 1 Einstein equations (2.1–2.4) in the situation where the isotropy representation of the principal orbit G/K splits into two inequivalent sub-representations and H/K is a sphere of dimension greater than 1. Because of the first integral (2.5), the Einstein equation can be thought of as a vector field on the 3-dimensional constant energy hypersurface defined by it.

THEOREM 2.13. [**19**] *There exists infinitely many pairwise non-isometric Einstein metrics of cohomogeneity* 1 *with positive scalar curvature on* $S^{n+1}, 4 \leq n \leq 8$.

This theorem provides for the first time infinitely many inhomogeneous Einstein metrics on standard spheres as well as the existence of more than one Einstein metric on even-dimensional spheres. The group G in these examples is $SO(p+1)\times SO(q+1)$ where $p+q=n$, $p,q \geq 2$ and the principal isotropy group $K = SO(p) \times SO(q)$. The two singular orbits are $S^p \times \{*\}$ and $\{*\} \times S^q$. Using this range of values of p and q, Böhm obtains one infinite sequence of pairwise non-isometric Einstein metrics on S^5 and S^6, *two* infinite sequences of non-isometric Einstein metrics on S^7 and S^8, and *three* such infinite sequences on S^9.

A new phenomenon is exhibited by these sequences of Einstein metrics. Let the Einstein constants be normalized to be equal to 1. For fixed p,q, the sequence of Einstein metrics converges in the Gromov-Hausdorff distance to the singular *Einstein* metric

$$dt^2 + \frac{p-1}{n-1}\sin(t)^2 g_{S^p} + \frac{q-1}{n-1}\sin(t)^2 g_{S^q}.$$

Away from the singular orbit, the sequence actually converges in the C^∞ topology. Notice that the volume of the limiting space is positive and the sectional curvatures blow up at the singularities. Also, the group action survives in the limit with the exception that the singular orbits are blown down to points. By contrast, a sequence of similarly normalized examples from Theorem 1.1 have bounded sectional curvatures and volumes tending to 0. Furthermore, if the diameter of the fibres tends to 0, which is automatic in the case of circle bundles, then the sequence of Einstein manifolds collapse (in the sense of Gromov) to the base with some product metric, not necessarily Einstein.

Because of the inexplicit nature of Böhm's solutions, one cannot yet decide whether or not the infinitely many Einstein metrics belong to different components of the Einstein moduli space. However, from the convergence to the singular Einstein space, it is possible to check that the Einstein metrics with different G's belong to different components of the moduli space and also belong to different components than the homogeneous Einstein metrics on spheres.

Besides low-dimensional spheres, Böhm has also constructed cohomogeneity 1 Einstein metrics on certain low-dimensional product manifolds.

THEOREM 2.14. [**19**] *There exists infinitely many non-isometric Einstein metrics of cohomogeneity* 1 *on* $\hat{M} = S^{p+1} \times Q^q$, *where* $5 \leq p+q+1 \leq 9, p>1, q>1$, *and* Q *is a non-flat compact isotropy irreducible homogeneous space* $\overline{G}/\overline{H}$.

In the above, the group G is $SO(p+1) \times \overline{G}$ and the principal isotropy group is $K = SO(p) \times \overline{H}$. The two singular orbits are both $Q = G/H$, with $H = SO(p+1) \times \overline{H}$. Note that $\overline{G}/\overline{H}$ is just the effective version of G/H.

Furthermore, Böhm was able to construct analytically an Einstein metric on $\mathbb{HP}^2\sharp(-\mathbb{HP}^2)$. Numerical solutions were obtained in **[94]** on the connected sum of two $\mathbb{HP}^n$ for a range of n.

We will now give a sketch of the methods employed by Böhm to obtain the above existence theorems.

In the situations of Theorems 2.13 and 2.14, the principal orbit is a product manifold whose isotropy representation consists of two inequivalent irreducible summands $\mathfrak{p}_1$ and $\mathfrak{p}_2$. Hence the metric $\hat{g}$ can be written as $dt^2 + f_1(t)^2 g_b|_{\mathfrak{p}_1} + f_2(t)^2 g_b|_{\mathfrak{p}_2}$, where g_b is an appropriately normalised background product Einstein metric on P. For Theorem 2.14, Böhm looks for solutions on an interval $[0,T]$ with boundary conditions $f_1(0) = 0 = f_1(T)$, $f_1'(0) = 1 = -f_1'(T)$, and $f_2(0) = f_2(T) = a > 0$, $f_2'(0) = 0 = f_2'(T)$. Geometrically, this means that reflection about the midpoint of the interval $[0,T]$ is an isometry and the principal orbit at the midpoint is totally geodesic. In Theorem 2.13, the boundary conditions used are instead $f_1(0) = 0 = f_2(T)$, $f_1'(0) = 1 = -f_2'(T)$, and $f_2(0) = a > 0, f_1(T) = b > 0$, $f_1'(T) = 0 = f_2'(0)$. In either case, the boundary conditions are precisely the smoothness conditions for the particular singular orbit type, and the initial value problem for Einstein metrics has a unique solution depending continuously on the single initial value a or b.

For any (local) solution emanating from a singular orbit it is first shown that the trace of the shape operator of the principal orbits is strictly decreasing and reaches zero before the maximal time of existence of the solution. Such a zero is called a *turning point*. Furthermore, all the critical points of the function $w = f_1/f_2$ of t are non-degenerate. Let N_a denote the number of critical points of w occurring before the turning point of the solution $f_a = (f_1, f_2)$ with initial value a. N_a is finite and remains constant as a is varied in an interval $[a_1, a_2] \subset \mathbb{R}_+$ provided that no a in the interval corresponds to a reflection symmetric solution, i.e., one which reaches a totally geodesic principal orbit ($f_1'(t^*) = f_2'(t^*) = 0$), which can therefore be extended to a global solution by reflection. On the other hand, if f_a passes through a reflection symmetric solution, then N_a jumps up or down by at most 1. Consequently, the change in N_a as a is varied can be used to detect and give a lower bound for reflection symmetric solutions.

In order to exploit this fact, Böhm shows that in the examples of the theorems above, N_a tends to $+\infty$ as $a \to 0$. It is here that the dimension restrictions in Theorems 2.13 and 2.14 enter crucially, together with the special properties of two-dimensional vector fields.

Recall that the first integral (2.5) implies that the Einstein equations can be viewed as a vector field on the three-dimensional constant energy manifold $E(w, w', f_2, f_2') = 0$. Böhm first shows that the spherical cone $(dt^2 + \sin(t)^2 g_b)$ of the product Einstein metric of the principal orbit is a local attractor for the integral curves of the Einstein vector field. Next, he uses special charts to study this vector field in detail and establishes certain rotational behaviour of the solutions. In a chart parametrized by w, f_2, f_2', it is shown that after a suitable blow-up, the Einstein vector field extends to a vector field V defined on a rectangular region in the boundary $f_2 \equiv 0$. V has two zeros: $(0,0)$, and z which corresponds to the spherical cone. Now z is a focal point (in the sense that the linearization of V at z has non-real eigenvalues) only if $n \leq 8$. In this case, the integral curve starting from $(0,0)$ eventually spirals around z. Here, one has to use Poincaré-Bendixson and also rule out the possibility of a limit cycle. From this behaviour of the integral curve, one can then deduce the limiting behaviour of N_a as $a \to 0^+$.

Theorem 2.14 now follows immediately from the above properties of N_a.

In order to prove Theorem 2.13, Böhm again uses the attracting property of the Einstein spherical cone. This time, he constructs a 2-dimensional slice whose origin is a point on the integral curve of the spherical cone. He shows that solutions emanating from the two singular orbits with small enough initial values a or b intersect this slice at a unique point. As a (resp. b) tends to 0, the locus of the intersection point is a clockwise (resp. anti-clockwise) spiral, both with the origin as limit point. The two spirals intersect in infinitely many points (in the slice), and each intersection represents an integral curve emanating from one singular orbit which continues to the other singular orbit. In this way, one obtains infinitely many Einstein metrics.

Finally, simple geometric arguments show that the metrics constructed cannot be homogeneous and cannot be isometric to each other.

In the case of $\mathbb{HP}^2\sharp(-\mathbb{HP}^2)$, the zero z of the vector field V is a node and one only has $N_a \geq 1$ as $a \to 0^+$.

Readers who are familiar with the many constructions of minimal submanifolds in spheres and other symmetric spaces using equivariant geometry will recognize that Böhm uses many of the same techniques. Of course, the Einstein equation is somewhat more complicated because one is dealing with a system rather than a single ODE.

A large family of complete, non-compact Einstein metrics of cohomogeneity 1 has very recently been found by Böhm [**21**] as a result of further study of the dynamic properties of the cohomogeneity 1 Einstein equations (2.1–2.4).

THEOREM 2.15. [**21**] *Let $m \geq 1$ and $k \geq 3$ be integers and $G_i/K_i, 1 \leq i \leq m$, be non-flat, compact isotropy irreducible spaces. Then $\mathbb{R}^k \times G_1/K_1 \times \cdots \times G_m/K_m$ has an m-dimensional family of complete Einstein metrics with negative scalar curvature as well as an $(m-1)$-dimensional family of complete Ricci flat metrics. All these metrics are of cohomogeneity* 1 *under the group $SO(k) \times G_1 \times \cdots \times G_m$.*

In certain cases, Böhm also finds finite subgroups of $SO(k) \times G_1 \times \cdots \times G_m$ which act freely on the product manifold, and in this way obtain families of Einstein metrics on the corresponding quotient manifolds.(Compare Theorem 4.1(i) below.)

E. Non-Existence. It follows from the analyses in [**16**] and [**93**] that the cohomogeneity 1 Einstein equations, specifically (3.2–3.4), can fail to have global smooth solutions, and hence there are closed simply connected manifolds of cohomogeneity 1 with respect to a fixed G-action which do not admit any G-invariant Einstein metrics. We present here one rather intriguing example, which was already mentioned in [**17**, p. 275].

EXAMPLE 2.2. As in §1, let P_b be the principal $U(1)$ bundle over $S^2 = \mathbb{CP}^1$ with Euler class $b \cdot \alpha$, where α is the generator of $H^2(S^2; \mathbb{Z})$ corresponding to the hyperplane bundle. P_b is really the lens space $U(2)/(U(1) \cdot \mathbb{Z}_b)$. The associated $\mathbb{CP}^1$

bundles are closed manifolds with an almost effective cohomogeneity 1 $U(2)$-action. (In fact, with the natural induced complex structure, these are the Hirzebruch surfaces.) For $|b| \geq 2$, it follows from **[16]** or **[93]** that there are no $U(2)$-invariant Einstein metrics. However, there are only 2 diffeomorphism types among the $\mathbb{CP}^1$ bundles: $S^2 \times S^2$ when b is even and $\mathbb{CP}^2 \sharp(-\mathbb{CP}^2)$ when b is odd. These two smooth manifolds admit respectively a homogeneous (the product metric) and a cohomogeneity 1 Einstein metric (the Page metric). This shows that the same manifold can have infinitely many cohomogeneity 1 actions by the same abstract group, but only some actions support invariant Einstein metrics.

Recently, Böhm has obtained a non-existence criterion for cohomogeneity 1 Einstein metrics on closed manifolds in terms of the orbit structure and the geometry of the principal orbit.

THEOREM 2.16. **[20]** *Let $\hat{M}$ be a closed G-manifold with cohomogeneity* 1 *and two singular orbits $Q_i = G/H_i$, $i = 1, 2$. Let G/K be the principal orbit type, with $K \subset H_i$. Suppose that $\mathfrak{h}_i = \mathfrak{k} \oplus \mathfrak{p}_i$ are Ad(K) invariant decompositions, and $\mathfrak{m}_1 \oplus \cdots \oplus \mathfrak{m}_l$ is the decomposition of the isotropy representation of G/K into Ad(K) invariant isotypic components. (Isotypic means a direct sum of equivalent irreducible representations.) If for some j, $\mathfrak{m}_j$ is Ad(K)-irreducible, $\mathfrak{m}_j \cap (\mathfrak{p}_1 \cup \mathfrak{p}_2) = \{0\}$, and the restriction of the trace-free part of the Ricci tensor of* **any** *G-invariant metric on G/K to the summand $\mathfrak{m}_j$ is negative definite, then there cannot be any smooth G-invariant Einstein metrics on $\hat{M}$.*

A large number of examples satisfying the hypotheses of the above theorem can be constructed **[20]** using compact homogeneous manifolds which do not admit any homogeneous Einstein metrics (see §4A). A simple example is the following. Let $G = SO(k+1) \times \overline{G}$, $H = SO(k+1) \times \overline{H}$, and $K = SO(k) \times \overline{H}$, where $\overline{G}/\overline{H}$ is $SO(2l)/(SO(l) \times U(1)), l \geq 32$. The G-manifold is $S^{k+1} \times (\overline{G}/\overline{H})$, and has no G-invariant Einstein metrics if $1 \leq k \leq l/3$.

3. Modified Kaluza-Klein Ansatz on Fibre Bundles.

In this section we consider a useful modification of the Kaluza-Klein ansatz. Let $\tilde{\pi} : \tilde{P} \longrightarrow M$ be a principal G-bundle and F a manifold on which G acts almost effectively with cohomogeneity 1. Let K denote a principal isotropy group of this action and let H (resp. H_1, H_2) denote the isotropy group(s) of the singular orbit(s). We will only refer to the situation having one singular orbit since analogous statements hold for the other singular orbit, if it is present. As in §2 we may assume that $K \subset H$. Recall also that H/K is diffeomorphic to a sphere S^k.

Let W be the manifold $\tilde{P} \times_G F$. Then W is the union of a 1-parameter family of hypersurfaces diffeomorphic to $P = \tilde{P} \times_G (G/K) = \tilde{P}/K$ which collapse onto $Q = \tilde{P} \times_G (G/H) = \tilde{P}/H$. P is an $H/K \cong S^k$ bundle over Q and this sphere bundle may be identified with the unit sphere bundle of the normal bundle of Q in W.

We can construct a metric $\hat{g}$ on W as follows. We choose a connection ϕ on the principal bundle $\tilde{P}$. This induces connections on $W \to M$ as well as on P and Q. Recall that a G-invariant metric on F can be written as $dt^2 + q_t$ where q_t is a 1-parameter family of homogeneous metrics on G/K defined on an interval $I = (0, T)$ where T is allowed to be $+\infty$. Let g_t^* be a 1-parameter family of metrics on M defined on the same interval. Using the connection ϕ, we let

$$\hat{g} = dt^2 + q_t + \tilde{\pi}^* g_t^*. \tag{3.1}$$

Of course there has to be boundary conditions at $t = 0$ and, in the compact case, at $t = T$ to ensure that $\hat{g}$ is smooth, just as in the cohomogeneity 1 case. Notice that for a fixed t, $g_t = q_t + \tilde{\pi}^* g_t^*$ makes $(P, g_t) \longrightarrow (M, g_t^*)$ into a Riemannian submersion with totally geodesic fibres. The modified Kaluza-Klein ansatz asks for $\hat{g}$ to be Einstein.

In order to write down the Einstein condition for $\hat{g}$, we need only observe that $W \setminus Q \cong I \times P$ is an equidistant family of hypersurfaces with unit normal field $N = \partial/\partial t$. Introducing the shape operators $\mathcal{L}_t$ of the hypersurfaces as in the cohomogeneity 1 case, we see that the Einstein condition for $\hat{g}$ is again given by (2.1–2.4) where the Ricci operator r_t of g_t can be computed using the theory of Riemannian submersions.

If we examine the arguments in Lemma 2.1 and Proposition 2.2 we find that Proposition 2.2 also holds in the present situation provided that for each t the scalar curvature of g_t^* is constant and the pointwise norm of the curvature form Ω is a constant function on $\tilde{P}$.

We will first discuss a special case that gives rise to large families of Einstein hermitian metrics and also unifies and generalizes many known examples. In particular, we obtain Einstein metrics on certain Fano manifolds when Kähler-Einstein metrics are obstructed (see Theorems 3.3 and 3.4).

Let (M_j, J_j, g_j^*) be Kähler-Einstein Fano manifolds as in §1 and let P_χ be a principal $U(1)$ bundle over $M = M_1 \times \cdots \times M_m$, where its Euler class $\chi = \sum_j b_j \pi_j^* \alpha_j \in H^2(M; \mathbb{Z})$. For F we take $\mathbb{C}$ or $S^2 = \mathbb{CP}^1$ or $\mathbb{RP}^2$, on which S^1 acts by complex multiplication in the first two cases. In the last case, the circle acts by the induced action on the $\mathbb{Z}/2$ quotient, so that the singular orbits are a point and a circle. Then W is a complex line bundle (resp. $\mathbb{CP}^1$, $\mathbb{RP}^2$ bundle) over M. As before, the connection ϕ on P_χ will be chosen so that $\frac{\sqrt{-1}}{2\pi} d\phi$ is the pull-back of a 2-form harmonic with respect to the product metric on M. We let

$$\hat{g} = dt^2 + h(t)^2 \langle\ ,\ \rangle + \sum_{j=1}^{m} f_j(t)^2 \pi_j^* g_j^*,$$

where $h, f_1, \cdots, f_m$ are smooth positive functions on I and $\langle\ ,\ \rangle$ is the metric on S^1 so that it has length 2π. We shall denote the real dimension of M_j by $2n_j$.

The Einstein equation is the following system.

$$-\frac{h''}{h} - \sum_{j=1}^{m} 2n_j \frac{f_j''}{f_j} = \Lambda \tag{3.2}$$

$$-\frac{h''}{h} - \sum_{j=1}^{m} 2n_j \frac{h' f_j'}{h f_j} + \sum_{j=1}^{m} n_j \frac{b_j^2}{2} \frac{h^2}{f_j^4} = \Lambda \tag{3.3}$$

$$-\frac{f_i''}{f_i} - \frac{h' f_i'}{h f_i} - \sum_j 2n_j \frac{f_i' f_j'}{f_i f_j} + \left(\frac{f_i'}{f_i}\right)^2 - \frac{b_i^2}{2} \frac{h^2}{f_i^4} + \frac{r_i^*}{f_i^2} = \Lambda \tag{3.4}$$

where r_i^* is the Ricci endomorphism of g_i^*, which is $p_i \cdot \mathbf{I}$ in our situation.

For the detailed analysis of these equations we refer the reader to [**45, 118**]. We will, however, make a few remarks about the special features of this system. First, equating the first two equations gives the relation

$$\sum_{j=1}^{m} 2n_j \left(\frac{f_j''}{f_j} - \frac{h' f_j'}{h f_j} + \frac{b_j^2}{4} \frac{h^2}{f_j^4} \right) = 0. \tag{3.5}$$

Let

$$\mu_i = \frac{f_i''}{f_i} - \frac{h' f_i'}{h f_i} + \frac{b_i^2}{4} \frac{h^2}{f_i^4}. \tag{3.6}$$

It turns out that explicit solutions can be obtained by setting all μ_i to be identically zero. To see this, we change variables by letting $dr = h(t)dt$ and defining $A_0(r) = h(t)^2$ and $A_i(r) = f_i(t)^2$. Then the Einstein equations become

$$\frac{1}{2} A_0'' + \frac{1}{2} A_0'(\log v)' + A_0 \sum_{j=1}^{m} n_j \left(\frac{A_j''}{A_j} - \frac{1}{2} \left(\frac{A_j'}{A_j} \right)^2 \right) = -\Lambda, \tag{3.7}$$

$$\frac{1}{2} A_0'' + \frac{1}{2} A_0'(\log v)' - \frac{A_0}{2} \sum_{j=1}^{m} n_j \frac{b_j^2}{A_j^2} = -\Lambda, \tag{3.8}$$

$$\frac{A_0' A_i'}{2A_i} + \frac{A_0}{2} \left(\frac{b_i^2}{A_i^2} + \frac{A_i''}{A_i} - \left(\frac{A_i'}{A_i} \right)^2 + (\log v)' \frac{A_i'}{A_i} \right) - \frac{r_i^*}{A_i} = -\Lambda, \tag{3.9}$$

where

$$v = \prod_{j=1}^{m} f_j^{2n_j} = \prod_j A_j^{n_j}.$$

There are two types of solutions of the equation $\mu_i = 0$. Either

$$A_i(r) = \pm (b_i r + a_i),$$

or

$$A_i(r) = \lambda_i (r + c_i)^2 - \frac{1}{4} \frac{b_i^2}{\lambda_i},$$

where a_i and λ_i are constants of integration. We shall refer to these two cases respectively as the linear and quadratic cases. Observe that (3.9) is a *linear* equation in A_0, so it can be integrated, and A_0 can be expressed in terms of the functions A_i.

In order to have explicit local solutions of (3.7–3.9), there will be consistency conditions to satisfy so that the m equations in (3.9) determine the same function A_0. One can then see that all of (3.7–3.9) hold. Using these local solutions, we can analyse the boundary conditions that will ensure that the solutions extend to a smooth metric on W. When W is non-compact, there are additional conditions which guarantee a complete metric.

In the cases where $F = \mathbb{C}$ or S^2, there is a complex structure J_h on W obtained by lifting the product complex structure of the base to the horizontal tangent spaces using the connection ϕ and defining $J_h(N)$ to be $h^{-1}U$, where U is the infinitesimal generator of the S^1 action, i.e., $\phi(U) = \sqrt{-1}$. Then $\hat{g}$ is hermitian with respect to J_h. In the compact and Ricci-flat cases, J_h is equivalent to the natural complex structure induced from the base and fibres. Furthermore, the Kähler condition for

$\hat{g}$ becomes simply $A_i' = -b_i$, $1 \le i \le m$. Therefore, the linear case of the explicit solutions corresponds to Kähler metrics modulo orientation.

If $\hat{g}$ is Kähler, then the right S^1 action on $I \times P$ has a momentum map and it is easy to see that it corresponds to an anti-derivative of h. Thus the change of variable used above to simplify the equations (3.2–3.4) is not at all ad hoc. It gives an explicit representation of the Einstein metric $\hat{g}$ in terms of rational functions. Also, there is a geometrical interpretation of the condition $\mu_i = 0, 1 \le i \le m$. It can be shown (see [**118**, §7]) that it is equivalent to

$$-(J_h \cdot \hat{R})(X,Y,Z,V) := \hat{R}(J_hX, J_hY, J_hZ, J_hV) = \hat{R}(X,Y,Z,V),$$

for all tangent vectors X, Y, Z, V of W, where $\hat{R}$ in the above is the Riemann curvature tensor of $\hat{g}$.

Thus far in the case under consideration, we have $\tilde{P} = P_\chi$ and $Q = M$, and we will refer to the situation as an S^1 collapse. On the other hand, it is possible to have S^k collapses with $k > 1$. In that case, one of the factors of M must be $\mathbb{CP}^{l-1}$ with $k = 2l - 1$. Our convention is then that the S^1 collapse case is identified with the case when $l = 1$ and one of the factors of M reduces to a point.

We will now give precise statements of the existence theorems and describe the special cases which were previously known. We begin with the linear (Kähler) case.

THEOREM 3.1. [**75, 76**] *Let $m \ge 3$, $(M_1, J_1^*) = (\mathbb{CP}^{l_1-1}, can)$ and $(M_m, J_m^*) = (\mathbb{CP}^{l_m-1}, can)$. Suppose that $b_1 = -b_m = -1$ and $b_2, \cdots, b_{m-1}$ are non-zero integers. Suppose that further that $l_1 b_j > -p_j$ and $p_j > l_m b_j$ for all j, $2 \le j \le m-1$. Then there is a Kähler Einstein metric with positive scalar curvature on $[P_\chi \times_{U(1)} \mathbb{CP}^1]/\sim$, (where $\sim$ means collapsing $M \times \{0\}$ onto $M_2 \times \cdots \times M_m$ and/or $M \times \{\infty\}$ onto $M_1 \times \cdots \times M_{m-1}$) iff the Futaki integral*

$$\int_{-l_1}^{l_m} \prod_{j=1}^{m} \left(\frac{p_j}{b_j} - x \right)^{n_j} x\,dx = 0.$$

(Note that $n_1 = l_1 - 1$ and $n_m = l_m - 1$. Hence when $l_1 = 1$ or $l_m = 1$, then the corresponding factors are identically 1 in the above integral.) Actually, Theorem 3.1 is a version of the existence theorem of Koiso-Sakane adapted to the present framework in order to facilitate comparison with Theorem 3.4 below. The general form of their theorem [**75**, Theorem 4.2] deals with compactifications of hermitian line bundles over a Kähler-Einstein Fano manifold such that the eigenvalues of the curvature form of the line bundles are constant with respect to the Ricci form of the base.

THEOREM 3.2. [**117, 118, 45**] *Let $m \ge 2$ and $(M_1, J_1^*) = (\mathbb{CP}^{l_1-1}, can)$, $l_1 \ge 1$. Assume that χ is determined by integers $b_1 = -1$ and $b_2, \cdots, b_m$.*

- (i) *If $b_j l_1 = -p_j$ for all $j \ge 2$, then there is an $(m-1)$-parameter family of complete Ricci-flat Kähler metrics on $[P_\chi \times_{U(1)} \mathbb{C}]/\sim$, where $\sim$ means collapsing $M \times \{0\}$ onto $M_2 \times \cdots \times M_m$.*
- (ii) *If $-b_j l_1 > p_j$ for all $j \ge 2$, then there exists a complete Kähler-Einstein metric with negative scalar curvature and infinite volume on $[P_\chi \times_{U(1)} D]/\sim$, where D is an open disk containing 0 in $\mathbb{C}$ and $\sim$ is as in (i).*

The $m = 2, l_1 = 1$ case is due to Bérard Bergery. We turn next to the quadratic case and begin with compact examples.

THEOREM 3.3. [**117, 118, 45**] *Let* $b_1, \cdots, b_m$, $m \geq 2$, *be integers satisfying*

$$|b_1| = 1, \quad 0 < l_1|b_j| < p_j, \quad j > 1$$

where $(M_1, J_1^*) = (\mathbb{CP}^{l_1-1}, can)$, $l_1 \geq 1$. *Then there is an Einstein metric with positive scalar curvature on* $[P_\chi \times_{U(1)} \mathbb{CP}^1]/\sim$ *where* $\sim$ *means collapsing* $M \times \{0\}$ *and* $M \times \{\infty\}$ *down to* $M_2 \times \cdots \times M_m$. *The Einstein metric is hermitian with respect to* J_h *and has* $\mathbb{Z}/2$ *symmetry about the equator of* $\mathbb{CP}^1$.

When $m = 2$ and $l_1 = 1$, we obtain the well-known Einstein metrics of Bérard Bergery [**16**] and Page and Pope [**93**]. Taking M_2 further to be $\mathbb{CP}^1$ we recover the Page metric [**91**] on $\mathbb{CP}^2\sharp(-\mathbb{CP}^2)$, which has the distinction of being the first compact *inhomogeneous* Einstein metric of positive scalar curvature discovered. Note that in these cases $\mu_i = 0$ is not an additional condition.

THEOREM 3.4. [**117, 118, 45**] *Let* $m \geq 3$ *and* $(M_1, J_1^*) = (\mathbb{CP}^{l_1-1}, can)$, $(M_m, J_m^*) = (\mathbb{CP}^{l_m-1}, can)$, *with* $l_1 \geq 1, l_m \geq 1$. *Suppose that there are integer* $b_1, \cdots, b_m$, *and* $\varepsilon_j = \pm 1, 1 \leq j \leq m$, *with the following properties:*

(i) $|b_1| = 1 = |b_m|$.
(ii) *For* $2 \leq i \leq m-1$, *if* $\varepsilon_i = 1$, *then* $0 < l_1|b_i| < p_i$ *and if* $\varepsilon_i = -1$, *then* $0 < l_m|b_i| < p_i$.
(iii) $\varepsilon_1 = 1$, $\varepsilon_m = -1$, *and when* $l_1 = l_m = 1$, *then at least one of the* ε_i, $2 \leq i \leq m-1$, *is positive.*
(iv) *The integral*

$$\int_{-l_1}^{l_m} \prod_{j=1}^{m} \left(\frac{p_j}{|b_j|} + \varepsilon_j x \right)^{n_j} x dx < 0.$$

Then there exists an Einstein metric with positive scalar curvature on $[P_\chi \times_{U(1)} \mathbb{CP}^1]/\sim$, *where* $\sim$ *means collapsing* $M \times \{0\}$ *onto* $M_2 \times \cdots \times M_m$ *and/or collapsing* $M \times \{\infty\}$ *onto* $M_1 \times \cdots \times M_{m-1}$. *The Einstein metric is hermitian with respect to the complex structure* J_h *and does not have* $\mathbb{Z}/2$ *symmetry with respect to the equator of* S^2.

The Einstein metrics in this theorem should perhaps be grouped with the Kähler-Einstein metrics in Theorem 3.1. This is because the integral condition in (iv) complements that in Theorem 3.1, which expresses the vanishing of the Futaki invariant evaluated on the (real) holomorphic vector field $h\partial/\partial t$. For example, over $\mathbb{CP}^1 \times \mathbb{CP}^n, n > 1$, the $\mathbb{CP}^1$ bundles with $(b_1, b_2) = (-1, k), 0 < k < n+1$, do not admit any Kähler Einstein metric, but they all admit an Einstein metric by the above theorem.

As for non-Kähler Einstein metrics with non-positive Einstein constant, we have

THEOREM 3.5. [**117, 118, 45**] *Let* $(M_1, J_1^*) = (\mathbb{CP}^{l_1-1}, can)$ *with* $l_1 \geq 1$ *and* $b_1, \cdots, b_m$, $m \geq 2$, *be non-zero integers.*

(i) *If* $l_1|b_j| < p_j$ *for* $j > 1$ *and* $|b_1| = 1$ *whenever* $l_1 > 1$, *then there exists a complete Ricci flat* J_h*-hermitian metric on* $[P_\chi \times_{U(1)} \mathbb{C}]/\sim$, *where* $\sim$ *denotes collapsing* $M \times \{0\}$ *onto* $M_2 \times \cdots \times M_m$.
(ii) *If* $|b_1| = 1$ *when* $l_1 > 1$, *then there exists a* 1*-parameter family of non-homothetic complete Einstein metrics with negative constant on* $[P_\chi \times_{U(1)} D]/\sim$, *where* $D \subset \mathbb{C}$ *is an open disk about* 0 *and* $\sim$ *denotes collapsing*

$M \times \{0\}$ onto $M_2 \times \cdots \times M_m$. These Einstein metrics are J_h hermitian and have infinite volume.

In all the above results (3.2)–(3.5), the versions without blow-downs were first obtained in [**117**]. Complete proofs together with a study of the associated hermitian geometry and analogues in Einstein-Weyl geometry can be found in [**118**]. The present versions were announced in the preprint version of [**45**]. The proofs consist of combining the analysis in [**118**] with the study of the blow-down situation in the cohomogeneity 1 Kähler case in [DaWa].

Just as in [**31**] and [**16**], it is also possible to construct both Kähler and non-Kähler Einstein metrics with negative constant on the total spaces of complex line bundles over a product of compact Kähler-Einstein manifolds with negative first Chern class and/or compact Ricci-flat Hodge manifolds. For a factor M_j of the first type, as in §1, we shall write the first Chern class as $-p_j \cdot \alpha_j$, where α_j is indivisible, p_j is positive, and we normalize the Kähler-Einstein metric to have constant $-p_j$. For a Ricci-flat factor, we shall assume that the Kähler class is of the form $2\pi\alpha_j$, i.e., the Kähler-Einstein metric is Hodge.

THEOREM 3.6. *Let $M = M_1 \times \cdots \times M_m$ be a product of compact Kähler-Einstein manifolds of the above types and P_χ be the principal $U(1)$-bundle over M with Euler class $\chi = \sum_j b_j \pi_j^* \alpha_j$, $b_j \neq 0$. Let W_χ denote the associated complex line bundle with the induced complex structure.*

(i) *If $b_j < 0$ for all j, then there exists a complete Kähler-Einstein metric with negative constant on a disk subbundle of W_χ.*
(ii) *For any choice of the b_j, there exists a 1-parameter family of non-homothetic hermitian but non-Kähler Einstein metrics with negative constant on a disk subbundle of W_χ.*

Note that in the above theorem, the base M can contain factors of both types.

The modified Kaluza-Klein ansatz can also be used to construct Einstein metrics with special holonomy.

THEOREM 3.7. [**29, 57**] *There are complete Einstein metrics of type (3.1) with the indicated holonomy on the following bundles:*

(i) *the bundle of anti-self-dual 2-forms over S^4 or $\mathbb{CP}^2$; holonomy type G_2*
(ii) *the bundle of real spinors over S^3; holonomy type G_2*
(iii) *the negative spin bundle over S^4 regarded as a self-dual manifold; holonomy type $Spin(7)$.*

In (i) above, $\tilde{P}$ is an $SO(3)$-bundle and $SO(3)$ acts on $F = \Lambda^2_-(\mathbb{R}^4) \cong \mathbb{R}^3$ in the usual manner. Likewise in (iii), $\tilde{P}$ is an $SU(2)$-bundle with $SU(2)$ acting in the usual way on $F = \mathbb{C}^2$, which is also a cohomogeneity 1 action. The fact that the base manifolds are self-dual and Einstein is significant because these properties provide $\tilde{P}$ with the Yang-Mills connection ϕ required in the construction. In (ii), $\tilde{P}$ is the trivial $SU(2)$ bundle over S^3 and $SU(2) = Spin(3)$ acts on $\mathbb{H}$ by quaternion multiplication. The bundles in (ii) and (iii) are both topologically trivial. Furthermore, all the metrics in the above theorem are explicit and, since the base manifolds are homogeneous, are of cohomogeneity 1 as well. Many incomplete metrics with G_2 or $Spin(7)$ holonomy were also found in [**29**]. We refer the reader to the article by Joyce for further information about the search for Einstein metrics with these holonomy groups.

Finally, we mention some numerical solutions of the Einstein system of the modified Kaluza-Klein ansatz.

In [**94**], the authors considered the situation where $G = Sp(1)$ and $\tilde{P} = S^{4n+3}$ is the total space of the Hopf bundle over $\mathbb{HP}^n$. The fibre F is $S^4 = \mathbb{H} \cup \{\infty\}$ or $\mathbb{H}$, on which $Sp(1)$ acts as the unit quaternions. In the first case, W is $\mathbb{HP}^{n+1} \sharp (-\mathbb{HP}^{n+1})$. Numerical evidence for an Einstein metric of type (3.1) was given. As was pointed out in §3, the case $n = 1$ has now been analytically established by Böhm [**19**]. He also produced numerical evidence for a second solution when $n \geq 2$. In the second case, W is either $\mathbb{H}^{n+1}$ or a non-trivial quaternionic line bundle over $\mathbb{HP}^n$. Page and Pope produced numerical solutions with negative and zero Einstein constants. Recently, Böhm [**21**] gave a proof of these numerical results as well as the corresponding result for the Hopf bundle over the Cayley plane.

A second situation was studied by Gibbons, Page, and Pope in [**57**]. Here, $G = SO(3)$ and $\tilde{P} \to M$ is the canonical $SO(3)$ bundle of a quaternionic Kähler manifold with positive scalar curvature. The fibre F is either $S^3 = \mathbb{R}^3 \cup \{\infty\}$ or $\mathbb{R}^3$ with G acting in the usual way as rotations. The hypersurfaces P are the quaternionic-Kähler twistor spaces of M. In the first case, numerical solutions exist when the dimension of M is small, e.g., when $M = S^4$ or $\mathbb{CP}^2$. Again, Böhm found a second numerical solution for $M = \mathbb{HP}^n$ for a certain n. In the second case, numerical solutions were also found, which have again recently been proved in [**21**] for all values of $n \geq 3$.

4. Homogeneous Einstein Metrics.

The Einstein condition for a homogeneous metric is a system of algebraic equations for which one seeks a real solution satisfying some positivity condition reflecting the positive definiteness of the metric which the solution represents. By the theorem of Alekseevsky-Kimel'fel'd [**6**], Ricci-flat homogeneous spaces are flat. A homogeneous Einstein manifold with positive Einstein constant must be compact with finite fundamental group by Bonnet-Myers, while an Einstein manifold with negative Einstein constant must be noncompact by Bochner's theorem.

A. Positive Einstein Constant. We begin with a qualitative picture. Homogeneous Einstein metrics on a compact homogeneous space G/K are precisely the critical points of the scalar curvature function on the space of G-invariant metrics with a fixed value for the volume. This fact was first exploited in [**67**] for left-invariant metrics on Lie groups.

THEOREM 4.1. [**123**] *Let G/K be an effective, compact homogeneous space with G and K compact and connected. Let S denote the function that assigns to each G-invariant metric of volume 1 its scalar curvature.*

(i) *S is bounded from below iff the universal cover of G/K has the form $\mathbb{R}^k \times (G_1/K_1) \times \cdots \times (G_l/K_l)$ where G_j/K_j are isotropy irreducible. In this case, S is proper iff $k = 0$. When $k = 0$, there is a unique critical point and S is bounded below by a positive constant. The critical point corresponds to the product Einstein metric. If $k \geq 1$, then S has a critical point iff G/K is a torus.*

(ii) *S is bounded from above and proper iff $\mathfrak{k}$ is a maximal subalgebra (by inclusion) in $\mathfrak{g}$. In this case, S has a global maximum which is therefore an Einstein metric.*

((iii) *S is bounded from above but is not proper iff $K \cdot U(1)$ is a subgroup of G and $G/(K \cdot U(1))$ is a compact irreducible hermitian symmetric space other than $SO(n+2)/(SO(n)SO(2))$.*

Isotropy irreducible spaces will be defined and described presently. Compact homogeneous spaces with $\mathfrak{k}$ maximal in $\mathfrak{g}$ are quite numerous and Theorem 4.1 therefore gives rise to many homogeneous Einstein manifolds. Roughly speaking, there are as many maximal subalgebras as there are irreducible representations. To explain this we recall some classification theorems of Dynkin in [**52**]. Let K be a compact simple Lie group, $\rho : K \longrightarrow SU(N)$ be an irreducible finite-dimensional unitary representation of K, and τ be the usual complex 1-dimensional representation of the circle. What Dynkin proved in the unitary case can be formulated (see [**121**]) more succintly as follows. $\rho(K)$ is maximal in $SU(N)$ unless $\rho(K) \subset SU(n) \subset SU(N)$ in which (a) the realification of $\rho \otimes \tau : K \times U(1) \longrightarrow U(n)$ is the isotropy representation of an irreducible hermitian symmetric space, and (b) the second inclusion belongs to a distinguished subset of the (irreducible) exterior powers of the vector representation of $SU(n)$. For the symplectic and orthogonal cases, see [**121**].

When K is closed but not necessarily connected, and $\mathfrak{k}$ is maximal in $\mathfrak{g}$, then 4.1(ii) still implies that G/K admits a G-invariant Einstein metric. This is because the $Ad(K)$-invariant inner products on the tangent space at the coset (K) form a closed subset of the $Ad(K_0)$-invariant inner products, where K_0 is the identity component of K.

Theorem 4.1 implies that when $\mathfrak{k}$ is not maximal in $\mathfrak{g}$ then generically the scalar curvature function S would be unbounded from above and from below. In this case, S can fail to have any critical points. For example, let $G = SU(4)$ and $SU(2)$ be the subgroup given by the irreducible 4-dimensional (symplectic) representation. Then $M = SU(4)/SU(2)$ does not have any homogeneous Einstein metric. Other families of similar examples can be found in [**123**]. A more complicated family was constructed recently by Park and Sakane [**96**]. The existence of compact homogeneous spaces which do not admit homogeneous Einstein metrics shows that in higher dimensions, Hamilton's Ricci flow with an initial metric of positive Ricci tensor need not converge to an Einstein metric, even if it exists for all time. (Hamilton's flow preserves the symmetries of the initial metric.)

From the point of view of Riemannian geometry, isometric Einstein metrics are always identified. In the homogeneous situation, the normalizer $N(K)$ of K in G acts on G/K via the adjoint action and hence induces an action of $N(K)/K$ on the space of invariant metrics. The moduli space of homogeneous Einstein structures can therefore be regarded as the quotient of the set of G-invariant Einstein metrics of volume 1 by this action. Since the space of volume 1 G-invariant Einstein metrics is diffeomorphic to Euclidean space, its quotient by the $N(K)/K$ action is contractible. (This is a special case of a general theorem of R. Oliver that the orbit space of a topological action of a compact Lie group on a contractible space is contractible.) Put another way, the domain of the scalar curvature function S does not acquire topology in passing to the quotient by $N(K)/K$.

An open general question for homogeneous Einstein manifolds with positive constant is the following. Suppose that G/K is a compact homogeneous manifold with finite fundamental group. Is it true that in the quotient of the space of G-invariant metrics with volume 1 by the above action of $N(K)/K$, there are only finitely many Einstein structures ?

Moving now to specific examples, we begin with *isotropy irreducible* spaces. These are connected homogeneous spaces G/K where K is compact and acts irreducibly on the tangent space. As was observed in [**127**], these spaces are Einstein since any $Ad(K)$-invariant inner product on the tangent space at the coset (K) must have Ricci tensor (also $Ad(K)$-invariant) proportional to itself by Schur's lemma. The main problem is therefore the classification of such spaces.

Among the isotropy irreducible spaces are the irreducible symmetric spaces, which were classified by Cartan in [**33, 34**]. In view of the importance of holonomy in the study of Einstein metrics, we might suggest [**126**, §2] as a modern version of Cartan's first method of classification, which is based on classifying the holonomy representation. The classification of the remaining isotropy irreducible spaces involves two steps. First, one classifies the *strongly isotropy irreducible* spaces, which are those for which the identity component of K already acts irreducibly on the tangent space. This classification is due independently to Manturov [**84, 85, 86**] and J. Wolf [**127**]. It should be noted that Wolf in addition made an extensive study of the geometry of these spaces. Another classification of these spaces appeared in [**78**] much later. There is a conceptual relation between the strongly isotropy irreducible quotients of the classical groups and irreducible symmetric spaces which allows one to deduce the classification of the former from that of the latter. This relationship was noticed by C. T. C. Wall [**127**, pp. 147-148], and is proved in two different ways in [**126**] and [**62**].

The isotropy irreducible spaces which are not strongly isotropy irreducible were classified in [**125**] using the classification of normal homogeneous Einstein quotients of compact connected simple groups in [**122**]. It was shown in [**122**] that if G/K is a homogeneous space with G compact, connected, and simple, then the Killing form metric is Einstein iff the Casimir operator of the isotropy representation is a multiple of the identity. This fact formed the basis of the classification in [**122**], which is relevant to the isotropy irreducible case because the key case in this latter classification is the situation of a compact and simple G, with the Killing form inducing an Einstein metric on G_0/K_0, where G_0 and K_0 are the identity components of G and K respectively. Instead of describing the classification in [**125**] in detail, we will illustrate the phenomenon using an example.

EXAMPLE 4.1. Let $G = SO(nk), K = SO(k) \times \cdots \times SO(k)$ (n times). Then the isotropy representation of G/K is easily seen to be

$$\sum_{1\le i<j\le n} \mathbf{1}\otimes\cdots\otimes\mathbf{1}\otimes\rho_k\otimes\mathbf{1}\otimes\cdots\otimes\mathbf{1}\otimes\rho_k\otimes\mathbf{1}\otimes\cdots\otimes\mathbf{1},$$

where the ρ_k belongs to the ith and jth factor in each summand. Hence G/K is not strongly isotropy irreducible. Although the irreducible summands are inequivalent, they become equivalent under outer automorphisms of K. Because the Casimir constants are equal, the Killing form metric is Einstein. On the other hand, the outer automorphisms which interchange the $SO(k)$ factors actually extend to automorphisms of G. Therefore, if we enlarge G to include these automorphisms, we obtain an isotropy irreducible space. Note that depending on which automorphisms are included there could be different ways of enlarging G to give an isotropy irreducible space.

In general, therefore, one has to select from the spaces classified in [**122**] those whose isotropy representations are permuted transitively by automorphisms of K and carry out the corresponding analysis of the extensions of the necessary automorphisms to G.

The Einstein condition for the Killing form metric on G/K where G is semisimple, compact, connected but *non-simple* has been studied recently in [**104, 105, 106, 108, 109, 95, 89**].

In [**104, 108**], Rodionov proved that if G is compact, connected, semisimple but nonsimple, and K is a closed *simple* subgroup such that G/K is simply connected and effective, then $G = K \times \cdots \times K$ and K is embedded diagonally. So G/K is actually isotropy irreducible but not strongly isotropy irreducible unless $G = K \times K$, in which case we get a symmetric space. The automorphisms which one has to add to get isotropy irreducibility could be any subgroup of the permutation group which permutes the K factors in G transitively.

Several new infinite families of homogeneous spaces G/K whose Killing form metric is Einstein were produced in [**89**] and [**109**]. We mention here one of the families in these references.

EXAMPLE 4.2. [**109, 89**] Let $G = Sp(m+n) \times Sp(n)^p$ and $K = Sp(m) \times Sp(n)$, where $m, n \geq 1$. Suppose also that the embedding of K in G is given by composition of

$$id \times \Delta : Sp(m) \times Sp(n) \subset Sp(m) \times Sp(n)^{p+1},$$

where Δ denotes the diagonal map, and the obvious embedding

$$(Sp(m) \times Sp(n)) \times Sp(n)^p \subset Sp(m+n) \times Sp(n)^p.$$

Then it was shown in [**109**] that the Einstein condition is satisfied for the Killing form metric on G iff m, n, p satisfy the diophantine equation $2n^2 + (4-p)n + 2 - p = 2m(p(n+1)+m)$. This equation was completely solved in [**89**] and there are infinitely many solutions.

In the study of homogeneous manifolds it sometimes happens that several Lie groups act transitively on the same manifold. If a subgroup $H \subset G$ acts transitively on G/K, then the isotropy group of the H-action is $K \cap H$. Therefore the dimension of the space of invariant metrics could very well increase. When this happens, there may be more Einstein metrics. This phenomenon is of special interest if the underlying manifold is important in many geometrical situations. The irreducible symmetric spaces fit this criterion. In this regard, Ziller determined all the homogeneous Einstein metrics on the compact rank 1 symmetric spaces [**130**]. The non-symmetric Einstein metrics on these manifolds all come from Hopf fibrations by scaling the fibres differently. They are

1. the Jensen metric on S^{4n+3} from the fibration $Sp(n+1)/Sp(n) \longrightarrow Sp(n+1)/(Sp(n) \times Sp(1))$,
2. the metric found in [**24**] on S^{15} from the fibration $Spin(9)/Spin(7) \longrightarrow Spin(9)/Spin(8)$,
3. the Ziller metric on $\mathbb{CP}^{2n+1}$ from the twistor fibration $Sp(n+1)/(Sp(n) \times U(1)) \longrightarrow Sp(n+1)/(Sp(n) \times Sp(1))$.

THEOREM 4.2. [**70**] *Let G/K be a compact irreducible symmetric space of rank > 1 such that G is the identity component of the isometry group of the symmetric metric. Assume that $G/K \neq (K \times K)/K$. Then there is a non-symmetric homogeneous Einstein metric with respect to a transitive subgroup $H \subset G$ precisely in the following cases:*

(i) $G = SO(2n), K = U(n), H = SO(2n-1), \; H \cap K = U(n-1), \; n \geq 4,$
(ii) $G = SO(7), K = SO(2) \times SO(5), H = G_2, \; H \cap K = U(2),$
(iii) $G = SO(8), K = SO(3) \times SO(5), H = Spin(7), \; H \cap K = SO(4).$

There are respectively 1, 2, *and* 2 *non-symmetric Einstein metrics.*

The non-symmetric Einstein metric in Theorem 4.2(i) was first found in [**123**]. The case of group manifolds was studied earlier in [**67, 68**] and more extensively in [**40**]. While a complete classification has not been achieved, there are numerous examples of left-invariant Einstein metrics. For example, every compact simple Lie group of dimension greater than 3 has a left-invariant Einstein metric other than the Killing form metric. $SO(2n)$ and $SO(2n+1)$ have at least $3n-2$ distinct left-invariant Einstein metrics.

Another important family of homogeneous spaces consists of the coadjoint orbits of compact connected semisimple Lie groups G, also known as the generalized flag manifolds. Each coadjoint orbit is of the form $G/C(T)$ where $C(T)$ is the centralizer of a torus in G. It has a natural invariant complex structure and the first Chern class is positive. The existence of a homogeneous Kähler-Einstein metric is due to Koszul. Both [**17**, chapter 8] and [**7**] are excellent references for these and other classical facts about coadjoint orbits. The question therefore arises if there are any other homogeneous Einstein metrics on the generalized flag manifolds.

The classification of [**122**] shows that the Killing form metric is Einstein for the generalised flag manifolds $SU(nk)/S(U(k)\times\cdots\times U(k))$, $Sp(3n-1)/(Sp(n)\times U(2n-1))$, $SO(3n+2)/(SO(n)\times U(n+1))$, $E_6/(Spin(8)\cdot U(1)\cdot U(1))$, and G/T where T is a maximal torus in a compact semisimple connected Lie group whose local factors are of type A_l, D_l, E_6, E_7 or E_8. In [**9**] and [**72**] all the homogeneous Einstein metrics on certain families of coadjoint orbits were determined by explicitly solving the Einstein equation, e.g., $SU(p+q+r)/S(U(p)U(q)U(r))$ and $SO(2n)/(U(n-1)U(1))$. In addition, Arvanitoyeorgos found at least $\frac{n!}{2}+n+1$ *solutions* of the Einstein equation on $SU(n)/T$ when $n \geq 4$, of which $n+1$ are non-Kähler. It should be pointed out that in [**9**] and [**72**], the distinct solutions sometimes represent isometric metrics. For example, the $\frac{n!}{2}$ solutions in the $SU(n)/T$ case are just the Kähler-Einstein metrics for the $n!$ distinct invariant complex structures (which correspond to the natural complex structures induced by different embeddings of $SU(n)/T$ as coadjoint orbits). The action of the Weyl group (the symmetric group) identifies them as Riemannian metrics. Similarly, n of the other solutions are also isometric. The remaining solution is the Killing form metric.

Homogeneous Einstein metrics with positive sectional curvature have been studied. One takes the classification of homogeneous spaces admitting positive sectional curvature (due to Berger, Bérard Bergery, Wallach, and B. Wilking) and examines which of the positively curved homogeneous metrics are Einstein as well. For the compact symmetric spaces of rank 1, Ziller [**130**] found that all homogeneous Einstein metrics have positive sectional curvature. In [**122**, 5.4] it is shown that in Berger's classification of normal homogeneous manifolds with positive sectional curvature, $Sp(2)/SU(2)$ is the only space other than the rank 1 symmetric spaces that has a positively curved normal homogeneous Einstein metric. The situation with the Aloff-Wallach spaces was studied in [**77**] (cf remarks about Theorem 1.3(i)). Although homogeneous Einstein metrics with positive curvature were found on an infinite family, a complete classification has not been achieved. The remaining cases of $F_4/Spin(8), SU(5)/(Sp(2)\cdot U(1)), Sp(3)/(Sp(1)Sp(1)Sp(1))$, and $SU(3)/T^2$ were studied by Rodionov in [**104**].

The classification of compact homogeneous Einstein manifolds of low dimension has also been attempted. A classical result of G. Jensen [**66**] is that all four-dimensional homogeneous Einstein manifolds are symmetric. Compact five-dimensional homogeneous Einstein manifolds with positive scalar curvature were classified in [**5, 107**]. Besides symmetric metrics and product metrics, there is an infinite family of homogeneous Einstein metrics on $S^2 \times S^3$. These are precisely the circle bundles over $\mathbb{CP}^1 \times \mathbb{CP}^1$ in Theorem 1.1. While they are all diffeomorphic, their homogeneous structures are all distinct. See remark (1) after Theorem (1.1).

While a complete classification has not been obtained, the dimension 7 case has been extensively studied by mathematicians as well as by theoretical physicists in connection with 11-dimensional supergravity theory. See [**50**, pp. 63-64] for more references. The circles bundles over $\mathbb{CP}^1 \times \mathbb{CP}^2$, $\mathbb{CP}^1 \times \mathbb{CP}^1 \times \mathbb{CP}^1$, and $SU(3)/T^2$ encountered in Theorems 1.1 and 1.3 are homogeneous Einstein manifolds of dimension 7. Indeed, every simply connected compact homogeneous 7-manifold admits a homogeneous Einstein metric with positive scalar curvature, by [**37**].

Finally, a well-known technical difficulty in studying homogeneous Einstein metrics is the presence of multiplicities in the decomposition of the isotropy representation into irreducible representations. In [**71**] examples of this phenomena are studied. They include $Spin(8)/G_2 \cong S^7 \times S^7$, $Spin(7)/SU(3) \cong S^7 \times S^6$, $Spin(8)/U(3) \cong S^7 \times G_2^+(\mathbb{R}^8)$, and the Stiefel manifolds $SO(n+1)/SO(n-1)$.

Kerr classified all the G-homogeneous Einstein metrics on these spaces, although some of the Einstein metrics were already known. In the first three cases, modulo the action of $N(K)/K$, there is at least one Einstein metric other than the product Einstein metric. In the case of the Stiefel manifolds, there is a unique invariant Einstein metric, a fact already proved in [**14**] and later reproved in [**10**]. The existence of this Einstein metric, however, already follows from Kobayashi's theorem [**74**], since the Stiefel manifold under consideration is a circle bundle over the corresponding oriented Grassmann of 2-planes. In all of the above cases except $Spin(8)/G_2$, the group $N(K)/K$ is a circle and Kerr uses this action to eliminate one of the off-diagonal components of the invariant metrics. It turns out that apart from $Spin(8)/U(3)$ all the non-product Einstein metrics, up to isometry, can be found among G-invariant metrics which are diagonal with respect to a fixed decomposition of the isotropy representation.

B. Negative Einstein Constant. Let $(M = G/K, g)$ be a homogeneous Einstein manifold with negative scalar curvature. Then it follows from Bochner's theorem that both M and G are non-compact. The well-known unsolved conjecture of D. V. Alekseevsky asserts that K must be a maximal compact subgroup of G. If this conjecture is true, then using the Levi and Iwasawa decompositions, it follows that (M, g) is isometric to a left-invariant metric on some solvable Lie group S. Indeed, all known examples are isometric to left-invariant metrics on simply connected solvable Lie groups. If S is unimodular, then a theorem of Dotti-Miatello [**49**] shows that there are no left-invariant Einstein metrics, generalizing an earlier result of Milnor [**88**] for the nilpotent case.

The most classical examples of homogeneous Einstein manifolds with negative scalar curvature are the symmetric spaces of non-compact type and the non-compact homogeneous Kähler-Einstein manifolds, which are shown to be precisely the bounded homogeneous domains with the Bergmann metric in [**39**] and in unpublished work of Koszul. For the latter class of examples, Piatetskii-Shapiro [**101**] has shown that there exists continuous Einstein moduli on the bounded homogeneous domains. The remaining case for which the holonomy is not generic is that of homogeneous quaternionic-Kähler manifolds, whose classification was begun by Alekseevsky [**2**] and completed by Cortés [**38**].

We will therefore consider below the non-unimodular case with generic holonomy. Also, we will confine ourselves to the simply connected case and hence will describe results in terms of Lie algebras. We begin with several definitions, following [**128**] and [**61**].

A *metric solvable Lie algebra* is a pair $(\mathfrak{s}, g)$, where $\mathfrak{s}$ is a finite-dimensional solvable Lie algebra and g is a left-invariant metric. A solvable Lie algebra is *completely solvable* if for all $X \in \mathfrak{s}$ the eigenvalues of ad_X are real. A metric solvable Lie algebra $(\mathfrak{s}, g)$ is called *standard* if the g-orthogonal complement of the derived algebra $[\mathfrak{s}, \mathfrak{s}]$ is an abelian subalgebra $\mathfrak{a}$ of $\mathfrak{s}$. The left-invariant metric g is then referred to as a standard left-invariant metric. Finally, a standard metric solvable Lie algebra is of *Iwasawa type* if in addition (a) for all $0 \neq X \in \mathfrak{a}$, $ad_X \neq 0$ and is a symmetric linear operator with respect to g, and (b) there exists some $X_0 \in \mathfrak{a}$ such that $ad_{X_0}| \, [\mathfrak{s}, \mathfrak{s}]$ has positive eigenvalues.

There are many examples of simply connected Einstein solvmanifolds with non-positive sectional curvature. See [**3, 47, 41, 22**] for earlier results. Recently, Wolter [**128**] constructed two infinite families of such examples using Heisenberg groups. These examples include the examples of Deloff. Wolter's construction uses a general sufficient condition for a left-invariant metric on a solvable Lie algebra of Iwasawa type to be Einstein. This condition is then verified in special cases to give Einstein solvmanifolds with non-positive sectional curvature. In [**80**], Lanzendorf classified all the left-invariant Einstein metrics (with non-positive sectional curvature) that can be obtained by Wolter's construction, and obtained some new examples.

In [**53**], an infinite family of quite explicit examples was constructed as follows. Let K be a compact connected Lie group with an almost faithful irreducible representation ρ on $\mathbb{R}^n$. Let $\mathfrak{g} = \mathfrak{k} \oplus \mathbb{R}^n \oplus \mathbb{R}A$ be equipped with an inner product g such that the three summands are orthogonal, $g(A, A) = 1$, on $\mathbb{R}^n$ it is an arbitrary K-invariant inner product, and on $\mathfrak{k}$ it is $-\frac{1}{4n}$ times the trace form of ρ. Define the Lie bracket on $\mathfrak{g}$ by using the Lie bracket of $\mathfrak{k}$ and by declaring that $ad(A)$ acts as the identity on $\mathbb{R}^n$ and twice the identity on $\mathfrak{k}$, that $[\mathfrak{k}, \mathbb{R}^n] = 0$, and that $[\mathbb{R}^n, \mathbb{R}^n] \subset \mathfrak{k}$ so that for $X, Y \in \mathbb{R}^n$, $Z \in \mathfrak{k}$, one has $g([X, Y], Z) = g(\rho(Z)X, Y)$. Then g gives a left-invariant Einstein metric on the associated simply connected solvable Lie group.

The above examples as well as all other known examples are of standard type.

Recently, J. Heber, building on the work in [**128**], has made a very systematic study of Einstein solvmanifolds of standard type. We will only describe some of the results in [**61**], referring the reader to that paper for details as well as other results.

Heber has also constructed in [**61**] many examples of solvable Lie algebras which do not admit any left-invariant Einstein metrics. While nonstandard left-invariant Einstein metrics are not known to exist, there is a dichotomy and a uniqueness theorem for standard Einstein metrics.

THEOREM 4.3. [**61**] *Let $\mathfrak{s}$ be a solvable Lie algebra. If $\mathfrak{s}$ admits a standard left-invariant Einstein metric, then it cannot admit a non-standard left-invariant Einstein metric. Furthermore, if $\mathfrak{s}$ admits two standard left-invariant Einstein metrics g_1 and g_2, then there is a positive constant c and an isometry ϕ such that $g_1 = c\phi^* g_2$. If $\mathfrak{s}$ is in addition completely solvable, then ϕ is an automorphism of $\mathfrak{s}$.*

Heber has also characterized when an Einstein left-invariant metric must be of standard type. The characterization is an open algebraic condition, which is satisfied, for example, if $\mathfrak{s}$ is completely solvable, or if the Killing form is either non-negative semidefinite or have signature at most 1. According to Azencott and Wilson [**12**], $(\mathfrak{s}, g)$ is standard as well if g has non-positive sectional curvature.

The first step towards classifying standard left-invariant Einstein metrics is the theorem in [**61**] that such a metric solvable Lie algebra $(\mathfrak{s}, g)$ is isometric to one of Iwasawa type after possibly altering the Lie bracket. Combining [**4**, Corollary 1.10] with this theorem, one obtains the following interesting fact.

THEOREM 4.4. [**4, 61**] *A simply connected standard Einstein solvmanifold with negative scalar curvature has a quotient of finite volume iff it is symmetric.*

Returning to the situation of the associated (solvable) algebra of Iwasawa type of a standard, Einstein, metric solvable Lie algebra, a canonical element $X_0 \in \mathfrak{a}$ can now be chosen such that ad_{X_0} acting on $[\mathfrak{s}, \mathfrak{s}]$ has eigenvalues which are positive integers with no common divisors. The ordered list of eigenvalues together with their multiplicities is called the *eigenvalue type* of $(\mathfrak{s}, g)$. For a fixed dimension, only finitely many eigenvalue types can occur. Heber then gives an outline of the steps towards a complete classification of standard Einstein metrics. In particular, he gives a reduction theorem which allows the description of Einstein solvable Lie algebras of Iwasawa type for which $\dim \mathfrak{a} > 1$ in terms of those with $\dim \mathfrak{a} = 1$.

Now let n be a positive integer and define $\mathcal{M}^n$ to be space of all Einstein metric solvable Lie algebras of dimension n and scalar curvature -1 modulo the action of the diffeomorphism group of the underlying $\mathbb{R}^n$. Let $\mathcal{M}^n_{st}$ be the subspace of Einstein algebras of standard type. Both spaces are regarded as subspaces of the space of all metric solvable Lie algebras of dimension n modulo diffeomorphisms, equipped with the C^∞ topology.

THEOREM 4.5. [**61**] *Let $\mathcal{M}^n_\lambda \subset \mathcal{M}^n_{st}$ be the subset of standard Einstein metric solvable Lie algebras with a fixed eigenvalue type λ. Then*

(i) *$\mathcal{M}^n_{st}$ is a finite disjoint union of spaces $\mathcal{M}^n_\lambda$ which are homeomorphic to compact, semi-algebraic analytic subsets of some auxiliary Euclidean space,*
(ii) *each $\mathcal{M}^n_\lambda$ is open in $\mathcal{M}^n$ in the C^∞ topology.*

Finally, Heber has computed the dimensions of those spaces $\mathcal{M}^n_\lambda$ which contain an irreducible non-compact symmetric space of rank 1.

THEOREM 4.6. [**61**] *Let $\mathcal{M}^n_\lambda$ contain a rank 1 symmetric space of non-compact type. In the cases of real or complex hyperbolic space, $\mathcal{M}^n_\lambda$ consists of only one point.*

For quaternionic hyperbolic space of real dimension $n = 4(m+1)$, *if* $m = 1$ *then the symmetric metric is an isolated point in* $\mathcal{M}^n_\lambda$, *while for* $m \geq 2$, *a neighborhood of the symmetric metric in* $\mathcal{M}^n_\lambda$ *has dimension* $8m^2 - 6m - 8$. *For the hyperbolic Cayley projective plane, the corresponding dimension is 84.*

There is actually a geometrical description of a neighborhood of the symmetric space in $\mathcal{M}^n_\lambda$ in terms of spaces of orbits in certain representations. See [**61**] for further details.

References

[1] A. V. Alekseevsky & D. Alekseevsky, Riemannian G-manifolds with one-dimensional orbit space, Advances in Soviet Math., **8**, (1992), 1-31.

[2] D. V. Alekseevsky, Classification of quaternionic spaces with a transitive solvable group of motions, Math. USSR-Izv., **9**, (1975), 297-339.

[3] D. V. Alekseevsky, Homogeneous Riemannian spaces of negative curvature, Math. USSR-Sbornik, **96**, (1975), 93-117.

[4] D. V. Alekseevsky & V. Cortés, Isometry groups of homogeneous quaternionic Kähler manifolds, Preprint 1995.

[5] D. Alekseevsky, I. Dotti & C. Ferraris, Homogeneous Ricci positive 5-manifolds, Pacific J. Math., **175**, (1996), 1-12.

[6] D. Alekseevsky & B. N. Kimel'fel'd, Structure of homogeneous Riemannian spaces with zero Ricci curvature, Func. Anal. Appl., **9**, (1975), 97-102.

[7] D. V. Alekseevsky & A. M. Perelomov, Invariant Kähler-Einstein metrics on compact homogeneous spaces, Func. Anal. Appl., **20**, (1986), 171-182.

[8] D. V. Alekseevsky & F. Podestà, Compact cohomogeneity one Riemannian manifolds of positive Euler characteristic and quaternionic Kähler manifolds, in *Geometry, Topology, Physics, Proceedings of the First USA-Brazil Workshop*, Campinas, Brazil, 1996, de Gruyter, Berlin, (1997), 1-33.

[9] A. Arvanitoyeorgos, New invariant Einstein metrics on generalized flag manifolds, Transac. A. M. S., **337**, (1993), 981-995.

[10] A. Arvanitoyeorgos, $SO(n)$-invariant Einstein metrics on Stiefel Manifolds, *Differential Geometry and Applications, 1995 Brno*, (1996), 1-5.

[11] M. Atiyah & N. Hitchin, *The geometry and dynamics of magnetic monopoles*, Princeton University Press, (1988).

[12] H. Azencott & E. Wilson, Homogeneous manifolds with negative curvature I, Transac. A. M. S., **215**, (1976), 323-362; II: Mem. A.M.S., **8**, (1976).

[13] A. Back, Local theory of equivariant Einstein metrics and Ricci realizability on Kervaire spheres, Preprint 1986.

[14] A. Back & W. Y. Hsiang, Equivariant geometry and Kervaire Spheres, Transac. A. M. S., **304**, (1987), 207-227.

[15] V. Belinskii, G. Gibbons, D. Page & C. Pope, Asymptotically Euclidean Bianchi IX metrics in quantum gravity, Phys. Lett., **76B**, (1978), 433-435.

[16] L. Bérard Bergery, Sur de nouvelles variétés riemanniennes d'Einstein, Publications de l'Institut E.Cartan, n^o4, (Nancy), (1982), 1-60.

[17] A. Besse, *Einstein Manifolds*, Ergebnisse der Mathematik und ihrer Grenzgebiete, 3. Folge, Band 10, Springer-Verlag, (1987).

[18] R. Bielawski, Invariant hyperkähler metrics with a homogeneous complex structure, Math. Proc. Camb. Phil. Soc., **122**, (1997), 473-482.

[19] C. Böhm, Inhomogeneous Einstein metrics on low-dimensional spheres and other low-dimensional spaces, Invent. Math., **134**, (1998), 145-176.

[20] C. Böhm, Non-existence of cohomogeneity 1 Einstein metrics, Preprint 1997.

[21] C. Böhm, Non-compact cohomogeneity one Einstein manifolds, Preprint 1998.

[22] J. Boggino, Generalized Heisenberg groups and solvmanifolds naturally associated, Rend. Sem. Mat. Univ. Politec. Torino, **43**, (1985), 529-547.

[23] J.-P. Bourguignon, A mathematician's visit to the Kaluza-Klein theory , *Conference on Partial Differential Equations and Geometry* (Torino, 1988), Rend. Sem. Mat. Univ. Politec. Torino, Fas. Spez , (1989), 143-163.

[24] J.-P. Bourguignon & H. Karcher, Curvature operators: pinching estimates and geometric examples, Ann. scient. Éc. Norm. Sup., 4e série, **t.11**, (1978), 71-92.

[25] C. Boyer, K. Galicki & B. Mann, Quaternionic reduction and Einstein metrics, Comm. Anal. Geom., **1**, (1993), 229-279.

[26] C. Boyer, K. Galicki & B. Mann, The geometry and topology of 3-Sasakian manifolds, Jour. reine angew. Math., **455**, (1994), 183-220.

[27] C. Boyer, K. Galicki & B. Mann, On strongly inhomogeneous Einstein manifolds, Bull. Lond. Math. Soc., **28**, (1996), 401-408.

[28] C. Boyer, K. Galicki, B. Mann & E. Rees, Compact 3-Sasakian 7-manifolds with arbitrary second Betti number, Invent. Math., **131**, (1998), 321-344.

[29] R. Bryant & S. Salamon, On the construction of some complete metrics with exceptional holonomy, Duke Math. Jour., **58**, (1989), 829-850.

[30] F. M. Cabrera, M. D. Monar & A. Swann, Classification of G_2-structures, J. Lond. Math. Soc., **53**, (1996), 407-416.

[31] E. Calabi. Métriques kähleriennes et fibrés holomorphes, Ann. Sci. Éc. Norm. Sup., 4e série, **12**, (1979), 269-294.

[32] M. Capria & S. Salamon, Yang-Mills fields on quaternionic spaces, Nonlinearity, **1**, (1988), 517-530.

[33] E. Cartan, Sur une classe remarquable d'espaces de Riemann, Bull. Soc. Math. France, **54**, (1926), 214-264; **55**, (1927), 114-134.

[34] E. Cartan, Sur la structure des groupes de transformation finis et continus, Thèse, Paris, (1894).

[35] L. Castellani, R. D'Auria & P. Fré, $SU(3) \Theta SU(2) \Theta U(1)$ from $D = 11$ supergravity, Nuclear Physics, **239 B**, (1984), 610-652.

[36] L. Castellani & L. J. Romans, $N = 3$ and $N = 1$ supersymmetry in a new class of solutions for $D = 11$ supergravity, Nuclear Physics, **238 B**, (1984), 683-701.

[37] L. Castellani, L. J. Romans & N. P. Warner, A classification of compactifying solutions for $D = 11$ supergravity, Nuclear Phys., **241 B**, (1984), 429-462.

[38] V. Cortés, Alekseevskian spaces, Diff. Geom. Appl., **6**, (1996), 129-168.

[39] J. D'Atri & I. Dotti-Miatello, A characterization of bounded symmetric domains by curvature, Trans. A. M. S., **276**, (1983), 531-540.

[40] J. D'Atri & W. Ziller, Naturally reductive metrics and Einstein metrics on compact Lie groups, Mem. A. M. S., **215**, (1979).

[41] E. Damek, The curvature of a semi-direct extension of a type H nilpotent group, Coll. Math., **53**, (1987), 249-253.

[42] A. Dancer & I. Strachan, Kähler-Einstein metrics with $SU(2)$ action, Math. Proc. Camb. Phil. Soc., **115** (1994), 513-525.

[43] A. Dancer & A. Swann, Hyperkähler metrics of cohomogeneity one, J. Geom. Phys., **21** (1997), 218-230.

[44] A. Dancer & A. Swann, Quaternionic Kähler manifolds of cohomogeneity one, Preprint 1998.

[45] A. Dancer & M. Wang, Kähler-Einstein manifolds of cohomogeneity one and bundle constructions for Einstein Hermitian metrics, Preprint 1997; Kähler-Einstein manifolds of cohomogeneity one, to appear in Math. Ann.

[46] R. D'Auria, P. Fré & P. van Nieuwenhuisen, $N = 2$ matter coupled supergravity from compactification on a coset G/H possessing an additional Killing vector, Phys. Lett. B, **136**, (1984), 347-353.

[47] E. Deloff, Naturally reductive metrics and metrics with volume preserving geodesic symmetries on NC algebras, Dissertation, Rutgers University, (1979).

[48] D. DeTurck & J. L. Kazdan, Some regularity theorems in Riemannian geometry, Ann. Sci. École Norm. Sup., 4e série, **14**, (1981), 249-260.

[49] I. Dotti-Miatello, Ricci curvature of left-invariant metrics on solvable unimodular Lie groups, Math. Zeit., **180**, (1982), 257-263.

[50] M. J. Duff, B. E. W. Nilsson & C. Pope, Kaluza-Klein supergravity, Phys. Reports, **130**, (1986), 1-142.

[51] E. B. Dynkin, Semisimple subalgebras of semisimple Lie algebras, Translations of the A.M.S., Series 2, **6**, (1957), 111–244.

[52] E. B. Dynkin, Maximal Subgroups of the Classical Groups, Translations of the A.M.S., Series 2, **6**, (1957), 245–378.
[53] P. Eberlein & J. Heber, Quarter-pinched homogeneous spaces of negative curvature, Int. J. Math., **7**, (1996), 441-500.
[54] T. Eguchi & A. Hanson, Asymptotically flat self-dual solutions to Euclidean gravity, Phys. Lett., **74B**, (1978), 249-251.
[55] J. Eschenburg & M. Y. Wang, The initial value problem for cohomogeneity one Einstein metrics, to appear in J. Geom. Analysis.
[56] T. Friedrich & I. Kath, 7-dimensional compact Riemannian manifolds with Killing spinors, Comm. Math. Phys., **133**, (1990), 543-561.
[57] G. Gibbons, D. Page & C. Pope, Einstein metrics on $S^3, \mathbb{R}^3$ and $\mathbb{R}^4$ bundles, Comm. Math. Phys., **127**, (1990), 529-553.
[58] G. Gibbons & C. Pope, The positive action conjecture and asymptotically Euclidean metrics in quantum gravity, Comm. Math. Phys., **66**, (1979), 267-290.
[59] V. Guillemin & M. Stenzel, Grauert tubes and the homogeneous Monge-Ampère equation, J. Diff. Geom., **34**, (1992), 561-570; II., ibid., **35**, (1992), 627-641.
[60] S. Hawking, Gravitational instantons, Phys. Lett., **60A**, (1977), 81-83.
[61] J. Heber, Non-compact homogeneous Einstein spaces, Invent. Math., **133**, (1998), 279-352.
[62] E. Heintze & W. Ziller, Isotropy irreducible spaces and s-representations, Diff. Geom. Appl., **6**, (1996), 181-188.
[63] N. Hitchin, Kählerian twistor spaces, Proc. Lond. Math. Soc., **43**, (1981), 133-150.
[64] N. Hitchin, Twistor spaces, Einstein metrics and isomonodromic deformations, Jour. Diff. Geom., **42**, (1995), 30-112.
[65] W. Y. Hsiang & H. B. Lawson, Minimal submanifolds of low cohomogeneity, Jour. Diff. Geom., **5**, (1971), 1-38.
[66] G. Jensen, Homogeneous Einstein spaces of dimension 4, Jour. Diff. Geom., **3**, (1971), 309-349.
[67] G. Jensen, The scalar curvature of left invariant Riemannian metrics, Indiana U. Math. J., **20**, (1971), 1125-1144.
[68] G. Jensen, Einstein metrics on principal fibre bundles, J. Diff. Geom., **8**, (1973), 599-614.
[69] G. Jensen & M. Rigoli, Einstein metrics on circle bundles, Proc. Symp. Pure Math., Vol. 54, Part 2, (1993), 331-336.
[70] M. Kerr, Some new homogeneous Einstein metrics on symmetric spaces, Trans. A. M. S., **348**, (1996), 153-172.
[71] M. Kerr, New examples of homogeneous Einstein metrics, Michigan J. Math., **45**, (1988), 115-134.
[72] M. Kimura, Homogeneous Einstein metrics on certain Kähler C-spaces, *Recent topics in Differential and Analytic Geometry*, Adv. Studies in Pure Math., **18-I**, (1990), 303-320.
[73] S. Kobayashi, On compact Kähler manifolds with positive definite Ricci tensor, Ann. Math., **74**, (1961), 570-574.
[74] S. Kobayashi, Topology of positively pinched Kähler manifolds, Tôhoku Math. J., **15**, (1963), 121-139.
[75] N. Koiso & Y. Sakane, Nonhomogeneous Kähler-Einstein metrics on compact complex manifolds, in *Curvature and Topology of Riemannian Manifolds*, (Katata, 1985), Lecture Notes in Math., Vol. 1201, Springer-Verlag, (1986), 165-179.
[76] N. Koiso & Y. Sakane, Non-homogeneous Kähler-Einstein metrics on compact complex manifolds II, Osaka J. Math., 25 (1988), 933-959.
[77] O. Kowalski & Z. Vlásek, Homogeneous Einstein metrics on Aloff-Wallach spaces, Diff. Geom. & Appl., **3**, (1993), 157-167.
[78] M. Krämer, Eine Klassifikation bestimmter Untergruppen kompakter zusammenhängender Liegruppen, Comm. Alg., **3**, (1975), 691-737.
[79] M. Kreck & S. Stolz, A diffeomorphism classification of 7-dimensional homogeneous Einstein manifolds with $SU(3) \Theta SU(2) \Theta U(1)$ symmetry, Ann. Math., **127**, (1988), 373-388.
[80] M. Lanzandorf, Einstein metrics with nonpositive sectional curvature on extensions of Lie Algebras of Heisenberg type, Geom. Ded., **66**, (1997), 187-202.
[81] C. LeBrun, Counterexamples to the generalized positive action conjecture, Comm. Math. Phys., **118**, (1988), 591-596.

[82] L. Lempert & R. Szöke, Global solution of the homogeneous complex Monge-Ampère equation and complex structures on the tangent space of Riemannian manifolds, Math. Ann., **290**, (1991), 689-712.

[83] B. Malgrange, Sur les points singuliers des équations différentielles, L'Enseignment Math., **20**, (1974), 147-176.

[84] O. V. Manturov, Homogeneous asymmetric Riemannian spaces with an irreducible group of motions, Dolk. Akad. Nauk. SSSR, **141**, (1961), 792-795.

[85] O. V. Manturov, Riemannian spaces with orthogonal and symplectic groups of motions and an irreducible group of rotations, Dolk. Akad. Nauk., SSSR, **141**, (1961), 1034-1037.

[86] O. V. Manturov, Homogeneous Riemannian manifolds with irreducible isotropy group, Trudy Sem. Vector Tenzor Anal., **13**, (1966), 68-145.

[87] P. Melvin & J. Parker, 4-manifolds with large symmetry groups, Topology, **25**, (1986), 71-83.

[88] J. Milnor, Curvatures of left invariant metrics on Lie groups, Adv. in Math., **21**, (1976), 293-329.

[89] Y. Nikonorov & E. Rodionov, Standard homogeneous Einstein manifolds and Diophantine equations, Arch. Math. (Brno), **32**, (1996), 123-136.

[90] T. Nitta, Connections for vector bundles over quaternionic Kähler manifolds, Proc. Japan Acad., **63A**, (1987), 23-25.

[91] D. Page, A compact rotating gravitational instanton, Phys. Lett., **79B** (1979), 235-238.

[92] D. Page & C. Pope, New squashed solutions of $d = 11$ supergravity, Phys. Lett. **147B**, (1984), 55-60.

[93] D. Page & C. Pope, Inhomogeneous Einstein metrics on complex line bundles, Classical and Quantum Gravity, **4**, (1987), 213-225.

[94] D. Page & C. Pope, Einstein metrics on quaternionic line bundles, Classical and Quantum Gravity, **3**, (1986), 249-259.

[95] J. Park, Einstein normal homogeneous Riemannian manifold, Proc. Jap. Acad., **72**, (1996), 197-198.

[96] J. Park & Y. Sakane, Invariant Einstein metrics on certain homogeneous spaces, Tokyo J. Math., **20**, (1997), 51-61.

[97] J. Parker, 4-dimensional G-manifolds with 3-dimensional orbits, Pac. J. Math., **125**, (1986), 187-204.

[98] H. Pedersen, Eguchi-Hanson metrics with cosmological constant, Classical and Quantum Gravity, **2**, (1985), 579-587.

[99] H. Pedersen, Einstein metrics, spinning top motions and monopoles, Math. Ann., **274**, (1986), 35-59.

[100] P. Petersen & S. Zhu, $U(2)$-invariant Einstein metrics on 4-manifolds, Indiana U. Math. J., **44**, (1995), 451-465.

[101] I. Piatetskii-Shapiro, *Geometry of classical domains and automorphic forms*, Gordon and Breach, N. Y., (1969).

[102] F. Podestà & A. Spiro, Kähler manifolds with large isometry group, Preprint 1997.

[103] E. D. Rodionov, On a new family of homogeneous Einstein manifolds, Arch. Math. (Brno), **28**, (1992), 199-204.

[104] E. D. Rodionov, Homogeneous Riemannian manifolds with Einstein metrics, Dissertation, Novosibirsk, 1994.

[105] E. D. Rodionov, Simply connected compact standard homogeneous Einstein manifolds, Siberian Math. J., **33**, (1992), 641-653.

[106] E. D. Rodionov, Standard homogeneous Einstein manifolds, Dokl. Math. USSR, **47**, (1993), 37-40.

[107] E. D. Rodionov, Simply connected compact five-dimensional homogeneous Einstein manifolds, Siberian Math. J., **35**, (1994), 163-168.

[108] E. D. Rodionov, The structure of standard homogeneous Einstein manifolds with a simple isotropy group I, Siberian Math. J., **37**, (1966), 151-167.

[109] E. D. Rodionov, The structure of standard homogeneous Einstein manifolds with a simple isotropy group II, Siberian Math. J., **37**, (1966), 542-551.

[110] Y. Sakane, Examples of compact Einstein Kähler manifolds with positive Ricci tensor, Osaka J. Math., **23**, (1986), 585-617.

[111] Y. Sakane, Homogeneous Einstein metrics on a principal circle bundle I: *Complex Geometry* (Osaka 1990), Lecture Notes in Pure and Applied math., **143**, Dekker, N. Y., (1993), 161-178 ; II : *Differential Geometry* (Shanghai 1991), World Scientific Publishing, (1993), 177-186.

[112] M. Stenzel, Ricci-flat metrics on the complexification of a compact rank one symmetric space, Manu. Math., **80**, (1993), 151-163.

[113] E. Straume, Compact connected Lie transformation groups on spheres with low cohomogeneity I, Mem. A. M. S., **119**, (569), (1996).

[114] A. Swann, Hyperkähler and quaternionic Kähler geometry, Math. Ann., **289**, (1991), 421-450.

[115] R. Szöke, Complex structures on tangent bundles of Riemannian manifolds, Math. Ann., **291**, (1991), 409-428.

[116] J. Wang, Einstein metrics on principal circle bundles, Diff. Geom. and its Appl., **7**, (1997), 377-388.

[117] J. Wang, Einstein metrics on bundles, Ph.D. thesis, McMaster Univ., 1996.

[118] J. Wang & M. Wang, Einstein metrics on S^2 bundles, Math. Ann., **310**, (1998), 497-526.

[119] M. Wang, Some examples of homogeneous Einstein manifolds in dimension seven, Duke Math. J., **49**, (1982), 23-28.

[120] M. Wang, Einstein metrics and quaternionic Kähler manifolds, Math. Zeit., **210**, (1992), 305-326.

[121] M. Wang & W. Ziller, On the isotropy representation of a symmetric spaces, Rend. Sem. Mat. Univers. Politecn. Torino, Fasc. Speciale, (1985), 253–261.

[122] M. Wang & W. Ziller, On normal homogeneous Einstein manifolds, Ann. Scient. Éc. Norm. Sup., 4e série, **t.18**, (1985), 563–633.

[123] M. Wang & W. Ziller, Existence and non-existence of homogeneous Einstein metrics, Invent. Math., **84**, (1986), 177-194.

[124] M. Wang & W. Ziller, Einstein metrics on principal torus bundles, J. Diff. Geom., **31**, (1990), 215-248.

[125] M. Wang & W. Ziller, On isotropy irreducible Riemannian manifolds, Acta Math., **166**, (1991), 223–261.

[126] M. Wang & W. Ziller, Symmetric spaces and strongly isotropy irreducible spaces, Math. Ann., **296**, (1993), 285-326.

[127] J. Wolf, The geometry and structure of isotropy irreducible homogeneous spaces, Acta Math., **120**, (1968), 59-148; correction: Acta Math., **152**, (1984), 141–142.

[128] T. H. Wolter, Einstein metrics on solvable groups, Math. Zeit., **206**, (1991), 457-471.

[129] S. T. Yau, On the Ricci curvature of a compact Kähler manifold and the complex Monge-Ampère equation I, Comm. Pure App. Math., **31**, (1978), 339-411.

[130] W. Ziller, Homogeneous Einstein metrics on spheres and projective spaces, Math. Ann., **259**, (1982), 351-358.

DEPARTMENT OF MATHEMATICS AND STATISTICS, MCMASTER UNIVERSITY, HAMILTON, ONTARIO L8S 4K1, **Canada**

E-mail address: wang@mcmail.mcmaster.ca

Part III: Relativity Revisited

General Relativity

K.P.Tod

1. Introduction

1.1. Aims. In this essay, my brief is to describe some current research in general relativity which would be of interest to mathematicians working elsewhere in geometry. To achieve this, I shall need first to review a range of background material in modern general relativity, corresponding roughly to a second or graduate-level course. For reasons of space, I shall need to assume that the reader has had a first course in the subject. After the review, the choice of topics is my own.

1.2. A way in. One way into relativity for a mathematical audience is to compare and contrast Riemannian and Lorentzian geometry — what changes when the signature of the metric changes? What familiar things cease to be of interest and what new things become of interest? One may classify topics of interest in relativity by their relation to Riemannian geometry into one of three classes:

- **direct uses of Riemannian geometry.** e.g. space-like surfaces are intrinsically Riemannian, therefore so are questions to do with the Initial Value Problem; the classification of black holes is concerned with time-independent solutions, where the field equations become elliptic; the first proof of the Positive Mass theorem uses the methods of Riemannian geometry;
- **Lorentzian problems motivated by analogy with Riemannian ones.** e.g. there are Lorentzian Splitting theorems, motivated by analogy with the Riemannian ones, but with their own physical interpretation; with an indefinite metric, positive sectional curvature is not a helpful notion but certain conditions of positivity of the Ricci tensor are of crucial importance, and play an analogous role in forcing the existence of conjugate points on geodesics;
- **complete novelties.** e.g. anything explicitly hyperbolic, so existence theory for the Einstein equations; singularity theorems and cosmic censorship.

1.3. One difference. It is instructive to pursue one answer to the question 'what changes when the signature changes?', namely the answer 'the Hopf-Rinow theorem;' cf. e.g. [**9**]. In Riemannian geometry, the manifold becomes a metric space with the distance defined by the metric tensor, and the open sets in the manifold are determined by the metric. The Hopf-Rinow theorem asserts that the manifold is complete as a metric space if and only if it is geodesically complete. Further, in this case, there will be a geodesic connecting any two points which achieves the (minimum) distance between them. When the metric tensor is indefinite, none of this works.

For the topology, one can seek instead to define the open sets by causal relations and one is led into a study of causal spaces, which represent an important layer of structure between the topological and the metric in relativity. For the completeness, one can distinguish a whole range of (independent) geodesic completenesses, and completeness for other types of curve. There is also the important condition of global hyperbolicity in relativity which implies the existence of maximal curves in appropriate circumstances. (Note 'maximal' rather than 'minimal': time-like geodesics locally maximise distance; it is always possible to join points by 'short' curves by making them nearly null.)

1.4. Physical arguments. Relativity is a theory of gravity, and an extremely accurate one; cf. e.g. the theory of binary pulsars [**95**, p.230]. This means on the one hand that physical concerns and heuristic arguments have a proper place in the subject, and on the other that physical insight can lead one to results which can then be proved to the most rigorous standards — physical insight can coincide with what is true in the theory. However as a mathematician, one may not want to delve too deeply into the physical aspects of the theory. There is a standard way to achieve this aim:

1.5. The Einstein inequalities. Recall the Einstein equations in the form

$$G_{ab} = 8\pi G T_{ab} \tag{1.1}$$

where G_{ab} is the Einstein tensor of some Lorentzian metric and T_{ab} is the energy-momentum tensor of some matter source. (Relativists commonly, though by no means invariably, use indices. In this article, where necessary, I shall use the *abstract index convention* of Penrose [**96**]. This allows one to use all the notations of local tensor calculus, so that one does not need to devise notational synonyms for tensor operations, while remaining perfectly invariant.)

The left-hand side of (1.1) is the mathematical side (the 'marble palace' of Einstein) and the right-hand side is the physical side (the 'wooden shed'). In many situations, one may regard (1.1) as producing a set of inequalitites by requiring of the right-hand side only that it have some positivity properties, and ignoring its details. The physical input of general relativity into geometry is then confined to demanding these positivity properties of the left-hand-side. These positivity properties are the various energy conditions: they express different conditions of positive energy locally, and most of what we shall see below is premised on one or another energy condition.

1.6. Conjugate point arguments. It is the energy conditions which make gravity attractive. One consequence of this attractiveness is that, given a large amount of mass in a small region, gravity may overwhelm the forces holding the matter up and bring about a gravitational collapse to a singularity. The mathematical counterpart of this physical argument is that energy conditions eventually lead to the existence of conjugate points on geodesics provided the geodesics can be extended to arbitrary values of affine parameter. Then these conjugate points can be shown to be inconsistent with other physical hypotheses which encode the fact of collapse, from which one is led to geodesic incompleteness. This is a paradigm *conjugate point argument.* A consequence of it is that relativists are obliged to consider manifolds which are geodesically incomplete or singular in other ways. Many of the arguments in sections 3 to 7 are conjugate point arguments in this sense.

1.7. Causality. One of the other ways to be singular is to have a closed time-like curve (or CTC). If such a thing existed in a space-time, then one could travel along it into one's past, when various, usually murderous, paradoxes could be generated. There is a whole range of causality pathologies which one might seek to forbid for physical reasons. Now a *compact* Lorentzian manifold necessarily has a CTC (in fact many, joining any point to any point), which is why these have traditionally held less interest for relativists.

1.8. Positive energy. An interesting problem historically has been how to derive global positive energy, as measured 'at infinity' and containing non-local contributions from the gravitational field, from an assumption of positive energy locally, expressed by an energy condition. (This problem is difficult because one expects gravitational energy to exist and so to contribute to total energy, but not to be the integral of any local quantity.) There are now three different ways to derive this result, two which work on space-like surfaces and are therefore 'elliptic' and a newer four-dimensional way. The result, the Positive Energy theorem, has subsequently been used to prove new results and strengthen old ones.

1.9. Cosmic censorship. Arguably the biggest unsolved problem in relativity is to prove or disprove the cosmic censorship hypothesis. In a weak form, this is the hypothesis that, while the formation of singularities in certain circumstances is inevitable, these singularities are hidden inside black holes and cannot be seen from large distances. In a strong form, the hypothesis is that only particular kinds of singularities can ever arise in an evolution of regular data. Either form is a hard problem, made harder by a physical consideration: these are supposed to be statements about the world so that one is interested in *generic* or *stable* sets of circumstances arising with *reasonable* matter, and all the italicised words are problematic.

1.10. Contents. This essay is organized as follows:

In §2, we describe the landscape of general relativity as it is now. The development here is inevitably condensed almost to the telegraphic but it sketches what is needed to locate the later sections.

In §3, we review various topological issues in relativity. These include those mentioned above, ideas related to dynamic topology, changing with time, and the recent notion of topological censorship, which is analogous to cosmic censorship.

In §4, we describe the Lorentzian Splitting Theorems and related material and in §5 we review what is known about existence for solutions of the Einstein equations.

In §6, we review work on the Black Hole Uniqueness theorems, where there has been a resurgence of interest recently, and finally in §7 we review work on the evidence for and against the Cosmic Censorship Hypothesis.

ACKNOWLEDGMENTS 1.10.1. In composing this review, I have benefitted from discussions with many people among whom I would like to mention Lars Andersson,

Piotr Chruściel, Helmut Friedrich, Lionel Mason, Vince Moncrief, Ted Newman, Roger Penrose, Alan Rendall and Bernd Schmidt.

1.11. Further reading. A review such as this, to be successful, needs to lead the reader onward and beyond itself. Thus a good text for Section 2 is [**137**]; more details in particular directions will be found in [**60**], which after 24 years is still the place to start, and in [**9**]. A useful resource in the near future will be the 'Living reviews' on various topics in relativity maintained by the Albert Einstein Institute in Potsdam at http://www.aei-potsdam.mpg.de, and much of the topical material discussed in this review first appeared in the gr-qc archive at http://xxx.lanl.gov/ or one of its mirrors.

2. Background Material

We will use conventions as in [**96**], so that the signature of the metric is $(+,-,-,-)$ and indices are abstract.

2.1. Infinity for flat space. We need a definition of isolated source in general relativity, which must convey the idea of asymptotic flatness at large distances. The idea is to define an infinity for flat space as a boundary, so that one may later define a space to be asymptotically flat if it has the same kind of infinity as flat space.

To this end, call flat space M and consider the metric of M in spherical polar coordinates:

$$ds^2 = dt^2 - dr^2 - r^2(d\theta^2 + \sin^2\theta d\phi^2) \tag{2.1}$$

Radially in- and out-going null geodesics have respectively

$$v = t + r = \text{ constant; } u = t - r = \text{ constant; } -\infty < u \leq v < \infty \tag{2.2}$$

Introduce u and v as coordinates in (2.1) to find:

$$ds^2 = dudv - \frac{1}{4}(v-u)^2(d\theta^2 + \sin^2\theta d\phi^2) \tag{2.3}$$

We shall add a boundary to M by first adding a point to the end of each radial null geodesic. This is achieved by introducing coordinates p and q via

$$u = \tan p;\ v = \tan q;\ \text{so that } -\frac{\pi}{2} < p \leq q < \frac{\pi}{2}$$

when (2.3) becomes

$$ds^2 = \frac{1}{4}\sec^2 p \sec^2 q[4dpdq - \sin^2(p-q)(d\theta^2 + \sin^2\theta d\phi^2)] \tag{2.4}$$

In this form, the metric can be conformally-rescaled to give a new metric on a larger manifold than M:

$$d\hat{s}^2 = \Omega^2 ds^2 = 4dpdq - \sin^2(p-q)(d\theta^2 + \sin^2\theta d\phi^2) \tag{2.5}$$

where

$$\Omega = 2\cos p \cos q$$

and now we can extend the range of p and q to include the end-points $\pm\frac{\pi}{2}$. The rescaled metric (2.5) is the product metric on $\mathbb{R} \times S^3$ as one may see by introducing one last set of coordinates via

$$T = p + q;\ R = p - q$$

when

$$ds^2 = dT^2 - dR^2 - \sin^2 R(d\theta^2 + \sin^2\theta d\phi^2) \tag{2.6}$$

In this context, this product metric is known as 'the Einstein static cylinder', having been at one time proposed as a cosmological model by Einstein.

We have found that flat space M is conformally related to the part of the Einstein static cylinder lying in the range

$$T + R > -\pi; \; T - R < \pi \tag{2.7}$$

The conformal structure of M extends to the boundary of this region in $\mathbb{R} \times S^3$, which is the locus where Ω from (2.5) vanishes. The boundary consists of the past null cone $\mathcal{I}^+$ (pronounced 'scri-plus') of the point i^+, which is also the future null cone of the point i^0, together with the past null cone $\mathcal{I}^-$ of i^0, which is also the future null cone of the point i^- (see figure 1). These symbols are conventional and

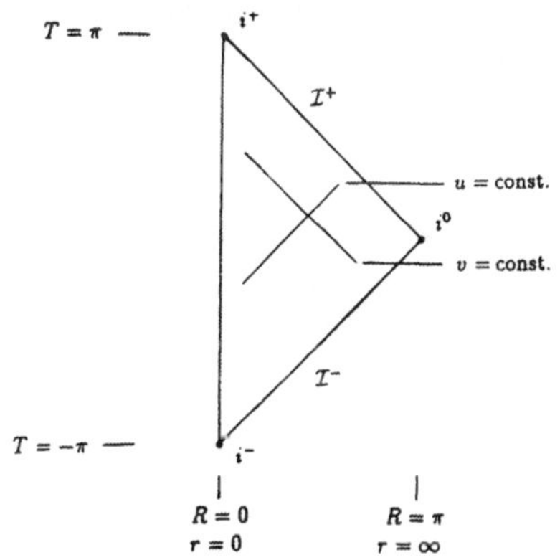

FIGURE 1. The (t, r)-half-plane of Minkowski space in the (T, R)-space of $\mathbb{R} \times S^3$; i^0 is antipodal to the origin on S^3.

are associated with the following names:

TERMINOLOGY 2.1.1.

- $\mathcal{I}^+$ is future null infinity;
- $\mathcal{I}^-$ is past null infinity;
- i^+ is future time-like infinity;
- i^- is past time-like infinity;
- i^0 is space-like infinity.

All null geodesics have a past end-point on $\mathcal{I}^-$ and a future one on $\mathcal{I}^+$; all time-like geodesics run from i^- to i^+; all space-like geodesics run from i^0 back to i^0. We may sometimes use $\mathcal{I}$ to mean the union $\mathcal{I}^+ \cup \mathcal{I}^-$.

We have defined this boundary using coordinates but invariant descriptions are possible.

2.2. Asymptotic simplicity. We use the work of the previous section to give a definition intended to capture the notion of asymptotic flatness.

DEFINITION 2.2.1. A space-time M with metric g is *asymptotically simple* if there is a smooth manifold $\hat{M}$ with boundary $\mathcal{I} = \partial\hat{M}$ and metric $\hat{g}$ and a scalar field Ω such that

- $M = \text{int } \hat{M}$;
- $\hat{g} = \Omega^2 g$ in M;
- Ω and $\hat{g}$ are smooth everywhere in $\hat{M}$;
- $\Omega > 0$ in M; $\Omega = 0$ and $d\Omega \neq 0$ on $\mathcal{I}$; and
- every null geodesic in M acquires a future and a past end-point on $\mathcal{I}$.

The last condition is needed to avoid trivial satisfaction of the conditions with $\mathcal{I}$ empty, but is too strong in practice since even the extended Schwarzschild solution will fail to be asymptotically simple. Thus one defines:

DEFINITION 2.2.2. A space-time M with metric g is *weakly asymptotically simple* (or WAS) if there is an asymptotically simple M' and a neighbourhood U of $\mathcal{I}$ in the corresponding $\hat{M}$ such that $U \cap M'$ is isometric to a subset of M.

2.3. Causal relations. Causal relations define a layer of structure prior to the smooth in a space-time. This section consists largely of definitions, made to introduce a convenient language.

DEFINITION 2.3.1. A Lorentzian manifold M is *time-orientable* if it is possible to make a consistent choice of future-light-cone at every point; M is *space-orientable* if it is possible to make a consistent choice of a right-handed triad of space-like vectors at every point.

If M is time and space orientable, then M is orientable but not conversely. If M admits spinors then M is orientable in all three senses.

DEFINITION 2.3.2. For points p and q in a time-orientable M define the relations:

$p \ll q$ (read 'p chronologically precedes q') iff there is a future-directed (non-empty) time-like path from p to q;

$p \prec q$ (read 'p causally precedes q') iff there is a future-directed (possibly empty) path from p to q which is everywhere non-space-like (i.e. is time-like or null at each point; call this a *causal path*).

DEFINITION 2.3.3. We define the sets:

- $I^+(p) = \{q | p \ll q\}$ the chronological future of p;
- $I^-(p) = \{q | p \gg q\}$ the chronological past of p;
- $J^+(p) = \{q | p \prec q\}$ the causal future of p;
- $J^-(p) = \{q | p \succ q\}$ the causal past of p.

'Time-like' is an open condition, whence it follows that $I^+(p)$ and $I^-(p)$ are open, but $J^+(p)$ and $J^-(p)$ are not necessarily closed (though they will be in Minkowski space).

In terms of these notions one can frame various causality conditions:

DEFINITION 2.3.4. M satisfies the *chronology condition* if it contains no closed time-like curves, equivalently if for no $p \in M$ is it true that $p \in I^+(p)$ or $p \in I^-(p)$.

DEFINITION 2.3.5. M satisfies the *causality condition* if it is never true that $p \prec q \prec p$ for distinct p and q.

A range of stronger conditions restricting causal pathologies is available. A useful one, needed in 3.1.1, which excludes almost closed causal paths is:

DEFINITION 2.3.6. M is *strongly causal* at p if there is a neighbourhood of p which no non-space-like path intersects more than once.

The strongest condition normally encountered is the following:

DEFINITION 2.3.7. M is *globally hyperbolic* if the strong causality condition holds everywhere and, for any $p, q \in M$, the set $J^+(p) \cap J^-(q)$ is compact.

Global hyperbolicity is related to Cauchy developments, so we need to define these:

DEFINITION 2.3.8. An *achronal set* S is one for which $I^+(S) \cap S = \emptyset$.

DEFINITION 2.3.9. The *future Cauchy development* or *future domain of dependence* $D^+(S)$ of an achronal set S in a space-time M is the set of $p \in M$ such that every past-inextendible non-space-like path through p intersects S.

DEFINITION 2.3.10. The *future Cauchy horizon* of S is the future boundary of $D^+(S)$, that is the set $H^+(S) = \overline{D^+(S)} - I^-(D^+(S))$ (writing $\bar{U}$ for the closure of U).

One defines $D^-(S)$ and $H^-(S)$ analogously, and then $D(S) = D^+(S) \cup D^-(S)$. The relation with global hyperbolicity is provided by the result:

PROPOSITION 2.3.11. [**60**, Prop 6.6.3] *If S is a closed achronal set then int $D(S)$, if non-empty, is globally hyperbolic.*

PROPOSITION 2.3.12. *An achronal set S is a Cauchy surface for M if $M = D(S)$.*

Thus if a space-time M has a Cauchy surface, then it is globally hyperbolic. We shall encounter a converse in 3.3.1.

An aspect of the role of global hyperbolicity as a completeness condition is provided by the result:

PROPOSITION 2.3.13. [**60**, Prop 6.7.1] *If p, q lie in a globally hyperbolic set U with $q \in J^+(p)$ then there is a non-space-like geodesic from p to q whose length is greater than or equal to the length of any other non-space-like curve from p to q.*

Finally in this section, we note that there is an invariant characterisation of $\mathcal{I}^-$ and $\mathcal{I}^+$ in terms of causal structure, so that these can be added as future and past *causal boundaries*.

2.4. The Schwarzschild solution.

SPACE-LORE 2.4.1. The Schwarzschild solution is characterised by Birkhoff's theorem [**60**] as *the* spherically-symmetric vacuum solution. It is weakly asymptotically simple, and *static*, which means that it admits a hypersurface-orthogonal Killing vector which is time-like at large distances (one reserves the term *stationary* for a solution with a time-like Killing vector which is not hypersurface-orthogonal). The solution depends on a single parameter which can be identified as the mass (see §2.11).

EXAMPLE 2.4.2 (Extending the Schwarzschild solution). In a first course on general relativity, the Schwarzschild solution is usually exhibited in coordinates as

$$ds^2 = (1 - 2\frac{m}{r})dt^2 - (1 - 2\frac{m}{r})^{-1}dr^2 - r^2(d\theta^2 + \sin^2\theta d\phi^2) \tag{2.8}$$

where the coordinate ranges are $-\infty < t < \infty$, $2m < r < \infty$.

This form of the metric is singular at $r = 2m$ but this is only a coordinate singularity. A first exercise is to solve the geodesic equations for radial null geodesics, when one readily finds that these geodesics run off the coordinate patch by arriving at $r = 2m$ at finite values of affine parameter, but infinite values of t. The strategy is now to mimic the process leading to equation (2.2), introducing coordinates u and v constant on out- and in-going radial null geodesics respectively, to arrive at an extended form of the metric:

$$ds^2 = \frac{32m^3}{r}\exp(-\frac{r}{2m})dudv - r^2(d\theta^2 + \sin^2\theta d\phi^2) \tag{2.9}$$

where $uv = -(\frac{r}{2m} - 1)\exp(\frac{r}{2m})$.

The metric is no longer 'time-independent', the Killing vector K^a which was $\partial/\partial t$ has become

$$\frac{\partial}{\partial t} = \frac{1}{4m}(v\frac{\partial}{\partial v} - u\frac{\partial}{\partial u}). \tag{2.10}$$

We may represent the manifold on which the metric is defined by its *Carter-Penrose diagram,* figure 2.

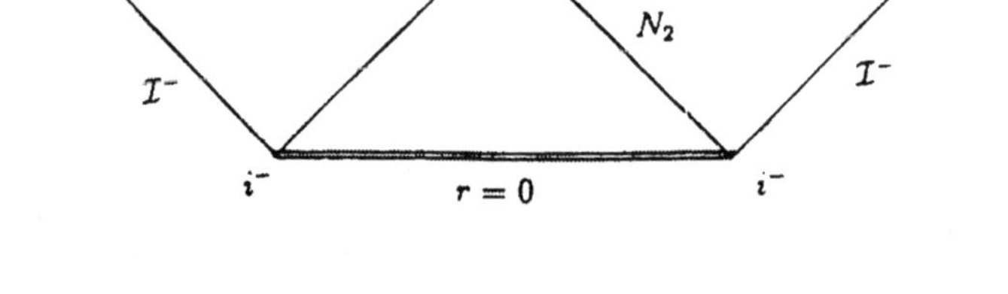

FIGURE 2. Carter-Penrose diagram of maximally analytically extended Schwarzschild solution; each point represents a 2-sphere; null-lines are at 45°; note initial and final $r = 0$ singularities, two asymptotic regions and two Killing horizons at $r = 2m$.

In figure 2 each point represents a 2-sphere of symmetry, and null directions are at 45°. The surprise about the diagram is the presence of two singularities, one in the past (at the bottom) and one in the future, and two asymptotic regions. The picture includes the two distinct $\mathcal{I}^+$'s and two distinct $\mathcal{I}^-$'s, where the conformal structure is regular. The conformal structure is singular at the points $i^\pm$, but also, perhaps surprisingly, at i^0.

The Killing vector (2.10) is time-like and future-pointing near the righthand asymptotic region and time-like, past-pointing near the left-hand one (choosing again the time-orientation which has the future towards the top of the page). The Killing vector becomes null on the pair of null hypersurfaces $N_1 : u = 0$ and $N_2 : v = 0$; each of these is a *Killing horizon:*

DEFINITION 2.4.3. A *Killing horizon* is a null hypersurface with a null Killing vector K^a tangent to the (null, geodesic) generators.

DEFINITION 2.4.4. The Killing horizon has a *surface gravity* κ defined by

$$\nabla_a(K^b K_b) = -2\kappa K_a.$$

Under quite general conditions the surface gravity is constant on the Killing horizon. For the Schwarzschild solution, $\kappa = 1/4m$.

DEFINITION 2.4.5. A Killing horizon is *degenerate* if it has zero surface gravity.

In the extended Schwarzschild case, there are two Killing horizons, which intersect in the *bifurcation surface* at $u = v = 0$.

A Killing horizon often defines an *event-horizon:*

DEFINITION 2.4.6. In a weakly asymptotically simple space-time, the event horizon (strictly, the future event horizon) is $\partial J^-(\mathcal{I}^+)$ if this is non-empty.

Thus if there is an event horizon, then it separates points from which there is a causal path to $\mathcal{I}^+$ from those where there is no such path i.e. it bounds the region from which one can 'escape' to infinity. In the extended Schwarzschild manifold, N_1 defines the event horizon for the $\mathcal{I}^+$ to the right.

We noted above that any point in figure 2 defines a 2-sphere. Furthermore, the area of the 2-sphere is $4\pi r^2$. Now consider a point in the top triangle, that is one with $u < 0$, $v > 0$, $r < 2m$; if the corresponding 2-sphere is moved in any direction normal to itself and into its own future then it will move to a smaller value of r and so its area will decrease (strictly speaking, one needs to calculate something to prove this). We define:

DEFINITION 2.4.7. A space-like 2-surface is said to be *trapped* if its area locally decreases in every future-pointing normal direction.

Now consider a line like γ running across the Carter-Penrose diagram from one i^0 to the other (and not necessarily through the bifurcation surface). This defines a spherically-symmetric space-like surface which is a Cauchy surface for the space-time. At the minimum value of r there will be a (stable) minimal surface, in the usual sense, but every sphere of constant r less than $2m$ will be trapped, while the spheres $r = 2m$ are *marginally-trapped* in that the area is non-increasing in every future-normal direction, and is strictly decreasing in all directions except one of the two null normal directions.

TERMINOLOGY 2.4.8. This Cauchy surface has the character of a *worm-hole* in that it connects two asymptotically flat regions through a minimal surface.

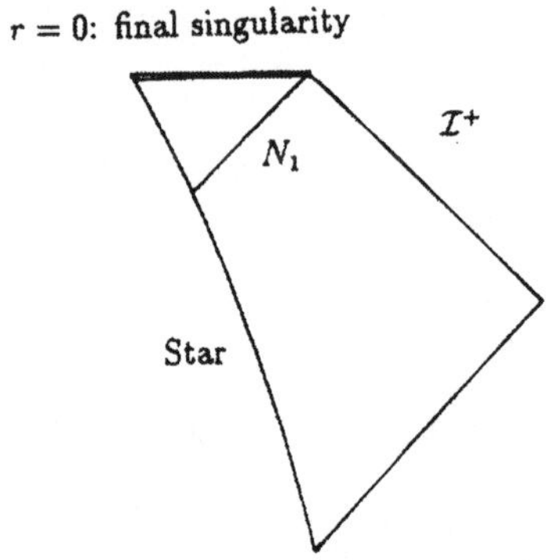

FIGURE 3. Carter-Penrose diagram of collapse of a star to a black hole; the solution outside the star is Schwarzschild and N_1 is the event horizon; the final singularity is formed in the collapse.

However it is not possible in the Schwarzschild manifold to follow a causal path through the worm-hole from one asymptotic region to the other (this can be seen from figure 2, which correctly shows causal relations).

On a Cauchy surface through the bifurcation surface, the bifurcation surface itself is both minimal and marginally-trapped with respect to both its null normals. This is a rather degenerate situation.

The collapse of a spherically symmetric body, say a star, surrounded by vacuum, to a singularity may be represented by a Carter-Penrose diagram, figure 3, consisting of the outer region of figure 2 joined across the surface of the star to another solution with matter. The matter solution cuts off the 'unphysical' past singularity. Now the null hypersurface N_1 defines the event horizon as the boundary of a black hole. A singularity forms in this collapse but it cannot be seen from infinity, that is to say no future causal path connects it to $\mathcal{I}^+$ - it is 'censored'.

An important property of the Schwarzschild solution is the following:

CONDITION 2.4.9. For any $p \in \mathcal{I}^-$, $I^+(p)$ contains all of $\mathcal{I}^+$.

This surprising result is a consequence of the phenomenon of time-delay in the passage of light past a massive body (equivalently 'of time-delay in the solutions of the null-geodesic equation in the Schwarzschild metric'). It is characteristic of positive mass - it is not true in flat space or in the negative-mass Schwarzschild solution.

2.5. The Reissner-Nordstrom solution.

SPACE-LORE 2.5.1. The *Reissner-Nordstrom solution* is characterised as the spherically-symmetric electrovac solution, which is to say a solution of the Einstein equations for which the energy-momentum tensor is that for electromagnetism, in which the spheres of symmetry vary in size. Again it is static and weakly asymptotically simple.

The metric is usually encountered first in the form

$$ds^2 = V(r)dt^2 - (V(r))^{-1}dr^2 - r^2(d\theta^2 + \sin^2\theta d\phi^2) \tag{2.11}$$

where now $V = 1 - \frac{2m}{r} + \frac{e^2}{r^2}$; e, m real constants, and we shall suppose that

$$e^2 < m^2 \tag{2.12}$$

The parameter m may be identified with the mass at infinity, while e is the charge.

Because V has two zeroes, the extension is more complicated. The Carter-Penrose diagram is as in figure 4. The Killing vector $K^a = \partial/\partial t$ is time-like in the (infinitely many) asymptotic regions, and again near the (infinitely many) singularities, being space-like in an intermediate region. There are inner and outer Killing horizons (with different surface gravities) where the Killing vector is null, intersecting at bifurcation surfaces where it vanishes.

A curve like γ defines a spherically-symmetric achronal surface S which is now not a Cauchy surface: note that S has a Cauchy horizon, because of the time-like character of the singularities.

Also because of the time-like character of the singularities, it is possible to follow a causal path from one asymptotic region down to small values of r then into the future and into a second asymptotic region. By identifying the diagram with a periodicity vertically one may therefore introduce closed time-like curves. (We shall see this again in §3.5.)

In the case $e^2 = m^2$, the zeroes of V coincide. The Carter-Penrose diagram simplifies to figure 5. The Killing vector is time-like everywhere except on the Killing horizon where it becomes null. The bifurcation surface has disappeared from the picture and the Killing horizon has become degenerate (these phenomena are related). There are internal points 'at infinity' in that the Riemannian distance on a (space-like) hypersurface of constant t from a value of r greater than m down to $r = m$ is infinite. Thus a constant t hypersurface is asymptotically flat at large distances, but asymptotic to an infinite cylinder as r tends to m. (Degenerate horizons will lead to problems in §6.)

By matching to a collapsing spherically-symmetric charged body, the Reissner-Nordstrom solution can be interpreted as a black hole solution.

2.6. The Majumdar-Papapetrou solutions. The Majumdar-Papapetrou solutions are electrovac solutions generalising the Reissner-Nordstrom solution with $e^2 = m^2$. They may be written

$$ds^2 = V^2dt^2 - V^{-2}(dx^2 + dy^2 + dz^2) \tag{2.13}$$

where V is harmonic

$$\frac{\partial^2 V}{\partial x^2} + \frac{\partial^2 V}{\partial y^2} + \frac{\partial^2 V}{\partial z^2} = 0$$

In the special case

$$V = 1 + \sum_{i=1}^{n} \frac{m_i}{|r - r_i|} \tag{2.14}$$

the solution represents a superposition of charged black holes, the i-th having mass m_i and charge e_i satisfying $e_i^2 = m_i^2$, the same sign taken for all. The locations are

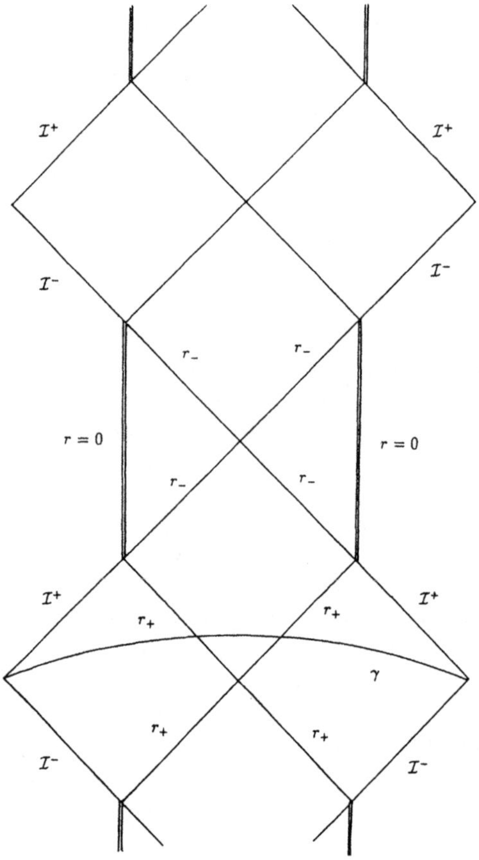

FIGURE 4. Carter-Penrose diagram of the maximally analytically extended Reissner-Nordstrom solution with $e^2 < m^2$; note the $r = 0$ singularity has become time-like and there are infinitely many asymptotic regions, also infinitely many Killing horizons at $r = r_+$ and $r = r_-$; note also the occurrence of a Cauchy horizon at $r = r_-$ for the surface γ.

freely specifiable since, physically speaking, the mutual gravitational attractions are balanced by the electrostatic repulsions. All the black holes have degenerate horizons.

2.7. Homogeneous and isotropic cosmologies. We shall need these in §5.

TERMINOLOGY 2.7.1. The *Robertson-Walker* (or FRW) metric is the metric

$$ds^2 = dt^2 - (R(t))^2 d\sigma_k^2 \tag{2.15}$$

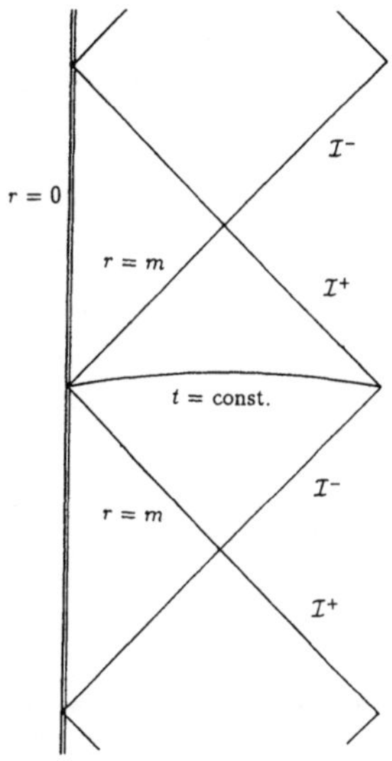

FIGURE 5. Carter-Penrose diagram of the maximally analytically extended Reissner-Nordstrom solution with $e^2 = m^2$; the Killing horizons of figure 4 merge in pairs and become degenerate at $r = m$; the surfaces of contant t have points at infinite distance at $r = m$.

where $R(t)$ is the *scale factor* and the spatial part $d\sigma_k^2$ is the metric of a 3-space of constant curvature k.

With any scale factor, this metric has a 6-dimensional isometry group transitive on the surfaces of constant time t, hence the title of this section. The metric is always conformally-flat and so is conformal to part of the Einstein static cylinder. Particular examples are:

EXAMPLE 2.7.2. The *de Sitter space*, for which $k = 1$ and $R(t) = a\cosh(t/a)$, or $k = 0$ and $R(t) = \exp(t/a)$.

EXAMPLE 2.7.3. The (universal cover of the) *anti-de Sitter space*, for which $k = -1$ and $R(t) = a\cos(t/a)$.

These are in fact Lorentzian symmetric spaces. De Sitter space is conformal to the region $0 < T < \pi$ on the Einstein static cylinder, so it has a space-like $\mathcal{I}^-$ at $T = 0$ and a space-like $\mathcal{I}^+$ at $T = \pi$. Anti-de Sitter space is conformal to the region $R < \pi/2$ on the Einstein cylinder and has a time-like $\mathcal{I}$ at $R = \pi/2$. It is a consequence of the definition of weak asymptotic simplicity 2.2.2 that, if the Einstein equations in the form

$$R_{ab} - \lambda g_{ab} = 0$$

hold near $\mathcal{I}$, then $\mathcal{I}$ is time-like, space-like or null according as the cosmological constant λ is greater than, less than or equal to zero (there is a choice of convention in the sign of the Ricci tensor R_{ab} which can confuse this issue; see §5.4).

2.8. Energy conditions. With the Einstein equations as in (1.1):

CONDITION 2.8.1. The stress-energy tensor T_{ab} is said to satisfy the *weak energy condition* if $T_{ab}t^a t^b \geq 0$ for every time-like vector t^a.

CONDITION 2.8.2. The stress-energy tensor T_{ab} is said to satisfy the *strong energy condition* if $T_{ab}t^a t^b \geq \frac{1}{2}T^c_c g_{ab}t^a t^b$ for every time-like vector t^a.

CONDITION 2.8.3. The stress-energy tensor T_{ab} is said to satisfy the *dominant energy condition* if $T_{ab}t^a$ is a non-spacelike, future-pointing vector for every future-pointing time-like vector t^a.

These conditions can all be regarded as reasonable conditions on (classical) matter. From them and the Einstein equations one deduces:

CONDITION 2.8.4. If T_{ab} satisfies the weak energy condition, then the Ricci tensor R_{ab} satisfies the *null convergence condition*: $R_{ab}n^a n^b \geq 0$ for every null vector n^a.

CONDITION 2.8.5. If T_{ab} satisfies the strong energy condition then R_{ab} satisfies the *time-like convergence condition*: $R_{ab}t^a t^b \geq 0$ for every time-like vector t^a.

The dominant energy condition is the one needed in the first two proofs of the positive energy theorem; the others are relevant to the existence of conjugate points, which we turn to next.

2.9. Geodesic deviation. We need some formalism here. Suppose γ is a time-like geodesic with unit future-pointing tangent vector T^a. Write $D = T^a\nabla_a$ for the directional derivative along γ and s for proper time along γ, and let $e^a_i = \{e^a_1, e^a_2, e^a_3\}$ be an orthonormal basis of vectors orthogonal to T^a and parallely-propagated along γ. A Jacobi field X^a is a vector field defined at points of γ and satisfying the geodesic deviation equation. If we assume that X^a is orthogonal to T^a and expand it in the triad e^a_i then geodesic deviation is the equation

$$D^2 X^i = \Phi^i_j X^j \tag{2.16}$$

where

$$X^a = X^i e^a_i; \text{ and } \Phi^j_i e^a_j = -R_{bcd}{}^a T^b T^d e^c_i. \tag{2.17}$$

We wish to consider simultaneously all Jacobi fields vanishing at a point p taken as $s = 0$. These can be represented by the columns of a matrix $A = (A^j_i)$ satisfying

$$D^2 A = \Phi A \tag{2.18}$$

where we adopt a matrix notation and write $\Phi = (\Phi^j_i)$.

Introduce the matrices M and Σ and the scalar θ by

$$\begin{aligned} DA &= MA \\ M &= \Sigma + \frac{1}{3}\theta I \end{aligned} \tag{2.19}$$

where Σ is trace-free and I is the identity. Then (2.18) implies

TERMINOLOGY 2.9.1. the *Raychaudhuri equation*:

$$D\theta + \theta^2 + tr(\Sigma^2) = tr\Phi \tag{2.20}$$

and a (nameless) propagation equation for Σ:

$$D\Sigma + \frac{2}{3}\theta\Sigma + \Sigma^2 - \frac{1}{3}Itr(\Sigma^2) = \Phi - \frac{1}{3}Itr(\Phi) \tag{2.21}$$

Note that $tr(\Phi)$ in (2.20) is, by (2.17), equal to $-R_{ab}T^aT^b$ which is nonpositive if we have the time-like convergence condition, so $D\theta + \theta^2$ in (2.20) is non-positive. Now a point q is conjugate to a point p iff there is a (non-trivial) Jacobi field vanishing at p and at q. This will happen iff $\det A$ satisfying (2.18) vanishes at q, but from (2.19)

$$\theta = trM = D(\log \det A) \tag{2.22}$$

Thus q is conjugate to p iff θ is infinite at q. The idea is to prove from (2.20) and (2.21) that this is inevitable: by (2.20) θ will become infinite along γ if it once becomes negative, and by (2.21) if Φ is non-zero somewhere on γ, then that will produce Σ which will enter (2.20) to reduce θ. This can be made precise:

PROPOSITION 2.9.2. [**60**, Prop 4.4.2] *Given*

(i) *the time-like convergence condition;*

(ii) *the generic condition:* $R_{abcd}T^aT^c \neq 0$ *at some point of each time-like geodesic;*

(iii) *time-like geodesic completeness;*

then every time-like geodesic contains a pair of conjugate points.

A similar formalism can be developed for geodesic deviation along null geodesics with one slight difference: one concentrates on Jacobi fields representing infinitesimally-neighbouring geodesics 'abreast' of the fiducial one, which is to say lying in a null hypersurface with it. This entails that the matrix A in this case is 2×2 rather than 3×3. The proposition analogous to 2.9.2 can be proved:

PROPOSITION 2.9.3. [**60**, Prop 4.4.5] *Given*

(i) *the null convergence condition;*

(ii) *the generic condition:* $T_{[a}R_{b]ef[c}T_{d]}T^eT^f \neq 0$ *at some point of each null geodesic;*

(iii) *null geodesic completeness;*

every null geodesic contains a pair of conjugate points.

The role of the generic condition is to constrain the relevant term for the modification of (2.21).

The significance of conjugate points is their relation to maximising properties of geodesics. One has:

LEMMA 2.9.4. (i) *a time-like geodesic curve γ from p to q is maximal iff there is no point conjugate to p along γ in (p, q);*

(ii) *if p and q lie on a null geodesic γ and there is a point r conjugate to p between them, then there is a time-like curve from p to q.*

As an application of (ii) used below, consider the boundary of the future of p, $\partial I^+(p)$; near p this is ruled by the null geodesics generating the null cone at p; if one of these generators meets a point r conjugate to p, then, by (ii), beyond r it lies inside $I^+(p)$ and no longer on $\partial I^+(p)$. This observation is the key ingredient in the proof of 2.12.1.

2.10. The Cauchy problem for general relativity. Here the problem is to express the Einstein equations as the evolution of something, and then to prove existence and uniqueness of solutions. The idea is to decompose tensorial quantities with respect to a foliation by hypersurfaces of constant 'time', t say, in the knowledge that the choice of this foliation usually has a great deal of arbitrariness in it. The variables are the first and second fundamental forms of the 3-surfaces of constant t, say h_{ij} and K_{ij}, where the indices are abstract but 3-dimensional, together with whatever matter variables are needed. One needs the Gauss and Codazzi equations to relate 3-dimensional and 4-dimensional tensors.

Suppose that the normal to the 3-surfaces is N^a and take the Einstein equations to be

$$G_{ab} = 8\pi T_{ab} \tag{2.23}$$

These decompose into 'constraints plus evolution'. The (time,time) component, using the Gauss equation twice-contracted, is

$${}^3R + K^2 - K_{ij}K^{ij} = 2G_{ab}N^aN^b = 16\pi T_{ab}N^aN^b \tag{2.24}$$

where ${}^3R$ is the 3-dimensional scalar curvature and $K = h_{ij}K^{ij}$. This is known as *the Hamiltonian constraint.* The (time, space) component, using the Codazzi equation once contracted is

$$D_jK_i^j - D_iK = G_{ai}N^a = 8\pi T_{ai}N^a \tag{2.25}$$

where D_i is the intrinsic 3-dimensional Levi-Civita derivative. This is the *momentum constraint.* These are four constraints: they are conditions on the data which must hold at each time and so in particular must hold initially.

The (space,space) components are the evolution equations, determining the time-derivative of K_{ij}, equivalently the second derivative of h_{ij}. There will also be matter evolution equations, and possibly matter constraints too.

The equations will not be strictly hyperbolic until the diffeomorphism invariance (or coordinate freedom) has been constrained. One then needs to verify that all the constraints are preserved by the evolution, which usually follows from the contracted Bianchi identities.

Notice from the Hamiltonian constraint that, if a 3-surface is *maximal*, which is to say that the trace $K = h^{ij}K_{ij}$ is zero, then either the weak energy condition or the dominant energy condition implies that the 3-dimensional Ricci scalar is non-negative.

2.11. Definitions of mass and positive energy theorems. In §2.4, we mentioned the 'mass' of the Schwarzschild solution. How is this defined? Without going into details, let us note that there is a definition of mass 'at infinity' on asymptotically flat hypersurfaces in asymptotically flat space-times. This is the ADM mass and is, roughly speaking, read off from the $O(1/r)$ terms in the metric. In an analogous way, one can define a mass at any (topologically spherical) section (or *cut*) of $\mathcal{I}^+$ or $\mathcal{I}^-$ in a weakly asymptotically simple space-time. This is the Bondi mass, and it decreases as the cut is moved into the future on $\mathcal{I}^+$, or into the past on $\mathcal{I}^-$. In both these cases, the mass is more properly called the energy as it is the time-like component of a 4-vector at infinity, the total energy-momentum.

In a stationary space-time with a Killing vector K_a one may associate a mass with any 2-surface by the Komar integral

$$m = \frac{1}{4\pi}\int *dK \tag{2.26}$$

The integrand is closed given the Einstein vacuum equations. On a sphere at large distances the Komar integral gives the Bondi or ADM mass (which are equal in a stationary space-time). In a vacuum space-time containing one or more black holes, the Komar integral gives the formula:

$$M = \frac{1}{4\pi} \sum_i \kappa_i A_i \tag{2.27}$$

in terms of the individual surface gravities κ_i and areas A_i of the black holes (of course, one does not expect there to be multiple static vacuum black hole solutions, but the extension of this formula to charged, rotating holes is a significant resource).

There has been a great deal of work with the aim of defining a mass or energy-momentum vector to be associated with an arbitrary 2-surface in an arbitrary space-time [**59, 94, 5, 129**]. Usually such a mass is called 'quasi-local' since one does not expect it to be the integral of a local density over a spanning 3-surface - gravitational mass-energy is not a local quantity - but one does require that it be determined by geometrical quantities at the 2-surface.

Given one of the definitions of total energy-momentum, one can seek to prove that the vector is time-like given some local energy condition, and vanishes only in flat space. We call such a result a Positive Energy Theorem.

The first proof that the ADM momentum is time-like and vanishes only for flat space given the dominant energy condition and an asymptotically flat maximal space-like hypersurface diffeomorphic to $\mathbb{R}^3$ was given by [**112**]. In a sequence of extensions, they subsequently dropped the condition of maximality, allowed the hypersurface to have an inner boundary which was minimal, and extended the result to the Bondi mass [**113, 114**]. They use methods of Riemannian geometry applied to the data for the space-time on the maximal hypersurface: they show that nonpositive mass together with non-negative Ricci scalar (which follows from the Hamiltonian constraint) permit the existence of a particular kind of minimal surface, which in turn forces the data to be data for flat space.

Under the assumptions of the dominant energy condition and the existence of an asymptotically flat space-like hypersurface diffeomorphic to $\mathbb{R}^3$, the same result was proved by Witten in a very different way [**145**]. He uses a 2-component spinor field and an identity, quadratic in the spinor field, which relates a component of the ADM energy-momentum to an integral over the space-like hypersurface. This integral is manifestly non-negative if the spinor field satisfies a linear equation, a modification of the 3-dimensional Dirac equation generally known now as the Witten or Sen-Witten equation. The problem is therefore reduced to the existence theory for the Witten equation. This has been established, and the Witten-style proof has been extended to permit inner boundaries and to prove positivity of the Bondi energy [**86, 75, 54, 109, 61**].

There is a third approach to the positive energy theorem [**97**]. First we need a definition:

DEFINITION 2.11.1. For a weakly asymptotically simple space-time M, define the *domain of outer communications* $D = I^+(\mathcal{I}^-) \cap I^-(\mathcal{I}^+)$.

These authors prove:

PROPOSITION 2.11.2. *In a WAS space-time M, if D is globally hyperbolic and every null geodesic in D possesses a pair of conjugate points then the ADM energy-momentum is future-pointing.*

The idea is to exploit the result noted in 2.4.9: causal properties of the point i^0 are quite different if the ADM mass is positive or negative; for positive mass and any point $p \in \mathcal{I}^-$, all of $\mathcal{I}^+$ is contained in $I^+(p)$; for negative mass, this is not true and there is a $q \in \partial I^+(p) \cap \mathcal{I}^+$; in this case one then shows that there is a null geodesic γ from p to q lying in the boundary $\partial I^+(p)$; but γ contains a pair of conjugate points and so cannot remain on the boundary $\partial I^+(p)$ by 2.9.4 (ii) yielding a contradiction. The existence of conjugate points follows from 2.9.3 given the Einstein equations, an energy condition and the generic condition.

2.12. Singularity Theorems. We saw in §2.4 how the Schwarzschild singularity may be seen to 'form' in gravitational collapse to a black hole. It was at one time argued that the formation of singularities was a very special circumstance, attributable possibly to the high degree of symmetry in the Schwarzschild solution. This position changed after the first singularity theorem appeared.

PROPOSITION 2.12.1. [**87**] *The following conditions on a space-time M cannot hold simultaneously:*

(i) *M has a non-compact Cauchy surface S;*
(ii) *M contains a closed trapped surface T;*
(iii) *M is null geodesically complete;*
(iv) *the null convergence condition holds in M.*

This is a 'singularity theorem' to the extent that geodesic incompleteness is taken as the criterion of singularity. We sketch the proof: by (ii) the outgoing null geodesics orthogonal to T are converging at T; (iii) and (iv) then enforce the appearance of a point conjugate to T along each such geodesic by a version of the argument leading to 2.9.3; beyond this conjugate point, the geodesic is in the interior of $I^+(T)$ by a modification of 2.9.4; thus the boundary $\partial I^+(T)$ is compact; this is incompatible with (i) - to see this, choose a smooth time-like vector field on M and use the integral curves of it to map $\partial I^+(T)$ continuously into S, which is not compact.

This was the first of the *conjugate point arguments* which have been crucial in mathematical relativity. There have been many more singularity theorems proved under different assumptions, for example different energy conditions, dropping global hyperbolicity, allowing causality violations, allowing compact spatial sections. The proofs typically derive contradictions from the simultaneous existence of conjugate points and some geometric condition implying collapse.

3. Topological Issues; Topological Censorship

3.1. The Alexandrov topology. As observed in §1, in a space-time M the spacetime metric does not define a topological metric. One may seek instead to define the open sets of the manifold by causal properties. The Alexandrov topology is the one generated by open sets of the form $I^+(p) \cap I^-(q)$; when does it coincide with the manifold topology (which will always be assumed to be Hausdorff)?

PROPOSITION 3.1.1. [**89**] *The following are equivalent:*

(i) *M is strongly causal*
(ii) *the Alexandrov topology agrees with the manifold topology;*
(iii) *the Alexandrov topology is Hausdorff.*

Clearly some causal condition is needed, and strong causality turns out to be the right one.

3.2. Compact Lorentzian manifolds. Historically, relativists have not been much interested in compact space-times. There are several reasons for this:

PROPOSITION 3.2.1. [8] *Any compact* M *contains closed time-like curves.*

PROOF. Take an open cover of M by sets $I^+(p)$ and contemplate a finite subcover. □

Next:

PROPOSITION 3.2.2. [8] *The 4-manifold* M *admits a Lorentzian metric iff* M *admits an everywhere time-like direction field. If* M *is compact this happens iff the Euler characteristic is zero, so in particular would imply that* M *is not simply-connected.*

Hawking and Ellis [**60**] interpret 3.2.2 as meaning that a compact spacetime is 'really' a non-compact space-time with identifications. Against this view is Tipler's 'No-return' theorem [**125**]: call a space-time M with a Cauchy surface S *time-periodic* if M admits an infinite cyclic group of isometries $G = \{\theta_i | i \in Z\}$ with $\theta_i(S) \cap \theta_j(S) = \emptyset$ for all i, j. Then:

PROPOSITION 3.2.3. *If* M *admits a compact Cauchy surface* S *and the generic and time-like convergence conditions hold in* M *then* M *cannot be time-periodic.*

PROOF. Note first that the generic and convergence conditions imply the existence of conjugate points on time-like geodesics; now one connects copies S_i and S_j of the Cauchy surface under the isometry by maximising time-like geodesics; take a limit, then the limit geodesic has conjugate points which contradicts the maximality. □

From Tipler's no-return theorem, [**82**] deduces another pathology of compact space-times:

PROPOSITION 3.2.4. *If* M *is compact and satisfies the null and time-like convergence and generic conditions, then* M *cannot admit a closed, embedded, edgeless, space-like hypersurface.*

The proof shows that, if it did, 3.2.3 would be violated in a suitable covering space.

3.3. Topology change. The idea that space-like hypersurfaces might have nontrivial topology which, furthermore, might change with time has long interested relativists. Typically, though, there are problems with topology change:

PROPOSITION 3.3.1. [**48**] *If* M *is globally hyperbolic then* M *admits a Cauchy surface* S *and* M *is homeomorphic (in fact diffeomorphic) to* $\mathbb{R} \times S$.

Thus the topology cannot change with time if M is globally hyperbolic. In the absence of conditions, however, topology can change:

PROPOSITION 3.3.2. [**103, 46**] *Any two compact (not necessarily connected) 3-manifolds* S *and* S' are Lorentz cobordant*: there is a compact* M*, whose boundary is the disjoint union* $S \coprod S'$*, and which admits a Lorentzian metric in which* S *and* S' *are space-like.*

But causality is necessarily violated if topology does change:

PROPOSITION 3.3.3. [**46**] *With M, S and S' as in 3.3.2, if M is time-oriented and contains no closed time-like curves then S and S' are diffeomorphic.*

The idea for 3.3.3 is to use the time-like direction field which M admits (by 3.2.2) to map S to S'.

Even giving up causality is not enough:

PROPOSITION 3.3.4. [**123, 124**] *With M, S and S' as in 3.3.2, if the null convergence and null generic conditions hold in M then S and S' are diffeomorphic and M is $\mathbb{R} \times S$.*

A different kind of difficulty with topology change was found by Gibbons and Hawking [**55**]. This is the problem of defining spinors on a topology changing space-time.

PROPOSITION 3.3.5. *There is a mod 2 invariant $u(S)$ of 3-manifolds such that, with M, S and S' as in 3.3.2, M will admit $SL(2,\mathbb{C})$ spinors iff $u(S) = u(S')$. Here $u(S)$ is the Kervaire invariant:*

$$u(S) = \dim_{\mathbb{Z}_2}(H_0(S;\mathbb{Z}_2) \oplus H_1(S;\mathbb{Z}_2)) \mod 2$$

Thus, for example, a Lorentzian metric can be defined on the topology-changing space-time M with $S = S^3$ and $S' = S^3 \amalg S^3$, but M will not admit spinors. Gibbons and Hawking argue that failing to admit spinors is a more serious defect in a space-time than having closed time-like curves.

3.4. Obstructions to spatial topology. Given that it is difficult to change spatial topology, are there are obstructions to having it at all? The answer is, "No, but ... "

PROPOSITION 3.4.1. [**144**] *Every closed 3-manifold occurs as a space-like hypersurface in a vacuum space-time; every closed 3-manifold minus a point occurs as an asymtotically flat initial data set for a vacuum space-time.*

The proof is by an explicit construction of a solution of the constraints for the vacuum field equations exhibited in §2.10.

However, if one seeks to impose the extra condition that the hypersurface is maximal then there is a problem: the Hamiltonian constraint implies that the (3-dimensional) Ricci scalar is positive. Thus:

PROPOSITION 3.4.2. [**144**] *Any closed oriented 3-manifold with a $K(\pi,1)$ as a prime factor admits no metric with $R > 0$ and only flat metrics with $R \geq 0$; thus there are many space-times (vacuum or with matter satisfying an energy condition) with no maximal slice.*

A simple explicit example of an asymptotically flat space-time containing no maximal surface due to Brill [**12**] contains an asymptotically flat space-like hypersurface which is topologically T^3 minus a point. It is constructed by joining part of the Schwarzschild solution to a piece of the $k = 0$ dust-filled FRW universe across a collapsing sphere, and then identifying the FRW part to a torus. This example in turn has been generalised by Bartnik [**4**] to give a space-time with spatial topology $T^3\#T^3$ which admits no space-like hypersurface of constant mean curvature for any value of the constant.

There is current interest in the existence of foliations by constant-mean-curvature or CMC hypersurfaces: see §5. There are at present no examples in the literature of vacuum space-times which admit no CMC hypersurfaces.

3.5. Topological censorship. We met cosmic censorship in §2.4 and will meet it again in §7. Topological censorship [**35**] is a related idea, that an asymptotically flat space-time may well have complicated topology close in, but this fact cannot be communicated to large distances. The starting point is the singularity theorem of Gannon [**45**], which needs a definition:

DEFINITION 3.5.1. A space-like hypersurface S in an asymptotically flat space is *regular near infinity* if it satisfies the following three conditions:

(i) $S = \bigcup_{i=1}^{\infty} W_i$, $W_i \subset W_{i+1}$ and each W_i is is a compact 3-manifold with boundary homeomorphic to a 2-sphere;
(ii) $S - \text{int}W_i$ is homeomorphic to $\partial W_i \times \mathbb{R}^+$;
(iii) the ingoing null geodesics normal to ∂W_i are converging everywhere on ∂W_i.

Note that (iii) is what you would expect on a large 2-sphere — this is not a 'trapped' condition. Then:

PROPOSITION 3.5.2. [**45**] *If a space-time M admits a Cauchy surface which is regular near infinity and not simply-connected, and if the time-like convergence condition is satisfied in M, then M is not null geodesically complete.*

PROOF. The idea is to consider, in the universal covering space $\bar{M}$ of M, a copy A of one of the large spheres ∂W_i lying on a copy $\bar{S}$ of S; the ingoing null geodesics normal to A define a submanifold N which is part of the boundary $\partial J^+(A)$; by the argument in 2.12.1 they leave the boundary after passing conjugate points if they are complete, so that N is compact and $A = \partial N$; now a time-like direction-field maps N down to S, but A cannot bound a compact 3-manifold in S. □

Topological censorship deals with a weakly asymptotically simple space-time M and causal curves from $\mathcal{I}^-$ to $\mathcal{I}^+$. Let γ be such a curve which lies in a simply-connected neighbourhood of $\mathcal{I} = \mathcal{I}^+ \cup \mathcal{I}^-$.

PROPOSITION 3.5.3. [**35**] *If M is WAS and globally hyperbolic and the null convergence condition holds in M then every causal curve from $\mathcal{I}^-$ to $\mathcal{I}^+$ is homotopic to γ.*

The idea is that, if Γ is a causal curve from $\mathcal{I}^-$ to $\mathcal{I}^+$ not homotopic to γ then, in the universal cover of M, Γ connects different asymptotic regions; to do this Γ must pass through a trapped surface T say on its way to $\mathcal{I}^+$; one derives a contradiction from a conjugate point argument applied to a null geodesic generator of the boundary of the future of T, $\partial\mathcal{I}^+(T)$, which meets $\mathcal{I}^+$.

The interpretation of 3.5.3 is that topological complexity close in in an asymptotically flat space-time satisfying an energy condition collapses 'too fast' for an observer outside to probe the topology, and in particular therefore, too fast for the observer to pass through any wormholes and escape safely. A result equivalent to 3.5.3 due to Galloway is:

PROPOSITION 3.5.4. [**41**] *If M is WAS, the null convergence condition holds in M, and the domain of outer communication $D = I^+(\mathcal{I}^-) \cap I^-(\mathcal{I}^+)$ is globally hyperbolic, then D is simply connected.*

In this form, the result will be seen to be relevant to the study of black holes in §6.

Finally, there is a version of topological censorship due to Galloway and Woolgar [**44**] which drops the condition of global hyperbolicity, replacing it with a form of cosmic censorship and a causal condition at i^0.

A testing example of a *traversable wormhole* was provided by Schein and Aichelburg [**111**]. Their electrovac solution can be interpreted as an exterior consisting of a 2-body Majumdar-Papapetrou solution containing two topologically-spherical charged shells, joined across the shells to an interior consisting of part of the extended Reissner-Nordstrom solution; the trick is that the two shells are in two different asymptotic regions in the Reissner-Nordstrom solution, one later than the other. The matching is done without violating energy conditions. Now it is possible to follow a causal curve through one shell at a certain time t_0, move forward in time in the Reissner-Nordstrom part but re-emerge into the Majumdar-Papapetrou exterior from the second shell at a time earlier than t_0: there are closed time-like curves through every point of the space-time; the wormhole is traversable but the energy conditions are not violated.

3.6. Signature change. Signature change, while not a topological issue, is related to the idea of topology change. The question is can the Einstein equations have solutions in which the signature of the metric changes from Riemannian to Lorentzian or vice-versa? The motivation for considering the possibility has come from the Hartle-Hawking 'No-boundary' proposal in quantum gravity [**58**]. There is a need for care because the metric must degenerate to change signature.

Gibbons and Hartle [**53**] consider the general theory, showing that the signature can only change across an umbilic (equivalently, a totally geodesic) space-like hypersurface S. Then the Hamiltonian constraint again constrains the topology of S as in 3.4.2. Ellis *et al* [**31**] present some explicit solutions of the Einstein equations with matter which do change signature.

4. Lorentzian Splitting Theorems; Related Matters

4.1. Yau's question. For this we first need a definition:

DEFINITION 4.1.1. A time-like *line* is an inextendible time-like geodesic which maximises the distance between any two of its points.

Yau [**147**] posed the problem, slightly rephrased here, of proving that a geodesically complete space-time M in which the time-like convergence condition holds and which contains a time-like line is isometrically the product of the line and a space-like hypersurface. This was proposed as an analogue of the Cheeger-Gromoll splitting theorem in Riemannian geometry.

The Lorentzian Splitting Theorem in this form was proved by Eschenburg [**32**], with the extra assumption that M is globally hyperbolic, by a modification of the Riemannian proof. Galloway [**40**] proved the theorem with the assumption of global hyperbolicity but dropping the assumption of time-like geodesic completeness. Then Newman [**83**] proved the theorem precisely in Yau's form, with the assumption of time-like geodesic completeness and without the assumption of global hyperbolicity. (See [**9**] for a more detailed account of this history.)

4.2. Geroch's suggestion. A related set of ideas is associated with the suggestion of Geroch [**46, 49**] that most closed universes should be flat or become singular. This was interpreted by Galloway and Horta [**43**] as 'spatially closed space-times should fail to be flat only under exceptional circumstances'. Geroch supported his contention with a singularity theorem, which we give in a modified form due to Bartnik [**4**]:

PROPOSITION 4.2.1. *Suppose the time-orientable space-time M has a compact Cauchy surface S and that the time-like convergence condition holds in M; suppose that there is at least one point $p \in S$ with no horizon in the sense that $M - (I^+(p) \cup I^-(p))$ is compact; then M is time-like geodesic incomplete or splits as a metric product.*

The 'no-horizon' condition means roughly that every observer can exchange communications with p. The idea of the proof is to move the surface S until it has everywhere negative or everywhere zero expansion; then use a conjugate point argument on the geodesic normals to S to prove incompleteness, or find a time-like line. Geroch assumed a stronger form of time-like convergence, namely that $R_{ab}t^a t^b \geq 0$ for all time-like t^a, and $R_{ab}t^a t^b = 0$ for some time-like t^a only if $R_{ab} = 0$. With this, the split case is actually flat.

For this section only, and following [**4**], call a space-time 'cosmological' if it is globally hyperbolic with a compact Cauchy surface and satisfies the time-like convergence condition. Then Bartnik [**4**] further conjectures that:

CONJECTURE 4.2.2. Any cosmological space-time is time-like geodesically incomplete or splits as a metric product.

One approach to this would be to find a maximal surface and use a conjugate point argument to prove incompleteness. Another would be to prove that a time-like line exists. The difficulty with the second strategy is that one can seek to construct the line as a limit, only to have the limit become null, a Lorentzian difficulty not existing in Riemannian geometry. With extra assumptions, the second route has been successfully followed by Eschenburg and Galloway [**33, 42**].

A related splitting theorem, due to Andersson *et al* [**2**], is concerned with warped products. They show that a globally hyperbolic space-time satisfying an energy condition with a negative cosmological constant (positive with their conventions) and having a finite but long enough time-like line is a warped product. With some more assumptions, they characterise anti-de Sitter space by this route.

5. Existence and Uniqueness

Questions of existence and uniqueness for the Einstein equations split into problems with the constraints, which are usually elliptic, and problems with evolution, which are hyperbolic. Solution of the constraints on constant-mean-curvature hypersurfaces, either compact, asymptotically flat or asymptotically hyperbolic, is well understood. Local-in-time solution of the evolution equations is also well understood. References to this material can be found in the 'Living review' *Existence theorems for the Einstein equations* by A.D.Rendall at http://www.aei-potsdam.mpg.de. For an earlier review, see [**34**]; for the situation with matter, see [**108**].

The big question now is global or long-time existence. In cosmological solutions, one expects initial and sometimes final singularities to form; in asymptotically flat solutions one expects gravitational collapse to be possible, resulting in singularities. Thus there is often no expectation that a solution obtained from Cauchy data will exist forever. Rather one hopes to investigate and perhaps constrain the kinds of singularities that are formed, and to prove existence up to the singularity. In particular, one would like to know whether Cauchy horizons ever arise in an evolution, or equivalently whether the maximal development of a set of data is globally hyperbolic; cf. e.g. [**20**]. Here we are getting close to cosmic censorship, discussed below in §7.

Results on long-time existence can be classified by the amount of symmetry a solution possesses:

5.1. Spatially-homogeneous cosmologies. These have isometry group transitive on space-like hypersurfaces (they are 'cohomogeneity-one') so the Einstein equations reduce to a system of ordinary differential equations. This system can often be solved; cf. e.g. [**135**]. Rendall [**105**] gives existence theorems for some symmetry types and matter models, which expand forever from an initial singularity, or expand and recollapse, with no Cauchy horizons. This is a 'large-data, long-time' theorem.

5.2. 1+1 reductions. Reductions with two commuting space-like symmetries lead to partial differential equations with one time and one space variable. These include Einstein-Rosen cylindrically-symmetric gravitational waves [**10, 146**] and Gowdy vacuum cosmologies [**80, 19, 24**], where the group orbits are compact . In both cases there are large-data, long-time existence theorems.

5.3. Spherical symmetry. Here collapse is possible. In a long series of papers, Christodoulou has investigated spherically-symmetric solutions with scalar-fields, which also lead to (1+1)-pdes; see [**16**] for references. He has produced a very complete picture, reviewed by Wald [**138**] and described at greater length in 7.2.3. From sufficiently small, asymptotically flat initial data, solutions last forever with a complete $\mathcal{I}^+$.

Rein et al. [**102**] established long-time existence for spherically-symmetric and other (1+1)-solutions of the Einstein-Vlasov equations, again for small data.

5.4. No symmetry. Christodoulou and Klainerman [**18**] have proved the global existence of solutions to the vacuum equations with data close to flat on an asymptotically flat initial hypersurface (see the chapter by Christodoulou in this volume). They find that, with generic (small) asymptotically flat data the conformal structure is not smooth at $\mathcal{I}^+$.

Friedrich, in a long series of papers, has studied the vacuum equations with cosmological constant. We noted in 2.7.3 that the nature of $\mathcal{I}^+$ depends on the sign of the cosmological constant (though note that Friedrich's conventions have the opposite sign for the cosmological constant). For positive cosmological constant, (negative with his conventions) Friedrich [**36**] shows that, with data on S^3 close to the data for de Sitter space, the solution exists globally, and is asymptotically simple. For negative cosmological constant (positive with his conventions), he poses an initial-boundary-value problem, with boundary data on a finite interval of the (time-like) $\mathcal{I}$, and initial data on a ball and proves existence of an asymptotically simple space-time generalising the anti-deSitter metric, with no assumption of smallness. In vacuum, he first considers 'hyperboloidal initial data' which is data on an asymptotically hyperbolic surface S spanning a cut of $\mathcal{I}^+$ [**36, 1**]. He proves

[**36**] that the solution exists to the future of S and that the smoothness of $\mathcal{I}^+$ is preserved by the evolution. Most recently [**38**], he considers the case of data for the vacuum equations on an asymptotically flat hypersurface S. What one wants to know here is what, if any, conditions on the data lead to hyperboloidal data, or equivalently to a smooth $\mathcal{I}^+$. This requires an intricate analysis of the geometry near i^0. Friedrich has a necessary condition on the data for the evolution to admit a smooth $\mathcal{I}^+$, but it is not yet known whether the condition is sufficient. For more details, see [**39**].

It is worth remarking that one knows already from the study of Maxwell's equations in flat space that, to obtain a solution which is smooth at $\mathcal{I}^+$, one needs to impose conditions on the data on an asymptotically flat hypersurface which are stronger than naive asymptotic flatness: crudely speaking with increasing k each 2^k-pole must fall off at a faster rate in r; equivalently the solution must be smooth at i^0. It seems reasonable that one should expect something similar in the vacuum equations.

There do exist radiating electrovac solutions which are smooth at $\mathcal{I}^+$. Cutler and Wald [**27**] give a spherically-symmetric solution of the constraints for an electrovac (actually 'magnetovac') solution. The data is asymptotically flat, and in fact is data for the Schwarzschild solution outside of a certain radius. These authors are able to show that the evolution therefore leads to hyperboloidal data which by Friedrich [**36**] evolves to have a complete $\mathcal{I}^+$.

5.5. Isotropic singularities. Another class of cosmological space-times where the solution is known 'up to' the singularity is the cosmologies with an isotropic singularity [**56, 130, 84**]. In fact, these are the other way round: data is given at the singularity, then local existence is proved; the data is unconstrained, but less data can be given than at a finite surface. Existence and uniqueness has been proved for some perfect fluid matter models [**26, 3**] and for the spatially homogeneous massless Einstein-Vlasov equations [**3**].

5.6. CMC foliations. Under certain circumstances, a space-time will admit a foliation by constant-mean-curvature space-like hypersurfaces, one for each value of the mean curvature in some range (e.g. the range $(-\infty, +\infty)$ in the $k = 1$ FRW solutions; the range $(-\infty, 0)$ in the $k = 0$ FRW solution). The value of the mean curvature is then a useful time coordinate, and 'global in CMC-time' can be thought of as the canonical existence problem. For spatially compact space-times, and given the right energy condition, such a foliation is unique if it exists [**77**]; see also [**6**] for the asymptotically flat case, and [**30**] for application to numerical relativity. We saw in 3.4.2 that there are space-times without such a foliation. There are examples where only part of the space-time is covered, and cases where the existence of the foliation is assured. See [**107**] for a recent review.

One may also seek foliations of a space-like hypersurface by constant mean curvature 2-surfaces (in fact 2-spheres). This is a wholly Riemannian problem. Huisken and Yau [**66**] use a mean curvature flow and the positive energy theorem to prove that a unique stable foliation exists in a neighbourhood of infinity on an asymptotically flat space-like hypersurface. The spheres approach a family of Euclidean spheres at large distances, all with the same centre which these authors interpret as a 'centre of mass'. See [**148**] for a similar result.

6. Black-Hole Uniqueness

6.1. The problem. A time-independent, asymptotically flat but not flat, vacuum space-time cannot be everywhere non-singular [**74**]; cf. [**21**]. Roughly speaking, the (non-zero) gravitational field needs a source. However, as we saw in the example of the Schwarzschild metric in §2.4, it can be non-singular everywhere outside a horizon. The problem of black hole uniqueness is the problem of first finding all time-independent solutions of the Einstein field equations which are asymptotically flat outside a horizon, and then showing that these are indeed all. The field equations are allowed to have one of a small number of matter sources corresponding to various fields. (This is hard to make precise and indeed the rules of the game evolve: the idea is that there are no sources in the sense of non-zero T_{ab} outside the hole except for fields generated by 'charges' attributable to the hole.)

SPACE-LORE 6.1.1 (Classical Results). The first results were Israel's characterisations [**68, 69**] of the Schwarzschild (respectively, Reissner-Nordstrom) solutions as the only static vacuum (respectively, electrovac) black holes. Then, at the end of a long chain of results, with contributions from Carter, Hawking, Robinson, Bunting and Mazur, the Kerr and Kerr-Newman solutions were characterised as the corresponding stationary black holes. A sequence of arguments shows first that one need only consider stationary, axisymmetric metrics with a single, topologically spherical hole, and then that the system of non-linear PDEs to which the Einstein equations reduce has the corresponding unique solutions, depending on a small number of constants. References can be found in the excellent recent monograph of Heusler [**63**].

SPACE-LORE 6.1.2 (No Hair). The Kerr-Newman solution depends on three constants, interpretable as the mass, charge and specific angular momentum. The black hole uniqueness theorem is often aphoristically stated as 'a black hole has no hair' [**79**], being characterised uniquely by its values of these constants.

There has been a resurgence of interest in black hole uniqueness. This is partly with the aim of proving the uniqueness theorems under weaker conditions, or of proving stronger theorems, and partly because new solutions have been found with other fields.

Under the first heading, see the critical account due to Chruściel [**21**]. According to this author, weak links in the proof of the theorem as it stood at his time of writing included:

- the proof that stationary black holes are topologically spherical; the horizon is a null hypersurface, but one thinks of the black hole as being a space-like cross-section of the horizon and so 2- dimensional;
- the proof that stationary black holes are axisymmetric if not static: this requires the metric to be analytic on the horizon; it also uses a physical argument about ergoregions;
- the assumption that the black holes are connected — i.e. that there is only one hole.

We first consider progress in these areas. We recall the definition of domain of outer communication (DOC) from 2.11.1, then a recent result due to Galloway [**41**] met in 3.5.4 is the following:

PROPOSITION 6.1.3. *If in an asymptotically flat space-time M the null convergence condition is satisfied and the DOC is globally hyperbolic, then the DOC is simply connected.*

Galloway notes that this is equivalent to the Friedman-Schleich-Witt topological censorship theorem 3.5.3 (the proof is similar; note that there is no assumption of stationarity). In the context of this section, 6.1.3 shows that all black holes are topologically spherical.

EXAMPLE 6.1.4. Chrusciel and Galloway [**22**] show by an example that a Cauchy horizon can be nowhere differentiable. Analyticity for stationary black holes is proved where the Killing vector is time-like so that the Einstein equations become elliptic. However, at the horizon no Killing vector is time-like so there is a real question of whether elliptic regularity holds 'up to the boundary'.

6.2. Multiple static black holes. The example of the Majumdar-Papapetrou solutions shows that time-independent solutions corresponding to multiple black holes are possible, that is the horizon need not be connected. The physical explanation of the Majumdar-Papapetrou solutions is that electrostatic repulsion balances gravitational attraction, so that this should not happen with vacuum solutions. Bunting and Masood-ul-Alam [**14**] show that, indeed, in the vacuum case there cannot be multiple, static black holes with all components non-degenerate. The proof, which is extremely elegant, is an application of the positive mass theorem with black holes; cf. e.g. [**54, 61**].

Ruback [**110**] extended their work to show that, in the electrovac case, there could not be multiple black holes with every horizon non-degenerate. Here the proof uses a positive energy theorem for charged black holes [**54**].

However, the Majumdar-Papapetrou black holes have all components degenerate. Heusler [**64**] shows that, if all components are degenerate and the charges of all holes have the same sign, then a multiple-black-hole static electrovac solution is necessarily in the Majumdar-Papapetrou family. There are still open questions about mixtures of degenerate and non-degenerate holes.

6.3. Multiple stationary black holes. One may ask if there exist solutions corresponding to multiple rotating black holes. The physical idea would be that there is a spin-spin repulsion which could balance gravitational attraction. There are 2-body solutions in the literature [**28, 65**] but the solutions are extremely complicated and hard to analyse.

In a series of papers, Weinstein [**139, 140, 141, 142**] has analysed the problem of multiple, rotating black holes. The solutions are stationary and axisymmetric, with the holes strung out along the axis. He shows that, for each n, there is a $4n-1$ parameter family of solutions containing n charged, rotating holes, where the parameters are related to mass, charge, and angular momentum of the n holes and their $n-1$ separations. The solutions are asymptotically flat and regular everywhere except that there may be conical singularities on the axis segments. Physically, the idea is that 'rods' or 'struts' may be needed on the axis to keep the black holes apart. It is not yet known whether the axis can be regular for suitable parameter values. The black holes are all non-degenerate.

6.4. Yang-Mills fields and other sources. One may seek black hole solutions in theories with various other matter fields. Very often, one finds that the

Kerr or Schwarzschild solutions are still the only regular black holes. See **[63]** for this with various scalar field and harmonic map sources.

The situation is different when the source is the Yang-Mills field. Bartnik and McKinnon **[7]** described a numerical study which found spherically symmetric, asymptotically flat, static solutions of the Einstein-Yang-Mills equations with a regular centre. This paper generated a great deal of excitement. Spherical solutions with black holes were subsequently found numerically by Bizon **[11]** and Volkov and Galtsov **[133]**, and a countably infinite family of solutions describing spherical black holes was found numerically by Künzle and Masood-ul-Alam **[73]**.

Proofs that the solutions really do exist were given by Smoller *et al* **[120]**, for the solutions with a regular centre, by Smoller and Wasserman **[118]** for a countably infinite family of solutions with a regular centre, and by Smoller, Wasserman and Yau **[121]** for black hole solutions. The problem is to show that solutions exist to the boundary-value problem for the coupled non-linear ODEs which the field equations reduce to. The infinite families are associated with a winding number.

Smoller and Wasserman **[119]** showed that the extreme Reissner-Nordstrom metric is the unique degenerate black hole among the $SU(2)$-Yang-Mills solutions. Brodbeck and Straumann **[13]** have shown that both the solutions with a regular centre and the black holes are unstable. This indicates that the solutions are probably not significant physically.

Other black hole solutions with matter sources related to the Yang-Mills field continue to appear in the literature, and may be found at the gr-qc archive.

7. Cosmic Censorship

7.1. Terminology. The term is due to Penrose **[88]**: 'Does there exist a "cosmic censor" who forbids the appearance of naked singularities, clothing each one in an absolute event horizon?'. This would now be called the weak cosmic censorship hypothesis, that singularities will form in gravitational collapse, but they will be hidden behind horizons, while the strong cosmic censorship hypothesis is the suggestion that space-time is globally hyperbolic. This distinction is also due to Penrose **[92]**; for example, weak cosmic censorship would allow time-like singularities to form inside horizons while strong cosmic censorship would not allow them anywhere.

7.2. Counterexamples. Attempts to disprove the cosmic censorship hypothesis (CCH) inevitably centre on finding counterexamples. A counterexample would be an evolution of regular data which results in an asymptotically flat space-time with a singularity in $J^-(\mathcal{I}^+)$. So far, proposed counterexamples have served mostly to hone the 'correct' statement of the CCH (a process which the unsympathetic may regard as moving the goal-posts).

CONDITION 7.2.1. The notion of a '*tame*' matter model has emerged **[108, 81]**. Suppose one has a putative counterexample to the CCH with some matter model; for this to be a real threat to the CCH the matter model should be 'tame' in the sense that it would not lead to the same kind of singularities without gravity - it is not reasonable to expect general relativity to remove a pathology from a matter model which produces singularities already in special relativity. Thus the shell-crossing singularities of the first counterexamples **[149, 150]** occur already with perfect fluids in flat-space: perfect fluids are not tame.

EXAMPLE 7.2.2. The Einstein-Vlasov equations are tame, because the Vlasov equation is linear, and also because the Newtonian limit has long-time existence [**98**]; cf. [**101**]. Shell-crossing singularities cannot occur in the spherically-symmetric Einstein-Vlasov equations [**102**], although shell-focussing singularities may arise [**30**]. Shapiro and Teukolsky [**117**] have presented numerical evidence of a violation of the weak CCH with solutions of the Einstein-Vlasov equations. However, it is difficult to be certain that the CCH is violated in their examples, and their initial distribution function is non-smooth; cf. [**104**].

EXAMPLE 7.2.3. Scalar fields are tame.

Christodoulou has investigated collapsing, spherically symmetric, massless scalar field configurations in a long and ongoing series of papers; for the references, see [**16**]. He gives data on a future light cone, centred at the origin, and shows [**16**] that there are choices of asymptotically flat initial data which evolve to solutions with a naked singularity. The singularity forms first at the origin and then propagates out to $\mathcal{I}^+$ along a singular null cone arriving at a finite (retarded) time. In a recent preprint [**17**], he obtains a very complete picture according to which one of three things happens:

(i) long-time existence with a complete $\mathcal{I}^+$;
(ii) a singularity forms, surrounded by a horizon, and again $\mathcal{I}^+$ is complete;
(iii) neither of the above;

and (iii) includes naked singularities, but the third case is non-generic: Christodoulou exhibits an arbitrarily small perturbation of the data converting (iii) to (ii).

7.3. Evidence for the CCH.

EXHIBIT 7.3.1 (stability of black holes). If the time-independent black hole solutions were unstable, then they could not be the (stable) endpoint of collapse and it is hard to see how the weak CCH could be true. However first the Schwarzschild solution [**134, 99, 72**] and later the Kerr solution [**143**] have been shown to be (linearly) stable.

EXHIBIT 7.3.2 (Existence Theorems). In addition to the work of Christodoulou described in 7.2.3, there are other existence theorems supporting various aspects of the CCH. Much of what is described in §5 can be interpreted in this light: for example, Christodoulou and Klainerman [**18**] prove the CCH for small data and vacuum, Friedrich [**36, 37**] proves it for vacuum plus cosmological constant and small data. Strong cosmic censorship is the claim that the maximal evolution of Cauchy data is a globally hyperbolic space-time, possibly with singularities but with no Cauchy horizons, probably with a requirement that the data be 'generic'. Proofs of strong cosmic censorship have been given in some restricted cases [**25, 105**]. Note that Cauchy horizons can exist in solutions of the Einstein equations (e.g. in the Reissner-Nordstrom solution; see figure 4) but the expectation is that they are necessarily unstable, [**57**], and therefore non-generic.

EXHIBIT 7.3.3 (Area Theorem for Black Holes). For the next pieces of evidence, we first need the following result [**60**, Prop 9.2.7]: Given the null convergence condition in a WAS, black-hole space-time M in which the weak CCH holds in the sense that $\mathcal{I}$ is in the domain of dependence $D(S)$ of a Cauchy surface S, then the area of any connected space-like cross-section of the event horizon $\partial J^-(\mathcal{I}^+)$ increases into the future.

The proof uses a conjugate point argument: if the area starts to decrease then it goes to zero in finite time and a naked singularity will appear. Chrusciel and Galloway [**23**] have emphasised that the present proof also assumes smoothness of the horizon.

Now one may attempt to devise gedanken experiments which reduce the area of a black hole and therefore violate the weak CCH, for example by firing in charged [**136**], dyonic [**116**] or spinning [**136, 126**] particles. The details of the particle trajectories turn out to foil these attempts.

7.4. The Penrose inequality.

EXHIBIT 7.4.1 (The Inequality). There is a whole cycle of ideas around this prediction of the CCH. The weak CCH, together with the null convergence condition, implies via the area theorem 7.3.2 an inequality between mass and area of a black hole: suppose a black hole forms in a gravitational collapse, and then settles down to a stationary or static one; in the process its area A will increase, but its Bondi mass m will decrease; when it has settled down, black hole uniqueness tells us that it will be a Kerr or Kerr-Newman solution and these quantities will satisfy the following inequality, which can readily be seen to be true for the Kerr family:

$$A \leq 16\pi m^2 \tag{7.1}$$

Thus this inequality must be true at all earlier times too.

In fact there is a whole range of inequalities like (7.1) in the literature where A may be the area of a trapped or marginally-trapped surface and m may be the Bondi, ADM or even quasi-local mass. Not all of these proposed inequalities are strictly speaking predictions of the CCH.

EXHIBIT 7.4.2 (Special Cases). Ludvigsen and Vickers [**75**], Tod [**127, 128**], Jezierski [**71**] and Malec and O'Murchadha [**76**] have proved versions of the inequality under different assumptions. Herzlich [**62**] proves something very like (7.1) using a modification of the Witten positive energy theorem on an umbilic space-like hypersurface S (as we saw in §2.4, a marginally-trapped surface is then a minimal surface; A is the area of the minimal surface and m is the ADM mass. Herzlich's inequality has a different constant from (7.1), related to a Sobolev constant).

Gibbons [**52**] recently completed a programme started by himself [**51**] and Penrose [**90**] to prove (7.1) when A is the area of a marginally trapped surface and m is Bondi mass at $\mathcal{I}^-$ in an idealised model of gravitational collapse. Here (7.1) is implied by other geometric inequalities, like Minkowski's inequality for convex bodies as recently generalised by Trudinger [**132**].

Huisken and Ilmanen [**67**] prove (7.1) in the same setting as Herzlich; they show that an inverse mean curvature flow proposed with the aim of proving the positive energy theorem by Geroch [**50**] and modified to prove (7.1) by Jang and Wald [**70**] does indeed work, evolving in from infinity to the (outermost) minimal surface.

EXHIBIT 7.4.3 (Converse). There are physical reasons for hoping for a converse to (7.1): that a black hole will form if matter of mass m is squeezed into a small region, for example one enclosed by area A. This is is hard to make precise. There is a result of this form due to Schoen and Yau [**115**], modified by O'Murchadha [**85**]. Rather than area, they have a sophisticated measure of the size of a region in terms of the largest torus which it can contain. The proof uses minimal surface techniques.

CONJECTURE 7.4.4 (Hoop Conjecture). Related to the Penrose inequality and its converse is the *hoop conjecture* of Thorne [**122**]. The idea is that a black hole will form when and only when matter of mass m is squeezed into a region whose every circumference C satisfies

$$C \leq 4\pi m \tag{7.2}$$

Part of the idea here is that there could be a collapse of a long, thin object to a singularity without the formation of a horizon, therefore with a violation of the CCH. The study of Shapiro and Teukolsky [**117**] was presented as just such a violation of the CCH and a vindication of the hoop conjecture. Note [**131**] that there are real difficulties with making (7.2) precise and some formulations of it are false.

References

[1] L. Andersson, P.T. Chruściel and H. Friedrich, *On the regularity of solutions to the Yamabe equation and the existence of smooth hyperboloidal initial data for Einstein's field equations.* Comm. Math. Phys. **149** (1992) 587–612.

[2] L. Andersson, G.J. Galloway and R. Howard, *A strong maximum principle for weak solutions of quasi-linear elliptic equations with applications to Lorentzian and Riemannian geometry.* Comm. Pure. App. Math. **51** (1998) 581–624.

[3] K. Anguige, **Isotropic cosmological singularities**. D.Phil. thesis, Oxford University, 1997.

[4] R. Bartnik, *Remarks on cosmological space-times and constant mean curvature hypersurfaces.* Comm. Math. Phys. **117** (1988) 615–624.

[5] R. Bartnik, New definition of quasi-local mass. Phys. Rev. Lett. **62** (1989) 2346

[6] R. Bartnik, P.T. Chruściel and N. O'Murchadha, On maximal surfaces in asymptotically flat space-times. Comm. Math. Phys. **130** (1990) 95–109.

[7] R. Bartnik and J. McKinnon, *Particle-like solutions of the Einstein-Yang-Mills equations.* Phys. Rev. Lett. **61** (1988) 141–143.

[8] R.W. Bass and L. Witten, *Remark on cosmological models.* Rev. Mod. Phys. **29** (1957) 452–453.

[9] J.K. Beem, P.E. Ehrlich and K.L. Easley, **Global Lorentzian Geometry**, Second Edition, Marcel Dekker inc: New York, 1996.

[10] B.K. Berger, P.T. Chruściel and V. Moncrief, *On "asymptotically flat" space-times with G_2-invariant Cauchy surfaces*, Ann. Phys. **237** (1995) 322–54.

[11] P. Bizon, *Colored black holes.* Phys. Rev. Lett. **64** (1990) 2844–2847.

[12] D. Brill, *On space-times without maximal surfaces*, in **Proc. 3rd Marcel Grossman Meeting**, ed. H. Ning, North Holland, Amsterdam, 1982.

[13] O. Brodbeck and N. Straumann, *Instability proof for the Einstein-Yang-Mills solitons and black holes with arbitrary gauge groups*, J. Math. Phys. **37** (1996) 1414–1433.

[14] G.L. Bunting and A.K.M. Masood-ul-Alam, *Nonexistence of multiple black holes in asymptotically Euclidean vacuum space-times*, Gen. Rel. Grav. **19** (1987) 147–154.

[15] B. Carter, *The general theory of the mechanical, electromagnetic and thermodynamical properties of black holes*, in **General Relativity**, eds S.W. Hawking and W. Israel, Cambridge University Press, 1979.

[16] D. Christodoulou, *Examples of naked singularity formation in the gravitational collapse of a scalar field.* Ann. Math. **140** (1994) 607–653.

[17] D. Christodoulou, *The instability of naked singularities in the gravitational collapse of a scalar field*, Ann. Math. **149** (1999) 183–217

[18] D. Christodoulou and S. Klainerman, **The Global Nonlinear Stability of Minkowski Space**. Princeton University Press, 1993.

[19] P.T. Chruściel, *On space-times with $U(1)\Theta U(1)$ symmetric compact Cauchy surfaces*, Ann. Phys. **202** (1990) 100–150.

[20] P.T. Chruściel, *On uniqueness in the large of solutions of Einstein's equations ("Strong cosmic censorship")*, Proc. Centre for Math. and Appns. ANU **27** (1991)

[21] P.T. Chruściel, "No hair" theorems — folklore, conjectures, results. Contemporary Mathematics **170** (1994) 23–49.

[22] P.T. Chruściel, Uniqueness of stationary, electro-vacuum black holes revisited. Helv. Phys. Acta **69** (1996) 529–552.

[23] P.T. Chruściel and G.J. Galloway. "Nowhere" differentiable horizons, e-print gr-qc/9611032, 1996.

[24] P.T. Chruściel, J. Isenberg and V. Moncrief, *Strong cosmic censorship in polarised Gowdy space-times*, Class. Quant. Grav. **7** (1990) 1671–1680

[25] P.T. Chruściel and A.D. Rendall, *Strong cosmic censorship in vacuum space-times with compact, locally homogeneous Cauchy surfaces.* Ann. Phys. **242** (1995) 349–385.

[26] C.M. Claudel and K. Newman, *The Cauchy problem for quasi-linear hyperbolic evolution problems with a singularity in the time.* Proc. Roy. Soc. Lond. **A 454** 1073–1107

[27] C. Cutler and R.M. Wald, *Existence of radiating Einstein-Maxwell solutions which are regular on all of* $\mathcal{I}^+$ *and* $\mathcal{I}^-$. Class. Quant. Grav. **6** (1989) 452–466

[28] W. Dietz and C. Hoenselaers, *Two mass solutions of Einstein's vacuum equations: the Double Kerr solution.* Ann. Phys. **165** (1985) 319–383

[29] D.M. Eardley, *Naked singularities in spherical gravitational collapse in Gravitation*, in **Astrophysics** eds B.Carter and J.B.Hartle, Plenum, 1987.

[30] D.M. Eardley and L. Smarr, *Time functions in numerical relativity: marginally bound dust collapse.* Phys. Rev. **D19** (1979) 2239–59

[31] G.F.R. Ellis, A. Sumeruk, D. Coule and C. Hellaby, *Change of signature in classical relativity.* Class. Quant. Grav. **9** (1992) 1535–1554

[32] J.-H. Eschenburg,The splitting theorem for space-times with strong energy conditions. J. Diff. Geom. **27** (1988) 477–491

[33] J.-H. Eschenburg and G.J. Galloway, *Lines in space-time.* Comm. Math. Phys. **148** (1992) 209–216

[34] A.E. Fischer and J.E. Marsden, *The initial value problem and the dynamical formulation of general relativity*, in **General Relativity: an Einstein Centennary Survey** eds S.W. Hawking and W. Israel, Cambridge University Press, (1973)

[35] J. Friedman, K. Schleich and D.W. Witt. *Topological censorship.* Phys. Rev. Lett. **71** (1993) 1486–89

[36] H. Friedrich, *On the existence of n-geodesically complete or future complete solutions of Einstein's field equations with smooth asymptotic structure.* Comm. Math. Phys. **107** (1986)587–609

[37] H. Friedrich, *Einstein equations and conformal structure: existence of anti-de-Sitter type space-times.* J. Geom. Phys. **17** (1995) 125–184

[38] H. Friedrich, *Gravitational fields near space-like and null infinity AEI.* Preprint 022, Albert Einstein Institute, Potsdam, 1996.

[39] H. Friedrich, *Einstein's equations and conformal structure*, in **The Geometric Universe** eds S. Huggett, L. Mason, K.P. Tod, S.T. Tsou and N.M.J. Woodhouse, Oxford Univ.Press, 1998.

[40] G.J. Galloway, *The Lorentzian splitting theorem without completeness assumptions.* J. Diff. Geom. **29** (1989) 373–387

[41] G.J. Galloway, *On the topology of the domain of outer communication.* Class. Quant. Grav. **12** (1995) L99–L101

[42] G.J. Galloway, *Some rigidity results for spatially closed space-times*, in **Mathematics of Gravitation 1** ed P. Chruściel, Banach Center Publications 41, 1997.

[43] G.J. Galloway and A. Horta, *Regularity of Lorentzian Busemann functions.* Trans. A.M.S. **348** (1996) 2063–2084.

[44] G.J. Galloway and E. Woolgar, *The cosmic censor forbids naked topology.* Class. Quant. Grav. **14** (1997) L1–L7

[45] D. Gannon, *Singularities in nonsimply connected space-times.* J. Math. Phys. **16** (1975) 2364–67.

[46] R.P. Geroch, *Singularities in closed universes.* Phys. Rev. Lett. **17** (1966) 445–447.

[47] R.P. Geroch, *Topology in general relativity.* J. Math. Phys. **8** (1969) 782–786.

[48] R.P. Geroch, *The domain of dependence.* J. Math. Phys. **11** (1970) 437–439

[49] R.P. Geroch, *Singularities*, in **Relativity** eds M.Carmeli, S.Fickler and L.Witten, Plenum Press, 1970.

[50] R.P. Geroch, *Energy extraction.* Ann. N.Y. Acad. Sci. **224** (1973) 108–117

[51] G.W. Gibbons, *The isoperimetric and Bogomolny inequalities for black holes*, in **Global Riemannian geometry** eds T.Willmore and N.Hitchin, Ellis Horwood, 1984.

[52] G.W. Gibbons, *Collapsing shells and the isoperimetric inequality for black holes.* Class.Quant.Grav. **14** (1997) 2905–2915.

[53] G.W. Gibbons and J.B. Hartle, *Real tunnelling geometries and the large scale topology of the universe.* Phys. Rev. **D42** (1990) 2458–68

[54] G.W. Gibbons, S.W. Hawking, G.T. Horowitz and M.J. Perry, *Positive mass theorems for black holes.* Comm. Math. Phys. **88** (1983) 259–308

[55] G.W. Gibbons and S.W. Hawking, *Selection rules for topology change.* Comm. Math. Phys. **148** (1992) 342–352

[56] S.W. Goode and J. Wainwright, *Isotropic singularities in cosmological models.* Class. Quant. Grav. **9** (1985) 445–455

[57] J.B. Hartle and S. Chandrasekhar, *On crossing the Cauchy horizon of a Reissner-Nordstrom black hole.* Proc. Roy. Soc. Lond. **A384** (1982) 301–315

[58] J.B. Hartle and S.W. Hawking, *The wave-function of the universe.* Phys. Rev. **D28** (1983) 2960–2975

[59] S.W. Hawking, *Gravitational radiation in an expanding universe.* J. Math. Phys. **9** (1968) 598–604

[60] S.W. Hawking and G.F.R. Ellis, **The Large-Scale Structure of Space-Time**, Cambridge University Press, 1973

[61] M. Herzlich, *The positive mass theorem for black holes revisited.* To appear in Jour. Geom. Phys.

[62] M. Herzlich, *A Penrose-like inequality for the mass of Riemannian asymptotically flat manifolds.* Comm. Math. Phys. **188** (1997) 121–133.

[63] M. Heusler, **Blackhole uniqueness theorems**. Cambridge University Press, 1996.

[64] M. Heusler, *On the uniqueness of the Majumdar-Papapetrou metric.* Class. Quant.Grav. **14** (1997) L129–L134.

[65] C. Hoenselaers, *The double Kerr solution: a survey*, in **Proceedings of the 4th M. Grossman meeting on general relativity** ed R.Ruffini, North Holland, 1986.

[66] G. Huisken and S.T. Yau, *Definition of centre of mass for isolated physical systems and unique foliations by stable spheres with constant mean curvature.* Invent. Math. **124** (1996) 281–311.

[67] G. Huisken and T. Ilmanen, *The Riemannian Penrose inequality.* Int. Math. Res. Not. **20** (1997) 1045–1058.

[68] W. Israel, *Event horizons in static vacuum space-times.* Phys. Rev. **164** (1967) 1776–1779.

[69] W. Israel, *Event horizons in static electrovac space-times.* Comm. Math. Phys. **8** (1968) 245–260.

[70] P.S. Jang and R.M. Wald, *The positive energy conjecture and the cosmic censorship hypothesis.* J. Math. Phys. **18** (1977) 41–44.

[71] J. Jezierski, *Positivity of mass for certain space-times with horizons.* Class. Quant. Grav. **6** (1989) 1535–1539.

[72] B.S. Kay and R.M. Wald, *Linear stability of Schwarzschild under perturbations which are non-vanishing on the bifurcation 2-surface.* Class. Quant. Grav. **4** (1987) 893–898.

[73] H.P. Künzle and A.K.M. Masood-ul-Alam, *Spherically symmetric static $SU(2)$ Einstein-Yang-Mills fields.* J. Math. Phys. **31** (1990) 928–935,

[74] A. Lichnerowicz, **Théories Relativistes de la Gravitation et de l'Électromagnétisme**, Masson et Cie, Paris, 1955.

[75] M. Ludvigsen and J. Vickers, *An inequality relating total mass and area of a trapped surface in general relativity.* J. Phys. **A16** (1983) 3349–3353.

[76] E. Malec and N. O'Murchadha, *Trapped surfaces and the Penrose inequality in spherically symmetric geometries.* Phys. Rev. **D49** (1994) 6931–6934.

[77] J.E. Marsden and F.J. Tipler, *Maximal hypersurfaces and foliations of constant mean curvature in general relativity.* Phys. Rep. **66** (1980) 109–139.

[78] P.O. Mazur, *Proof of uniqueness of the Kerr-Newman black hole solution.* J. Phys. **A15** (1982) 3173–3180.

[79] C. W. Misner, K. S. Thorne and J. A. Wheeler, **Gravitation**, W.H. Freeman, 1973.

[80] V. Moncrief, *Global properties of Gowdy space-times with T^3 topology.* Ann. Phys. **132** (1981) 87–107.

[81] V. Moncrief, *Space-time singularities and cosmic censorship*, in **Proc. 14th Conference on Gen. Rel. Grav.** ed M. Francaviglia et al, World Scientific, 1997.

[82] R.P.A.C. Newman, *Compact space-times and the no-return theorem*. Gen. Rel. Grav. **18** (1986) 1181–1186.

[83] R.P.A.C. Newman, *A proof of the splitting conjecture of S.T. Yau*. J. Diff. Geom. **31** (1990) 163–184.

[84] R.P.A.C. Newman, *On the structure of conformal singularities in classical general relativity I and II*. Proc. Roy. Soc. Lond. **A443** (1993) 473–492, 493–515.

[85] N. O'Murchadha, *How large can a star be?* Phys. Rev. Lett. **57** (1986) 2466–2469.

[86] T. Parker and C.H. Taubes, *On Witten's proof of the positive energy theorem*. Comm. Math. Phys. **84** (1982) 223–238.

[87] R. Penrose, *Gravitational collapse and space-time singularities*. Phys. Rev. Lett. **14** (1965) 57–59.

[88] R. Penrose, *Gravitational collapse: the role of General Relativity*. Riv. del N. Cim. **1** (1969) 252–276.

[89] R. Penrose, **Techniques of Differential Topology in Relativity**, SIAM, 1972.

[90] R. Penrose, *Naked singularities*. Ann. NY Acad. Sci. **224** (1973) 125–134.

[91] R.Penrose, *Singularities of space-times*, in **Theoretical Principles in Astrophysics and Relativity,** eds. N.R.Lebowitz *et al*. University of Chicago Press, 1978.

[92] R. Penrose, *Singularities and time-asymmetry* in **General Relativity: an Einstein Centennial Survey**, eds S.W.Hawking and W.Israel, Cambridge University Press, 1979.

[93] R. Penrose, *On Schwarzschild causality*, in **Essays in General Relativity**, ed F.J.Tipler, 1980.

[94] R. Penrose, *Quasi-local mass and angular momentum in general relativity*. Proc. Roy. Soc. Lond. **A381** (1982) 53–63.

[95] R. Penrose, **Shadows of the Mind**, Oxford Univ. Press, 1994.

[96] R. Penrose and W. Rindler, **Spinors and Spacetime**, volumes 1 and 2, Cambridge University Press, 1984.

[97] R. Penrose, R.D. Sorkin and E. Woolgar, *A positive mass theorem based on the focussing and retardation of null geodesics*. gr-qc/9301015, 1993.

[98] K. Pfaffelmoser, *Global classical solutions of the Vlasov-Poisson system in three dimensions for general initial data*. J. Diff. Eqns **95** (1992) 281–303.

[99] R. Price, *Nonspherical perturbations of relativistic gravitational collapse I and II*. Phys. Rev. **D5** (1972) 2419–2438, 2439–2454.

[100] G. Rein, *Cosmological solutions of the Vlasov-Einstein system with spherical, plane and hyperbolic symmetry*. Math. Proc. Cam. Phil. **119** (1996) 739–762.

[101] G. Rein , *Self-gravitating systems in Newtonian theory — the Vlasov-Poisson system*, in **Mathematics of Gravitation,** ed P.Chruściel, Banach Center Publications **41**, 1997.

[102] G. Rein, A.D. Rendall and J. Schaeffer, *A regularity theorem for solutions of the spherically symmetric Einstein-Vlasov system*. Comm. Math. Phys. **168** (1995) 467–478.

[103] B.L. Reinhart,*Cobordism and the Euler number*. Topology **2** (1963) 173–177.

[104] A.D. Rendall, *Cosmic censorship and the Vlasov equation*. Class. Quant. Grav. **9** (1992) L99–L101.

[105] A.D. Rendall, *Cosmic censorship for some spatially-homogeneous cosmological models*. Ann. Phys. **233** (1994) 82–96.

[106] A.D. Rendall, *Crushing singularities in space-times with spherical, plane and hyperbolic symmetry*. Class. Quant. Grav. **12** (1995) 1517–1535.

[107] A.D. Rendall, *Constant mean curvature foliations in cosmological spacetimes*. Helv. Phys. Acta **69** (1996) 490–500.

[108] A.D. Rendall, *Solutions of the Einstein equations with matter*, in **Proc.14th Conference on Gen. Rel. Grav.** ed M. Francaviglia et al. World Scientific, 1997.

[109] O. Reula and K.P. Tod, *Positivity of the Bondi energy*. J. Math. Phys. **25** (1984) 1004–1008.

[110] P. Ruback , *A new uniqueness theorem for charged black holes*. Class. Quant. Grav. **5** (1988) L155–L159.

[111] F. Schein and P.C. Aichelburg, *Traversable wormholes in geometries of charged shells*. Phys. Rev. Lett. **77** (1996) 4130–4133.

[112] R. Schoen and S.T. Yau, *On the proof of the positive mass conjecture in general relativity*. Comm. Math. Phys. **65** (1979) 45–76.

[113] R. Schoen and S.T. Yau, *Proof of the positive mass theorem II.* Comm. Math. Phys. **79** (1981) 231–260.

[114] R. Schoen and S.T. Yau, *Proof that the Bondi mass is positive.* Phys. Rev. Lett. **48** (1982) 369–371.

[115] R. Schoen and S.T. Yau, *The existence of a black hole due to condensation of matter.* Comm. Math. Phys. **90** (1983) 575–579.

[116] I. Semiz, *Dyon black holes do not violate cosmic censorship.* Class. Quant. Grav. **7** (1990) 353–359.

[117] S.L. Shapiro and S.A. Teukolsky, *Formation of naked singularities: the violation of cosmic censorship.* Phys. Rev. Lett. **66** (1991) 994–997.

[118] J. Smoller and A. Wasserman, *Existence of infinitely many smooth static global solutions of the Einstein-Yang-Mills equations. Comm. Math. Phys.* **151** (1993) 303–325.

[119] J. Smoller and A. Wasserman, *Uniqueness of zero surface gravity SU(2) Einstein-Yang-Mills black holes.* J. Math. Phys. **37** (1995) 1434–1484.

[120] J. Smoller, A. Wasserman, S.T. Yau and J.B. Mcleod, *Smooth static solutions of the Einstein-Yang-Mills equations.* Comm. Math. Phys. **143** (1991) 115–147.

[121] J. Smoller, A. Wasserman and S.T. Yau, *Existence of black hole solutions of the Einstein-Yang-Mills equations.* Comm. Math. Phys. **154** (1993) 137–401.

[122] K.S. Thorne, *Non-spherical gravitational collapse: a short review,* in Magic without magic: John Archibald Wheeler, ed J. Klauder, W.H. Freeman, 1972.

[123] F.J. Tipler, *Singularities and causality violation.* Ann. Physics **108**(1977) 1–36.

[124] F.J. Tipler, *Singularities and causality violation.* Gen. Rel. Grav. **10** (1979) 983–984.

[125] F.J. Tipler, *General relativity and the Eternal Return,* in Essays in general relativity ed F.J.Tipler, Academic Press, 1980.

[126] K.P. Tod, F. de Felice and M. Calvani, *Spinning test particles in the field of a black hole.* Nuovo Cim. **34B** (1976) 365–379.

[127] K.P. Tod, *Penrose's quasi-local mass and the isoperimetric inequality for static black holes.* Class. Quant. Grav. **2** (1985) L65–L68.

[128] K.P. Tod, *More on Penrose's quasi-local-mass.* Class.Quant.Grav. **3** (1986) 1169–1189.

[129] K.P. Tod, *Penrose's quasi-local mass,* in **Twistors in Mathematics and Physics**, eds T.N. Bailey and R.J. Baston, LMS Lecture Notes **156**, Cambridge Univ. Press, 1990.

[130] K.P. Tod, *Isotropic singularities.* Rend. Sem. Mat. Uni. Pol. Tor. **50** (1992) 69–93.

[131] K.P. Tod, *The hoop conjecture and the Gibbons-Penrose construction of trapped surfaces.* Class. Quant. Grav. **9** (1992) 1581–1591.

[132] N. Trudinger, *Isoperimetric inequalities for quermassintegrals.* Ann. Inst. H. Poincare **11** (1994) 411–425.

[133] D. V. Volkov and M. S. Galtsov, *Black holes in Einstein-Yang-Mills theory.* Sov. Jour. Nucl. Phys. **51** (1990) 747–753.

[134] C.V. Vishveshwara, *Stability of the Schwarzschild metric.* Phys. Rev. **D1** (1970) 2870–2879.

[135] J. Wainwright and G.F.R. Ellis, **Dynamical Systems in Cosmology**, Cambridge University Press, 1997.

[136] R.M. Wald, *Gedanken experiments to destroy a black hole.* Ann. Phys. **82** (1974) 548–556.

[137] R.M. Wald, **General Relativity**, Univ of Chicago Press, 1984.

[138] R.M. Wald, *Gravitational collapse and cosmic censorship.* gr-qc/9710068, 1997.

[139] G. Weinstein, *On rotating black holes in equilibrium in general relativity.* Comm. Pure Appl. Maths. **XLIII** (1990) 903–948.

[140] G. Weinstein, *The stationary axisymmetric two-body problem in general relativity.* Comm. Pure Appl. Maths **XLV** (1990) 1183–1203.

[141] G. Weinstein, *On the force between rotating coaxial black holes.* Trans. A.M.S. **343** (1994) 899–906.

[142] G. Weinstein, *Black hole stationary and axisymmetric solutions of the Einstein-Maxwell equations.* Comm. P.D.E. **21** (1996) 1389–1430.

[143] B. Whiting, *Mode stability of the Kerr black hole.* J. Math. Phys. **30** (1989) 1301–1305.

[144] D.M. Witt, *Vacuum space-times that admit no maximal slice.* Phys. Rev. Lett. **57** (1986) 1386–1389.

[145] E. Witten, *A new proof of the positive energy theorem.* Comm. Math. Phys. **80** (1981) 381–402.

[146] N.M.J. Woodhouse, *Integrability and Einstein's equations* in **Mathematics of Gravitation**, ed P.T.Chruściel, Banach Center Publications **41**, 1997.

[147] S.T. Yau, *Problem section* in **Seminar on differential geometry**, Ann. Math. Studs. **102**, Princeton University Press, 1982.

[148] R. Ye, *Foliation by constant mean curvature spheres on asymptotically flat manifolds* in **Geometric Analysis and Calculus of Variations**, ed J.Jost, Internat. Press, Cambridge, MA, 1996.

[149] P. Yodzis, H-J. Seifert and H. Muller zum Hagen, *On the occurrence of naked singularities in general relativity*. Comm. Math. Phys. **34** (1973) 135–148.

[150] P. Yodzis, H-J. Seifert and H. Muller zum Hagen, *On the occurrence of naked singularities in general relativity*. Comm. Math. Phys. **37** (1974) 29–40.

Mathematical Institute, 24–29 St. Giles, Oxford OX1 3LB, England
E-mail address: tod@maths.ox.ac.uk

The Stability of Minkowski Space-Time

Demetrios Christodoulou

1. Introduction

The *general theory of relativity*, discovered by Einstein in 1915 [**9, 10**], is a unified theory of space, time and gravitation. According to general relativity, the *space-time manifold* is a four-dimensional oriented differentiable manifold $\mathcal{M}$ which is endowed with a *Lorentzian metric* g, that is, a continuous assignment of g_p, a symmetric bilinear form of index 1 in $T_p\mathcal{M}$, at each $p \in \mathcal{M}$. The Lorentzian metric divides $T_p\mathcal{M} \setminus 0_p$ into three subsets, I_p, N_p, S_p, the set of *time-like, null, space-like* vectors at p, according as to whether the quadratic form g_p is respectively negative, zero, or positive. The subset N_p is a double cone $N_p^+ \bigcup N_p^-$, the *null cone* at p. The subset I_p is the interior of this cone, an open set consisting of two components I_p^+ and I_p^-, the *future* and *past* components respectively. The boundaries of these components are the corresponding components of N_p. The subset S_p is the exterior of the null cone, a connected open set.

A curve in $\mathcal{M}$ is called *causal* if its tangent vector at each point belongs to the set $I \bigcup N$ correponding to that point. We assume that $(\mathcal{M}, g)$ is *time oriented*, that is a continuous choice of future component of I_p at each $p \in \mathcal{M}$ can and has been made. A causal curve is then *future directed* or *past directed* according as to whether its tangent vector at a point belongs to the subset $I^+ \bigcup N^+$ or $I^- \bigcup N^-$ corresponding to that point. The *causal future* $\mathcal{J}^+(\mathcal{K})$ of a set $\mathcal{K} \subset \mathcal{M}$ is the set of points which can be reached by a future directed causal curve initiating at $\mathcal{K}$. Similarly $\mathcal{J}^-(\mathcal{K})$, the *causal past* of $\mathcal{K}$, is the set of points which can be reached by a past directed causal curve initiating at $\mathcal{K}$. The boundaries $\partial\mathcal{J}^+(\mathcal{K}) \setminus \mathcal{K}$ and $\partial\mathcal{J}^-(\mathcal{K}) \setminus \mathcal{K}$ are hypersurfaces generated by null geodesics, *null hypersurfaces*, with the past end points of the null geodesics generating $\partial\mathcal{J}^+(\mathcal{K}) \setminus \mathcal{K}$ and the future end points of those generating $\partial\mathcal{J}^-(\mathcal{K}) \setminus \mathcal{K}$ all lying in $\mathcal{K}$. The specification of $\mathcal{J}^+(p)$ and $\mathcal{J}^-(p)$ for every $p \in \mathcal{M}$ defines the *causal structure*, which is equivalent to the conformal geometry of $\mathcal{M}$.

A hypersurface $\mathcal{H}$ in $\mathcal{M}$ is called *space-like* if at each $x \in \mathcal{H}$ the restriction of g_x to $T_x\mathcal{H}$ is positive definite. We denote by $\overline{g}$ the induced metric or first fundamental form of $\mathcal{H}$:

$$\overline{g}_x = g_x\Big|_{T_x\mathcal{H}}$$

The pair $(\mathcal{H}, \overline{g})$ is then a Riemannian manifold. The orthogonal complement of $T_x\mathcal{H}$ in $T_x\mathcal{M}$ is a one dimensional linear subspace of $T_x\mathcal{M}$ contained in I_x. There is therefore a unique future directed unit time-like vector N_x whose span is this orthogonal complement, the *unit normal* to $\mathcal{H}$ at x. We denote by k the second fundamental form of $\mathcal{H}$. Its components in an arbitrary frame e_i, $i = 1,2,3$ in $\mathcal{H}$ are given by:

$$k_{ij} = g(\nabla_{e_i} N, e_j)$$

where we denote by ∇ the covariant derivative operator on $\mathcal{M}$ associated to g. A space-like hypersurface $\mathcal{H}$ in $\mathcal{M}$ is called a *Cauchy hypersurface* if $(\mathcal{H}, \overline{g})$ is complete and each causal curve in $\mathcal{M}$ intersects $\mathcal{H}_0$ at one and only one point. We assume that $(\mathcal{M}, g)$ posesses such a Cauchy hypersurface. This assumption essentially means that we consider only space-times arising from the evolution of initial data. Under this assumption we can define on $\mathcal{M}$ a *time function*, that is a differentiable function t such that at each $p \in \mathcal{M}$, $dt \cdot X > 0$ whenever $X \in I_p^+$. The level sets $\mathcal{H}_t$ of a time function constitute a foliation of $\mathcal{M}$ into space-like hypersurfaces. The *lapse function* of the foliation is defined by:

$$\phi = (-g^{\mu\nu}\partial_\mu t \partial_\nu t)^{-1/2}$$

It measures the normal separation of the leaves of the foliation. We also have the time-like future directed vectorfield whose components in an arbitrary frame are given by:

$$T^\mu = -\phi^2 g^{\mu\nu}\partial_\nu t$$

It is characterized by the fact that its integral curves are orthogonal to the foliation and are parametrized by t. The one parameter group of diffeomorphisms generated by T maps the hypersurfaces $\mathcal{H}_t$ onto each other. We call T the *time translation* vectorfield corresponding to the time function t. The space-time manifold $\mathcal{M}$ is represented by the product $\Re \times \mathcal{H}_0$, where we identify $p \in \mathcal{M}$ with the pair (t, x) and the integral curve of T through p intersects $\mathcal{H}_0$ at x. In this representation we have:

$$T = \frac{\partial}{\partial t}$$

and the space-time metric is given by:

$$g = -\phi^2 dt^2 + \overline{g}$$

If e_i, $i = 1,2,3$ is a local frame in $\mathcal{H}_0$ we propagate it to a local frame in each $\mathcal{H}_t$ according to:

$$[T, e_i] = 0$$

The components of the first fundamental form of $\mathcal{H}_t$ then satisfy the first variation equations:

$$\frac{\partial \overline{g}_{ij}}{\partial t} = 2\phi k_{ij} \tag{1.1}$$

2. The Einstein Vacuum Equations

In general relativity the connection of the Lorentzian metric g is identified with the gravitational force, while its curvature, which produces geodesic deviation, is identified with the tidal force. Einstein's basic physical insight in discovering the theory was the fact that the gravitational force can be locally elliminated by going to a freely falling frame, just as the connection coefficients can be made to vanish along a geodesic by going to cylindrical normal coordinates (*equivalence principle*).

The laws of general relativity are the *Einstein equations* [**10**] linking the space-time curvature to the matter content:

$$G_{\mu\nu} = 2T_{\mu\nu} \tag{2.1}$$

Here $G_{\mu\nu}$ is the *Einstein tensor*, given by:

$$G_{\mu\nu} = R_{\mu\nu} - (1/2)Rg_{\mu\nu} \tag{2.2}$$

with $R_{\mu\nu}$ the Ricci tensor and R the scalar curvature of the metric $g_{\mu\nu}$, while $T_{\mu\nu}$ is the *energy-momentum tensor* of matter. The twice contracted Bianchi identities,

$$\nabla^\nu G_{\mu\nu} = 0, \tag{2.3}$$

then imply the *energy-momentum conservation laws*:

$$\nabla^\nu T_{\mu\nu} = 0, \tag{2.4}$$

Thus general relativity incorporates the equations of motion of classical mechanics.

In the absense of matter equations (2.1) reduce to the *Einstein vacuum equations* for the space-time manifold:

$$R_{\mu\nu} = 0 \tag{2.5}$$

In the present article we shall confine our attention to this case.

The principal part of the Ricci tensor is:

$$(1/2)g^{\alpha\beta}(\partial_\mu\partial_\alpha g_{\beta\nu} + \partial_\nu\partial_\alpha g_{\beta\mu} - \partial_\mu\partial_\nu g_{\alpha\beta} - \partial_\alpha\partial_\beta g_{\mu\nu})$$

For a given metric g the symbol σ_ξ at a point $p \in \mathcal{M}$ and a covector $\xi \in T^*_p\mathcal{M}$ is the linear operator on $S_2(T_p\mathcal{M})$, the space of symmetric bilinear forms $\dot{g}$ in $T_p\mathcal{M}$ (variations of g), obtained by the replacement:

$$\partial_\mu g_{\alpha\beta} \mapsto \xi_\mu \dot{g}_{\alpha\beta}$$

This gives:

$$(\sigma_\xi \cdot \dot{g})_{\mu\nu} = (1/2)(\xi_\mu\xi^\alpha\dot{g}_{\alpha\nu} + \xi_\nu\xi^\alpha\dot{g}_{\alpha\mu} - g^{\alpha\beta}\dot{g}_{\mu\nu} - \xi_\mu\xi_\nu g^{\alpha\beta}\dot{g}_{\alpha\beta})$$

which we can write as:

$$\sigma_\xi \cdot \dot{g} = (1/2)(\xi \otimes i_\xi\dot{g} + i_\xi\dot{g} \otimes \xi - (\xi,\xi)\dot{g} - \xi \otimes \xi \operatorname{tr}\dot{g})$$

Here $i_\xi\dot{g}$ is the covector obtained by contracting $\dot{g}$ with the vector corresponding to ξ:

$$(i_\xi\dot{g})_\nu = \xi^\mu\dot{g}_{\mu\nu}$$

and we denote $g^{\alpha\beta}\xi_\alpha\xi_\beta = (\xi,\xi)$. We see that for any given ξ and any other covector ζ, the variation

$$\dot{g} = \xi \otimes \zeta + \zeta \otimes \xi$$

belongs to the null space $N(\sigma_\xi)$ of the symbol at ξ. Thus the Einstein equations seem at first sight to be a degenerate differential system, $N(\sigma_\xi)$ being non-zero for any $\xi \in T^*_p\mathcal{M}$. This is due to the fact that the equations are generally covariant; proper account must be taken of the geometric equivalence of metrics related by a diffeomorphism. Since any one parameter group of diffeomorphisms is generated by a vectorfield ζ and the infinitesimal action of the group on the space of metrics is the Lie derivative

$$(\mathcal{L}_\zeta g)_{\mu\nu} = \nabla_\mu\zeta_\nu + \nabla_\nu\zeta_\mu$$

the symbol of which is $\xi\otimes\zeta+\zeta\otimes\xi$, we consider the quotient Q_ξ of $S_2(T_p\mathcal{M})$ by the following equivalence relation: $\dot{g}_1 \sim \dot{g}_2$ if and only if $\dot{g}_1 - \dot{g}_2 = \xi\otimes\zeta+\zeta\otimes\xi$ for some covector ζ at p. The symbol σ_ξ when reduced to Q_ξ is then seen to have zero null space when ξ is not null. Moreover, when ξ is a non-zero null covector, choosing a null conjugate to $\underline{\xi}$ to ξ, i.e. another null covector in the same component of the dual null cone at p such that $(\xi, \underline{\xi}) = -2$, we can identify Q_ξ with the space of all $\dot{g} \in S_2(T_p\mathcal{M})$ such that $i_{\underline{\xi}}\dot{g} = 0$. Then $N(\sigma_\xi)$ is seen to be the subspace of Q_ξ consisting of those $\dot{g}$ which also verify $i_\xi \dot{g} = 0$ and $\mathrm{tr}\dot{g} = 0$. Therefore, when ξ is null $N(\sigma_\xi)$ can be identified with $\hat{S}_2(\Pi)$, the space of trace-free symmetric bilinear forms on Π, the space-like plane which is the intersection of the null spaces of ξ and $\underline{\xi}$. Thus, the two dimensional space $\hat{S}_2(\Pi)$ represents the space of *dynamical degrees of freedom of the gravitational field at a point* (gravitational waves).

In terms of the foliation induced by a time function t the Einstein vacuum equations become:

$$\overline{R} - |k|^2 + (\mathrm{tr}k)^2 = 0 \tag{2.6}$$

$$\overline{\nabla}^j k_{ij} - \overline{\nabla}_i \mathrm{tr}k = 0 \tag{2.7}$$

$$\frac{\partial k_{ij}}{\partial t} = \overline{\nabla}_i \overline{\nabla}_j \phi - (\overline{R}_{ij} + k_{ij}\mathrm{tr}k - 2k_{im}k^m_{\ j})\phi \tag{2.8}$$

Equations (2.6) and (2.7) correspond to the Gauss and Codazzi equations respectively, while equations (2.8) represent the second variation equations and must be considered in conjunction with equations (1.1). Here $\overline{\nabla}$ is the covariant derivative operator, $\overline{R}_{ij}$ the Ricci tensor and $\overline{R}$ the scalar curvature on $\mathcal{H}_t$, defined by $\overline{g}$. Note that ϕ is left completely undetermined by the above equations, a freedom which corresponds to the complete arbitrariness in chosing the time function. Remark that by virtue of the identities (2.3), if $\overline{g}, k$ satisfy equations (1.1,2.8), then equations (2.6,2.7) are satisfied on any $\mathcal{H}_t$ provided that they are satisfied on $\mathcal{H}_0$. Therefore they can be regarded as constraints on given initial conditions for $\overline{g}, k$. Accordingly, an *initial data set* for the Einstein vacuum equations is defined to be a triplet $(\mathcal{H}_0, \overline{g}_0, k_0)$ consisting of a complete three dimensional Riemannian manifold $(\mathcal{H}_0, \overline{g}_0)$ equipped with a 2-covariant symmetric tensorfield k_0, satisfying the constraint equations (2.6,2.7). By a *development* of such an initial data set we mean a Lorentzian manifold $(\mathcal{M}, g)$ satisfying the Einstein vacuum equations (2.5) and an embedding of $\mathcal{H}_0$ as a Cauchy hypersurface in $\mathcal{M}$ such that $\overline{g}_0$ and k_0 are the induced first and second fundamental forms respectively.

3. Asymptotic Flatness

The central mathematical problem of the theory is the study of the developments of general asymptotically flat initial data sets. These represent isolated gravitating physical systems. By an *asymptotically flat initial data set* we mean an initial data set $(\mathcal{H}, \overline{g}, k)$ such that the complement of a compact set in $\mathcal{H}$ is diffeomorphic to the complement of a closed ball in $\Re^3$ and there exists a coordinate system in this complement relative to which the metric components $\overline{g}_{ij}$ approach δ_{ij} and those of k approach zero, sufficiently rapidly for the notions of *total energy*, *linear momentum* and *angular momentum* to be well defined and finite. The Arnowitt , Deser and Misner [**1**] definitions of these notions are, respectively,

$$E = \frac{1}{4}\lim_{r\to\infty}\int_{S_r}\sum_{i,j}(\partial_i \overline{g}_{ij} - \partial_j \overline{g}_{ii})dS_j \tag{3.1}$$

$$P^i = -\frac{1}{2}\lim_{r\to\infty}\int_{S_r}\sum_j (k_{ij} - \overline{g}_{ij}\mathrm{tr}k)dS_j \tag{3.2}$$

$$J^i = -\frac{1}{2}\lim_{r\to\infty}\int_{S_r}\sum_{j,m,n} \epsilon_{ijm}\, x^j(k_{mn} - \overline{g}_{mn}\mathrm{tr}k)dS_n \tag{3.3}$$

where S_r is the coordinate sphere of radius r and dS_i are the components of its oriented area element.

The notions of total energy, linear and angular momentum are in particular well defined and finite if there is a coordinate system in a neighborhood of infinity in which

$$g_{ij} = (1 + M_0/2\pi r)\delta_{ij} + o_2(r^{-3/2}),\ k_{ij} = o_1(r^{-5/2}) \text{ as } r \to \infty \tag{3.4}$$

Here, a function f is said to be $o_n(r^{-\alpha})$ as $r \to \infty$ if f is C^n and $\partial^m f = o(r^{-m-\alpha})$ as $r \to \infty$, for any $m = 0, ..., n$ where ∂^m denotes all partial derivatives of order m. Initial data sets verifying (3.4) we call *strongly asymptotically flat.* The leading term

$$(1 + M_0/2\pi r)\delta_{ij}$$

in the expansion of the metric of a strongly asymptotically flat initial data set we call the *Schwarzschild part* of the metric. The conditions (3.4) imply:

$$E = M_0,\ P^i = 0$$

Thus a strongly asymptotically flat initial data set defines a *center of mass frame.* The *posivite mass theorem* first proved by R. Schoen and S.T. Yau [**14**] and later, by a different method, by E. Witten [**15**], states that $M_0 \geq 0$ with equality if and only if the initial data set is embedded in the flat Minkowski space-time.

The total energy, the linear momentum and the angular momentum are *conserved quantities.* That is, given a time function t whose lapse function ϕ tends to 1 at infinity each level set, then if the zero level set $\mathcal{H}_0$ defines an asymptotically flat initial data set, so do all the level sets $\mathcal{H}_t$ and the values of each of these quantities are the same for all the $\mathcal{H}_t$.

4. The Maximal Time Function

In a space-time arising from asymptotically flat initial conditions we can define a unique *maximal time function t.* This is defined by the condition that its level sets $\mathcal{H}_t$ are complete space-like hypersurfaces of maximal volume on which ϕ tends to 1 at infinity and $P^i = 0$. The maximality condition is expressed by:

$$\mathrm{tr}k = 0 \tag{4.1}$$

Relative to the maximal time function the constraint equations (2.6,2.7) reduce to:

$$\overline{R} = |k|^2 \tag{4.2}$$

$$\overline{\nabla}^j k_{ij} = 0 \tag{4.3}$$

while the evolution equations (1.1,2.8) reduce to:

$$\frac{\partial \overline{g}_{ij}}{\partial t} = 2\phi k_{ij} \tag{4.4}$$

$$\frac{\partial k_{ij}}{\partial t} = \overline{\nabla}_i\overline{\nabla}_j\phi - (\overline{R}_{ij} - 2k_{im}k^m_{\ j})\phi \tag{4.5}$$

Furthermore, taking the trace of (4.4) and imposing (4.1) we obtain the following elliptic equation for the lapse function:

$$\overline{\triangle}\phi = |k|^2\phi \tag{4.6}$$

A complete maximal space-like hypersurface in Minkowski space-time is necessarily a hyperplane. Thus if the initial data set $(\mathcal{H}_0, \overline{g}_0, k_0)$ satisfies the maximality condition $\mathrm{tr}k_0 = 0$, it has trivial development if and only if $(\mathcal{H}_0, \overline{g}_0)$ is the Euclidean space and $k_0 = 0$. In the following we shall restrict ourselves to strongly asymptotically flat initial data sets satisfying the maximality condition. An appropriate version of the local existence theorem gives us a development $\mathcal{M}$ represented by the product $\mathcal{I} \times \mathcal{H}_0$, where $\mathcal{I}$ is an interval containing 0 and the projection to the first factor is the maximal time function. We remark here that $\mathcal{I} = \Re$ does not imply that the development is geodesically complete, for we may have $\inf_{\mathcal{M}} \phi = 0$.

5. Statement of The Problem

The simplest solution of the Einstein vacuum equations is of course the flat Minkowski space-time of special relativity, introduced by Minkowski in 1908 [**13**] as the geometric framework of that theory, in a work which was instrumental in the transition from Einstein's formulation of special relativity of 1905 [**8**] to his discovery of the general theory in 1915 [**9**]. Minkowski space-time is the manifold $\Re^4$ together with the metric η whose components form the diagonal matrix with entries -1, 1, 1, 1.

The problem which we shall discuss in the present article is the problem of the global stability of Minkowski space-time in the framework of general relativity. That is, whether any asymptotically flat initial data set which is sufficiently close to a trivial one has a development which is a geodesically complete space-time approaching the Minkowski space-time at infinity along any geodesic. This question has been answered in the affirmative in my joint work with Sergiu Klainerman [**7**] when asymptotic flatness of the initial data set is meant in the strong sense defined above and an appropriate notion of closeness is required.

In the following we shall discuss the main ideas and methods of the proof, after a brief exposition of general methods of treating problems of global stability of the trivial solution for field theories in Minkowski space-time and a discussion of the peculiar difficulties present in the problem at hand and the obstacles that had to be overcome.

6. Field Theories in a Given Spacetime

Consider a field theory in a given space-time $(\mathcal{M}, g)$ whose field equations are derivable from an *action* $\mathcal{A}$. For any domain $\mathcal{D}$ with compact closure in $\mathcal{M}$ the action in $\mathcal{D}$ is the integral:

$$\mathcal{A}[\mathcal{D}] = \int_{\mathcal{D}} L d\mu_g \tag{6.1}$$

where L is the *Lagrangian*. The field equations of the theory express the condition that for any such domain $\mathcal{D}$ the action is stationary with respect to variations of the field with support in $\mathcal{D}$. On the other hand, variations of the action, supported in $\mathcal{D}$, with respect to the underlying metric, give rise to the energy-momentum tensor through the formula:

(6.2) $$\dot{\mathcal{A}}[\mathcal{D}] = -\frac{1}{2}\int_{\mathcal{D}} T^{\mu\nu}\dot{g}_{\mu\nu}d\mu_g$$

By its definition $T^{\mu\nu}$ is symmetric. If $\mathcal{A}$ is invariant under diffeomorphisms of $\mathcal{M}$ reducing to the identity outside $\mathcal{D}$, then the field equations imply that $T^{\mu\nu}$ is divergence-free:

(6.3) $$\nabla^\nu T_{\mu\nu} = 0$$

This is in accordance with (2.4), so the theory is compatible with general relativity.

Now suppose that X is a vectorfield generating a one parameter group of isometries of $(\mathcal{M}, g)$ (Killing vectorfield). Then the 1-form

(6.4) $$P_\mu = -T_{\mu\nu}X^\nu$$

is divergence-free

(6.5) $$\nabla^\mu P_\mu = 0$$

or, equivalently, the dual 3-form ${}^\star P$ is closed:

(6.6) $$d{}^\star P = 0$$

It follows that the integral of ${}^\star P$ on two homologous hypersurfaces is the same and the integral

$$\int_{\mathcal{H}} {}^\star P$$

on a Cauchy hypersurface $\mathcal{H}$ is a conserved quantity, that is, its value is the same for all Cauchy hypersurfaces. This is essentially what is called *Noether's Principle.* Moreover if the action is invariant under conformal transformations of the metric then the energy-momentum tensor is trace-free and these considerations extend to the case where X generates a one parameter group of conformal isometries of $(\mathcal{M}, g)$ (conformal Killing vectorfield). An important requirement on a physical theory is that the energy-momentum tensor should satisfy the positivity condition:

$$T(X_1, X_2) \geq 0$$

for any pair X_1, X_2 of time-like future directed vectors at a point. Then, provided that the vector multiplier X above is time-like future directed, the quantity

$$\int_{\mathcal{H}} {}^\star P = \int_{\mathcal{H}} T(X, N)d\mu_{\bar{g}}$$

is non-negative, N being the unit normal to $\mathcal{H}$. As its value is the same as that on the Cauchy hypersurface on which the initial data is given, it provides an estimate for the solution in terms of the initial data.

Furthermore, if we suppose, as is natural, that the Lagrangian posesses the symmetries of the underlying metric, the pullback by an isometry of a solution is also a solution of the field equations. Moreover, if the field equations are linear then the difference of two solutions is also a solution. It follows that given a vectorfield which generates a one parameter group of isometries of the space-time, the Lie derivative of a solution with respect to this vectorfield is also a solution of the same equations, being the limit of a difference quotient of solutions. In the case of a conformally invariant action, the same is true for the Lie derivative with respect to a vectorfield generating conformal space-time isometries. Thus in the linear case the previous construction applies to Lie derivatives as well, in fact to iterated

Lie derivatives of arbitrary order, giving a series of positive conserved quantities controlling the solutions. In fact, once enough such quantities of sufficiently high order are obtained, the Sobolev inequalities imply uniform decay estimates of the solutions at infinity.

In the non-linear case, the Lie derivative of a solution is no longer a solution of the same field equations. An analogous construction does give energy tensors corresponding to the Lie derivatives, but their divergence no longer vanishes. The positive quantities obtained using suitable vector multipliers as before, are consequently not conserved. The difference of the values corresponding to two Cauchy hypersurfaces is the integral of error terms over the space-time region bounded by the hypersurfaces. Nevertheless, if we have enough quantities of sufficiently high order at our disposal then the integral of the error terms may be estimated, using Sobolev-type inequalities, in terms of the quantities themselves. Thus one arrives at a closed system of ordinary differential inequalities which controls the growth of these quantities in time and implies that they remain bounded for all time provided that their initial values are sufficiently small. This yields a global existence theorem for small initial data.

In the case that the underlying space-time is the Minkowski space-time, there is a large group of conformal isometries available, consisting of the space-time translations, the space-time rotations (Lorentz group), the scaling, and the inverted space-time translations, generated by the vectorfields:

$$T_\mu = \partial_\mu; \ \mu = 0, 1, 2, 3 \tag{6.7}$$

$$\Omega_{\mu\nu} = x_\mu \partial_\nu - x_\nu \partial_\mu; \ \mu, \nu = 0, 1, 2, 3 \tag{6.8}$$

$$S = x^\mu \partial_\mu \tag{6.9}$$

$$K_\mu = -2x_\mu S + (x, x)\partial_\mu; \ \mu = 0, 1, 2, 3 \tag{6.10}$$

respectively. Here,

$$x_\mu = \eta_{\mu\nu} x^\nu, \ (x, x) = \eta_{\mu\nu} x^\mu x^\nu$$

Of these only the time translations and the inverted time translations are generated by everywhere time-like future directed vectorfields, T_0 and K_0 respectively, and are thus suitable for use as multipliers. Lie derivatives can be taken with respect to all generating vectorfields.

The general method outlined above grew as a synthesis of the *conformal method* which I introduced in the case of the Yang-Mills equations [**4**] and later applied it to quasilinear hyperbolic systems of scalar equations [**5**], and the *commutator method* introduced by Klainerman [**12**] in the study of non-linear perturbations of the wave equation. The conformal method corresponded to a special case of the method just outlined, namely the case where Lie derivatives are taken only with respect to inverted space-time translations and only the inverted time translation is used as a multiplier, the integrations being carried over space-like hyperboloids, while Klainerman's commutator method corresponded to the case where Lie derivatives are taken only with respect to the Lorentz group and scaling and only the usual time translation is used as a multiplier, the integrations being carried over space-like hyperplanes.

7. Weyl Fields and Bianchi Equations

If one tries to apply the general method just outlined to the problem of the global stability of the Minkowski space-time in general relativity, one quickly reaches an impasse for the following two reasons. First, the energy-momentum tensor in the case of gravitation, defined as in (6.2) above, but relative to the Einstein-Hilbert action:

$$\mathcal{A}[\mathcal{D}] = -\frac{1}{4}\int_{\mathcal{D}} R d\mu_g$$

vanishes, as this expresses the field equations of gravitation, namely the Einstein vacuum equations. And, second, space-time in general relativity posesses in general no symmetries, hence the conformal isometry group is trivial and the vectorfields required in the construction do not exist. At this point two main ideas were introduced which overcame these obstacles.

The first idea was that instead of the Einstein equations we should concentrate our attention on the Bianchi identities

$$\nabla_{[\alpha}R_{\beta\gamma]\delta\epsilon} = 0 \tag{7.1}$$

(here [ ] stands for cyclic permutation), considering them as equations for the curvature. This leads us to introduce the concept of a *Weyl field* $W_{\alpha\beta\gamma\delta}$, in a given space-time, a 4-covariant tensorfield posessing the algebraic properties of the Weyl or conformal curvature tensor. The natural field equations for a Weyl field are the *Bianchi equations*, identical in form to the Bianchi identities:

$$\nabla_{[\alpha}W_{\beta\gamma]\delta\epsilon} = 0 \tag{7.2}$$

We can write these simply as:

$$DW = 0 \tag{7.3}$$

A particular case of a Weyl field is, of course, the Riemann curvature tensor of a metric satisfying the Einstein vacuum equations, but the situation considered here is more general as there need be no connection between a Weyl field and the underlying space-time metric. In a four dimensional space-time the dual ${}^*W$ of a Weyl field W is also a Weyl field and if W satisfies the Bianchi equations so does ${}^*W$. The operator D although formally identical to the exterior derivative, is not an exterior differential operator and $D^2 \neq 0$. As a consequence, the Bianchi equations imply an algebraic condition:

$$R_\mu{}^{\alpha\beta\gamma}\,{}^*W_{\nu\alpha\beta\gamma} - R_\nu{}^{\alpha\beta\gamma}\,{}^*W_{\mu\alpha\beta\gamma} = 0$$

The Bianchi equations are conformally covariant. If f is a conformal isometry of $(\mathcal{M}, g)$, that is $f^*g = \Omega^2 g$ for some positive function Ω, and W is a solution of the Bianchi equations then so is $\Omega^{-1}f^*W$.

To a Weyl field we can associate a tensorial quadratic form, a 4-covariant tensorfield which is fully symmetric and trace-free. This tensorfield is a generalization of one found previously by Bel and Robinson [**3**] so we call it the *Bel-Robinson tensor*. It is given by:

$$Q_{\alpha\beta\gamma\delta} = (1/2)(W_{\alpha\rho\gamma\sigma}W_\beta{}^\rho{}_\delta{}^\sigma + {}^*W_{\alpha\rho\gamma\sigma}{}^*W_\beta{}^\rho{}_\delta{}^\sigma) \tag{7.4}$$

and satisfies the following positivity condition:

$$Q(X_1, X_2, X_3, X_4) \geq 0$$

for any tetrad of time-like future directed vectors at a point, with equality if and only if W vanishes at that point. Furthermore, if W satisfies the Bianchi equations then Q is divergence-free:

$$\nabla^\alpha Q_{\alpha\beta\gamma\delta} = 0 \tag{7.5}$$

It follows that given three vector fields X_1, X_2, X_3, each generating a one parameter group of conformal isometries of $(\mathcal{M}, g)$, some or all of which possibly coincident, then the 1-form

$$P = -Q(\cdot, X_1, X_2, X_3) \tag{7.6}$$

is divergence-free, consequently the integral

$$\int_{\mathcal{H}} {}^*P$$

on a Cauchy hypersurface $\mathcal{H}$ is a conserved quantity, which is positive definite in the case that the X_1, X_2, X_3 are all time-like future directed.

Given a Weyl field W and a vector field X the usual Lie derivative $\mathcal{L}_X W$ of W with respect to X is not in general a Weyl field. However we can define a modified Lie derivative $\hat{\mathcal{L}}_X W$ which is a Weyl field:

$$\begin{aligned} \hat{\mathcal{L}}_X W_{\alpha\beta\gamma\delta} &= \mathcal{L}_X W_{\alpha\beta\gamma\delta} - (1/8)\mathrm{tr}\pi W_{\alpha\beta\gamma\delta} \\ &\quad -(1/2)(\hat{\pi}_\alpha{}^\mu W_{\mu\beta\gamma\delta} + \hat{\pi}_\beta{}^\mu W_{\alpha\mu\gamma\delta} + \hat{\pi}_\gamma{}^\mu W_{\alpha\beta\mu\delta} + \hat{\pi}_\delta{}^\mu W_{\alpha\beta\gamma\mu}) \end{aligned} \tag{7.7}$$

Here $\pi_{\mu\nu} = \mathcal{L}_X g_{\mu\nu}$ and $\hat{\pi}$ is the *deformation tensor* of X, namely the trace-free part of π. The modified Lie derivative commutes with the Hodge dual:

$$\hat{\mathcal{L}}_X {}^*W = {}^*\hat{\mathcal{L}}_X W \tag{7.8}$$

As a consequence of the linearity and the conformal invariance of the Bianchi equations, if W is a solution of these equations and X is a vector field generating a one parameter group of conformal isometries f_t, then

$$\hat{\mathcal{L}}_X W = \frac{d}{dt}(\Omega_t^{-1} f_t^* W)\Big|_{t=0}$$

is also a solution of the same equations. Therefore the considerations regarding conserved quantities can be applied to the Weyl field $\hat{\mathcal{L}}_X W$ as well.

8. The Optical Function

The second main idea of the proof of the global stability of Minkowski space-time was in overcoming the obstacle that a general metric in fact posesses only a trivial conformal isometry group. The idea originates in the observation that a space-time which arises from asymptotically flat initial conditions should itself be asymptotically flat, approaching the Minkowski space-time at infinity. Thus we have a group acting at infinity as a conformal isometry. The problem is how to extend this action to the whole space-time in such a way that the deviation from conformal isometry is globally small and approaching zero at infinity sufficiently rapidly. The crux of the idea was the solution of this problem by means of a geometric construction. It turns out that we can only define the action of the subgroup of the Minkowskian conformal group consisting of the time translations, the scaling, the inverted time translations and the spatial rotation group $O(3)$ leaving the total energy-momentum vector invariant, however this subgroup suffices to derive a complete system of estimates. First, the action of the group of time translations is the simplest to define, for, we have a unique maximal time function. The corresponding time translation vector field T generates the action, mapping the maximal hypersurfaces of vanishing linear momentum $\mathcal{H}_t$ onto each other. The action of the other groups is defined with the help of an *optical function* u.

This is a function whose level sets $\mathcal{C}_u$ are null hypersurfaces, defined as follows. We start with a surface $\mathcal{S}_{0,0}$ diffeomorphic to S^2 on $\mathcal{H}_0$ and we define the level set $\mathcal{C}_0$ to be the outer component of $\partial\mathcal{J}^+(\mathcal{S}_{0,0})$, an outgoing null hypersurface. The surface $\mathcal{S}_{0,0}$ must be chosen so that the null geodesics generating the latter have no future end points. We would like then to define the level sets $\mathcal{C}_u$, $u \neq 0$, to be other outgoing null hypersurfaces such that, if we consider the surfaces $\mathcal{S}_{t,u} = \mathcal{H}_t \bigcap \mathcal{C}_u$, the restriction to $\mathcal{S}_{t,u}$ of minus the signed distance function along $\mathcal{H}_t$ from $\mathcal{S}_{t,0}$ tends to u as $t \to \infty$. However, this definition can be implemented only after global existence has already been proven.

In the course of the proof, a continuity argument, we have a final maximal hypersurface $\mathcal{H}_{t_*}$. We would like then to define u on $\mathcal{H}_{t_*}$ to be minus the signed distance function along $\mathcal{H}_{t_*}$ from $\mathcal{S}_{t_*,0}$. However, the definition is inappropriate because this is only as smooth as the metric, two orders of differentiability smoother than the curvature, even though $\mathcal{S}_{t_*,0}$ itself is of the maximal smoothness allowed, one order smoother than the metric. With such a loss of smoothness we would not arrive at a closed system of estimates. We instead define u on $\mathcal{H}_{t_*}$ by imposing certain equation for the lapse function a of the foliation of $\mathcal{H}_{t_*}$ generated by u:

$$a = (\overline{g}^{ij}\partial_i u \partial_j u)^{-1/2} \tag{8.1}$$

As the lapse function measures the normal separation of the leaves of the foliation, the equation for a, to be given below, can be thought of as an equation of motion for a surface on a the three dimensional Riemannian manifold. The given surface $\mathcal{S}_{t_*,0}$, which is to be the zero level set of u on $\mathcal{H}_{t_*}$, plays the role of an initial condition. To write the equation for a in a form which is as simple as possible we shall neglect the terms contributed by the second fundamental form of $\mathcal{H}_{t_*}$. Then a satisfies on each surface $\mathcal{S}_{t_*,u}$, level set of u on $\mathcal{H}_{t_*}$ the equation:

$$\not{\Delta} \log a = f - \overline{f}, \quad \overline{\log a} = 0 \tag{8.2}$$

where f is the function:

$$f = K - \frac{1}{4}(\mathrm{tr}\theta)^2 \tag{8.3}$$

Here K is the Gauss curvature of $\mathcal{S}_{t_*,u}$ and θ is the second fundamental form of $\mathcal{S}_{t_*,u}$ relative to $\mathcal{H}_{t_*}$. Also, $\not{\nabla}$ is the covariant derivative operator on $\mathcal{S}_{t_*,u}$ associated to the induced metric γ. Finally, we denote by an overline the mean value of a function on $\mathcal{S}_{t_*,u}$. To see why the function u constructed by solving (8.2,8.3) has the required smoothness properties, recall the trace of the second variation equations of the foliation of a three dimensional Riemannian manifold induced by a function u:

$$\frac{\partial \mathrm{tr}\theta}{\partial u} = \not{\Delta} a + \frac{1}{2}a(\overline{R} + |\theta|^2 + (\mathrm{tr}\theta)^2 - 2K)$$

Since we are neglecting the second fundamental form of $\mathcal{H}_{t_*}$ we have, in accordance with (4.3), $\overline{R} = 0$; therefore, by virtue of (8.2) this reduces to:

$$\frac{1}{a}\frac{\partial \mathrm{tr}\theta}{\partial u} = \frac{1}{2}|\hat{\theta}|^2 + \frac{1}{2}(\mathrm{tr}\theta)^2 + |\not{\nabla}\log a|^2 - \overline{f} \tag{8.4}$$

Here we denote by $\hat{\theta}$ the trace-free part of θ. The gain in smoothness is evident from the fact that the curvature terms have been elliminated in favor of terms which are one order smoother. The propagation equation (8.4) is considered in conjunction with the Codazzi equations:

$$\overline{\nabla}^B\hat{\theta}_{AB} - \frac{1}{2}\overline{\nabla}_A \mathrm{tr}\theta = \overline{R}_{A3} \tag{8.5}$$

an elliptic equation for $\hat{\theta}$ on each $S_{t_*,u}$, and with the Gauss equation:

$$K - \frac{1}{4}(\mathrm{tr}\theta)^2 + \frac{1}{2}|\hat{\theta}|^2 = -\overline{R}_{33} \tag{8.6}$$

to complete the smoothness argument. Here e_A, $A = 1,2$ is an arbitrary local frame in $\mathcal{S}_{t_*,u}$, complemented by e_3, the unit outward normal to $\mathcal{S}_{t_*,u}$ in $\mathcal{H}_{t_*}$.

Once the surfaces $\mathcal{S}_{t_*,u}$ have been constructed, the null hypersurfaces $\mathcal{C}_u$ are defined to be the inner components of $\partial\mathcal{J}^-(\mathcal{S}_{t_*,u})$ and the construction of the optical function is complete.

9. Vector Fields and the Controlling Quantity

The surfaces $\mathcal{S}_{t,u}$ define a two parameter foliation of the space-time slab bounded by $\mathcal{H}_0$ and $\mathcal{H}_{t_*}$. Let $r(t,u)$ be the *area radius* of $\mathcal{S}_{t,u}$, defined by:

$$r(t,u) = \sqrt{\frac{\mathrm{Area}(\mathcal{S}_{t,u})}{4\pi}} \tag{9.1}$$

We then define the function

$$\underline{u} = u + 2r \tag{9.2}$$

Let L and $\underline{L}$ be respectivelly the outgoing and incoming null normals to $\mathcal{S}_{t,u}$ whose component along T is equal to T. We then have:

$$T = \frac{1}{2}(L + \underline{L}) \tag{9.3}$$

and we define the generator of scalings by:

$$S = \frac{1}{2}(\underline{u}L + u\underline{L}) \tag{9.4}$$

and the generator of inverted time translations by:

$$K = \frac{1}{2}(\underline{u}^2 L + u^2\underline{L}) \tag{9.5}$$

To define the action of the rotation group $O(3)$ on $\mathcal{H}_{t_*}$, we consider the vector field on $\mathcal{H}_{t_*}$ whose components in an arbitrary frame in $\mathcal{H}_{t_*}$ are given by:

$$U^i = a^2\overline{g}^{ij}\partial_j u \tag{9.6}$$

The integral curves of U are orthogonal to the foliation induced by u on $\mathcal{H}_{t_*}$ and are parametrized by u. The one parameter group of diffeomorphisms generated by U maps the surfaces $\mathcal{S}_{t_*,u}$ onto each other. The induced metric γ on $\mathcal{S}_{t_*,u}$ rescaled by the factor r^{-2} tends along the flow of U to a metric of Gauss curvature equal to 1 as $u \to -\infty$. We can thus attach the standard sphere S^2 at infinity on $\mathcal{H}_{t_*}$. We have the standard action of $O(3)$ on S^2 by isometries. The action is then extended to $\mathcal{H}_{t_*}$ by conjugation: Given an element $O \in O(3)$ and a point $p_* \in \mathcal{S}_{t_*,u}$, there is a point $q \in S^2$, the ideal point at parameter value $-\infty$ along the integral curve of U through p_* at parameter value u. The action of $O(3)$ on S^2 gives us the point

$Oq \in S^2$. The point $Op_* \in \mathcal{S}_{t_*,u}$ is then defined to be the point at parameter value u along the integral curve of U leading to the ideal point Oq at parameter value $-\infty$.

The action of $O(3)$ is then extended to the space-time slab using the vector field L. The integral curves of L are the null geodesic generators of the hypersurfaces $\mathcal{C}_u$ and are parametrized by t. The one parameter group of diffeomorphisms generated by L maps the surfaces $\mathcal{S}_{t,u}$ corresponding to the same value of u but different values of t onto each other. Given an element $O \in O(3)$ and a point $p \in \mathcal{S}_{t,u}$, to obtain the point Op we follow the integral curve of L through p at parameter value t to the point $p_* \in \mathcal{S}_{t_*,u}$ at parameter value t_*. The action of $O(3)$ on $\mathcal{H}_{t_*}$ just defined gives us the point $Op_* \in \mathcal{S}_{t_*,u}$. The point $Op \in \mathcal{S}_{t,u}$ is then defined to be the point at parameter value t along the integral curve of L through Op_* at parameter value t_*.

The three rotation vector fields ${}^{(a)}\Omega$, $a = 1,2,3$, generating the above action satisfy:

$$[{}^{(a)}\Omega, L] = 0 \tag{9.7}$$

$$g({}^{(a)}\Omega, L) = g({}^{(a)}\Omega, T) = 0 \tag{9.8}$$

and, of course, the commutation relations of the Lie algebra of $O(3)$:

$$[{}^{(a)}\Omega, {}^{(b)}\Omega] = \in_{abc}\, {}^{(c)}\Omega \tag{9.9}$$

The group orbits are the surfaces $S_{t,u}$.

By the above construction the deformation tensors of the generating vector fields depend entirely on the geometric properties of the hypersurfaces $\mathcal{C}_u$ and $\mathcal{H}_t$.

Once the vector fields are defined we consider the 1-form P, given by

$$P = P_0 + P_1 + P_2 \tag{9.10}$$

where:

$$
\begin{aligned}
P_0 &= -Q(R)(\cdot, \overline{K}, T, T) \\
P_1 &= -Q(\hat{\mathcal{L}}_O R)(\cdot, \overline{K}, \overline{K}, T) - Q(\hat{\mathcal{L}}_T R)(\cdot, \overline{K}, \overline{K}, \overline{K}) \\
P_2 &= -Q(\hat{\mathcal{L}}_O^2 R)(\cdot, \overline{K}, \overline{K}, T) - Q(\hat{\mathcal{L}}_O \hat{\mathcal{L}}_T R)(\cdot, \overline{K}, \overline{K}, \overline{K}) \\
&- Q(\hat{\mathcal{L}}_S \hat{\mathcal{L}}_T R)(\cdot, \overline{K}, \overline{K}, \overline{K}) - Q(\hat{\mathcal{L}}_T^2 R)(\cdot, \overline{K}, \overline{K}, \overline{K})
\end{aligned}
\tag{9.11}
$$

and

$$\overline{K} = K + T$$

while O stands for the collection ${}^{(a)}\Omega$, $a = 1,2,3$. Here $Q(W)$ is the Bel-Robinson quadratic form associated to the Weyl field W and R stands for the space-time curvature, the original Weyl field. We then define the *controlling quantity*:

$$E = \max\{E_1, E_2\} \tag{9.12}$$

where

$$E_1 = \sup_t \int_{\mathcal{H}_t} {}^\star P, \; E_2 = \sup_u \int_{\mathcal{C}_u} {}^\star P \tag{9.13}$$

and everything is restricted to the space-time slab ${}^{(t_*)}\mathcal{M} = \bigcup_{t \in [0,t_*]} \mathcal{H}_t$ under consideration.

10. The Continuity Argument

The values of the integral of $^*P$ on two homologous hypersurfaces are not the same, for the vector fields T, S, K and O are not exact conformal Killing vector fields. The difference of these values is the integral of error terms, linear in the deformation tensors of the vector fields and quadratic in the Weyl fields, over the space-time region bounded by the hypersurfaces. The crucial point and success of the geometric construction is the fact that these error integrals can be bounded in terms of the controlling quantity itself.

The proof of the stability theorem is by the method of continuity and it involves a complex bootstrap argument. Starting with a strongly asymptotically flat initial data set satisfying the maximality condition, and using an appropriate version of the local existence theorem we can assume that the space-time is maximally extended up to a value t_* of the maximal time function. This value is defined to be the maximal one such that certain geometric quantities defined by the hypersurfaces $\mathcal{H}_t$ and $\mathcal{C}_u$ remain bounded by a small positive number ε_0. These quantities include, in particular,

$$\sup_{t,u} \sup_{\mathcal{S}_{t,u}} \left| r^2 K - 1 \right|$$

which controlls the isoperimetric constant of the surfaces $\mathcal{S}_{t,u}$, on which the Sobolev inequalities depend. They also include:

$$\sup_{t} \sup_{\mathcal{H}_t} (1 - \phi)$$

(note that by the maximum principle applied to (4.6): $\phi \leq 1$). It then follows that a cetrain norm of the deformation tensors of the vector fields T, S, K and O in the space-time slab bounded by $\mathcal{H}_0$ and $\mathcal{H}_{t_*}$ is less than another small positive constant ε_1. Using this bound for the deformation tensors, as well as the Sobolev inequalities, we are able to estimate the integral of the error terms over the space-time slab by $c\varepsilon_1 E$ and thus arrive at an inequality of the form:

$$E \leq c(D + \varepsilon_1 E)$$

where D stands for initial data. When ε_1 is chosen sufficiently small, which is achieved by choosing ε_0 suitably small, this implies $E \leq cD$. On the other hand we are able to show that the aforementioned geometric quantities associated to the hypersurfaces $\mathcal{H}_t$ and $\mathcal{C}_u$ are bounded by cE. Thus if D is suitably small this bound does not exceed $\varepsilon_0/2$, which by contituity contradicts the maximality of t_*, unless of course $t_* = \infty$, in which case, in view of the fact that ϕ has a positive lower bound, we have geodesic completeness and the theorem is proved.

We remark that the estimate of the error terms would fail if it were not for the fact that the worst error terms vanish due to a simple algebraic identity: if A, B, C are any three symmetric trace-free two dimensional matrices then $\mathrm{tr}(ABC) = 0$. The reason why such matrices appear can be traced back to the symbol of the Einstein equations; they represent the dynamical degrees of freedom of the gravitational field.

The smallness condition on the initial data which is required in the proof of the theorem is the following. Take a point $p \in \mathcal{H}_0 = \mathcal{H}$ and a positive real number

λ. Let d_p be the distance function on $\mathcal{H}$ from p. Set:

$$\begin{aligned} D(p,\lambda) &= \sup_{\mathcal{H}}\{\lambda^{-2}(d_p^2+\lambda^2)^3|\overline{Ric}|^2\} \\ &+ \lambda^{-3}\left\{\int_{\mathcal{H}}\sum_{l=0}^{3}(d_p^2+\lambda^2)^{l+1}|\overline{\nabla}^l k|^2 d\mu_{\overline{g}}\right. \\ &+ \left.\int_{\mathcal{H}}\sum_{l=0}^{1}(d_p^2+\lambda^2)^{l+3}|\overline{\nabla}^l B|^2 d\mu_{\overline{g}}\right\} \end{aligned} \tag{10.1}$$

Here, $|\overline{Ric}|^2 = \overline{R}^{ij}\overline{R}_{ij}$, $\overline{\nabla}^l$ denotes the covariant derivative of order l, and B is the Bach tensor or conformal curvature of $(\mathcal{H},\overline{g})$, a symmetric trace-free 2-covariant tensorfield given by:

$$B_{ij} = \frac{1}{2}(\in_i{}^{ab}\,\overline{\nabla}_a\hat{\overline{R}}_{jb} + \in_j{}^{ab}\,\overline{\nabla}_a\hat{\overline{R}}_{ib}) \tag{10.2}$$

with $\hat{\overline{R}}_{ij}$ the traceless part of $\overline{R}_{ij}$. Then it is the dimensionless invariant

$$\inf_{p\in\mathcal{H},\lambda>0} D(p,\lambda)$$

which must be sufficiently small.

11. The Geometry of Maximal and Null Hypersurfaces

The most difficult and complex step in the proof of the stability theorem is the step demonstrating that if the geometric quantities defined by the hypersurfaces $\mathcal{H}_t$ and $\mathcal{C}_u$ do not exceed ε_0 they are in fact bounded by cE.

The instrinsic and extrinsic geometry of a maximal hypersurface $\mathcal{H}_t$ is is controlled by the elliptic system:

$$\overline{R}_{ij} - k_{im}k^m_j = E_{ij} \tag{11.1}$$

$$\overline{\nabla}_i k_{jm} - \overline{\nabla}_j k_{im} = \in_{ij}^{\ n} H_{mn},\ \overline{\nabla}^j k_{ij} = 0,\ \mathrm{tr}k = 0 \tag{11.2}$$

Here E_{ij} and H_{ij} stand for the *electric* and *magnetic* parts of the space-time curvature respectively, symmetric trace-free 2-covariant tensorfiels on $\mathcal{H}_t$, defined in terms of an arbitray frame e_i, $i = 1,2,3$ in $\mathcal{H}_t$ by:

$$E_{ij} = R(e_i,\hat{T},e_j,\hat{T}),\ H_{ij} = -{}^\star R(e_i,\hat{T},e_j,\hat{T}) \tag{11.3}$$

where $\hat{T}$ is the unit normal to $\mathcal{H}_t$. These are directly controlled by the quantity E. The estimates however involve the foliation of $\mathcal{H}_t$ given by the surfaces $\mathcal{S}_{t,u}$, the level sets of the restriction to $\mathcal{H}_t$ of the optical function u, and some control of the properties of this foliation, provided by the a priori assumption that the geometric quantities do not exceed ε_0, is needed in order to proceed.

The intrinsic geometry of a null hypersurface $\mathcal{C}_u$ is described in terms of the foliation of $\mathcal{C}_u$ given by the surfaces $\mathcal{S}_{t,u}$. If we denote by $e_0 = \phi^{-1}T$ the unit normal to $\mathcal{H}_t$, then e_+ and e_-, respectively the outgoing and incoming null normals to to $\mathcal{S}_{t,u}$, whose component along e_0 is equal to e_0, are given by:

$$e_+ = e_0 + e_3,\ e_- = e_0 - e_3 \tag{11.4}$$

where $e_3 = -a^{-1}U$ is the unit outward normal to $\mathcal{S}_{t,u}$ in $\mathcal{H}_t$. We have:

$$e_+ = \phi^{-1}L \tag{11.5}$$

As e_+ is tangent to $\mathcal{C}_u$, χ, the second fundamental form of $\mathcal{S}_{t,u}$ relative to e_+ is an aspect of the intrinsic geometry of $\mathcal{H}_u$. Its components in an arbitrary local frame e_A, $A = 1, 2$ in $\mathcal{S}_{t,u}$ are given by:

(11.6) $$\chi_{AB} = g(\nabla_{e_A} e_+, e_B)$$

The second fundamental form of $\mathcal{S}_{t,u}$ relative to e_-, which is transverse to $\mathcal{H}_u$, we denote by $\underline{\chi}$:

(11.7) $$\underline{\chi}_{AB} = g(\nabla_{e_A} e_-, e_B)$$

We have:

(11.8) $$\chi = \theta + \eta, \ \underline{\chi} = -\theta + \eta$$

where θ is the second fundamental form of $\mathcal{S}_{t,u}$ relative to $\mathcal{H}_t$ and η is the restriction of k to $\mathcal{S}_{t,u}$. As we have already discussed how k is estimated we shall describe below how estimates for χ are obtained; the intrinsic geometry of $\mathcal{S}_{t,u}$ is controlled by the Gauss equation:

(11.9) $$K + \frac{1}{4}\mathrm{tr}\chi\mathrm{tr}\underline{\chi} - \frac{1}{2}\hat{\chi} \cdot \hat{\underline{\chi}} = -\rho$$

where

(11.10) $$\rho = \frac{1}{4}R(e_-, e_+, e_-, e_+)$$

and we denote by $\hat{\chi}, \hat{\underline{\chi}}$ the traceless parts of $\chi, \underline{\chi}$, respectively. The function $\mathrm{tr}\chi$ satisfies along the integral curves of L (which are parametrized by t) the propagation equation:

(11.11) $$\frac{1}{\phi}\frac{\partial \mathrm{tr}\chi}{\partial t} = \nu\mathrm{tr}\chi - \frac{1}{2}(\mathrm{tr}\chi)^2 - |\hat{\chi}|^2$$

Here,

(11.12) $$\nu = g(\nabla_{e_+} e_0, e_3) = \nabla_3 \log\phi + \delta, \ \delta = k_{33}$$

Note that by virue of the Einstein vacuum equations no curvature term appears on the right hand side of (11.11). The propagation equation (11.11) is considered in conjunction with the Codazzi equation:

(11.13) $$\nabla\!\!\!/^B \hat{\chi}_{AB} - \frac{1}{2}\nabla\!\!\!/_A \mathrm{tr}\chi = \epsilon^B \hat{\chi}_{AB} - \frac{1}{2}\epsilon\mathrm{tr}\chi - \beta_A$$

(11.14) $$\epsilon_A = k_{A3}, \ \beta_A = \frac{1}{2}R(e_A, e_+, e_-, e_+)$$

an elliptic equation for $\hat{\chi}$ on each $\mathcal{S}_{t,u}$, to obtain the required optimal estimates for χ, one order of differentiability smoother than the space-time curvature.

The foliation of space-time given by the null hypersurfaces $\mathcal{C}_u$ are described in terms of the foliation of each $\mathcal{H}_t$ given by the surfaces $\mathcal{S}_{t,u}$. The properties of the latter include, besides what we have already discussed, the lapse function a given, on each $\mathcal{H}_t$, by (8.1). The estimation of $\log a$ is the most subtle part of the argument. It is accomplished by introducing the *mass aspect* function:

(11.15) $$\mu = -\nabla\!\!\!/ \cdot \zeta + K + \frac{1}{4}\mathrm{tr}\chi\mathrm{tr}\underline{\chi}$$

where

(11.16) $$\zeta = \nabla\!\!\!/ \log a - \epsilon$$

The function μ turns out to satisfy along the integral curves of L the propagation equation:

$$\begin{aligned}\frac{1}{\phi}\frac{\partial\mu}{\partial t}+\mu\mathrm{tr}\chi &= 2\hat{\chi}\cdot(\nabla\!\!\!/\hat{\otimes}\zeta)-2\zeta\cdot\beta\\ &-\frac{1}{2}\mathrm{tr}\chi(\nabla\!\!\!/\cdot\lambda+|\lambda|^2+\frac{1}{2}\hat{\chi}\cdot\underline{\hat{\chi}}-\rho)\\ &+(\zeta-\lambda)\cdot(\nabla\!\!\!/\mathrm{tr}\chi-\epsilon\mathrm{tr}\chi)\\ &-\frac{1}{4}\mathrm{tr}\underline{\chi}|\hat{\chi}|^2+\zeta\cdot\hat{\chi}\cdot\lambda\end{aligned}\tag{11.17}$$

Here,

$$\lambda_A=g(\nabla_{e_+}e_0,e_A)=\nabla\!\!\!/_A\log\phi+\epsilon_A\tag{11.18}$$

and we denote by $\nabla\!\!\!/\hat{\otimes}\zeta$ the 2-covariant symmetric trace-free tensorfield on $\mathcal{S}_{t,u}$ given by:

$$(\nabla\!\!\!/\hat{\otimes}\zeta)_{AB}=\frac{1}{2}(\nabla\!\!\!/_A\zeta_B+\nabla\!\!\!/_B\zeta_A-\gamma_{AB}\nabla\!\!\!/\cdot\zeta)$$

What is remarkable here is that, by virtue of the Einstein vacuum equations, the right hand side of (11.17) does not contain terms involving the first derivatives of the curvature. This fact allows us to consider the propagation equation (11.17) in conjuction with the definition (11.15), which is equivalent to:

$$\Delta\!\!\!/\log a=-\mu+\nabla\!\!\!/\cdot\epsilon+K+\frac{1}{4}\mathrm{tr}\chi\mathrm{tr}\underline{\chi}\tag{11.19}$$

an elliptic equation for $\log a$ on each $\mathcal{S}_{t,u}$, to obtain the required optimal estimates for $\log a$, two orders of differentiability smoother than the space-time curvature.

We remark that equation (8.2) on $\mathcal{H}_{t_*}$, when the terms contributed by the second fundamental form of $\mathcal{H}_{t_*}$ are no longer neglected, takes in terms of the function μ the form, simply:

$$\mu=\overline{\mu}\tag{11.20}$$

where $\overline{\mu}$ denotes the mean value of μ on each $\mathcal{S}_{t,u}$.

12. Asymptotic Behaviour

Once the proof of the stability theorem is completed we show that the optical function ${}^{(t_*)}u$ defined during the course of the proof in the slab ${}^{(t_*)}\mathcal{M}$, converges as $t_*\to\infty$ to a global optical function u. For each $t\geq 0$, the 0-level set of ${}^{(t_*)}u$ is the part of C_0, the 0-level set of u, contained in the slab ${}^{(t_*)}\mathcal{M}$. Thus the restrictions of ${}^{(t_*)}L$, L to $C_0\bigcap{}^{(t_*)}\mathcal{M}$ coincide. We shall describe in the remainder of this article the asymptotic behaviour of the solutions. The derivation of these results is found in the last chapter of "".

Let us denote by ${}^{(t_*)}\omega_t$ and ω_t the one parameter groups of transformations generated by ${}^{(t_*)}L$ and L respectively. Let us also denote by ${}^{(t_*)}\psi_s$ and ψ_s the one parameter groups of transformations generated by ${}^{(t_*)}U$ and U respectively. Given a diffeomorphism χ of S^2 onto the surface $\mathcal{S}_{0,0}$ we define the one parameter family $\varphi_{t,0}$ of diffeomorphisms of S^2 onto $\mathcal{S}_{t,0}$ by:

$$\varphi_{t,0}=\omega_t\circ\chi$$

We then define the one parameter family ${}^{(t_*)}\varphi_{t_*,s}$ of diffeomorphisms of S^2 onto ${}^{(t_*)}\mathcal{S}_{t_*,s}$ by:

$$ {}^{(t_*)}\varphi_{t_*,s} = \psi_s \circ \varphi_{t_*,0} $$

Finally we define the two parameter family ${}^{(t_*)}\varphi_{t,s}$, $t \in [0, t_*]$, of diffeomorphisms of S^2 onto ${}^{(t_*)}\mathcal{S}_{t,s}$ by:

$$ {}^{(t_*)}\varphi_{t,s} = \omega_{t-t_*} \circ {}^{(t_*)}\varphi_{t_*,s} $$

We then show that as $t_* \to \infty$, ${}^{(t_*)}\varphi_{t,s}$ converges for each t and s to a diffeomorphism of S^2 onto $\mathcal{S}_{t,s}$.

We call a n-covariant tensorfield w on $\mathcal{M}$ $\mathcal{S}_{t,u}$-tangent if at each $p \in \mathcal{M}$ and for any n-tuplet $X_1, ..., X_n$ of vectors at $p \in \mathcal{S}_{t,u}$ we have:

$$ w(X_1, ..., X_n) = w(\Pi X_1, ..., \Pi X_n) $$

where Π is the orthogonal projection to $T_p\mathcal{S}_{t,u}$. Given any such tensorfield we define:

$$ \tilde{w}_{t,u} = \varphi^*_{t,u}(r^{-n} w) $$

Then $\tilde{w}_{t,u}$ is a n-covariant tensorfield on S^2, for each t and u. We say that on C_u w tends to a limit $W(u)$ as $t \to \infty$, and we write:

$$ \lim_{C_u, t\to\infty} w = W(u) $$

if:

$$ \lim_{t\to\infty} \tilde{w}_{t,u} = W(u) $$

on S^2. It then follows that:

$$ \frac{\partial W}{\partial u} = \lim_{C_u, t\to\infty} \Pi \mathcal{L}_U w $$

The induced metric γ on $\mathcal{S}_{t,u}$ tends in this sense to a metric $\overset{\circ}{\gamma}$ on S^2, which is independent of u and of Gauss curvature equal to 1. Therefore $(S^2, \overset{\circ}{\gamma})$ can be identified with the unit sphere in $\Re^3$. Also,

(12.1) $$ \lim_{C_u, t\to\infty} \phi = \lim_{C_u, t\to\infty} a = 1 $$

and:

(12.2) $$ \lim_{C_u, t\to\infty} r\mathrm{tr}\chi = - \lim_{C_u, t\to\infty} r\mathrm{tr}\underline{\chi} = 2 $$

Moreover,

(12.3) $$ \lim_{C_u, t\to\infty} r^2 \hat{\chi} = \Sigma(u), \quad \lim_{C_u, t\to\infty} r\underline{\hat{\chi}} = \Xi(u) $$

where Σ and Ξ are symmetric trace-free 2-covariant tensorfields on S^2 depending on u and related by:

(12.4) $$ \frac{\partial \Sigma}{\partial u} = -\frac{1}{2}\Xi $$

Also,

(12.5) $$ \Xi = o(|u|^{-3/2}) \text{ as } |u| \to \infty $$

Note that according to (12.2,12.3),

(12.6) $$ \lim_{C_u, t\to\infty} \frac{\hat{\theta}}{\mathrm{tr}\theta} = -\frac{1}{4}\Xi(u) $$

so the surface $\mathcal{S}_{t,u}$ for fixed u does not become umbilical relative to $\mathcal{H}_t$ as $t \to \infty$.

The space-time curvature decomposes relative to the surfaces $\mathcal{S}_{t,u}$ into the $\mathcal{S}_{t,u}$-tangent 2-covariant symmetric trace-free tensorfields α, $\underline{\alpha}$, whose components in an arbitrary local frame e_A, $A = 1, 2$ in $\mathcal{S}_{t,u}$ are given by:

$$\alpha_{AB} = R(e_A, e_+, e_B, e_+), \ \underline{\alpha}_{AB} = R(e_A, e_-, e_B, e_-) \tag{12.7}$$

the $\mathcal{S}_{t,u}$-tangent 1-forms β, $\underline{\beta}$, with components:

$$\beta_A = \frac{1}{2} R(e_A, e_+, e_-, e_+), \ \underline{\beta}_A = \frac{1}{2} R(e_A, e_-, e_-, e_+) \tag{12.8}$$

and the functions ρ and σ, defined by:

$$\rho = \frac{1}{4} R(e_-, e_+, e_-, e_+), \ \sigma\varepsilon(e_A, e_B) = \frac{1}{2} R(e_A, e_B, e_-, e_+) \tag{12.9}$$

where ε is here the area 2-form of $\mathcal{S}_{t,u}$. We have:

$$\begin{aligned} &\lim_{C_u, t \to \infty} r^{7/2} \alpha = 0, && \lim_{C_u, t \to \infty} r \underline{\alpha} = A(u) \\ &\lim_{C_u, t \to \infty} r^{7/2} \beta = 0, && \lim_{C_u, t \to \infty} r^2 \underline{\beta} = B(u) \\ &\lim_{C_u, t \to \infty} r^3 \rho = P(u), && \lim_{C_u, t \to \infty} r^3 \sigma = Q(u) \end{aligned} \tag{12.10}$$

where A is a symmetric trace-free 2-covariant tensorfield, B is a 1-form and P and Q are functions on S^2, all depending on u and having the decay properties:

$$\begin{aligned} &A = o(|u|^{-5/2}), && B = o(|u|^{-3/2}) \\ &P - \overline{P} = o(|u|^{-1/2)}), && Q = o(|u|^{-1/2)}) \\ & && \text{as } |u| \to \infty \end{aligned} \tag{12.11}$$

while:

$$\begin{aligned} &\overline{P} = o(|u|^{-1/2}) && \text{as } u \to \infty \\ &\overline{P} + \frac{M_0}{2\pi} = o(|u|^{-1/2}) && \text{as } u \to -\infty \end{aligned} \tag{12.12}$$

Here M_0 is the ADM mass. Moreover A and B are related to Ξ according to:

$$\frac{\partial \Xi}{\partial u} = -\frac{1}{2} A \tag{12.13}$$

and (relative to an arbitrary local frame in S^2)

$$\overset{\circ}{\nabla}{}^B \Xi_{AB} = B_A \tag{12.14}$$

The following result shows that the ADM mass enters the asymptotic expansion of the area radius of the sections $\mathcal{S}_{t,u}$ of a null hypersurface C_u as $t \to \infty$:

$$r(t, u) = t - \frac{M_0}{2\pi} \log t + O(1) \text{ :at constant } u \text{ as } t \to \infty \tag{12.15}$$

The *Hawking mass* $m(t, u)$ contained by a surface $\mathcal{S}_{t,u}$ is defined by [**11**]:

$$m(t, u) = 2\pi r \left(1 + \frac{1}{16\pi} \int_{\mathcal{S}_{t,u}} \mathrm{tr}\chi \mathrm{tr}\underline{\chi} \right) \tag{12.16}$$

Note that:

$$\overline{\mu} = \frac{m}{2\pi r^3} \tag{12.17}$$

The *Bondi mass* $M(u)$ contained in $\mathcal{C}_u$ is defined by:

$$M(u) = \lim_{t\to\infty} m(t,u) \tag{12.18}$$

One of the achievements of our work was the rigorous derivation of the formula:

$$\frac{\partial M}{\partial u} = -\frac{1}{8}\int_{S^2} |\Xi|^2 d\mu_{\overset{\circ}{\gamma}} \tag{12.19}$$

due to Bondi [**2**]. Morever, we obtain:

$$\lim_{u\to-\infty} M(u) = M_0, \quad \lim_{u\to\infty} M(u) = 0 \tag{12.20}$$

Our final result has to do with the difference of the limits Σ^+, Σ^-, of Σ as $u\to\infty$, $u\to-\infty$, respectively. This difference is determined by the equation:

$$\overset{\circ}{\nabla}{}^B (\Sigma^+_{AB} - \Sigma^-_{AB}) = \overset{\circ}{\nabla}_A \Phi \tag{12.21}$$

where Φ is the solution of:

$$\overset{\circ}{\Delta}\Phi = -2(F - \overline{F}), \ \overline{\Phi} = 0 \tag{12.22}$$

and F is the function on S^2 defined by:

$$F = \frac{1}{8}\int_{-\infty}^{\infty} |\Xi(u)|^2 du \tag{12.23}$$

In view of (12.19), $F/4\pi$ is the total energy radiated to infinity in a given direction, per unit solid angle. The integrability condition of (12.21,12.22), is that F is L^2-orthogonal to the 1st eigenspace of $\overset{\circ}{\Delta}$:

$$F_{(1)} = 0 \tag{12.24}$$

Now the L^2-inner products of F with the three Cartesian coordinate functions x^i, i=1,2,3, on $S^2 \subset \Re^3$, which form an orthogonal basis for the 1st eigenspace of $\overset{\circ}{\Delta}$ represent the components of the total linear momentum radiated to infinity. Since the initial and final states both have zero linear momentum, (12.24) expresses here the law of conservation of linear momentum.

The solution of (12.21,12.22), evaluated at an arbitrary pair X, Y of vectors in $\Re^3$, tangent to S^2 at an arbitrary point ξ, is given by:

$$(\Sigma^+ - \Sigma^-)(X,Y) = -\frac{1}{2\pi}\int_{|\xi'|=1} (F - F_{[1]})(\xi')\frac{(X,\xi')(Y,\xi') - (1/2)(X,Y)|\Pi\xi'|^2}{1-(\xi,\xi')} d\mu_{\overset{\circ}{\gamma}}(\xi') \tag{12.25}$$

Here the subscript [1] denotes the projection on the sum of the 0th and 1st eigenspaces of $\overset{\circ}{\Delta}$, the projection on the 0th eigenspace being the mean value, $(,)$ denotes inner product in $\Re^3$ and Π denotes projection to the plane orthogonal to ξ. Now, by (12.4) we have:

$$\Sigma^+ - \Sigma^- = -\frac{1}{2}\int_{-\infty}^{\infty} \Xi(u) du \tag{12.26}$$

In view of (12.26,12.23), equation (12.25) constitutes a non-linear relationship satisfied by Ξ on $\Re \times S^2$. The non-linearity of Einstein's equations is therefore partially retained even at infinity!

It turns out that the difference $\Sigma(u) - \Sigma^-$ is directly related to the instantaneous displacements of faraway test masses with respect to a reference test mass, relative to which they are initially at rest. The difference $\Sigma^+ - \Sigma^-$, thus yields a permanent displacement of the test masses, a non-linear effect, which is observable in principle (see **[6]** for the details).

References

[1] R. Arnowitt, S. Deser, C.W. Misner, *The dynamics of general relativity*, pp. 227–265 in **Gravitation: An Introduction to Current Research**, L. Witten (ed.), Wiley, 1962.

[2] H. Bondi, M.G.J. van der Burg, A.W.K. Metzner, *Gravitational waves in general relativity, VII*, Proc. Roy. Soc. Lond. **A269** (1962), 21–52.

[3] L. Bel, *Introduction d'un tenseur du quartieme ordre*, CR. Acad. Sci. Paris **248** (1959), 1094–1096.

[4] D. Christodoulou, *Solutions globales des equations de champ de Yang-Mills*, C.R. Acad. Sci. Paris **A293** (1981), 139–141.

[5] D. Christodoulou, *Global solutions of nonlinear hyperbolic equations for small initial data*, Commun. Pure & Appl. Math. **39** (1986), 267–281.

[6] D. Christodoulou, *The nonlinear nature of gravitation and gravitational wave experiments*, Phys. Rev. Letters **67** (1991), 1486–1489.

[7] D. Chistodoulou, S. Klainerman, **the Global Nonlinear Stability of the Minkowski Space**, Princeton Mathematical Series **41**, Princeton University Press, 1993.

[8] A. Einstein, *Zur Elektrodynamik bewegter Körper*, Annalen der Physik **17** (1905), 891–921.

[9] A. Einstein, *Zur allgemeinen Relativitätstheorie*, Preuss. Akad. Wiss. Berlin, Sitzber. **44** (1915) 778–786; 799–801.

[10] A. Einstein, *Die Feldgleichungen der Gravitation*, Preuss. Akad. Wiss. Berlin, Sitzber. **50** (1915) 844–847.

[11] S.W. Hawking, *Gravitational radiation in an expanding universe*, J. Math. Phys. **9** (1968), 598–604.

[12] S. Klainerman, *Uniform decay estimates and the Lorentz invariance of the classical wave equation*, Commun. Pure & Appl. Math. **38** (1985), 321–332.

[13] H. Minkowski, *Raum und Zeit*, Address at the 80th Assembly of German Natural Scientists and Physicians, Cologne (1908), translated in *The Principle of Relativity*, pp. 73–91, Dover, 1952.

[14] R. Schoen, S.T. Yau, *On the proof of the positive mass conjecture in general relativity*, Commun. Math. Phys. **65** (1979), 45–76.

[15] E. Witten, *A new proof of the positive energy theorem* Commun. Math. Phys. **80** (1981), 381–402.

Department of Mathematics, Princeton University, Princeton, NJ 08544
E-mail address: demetri@math.princeton.edu

Einstein-Weyl Geometry

David M. J. Calderbank and Henrik Pedersen

A Weyl manifold is a conformal manifold equipped with a torsion free connection preserving the conformal structure, called a *Weyl connection.* It is said to be *Einstein-Weyl* if the symmetric trace-free part of the Ricci tensor of this connection vanishes. In particular, if the connection is the Levi-Civita connection of a compatible Riemannian metric, then this metric is Einstein. Such an approach has two immediate advantages: firstly, the homothety invariance of the Einstein condition is made explicit by focusing on the connection rather than the metric; and secondly, not every Weyl connection is a Levi-Civita connection, and so Einstein-Weyl manifolds provide a natural generalization of Einstein geometry.

The simplest examples of this generalization are the locally conformally Einstein manifolds. A Weyl connection on a conformal manifold is said to be *closed* if it is *locally* the Levi-Civita connection of a compatible metric; but it need not be a global metric connection unless the manifold is simply connected. Closed Einstein-Weyl structures are then locally (but not necessarily globally) Einstein, and provide an interpretation of the Einstein condition which is perhaps more appropriate for multiply connected manifolds. For example, $S^1 \times S^{n-1}$ admits flat Weyl structures, which are therefore closed Einstein-Weyl. These closed structures arise naturally in complex and quaternionic geometry.

Einstein-Weyl geometry not only provides a different way of viewing Einstein manifolds, but also a broader setting in which to look for and study them. For instance, few compact Einstein manifolds with positive scalar curvature and continuous isometries are known to have Einstein deformations, yet we shall see that it is precisely under these two conditions that nontrivial Einstein-Weyl deformations can be shown to exist, at least infinitesimally.

The Einstein-Weyl condition is particularly interesting in three dimensions, where the only Einstein manifolds are the spaces of constant curvature. In contrast, three dimensional Einstein-Weyl geometry is extremely rich [**16, 68, 72**], and has an equivalent formulation in twistor theory [**34**] which provides a tool for constructing self-dual four dimensional geometries. In section 10, we shall discuss a construction relating Einstein-Weyl 3-manifolds and hyper-Kähler 4-manifolds [**40, 29, 50, 79**]. Twistor methods also yield complete self-dual Einstein metrics of negative scalar curvature with prescribed conformal infinity [**48, 35**]. An important special case of this construction is the case of an Einstein-Weyl conformal infinity [**34, 61**].

Although Einstein-Weyl manifolds can be studied, along with Einstein manifolds, in a Riemannian framework, the natural context is Weyl geometry [**23**]. We take this point of view seriously, because of the insight it provides into the formulation of ideas and results about Einstein-Weyl manifolds. For this reason, and because the approach is less familiar, the first few pages of this essay are devoted to a brief presentation of some concepts invaluable in Weyl geometry, such as densities, Weyl derivatives and conformal metrics; these concepts, despite being as basic as tangent vectors, linear connections and Riemannian metrics, are not common currency.

In section 3, Einstein-Weyl manifolds are introduced. After giving a few examples, we present the initial results of the theory. A key role is played by the contracted Bianchi identity, which implies the constancy of the scalar curvature in the Einstein case. The implications in Einstein-Weyl geometry are more subtle and are described in Theorem 3.6: closed Einstein-Weyl structures have parallel scalar curvature, and the converse holds in the compact case [**27**] or when the dimension is not four [**12**]. We also give the definition of Einstein-Weyl manifolds in two dimensions [**11**] and we indicate throughout how general results apply to this case.

Many of the general theorems about compact Einstein-Weyl manifolds follow from the existence of a distinguished compatible metric, the Gauduchon metric [**24**]. These results are given in section 4 and imply that, apart from in the Einstein case, the isometry group of the Gauduchon metric on a compact Einstein-Weyl manifold is at least one dimensional [**72**]. We also observe that the sign of the scalar curvature is constant in four or more dimensions [**67, 12**], contrary to some previous claims. In section 10 we show that this need not hold in dimensions two and three.

In section 6 we give an extensive supply of examples of Einstein-Weyl manifolds. These examples are often obtained from Riemannian submersions, which we discuss, in section 5, within the more general framework of conformal submersions. In particular we give an ansatz aimed at a study of submersions between Einstein-Weyl manifolds, which includes as special cases both circle bundles over Kähler-Einstein manifolds [**66, 56**] and hyper-complex 4-manifolds (which are Einstein-Weyl) over Einstein-Weyl 3-manifolds [**17, 29**]. We discuss this latter case in section 10, where we use submersions to give a direct proof of the result, of Jones and Tod [**40**], concerning the construction of three dimensional Einstein-Weyl spaces from self-dual 4-manifolds with a conformal vector field. The Jones-Tod construction was used in [**68**] to obtain the full moduli of Einstein-Weyl structures near the round metric on the 3-sphere. More generally, in section 7 we study Einstein-Weyl moduli spaces near Einstein metrics [**65**].

The material in section 8 illustrates how additional conditions on Einstein-Weyl manifolds often lead to closed structures (see [**39**] for another instance of this), and also highlights the role of Weyl structures in complex and quaternionic geometry [**58, 59, 63**]. This is further amplified in following section on four dimensions, where Weyl geometry and complex geometry are intimately linked. However, as shown by Gauduchon and Ivanov [**28**], in the compact case the Einstein-Weyl condition again gives only closed structures. Also in section 9 we discuss the interactions between four dimensional Weyl geometry and twistor theory [**25, 66**], and give a local formula for the Bach tensor on an Einstein-Weyl manifold [**12**]. In the compact case, a similar formula was given in [**67**], where it was used to show that compact Einstein-Weyl manifolds with self-dual Weyl curvature are closed. This fact was obtained in a different way in [**27**] and may be combined with the results of [**66**] to give a classification of the compact self-dual examples. It is now known that, even locally, half conformally flat Einstein-Weyl structures are "half-closed" [**12**] and therefore Einstein or locally hyper-complex [**66**]. We briefly discuss some local examples taken from [**4**].

It is far from being true, however, that the only compact Einstein-Weyl manifolds in dimension four are the closed ones. After discussing topological constraints given by an analogue of the Hitchin-Thorpe inequality [**64**], we end section 9 by presenting the classification [**54**] of four dimensional Einstein-Weyl manifolds with symmetry group of dimension at least four.

In section 10, after discussing the twistor theory of Einstein-Weyl 3-manifolds and the Jones-Tod construction, we present some special classes of three dimensional Einstein-Weyl geometries [**17, 29, 50, 79**] and place them in a unified framework [**14**]. We also explain why the possible geometries on compact 3-manifolds are all obtained as quotients of $\mathbb{R}^4$ and review their classification [**72**]. Finally we give the analogous classification result in dimension two [**11**].

Acknowledgements

During the course of our work on Einstein-Weyl geometry in general, and this survey in particular, we have benefited from discussions with many people. We would like to thank Paul Gauduchon for discussing with us his work with Stefan Ivanov. We also are grateful to Claude LeBrun for advising us on questions of moduli, and to Andrew Swann for useful conversations on submersions. Finally, we thank Paul Tod for updating us on his work with Michael Eastwood on local obstructions (a question raised by Sergei Merkulov).

1. Density line bundles and Weyl derivatives

DEFINITION 1.1. Let V be a real n-dimensional vector space and w any real number. Then a *density of weight* w or w*-density* on V is defined to be a map $\rho\colon (\Lambda^n V)\smallsetminus 0 \to \mathbb{R}$ such that $\rho(\lambda\omega) = |\lambda|^{-w/n}\rho(\omega)$ for all $\lambda \in \mathbb{R}^\times$ and $\omega \in (\Lambda^n V)\smallsetminus 0$. The space of densities of weight w is denoted $L^w = L^w(V)$.

REMARKS. L^w naturally carries the representation $\lambda.\rho = |\lambda|^w\rho$ of the center of $\mathrm{GL}(V)$ or equivalently the representation $A.\rho = |\det A|^{w/n}\rho$ of $\mathrm{GL}(V)$. Note also:

- L^w is an oriented one dimensional linear space with dual space L^{-w}, and L^0 is canonically isomorphic to $\mathbb{R}$.
- The absolute value defines a map from $\Lambda^n V^*$ to L^{-n}. If V is oriented then the $(-n)$-densities can be identified with the volume forms.
- The densities of $L^{-1} \otimes V$ are canonically isomorphic to $\mathbb{R}$.

Now let M be any manifold. Then the *density line bundle* $L^w = L^w_{TM}$ of M is defined to be the bundle whose fiber at $x \in M$ is $L^w(T_xM)$. Equivalently it is the associated bundle $\mathrm{GL}(M) \times_{\mathrm{GL}(n)} L^w(n)$ where $\mathrm{GL}(M)$ is the frame bundle of M and $L^w(n)$ is the space of w-densities of $\mathbb{R}^n$.

One advantage of using densities is that they permit a simple geometric dimensional analysis to be carried out on tensors. Sections of $L = L^1$ are scalar fields with dimensions of length. More generally:

DEFINITION 1.2. The tensor bundle $L^w \otimes (TM)^j \otimes (T^*M)^k$ (and any subbundle, quotient bundle, element or section) will be said to have weight $w + j - k$, or dimensions of $[\text{length}]^{w+j-k}$.

It is quite common in the literature to call a section of such a bundle a tensor field of weight w, or perhaps $-w$, $w/2$, w/n ... various normalization are possible. In view (for instance) of the isomorphism $\Lambda^n T^*M \cong L^{-n}$ on an oriented manifold, such notions of weight would not permit a reasonable dimensional analysis.

On the other hand, the weight defined above can be interpreted invariantly as the representation of the center of $\mathrm{GL}(TM)$. It is additive under tensor product, compatible with contractions, and gives tangent vectors dimensions of length. Here "length" has been identified with weight $+1$, which is not the only reasonable choice. For instance in Fegan [**22**], the weight $+1$ is assigned to cotangent vectors.

NOTATION 1.3. When tensoring a vector bundle with some L^w, we shall often omit the tensor product sign.

The real line bundles L^w are oriented and hence trivializable. However, there is generally no preferred trivialization, and so we prefer to make such a choice explicit.

DEFINITION 1.4. A non-vanishing (usually positive) section of L^1 (or L^w for $w \neq 0$) will be called a *length scale* or *gauge* (of weight w).

It can be convenient in computation and examples to choose a length scale. Nevertheless, the following will be viewed as being more geometrically fundamental.

DEFINITION 1.5. A *Weyl derivative* is a covariant derivative D on L^1. It induces covariant derivatives on L^w for all w. The curvature of D is a real 2-form F^D which will be called the *Faraday curvature.* If $F^D = 0$ then D is said to be *closed*, and there exist local length scales μ with $D\mu = 0$. If such a length scale exists globally, then D is said to be *exact.*

Note that the Weyl derivatives form an affine space modeled on the space of 1-forms, while the spaces of closed and exact Weyl derivatives are modeled on the closed and exact 1-forms respectively.

A length scale μ induces an exact Weyl derivative D^μ such that $D^\mu \mu = 0$. Consequently we shall sometimes call an exact Weyl derivative a gauge, but note that $c\mu$ induces the same derivative for any $c \in \mathbb{R}^+$. If D is any other Weyl derivative then $D = D^\mu + \omega^\mu$ for the 1-form $\omega^\mu = \mu^{-1} D\mu$.

A *gauge transformation* on M is a positive function e^f which rescales a gauge $\mu \in \mathrm{C}^\infty(M, L^w)$ to give $e^{wf}\mu$. Gauge transformations also act on Weyl derivatives via $e^f \cdot D = e^f \circ D \circ e^{-f} = D - df$. However, we shall normally only consider the action on length scales, so that if, for a fixed Weyl derivative D and any length scale μ, we write $D = D^\mu + \omega^\mu = D^{e^f\mu} + \omega^{e^f\mu}$, then $\omega^{e^f\mu} = \omega^\mu + df$.

REMARKS. The theory of Weyl derivatives is a gauge theory with gauge group $\mathbb{R}^+$, and is a geometrization of classical electro-magnetism: the Faraday curvature represents the electro-magnetic field. Indeed this is the original gauge theory of "metrical relationships" introduced by Weyl [**80**]. As a model for electromagnetism, however, it was subsequently rejected in favor of a $\mathrm{U}(1)$ gauge theory. An unfortunate consequence of this is that Weyl derivatives have suffered a period of neglect in differential geometry, although there are several contexts in which they are useful.

EXAMPLE 1.6. Let Ω be a non-degenerate 2-form on M^n. Then Ω^m $(n = 2m)$ equips M with an orientation and a length scale (hence an exact Weyl derivative). Suppose instead that $\Omega \in \mathrm{C}^\infty(M, L^2\Lambda^2 T^*M)$ and that Ω^m is a constant nonzero section of the orientation line bundle $L^n\Lambda^n T^*M$. Now $d\Omega$ is no longer well defined: for each Weyl derivative D on L^1 one can define $d^D\Omega$, but if γ is a 1-form then $d^{D+\gamma}\Omega = d^D\Omega + 2\gamma\wedge\Omega$. However, for $2m > 2$, the non-degeneracy of Ω implies that there is a unique Weyl derivative such that $\mathrm{tr}_\Omega\, d^D\Omega = 0$. In four dimensions this forces $d^D\Omega = 0$, so that every weightless almost symplectic form is "symplectic" with respect to a unique Weyl derivative: it is symplectic in the usual sense iff the Weyl derivative is exact. This Weyl derivative is a manifestly scale invariant version of the Lee form [**75**], which appears naturally in Hermitian geometry, since the Kähler form of an orthogonal almost complex structure on a conformal manifold is a weightless non-degenerate 2-form.

Weyl derivatives also arise in (oriented) contact and CR geometry, where they are induced by complementary subspaces to the contact distribution. The exact Weyl derivatives correspond to global contact forms.

Finally, whenever a geometry has a preferred family of linear connections affinely modeled on the space of 1-forms, these linear connections are usually parameterized by Weyl derivatives. This occurs in quaternionic geometry, projective geometry and the example of interest here: conformal geometry.

2. Conformal geometry

The modern approach to gauge theory has provided much geometrical clarification by identifying it as a theory of connections rather than potentials and gauge transformations. Yet this approach has not filtered back to conformal geometry, where the gauge is constantly being fixed by a metric, and then transformations under rescaling are considered. Part of the problem is that the standard definition of a conformal manifold is a manifold equipped with an equivalence class of Riemannian metrics. The very notation, $[g]$, for the conformal structure leads one to fix the gauge.

A conformal structure may alternatively be defined as a reduction of the frame bundle to a principal $\mathrm{CO}(n)$-bundle, just as a Riemannian metric is equivalently an $\mathrm{O}(n)$-structure. However, this definition has the disadvantage that although the group of invariance of the geometry is clear, it is not made clear exactly *what* remains invariant, and so a Riemannian metric is usually introduced. As counterpoint to the tendency to do conformal geometry in a Riemannian framework, we would like to suggest that a conformal structure is more fundamental than a Riemannian structure by defining the latter in terms of the former.

One motivation for this is that the notion of a Riemannian metric is dimensionally incorrect, since the length of a tangent vector should be a length, not a number. One can only turn it into a number by choosing a length scale.

DEFINITION 2.1. (See e.g., Hitchin [**33**]) A *conformal structure* on a manifold M is an L^2 valued inner product on TM. More precisely it is a section $\mathrm{c} \in C^\infty(M, L^2S^2T^*M)$ which is everywhere positive definite. Furthermore, we shall always take it to be normalized in the sense that $|\det \mathrm{c}| = 1$. Equivalently c is a normalized metric on the *weightless tangent bundle* $L^{-1}TM$. The normalization condition makes sense because the densities of $L^{-1}TM$ are canonically trivial.

In physics, where dimensional analysis is part of the culture, the determinant of a metric is often set to unity: physical metrics assign a length, not a number, to a vector. On the other hand a Riemannian metric is not dimensionless, and so it is meaningless to normalize it. Instead it defines a preferred length scale.

DEFINITION 2.2. A *Riemannian structure* on M is a conformal structure c together with a length scale μ. The metric is $g = \mu^{-2}\mathrm{c} \in C^\infty(M, S^2T^*M)$ and we write (c_g, μ_g) for the corresponding conformal structure and length scale.

This decomposition of a Riemannian structure into two pieces is reflected in the linearized theory: the bundle S^2T^*M is not irreducible under the orthogonal group, but decomposes into a trace and a trace-free part.

An alternative definition of a Riemannian structure is a conformal structure together with an exact Weyl derivative. Such a definition does not distinguish between homothetic metrics, which is often appropriate in practice. The existence and uniqueness of the Levi-Civita connection inducing this exact Weyl derivative is then a special case of the following foundational result.

THEOREM 2.3 (The Fundamental Theorem of Conformal Geometry). **[80]** *On a conformal manifold M there is an affine bijection between Weyl derivatives and torsion free connections on TM preserving the conformal structure. More explicitly, the torsion free connection on TM is determined by the Koszul formula*

$$\begin{aligned}2\langle D_XY, Z\rangle &= D_X\langle Y,Z\rangle + D_Y\langle X,Z\rangle - D_Z\langle X,Y\rangle\\ &\quad + \langle [X,Y],Z\rangle - \langle [X,Z],Y\rangle - \langle [Y,Z],X\rangle,\end{aligned}$$

where $\langle X,Y\rangle \in C^\infty(M,L^2)$ denotes the conformal inner product of vector fields. (Note also that we shall write $|X|^2$ for $\langle X,X\rangle$.) The corresponding linear map sends a 1-form γ to the $\mathfrak{co}(TM)$-valued 1-form Γ defined by $\Gamma_X = \gamma(X)\mathrm{id} + \gamma \wedge X$, where $(\gamma \wedge X)(Y) = \gamma(Y)X - \langle X,Y\rangle\gamma$. Here γ is viewed as a vector field of weight -1 using the natural isomorphism $\sharp\colon T^*M \to L^{-2}TM$ given by the conformal structure.

Henceforth, we identify a Weyl derivative on a conformal manifold with the induced "Weyl connection" on the tangent bundle and all associated bundles. We also use the sharp isomorphism freely, only writing it explicitly to avoid ambiguity.

DEFINITION 2.4. A conformal structure c and a Weyl derivative D define a *Weyl structure* on M, making it into a *Weyl manifold*. For each $w \in \mathbb{R}$, $R^{D,w}$ will denote the *curvature* alt D^2 of D on $L^{w-1}TM$: it is a section of $\Lambda^2T^*M \otimes \mathfrak{co}(TM)$. We write $R^D = R^{D,1}$ for the curvature of the torsion free connection D on TM.

A basic fact in Weyl geometry is the existence of a weight -2 tensor r^D, called the *normalized Ricci endomorphism* of the Weyl structure, such that the curvature of D decomposes as follows:

$$R^{D,w}_{X,Y} = W_{X,Y} + wF^D(X,Y)\mathrm{id} - r^D(X) \wedge Y + r^D(Y) \wedge X.$$

Here W is the *Weyl curvature* of c, which is independent of D and is trace-free.

One way to establish this (and hence find r^D) is to study the way in which the curvature $R^{D,w}$ depends upon the choice of D. Since this is useful for other reasons, we state the result explicitly.

PROPOSITION 2.5. *Suppose D and $\tilde{D} = D+\gamma$ are Weyl derivatives on a conformal manifold (M^n, c). Then the curvatures of D and $\tilde{D}$ are related by the formula:*

$$\begin{aligned} R^{\tilde{D},w}_{X,Y} = R^{D,w}_{X,Y} &+ w\, d\gamma(X,Y)\mathrm{id} \\ &+ \left(D_X\gamma - \gamma(X)\gamma + \tfrac{1}{2}\langle\gamma,\gamma\rangle X\right) \vartriangle Y \\ &- \left(D_Y\gamma - \gamma(Y)\gamma + \tfrac{1}{2}\langle\gamma,\gamma\rangle Y\right) \vartriangle X. \end{aligned}$$

The proof is a matter of computing $d^D\Gamma + \Gamma \wedge \Gamma$ where Γ is related to γ by 2.3. The first term, $w\, d\gamma$, is simply the change in the Faraday curvature F^D on L^w, while the remainder is given in terms of the expression $D\gamma - \gamma \otimes \gamma + \frac{1}{2}\langle\gamma,\gamma\rangle\mathrm{id}$. In order to find a tensor r^D transforming in this way, define, for each $w \in \mathbb{R}$, a section of $L^{-2}\,\mathrm{End}\,TM$ by

$$\mathrm{Ric}^{D,w}(X) = \textstyle\sum_i R^{D,w}_{X,e_i} e_i,$$

where e_i is a weightless orthonormal basis. This Ricci endomorphism is not necessarily symmetric: its skew part turns out to be $(w - \frac{n-2}{2})F^D$, where F^D is viewed as the endomorphism $X \mapsto \sharp\iota_X F^D = \sharp F^D(X,.)$. The symmetric part of $\mathrm{Ric}^{D,w}$ is independent of w and hence so is the trace scal^D, which is a section of L^{-2} called the *scalar curvature* of D. Let $r^D_0 = \frac{1}{n-2}\,\mathrm{sym}_0\,\mathrm{Ric}^{D,w}$ be the (normalized) symmetric trace-free part, and define

$$r^D = r^D_0 + \tfrac{1}{2n(n-1)}\mathrm{scal}^D\mathrm{id} - \tfrac{1}{2}F^D.$$

PROPOSITION 2.6. *If D and $\tilde{D} = D + \gamma$ are Weyl derivatives on (M^n, c) then:*

$$\begin{aligned} r^{\tilde{D}}_0 &= r^D_0 - \mathrm{sym}_0\, D\gamma + \left(\gamma \otimes \gamma - \tfrac{1}{n}\langle\gamma,\gamma\rangle\mathrm{id}\right) \\ \mathrm{scal}^{\tilde{D}} &= \mathrm{scal}^D - 2(n-1)\,\mathrm{tr}\, D\gamma - (n-1)(n-2)\langle\gamma,\gamma\rangle \\ r^{\tilde{D}} &= r^D - \left(D\gamma - \gamma\otimes\gamma + \tfrac{1}{2}\langle\gamma,\gamma\rangle\mathrm{id}\right). \end{aligned}$$

This follows from 2.5 by taking traces, and also shows that W is independent of D.

PROPOSITION 2.7. *On any Weyl manifold of dimension $n > 2$,*

$$\mathrm{div}^D\left(r^D_0 - \tfrac{1}{2n}\mathrm{scal}^D\mathrm{id} + \tfrac{1}{2}F^D\right) = 0,$$

where $\mathrm{div}^D = \mathrm{tr}_c \circ D$ *and in particular,* $\mathrm{div}^D F^D = \sum_i (D_{e_i}F^D)(e_i,.)$.

This is a consequence of the differential Bianchi identity $d^D R^{D,0} = 0$.

The exterior divergence δ on sections of $L^{-n}\Lambda^k TM$ (multi-vector densities) is an invariant operator, just like the exterior derivative on forms. In fact, up to sign, these divergences form a complex formally adjoint to the deRham complex. Our convention is to define $\delta = \mathrm{tr}\, D$, the trace being taken with the first entry. On forms, div^D can therefore be identified with a twisted exterior divergence δ^D. Such twisted divergences no longer form a complex in general. One consequence of this is the following.

PROPOSITION 2.8. **[12]** *Let D be a Weyl derivative on a conformal n-manifold M. Then $(\delta^D)^2 F^D = -(n-4)|F^D|^2$. If $n \neq 4$ it follows that $\mathrm{div}^D F^D = 0$ iff $F^D = 0$.*

PROOF. F^D is a section of $\Lambda^2 T^*M \cong L^{n-4}L^{-n}\Lambda^2 TM$ and so the divergence has been twisted by D on L^{n-4}. The formula follows by direct computation using a trivialization of L^{n-4}. □

3. The Einstein-Weyl equation

We now come to the main definition of this essay.

DEFINITION 3.1. [**16, 34**] Let (M, c, D) be a Weyl manifold of dimension at least three. Then M is said to be *Einstein-Weyl* iff $r_0^D = 0$; equivalently, the symmetric trace-free part of the Ricci tensor vanishes.

EXAMPLES 3.2. We illustrate the three types of Einstein-Weyl manifold.

(i) M is Einstein-Weyl with D exact iff it is Einstein, in the sense that each length scale μ with $D\mu = 0$ defines an Einstein metric.
(ii) Suppose $M = S^1 \times S^{n-1} \cong (\mathbb{R}^n \setminus \{0\})/\mathbb{Z}$, where the $\mathbb{Z}$ action is generated by $x \mapsto 2x$. This action preserves the flat conformal structure and the flat Levi-Civita derivative on $\mathbb{R}^n$, but not the flat metric. Hence M has a natural flat Weyl structure, which is therefore Einstein-Weyl, but the Weyl derivative, although closed, is not exact [**66, 68**]. Note that $S^1 \times S^2$ and $S^1 \times S^3$ admit no Einstein metric [**3**], yet both are Einstein-Weyl in a simple way.
(iii) The simplest example of an Einstein-Weyl manifold with nonzero Faraday curvature is the following Weyl structure on the Berger sphere [**40**]:
$$g = d\theta^2 + \sin^2\theta d\phi^2 + a^2(d\psi + \cos\theta d\phi)^2$$
$$\omega = b(d\psi + \cos\theta d\phi).$$
Here $D = D^g + \omega$ and a, b are constants with $b^2 = a^2(1-a^2)$. This example is related to the Hopf fibration over S^2, and will be discussed again in section 6.

REMARK. In two dimensions, there is no symmetric trace-free Ricci tensor, and so the Einstein-Weyl condition is vacuous. A 2-manifold is usually said to be Einstein iff it has constant scalar curvature, since this follows from the contracted Bianchi identity in higher dimensions. There is a natural generalization in Einstein-Weyl geometry.

PROPOSITION 3.3. [**68, 27**] *Suppose M is Einstein-Weyl of dimension $n > 2$. Then* $D\mathrm{scal}^D - n\,\mathrm{div}^D F^D = 0$. *(As before, the trace is with the first entry of F^D.)*

This is immediate from 2.7, and suggests the following definition.

DEFINITION 3.4. A Weyl manifold (M, c, D) of dimension two is said to be *Einstein-Weyl* iff $D\mathrm{scal}^D - 2\,\mathrm{div}^D F^D = 0$.

Another justification for this definition is that a Weyl derivative D on a conformal 2-manifold defines an (almost) Möbius structure [**11**], and this Möbius structure is integrable (in other words, a complex projective structure) iff D is Einstein-Weyl.

The contracted Bianchi identity has several useful consequences.

PROPOSITION 3.5. [**12, 67**] *Let M be an n-dimensional Einstein-Weyl manifold. Then* $\Delta^D\mathrm{scal}^D = -n(n-4)|F^D|^2$, *where* $\Delta^D = \mathrm{tr}\, D^2$.

We also obtain the following result, essentially given in [**27, 31**], although by using 2.8 compactness assumptions can be avoided except in dimension four [**12**].

THEOREM 3.6. *If (M^n, D) is Einstein-Weyl, the following are equivalent:*

(i) *Either D is closed or $n = 4$, M is non-compact and F^D is harmonic.*
(ii) $\mathrm{div}^D F^D = 0$.
(iii) $D\mathrm{scal}^D = 0$.

(iv) *Either D is exact or* scal^D *is identically zero.*

PROOF. (ii) and (iii) are equivalent by 3.3, and clearly (iii) $\Longrightarrow$ (iv) $\Longrightarrow$ (ii) or (iii). The equivalence of (i) and (ii) follows from 2.8, together with the conformal invariance of the divergence on 2-forms in four dimensions, and the fact that an exact co-closed 2-form on a compact 4-manifold necessarily vanishes (write $F^D = d\gamma$ and integrate the section $|F^D|^2$ of L^{-4} by parts). $\square$

4. The Gauduchon gauge

For electro-magnetism, it is common to fix the gauge by requiring the potential to be divergence free. In Weyl geometry, there are several possible ways to interpret this. However, it is the following gauge that has become the most important, thanks to its global existence and the wealth of results that follow from it [**24, 27, 72**].

DEFINITION 4.1. Let (M, c, D) be a Weyl manifold. Then a length scale μ is called a *Gauduchon gauge* iff $D = D^\mu + \omega^\mu$ with $\mathrm{tr}_{\mathrm{c}}\, D^\mu \omega^\mu = 0$. The exact Weyl derivative D^μ will be called the *Gauduchon derivative* and ω^μ the *Gauduchon 1-form.*

Note that it is the *gauge* derivative being used to define the divergence and so *a priori* this condition is nonlinear, except in two dimensions where the divergence on 1-forms is conformally invariant. However, in higher dimensions the condition is easily linearized by using a length scale of weight $2 - n$.

PROPOSITION 4.2. *Suppose (M, c, D) is a Weyl manifold of dimension $n \geqslant 3$. Then a length scale λ of weight $2-n$ is a Gauduchon gauge iff* $\mathrm{div}\, D\lambda := \mathrm{tr}\, D^2\lambda = 0$.

This follows from the invariance of the divergence on $L^{-n}TM \cong L^{2-n}T^*M$.

On an oriented 3-manifold, a Gauduchon gauge is an "Abelian monopole": the Gauduchon gauge condition means that $*D\lambda$ is a closed 2-form, which is locally equivalent to $*D\lambda = d\theta$ for some 1-form θ. On an Einstein-Weyl 4-manifold, Proposition 3.5 shows that the scalar curvature scal^D defines a Gauduchon gauge wherever it is nonzero. More generally, there is the following theorem.

THEOREM 4.3. [**24**] *A compact Weyl manifold admits a Gauduchon gauge, unique up to homothety (i.e., the Gauduchon derivative is uniquely determined).*

PROOF. If $n = 2$ a Gauduchon gauge is a co-closed representative for the space of 1-forms γ such that $D - \gamma$ is exact (in particular $d\gamma = F^D$). The result in this case is therefore a consequence of the Hodge decomposition. Now suppose $n > 2$.

• The formal adjoint of $\mathrm{tr}\, D^2 \colon J^2L^w \to L^{w-2}$ is $\mathrm{tr}\, D^2 \colon J^2L^{-w+2-n} \to L^{-w-n}$. Now let Δ_D denote this Weyl Laplacian on functions, and Δ_D^* its formal adjoint on sections of L^{2-n}. By Proposition 4.2 a positive section λ of L^{2-n} defines a Gauduchon gauge iff $\Delta_D^*\lambda = 0$.

• Since Δ_D and Δ_D^* have the same principal symbol (after trivializing L^1), they have the same index, which is therefore zero, since they are adjoints. Consequently $\dim\ker\Delta_D^* = \dim\ker\Delta_D = 1$ by the maximum principle.

• No $\phi \in \ker\Delta_D^*$ may change sign: if it did, its integral in a gauge could take any real value and so in particular there would exist positive sections of L^{-2} orthogonal to ϕ. However, the image of Δ^D cannot contain such a positive section, since the Hopf maximum principle implies that super-solutions of Δ^D must be constant. Therefore any $\phi \in \ker\Delta_D^*$ is everywhere nonnegative or non-positive, and so (by the

Hopf maximum principle again) any nonzero ϕ is nowhere vanishing, whence $\ker \Delta^*_D$ consists precisely of the constant multiples of some positive section of L^{2-n}. □

The Gauduchon gauge is particularly powerful on compact Einstein-Weyl manifolds, because it is a *Killing gauge* in the sense that the Gauduchon 1-form is dual to a Killing field. This result of Tod [**72**] is closely related to the existence of a *Gauduchon constant* [**27**] generalizing the constant scalar curvature on an Einstein manifold.

THEOREM 4.4. *Let M be a compact Einstein-Weyl n-manifold and suppose that $D = D^g + \omega^g$ in the Gauduchon gauge. Then the section*

$$\kappa = \mathrm{scal}^g - (n+2)|\omega^g|^2 = \mathrm{scal}^D + n(n-4)|\omega^g|^2$$

of L^{-2} is constant and $\sharp\omega^g$ is a Killing field with respect to D^g. The Ricci endomorphism of D^g is given by

$$\mathrm{Ric}^g = \tfrac{1}{n}\mathrm{scal}^D\mathrm{id} + (n-2)\big(\langle\omega^g,\omega^g\rangle\mathrm{id} - \omega^g\otimes\omega^g\big). \tag{4.1}$$

Conversely suppose that M is Riemannian with Levi-Civita derivative D^g and that ω^g is a 1-form such that $\sharp\omega^g$ is a Killing field and the Ricci tensor of D^g is of the above form, where $\mathrm{scal}^D = \mathrm{scal}^g - (n-1)(n-2)|\omega^g|^2$. Then $D = D^g \pm \omega^g$ is Einstein-Weyl with Gauduchon derivative D^g. (In two dimensions it is also necessary to suppose that $\mathrm{scal}^g - 4|\omega^g|^2$ is constant with respect to D^g.)

The proof of this theorem involves the contracted Bianchi identity for D^g. In general, let μ be a gauge on a conformal n-manifold. Then for $n > 2$, $r^\mu_0 - \frac{1}{2n}\mathrm{scal}^\mu\mathrm{id}$ is divergence free. If M is Einstein-Weyl, r^μ_0 may also be defined by the difference $\mathrm{sym}_0(D^\mu)^2 - \mathrm{sym}_0 D^2$, and this definition works in dimension two: the contracted Bianchi identity for D^μ is then a consequence of the two dimensional Einstein-Weyl equation. From these observations, the following identities are obtained.

PROPOSITION 4.5. *Let M be Einstein-Weyl and let μ be any gauge. Then*

$$r^\mu_0 - \tfrac{1}{2n}\mathrm{scal}^\mu\mathrm{id} = \mathrm{sym}_0 D^\mu\omega^\mu - \omega^\mu\otimes\omega^\mu + \tfrac{1}{n}\big(\langle\omega^\mu,\omega^\mu\rangle - \tfrac{1}{2}\mathrm{scal}^\mu\big)\mathrm{id}$$

is divergence free with respect to μ and consequently:

$$\begin{aligned}
&\mathrm{div}^\mu(\mathrm{sym}_0 D^\mu\omega^\mu)\\
&\quad= 2\langle\mathrm{sym}_0 D^\mu\omega^\mu,\omega^\mu\rangle + \tfrac{n+2}{n}(\mathrm{div}^\mu\omega^\mu)\omega^\mu + \tfrac{1}{2n}D^\mu\big(\mathrm{scal}^\mu - (n+2)|\omega^\mu|^2\big)\\
&\mathrm{div}^\mu\Big(\langle\mathrm{sym}_0 D^\mu\omega^\mu,\omega^\mu\rangle - \tfrac{1}{2n}\big(\mathrm{scal}^\mu + (n-2)|\omega^\mu|^2\big)\omega^\mu\Big)\\
&\quad= 2|\mathrm{sym}_0 D^\mu\omega^\mu|^2 - \tfrac{1}{2n}\big(\mathrm{scal}^\mu - (n+2)|\omega^\mu|^2\big)\,\mathrm{div}^\mu\omega^\mu.
\end{aligned}$$

On a compact manifold, taking μ to be a Gauduchon gauge and integrating the second of these identities immediately gives the main part of Theorem 4.4. The rest of the theorem is now straightforward.

If (M, D) is an Einstein-Weyl manifold with Killing gauge $D = D^g + \omega^g$ then $2D^g\omega^g = F^D$ and $D^g\langle\omega^g,\omega^g\rangle = -F^D(\omega^g,.)$. Consequently the contracted Bianchi identity 3.3 and the constancy of κ imply:

$$2\,\mathrm{tr}(D^g)^2\omega^g = \mathrm{div}^g F^D = -\tfrac{2}{n}\mathrm{scal}^D\omega^g \tag{4.2}$$

$$\Delta^g|\omega^g|^2 + \tfrac{2}{n}\mathrm{scal}^D|\omega^g|^2 = |F^D|^2 \tag{4.3}$$

$$\Delta^g\mathrm{scal}^D - 2(n-4)|\omega^g|^2\mathrm{scal}^D = -n(n-4)|F^D|^2 \tag{4.4}$$

We now collect some geometrical consequences.

THEOREM 4.6. [**12, 27, 37, 67, 68, 72**] *Let M^n be a* compact *Einstein-Weyl manifold with $D = D^g + \omega^g$ in the Gauduchon gauge. Then*

(i) *If D is not exact, then the isometry group of the Gauduchon metric is at least one dimensional.*

(ii) *Contracting* (4.2) *with ω^g and integrating gives:*

$$\int_M |F^D|^2 = \frac{2}{n} \int_M \mathrm{scal}^D |\omega^g|^2.$$

Consequently, if $\mathrm{scal}^D \leqslant 0$ *then D is closed.*

(iii) *D closed $\Longrightarrow$ $D^g \omega^g = 0$ and, if D is not exact, $|\omega^g|^{-1}$ is a Gauduchon gauge.*

(iv) *If* $\mathrm{scal}^D > 0$ *then* $\mathrm{Ric}^g > 0$, *while if* $\mathrm{scal}^D \geqslant 0$ *then* $\mathrm{Ric}^g \geqslant 0$, *and* scal^g *is strictly positive if $n \geqslant 4$ or $n = 3$ and $\kappa \neq 0$.*

(v) *The Hopf maximum principle applied to* (4.4) *implies that if $n \geqslant 4$ and* scal^D *is not everywhere positive, then it is constant in the Gauduchon gauge.*

Theorems 3.6 and 4.6 together give the following rough classification result.

THEOREM 4.7. *If M is compact Einstein-Weyl, one of the following holds:*

(i) scal^D *is negative and D is exact.*

(ii) scal^D *is identically zero, D is closed and if D is not exact, M admits a metric of positive scalar curvature (zero scalar curvature in two dimensions).*

(iii) scal^D *is positive and M admits a metric of positive Ricci curvature.*

(iv) scal^D *is of non-constant sign,* $\dim M \leqslant 3$, $\kappa \leqslant 0$ *and F^D is nonzero.*

We also obtain topological consequences of the Einstein-Weyl condition.

THEOREM 4.8. [**27, 66**] *Let M be a compact Einstein-Weyl manifold. Then if* scal^D *is positive, M has finite fundamental group. Also if D is not exact and M is a spin manifold, then the $\hat{A}$-genus of M vanishes.*

THEOREM 4.9. [**27, 67**] *Let M^n be a compact Einstein-Weyl manifold* $(n > 2)$ *with D closed but not exact. Then the parallel 1-forms on M are precisely the multiples of the Gauduchon 1-form and so the first Betti number of M is one. Also, the universal cover of M is $\mathbb{R} \times \Sigma$ where Σ is a simply connected Einstein manifold of positive scalar curvature. If $n \leqslant 4$ then $\Sigma = S^{n-1}$ and D is flat.*

PROOF. These results all follow easily from the formula (4.1) for the Ricci endomorphism of the Gauduchon gauge, together with the fact that ω^g is D^g-parallel. The first part can be proven either by a Bochner argument [**67**] or as a consequence of the second part [**27**]. The flatness of D for $n = 3$ is immediate from its Ricci-flatness, while for $n = 4$ it follows because $\mathbb{R} \times S^3$ is conformally flat. □

The flat non-exact compact Weyl manifolds, or manifolds of type $S^1 \times S^{n-1}$, therefore exhaust the closed Einstein-Weyl manifolds in dimension less than or equal to four. A detailed study of the four dimensional case can be found in [**27**].

5. Conformal submersions

As in Einstein geometry, many examples of Einstein-Weyl manifolds arise from submersions. Although we shall mainly focus on Riemannian submersions with totally geodesic fibers [**3**], we would like to place these in a conformal context.

DEFINITION 5.1. Let $\pi\colon M \to B$ be a smooth surjective map between conformal manifolds and let the *horizontal bundle* $\mathcal{H}$ be the orthogonal complement to the vertical bundle $\mathcal{V}$ of π in TM. Then π will be called a *conformal submersion* iff for all $x \in M$, $d\pi_x|_{\mathcal{H}_x}$ is a nonzero conformal linear map.

It is not at all necessary to restrict attention to submersions in the following. The base could, for instance, be an orbifold, or be replaced altogether by the horizontal geometry of a foliation (see [**60**]). However, since we are primarily interested in the local geometry, we shall, for convenience of exposition, take the base to be a manifold. A bundle $\mathcal{H}$ complementary to $\mathcal{V}$ is often called a connection on π.

PROPOSITION 5.2. *If $\pi\colon M \to B$ is a submersion onto a conformal manifold B, then conformal structures on M making π into a conformal submersion correspond bijectively to triples $(\mathcal{H}, c^{\mathcal{V}}, \rho)$, where $\mathcal{H}$ is a connection on π, $c^{\mathcal{V}}$ is a conformal structure on the fibers, and $\rho\colon \pi^* L^1_B \cong L^1_{\mathcal{H}} \to L^1_{\mathcal{V}}$ is a (positive) isomorphism.*

The final ingredient ρ in this construction will be called a *relative length scale*, since it allows vertical and horizontal lengths to be compared. The freedom to vary ρ generalizes the so called "canonical variation" of a Riemannian submersion, in which the fiber metric is rescaled, while the base metric remains constant.

DEFINITION 5.3. Let $\pi\colon M \to B$ be a conformal submersion and D a Weyl derivative on M. Then, following O'Neill [**57**], we define fundamental forms A^D, II^D by $A^D(X,Y) = (D_X Y)^{\mathcal{V}}$ for $X, Y \in \mathcal{H}$ and $\mathrm{II}^D(U,V) = (D_U V)^{\mathcal{H}}$ for $U, V \in \mathcal{V}$, where $(\ldots)^{\mathcal{V}}$ and $(\ldots)^{\mathcal{H}}$ denote the vertical and horizontal components.

A remarkable feature of conformal submersions is the existence of a preferred Weyl derivative, much like the Bott connection of a foliation.

PROPOSITION 5.4. *Suppose M is conformal and $TM = \mathcal{V} \oplus_\perp \mathcal{H}$ with $\mathcal{V}, \mathcal{H}$ nontrivial. Then if D is any Weyl derivative, $U \mapsto \mathrm{tr}_{\mathcal{H}} DU$ and $X \mapsto \mathrm{tr}_{\mathcal{V}} DX$ are tensorial for $U \in \mathcal{V}$ and $X \in \mathcal{H}$, and there is a unique $D = D^0$ such that $\mathcal{V}$ and $\mathcal{H}$ are minimal, in the sense that these mean curvature tensors are zero.*

(The last part follows by comparing the mean curvature tensors of D and $D + \gamma$.)

For a conformal submersion, D^0 will be called the *minimal Weyl derivative*, and the corresponding fundamental forms will be denoted II^0 and A^0. The integrability of $\mathcal{V}$ implies that for any D, II^D is symmetric in U, V (it is just the second fundamental form of the fibers), and so II^0 is symmetric and trace free. On the other hand, the conformal property of π implies that the symmetric part of $\langle A^D(X,Y), U\rangle = -\langle D_X U, Y\rangle$ is a pure trace, and so A^0 is skew in X, Y. If D^0 is exact, then in this gauge, the submersion is Riemannian and the fibers are minimal submanifolds.

The O'Neill formulae [**57, 30**] carry over to the conformal setting without substantial change, but here we restrict attention to the case of one dimensional fibers. A foliation of a conformal manifold with oriented one dimensional leaves is equivalently given by the weightless unit vector field tangent to the leaves. In this case the properties of D^0 can be reinterpreted as follows.

PROPOSITION 5.5. *Let ξ be a weightless unit vector field on a conformal manifold. Then the minimal Weyl derivative of the corresponding foliation is characterized by $D^0_\xi \xi = 0$ and $\mathrm{tr}\, D^0 \xi = 0$ and the foliation is (locally) a conformal submersion iff $D^0\xi$ is skew. D^0 is exact iff there is a conformal vector field K with*

$K = |K|\xi$, in which case $D^0|K| = 0$ and so D^0 is the Levi-Civita derivative of $g = |K|^{-2}c$ and K is a unit Killing field.

In other words, if a conformal submersion is given by the flow of a non-vanishing conformal vector field K then D^0 is the *constant length gauge* of K.

Let $\pi\colon M^{n+1} \to B^n$ be a conformal submersion with one dimensional fibers and D^0 exact (this is equivalently a Riemannian submersion with totally geodesic fibers and D^0 is the Levi-Civita derivative). Then D^0 is well defined on the base, and any other Weyl derivative on B is of the form $D^0 + \omega$ for some 1-form ω. On the total space M, we now consider the Weyl derivative $D = D^0 + \frac{n-2}{n-1}\pi^*\omega + \lambda\xi$ where ξ is the weightless (co)tangent vector to the fibers and λ is a section of L^{-1}.

Using the well known submersion formulae for the Ricci tensor of D^0 [**3**], together with the formulae in 2.6 we obtain the following.

PROPOSITION 5.6. *Let $D = D^0 + \frac{n-2}{n-1}\pi^*\omega + \lambda\xi$. Then:*

$$\begin{aligned}
\operatorname{sym}\operatorname{Ric}^D_M(X,Y) &= \operatorname{sym}\operatorname{Ric}^{D^0+\omega}_B(X,Y) - 2\langle A^0_X, A^0_Y\rangle - \big(D^0_\xi\lambda + (n-1)\lambda^2\big)\langle X,Y\rangle \\
&\quad - \tfrac{n-2}{n-1}\,\omega(X)\omega(Y) + \tfrac{1}{n-1}\big(\operatorname{div}^0\omega + (n-2)|\omega|^2\big)\langle X,Y\rangle \\
\operatorname{sym}\operatorname{Ric}^D_M(\xi,X) &= \textstyle\sum_i \langle (D^0_{e_i}A^0)(e_i,X),\xi\rangle - \tfrac12(n-1)D^0_X\lambda \\
&\quad + (n-2)\big(\omega(X)\lambda - \langle A^0(X,\omega),\xi\rangle\big) \\
\operatorname{sym}\operatorname{Ric}^D_M(\xi,\xi) &= |A^0|^2 - nD^0_\xi\lambda \\
&\quad - \tfrac{n-2}{n-1}\big(\operatorname{div}^0\omega + (n-2)|\omega|^2\big)
\end{aligned}$$

where X,Y are horizontal, $e_1,..e_n,\xi$ is a weightless orthonormal basis with ξ vertical, $\langle A^0_X, A^0_Y\rangle = \sum_i\langle A^0(X,e_i), A^0(Y,e_i)\rangle$ and $|A^0|^2 = \sum_i\langle A^0_{e_i}, A^0_{e_i}\rangle$.

The factor $\frac{n-2}{n-1}$ eliminates the difficult terms involving $D^0\omega$. It occurs naturally in the case of a hyper-complex 4-manifold over an Einstein-Weyl 3-manifold [**29**], which we shall discuss in section 10. In this section, though, we shall only treat the case $\omega = 0$, as considered by Pedersen and Swann [**66**].

THEOREM 5.7. [**66, 56**] *Let $\pi\colon M^{n+1} \to B^n$ be a Riemannian submersion, over an Einstein manifold B, with complete totally geodesic one dimensional fibers. Suppose that M admits an Einstein-Weyl structure of the form $D = D^0 + \lambda\xi$. Then*

(i) *$\operatorname{scal}^0_B \geqslant (n+2)|A^0|^2 + n(n-1)\lambda^2$ and $\operatorname{scal}^D_M \geqslant n|A^0|^2$, with equality (in both) iff λ is constant on the fibers (which necessarily holds if the fibers are compact).*

(ii) *In the D^0 gauge, A^0 defines a symplectic form on the open subset of B where it is nonzero, and so n is even unless A^0 is identically zero. If $|A^0|^2$ is a nonzero constant then B is almost Kähler and M is almost Sasakian.*

PROOF. By the submersion formulae, the Einstein-Weyl condition gives rise to the following three equations:

$$\begin{aligned}
\langle A^0_X, A^0_Y\rangle &= \tfrac1n|A^0|^2\langle X,Y\rangle \\
\textstyle\sum_i\langle (D^0_{e_i}A^0)(e_i,X),\xi\rangle &= \tfrac12(n-1)D^0_X\lambda \qquad (5.1)\\
n(n-1)(\lambda^2 - D^0_\xi\lambda) &= \operatorname{scal}^0_B - (n+2)|A^0|^2.
\end{aligned}$$

The last equation and the completeness of the fibers together imply that along each fiber, λ is either constant or a negative hyperbolic tangent with respect to

D^0. Hence $D^0_\xi\lambda$ is non-positive and the first part readily follows. The first equation implies that $(X,Y)\mapsto\langle A^0(X,Y),\xi\rangle$ is either zero or non-degenerate at each point. Also, if μ is a D^0-parallel length scale, then $\langle A^0(X,Y),\mu^{-1}\xi\rangle=-\frac{1}{2}d(\mu^{-1}\xi)$, which is a closed basic 2-form $\pi^*\Omega$. If A^0 is nonzero, the metric $\mu|A^0|\mathrm{c}$ on B is almost Hermitian with Kähler form Ω and so if $D^0|A^0|=0$ then B is almost Kähler and M is almost Sasakian. □

Examining this theorem more closely, we see that the Einstein-Weyl equations on M have in fact been encoded on B, suggesting that there should be an inverse construction. In fact one parameter families of Einstein-Weyl structures can be found on S^1-bundles in this way.

Suppose $\pi\colon M\to B$ is a fibration over an almost Kähler-Einstein manifold of positive scalar curvature and that it has a connection $\mathcal{H}$ with curvature $k\pi^*\Omega\otimes U$, where Ω is the Kähler form on B, U is a non-vanishing vertical vector field and k is constant. If for some choice of relative length scale, M becomes a Riemannian submersion with totally geodesic fibers and U constant, then the same holds for any constant multiple of this relative length scale, giving a one parameter family of metrics $g_t=\pi^*g_B+t^2g(U,.)^2$ called the *canonical variation.* The equations (5.1) with constant λ may be satisfied provided $\mathrm{scal}^0_B\geqslant(n+2)|A^0|^2_t$. If $A^0=0$ this holds for all t, while for $A^0\neq0$, it is only possible for $0<t\leqslant t_0$ where g_{t_0} is an Einstein metric.

THEOREM 5.8. **[66]** *Let B be a Kähler-Einstein manifold of positive scalar curvature and let M be a principal S^1-bundle with connection whose curvature is a multiple of the Kähler form. Then M admits a one parameter family of Einstein-Weyl structures.*

These results fit in with the idea that Einstein-Weyl geometry is a natural deformation of Einstein geometry, which we shall discuss again in section 7.

6. Examples

Examples of Einstein-Weyl structures on S^1-bundles include the following.

1. The basic example of a nontrivial S^1-bundle over a Kähler-Einstein base is the Hopf fibration $S^3\to S^2$. If σ is a left invariant 1-form on S^3 then the bi-invariant (round) metric is $g=\pi^*g_{S^2}+\sigma^2$ where π is the Riemannian submersion generated by the Killing field dual to σ. If we now consider the U(2) invariant Berger metric $g_a=\pi^*g_{S^2}+a^2\sigma^2$, we find that for $0<a<1$ there is a unique b up to sign such that $D=D^{g_a}+b\sigma$ is Einstein-Weyl. This is the example given in 3.2 and it easily generalizes to the higher dimensional Hopf fibration $S^{2n+1}\to\mathbb{C}P^n$. The Einstein-Weyl structures are parameterized by a point (a,b) on an ellipse, where the two points on the axis of symmetry $b=0$ are respectively degenerate and Einstein **[65]**.

2. The unit tangent bundle T_1S^n of S^n is an S^1-bundle over the Grassmannian $\widetilde{\mathrm{Gr}}_2(\mathbb{R}^{n+1})$ of oriented 2-planes in $\mathbb{R}^{n+1}$. Since $\widetilde{\mathrm{Gr}}_2(\mathbb{R}^{n+1})$ is Kähler-Einstein, T_1S^n admits a one parameter family of Einstein-Weyl structures.

3. The twistor space Z of a quaternionic manifold M possesses a natural S^1-bundle S. If M is quaternionic Kähler with positive scalar curvature, then Z is Kähler-Einstein and S is a 3-Sasakian manifold admitting a one parameter family of Einstein-Weyl structures which fiber over M with Berger 3-spheres as fibers.

Riemannian submersions have also been used [67] to construct Einstein-Weyl structures on S^2- or $\mathbb{R}P^2$-bundles over compact Kähler-Einstein manifolds of positive scalar curvature. For instance, we find Einstein-Weyl structures with $\mathrm{scal}^D > 0$ on $P(\mathcal{O}(k) \oplus \mathcal{O})$ over $\mathbb{C}P^n$ for $|k| \leqslant n$ (i.e., $S^2 \times S^2$ or $\mathbb{C}P^2 \# \overline{\mathbb{C}P}^2$ when $n = 1$).

In all these bundle constructions the base manifold may be taken to be a *product* $M_1 \times \cdots \times M_m$ of Kähler-Einstein manifolds (M_i, g_i) with $c_1(M_i)$ positive and proportional to an indivisible class α_i:

1. Let $\pi\colon P \to M_1 \times \cdots \times M_m$ be a principal T^r-bundle with characteristic classes $\beta_i = \sum_{j=1}^m b_{ij}\pi_j^*\alpha_j$, for $i = 1, \ldots, r \leqslant m+1$, where$(b_{ij})$ is a matrix of integers of rank at least $r - 1$. Then there is a family of Einstein-Weyl structures (g, ω) on P such that π is a Riemannian submersion with flat totally geodesic fibers, the metric on B is of the form $x_1g_1 + \cdots + x_mg_m$, and the 1-form ω is vertical.

More explicitly, for $r = 1$, let θ be the principal connection and set $\omega = f\theta$ for some function f. Let g be the metric $x_1\pi^*g_1 + \cdots + x_n\pi^*g_n + \theta^2$. Then, using the Riemannian submersion formula 5.6, the Einstein-Weyl equation forces the function f to be constant. A fixed point argument modeled on that of Wang and Ziller [78] shows the existence of a solution. For general r the technical condition on the rank of (b_{ij}) turns out to be equivalent to the necessary condition $b_1(P) \leqslant 1$ for the existence of an Einstein-Weyl solution [66].

2. Similarly, there are solutions on S^2- or $\mathbb{R}P^2$-bundles over $M_1 \times \cdots \times M_m$. For integers q_i, the cohomology class $q_1\alpha_1 + \cdots + q_m\alpha_m$ is the Euler class of a principal circle bundle P with curvature $\Omega = \sum_i q_i\Omega_i$ where Ω_i is the Kähler form on M_i. Then there are Einstein-Weyl structures on $M = P \times_{S^1} S^2$, of the form

$$h = dt^2 + f(t)^2\sigma^2 + \textstyle\sum_i h_i(t)^2 g_i$$
$$\omega = A\,dt + Bf\sigma,$$

where $d\sigma$ is the pull-back of Ω [77].

Another important reservoir of examples is provided by looking for Einstein-Weyl manifolds with a high degree of symmetry [52, 54]. The natural group of symmetries on a Weyl manifold is the group of automorphisms preserving both the conformal structure c (the conformal transformations) and the Weyl connection D (the affine transformations). One may argue that it is equally natural to consider projective transformations but on a Weyl manifold conformal projective transformations are automatically affine: if D^1 and D^2 are projectively equivalent then $D^1_X - D^2_X = \alpha(X)\mathrm{id} + \alpha \otimes X$, whereas if D^1 and D^2 are both compatible with the conformal structure, then $D^1_X - D^2_X$ has to be a section of $\mathfrak{co}(TM)$ for all X, which forces $\alpha = 0$. Furthermore, we can generally assume that the symmetry groups preserve the Gauduchon metrics:

PROPOSITION 6.1. *Let G act by symmetries on a compact Weyl manifold M. Then G preserves the Gauduchon derivative D^g and so acts by homotheties of each Gauduchon metric. If G is compact then it acts by isometries. (If M is not the n-sphere then G is compact by the theorems of Obata and Lelong-Ferrand* [46].)

PROOF. The Gauduchon gauge D^g satisfies $\mathrm{tr}\, D^g\omega^g = 0$. For each $a \in G$ we have $D = a^*D = a^*D^g + a^*\omega^g$. The pull-back of an exact Weyl derivative is exact and $\mathrm{tr}(a^*D^g)a^*\omega^g = a^*\,\mathrm{tr}\, D^g\omega^g = 0$, so uniqueness implies that $a^*D^g = D^g$ and $a^*\omega^g = \omega^g$. The action of G on the homothety class of Gauduchon metrics is thus described by a homomorphism $\rho\colon G \to \mathbb{R}$. If G is compact, ρ is constant. $\square$

COROLLARY 6.2. *Let G be the symmetry group of a compact Weyl manifold M with Weyl derivative $D = D^g + \omega^g$ in the Gauduchon gauge. Then G preserves ω^g and hence for each $x \in M$ either $\omega^g_x = 0$ or the isotropy representation at x has a trivial summand tangent to the orbit.*

This greatly restricts the possible non-exact homogeneous Weyl structures. If one then imposes the Einstein-Weyl equation, only a few examples are known. For instance, there are the S^1-bundles S^{2n+1} and T_1S^n with $\mathrm{scal}^D > 0$ given above, and examples of type $S^1 \times S^{n-1}$ with $\mathrm{scal}^D = 0$. Together with an additional family on S^{4n+3}, these are the only homogeneous examples on symmetric spaces [**43**].

We turn now to some explicit examples of compact Einstein-Weyl n-manifolds M of co-homogeneity one under a group of symmetries G. The principal orbits are therefore homogeneous submanifolds G/H of dimension $n-1$ and M/G is either a closed interval or a circle. In the latter case M will not have finite fundamental group and therefore (for $n \geqslant 4$) the Weyl derivative is closed. Consequently we restrict attention to the case $M/G = [0, \ell]$.

Motivated by the possible actions in low dimensions (in particular see Theorem 9.13 for a classification result in dimension four) we look for Einstein-Weyl manifolds of co-homogeneity one under $\mathrm{SO}(n)$, $S^1 \times \mathrm{SO}(n-1)$ or $\mathrm{U}(2m)$ $(n = 2m)$ with principal orbits covered by S^{n-1}, $S^1 \times S^{n-2}$ or S^{2m-1} respectively.

The case of $\mathrm{SO}(n)$ gives only Einstein manifolds by 6.2, since the principal orbits are isotropy irreducible. In the other cases, non-closed Einstein-Weyl structures may be constructed explicitly via the solution of ODEs with boundary conditions.

Consider, for example, $S^1 \times \mathrm{SO}(n-1)$ symmetry. First of all we study the possible $S^1 \times \mathrm{SO}(n-1)$-invariant Einstein-Weyl structures on $(0, \ell) \times S^1 \times S^{n-2}$. In the Gauduchon gauge they are easily seen to be of the form:

$$g = dt^2 + f(t)^2 d\theta^2 + h(t)^2 g_{\mathrm{can}}$$
$$\omega = Af(t)^2 d\theta$$

where g_{can} is the round metric on S^{n-2} of sectional curvature one. Here f and h are smooth functions on $[0, \ell]$ with $f, h > 0$ on $(0, \ell)$ and A is constant. With this Ansatz the Einstein-Weyl equation becomes:

$$-\frac{f''}{f} - (n-2)\frac{h''}{h} = \Lambda$$
$$-\frac{f''}{f} - (n-2)\frac{f'h'}{fh} + (n-2)A^2 f^2 = \Lambda$$
$$-\frac{h''}{h} - (n-3)\frac{h'^2}{h^2} - \frac{f'h'}{fh} + \frac{n-3}{h^2} = \Lambda.$$

At the boundary points $0, \ell$ we seek subgroups K such that $\mathrm{SO}(n-2) < K \leqslant S^1 \times \mathrm{SO}(n-1)$ with $K/\mathrm{SO}(n-2)$ a sphere. For instance, $M = S^n$ is obtained if we take $K_1 = \mathrm{SO}(n-1)$ at $t = 0$ and $K_2 = S^1 \times \mathrm{SO}(n-2)$ at $t = \ell$. The boundary conditions at $t = 0$ are then seen to be $f > 0$, $f', h, h'' = 0$, $h' = 1$ while at $t = \ell$, we have $h > 0$, $f, f'', h' = 0$, $f' = -1$. Solutions matching these boundary conditions can be found explicitly. In particular, when $n = 4$ we find the following solutions on S^4.

$$
(6.1)\quad
\begin{aligned}
g &= \frac{1 - a\cot a}{y\cot y - a\cot a}dy^2 + \frac{4(1-a\cot a)(y\cot y - a\cot a)}{(a + a\cot^2 a - \cot a)^2}d\theta^2 + \sin^2 y\, g_{\mathrm{can}}\\
\omega &= \frac{2(y\cot y - a\cot a)}{a + a\cot^2 a - \cot a}d\theta.
\end{aligned}
$$

Here $\sin y = h(t)$ and $(y,\theta) \in (0,a) \times (0,2\pi)$, where $0 \leqslant a < \pi$ and $a = 0$ corresponds to the standard Einstein metric on S^4.

In this way we find one parameter families of $S^1 \times \mathrm{SO}(n-1)$ symmetric Einstein-Weyl structures on S^n and $S^2 \times S^{n-2}$. Relaxing the boundary conditions gives solutions on line bundles over compact manifolds [**52**].

A similar calculation leads to families of solutions with $\mathrm{U}(m)$ symmetry on S^{2m}, $\mathbb{C}P^m$ and $P(\mathcal{O}(k) \oplus \mathcal{O})$ $(0 < |k| < m)$ over $\mathbb{C}P^{m-1}$. This last case fits into the framework of Einstein-Weyl structures on S^2-bundles discussed earlier (the Fubini-Study metric on $\mathbb{C}P^{m-1}$ being the Kähler-Einstein base).

One motivation for studying these highly symmetric examples is that the principal orbits provide an interesting family of submanifolds [**65**]. For instance, in the case of S^n with principal orbits $S^1 \times S^{n-2}$ there exists $t_0 \in [0,\ell]$ such that the corresponding $S^1 \times S^{n-2}$ is minimal in the Gauduchon metric of S^n. This generalizes the Clifford torus in the round 3-sphere. Likewise S^{2m} with the Gauduchon metric has a totally geodesic equator S^{2m-1} $(t = \frac{\ell}{2})$.

It should be pointed out, however, that the induced structures on the submanifolds $S^1 \times S^{n-2}$ in S^{n-1} and and S^{2m-1} in S^{2m} are *not* Einstein-Weyl. Of course, both $S^1 \times S^{n-2}$ and S^{2m-1} are Einstein-Weyl with respect to other Weyl structures and in fact these structures *do* sit as minimal hypersurfaces in *some* Einstein-Weyl space due to the following theorem which is inspired by the work of Koiso [**44**].

THEOREM 6.3. [**65**] *Let (M, c, D) be a real analytic Weyl manifold with an analytic symmetric bilinear form β taking values in a real line bundle over M. Then, there is a germ unique Einstein-Weyl space $(\bar{M}, \bar{\mathrm{c}}, \bar{D})$ in which (M, c, D) is embedded as a hypersurface with second fundamental form β. In particular the embedding could be minimal or totally geodesic.*

7. Moduli spaces of Einstein-Weyl structures

A possible motivation for studying Einstein-Weyl geometry in arbitrary dimensions is that Einstein manifolds with Killing fields often admit continuous families of Einstein-Weyl structures, as discussed in section 5. Since such Einstein manifolds are often rigid [**3**], the Einstein-Weyl condition may provide nontrivial deformations which would otherwise be lacking. One might then hope to get new Einstein metrics by going to the boundary of the Einstein-Weyl moduli space. So far, though, only known Einstein metrics have been obtained in this way.

Let M be compact and let the diffeomorphism group $\mathrm{Diff}(M)$ act on Weyl structures (c, D) by pull-back. Since the quotient space is not a manifold, we need to fix a slice to this action. One way of doing this is to describe Weyl structures in the Gauduchon gauge and use the Ebin slice [**21**] near a suitable Gauduchon metric g_0. The homothety factor of this metric may be fixed by specifying the Gauduchon constant κ. To do this, note that for $n \geqslant 4$, an Einstein-Weyl structure with $\kappa \leqslant 0$ is either Einstein or belongs to the known family of four dimensional manifolds of type $S^1 \times S^3$. In dimensions two and three there is a classification of Einstein-Weyl geometries on compact manifolds (see section 10), and so we may focus here on the case of positive Gauduchon constant and fix $\kappa = 1$. The Ebin slice $\mathcal{S}(g_0)$ is now given infinitesimally by $\operatorname{div}^{g_0} \dot{g} = 0$, where $\dot{g}$ is a tangent vector at g_0 to the space of metrics. This fixes the action of Diff(M) up to isometries of g_0.

DEFINITION 7.1. Suppose (g_0, ω_0) is an Einstein-Weyl solution in the Gauduchon gauge on M and let $\mathcal{S}(g_0)$ be the Ebin slice. The subset $\widetilde{\mathcal{M}}$ of $\mathcal{S}(g_0) \times \Omega^1 M$ given by $\operatorname{div}^g \omega = 0$, $\kappa = 1$ and $\operatorname{ric}^g = \frac{1}{n}\operatorname{scal}^D g + (n-2)(|\omega|^2 g - \omega \otimes \omega)$ is called the *premoduli space* of Einstein-Weyl structures around (g_0, ω_0). The local *moduli space* $\mathcal{M}$ near (g_0, ω_0) is the quotient of $\widetilde{\mathcal{M}}$ by the isometry group of g_0.

We now study the Einstein-Weyl moduli space near an Einstein metric.

THEOREM 7.2. [**65**] *Suppose (g_t, ω_t) is a smooth curve in $\tilde{\mathcal{M}}$ with $(g_0, \omega_0) = (g, 0)$, so that g is Einstein with* $\operatorname{scal}^g = 1$, *and let $(\dot{g}, \dot{\omega})$ be the tangent at $t = 0$.*

Then $\dot{\omega}$ is a Killing field of g and $\dot{g}$ satisfies the linearized Einstein equation. This equation is elliptic and the space of Einstein-Weyl deformations is finite dimensional. In particular, if the Einstein metric has no infinitesimal Einstein deformations then (M, g) has at most an m-dimensional family of infinitesimal Einstein-Weyl deformations, where m is the rank of the isometry group of g.

PROOF. The Killing condition $\operatorname{sym} D^{g_t} \omega_t = 0$ implies that $\operatorname{sym} D^g \dot{\omega} = 0$, so $\dot{\omega}$ is dual to a Killing field. Next, differentiating the equation defining the Gauduchon constant and using $\dot{\kappa} = 0$ gives $\dot{\operatorname{scal}}^D = 0$ as the derivative of the quadratic term involving ω_t vanish. Differentiating the Einstein-Weyl equation

$$\operatorname{ric}^{g_t} = \tfrac{1}{n}\operatorname{scal}^{D^t} g_t + (n-2)(|\omega_t|^2 g_t - \omega_t \otimes \omega_t)$$

now gives the linearized Einstein equation since $\omega_t = 0$ and $\operatorname{scal}^{D^t} = 1$ at $t = 0$. Together with the infinitesimal Ebin slice condition, this equation is known to be elliptic [**3**], which gives the finite dimensionality.

If the Einstein metric has no infinitesimal deformations then there remain only the deformations of ω, namely the Killing fields of g modulo isometries. The dimension of a generic orbit is the co-rank of the isometry group (since the stabilizer is a maximal torus), which gives the bound on the dimension of the moduli space. □

REMARK. More generally, following [**18**], it has been shown [**68, 67**] that the Einstein-Weyl equation is elliptic in harmonic coordinates, at least once supplemented by the Bianchi identities 2.7, 3.5. Consequently, Einstein-Weyl manifolds are real analytic, and on compact manifolds the moduli space is finite dimensional.

We will now give some examples which show that at least some of these deformations may be integrated to give a nontrivial moduli space.

1. Theorem 7.2 implies in particular that the number of infinitesimal Einstein-Weyl deformations of the standard n-sphere is equal to the rank $\lfloor (n+1)/2 \rfloor$ of $\mathrm{SO}(n+1)$.

On S^3 the number of infinitesimal deformations is two and all these have been integrated [**68**], using a relationship between three dimensional Einstein-Weyl manifolds and four dimensional self-dual manifolds, which we shall describe in section 10.

On S^4, as we discussed in section 6, there is a one parameter family with $S^1 \times \mathrm{SO}(3)$ symmetry and a one parameter family with $\mathrm{U}(2)$ symmetry, but so far a two parameter family of solutions integrating all the infinitesimal deformations has not been found.

2. Further examples where the rank of the isometry group agrees with the dimension of the space of known deformations can be found amongst the Einstein-Weyl structures on r-torus bundles over products of $m \geqslant r$ Kähler-Einstein manifolds, as described in section 6. The family of solutions is r-dimensional and the Einstein-Weyl structures are close to the Einstein metrics found by Wang and Ziller [**78**]. In particular, if the base manifold is a product of Kähler-Einstein manifolds without a continuous family of isometries (such as those found by Tian and Yau [**71**] on k-fold blow-ups of $\mathbb{C}P^2$ for $4 \leqslant k \leqslant 8$) then the isometry group of the Einstein metric on the torus bundle is the torus T^r itself, which has rank r. Note however, that the Einstein metrics on these T^r-bundles are not known to be rigid, so the full moduli space of Einstein-Weyl deformations could be larger.

3. Let (M, g_0) be a locally symmetric Einstein manifold of compact type and let $\prod_{a=1}^N M_a$ be the irreducible decomposition of the universal Riemannian covering manifold $\widetilde{M}$. Consider the following lists of compact symmetric manifolds.

$A.\quad \dfrac{\mathrm{SU}(p+q)}{S(\mathrm{U}(p)\times \mathrm{U}(q))}\quad (p \geqslant q \geqslant 2),\quad \dfrac{E_6}{F_4},\quad \dfrac{\mathrm{SU}(\ell)}{\mathrm{SO}(\ell)},\quad \dfrac{\mathrm{SU}(2\ell)}{\mathrm{Sp}(\ell)},\quad \mathrm{SU}(\ell)\quad (\ell \geqslant 3);$

$B.\quad \dfrac{G_2}{SO(4)},\quad G_2,$ or a Hermitian symmetric space of dimension $\geqslant 4$;

$C.\quad S^2.$

If $N = 1$ and M_a is not on list A, or $N = 2$ and M_a is not on lists A–B, or $N = 3$ and M_a is not on the lists A–C, then (M, g_0) has no infinitesimal Einstein deformations [**45**]. Thus, for example, $S^2 \times S^2$ can have at most a two parameter family of Einstein-Weyl solutions near the Einstein metric. An explicit one parameter family can be found using the constructions of section 6 [**54**].

As an example of a moduli of Einstein-Weyl structures away from an Einstein metric, we should mention the moduli of flat Weyl structures on the manifold $S^1 \times S^{n-1}$. In four dimensions, all Einstein-Weyl structures on $S^1 \times S^3$ are flat, as we shall see in section 9.

8. Complex and quaternionic structures

DEFINITION 8.1. A conformal manifold (M, c) will be called *Kähler Weyl* iff it is equipped with an almost complex structure $J \in \mathrm{C}^\infty(M, \mathfrak{so}(TM))$ and a Weyl connection D such that $DJ = 0$.

REMARK. These manifolds were called "Hermitian Weyl" in [**63**], since representative metrics for c are generally only Hermitian. From the perspective adopted here, however, properties of representative metrics are less relevant, and so, since $DJ = 0$ is a Kähler condition on D, we would like to advocate this change of terminology, which is also consonant with the term "locally (conformally) Kähler".

Since D is torsion-free, $DJ = 0$ implies that J is integrable. It also implies that $d^D\Omega = 0$ where Ω is the weightless Kähler form associated to J using c. Therefore $0 = (d^D)^2\Omega = 2F^D \wedge \Omega$ and so F^D vanishes in dimension $2m > 4$ [**63, 75**]. It then follows that if μ is a parallel local length scale, the metric $g = \mu^{-2}\mathrm{c}$ is a local Kähler metric. If D is exact, such a length scale exists globally and M is Kähler.

Conversely, a locally Kähler manifold is Kähler Weyl: the complex structure is the one given and the Weyl connection is locally the Levi-Civita connection of the compatible local Kähler metrics.

Since we shall discuss the four dimensional case in section 9, we confine ourselves here to the case that D is closed. Therefore if $D = D^g + \omega^g$ in the Gauduchon gauge, then ω^g is harmonic with respect to D^g. If also M is compact and Einstein-Weyl and D is not exact, then ω^g is a nontrivial D^g-parallel 1-form (see 4.6) and we may take $g = |\omega^g|^2 c$ as the Gauduchon metric.

One easily sees that $\sharp\omega^g$, $J\sharp\omega^g$ are commuting holomorphic Killing fields. Let $\mathcal{B}$ be the foliation generated by $\sharp\omega^g$ and let $\mathcal{E}$ be generated by $\sharp\omega^g$ and $J\sharp\omega^g$.

PROPOSITION 8.2. **[63]** *Let M be a compact Kähler Einstein-Weyl manifold of dimension $n = 2m > 4$ which is not exact. If the leaves of $\mathcal{B}$ and $\mathcal{E}$ are compact then there is a commutative diagram*

$$\begin{array}{ccc} M & \longrightarrow & P \\ & \searrow & \downarrow \\ & & N \end{array}$$

where $P = M/\mathcal{B}$ and $N = M/\mathcal{E}$ are Einstein orbifolds with positive scalar curvature and N is Kähler.

PROOF. Let ν be the 1-form $-\omega^g \circ J$. Then $h = g - (\omega^g)^2 - \nu^2$ descends to a Hermitian metric on N with Kähler form $\Omega_g + \omega^g \wedge \nu = d\nu$, where Ω_g is the Kähler form of J with respect to g. Since $d\omega^g = 0$, the Einstein-Weyl equation on M implies that $g_P = h + \nu^2$ is an Einstein metric on P with scalar curvature $(n-1)(n-2)$. Similarly, the submersion formulae for $P \to N$ show that N is also Einstein. □

The example to keep in mind is $M = S^1 \times S^{2n-1} = (\mathbb{C}^n \setminus \{0\})/\mathbb{Z}$ where the $\mathbb{Z}$ action is generated by $x \mapsto 2x$. Then $P = S^{2n-1}$ and $N = \mathbb{C}P^{n-1}$.

Conversely, if N is a Kähler-Einstein manifold, then the Calabi metric on $\mathcal{L}\backslash 0$, where $\mathcal{L}$ is a maximal root of the canonical bundle of N, gives a Kähler Einstein-Weyl structure on the universal cover of M **[63]**.

Next we turn to the quaternions.

DEFINITION 8.3. A conformal manifold (M, c) of dimension $n > 4$ will be called *quaternion Kähler Weyl* iff it is equipped with a rank 3 sub-bundle $\mathcal{Q} \leqslant \mathfrak{so}(TM)$ pointwise isomorphic to $\operatorname{Im}\mathbb{H} = \mathfrak{sp}(1)$, and a Weyl connection D preserving $\mathcal{Q}$. It is (locally) *hyper-Kähler Weyl* iff the induced covariant derivative on $\mathcal{Q}$ is (locally) trivial. (We discuss the four dimensional case in section 9.)

Since D is torsion free, a quaternion Kähler Weyl manifold is quaternionic and a (locally) hyper-Kähler Weyl manifold is (locally) hyper-complex.

PROPOSITION 8.4. **[63]** *Let M be a conformal manifold with* $\dim M > 4$. *Then M is quaternion Kähler Weyl iff it is locally quaternion Kähler, in which case it is closed Einstein-Weyl. A non-exact quaternion Kähler Weyl manifold is locally hyper-Kähler Weyl, and any locally hyper-Kähler Weyl manifold is locally hyper-Kähler.*

PROOF. If M is quaternion Kähler Weyl then the weightless 4-form Ω of the quaternionic structure satisfies $d^D\Omega = 0$, so $F^D \wedge \Omega = 0$ and therefore $F^D = 0$ if $\dim M > 4$. Parallel local length scales are therefore Einstein, and so D is Einstein-Weyl. If D is not exact then scal^D must vanish by Theorem 3.6. □

Assuming D is not exact, we can again use the Gauduchon metric with $|\omega^g| = 1$ and consider the foliation $\mathcal{B}$, as in the complex case. Also, let $\mathcal{D}$ be the foliation given by the quaternionic span of $\sharp\omega^g$. Then Proposition 8.2 has a quaternionic analogue. We concentrate on the following results of Ornea and Piccinni [**59**].

PROPOSITION 8.5. *Let M be a compact quaternion Kähler Weyl manifold such that the foliations $\mathcal{B}$ and $\mathcal{D}$ have compact leaves. Then there is a finite hyper-Kähler Weyl covering $\tilde{M}$ of M and a commutative diagram*

$$\begin{array}{ccccc} \tilde{M} & \stackrel{S^1}{\longrightarrow} & \tilde{P} & \stackrel{S^3/H}{\longrightarrow} & \tilde{N} \\ \downarrow & & \downarrow & & \downarrow \\ M & \stackrel{S^1}{\longrightarrow} & P & \stackrel{S^3/G}{\longrightarrow} & N \end{array}$$

with finite coverings as vertical arrows and Riemannian submersions over orbifolds as horizontal arrows. The orbifolds P and $\tilde{P}$ carry respectively local and global 3-Sasakian structures, while N and $\tilde{N}$ are quaternion Kähler orbifolds with positive scalar curvature. The fibers of $P \to N$ and $\tilde{P} \to \tilde{N}$ are spherical space forms, respectively locally and globally homogeneous. On M there is a global integrable compatible complex structure and $M/\mathcal{E}$ is the twistor space of N (see Proposition 8.2).

PROOF. Let $M \stackrel{\pi}{\longrightarrow} P$ be a flat S^1-bundle with connection ω^g. If $(\phi_\alpha, \xi_\alpha)$ is a locally defined 3-Sasakian structure on P, a quaternionic structure on M may be defined by

$$I_\alpha Y = -\phi_\alpha Y - g(\xi_\alpha, Y)\sharp\omega^g, \quad I_\alpha \sharp\omega^g = \xi_\alpha$$

and this is compatible with the metrics $g_M = \pi^* g_P + (\omega^g)^2$.

Now since all the leaves of $P \to N$ are spherical space forms S^3/G, each leaf has a global Sasakian structure induced by a conjugate complex structure on $S^3 \subseteq \mathbb{C}^2$. The unit vector field also belongs to the locally 3-Sasakian distribution in TP. Thus P has a global Sasakian structure which may be lifted to a complex structure J on M using the formulae above. Then $\sharp\omega^g$ and $J\sharp\omega^g$ generate the foliation $\mathcal{E}$ and $M/\mathcal{E}$ is the twistor space of N. $\square$

REMARKS. 1. The relation to 3-Sasakian geometry leads to a classification of compact homogeneous hyper-Kähler Weyl manifolds using the classification of homogeneous 3-Sasakian manifolds [**10, 58**].

2. Similarly, results on Betti numbers of 3-Sasakian geometry imply topological constraints and relations on M and N above [**58, 59**]. One such constraint is $b_1(M) = 1$, which is also known to hold for closed non-exact Einstein-Weyl manifolds for other reasons.

EXAMPLE 8.6. Let G be a discrete subgroup of $\mathrm{GL}(1,\mathbb{H})\mathrm{Sp}(1) = \mathrm{CO}^+(4)$ and let $M = (\mathbb{H}^n \setminus \{0\})/G$, where G is acting diagonally. Equip M with the metric $g = \left(\sum_\alpha q_\alpha \bar{q}_\alpha\right)^{-1} \sum_\alpha dq_\alpha \otimes d\bar{q}_\alpha$. Then (M,g) is a quaternion Kähler Weyl manifold (the Weyl connection coincides locally with the Levi-Civita connection of a quaternion Kähler metric). In fact $G \leqslant \mathrm{GL}(n,\mathbb{H})\mathrm{Sp}(1)$ and by choosing G so it is contained in $\mathrm{GL}(n,\mathbb{H})$, M becomes hyper-Kähler Weyl. If G is chosen inside $\mathrm{GL}(n,\mathbb{H})\mathrm{U}(1) \subseteq \mathrm{GL}(2n,\mathbb{C})$ there is a global integrable complex structure on M [**58**].

9. Einstein-Weyl geometry in four dimensions

On an oriented conformal 4-manifold, the bundle $\Lambda^2 T^*M$ decomposes as the direct sum of the bundles of self-dual and anti-self-dual 2-forms, $\Lambda^2_+ T^*M$ and $\Lambda^2_- T^*M$. For any Weyl derivative D there is a corresponding decomposition $F^D = F^D_+ + F^D_-$ of the Faraday curvature. The same is true of the bundle of Weyl tensors, and so the Weyl curvature W of the conformal structure splits into two components W^+ and W^-. When one of these components vanishes, it is well known [**2, 3**] that the conformal geometry of M may be studied in terms of the holomorphic geometry of an associated complex manifold or "twistor space".

DEFINITION 9.1. The *twistor space* Z of M is the bundle of negatively oriented orthogonal almost complex structures on TM, which is a sphere bundle in $L^2\Lambda^2_- T^*M$. The fibers are the real *twistor lines* of Z.

There is also a spinorial representation of the twistor space. Equipping M (at least locally) with a spin structure, there are weightless spinor bundles V_+, V_- (complex symplectic with rank 2) such that the complexified tangent bundle $\mathbb{C}TM$ is isomorphic to $L^1 \otimes (V_+ \otimes_{\mathbb{C}} V_-)$.

The twistor space Z is then isomorphic to $P(L^w V_-)$ for any w, although the tautological line bundle over this projectivized bundle will depend on the weight. Because the twistor operator is conformally invariant on $L^{1/2}V_- \cong (L^{-1/2}V_-)^*$, the following choice of projective structure is the "right" one.

NOTATION 9.2. If M is a spin manifold, then the twistor space Z of M will be identified with $P(L^{-1/2}V_-)$ and $\mathcal{O}(-1)$ will denote the corresponding tautological line bundle. Note that for m even, $\mathcal{O}(m)$ makes sense globally even if M is not spin. In particular the canonical bundle of Z is $K_Z \cong \mathcal{O}(-4)$. Let $L_{\mathbb{C}}$ denote the complexified pull-back of $L^1 \to M$ to Z. The Euler sequence on each fiber then implies that the vertical tangent bundle $\mathcal{V}$ of $Z \to M$ is $L_{\mathbb{C}}^{-1}\mathcal{O}(2) = L_{\mathbb{C}}^{-1} K_Z^{-1/2}$.

Let us now compare the standard twistor theory on a conformal manifold with twistor theory on a Weyl manifold.

THEOREM 9.3. *Let M be an oriented conformal manifold with twistor space Z.*

(i) [**2, 33**] *Z carries a natural almost complex structure, which is integrable if and only if $W^- = 0$. The complex line bundles $\mathcal{O}(m)$ are then holomorphic, and their (real) holomorphic sections correspond to solutions of conformally invariant differential equations on M* [**22**].

(ii) [**25, 26, 27, 66**] *Suppose now that $W^- = 0$ and let D be a Weyl derivative on M. Then for $w \neq 0$, the complex line bundle $L^w_{\mathbb{C}}$ over Z carries a pre-holomorphic structure depending on D which is integrable if and only if $F^D_- = 0$. Holomorphic sections of $L^w_{\mathbb{C}} \otimes \mathcal{O}(m)$ for $w \neq 0$ then correspond to solutions of differential equations on M depending on D.*

In particular if $W^- = F^D_- = 0$, then D defines a holomorphic structure on $\mathcal{V} = L_{\mathbb{C}}^{-1}\mathcal{O}(2)$ and the holomorphic sections correspond to solutions of a twistor-type equation on the bundle of anti-self-dual endomorphisms of TM. Gauduchon [**25, 26**] has used this equation to prove that a compact self-dual conformal manifold which admits a gauge of negative scalar curvature does not admit a global anti-self-dual complex structure. In [**27**] he also uses twistor theory to analyze conformal vector fields on Weyl manifolds, and to study manifolds of type $S^1 \times S^3$.

The above theorem suggests that we will only be able to obtain a twistor interpretation of the Einstein-Weyl condition when $W^- = F^D_- = 0$. Following the constructions in the metric case [**3**] *mutatis mutandis*, we obtain:

THEOREM 9.4. [**27, 66**] *Let M be an Einstein-Weyl 4-manifold with $W^- = F^D_- = 0$. Then there is a twisted 1-form $\theta \in H^0(Z, \Omega^1 \otimes \mathcal{V})$ which is holomorphic iff M is Einstein-Weyl. Furthermore, $\theta \wedge d\theta \in H^0(Z, L_{\mathbb{C}}^{-2})$ is a nonzero multiple of the pull-back of* scal^D. *In particular θ defines a holomorphic map $Z \to \mathbb{C}P^1$ iff the symmetrized Ricci endomorphism of D is identically zero.*

On a compact 4-manifold M, any Weyl derivative D with self-dual Faraday curvature is closed (since self-dual exact 2-forms must vanish). If M is also Einstein-Weyl then D is flat by 4.9. The self-duality of F^D therefore seems very restrictive, but is inevitable in view of the following result.

THEOREM 9.5. [**12**] *Let M, D be an Einstein-Weyl 4-manifold with self-dual Weyl tensor. Then F^D is also self-dual.*

COROLLARY 9.6. [**27, 67**] *Let M, D be a compact Einstein-Weyl 4-manifold with self-dual Weyl tensor. Then D is closed.*

This corollary may be established directly by using the classification of 4-manifolds admitting self-dual metrics with positive Ricci curvature [**27**], or by using the vanishing of the Bach tensor [**67**].

Even without assuming compactness, Theorem 3.6 shows that if D is Einstein-Weyl with $F^D_- = 0$, then $D\mathrm{scal}^D = 0$ and so either scal^D is identically zero or M is Einstein. In the self-dual case this can be seen on the twistor space as follows [**66**].

Let S be the divisor of $\theta \wedge d\theta$. Then S is either empty or it meets every twistor line [**70**]. However, $L_{\mathbb{C}}^{-2}$ is a pull-back from M and so it is trivial on twistor lines. Therefore, if S is non-empty, it must contain all twistor lines and therefore be all of Z, meaning that $\theta \wedge d\theta$ is identically zero. Otherwise, $\theta \wedge d\theta$ is nowhere vanishing, which implies the existence of a section of $\Omega^1 \otimes K_Z^{-1/2}$ and therefore, by standard twistor theory, an Einstein gauge of nonzero scalar curvature.

Consequently, all non-exact self-dual Einstein-Weyl 4-manifolds are scalar flat with self-dual Faraday curvature. The twistor space fibers over $\mathbb{C}P^1$ and D defines a flat connection on the bundle of anti-self-dual complex structures. Conversely if D is flat on this bundle, W is self-dual and $F^D \wedge \Omega_J = 0$ for each anti-self-dual complex structure J, so F^D is also self-dual. In fact we have the following equivalence.

PROPOSITION 9.7. [**66**] *A Weyl 4-manifold is Einstein-Weyl with $W^- = F^D_- = \mathrm{scal}^D = 0$ iff it is locally hyper-complex with Obata connection D.*

The anti-self-dual complex structures on a 4-manifold M form a bundle isomorphic to the imaginary quaternions and any Weyl derivative D preserves this bundle. By analogy with Theorem 8.4, we say that M is *quaternion Kähler Weyl* iff it Einstein-Weyl with $W^- = F^D_- = 0$ and locally *hyper-Kähler Weyl* if also $\mathrm{scal}^D = 0$.

The compact examples may be classified as follows [**66, 67**].

THEOREM 9.8. *A compact self-dual Einstein-Weyl 4-manifold is isometric to S^4, $\mathbb{C}P^2$ or an Einstein manifold of negative scalar curvature, or is covered by a flat torus, a $K3$ surface or a coordinate quaternionic Hopf surface* [**3, 8**].

One might hope to obtain more examples by replacing self-duality by the vanishing of the *Bach tensor*. This is a symmetric trace-less bilinear form B obtained from the Weyl tensor by applying a conformally invariant second order differential operator. It arises on compact manifolds as the gradient of the functional $\mathrm{c} \mapsto \int_M |W^{\mathrm{c}}|^2$. In terms of an arbitrary Weyl derivative

$$B(X,Y) = \textstyle\sum_i (D_{e_i} C^D_{e_i,X} Y - \langle W_{e_i,X}\, r^D(e_i), Y\rangle)$$

where C^D is the *Cotton-York* tensor of D, which is a vector valued 2-form defined by $C^D_{X,Y} := D_X r^D(Y) - D_Y r^D(X) = -\sum_i D_{e_i} W_{X,Y} e_i$ (by the second Bianchi identity). The Bach tensor may also be computed by applying the same formula to W^+ or W^- and doubling it (see [**12**]) and it therefore vanishes if c is (anti)self-dual. If M is Einstein-Weyl then we can compute B with the help of Proposition 3.3 to obtain the following result.

THEOREM 9.9. [**12, 67**] *Let M be an Einstein-Weyl 4-manifold. Then*

$$B(X,Y) = \tfrac{1}{24}\big(D^2_{X,Y}\mathrm{scal}^D + D^2_{Y,X}\mathrm{scal}^D\big) - \langle F^D_+(X), F^D_-(Y)\rangle - \langle F^D_-(X), F^D_+(Y)\rangle$$

and so if $B = 0$ and $\mathrm{scal}^D = 0$ then F^D is (anti)self-dual [**12**]. *When M is compact the formula for B in the Gauduchon gauge $D = D^g + \omega^g$ becomes*

$$B(X,Y) = \tfrac{2}{3}\mathrm{scal}^D(\omega^g \otimes_0 \omega^g)(X,Y) - \langle F^D_+(X), F^D_-(Y)\rangle - \langle F^D_-(X), F^D_+(Y)\rangle$$

and it follows that if $B = 0$ then D is closed [**67**].

We now turn to four dimensional Kähler Weyl geometry, which is richer than the higher dimensional case of the previous section. Indeed if (c, J) is *any* conformal Hermitian structure on a 4-manifold M, it follows from Example 1.6 that there is a unique Weyl derivative D with $DJ = 0$ [**75**]. We choose the orientation so that J is anti-self-dual. Therefore $\langle F^D_-, J\rangle = 0$, and since $[R^D, J] = 0$, r^D_0 is J-invariant and also (see [**1**])

$$W^- = \tfrac{1}{4}\mathrm{scal}^D\big(\tfrac{1}{3}\mathrm{id}_{\Lambda^2_-} - \tfrac{1}{2}\Omega_J \otimes \Omega_J\big) - \tfrac{1}{2}(JF^D_- \otimes \Omega_J + \Omega_J \otimes JF^D_-).$$

In particular, W^- vanishes iff F^D_- and scal^D both vanish, and so a compact self-dual Hermitian 4-manifold is locally scalar flat Kähler. (See [**7**], and also [**69**] for a twistor proof.)

Despite the wide generality of Kähler Weyl geometry, the Einstein-Weyl condition is much more restrictive. The following result is due to Gauduchon and Ivanov [**28**], although we sketch a different proof.

THEOREM 9.10. *Let M be a compact Kähler Einstein-Weyl 4-manifold. Then the Weyl derivative D is closed.*

PROOF. By 4.4, $D = D^g + \omega^g$ with $K = \sharp\omega^g$ a Killing field, which implies

$$\langle R^g_{JK,K} e_i, J e_i\rangle = \langle D^g_{JK}(D^g K)_{e_i}, J e_i\rangle = -\langle D^g_{e_i} K, (D^g_{JK} J) e_i\rangle.$$

This vanishes, since $D^g_X J + [\omega^g \wedge X, J] = D_X J = 0$ and when $X = JK$ we have $[\omega^g \wedge JK, J] = 0$. Now $\langle R^D_{JK,K} e_i, J e_i\rangle$ is also zero, since $R^D_{X,Y}$ commutes with J. Comparing these using 2.5, we find that $\mathrm{scal}^D |\omega^g|^2 = 0$ and so either D is exact or $\mathrm{scal}^D = 0$. In the latter case D is closed by 3.6. □

This conclusion continues to hold even if the complex structure is not integrable: Kamada [**42**] shows that a compact almost Kähler Einstein-Weyl manifold with non-negative scalar curvature is in fact Kähler Einstein-Weyl.

In contrast to these negative results we now present some non-compact examples and examples which are not (anti)self-dual. One interesting class of examples are those of Bianchi type IX, i.e., admitting a (possibly local) SU(2) action with three dimensional orbits. Apart from the Einstein case, the solutions are all diagonal, biaxial and conformally Kähler [**5, 53**] and so there is actually a (local) U(2) action. Madsen [**53**] obtains the (anti)self-dual examples, and the general solutions can be found in [**4, 54**], although the latter reference is concerned with the compact case, to which we shall return at the end of this section.

We write the solutions in the form $D = D^g + \omega$, where g is a Kähler metric, and we have reorganized Bonneau's parameters to simplify and unify the various cases. We use the coordinates of Madsen; in particular, σ_i are invariant 1-forms with $d\sigma_1 = \sigma_2 \wedge \sigma_3$ etc.

$$\begin{aligned}
(9.1)\quad g &= V(\rho)^{-1}d\rho^2 + \tfrac{1}{4}\rho^2(\sigma_1^2 + \sigma_2^2 + V(\rho)\sigma_3^2)\\
\omega &= -\frac{\rho(b+2c\rho^2)}{a+b\rho^2+c\rho^4}d\rho \pm \tfrac{1}{2}\rho^2 V(\rho)\frac{\sqrt{4ac-b^2}}{a+b\rho^2+c\rho^4}\sigma_3\\
V(\rho) &= \frac{(a+b\rho^2+c\rho^4)\big(a+c\rho^4+\lambda(a-c\rho^4)\big)}{2ac\rho^4}\\
&\quad + \mu\frac{a+b\rho^2+c\rho^4}{\rho^4(b+2c\rho^2)}\left[1 - \frac{4c(a-c\rho^4)}{4ac-b^2}\left(1 - \frac{b+2c\rho^2}{\sqrt{4ac-b^2}}\operatorname{arccot}\frac{b+2c\rho^2}{\sqrt{4ac-b^2}}\right)\right]
\end{aligned}$$

The parameters are constrained by $4ac \geqslant b^2$ and the requirement that $V(\rho)$ should be somewhere positive. The solution is homogeneous in (a, b, c) and is also invariant under the transformation $\rho \mapsto k\rho$, $a \mapsto k^2 a$, $c \mapsto k^{-2}c$, $\mu \mapsto k^2\mu$, $g \mapsto k^2 g$, so there are only three independent parameters. The Einstein case occurs when $b^2 = 4ac$: note that $1 - x \operatorname{arccot} x \approx 1/(3x^2)$ for x large, and so the μ term has a well defined limit. In this situation, and also when $\mu = 0$, $V(\rho)$ is of the form

$$V(\rho) = 1 + A_+\rho^2 + B_+\rho^4 + \frac{A_-}{\rho^2} + \frac{B_-}{\rho^4}$$

where $A_+A_- = 4B_+B_-$ in the Einstein case and $A_+^2B_- + A_-^2B_+ = A_+A_-$ in the case $\mu = 0$. The latter constraint is simply the condition for V to factorize as $(a + b\rho^2 + c\rho^4)\big((1+\lambda)a + (1-\lambda)c\rho^4\big)/2ac\rho^4$. The scalar curvature of the Kähler metric is $-24(A_+ + 2B_+\rho^2)$, but note that replacing ρ by $1/\rho$ and rescaling by ρ^4 gives a metric of the same form with A_+ and B_+ interchanged with A_- and B_-. Consequently these conformal metrics admit Kähler complex structures of both orientations. The $\mu = 0$ Einstein-Weyl structures satisfy $\mathrm{scal}^D = 0$, giving Ricci flat metrics when $b^2 = 4ac$. Also they are conformally Einstein for $\lambda = \pm 1$, when the Weyl curvature is (anti)self-dual. If we fix the orientation by $W^- = 0$, these self-dual conformal structures each admit:

- a compatible Einstein metric [**62**]
- a compatible scalar flat Kähler metric [**49**]
- a compatible self-dual Kähler metric [**4**]
- a compatible hyper-complex structure [**53**].

On the other hand, for $\mu = 0$, $\lambda \neq \pm 1$, $4ac > b^2$, we have examples of scalar flat Einstein-Weyl structures where F^D is *not* (anti)self-dual [**4**].

We turn now to the search for more compact examples and begin by noting that there are topological constraints on compact 4-manifolds admitting Einstein-Weyl structures, given by an analogue of the Hitchin-Thorpe inequality [**32**]. Related to this is a a generalization of the Lafontaine inequality [**47**], and also the fact that four dimensional Einstein-Weyl manifolds minimize a quadratic total curvature functional. These constraints were previously established using the Gauduchon gauge [**64, 65**], but we sketch here how they can be obtained in Weyl geometry. One advantage of this approach is that we find a quadratic total curvature functional minimized by all Einstein-Weyl structures, not just the closed ones.

The key idea is that a Weyl connection is a metric connection on $L^{-1}TM$. Since L^{-1} is a trivializable bundle, the Euler characteristic of M is given by the integral of a multiple of the Pfaffian of $R^{D,0}$. This may be computed by viewing $R^{D,0}$ as a weight -2 endomorphism of Λ^2T^*M and splitting into self-dual and anti-self-dual parts. The Pfaffian of $R^{D,0}$ reduces to $\langle R^{D,0}, *R^{D,0}*\rangle$. In block diagonal form $R^{D,0}$ may be written $\begin{bmatrix} A_+ & B \\ B & A_- \end{bmatrix}$, where $A_\pm$ is given by the action of $W^\pm$, scal^D and $F^D_\pm$, whereas B is given by the action of r^D_0. Hence the Pfaffian integrand is $|R^{D,0}|^2$ with the r^D_0 term negated. A straightforward computation of $\sum_{i<j,k<l} \langle R^{D,0}_{e_i,e_j}e_k, e_l\rangle^2$ gives:

$$|R^{D,0}|^2 = |W^+|^2 + |W^-|^2 + |F^D_+|^2 + |F^D_-|^2 + \frac{1}{24}\left(\mathrm{scal}^D\right)^2 + 2|r^D_0|^2,$$

where $S^2_0T^*M$ is given the tensor product norm, and Λ^2T^*M its usual norm.

It follows that we have the following integral formulae for the Euler characteristic, the signature and the trivial characteristic of L^1:

$$2\chi(M) = \frac{1}{4\pi^2}\int_M |W^+|^2 + |W^-|^2 + |F^D_+|^2 + |F^D_-|^2 + \frac{1}{24}\left(\mathrm{scal}^D\right)^2 - 2|r^D_0|^2$$

$$3\tau(M) = \frac{1}{4\pi^2}\int_M |W^+|^2 - |W^-|^2, \qquad 0 = \frac{1}{4\pi^2}\int_M |F^D_+|^2 - |F^D_-|^2.$$

THEOREM 9.11. *Let M be a compact 4-manifold. Then the quadratic total curvature functional $\int_M |R^{D,0}|^2$ is minimized by Einstein-Weyl structures and also by half conformally flat, scalar flat, closed Weyl structures. If D is Einstein-Weyl then*

$$2\chi(M) \geqslant 3|\tau(M)| + \frac{1}{2\pi^2}\int_M |F^D_\pm|^2$$

with equality iff $\mathrm{scal}^D = 0$ and W is (anti)self-dual. Similarly, if M, D is a Weyl manifold with $\mathrm{scal}^D = 0$ and W (anti)self-dual, then the reverse inequality holds, with equality iff M is Einstein-Weyl.

It follows from this [**64, 67**], that if M is a torus or $K3$ surface then M admits no non-exact Einstein-Weyl structures, and $M\#M$ admits no Einstein-Weyl structures at all. Also $k\mathbb{C}P^2$ can only be Einstein-Weyl for $k \leqslant 3$. Finally, any Einstein-Weyl structure on $S^1 \times S^3$ is closed and therefore flat by 4.9.

We end this section with the classification of compact Einstein-Weyl 4-manifolds with large symmetry group. First let us consider the homogeneous case.

THEOREM 9.12. [**54**] *A compact homogeneous Einstein-Weyl 4-manifold is either finitely covered by $S^1 \times S^3$ with its standard Einstein-Weyl structure or is a homogeneous Einstein manifold.*

PROOF. Assume D is not exact and let $M = G/H$ where G is the symmetry group. Theorem 9.8 implies that the only conformally flat Einstein-Weyl structures on S^4 are the Einstein metrics, and so (as noted in 6.1) we may assume G is compact. Let $\mathfrak{m}$ be an Ad_H invariant complement to $\mathfrak{h}$ in $\mathfrak{g}$. Then $\mathfrak{m} = \ker\omega^g \oplus (\ker\omega^g)^\perp$ where ω^g is the Gauduchon 1-form. Therefore $\mathfrak{h} < \mathfrak{o}(3) \oplus \mathfrak{o}(1)$ so the rank of $\mathfrak{h}$ is at most 1 and $\dim\mathfrak{g}$ at most 7. The classification of compact Lie groups now implies that we only need to consider a few cases which either gives M finitely covered by $S^1 \times S^3$ or $b_1(M) \geqslant 2$. But we have seen that $b_1(M) \leqslant 1$ for non-exact Einstein-Weyl manifolds with equality iff M is flat (see 4.9 and 9.11). Indeed, the manifolds of type $S^1 \times S^3$ exhaust the compact closed Einstein-Weyl manifolds. □

Inspired by the work of Bérard-Bergery **[3]** on Einstein manifolds with large symmetry group, we now consider the following situation.

THEOREM 9.13. **[54]** *Let G be the symmetry group of a compact four dimensional inhomogeneous Einstein-Weyl manifold with non-closed structure and assume that* $\dim G \geqslant 4$. *Then the Einstein-Weyl structure is of co-homogeneity one and it is defined on S^4, $\mathbb{C}P^2$, $S^2 \times S^2$, $\mathbb{C}P^2\#\overline{\mathbb{C}P}^2$ or some of their finite quotients. The solutions in each case come in one dimensional families.*

PROOF. If M is not homogeneous then as G preserves the metric on the principal orbit P^n, and so we must have $4 \leqslant \dim G \leqslant \frac{1}{2}n(n+1)$ and hence $n = \dim P = 3$. There are now only the following cases to consider:

- $\mathrm{SO}(4)$ with principal orbit $S^3 = \mathrm{SO}(4)/\mathrm{SO}(3)$
- $S^1 \times \mathrm{SO}(3)$ with orbit $S^1 \times S^2 = S^1 \times \mathrm{SO}(3)/\mathrm{SO}(2)$
- $\mathrm{U}(2)$ with orbit $S^3 = \mathrm{U}(2)/\mathrm{U}(1)$

or finite quotients of these. We have studied Einstein-Weyl manifolds with this kind of symmetry in section 6: the Einstein-Weyl equation reduces to a collection of ODEs over a closed interval or circle, the latter case yielding only closed structures. When $M/G = [0,\ell]$, it is convenient to write $M = [\,G/K_1\,|\,G/H\,|\,G/K_2\,]$ for the manifold with principal orbit G/H and special orbits G/K_i, $i = 1, 2$ at the endpoints.

For each symmetry group we classify the possible diffeomorphism types using Lie theory and the known topological constraints on Einstein-Weyl geometry. Then we impose the appropriate boundary conditions on the ODEs and solve explicitly. The case of $\mathrm{SO}(4)$ symmetry yields only closed Einstein-Weyl structures so let us consider $S^1 \times \mathrm{SO}(3)$ symmetry.

Some of the topologies here do not carry any Einstein-Weyl solutions. Firstly, if M/G is a circle then M is finitely covered by $T^2 \times S^2$ which cannot be Einstein-Weyl. When M/G is an interval with special orbits $\mathbb{R}P^1 \times S^2$ we have:

$$\begin{aligned} M &= [\,\mathbb{R}P^1 \times S^2\,|\,S^1 \times S^2\,|\,\mathbb{R}P^1 \times S^2\,] \\ &= [\,\mathbb{R}P^1\,|\,S^1\,|\,\mathbb{R}P^1\,] \times S^2 \\ &= ([\,\mathbb{R}P^1\,|\,S^1\,|\,\mathrm{pt}\,]\#[\,\mathrm{pt}\,|\,S^1\,|\,\mathbb{R}P^1\,]) \times S^2 \\ &= K^2 \times S^2, \end{aligned}$$

where K^2 is the Klein bottle. However, K^2 is double-covered by T^2 and $T^2 \times S^2$ is not Einstein-Weyl. Also, not all finite quotients of $S^1 \times S^2$ are possible principal orbits. For instance:

$$\begin{aligned} M &= [\,\mathbb{R}P^1 \times \mathbb{R}P^2 \,|\, S^1 \times \mathbb{R}P^2 \,|\, \mathbb{R}P^1 \times \mathbb{R}P^2\,] \\ &= K^2 \times \mathbb{R}P^2 \end{aligned}$$

which again cannot be Einstein-Weyl. The remaining cases give one parameter families on S^4, $S^2 \times S^2$ and some finite quotients, such as $\mathbb{R}P^2 \times S^2$. The family on S^4 was given in (6.1).

The U(2) symmetric examples are obtained from the family given in (9.1). For S^4, $\mathbb{C}P^2$, and $\mathbb{C}P^2\#\overline{\mathbb{C}P}^2$, the boundary value problem leads to one parameter families of solutions, and these and these descend to the finite quotients $\mathbb{R}P^4$ and $\mathbb{C}P^2\#\mathbb{R}P^4$.

We refer to [**54**] for the full details of all the cases, but note that this reference contains some errors in the U(2) case, corrected by Bonneau [**6**]. □

10. Einstein-Weyl geometry in three dimensions

In three dimensions, there is also a twistor theory of Einstein-Weyl manifolds, but unlike the four dimensional case, where twistor methods are limited to the self-dual structures, in three dimensions "mini-twistor theory" applies to all Einstein-Weyl spaces. Indeed this was the case first studied, by Cartan [**16**], who showed that the Einstein-Weyl equation is the integrability condition for the existence, in a complex three dimensional Weyl manifold, of a two parameter family of totally geodesic null hypersurfaces. Consequently, the space of oriented geodesics in a real three dimensional Einstein-Weyl manifold is a complex surface. Hitchin showed that this surface contains projective lines with normal bundle $\mathcal{O}(2)$ and conversely, that given such a complex surface (with a real structure), the real points in the Kodaira moduli space of these lines form a three dimensional Einstein-Weyl manifold [**34**]. In other words there is a twistor construction, the *Hitchin correspondence*, for three dimensional Einstein-Weyl manifolds, in terms of a class of complex surfaces called *mini-twistor spaces*. The conformal structure of the Einstein-Weyl space is given by the condition for nearby "mini-twistor lines" to intersect to second order, and the Weyl derivative can be obtained via a construction of projective structures on moduli spaces [**55**].

For example, the quadric surface $\mathbb{P}^1 \times \mathbb{P}^1$, together with the plane sections, generates the Einstein space S^3 or H^3 depending on the real structure, and the mini-twistor space of $\mathbb{R}^3$ is the punctured cone $T\mathbb{P}^1$, together with its sections over $\mathbb{P}^1$. The following result shows that other mini-twistor spaces are more complicated.

PROPOSITION 10.1. *A mini-twistor space which is an open set of a compact surface generates the Einstein-Weyl geometry of a space of constant curvature. The compact surface can be taken to be the cone or the quadric surface.*

Despite this, we can construct mini-twistor spaces locally by taking blow-ups and branched covers. For instance, a $(1,n)$-curve in $\mathbb{P}^1 \times \mathbb{P}^1$ is rational with normal bundle $\mathcal{O}(2n)$. If we take a branched n-fold covering, then in the covering the normal bundle is $\mathcal{O}(2)$ and we have a mini-twistor space [**61**], although the covering cannot extend to all of the quadric.

There are close connections between mini-twistor theory and twistor theory in four dimensions. In [**40**], Jones and Tod observed that, given a self-dual conformal 4-manifold M with a conformal vector field K, the quotient of the twistor space Z

of M by the induced holomorphic vector field is a mini-twistor space. They then wrote down a Weyl structure on the orbit space $B = M/K$ and showed that this agreed with the Weyl structure coming from the Hitchin correspondence. In other words, the quotient of a self-dual conformal manifold by a conformal vector field is Einstein-Weyl.

Although such a result would have been difficult to find without twistor theory, the twistor theoretic proof that the Jones-Tod Weyl structure is Einstein-Weyl is rather indirect. More direct arguments, sometimes only in special cases, have been given in [**14, 29, 41, 50**] and we would like to sketch the approach of [**14**], which has the advantage that it extends to a more general class of conformal submersions [**13**], although we shall treat only conformal vector fields here.

Let M be a self-dual conformal manifold with a conformal vector field K, and by restricting to an open set if necessary, assume K is nowhere vanishing. Then $|K|$ is a length scale on M and induces an exact Weyl derivative D^0, the constant length gauge of K. One can compute D^0 in terms of an arbitrary Weyl derivative D by the formula

$$D^0 = D - \frac{\langle DK, K\rangle}{\langle K, K\rangle} = D - \frac{1}{4}\frac{(\operatorname{tr} DK)K}{\langle K, K\rangle} + \frac{1}{2}\frac{(d^D K)(K, .)}{\langle K, K\rangle}.$$

Note that $\langle D^0 K, .\rangle$ is a weightless 2-form. The crucial observation is that there is a unique Weyl derivative D^{sd} on M such that $\langle D^{sd} K, .\rangle$ is a weightless *self-dual* 2-form. One way to see this is to observe that $\omega = (*d^D K)(K, .)/\langle K, K\rangle$ is a 1-form independent of the choice of D and define:

$$D^{sd} = D^0 - \frac{1}{2}\omega = D - \frac{1}{4}\frac{(\operatorname{tr} DK)K}{\langle K, K\rangle} + \frac{1}{2}\frac{(d^D K)(K, .) - (*d^D K)(K, .)}{\langle K, K\rangle}.$$

Since D is arbitrary, we may take $D = D^{sd}$ to see that $(D^{sd}K - *D^{sd}K)(K, .) = 0$ from which it is immediate that $D^{sd}K = *D^{sd}K$ since an anti-self-dual 2-form is uniquely determined by its contraction with a nonzero vector field.

Next recall that for any vector field K and torsion free connection D on TM, $(\mathcal{L}_K D)_X = D_X DK - R^D_{X,K}$. There is an analogous formula for Weyl derivatives.

PROPOSITION 10.2. *Let X be a vector field, μ a section of L^w and D a Weyl derivative on M^n. Then $\mathcal{L}_X \mu = D_X \mu - \frac{w}{n}(\operatorname{div}^D X)\mu$ and so the Lie derivative of the Weyl derivative on L^1 is: $(\mathcal{L}_K D)_X = \frac{1}{n}\partial_X(\operatorname{div}^D K) - F^D(X, K)$.*

Now if K is conformal then the Lie derivative (along K) of a Weyl connection D on TM is given by the linearized Koszul formula applied to the Lie derivative of D on L^1. Hence

$$D_X DK = R^D_{X,K} + \gamma_K(X)\mathrm{id} + \gamma_K \vartriangle X,$$

where $\gamma_K = \frac{1}{n}d(\operatorname{tr} DK) + F^D(K, .)$. (This formula also appears in [**27**].)

Applying this with $D = D^{sd}$ and decomposing the curvature gives:

$$D^{sd}_X D^{sd} K = W_{X,K} + r^{sd}(K) \vartriangle X - r^{sd}(X) \vartriangle K + F^{sd}(K, .) \vartriangle X.$$

Now $D^{sd}_X D^{sd} K$ and $W_{X,K}$ are both self-dual 2-forms and hence so is the sum of the remaining terms. This implies that if $\langle X, K\rangle = \langle Y, K\rangle = 0$ then

$$r^{sd}(X, Y)\langle K, K\rangle + r^{sd}(K, K)\langle X, Y\rangle = *\big(K \wedge (r^{sd} + F^{sd})(K) \wedge X \wedge Y\big).$$

Symmetrizing in X, Y, we see that the horizontal part of the symmetric Ricci endomorphism of D^{sd} is a multiple of the identity. It now looks as if $D^{sd} = D^0 - \frac{1}{2}\omega$

might be the Einstein-Weyl structure we seek. In fact this is not the case: instead it is $D^0 - \omega$ which is Einstein-Weyl on B.

THEOREM 10.3. [**40**] *Suppose M is a self-dual 4-manifold and K a conformal vector field such that $B = M/K$ is a manifold. Let D^0 be the constant length gauge of K and $\omega = 2(*D^0K)(K,.)/\langle K,K\rangle$. Then $D = D^0 - \omega$ is Einstein-Weyl on B and D^0 is a Gauduchon gauge.*

Conversely, if (B,D) is an Einstein-Weyl 3-manifold and $\mathrm{w} \in C^\infty(B,L^{-1})$ is a non-vanishing solution of the monopole equation $d{*}D\mathrm{w} = 0$ then there is a self-dual 4-manifold M with symmetry over B such that $*D\mathrm{w}$ is the curvature of the connection defined by the horizontal distribution.

PROOF. The conformal structure and Weyl derivative descend to B because K is Killing in the constant length gauge and ω is a basic 1-form. The first submersion formula in 5.6 relates the Ricci curvature of D on B to that of D^{sd} on M:

$$\operatorname{sym}\mathrm{Ric}^D_B(X,Y) = \operatorname{sym}\mathrm{Ric}^{sd}_M(X,Y) + 2\langle D^0_X K, D^0_Y K\rangle + \tfrac{1}{2}\omega(X)\omega(Y) + \mu^{-2}\langle X,Y\rangle$$

for some section μ of L^1. We have shown that $\operatorname{sym}\mathrm{Ric}^{sd}(X,Y)$ is a multiple of $\langle X,Y\rangle$. Since $D^0_K K = 0$, ω vanishes on the plane spanned by D^0K, and so by comparing the lengths of ω and D^0K one verifies that $2\langle D^0_X K, D^0_Y K\rangle + \tfrac{1}{2}\omega(X)\omega(Y)q$ is also a multiple of $\langle X,Y\rangle$, and hence B is Einstein-Weyl. Now D^0K is a closed 2-form with respect to D^0 on M, so ω is co-closed with respect to D^0 on B and D^0 is a Gauduchon gauge. Finally one sees that no information is lost in this construction. Indeed if $*D\mathrm{w} = d\theta$ (locally) then the metric $g_M = \pi^*\mathrm{w}^2 c_B + (dt+\theta)^2$ is self-dual and $\partial/\partial t$ is a unit Killing field. (More invariantly, let G be the group of D^0-parallel sections of L^1 under addition so that M is a principal G-bundle. Then the monopole equation $*D\mathrm{w} = \Omega$, with Ω closed, couples a relative length scale $\mathrm{w} : L^1 \to M \times_G \mathfrak{g}$ to the curvature Ω of a principal connection on M.) □

Two special cases of this construction have received particular attention. The first is the case of a scalar flat Kähler 4-manifold with a Killing field. In this case, the Einstein-Weyl structure on B, which we call a *LeBrun-Ward* geometry [**50, 79**], is given locally by

$$\begin{aligned} g &= e^u(dx^2 + dy^2) + dz^2 \\ \omega &= -u_z dz \end{aligned} \tag{10.1}$$

where $D = D^g + \omega$ and where u satisfies the Toda equation

$$u_{xx} + u_{yy} + (e^u)_{zz} = 0.$$

Consequently these Einstein-Weyl geometries are also said to be *Toda*. Examples can be found in [**15, 73, 79**].

Corresponding to a solution of the monopole equation $d{*}D\mathrm{w} = 0$ on B, is the scalar-flat Kähler manifold M given by the metric

$$g = e^u\mathrm{w}(dx^2 + dy^2) + \mathrm{w}dz^2 + \mathrm{w}^{-1}(dt+\theta)^2$$

and $\partial/\partial t$ is a Killing field. In this gauge, the monopole equation turns out to be equivalent to the linearized Toda equation

$$\mathrm{w}_{xx} + \mathrm{w}_{yy} + (e^u\mathrm{w})_{zz} = 0.$$

It follows that $\mathrm{w} = u_z$ is a distinguished monopole on B; if this monopole is used to construct M, then M is found to be hyper-Kähler [**9, 50**].

The LeBrun-Ward spaces may be characterized invariantly as the Einstein-Weyl spaces locally fibering as a conformal submersion with geodesic one dimensional fibers and integrable horizontal distribution (i.e., they admit a shear-free, twist-free congruence of geodesics). In the above description, these geodesics are the curves of constant (x, y) [**73**].

The extra data on the mini-twistor space S given by this Toda structure is a real holomorphic section of $K_S^{-1/2}$ and the particular form in (10.1) is obtained by choosing a holomorphic coordinate $x + iy$ on the corresponding divisor.

Mini-twistor theory can be used to prove some of these claims [**50, 51**]. The monopole solution w is given, via the mini-twistor Ward correspondence, by a holomorphic line bundle $\mathcal{L} \to S$ with $c_1(\mathcal{L}) = 0$. If N denotes the normal bundle to the lifted mini-twistor lines, then the obstruction to the splitting of

$$0 \to \mathcal{O} \to N \to \mathcal{O}(2) \to 0$$

over a twistor line $\mathbb{C}P^1_x$ is an element of $H^1(\mathbb{C}P^1_x, \mathcal{O}(-2))$ and may be identified with $\mathrm{w}(x)$. Therefore, for $\mathrm{w}(x) > 0$, $N \cong \mathcal{O}(1) \oplus \mathcal{O}(1)$ and so $\mathcal{L}\backslash 0$ is a twistor space. The two orientations of the distinguished family of geodesics correspond to two curves $\mathcal{C}, \overline{\mathcal{C}}$ in S and $x + iy$ is a complex coordinate on $\mathcal{C}$. We shall now show that the line bundle represented by the divisor $\mathcal{C} + \overline{\mathcal{C}}$ is $K_S^{-1/2}$.

Choose a monopole (w, θ) and consider the twistor space Z of the corresponding scalar-flat Kähler metric. The vector field $\partial/\partial t$ lifts to Z so we may assume Z is a line bundle over S. Let $\mathcal{D} \subseteq Z$ be the section of $Z \xrightarrow{\pi} M$ corresponding to the complex structure on M. The projection $Z \to S$ maps a complex structure J at a point of M to the geodesic in B in the direction $J\frac{\partial}{\partial t}$. The image of $\mathcal{D}$ is therefore $\mathcal{C}$ and $\overline{\mathcal{D}}$ maps to $\overline{\mathcal{C}}$. From [**69**] we know that $[\mathcal{D} + \overline{\mathcal{D}}] = K_Z^{-1/2}$ and so, since the vertical tangent bundle of $Z \to S$ is trivial, it follows that $[\mathcal{C} + \overline{\mathcal{C}}] = K_S^{-1/2}$.

The second special case is the case of hyper-complex 4-manifolds with tri-holomorphic conformal vector fields. These were studied in connection with local heterotic geometries by Chave, Tod and Valent in [**17**]—see also [**74**]. In [**29**], Gauduchon and Tod showed that the Einstein-Weyl quotients arising in this situation are characterized by the presence of what might be called a "scalar curvature monopole": the scalar curvature is nonnegative and if $\kappa^2 = \frac{1}{6}\mathrm{scal}^D$ then κ satisfies the special monopole equation $*D\kappa = \frac{1}{2}F^D$. Together with the Einstein-Weyl equation, this is equivalent to the flatness of the connection $D - \kappa *1$ on $L^{-1}TM$ and the parallel weightless unit vector fields are shear-free divergence-free geodesic congruences. We call these Einstein-Weyl spaces *Gauduchon-Tod* geometries or say that they are *hyper-CR*. Their mini twistor spaces fiber over $\mathbb{C}P^1$ and the only compact examples, apart from the manifolds of constant nonnegative curvature, are $S^1 \times S^2$, the Berger spheres [**29**], and some finite quotients of these.

The total space M of an arbitrary monopole over a Gauduchon-Tod geometry carries a hyper-complex structure, and this provides an example of the Ansatz we have given in 5.6. If the scalar curvature monopole itself is used, then M turns out to be hyper-Kähler with a tri-holomorphic homothetic vector field.

There are clearly close parallels between these two cases.

- The LeBrun-Ward structures arise as quotients of scalar flat Kähler manifolds by a holomorphic Killing field. They have a special monopole (given in terms of a solution to the Toda field equation) which may be used to construct a hyper-Kähler manifold with a holomorphic Killing field.

- Hyper-complex manifolds with tri-holomorphic conformal Killing fields give rise to Gauduchon-Tod structures on the space of orbits. Again there is a special monopole (namely the scalar curvature monopole) leading to a hyper-Kähler metric, this time with a tri-holomorphic homothetic vector field.

The two situations may be unified and generalized by considering a self-dual 4-manifold with an anti-self-dual complex structure and a holomorphic conformal vector field. It can be shown [**14**] that the complex structure induces a shear-free geodesic congruence on the quotient Einstein-Weyl geometry B. The twist κ and divergence τ of this congruence turn out to be "special" monopoles on B. The κ monopole, if nonzero, gives a scalar flat Kähler 4-manifold over B with a holomorphic conformal vector field, while the τ monopole, if nonzero, gives a hyper-complex 4-manifold over B with a holomorphic conformal vector field. Hyper-Kähler manifolds are obtained when κ and τ are linearly dependent.

We have seen that hyper-Kähler 4-manifolds with special conformal vector fields give rise to interesting Einstein-Weyl geometries. It is natural to ask which geometries arise as (local) quotients of $\mathbb{R}^4$. Now, $\mathbb{R}^4$ is conformal to S^4 (minus a point) and so this question has been answered by Pedersen and Tod in [**68**]. Viewing S^4 as the light-cone in $\mathbb{R}^{5,1}$, conformal vector fields correspond to elements of the Lie algebra $\mathfrak{so}(5,1)$. There are no globally non-vanishing conformal vector fields and so, since conjugate elements of $\mathfrak{so}(5,1)$ will produce equivalent quotients, we may conjugate into a normal form in which they vanish at ∞ and then stereographically project. There are essentially three distinct cases: the hyperbolic elements (with a nontrivial infinitesimal dilation); the elliptic elements (generating rotations); and the parabolic elements (generating transrotations). The corresponding Einstein-Weyl geometries are given explicitly in [**68**], as Cases (1, $a \neq 0$), (1, $a = 0$) and (2) respectively. The generic case is the hyperbolic case, which gives a two dimensional moduli space of Einstein-Weyl structures near the Einstein metric on S^3.

In fact, these quotients of $\mathbb{R}^4$ exhaust the possible geometries on compact Einstein-Weyl manifolds.

THEOREM 10.4. *Let B, D be an Einstein-Weyl 3-manifold with Killing gauge $D = D^g + \omega^g$. Then B is locally isomorphic, as a Weyl manifold, to the quotient of an open subset of $\mathbb{R}^4$ by a conformal vector field with its induced Einstein-Weyl structure.*

PROOF. Let M be the total space of the monopole given by D^g. Then, by the inverse Jones-Tod construction, M is a self-dual conformal 4-manifold. However, since D^g is a Killing gauge, $D^g - \omega^g$ is also Einstein-Weyl. Now if $*\omega^g = dA$ then $*(-\omega^g) = d(-A)$ and $dt + (-A) = -(d(-t) + A)$. Hence changing the sign of ω^g does not alter the conformal structure on M, only the orientation. Therefore M is both self-dual and anti-self-dual, and thus conformally flat. The local isomorphisms are now given by conformal charts on M. □

This theorem was originally established by Tod [**72**] as a consequence of his classification of the possible local geometries on compact Einstein-Weyl 3-manifolds. He did this by solving the Einstein-Weyl equation in the Gauduchon gauge, using the fact that B fibers locally over a surface since ω^g is a Killing field. The freedom in the choice of isothermal coordinates on this surface may be used to reduce the Einstein-Weyl equation to an ODE, which is readily integrated. It is perhaps worth remarking that the additional symmetry which arises comes from the Faraday 2-form: generically $*F^D$ and ω^g are dual to linearly independent Killing fields. These generic solutions are of the form [**72**]:

$$g = P(v)^{-1}dv^2 + P(v)dy^2 + v^2(dt + Cv^{-2}dy)^2$$
$$\omega = 2\lambda v^2(dt + Cv^{-2}dy),$$

where $$P(v) = -\lambda^2 v^4 + Av^2 + B - C^2 v^{-2}$$

and λ, A, B, C are arbitrary constants. The isothermal coordinates (x, y) can be found by solving the equation $v'(x) = P(v)$. Another change of coordinates relates these geometries to the quotients of S^4 in [**68**]. The parameters λ, A, B, C above are related to the parameters a, b, c in [**68**] by:

$$A = -a^2 + b^2 + c^2, \quad \lambda^2 B = a^2b^2 + a^2c^2 - b^2c^2, \quad \lambda^4 C^2 = a^2b^2c^2.$$

Examining Tod's argument, we find that the Gauduchon constant is $-6A$ and so $\mathrm{scal}^D = -6A + 3|\omega|^2 = 6(a^2 - b^2 - c^2 + 2\lambda^2 v^2)$. Also, the range of $\lambda^2 v^2$ when $P(v) \geqslant 0$ is the interval $[b^2, c^2]$. Therefore, for $|b^2 - c^2| > a^2$, the scalar curvature has non-constant sign [**12**].

In particular, there are Einstein-Weyl geometries globally defined on S^3 with scalar curvature of non-constant sign, contrary to remarks made in [**67, 68**]. Such examples are "far" from the Berger spheres, which are given by $b^2 = c^2$ (and $a^2 \neq 0$), but include some examples in the one parameter family given in [**52**].

We would also like to emphasize that, although most of the solutions above are globally defined on S^3, Tod's result [**72**] claims only to classify the local forms of solutions which can exist on compact manifolds. It should be possible to work out which compact 3-manifolds carry which local forms using the co-homogeneity one torus action given by the Killing fields $\partial/\partial y$ and $\partial/\partial t$, but care needs to be taken when considering the possible flows of $\partial/\partial t$ in this torus.

We now briefly treat the two dimensional case, where matters are simplified by the fact that the only compact 2-manifolds admitting metrics with Killing fields are S^2 and $S^1 \times S^1$. Hence only these manifolds can admit non-exact Einstein-Weyl structures.

In the Gauduchon gauge (g, ω), $\sharp_g\omega$ is holomorphic, so we may locally choose a complex coordinate $x + it$ such that $\sharp_g\omega = \partial/\partial t$. The Einstein-Weyl equation 3.4 immediately reduces to an ODE for a function of x, and we find [**11**]

$$g = P(v)^{-1}dv^2 + v^2 dt^2$$
$$\omega = Av^2\, dt,$$

where $$P(v) = -A^2v^4 + Bv^2 + C$$

and A, B, C are arbitrary constants. This time $v'(x)^2 = P(v)v^2$, but it is perhaps simpler to introduce a new coordinate r by $v'(r)^2 = P(v)$. The metric is now

$$g = dr^2 + v(r)^2 dt^2$$

and $v(r)$ is an elliptic function since P is a quartic polynomial. In terms of Jacobian elliptic functions (assuming $P(v)$ is somewhere positive),

$$v(r) = \begin{cases} \lambda\,\mathrm{cn}(\mu r + \alpha, k) \quad \text{or} \quad \lambda\,\mathrm{sd}(\mu r + \alpha, k) \quad \text{if } C > 0 & (1) \\ \lambda\,\mathrm{dn}(\mu r + \alpha, k) \quad \text{or} \quad \lambda\,\mathrm{nd}(\mu r + \alpha, k) \quad \text{if } C < 0 & (2), \end{cases}$$

where α is a constant of integration and λ, μ, k are constants depending on A, B, C. The two forms given in each case are equivalent by period translation, but behave differently in the limit $k \to 1$ when the (real) period becomes infinite. The Gauduchon constant is $-2B$ which is always negative in (2), but is proportional to $1 - 2k^2$ in (1). If w ranges over a half-period of cn or sd, (1) gives a family of global solutions on S^2, whereas dn and nd are periodic and non-vanishing, and so the solutions in (2) are defined on $S^1 \times S^1$. In particular, there are non-closed Einstein-Weyl structures on $S^1 \times S^1$ in stark contrast to the situation on $S^1 \times S^{n-1}$ for $n \geqslant 4$.

11. Further horizons

Notwithstanding the pioneering work of Cartan and Hitchin, a detailed understanding of the nature of Einstein-Weyl spaces from a differential geometric point of view has only matured during the last fifteen years. We believe that there is now a good supply of concepts, examples and results about Einstein-Weyl geometry. Nevertheless many basic questions remain unanswered and there are interesting avenues still to be explored.

1. Is there a Lagrangian for the Einstein-Weyl equations? The calculations made to date suggest that the Einstein-Weyl equations may not be the Euler-Lagrange equations of a natural functional, but there are at least interesting functionals which have Einstein-Weyl spaces as a special class of minima (see [**36**] for the Euler-Lagrange equations of total curvature functionals).

2. Three dimensional Einstein-Weyl manifolds with a Gauduchon gauge correspond to four dimensional self-dual manifolds with symmetry, and these minimize the L^2-norm of the Weyl curvature. The critical points of this functional are the Bach flat manifolds. What is the symmetry reduction of this functional and of the Bach flatness condition?

3. The classification of compact three dimensional Einstein-Weyl manifolds needs to be completed.

4. So far, interesting interactions of *non-closed* Einstein-Weyl geometry with special conditions on *compact* 4-manifolds have not been found. Nevertheless there is a good supply of highly symmetric examples and so we would like to know what special properties they have.

5. We have seen that there are global obstructions to the existence of Einstein-Weyl structures, but the question of local existence of Einstein-Weyl structures compatible with a given conformal structure is a nontrivial problem. In [**19, 20**], Eastwood and Tod have shown that there are conformal structures which do not admit compatible Einstein-Weyl structures even locally. For instance, in three dimensions, the general left invariant metric on S^3 and Thurston's Sol geometry do not admit Einstein-Weyl structures. In four dimensions the product of two spheres of different sizes is not locally Einstein-Weyl. Nevertheless, local questions still remain. We would like to know, for instance, whether the scalar curvature of a Bach flat Einstein-Weyl structure is necessarily zero. A similar question can be asked about Kähler Einstein-Weyl structures.

6. The theory of submersions between Einstein-Weyl spaces with one dimensional fibers has not been studied when the total space is even dimensional, except in the case of self-dual 4-manifolds, when we have hyper-complex structures and self-dual

Einstein metrics with symmetry (the latter are conformally scalar flat Kähler). The Ansatz we have presented in 5.6 might be useful for generalizing these situations.

References

[1] V. Apostolov and P. Gauduchon, *The Riemannian Goldberg-Sachs theorem*, Int. J. Math. **8** (1997) pp. 421–439.

[2] M. F. Atiyah, N. J. Hitchin and I. M. Singer, *Self-duality in four dimensional Riemannian geometry*, Proc. Roy. Soc. London **A 362** (1978) pp. 425–461.

[3] A. L. Besse, *Einstein manifolds*, Ergeb. Math. Grenzgeb., vol. 10, Springer, Berlin, 1987.

[4] G. Bonneau, *Einstein-Weyl structures corresponding to diagonal Kähler Bianchi IX metrics*, Class. Quantum Grav. **14** (1997) pp. 2123–2135.

[5] G. Bonneau, *Einstein-Weyl structures and Bianchi metrics*, Class. Quantum Grav. **15** (1998) pp. 2415–2425.

[6] G. Bonneau, *Compact Einstein-Weyl four-dimensional manifolds,*, Class. Quantum Grav. **16** (1999) pp. 1057–1068.

[7] C. P. Boyer, *Conformal duality and compact complex surfaces*, Math. Ann. **274** (1986) pp. 517–526.

[8] C. P. Boyer, *A note on hyperHermitian four-manifolds*, Proc. Amer. Math. Soc. **102** (1988) pp. 157–164.

[9] C. P. Boyer and J. D. Finley, *Killing vectors in self-dual Euclidean Einstein spaces*, J. Math. Phys **23** (1982) pp. 1126–1130.

[10] C. P. Boyer, K. Galicki and B. Mann, *The geometry and topology of 3-Sasakian manifolds*, J. reine angew. Math. **455** (1994) pp. 183–220.

[11] D. M. J. Calderbank, *Möbius structures and two dimensional Einstein-Weyl geometry*, J. reine angew. Math. **504** (1998) pp. 37–53.

[12] D. M. J. Calderbank, *The Faraday 2-form in Einstein-Weyl geometry*, to appear in Math. Scand.

[13] D. M. J. Calderbank, *Weyl derivatives, conformal submersions and self-dual Einstein metrics*, Edinburgh Preprint MS-98-013 (1998).

[14] D. M. J. Calderbank and H. Pedersen, *Self-dual spaces, complex structures, Einstein-Weyl geometry and geodesics,*, Edinburgh Preprint (1998)

[15] D. M. J. Calderbank and K. P. Tod, *Einstein metrics, hyper-complex structures and the Toda field equation*, Edinburgh Preprint MS-98-011 (1998).

[16] E. Cartan, *Sur une classe d'espaces de Weyl*, Ann. Sci. École Norm. Sup. (3) **60** (1943) pp. 1–16.

[17] T. Chave, K. P. Tod and G. Valent, $(4,0)$ *and* $(4,4)$ *sigma models with a tri-holomorphic Killing vector*, Phys. Lett. **B 383** (1996) pp. 262–270.

[18] D. M. DeTurck and L. Kazdan, *Some regularity theorems in Riemannian geometry*, Ann. Sci. École Norm. Sup. (4) **14** (1981) pp. 249–260.

[19] M. G. Eastwood and K. P. Tod, *Local constraints on Einstein-Weyl geometries*, J. reine angew. Math. **491** (1997) pp. 183–198.

[20] M. G. Eastwood and K. P. Tod, *Local constraints on Einstein-Weyl geometries: the three dimensional case*, Preprint (1998)

[21] D. G. Ebin, *The manifold of Riemannian metrics*, Proc. Symp. Pure Math. **15** (1968) pp. 11–40.

[22] H. D. Fegan, *Conformally invariant first order differential operators*, Quart. J. Math. Oxford **27** (1976) pp. 371–378.

[23] G. B. Folland, *Weyl manifolds*, J. Diff. Geom. **4** (1970) pp. 145–153.

[24] P. Gauduchon, *La 1-forme de torsion d'une variété hermitienne compacte*, Math. Ann. **267** (1984) pp. 495–518.

[25] P. Gauduchon, *Structures de Weyl et théorèmes d'annulation sur une variété conforme autoduale*, Ann. Sc. Norm. Sup. Pisa **18** (1991) pp. 563–629.

[26] P. Gauduchon, *Weyl structures on a self-dual conformal manifold*, Proc. Symp. Pure Math. **54** (1993) Part 2, pp. 259–270.

[27] P. Gauduchon, *Structures de Weyl-Einstein, espaces de twisteurs et variétés de type* $S^1 \Theta S^3$, J. reine angew. Math. **469** (1995) pp. 1–50.

[28] P. Gauduchon and S. Ivanov, *Einstein-Hermitian surfaces and Hermitian Einstein-Weyl structures in dimension four*, Math. Z. **266** (1997) pp. 317–326.

[29] P. Gauduchon and K. P. Tod, *Hyperhermitian metrics with symmetry*, J. Geom. Phys. **25** (1998) pp. 291–304.

[30] A. Gray, *Pseudo-Riemannian almost product manifolds and submersions*, J. Math. Mech. **16** (1967) pp. 715–737.

[31] T. Higa, *Curvature tensors and curvature conditions in Weyl geometry*, Comment. Math. Univ. Sancti Pauli **43** (1994) pp. 139–153.

[32] N. J. Hitchin, *Compact four-dimensional Einstein manifolds*, J. Diff. Geom. **9** (1974) pp. 435–441.

[33] N. J. Hitchin, *Linear field equations on self-dual spaces*, Proc. Roy. Soc. London **A 370** (1980) pp. 173–191.

[34] N. J. Hitchin, *Complex manifolds and Einstein equations*, Twistor Geometry and Non-linear Systems (eds H. D. Doebner and T. D. Palev), Lecture Notes in Math. 970, Springer, Berlin, 1982, pp. 79–99.

[35] N. J. Hitchin, *Einstein metrics and the eta-invariant*, Boll. Un. Mat. Ital. B (7) **11** (1997) pp. 95–105.

[36] T. Ichiyama, H. Furuhata and H. Urakawa, *A conformal gauge invariant functional for Weyl structures and the first variation formula*, Preprint (1998).

[37] M. Itoh, *Compact Einstein-Weyl manifolds and the associated constant*, Osaka J. Math. **35** (1998) pp. 567–578.

[38] M. Itoh, *Affine locally symmetryic structures and finiteness theorems for Einstein-Weyl manifolds*, Preprint (1998).

[39] S. Ivanov, *Einstein-Weyl structures on certain compact conformal manifolds*, Preprint (1997)

[40] P. E. Jones and K. P. Tod, *Mini-twistor spaces and Einstein-Weyl spaces*, Class. Quantum Grav. **2** (1985) pp. 565–577.

[41] D. D. Joyce, *Explicit construction of self-dual* 4*-manifolds*, Duke Math. J. **77** (1995) pp. 519–552.

[42] H. Kamada, *Compact Einstein-Weyl four-manifolds with compatible almost complex structures*, Preprint (1998).

[43] M. Kerr, *Homogeneous Einstein-Weyl structures on symmetric spaces*, Ann. Global Anal. Geom. **15** (1997) pp. 437–445.

[44] N. Koiso, *Hypersurfaces of Einstein manifolds*, Ann. Sci. École Norm. Sup. (4) **14** (1981) pp. 433–443.

[45] N. Koiso, *Rigidity and infinitesimal deformability*, Osaka J. Math. **19** (1982) pp. 643–668.

[46] R. S. Kulkarni and U. Pinkall (Eds.), *Conformal Geometry*, Aspects of Mathematics, Max-Planck-Institut für Mathematik (Bonn), Friedr. Vieweg & Sohn, Braunschweig/Wiesbaden (1988).

[47] J. Lafontaine, *Remarques sur les variétés conformément plates*, Math. Ann. **259** (1982) pp. 313–319.

[48] C. R. LeBrun, H*-space with a cosmological constant*, Proc. Roy. Soc. London **A 380** (1982) pp. 171–185.

[49] C. R. LeBrun, *Counterexamples to the generalized positive action conjecture*, Comm. Math. Phys. **118** (1988) pp. 591–596.

[50] C. R. LeBrun, *Explicit self-dual metrics on* $\mathbb{C}P^2 \# \Delta\Delta\#\Delta P^2$, J. Diff. Geom. **34** (1991) pp. 223–253.

[51] C. R. LeBrun, *Self-dual manifolds and hyperbolic geometry*, Einstein metrics and Yang-Mills connections (Sanda, 1990), Lecture Notes in Pure and Appl. Math., vol. 145, Marcel Dekker, New York, 1993, pp. 99–131.

[52] A. B. Madsen, *Compact Einstein-Weyl manifolds with large symmetry group*, Ph.D. thesis, Odense University, 1995.

[53] A. B. Madsen, *Einstein-Weyl structures in the conformal classes of LeBrun metrics*, Class. Quantum Grav. **14** (1997) pp. 2635–2645.

[54] A. B. Madsen, H. Pedersen, Y. S. Poon and A. Swann, *Compact Einstein-Weyl manifolds with large symmetry group*, Duke Math. J. **88** (1997) pp. 407–434.

[55] S. Merkulov and H. Pedersen, *Projective structures on moduli spaces of compact complex hypersurfaces*, Proc. Amer. Math. Soc. **125** (1997) pp. 407–416.

[56] F. Narita, *Riemannian submersions and Riemannian manifolds with Einstein-Weyl structures*, Geometriae Dedicata **65** (1997) pp. 103–116.

[57] B. O'Neill, *The fundamental equations of a submersion*, Michigan Math. J. **13** (1966) pp. 459–469.

[58] L. Ornea and P. Piccinni, *Locally conformal Kähler structures in quaternionic geometry*, Trans. Amer. Math. Soc. **349** (1997) pp. 641–655.

[59] L. Ornea and P. Piccinni, *Compact hyperHermitian Weyl and quaternionic Hermitian Weyl manifolds*, Ann. Global Anal. Geom. **16** (1998) pp. 383–398.

[60] H. K. Pak, *One dimensional metric foliations in Einstein spaces*, Illinois J. Math. **36** (1992) pp. 594–599.

[61] H. Pedersen, *Einstein-Weyl spaces and* $(1,n)$*-curves in the quadric surface*, Ann. Global Anal. Geom. **4** (1986) pp. 89–120.

[62] H. Pedersen, *Einstein metrics, spinning top motions and monopoles*, Math. Ann. **274** (1986) pp. 35–39.

[63] H. Pedersen, Y. S. Poon and A. Swann, *The Einstein-Weyl equations in complex and quaternionic geometry*, Diff. Geom. Appl. **3** (1993) pp. 309–321.

[64] H. Pedersen, Y. S. Poon and A. Swann, *The Hitchin-Thorpe inequality for Einstein-Weyl manifolds*, Bull. London Math. Soc. **26** (1994) pp. 191–194.

[65] H. Pedersen, Y. S. Poon and A. Swann, *Einstein-Weyl deformations and submanifolds*, Int. J. Math. **7** (1996) pp. 705–719.

[66] H. Pedersen and A. Swann, *Riemannian submersions, four-manifolds and Einstein-Weyl geometry*, Proc. London Math. Soc. **66** (1993) pp. 381–399.

[67] H. Pedersen and A. Swann, *Einstein-Weyl geometry, the Bach tensor and conformal scalar curvature*, J. reine angew. Math. **441** (1993) pp. 99–113.

[68] H. Pedersen and K. P. Tod, *Three-dimensional Einstein-Weyl geometry*, Adv. Math. **97** (1993) pp. 74–109.

[69] M. Pontecorvo, *On twistor spaces of anti-self-dual hermitian surfaces*, Trans. Amer. Math. Soc. **331** (1992) pp. 653–661.

[70] Y. S. Poon, *Compact self-dual manifolds with positive scalar curvature*, J. Diff. Geom. **24** (1986) pp. 97–132.

[71] G. Tian and S. T. Yau, *Kähler-Einstein metrics on complex surfaces with* $c_1 > 0$, Comm. Math. Phys. **112** (1987) pp. 175–203.

[72] K. P. Tod, *Compact 3-dimensional Einstein-Weyl structures*, J. London Math. Soc. **45** (1992) pp. 341–351.

[73] K. P. Tod, *Scalar-flat Kähler and hyper-Kähler metrics from Painlevé-III*, Class. Quantum Grav. **12** (1995) pp. 1535–1547.

[74] K. P. Tod, *Local heterotic geometry and self-dual Einstein-Weyl spaces*, Class. Quantum Grav. **13** (1996) pp. 2609–2616.

[75] I. Vaisman, *On locally conformal almost Kähler manifolds*, Israel J. Math. **24** (1976) pp. 338–351.

[76] I. Vaisman, *Generalized Hopf manifolds*, Geom. Dedicata **13** (1982) pp. 231–255.

[77] J. Wang and M. Y. Wang, *Einstein metrics on* S^2*-bundles*, Math. Ann **310** (1998) pp. 497–526.

[78] M. Y. Wang and W. Ziller, *Einstein metrics on principal torus bundles*, J. Diff. Geom. **31** (1990) pp. 215–248.

[79] R. S. Ward, *Einstein-Weyl spaces and* SU(1) *Toda fields*, Class. Quantum Grav. 7 (1990) pp. L95–L98.

[80] H. Weyl, *Space, Time, Matter* (Translation of the fourth edition of *Raum, Zeit, Materie*, the first edition of which was published in 1918 by Springer, Berlin), Dover, New York (1952).

MATHEMATICS INSTITUTE, UNIVERSITY OF WARWICK, COVENTRY CV4 7AL, ENGLAND.
Current address: Department of Mathematics and Statistics, University of Edinburgh, King's Buildings, Mayfield Road, Edinburgh EH9 3JZ, Scotland.
E-mail address: davidmjc@maths.ed.ac.uk

DEPARTMENT OF MATHEMATICS AND COMPUTER SCIENCE, ODENSE UNIVERSITY, CAMPUSVEJ 55, DK-5230 ODENSE M, DENMARK.
E-mail address: henrik@imada.ou.dk

Made in United States
Orlando, FL
04 November 2026

90206113R00238